Springer Proceedings in Mathematics & Statistics

Volume 79

For further volumes:
http://www.springer.com/series/10533

Springer Proceedings in Mathematics & Statistics

This book series features volumes composed of selected contributions from workshops and conferences in all areas of current research in mathematics and statistics, including OR and optimization. In addition to an overall evaluation of the interest, scientific quality, and timeliness of each proposal at the hands of the publisher, individual contributions are all refereed to the high quality standards of leading journals in the field. Thus, this series provides the research community with well-edited, authoritative reports on developments in the most exciting areas of mathematical and statistical research today.

Ivan Cheltsov • Ciro Ciliberto • Hubert Flenner •
James McKernan • Yuri G. Prokhorov •
Mikhail Zaidenberg
Editors

Automorphisms in Birational and Affine Geometry

Levico Terme, Italy, October 2012

Springer

Editors

Ivan Cheltsov
School of Mathematics
University of Edinburgh
Edinburgh, United Kingdom

Ciro Ciliberto
Department of Mathematics
University of Rome Tor Vergata
Rome, Italy

Hubert Flenner
Faculty of Mathematics
Ruhr University Bochum
Bochum, Germany

James McKernan
Department of Mathematics
University of California San Diego
La Jolla, California, USA

Yuri G. Prokhorov
Steklov Mathematical Institute
Russian Academy of Sciences
Moscow, Russia

Mikhail Zaidenberg
Institut Fourier de Mathématiques
Université Grenoble I
Grenoble, France

ISSN 2194-1009 ISSN 2194-1017 (electronic)
ISBN 978-3-319-05680-7 ISBN 978-3-319-05681-4 (eBook)
DOI 10.1007/978-3-319-05681-4
Springer Cham Heidelberg New York Dordrecht London

Library of Congress Control Number: 2014941702

Mathematics Subject Classification (2010): 14L30, 14E07, 14R20, 14R10, 32M17

Printed on acid-free paper

Springer is part of Springer Science+Business Media (www.springer.com)

Preface

The conference brought together specialists from the birational geometry of projective varieties, affine algebraic geometry, and complex algebraic geometry. Topics from these areas include Mori theory, Cremona groups, algebraic group actions, and automorphisms. The ensuing talks and the discussions have highlighted the close connections between these areas. The meeting allowed these groups to exchange knowledge and to learn methods from adjacent fields. For detailed information about this conference, see http://www.science.unitn.it/cirm/GABAG2012.html.

The chapters in this book cover a wide area of topics from classical algebraic geometry to birational geometry and affine geometry, with an emphasis on group actions and automorphism groups.

Among the total of 27 chapters, eight give an overview of an area in one of the fields. The others contain original contributions, or mix original contributions and surveys.

In the birational part, there are chapters on Fano and del Pezzo fibrations, birational rigidity and superrigidity of Fano varieties, birational morphisms between threefolds, subgroups of the Cremona group, Jordan groups, real Cremona group, and a real version of the Sarkisov program.

In the part on classical projective geometry, there are chapters devoted to algebraic groups acting on projective varieties, automorphism groups of moduli spaces, and the algebraicity problem for analytic compactifications of the affine plane.

The topics in affine geometry include: different aspects of the automorphism groups of affine varieties, flexibility properties in affine algebraic and analytic geometry, automorphisms of affine spaces and Shestakov–Umirbaev theory, affine geometry in positive characteristic, the automorphism groups of configuration spaces and other classes of affine varieties and deformations of certain group actions on affine varieties.

Hopefully these proceedings will be of help for both specialists, who will find information on the current state of research, and young researchers, who wish to learn about these fascinating and active areas of mathematics.

Acknowledgments

The editors would like to thank the contributors of this volume. They are grateful to the CIRM at Trento, its directors Marco Andreatta and Fabrizio Catanese, its secretary Augusto Micheletti, and the whole staff for their permanent support in the organization of the conference and the distribution of this volume. Our thanks are due also to Marina Reizakis, Frank Holzwarth, and the staff of the Springer publishing house for agreeable cooperation during the preparation of the proceedings.

Edinburgh, UK — Ivan Cheltsov
Rome, Italy — Ciro Ciliberto
Bochum, Germany — Hubert Flenner
La Jolla, CA — James McKernan
Moscow, Russia — Yuri G. Prokhorov
Grenoble, France — Mikhail Zaidenberg

Contents

Part I
Birational Automorphisms

Singular del Pezzo Fibrations and Birational Rigidity

Hamid Ahmadinezhad

Abstract A known conjecture of Grinenko in birational geometry asserts that a Mori fibre space with the structure of del Pezzo fibration of low degree is birationally rigid if and only if its anticanonical class is an interior point in the cone of mobile divisors. The conjecture is proved to be true for smooth models (with a generality assumption for degree 3). It is speculated that the conjecture holds for, at least, Gorenstein models in degree 1 and 2. In this chapter, I present a (Gorenstein) counterexample in degree 2 to this conjecture.

2010 Mathematics Subject Classification: 14E05, 14E30 and 14E08

1 Introduction

All varieties in this chapter are projective and defined over the field of complex numbers. Minimal model program played on a uniruled threefold results in a Mori fibre space (Mfs for short). Such output for a given variety is not necessarily unique. The structure of the endpoints is studied via the birational invariant called *pliability* of the Mfs, see Definition 2. An Mfs is a $\mathbb{Q}$-factorial variety with at worst terminal singularities together with a morphism $\varphi: X \to Z$, to a variety Z of strictly smaller dimension, such that $-K_X$, the anti-canonical class of X, is φ-ample and

$$\operatorname{rank}\operatorname{Pic}(X) - \operatorname{rank}\operatorname{Pic}(Z) = 1.$$

H. Ahmadinezhad (✉)
Radon Institute, Austrian Academy of Sciences, Altenberger Str. 69, 4040 Linz, Austria
e-mail: hamid.ahmadinezhad@oeaw.ac.at

I. Cheltsov et al. (eds.), *Automorphisms in Birational and Affine Geometry*, Springer Proceedings in Mathematics & Statistics 79,
DOI 10.1007/978-3-319-05681-4_1,

Definition 1. Let $X \to Z$ and $X' \to Z'$ be Mfs. A birational map $f: X \dashrightarrow X'$ is *square* if it fits into a commutative diagram

$$\begin{array}{ccc} X & \overset{f}{\dashrightarrow} & X' \\ \downarrow & & \downarrow \\ Z & \overset{g}{\dashrightarrow} & Z' \end{array}$$

where g is birational and, in addition, the map $f_L: X_L \dashrightarrow X'_L$ induced on generic fibres is biregular. In this case we say that X/Z and X'/Z' are *square birational*. We denote this by $X/Z \sim X'/Z'$.

Definition 2 (Corti [10]). The *pliability* of an Mfs $X \to Z$ is the set

$$\mathscr{P}(X/Z) = \{\text{Mfs } Y \to T \mid X \text{ is birational to } Y\}/\sim$$

An Mfs $X \to Z$ is said to be *birationally rigid* if $\mathscr{P}(X/Z)$ contains a single element.

A main goal in the birational geometry of threefolds is to study the geometry of Mfs, and their pliability. Note that finite pliability, and in particular birational rigidity, implies non-rationality. There are three types of Mfs in dimension 3, depending on the dimension of Z. If $\dim(Z) = 1$, then the fibres must be del Pezzo surfaces. When the fibration is over $\mathbb{P}^1$, I denote this by $dP_n/\mathbb{P}^1$, where $n = K_\eta^2$ and η is the generic fibre.

It is known that a $dP_n/\mathbb{P}^1$ is not birationally rigid when the total space is smooth and $n \geq 4$. For $n > 4$ the threefold is rational (see for example [20]), hence non-rigid, and it was shown in [5] that $dP_4/\mathbb{P}^1$ are birational to conic bundles.

Understanding conditions under which a $dP_n/\mathbb{P}^1$, for $n \leq 3$, is birationally rigid is a key step in providing the full picture of MMP, and hence the classification, in dimension 3. Birational rigidity for the smooth models of degree $1, 2$ and 3 is well studied, see for example [22]. However, as we see later, while the smoothness assumption for degree 3 is only a generality assumption, considering the smooth case for $n = 1, 2$ is not very natural. Hence the necessity of considering singular cases is apparent.

In this chapter, I focus on a well-known conjecture (Conjecture 1) on this topic that connects birational rigidity of del Pezzo fibrations of low degree to the structure of their mobile cone. A counterexample to this conjecture is provided when the threefold admits certain singularities.

2 Grinenko's Conjecture

Pukhlikov in [22] proved that a general smooth $dP_3/\mathbb{P}^1$ is birationally rigid if the class of 1-cycles $mK_X^2 - L$ is not effective for any $m \in \mathbb{Z}$, where L is the class of a line in a fibre. This condition is famously known as the K^2-condition.

Definition 3. A del Pezzo fibration is said to satisfy K^2-condition if the 1-cycle K^2 does not lie in the interior of the Mori cone $\overline{\mathrm{NE}}$.

The birational rigidity of smooth $dP_n/\mathbb{P}^1$ for $d = 1, 2$ was also considered in [22] and the criteria for rigidity are similar to that for $n = 3$.

In a sequential work [13–19], Grinenko realised and argued evidently that it is more natural to consider K-condition instead of the K^2-condition.

Definition 4. A del Pezzo fibration is said to satisfy K-condition if the anticanonical divisor does not lie in the interior of the Mobile cone.

Remark 1. It is a fun, and not difficult, exercise to check that K^2-condition implies K-condition. And the implication does not hold in the opposite direction.

One of the most significant observations of Grinenko was the following theorem.

Theorem 1 ([17, 19]). *Let X be a smooth threefold Mfs, with del Pezzo surfaces of degree 1 or 2, or a general degree 3, fibred over $\mathbb{P}^1$. Then X is birationally rigid if it satisfies the K-condition.*

He then conjectured that this must hold in general, as formulated in the conjecture below, with no restriction on the singularities.

Conjecture 1 ([19], Conjecture 1.5 and [13], Conjecture 1.6). Let X be a threefold Mori fibration of del Pezzo surfaces of degree $1, 2$ or 3 over $\mathbb{P}^1$. Then X is birationally rigid if and only if it satisfies the K-condition.

It is generally believed that Grinenko's conjecture might hold if one only considers Gorenstein singularities.

Note that, it is not natural to only consider the smooth case for $n = 1, 2$ as these varieties very often carry some orbifold singularities inherited from the ambient space, the non-Gorenstein points. For example a del Pezzo surface of degree 2 is naturally embedded as a quartic hypersurface in the weighted projective space $\mathbb{P}(1, 1, 1, 2)$. It is natural that a family of these surfaces meets the singular point $1/2(1, 1, 1)$. See [1] for construction of models and the study of their birational structure.

Grinenko also constructed many nontrivial (Gorenstein) examples, which supported his arguments. The study of quasi-smooth models of $dP_2/\mathbb{P}^1$ in [1], i.e. models that typically carry a quotient singularity, also gives evidence that the relation between birational rigidity and the position of $-K$ in the mobile cone is not affected by the presence of the non-Gorenstein point. Below in Sect. 3.1 I give a counterexample to Conjecture 1 for a Gorenstein singular degree 2 del Pezzo fibration.

On the other hand, in [8], Example 4.4.4, it was shown that this conjecture does not hold in general for the degree 3 case (of course in the singular case) and suggested that one must consider the semi-stability condition on the threefold X in order to state an updated conjecture:

Conjecture 2 ([8], Conjecture 2.7). Let X be a $dP_3/\mathbb{P}^1$ which is semistable in the sense of Kollár[21]. Then X is birationally rigid if it satisfies the K-condition.

Although this type of (counter)examples are very difficult to produce, the expectation is that such example in degree 1 is possible to be produced. On the other hand, a notion of (semi)stability for del Pezzo fibrations of degree 1 and 2 seems necessary (as already noted in [9], Problem 5.9.1), in order to state an improved version of Conjecture 1, and yet there has been no serious attempt in this direction.

3 The Counterexample

The most natural construction of smooth $dP_2/\mathbb{P}^1$ is the following, due to Grinenko [19].

Let $\mathscr{E} = \mathscr{O} \oplus \mathscr{O}(a) \oplus \mathscr{O}(b)$ be a rank 3 vector bundle over $\mathbb{P}^1$ for some positive integers a, b, and let $V = \mathrm{Proj}_{\mathbb{P}^1} \mathscr{E}$. Denote the class of the tautological bundle on V by M and the class of a fibre by L so that

$$\mathrm{Pic}(V) = \mathbb{Z}[M] + \mathbb{Z}[L]$$

Assume $\sigma\colon X \to V$ is a double cover branched over a smooth divisor $R \sim 4M - 2eL$, for some integer e. The natural projection $p\colon V \to \mathbb{P}^1$ induces a morphism $\pi\colon X \to \mathbb{P}^1$, such that the fibres are del Pezzo surfaces of degree 2 embedded as quartic surfaces in $\mathbb{P}(1,1,1,2)$. This threefold X can also be viewed as a hypersurface of a rank two toric variety. Let T be a toric fourfold with Cox ring $\mathbb{C}[u, v, x, y, z, t]$, that is $\mathbb{Z}^2$-graded by

$$\begin{pmatrix} u & v & x & y & z & t \\ 1 & 1 & 0 & -a & -b & -e \\ 0 & 0 & 1 & 1 & 1 & 2 \end{pmatrix} \tag{1}$$

The threefold X is defined by the vanishing of a general polynomial of degree $(-2e, 4)$. It is studied in [19] for which values a, b and e this construction provides an Mfs, and then he studies their birational properties. This construction can also be generalised to non-Gorenstein models [1]. As before, let X be defined by the vanishing of a polynomial of degree $(-n, 4)$ but change the grading on T to

$$\begin{pmatrix} u & v & x & y & z & t \\ 1 & 1 & -a & -b & -c & -d \\ 0 & 0 & 1 & 1 & 1 & 2 \end{pmatrix} \tag{2}$$

where c and d are positive integers. When X is an Mfs, it is easy to check that X is Gorenstein if and only if $e = 2d$. See [1] for more details and construction.

The cone of effective divisors modulo numerical equivalence on T is generated by the toric principal divisors, associate with columns of the matrix above, and we have $\mathrm{Eff}(T) \subset \mathbb{Q}^2$. This cone decomposes, as a chamber, into a finite union of subcones

$$\mathrm{Eff}(T) = \bigcup \mathrm{Nef}(T_i)$$

where T_i are obtained by the variation of geometric invariant theory (VGIT) on the Cox ring of T. See [11] for an introduction to the GIT construction of toric varieties, and [7] for a specific treatment of rank two models and connections to Sarkisov program via VGIT. In [7] it was also shown how the toric 2-ray game on T (over a point) can, in principle, be realised from the VGIT. I refer to [9] for an explanation of the general theory of 2-ray game and Sarkisov program. In certain cases, when the del Pezzo fibration is a Mori dream space, i.e. it has a finitely generated Cox ring, its 2-ray game is realised by restricting the 2-ray game of the toric ambient space. In other words, in order to trace the Sarkisov link one runs the 2-ray game on T, restricts it to X and checks whether the game remains in the Sarkisov category, in which case a winning game is obtained and a birational map to another Mfs is constructed. See [1–4, 7, 8] for explicit constructions of these models for del Pezzo fibrations or blow ups of Fano threefolds. I demonstrate this method in the following example, which also shows that Conjecture 1 does not hold.

3.1 Construction of the Example

Suppose T is a toric variety with the Cox ring $\mathrm{Cox}(T) = \mathbb{C}[u, v, x, t, y, z]$, grading given by the matrix

$$A = \begin{pmatrix} u & v & x & t & y & z \\ 1 & 1 & 0 & -2 & -2 & -4 \\ 0 & 0 & 1 & 2 & 1 & 1 \end{pmatrix},$$

and let the irrelevant ideal $I = (u, v) \cap (x, y, z, t)$. In other words T is the geometric quotient with character $\psi = (-1, -1)$. Suppose $\mathcal{L}$ is the linear system of divisors of degree $(-4, 4)$ in T; I use the notation $\mathcal{L} = |\mathcal{O}_T(-4, 4)|$. This linear system is generated by monomials according to the following table:

deg of u, v coefficient	0	2	4	6	8	10	12
fibre monomials	x^3z	xy^3	yzt	xyz^2	xz^3	yz^3	z^4
	t^2	ty^2	xy^2z	tz^2	y^2z^2		
	x^2y^2	x^2yz	y^4	y^3z			
	xyt	xzt	x^2z^2				

Let g be a polynomial whose zero set defines a general divisor in $\mathcal{L}$ (I use the notation $g \in \mathcal{L}$). Then, for example, x^3z is a monomial in g with nonzero coefficient, and appearance of y^4 in the column indicated by 4 means that $a(u,v)y^4$ is a part of g, for a homogeneous polynomial a of degree 4 in the variables u and v, so that $a(u,v)y^4$ has bidegree $(-4,4)$.

Now consider a sublinear system $\mathcal{L}' \subset \mathcal{L}$ with the property that u^i divides the coefficient polynomials according to the table

power of u	1	2	3	4	5	6	8	9	12
monomial	x^2yz	xzt	yzt	x^2z^2	xyz^2	tz^2	xz^3	yz^3	z^4
		xy^2z	y^3z			y^2z^2			

and general coefficients otherwise. And suppose $f \in \mathcal{L}'$ is general. For example $u^{12}z^4$ is a monomial in f and no other monomial that includes z^4 can appear in f. Or, for instance, xyz^2 in the column indicated by 5 carries a coefficient u^5; in other words, we can only have monomials of the form $\alpha u^6 xyz^2$ or $\beta v u^5 xyz^2$ in f, for $\alpha, \beta \in \mathbb{C}$, and no other monomial with xyz^2 can appear in f. Denote by X the hypersurface in T defined by the zero locus of f.

3.2 Argumentation Overview

Note that the threefold X has a fibration over $\mathbb{P}^1$ with degree 2 del Pezzo surfaces as fibres. In the remainder of this section, I check that it is a Mfs and then I show that the natural 2-ray game on X goes out of the Mori category. This is by explicit construction of the game. It is verified, from the construction of the game, that $-K_X \notin \operatorname{Int}\overline{\operatorname{Mob}}(X)$. Hence the conditions of Conjecture 1 are satisfied for X. In Sect. 4, a new (square) birational model to X is constructed, for which the anticanonical divisor is interior in the mobile cone. Then I show that it admits a birational map to a Fano threefold, and hence it is not birationally rigid.

Lemma 1. *The hypersurface $X \subset T$ is singular. In particular* $\operatorname{Sing}(X) = \{p\}$, *where* $p = (0:1:0:0:0:1)$. *Moreover, the germ at this point is of type cE_6, and X is terminal.*

Proof. By Bertini theorem $\operatorname{Sing}(X) \subset \operatorname{Bs}(\mathcal{L}')$. Appearance of y^4 in $\mathcal{L}'$ with general coefficients of degree 4 in u and v implies that this base locus is contained in the loci $(y=0)$ or $(u=v=0)$. However, $(u=v=0)$ is not permitted as (u,v) is a component of the irrelevant ideal. Hence we take $(y=0)$. Also $t^2 \in \mathcal{L}'$ implies $\operatorname{Bs}(\mathcal{L}') \subset (y=t=0)$. The remaining monomials are x^3z, $u^4x^2z^2, u^8xz^3$ and $u^{12}z^4$, where b is a general quartic. In fact, appearance of the first monomial, i.e. x^3z, implies that $x=0$ or $z=0$. And $u^{12}z^4 \in \mathcal{L}'$ implies $z=0$ or $u=0$. Therefore

$$\operatorname{Bs}(\mathcal{L}') = (u=x=y=t=0) \cup (y=t=z=0).$$

The part $(y = t = z = 0)$ is the line $(u : v; 1 : 0 : 0 : 0)$, which is smooth because of the appearance of the monomial x^3z in f. And the rest is exactly the point p.

Note that near the point $p \in T$, I can set $v \neq 0$ and $z \neq 0$, to realise the local isomorphism to $\mathbb{C}^4$, which is

$$\operatorname{Spec}\mathbb{C}\left[u, v, x, y, z, t, \frac{1}{v}, \frac{1}{z}\right]^{\mathbb{C}^*\times\mathbb{C}^*} = \operatorname{Spec}\mathbb{C}\left[\frac{u}{v}, \frac{x}{v^4z}, \frac{t}{v^6z^2}, \frac{y}{v^2z}\right]$$

I denote the new local coordinates by u, x, y, t. Now looking at f, in this local chart, we observe that the only quadratic part is t^2, and the cubic part is x^3. The variable y appears in degree 4, after some completing squares with t and x and u has higher order. In particular, after some analytic changes we have

$$f_{\mathrm{loc}} \equiv t^2 + x^3 + y^4 + u \times (\text{higher order terms})$$

which is a cE_6 singularity.

In the following lemma, I study the 2-ray game played on $X/\{\mathrm{pt}\}$. This will be used to figure out the shape of $\mathrm{Mob}(X)$ and the position of $-K_X$ against it.

Lemma 2. *The 2-ray game of T restricts to a game on X.*

Proof. The 2-ray game on T goes as follows

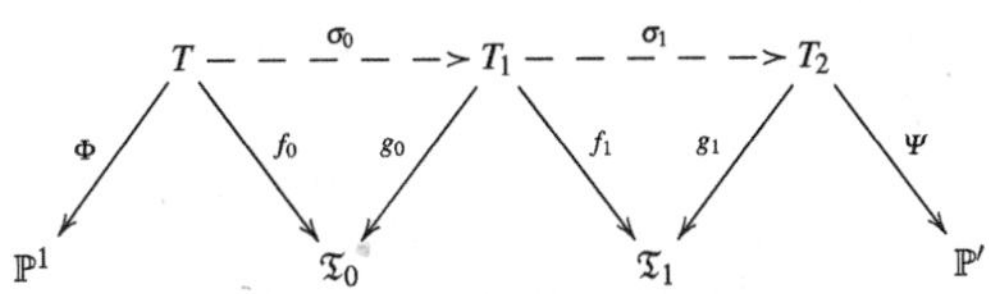

The varieties in this diagram are obtained as follows. The GIT chamber of T is indicated by the matrix A and it is

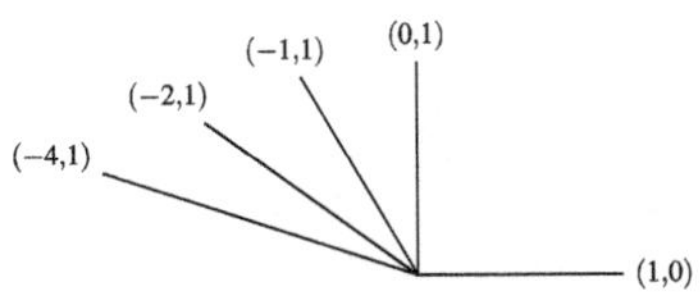

The ample cone of the variety T is the interior of the cone

$$\mathrm{Convex}\langle(1, 0), (0, 1)\rangle$$

and T_1 corresponds to the interior of the cone

$$\mathrm{Convex}\langle(0, 1), (-1, 1)\rangle,$$

i.e. this cone is the nef cone of T_1. In other words T_1 is a toric variety with the same coordinate ring and grading as T but with irrelevant ideal $I_1 = (u, v, x) \cap (t, y, z)$. Similarly T_2 corresponds to the cone generated by rays $(-1, 1)$ and $(-2, 1)$, i.e. the irrelevant ideal of T_2 is $I_2 = (u, v, x, t) \cap (y, z)$. On the other hand, the four varieties in the second row of the diagram correspond to the one-dimensional rays in the chamber. The projective line $\mathbb{P}^1$ corresponds to the ray generated by $(1, 0)$; monomial of degree $(n, 0)$ form the graded ring $\mathbb{C}[u, v]$. Similarly $\mathfrak{T}_0$ correspond to the ray generated by $(0, 1)$. In other words, $\mathfrak{T}_0$ is

$$\operatorname{Proj} \bigoplus_{n \geq 1} H^0(T, \mathcal{O}_T(0, n)) = \operatorname{Proj} \mathbb{C}[x, u^2t, uvt, v^2t, u^2y, \dots, v^4z],$$

which is embedded in $\mathbb{P}(1^9, 2^3)$ via the relations among the monomials above. The varieties $\mathfrak{T}_1$ and $\mathbb{P}'$ can also be explicitly computed in this way. In particular, $\mathbb{P}' = \mathbb{P}(1, 1, 2, 4, 6)$.

For the maps in the diagram we have that

1. $\Phi : T \to \mathbb{P}^1$ is the natural fibration.
2. $f_0 : T \to \mathfrak{T}_0$ is given in coordinate by

$$(u, v, x, t, y, z) \in T \longmapsto (x, u^2, uvt, v^2t, u^2y, \dots, v^4z) \in \mathfrak{T}_0 \subset \mathbb{P}(1^9, 2^3)$$

It is rather easy to check that away from $p = (1 : 0 : \cdots : 0) \in \mathfrak{T}_0$ the map f_0 is one-to-one. The pre-image of this point under f_0 is the set $(u = v = 0) \cup (t = y = z = 0)$. But $(u = v = 0)$ corresponds to a component of the irrelevant ideal, and hence it is empty on T. This implies that the line $(t = y = z = 0) \cong \mathbb{P}^1 \subset T$ is contracted to a point via f_0. In particular, $\mathfrak{T}_0$ is not $\mathbb{Q}$-factorial. Similarly $g_0 : T_1 \to \mathfrak{T}_0$ is the contraction of the surface $(u = v = 0)$, to the same point in $\mathfrak{T}_0$. In particular, σ_0 is an isomorphism in codimension 1.

Let us have a look at the local description of these maps. In $\mathfrak{T}_0$ consider the open set given by $x \neq 0$, a neighbourhood of p. This affine space is (with an abuse of notation for local coordinates)

$$\operatorname{Spec} \mathbb{C}[u^2t, uvt, v^2t, u^2y, \dots, v^4z].$$

And at the level of T (respectively T_1) using $\{x \neq 0\}$ I can get rid of the second grading (corresponding to the second row of the matrix A) and obtain a quasi-projective variety which is the quotient of $\mathbb{C}^5 - \{u = v = 0\}$ (respectively $\mathbb{C}^5 - \{t = y = z = 0\}$) by an action of $\mathbb{C}^*$ by

$$(\lambda; (u, v, t, y, z) \mapsto (\lambda u, \lambda v, \lambda^{-2}t, \lambda^{-2}y, \lambda^{-}4z)$$

I denote this by $(1, 1, -2, -2, -4)$ anti-flip, following [6, 23]. See these references for an explanation of this construction and notation. This all means that under σ_0 a copy of $\mathbb{P}^1$ is replaces by a copy of $\mathbb{P}(2, 2, 4)$.

3. Similarly, σ_1 is an isomorphism in codimension 1. Note that the action of $(\mathbb{C}^*)^2$ by A is invariant under $\mathrm{GL}(2,\mathbb{Z})$ action on A, so that multiplying it from the left by

$$\begin{pmatrix} 1 & 1 \\ 0 & 1 \end{pmatrix}$$

and setting $t = 1$, as in the previous case, shows that this map is of type $(1,1,1,-1,-3)$. So f_1 contracts a copy of $\mathbb{P}^2$ and g_1 extracts a copy of $\mathbb{P}(1,3)$.
4. As noted before, the irrelevant ideal of T_2 is $I_2 = (u,v,x,t) \cap (y,z)$, and $\Psi\colon T_2 \to \mathbb{P}' = \mathbb{P}(1,1,2,4,6)$ is the contraction of the divisor $E = (z=0)$ to a point in $\mathbb{P}'$. This map, similar to Φ, is given by

$$\bigoplus_{n\geq 1} H^0(T, \mathcal{O}_T(0,n))$$

where the degrees are now considered in the transformation of A by the action of

$$\begin{pmatrix} 1 & 2 \\ 0 & 1 \end{pmatrix}$$

from the left.

As already noted Φ restricted to X is just the degree 2 del Pezzo fibration. The restriction of f_0 contracts the same line, to the same point on a subvariety of $\mathfrak{T}_0$. Once we set $x = 1$, a linear form "z" appears in f, hence locally in the neighbourhood where the flip is happening one can eliminate this variable. Therefore, σ_0 restricts to $(1,1,-2,-2)$, which means a copy of $\mathbb{P}^1$ is contracted to a point and a copy of $\mathbb{P}(2,2) \cong \mathbb{P}^1$ is extracted. The restriction of σ_1 to the threefold X_1, i.e., the image of X under the anti-flip, is an isomorphism. This is because t^2 is a term in f, and hence the toric flip happens away from the threefold. Finally Ψ restricts to a divisorial contraction to X_1.

Remark 2. Note that it follows from some delicate version of Lefschetz principle that $\operatorname{rank}\operatorname{Pic}(X) = 2$. In fact, X is defined by a linear system of bi-degree $(-4,4)$, and does not belong to the nef cone of T, which is generated by $(1,0)$ and $(0,1)$. However, it is in the interior of the mobile cone of T. In particular, it is nef and big on both T_1 and T_2, which are isomorphic to T in codimension 1. Hence we have that

$$\operatorname{Pic}(X) \cong \operatorname{Pic}(T_1) \cong \operatorname{Pic}(T) \cong \mathbb{Z}^2$$

These isomorphisms follow a singular version of Lefschetz hyperplane theorem as in [12, Sect. 2.2], and in this particular case it holds because $(-4,4)$ represents an interior point in the cone of mobile divisors on T. A more detailed argument for this can be found in [1, Sect. 4.3] or [8, Proof of Proposition 32]. Also note that X

and X_1 are isomorphic, so it does not really matter which one to consider for these arguments.

Proposition 1. *The 2-ray game on X obtained in Lemma 2 does not provide a new Mfs model of X.*

Proof. This is quite clear now. The first reason for the failure of the game is the anti-flip $(1,1,-2,-2)$. As mentioned before, in this anti-flip a copy of $\mathbb{P}^1$ is contracted to a point and on the other side of the anti-flip a copy of $\mathbb{P}(2,2) \cong \mathbb{P}^1$ is extracted. In particular, the map replaces a smooth line by a line that carries singularities at each point of it. Note that the line itself is smooth (isomorphic to the projective line) but on the threefold it is singular (at each point). As terminal singularities are isolated, this game goes out of the Mori category. Another reason for the failure is that $-K_X \in \partial\overline{\mathrm{Mob}}(X)$, as explained below. This means that in the last map of the 2-ray game of X the contracted curves are trivial against the canonical divisor, hence not fulfilling the rules of Mori theory.

Remark 3. It is also a good point here to observe that the anti-canonical class of X has degree $(-2,1)$. This can be seen as follows. The variety T is toric, hence its anticanonical divisor is given by the sum of the toric principal divisors, in particular it has degree $(-6,5)$, note that I am still working with the matrix A and this bi-degree is nothing but the sum of the columns of A. By adjunction formula $K_X = (K_T + X)_{|X}$, which implies that $-K_X \sim \mathcal{O}(-2,1)$. On the other hand, the fact that the 2-ray game on X is essentially the restriction of the 2-ray game on T implies that $\overline{\mathrm{Mob}}(X)$ and $\overline{\mathrm{Mob}}(T)$, as convex cones in $\mathbb{Q}^2$, have the same boundaries. In fact, the decomposition of $\overline{\mathrm{Mob}}(X)$ into the union of nef cones is a sub-decomposition of $\overline{\mathrm{Mob}}(T)$. In particular

$$\overline{\mathrm{Mob}}(X) = \mathrm{Convex}\left\langle (1,0),(-2,1)\right\rangle,$$

which implies

$$-K_X \in \partial\overline{\mathrm{Mob}}(X).$$

And this shows that X satisfies conditions of Conjecture 1.

4 The Fano Variety Birational to X

Now consider the fibrewise transform $T \to \mathbb{F}$, given by

$$(u,v,x,t,y,z) \mapsto (u,v,u^4x,u^6t,u^3y,z)$$

where $\mathbb{F}$ is a toric variety with Cox ring $\mathrm{Cox}(\mathbb{F}) = \mathbb{C}[u,v,x,t,z,y]$, with irrelevant ideal $I_{\mathbb{F}} = (u,v) \cap (x,y,z,t)$, and the grading

$$A' = \begin{pmatrix} u & v & x & t & z & y \\ 1 & 1 & 0 & 0 & 0 & -1 \\ 0 & 0 & 1 & 2 & 1 & 1 \end{pmatrix}$$

Denote by X' the birational transform of X under this map. And suppose g is the defining equation of X'. The polynomial g can be realised from f by substituting x, y, z, t by u^4x, u^3y, z, u^6t, and then cancelling out u^{12} from it. In particular it is of the form

$$g = t^2 + x^3z + z^4 + x^2y^2 + a_4(u, v)y^4 + \{\text{other terms}\}$$

In fact the full table of monomials appearing in g is given by

deg of u, v coefficient	0	1	2	3	4
fibre monomials	x^3z	x^2yz	xy^2z	y^3z	y^4
	x^2z^2	xyz^2	y^2z^2		
	xz^3	yz^3	y^2t		
	z^4	yzt	uxy^3		
	t^2				
	xzt				
	z^2t				
	$u^2x^2y^2$				
	$uxyt$				

Note that this table does not generate a general member in the linear system $|\mathcal{O}_{\mathbb{F}}(0, 4)|$, and the missing monomials are x^4, tx^2 and x^3y. Also x^2y^2, xyt and xy^3 do not have general coefficients in u, v, as specified in the table above.

Theorem 2. *The threefold X' is smooth and it is birational to a Fano threefold. In particular, X', and hence X, are not birationally rigid.*

Proof. A similar, and easier, check to that of Lemma 1 shows that X' is smooth: First note that the base locus of the linear system is the line $(u : v; 1 : 0 : 0 : 0)$, given by $y = z = t = 0$. Then observe that any point on this line is smooth, guaranteed by appearance of the monomial $x^3z \in g$.

Now, let us play the 2-ray game on $\mathbb{F}$ and restrict it to X'. The 2-ray game on $\mathbb{F}$ proceeds, after the fibration to $\mathbb{P}^1$, by a divisorial contraction to $\mathbb{P}(1, 1, 1, 1, 2)$. This is realised by

$$\bigoplus_{n \geq 1} H^0(\mathbb{F}, \mathcal{O}_{\mathbb{F}}(0, n))$$

and is given in coordinates by

$$(u : v; x : t : z : y) \mapsto (x : z : uy : vy : t)$$

In particular, the divisor $E : (y = 0) \subset \mathbb{F}$ is contracted to the locus $\mathbb{P}(1, 1, 2) \subset \mathbb{P}(1, 1, 1, 1, 2)$. The restriction of this map to X' shows that the divisor $E_X = E \cap X$ is contracted to a quartic curve in $\mathbb{P}(1, 1, 2)$. The image of X' under this map is a quartic threefold in $\mathbb{P}(1, 1, 1, 1, 2)$, which is a Fano variety of index 2. Note that, similar to X, we can check that $-K_{X'} \sim \mathcal{O}_{X'}(1, 1)$. In particular, $-K_{X'}$ is ample, and of course interior in the mobile cone. This variety is also studied in [1] Theorem 3.3.

Acknowledgements I would like to thank Gavin Brown for his support during my Ph.D., where I learnt most of the techniques used in this chapter. I would also like to thank Ivan Cheltsov and Mikhail Grinenko for showing interest in this example. I am also grateful to the anonymous referees, whose valuable comments helped me shape the presentation of this paper.

References

1. H. Ahmadinezhad, On del Pezzo fibrations that are not birationally rigid. J. Lond. Math. Soc. **86**(1), 36–62 (2012)
2. H. Ahmadinezhad, On pliability of del Pezzo fibrations and Cox rings (2013) [arXiv:1304.4357]
3. H. Ahmadinezhad, A.-S. Kaloghiros, Non-rigid quartic 3-folds (2013) [arXiv:1310.5554]
4. H. Ahmadinezhad, F. Zucconi, Circle of Sarkisov links on a Fano threefold (2013) [arXiv:1304.6009]
5. V.A. Alekseev, On conditions for the rationality of three-folds with a pencil of del Pezzo surfaces of degree 4. Mat. Zametki **41**(5), 724–730, 766 (1987) [MR 898133 (88j:14048)]
6. G. Brown, Flips arising as quotients of hypersurfaces. Math. Proc. Camb. Philos. Soc. **127**(1), 13–31 (1999) [14E05 (14E30 14L24 14L30)]
7. G. Brown, F. Zucconi, Graded rings of rank 2 Sarkisov links. Nagoya Math. J. **197**, 1–44 (2010) [MR 2649280 (2011f:14020)]
8. G. Brown, A. Corti, F. Zucconi, Birational geometry of 3-fold Mori fibre spaces, in *The Fano Conference*, pp. 235–275 (University of Torino, Turin, 2004) [MR 2112578 (2005k:14031)]
9. A. Corti, Singularities of linear systems and 3-fold birational geometry, in *Explicit Birational Geometry of 3-Folds*. London Mathematical Society Lecture Note Series, vol. 281 (Cambridge University Press, Cambridge, 2000), pp. 259–312
10. A. Corti, M. Mella, Birational geometry of terminal quartic 3-folds. I. Am. J. Math. **126**(4), 739–761 (2004)
11. D.A. Cox, The homogeneous coordinate ring of a toric variety. J. Algebr. Geom. **4**(1), 17–50 (1995) [MR 1299003 (95i:14046)]
12. M. Goresky, R. MacPherson, in *Stratified Morse Theory*. Ergebnisse der Mathematik und ihrer Grenzgebiete (3) [Results in Mathematics and Related Areas (3)], vol. 14 (Springer, Berlin, 1988)
13. M. Grinenko, Birational properties of pencils of del Pezzo surfaces of degrees 1 and 2. Mat. Sb. **191**(5), 17–38 (2000)
14. M. Grinenko, On the birational rigidity of some pencils of del Pezzo surfaces. Algebr. Geom. **10**; J. Math. Sci. (New York) **102**(2), 3933–3937 (2000)

15. M. Grinenko, On fiberwise surgeries of fibrations on a del Pezzo surface of degree 2. Uspekhi Mat. Nauk **56**(4(340)), 145–146 (2001)
16. M. Grinenko, On fibrations into del Pezzo surfaces. Mat. Zametki **69**(4), 550–565 (2001)
17. M. Grinenko, Birational properties of pencils of del Pezzo surfaces of degrees 1 and 2. II. Mat. Sb. **194**(5), 31–60 (2003)
18. M. Grinenko, Non-rationality of a three-dimensional Fano variety of index 2 and degree 1 (2003) [math.AG/0304321]
19. M. Grinenko, Fibrations into del Pezzo surfaces. Uspekhi Mat. Nauk **61**(2(368)), 67–112 (2006) [MR 2261543 (2007g:14009)]
20. V.A. Iskovskikh, On the rationality problem for three-dimensional algebraic varieties fibered over del Pezzo surfaces. Trudy Mat. Inst. Steklov. **208** (1995), no. Teor. Chisel, Algebra i Algebr. Geom. 128–138. Dedicated to Academician Igor' Rostislavovich Shafarevich on the occasion of his seventieth birthday (Russian) [MR 1730261 (2000i:14019)]
21. J. Kollár, Polynomials with integral coefficients, equivalent to a given polynomial. Electron. Res. Announc. Am. Math. Soc. **3**, 17–27 (electronic) (1997) [MR 1445631 (98g:11076)]
22. A.V. Pukhlikov, Birational automorphisms of three-dimensional algebraic varieties with a pencil of del Pezzo surfaces. Izv. Ross. Akad. Nauk Ser. Mat. **62**(1), 123–164 (1998) [MR 1622258 (99f:14016)]
23. M. Reid, *What is a flip?* unpublished manuscript of Utah seminar (1992)

Additive Actions on Projective Hypersurfaces

Ivan Arzhantsev and Andrey Popovskiy

Abstract By an additive action on a hypersurface H in $\mathbb{P}^{n+1}$ we mean an effective action of a commutative unipotent group on $\mathbb{P}^{n+1}$ which leaves H invariant and acts on H with an open orbit. Brendan Hassett and Yuri Tschinkel have shown that actions of commutative unipotent groups on projective spaces can be described in terms of local algebras with some additional data. We prove that additive actions on projective hypersurfaces correspond to invariant multilinear symmetric forms on local algebras. It allows us to obtain explicit classification results for non-degenerate quadrics and quadrics of corank one.

2010 Mathematics Subject Classification: Primary 13H, 14L30; Secondary 14J50, 15A69

I. Arzhantsev
Faculty of Mechanics and Mathematics, I.A. Department of Higher Algebra, Moscow State University, Leninskie Gory 1, GSP-1, Moscow, 119991, Russia

School of Applied Mathematics and Information Science, National Research University Higher School of Economics, Bolshoi Trekhsvyatitelskiy 3, Moscow, 109028, Russia
e-mail: arjantse@mccme.ru

A. Popovskiy (✉)
Faculty of Mechanics and Mathematics, A.P. Department of Higher Algebra, Moscow State University, Leninskie Gory 1, GSP-1, Moscow, 119991, Russia
e-mail: PopovskiyA@gmail.com

I. Cheltsov et al. (eds.), *Automorphisms in Birational and Affine Geometry*, Springer Proceedings in Mathematics & Statistics 79,
DOI 10.1007/978-3-319-05681-4_2,

1 Introduction

Let $\mathbb{K}$ be an algebraically closed field of characteristic zero and $\mathbb{G}_a$ the additive group $(\mathbb{K}, +)$. Consider the commutative unipotent affine algebraic group $\mathbb{G}_a^n$. In other words, $\mathbb{G}_a^n$ is the additive group of an n-dimensional vector space over $\mathbb{K}$. The first aim of this paper is to survey recent results on actions of $\mathbb{G}_a^n$ with an open orbit on projective algebraic varieties. To this end we include a detailed proof of the Hassett–Tschinkel correspondence, discuss its corollaries, interpretations, and related examples. Also we develop the method of Hassett and Tschinkel to show that the generically transitive actions of the group $\mathbb{G}_a^n$ on projective hypersurfaces correspond to invariant multilinear symmetric forms on finite-dimensional local algebras. This leads to explicit classification results for non-degenerate quadrics and quadrics of corank one.

By an additive action on a variety X we mean a faithful regular action of the group $\mathbb{G}_a^n$ on X such that one of the orbits is open in X. The study of such actions was initiated by Brendan Hassett and Yuri Tschinkel [11]. They showed that additive actions on the projective space $\mathbb{P}^n$ up to equivalence are in bijection with isomorphism classes of local algebras of $\mathbb{K}$-dimension $n + 1$. In particular, the number of additive actions on $\mathbb{P}^n$ is finite if and only if $n \leq 5$.

Additive actions on projective subvarieties $X \subseteq \mathbb{P}^m$ induced by an action $\mathbb{G}_a^n \times \mathbb{P}^m \to \mathbb{P}^m$ can be described in terms of local $(m+1)$-dimensional algebras equipped with some additional data. This approach was used in [2, 15] to classify additive actions on projective quadrics. Elena Sharoiko proved in [15, Theorem 4] that an additive action on a non-degenerate quadric $Q \subseteq \mathbb{P}^{n+1}$ is unique up to equivalence. Recently Baohua Fu and Jun-Muk Hwang [9] used a different technique to show the uniqueness of additive action on a class of Fano varieties with Picard number 1. This result shows that an abundance of additive actions on the projective space should be considered as an exception.

A variety with a given additive action looks like an "additive analogue" of a toric variety. Unfortunately, it turns out that two theories have almost no parallels, see [2, 11].

Generalized flag varieties G/P of a semisimple algebraic group G admitting an additive action are classified in [1]. Roughly speaking, an additive action on G/P exists if and only if the unipotent radical P^u of the parabolic subgroup P is commutative. The uniqueness result in this case follows from [9]. In particular, it covers the case of Grassmannians and thus answers a question posed in [2]. Another proof of the uniqueness of additive actions on generalized flag varieties is obtained by Rostislav Devyatov [7]. It uses nilpotent multiplications on certain finite-dimensional modules over semisimple Lie algebras.

Evgeny Feigin proposed a construction based on the PBW-filtration to degenerate an arbitrary generalized flag variety G/P to a variety with an additive action, see [8] and further publications.

In [5], Ulrich Derenthal and Daniel Loughran classified singular del Pezzo surfaces with additive actions; see also [6]. By the results of [4], Manin's Conjecture is true for such surfaces.

In this paper we prove that additive actions on projective hypersurfaces of degree d in $\mathbb{P}^{n+1}$ are in bijection with invariant d-linear symmetric forms on $(n+2)$-dimensional local algebras. The corresponding form is the polarization of the equation defining the hypersurface. As an application, we give a short proof of uniqueness of additive action on non-degenerate quadrics and classify additive actions on quadrics of corank one. The case of cubic projective hypersurfaces is studied in the recent preprint of Ivan Bazhov [3].

The paper is organized as follows. In Sect. 2 we define additive actions and consider the problem of extension of an action $\mathbb{G}_a^n \times X \to X$ on a projective hypersurface X to the ambient space $\mathbb{P}^{n+1}$. The Hassett–Tschinkel correspondence is discussed in Sect. 3. Section 4 is devoted to invariant multilinear symmetric forms on local algebras. Our main result (Theorem 2) describes additive actions on projective hypersurfaces in these terms. Also we give an explicit formula for an invariant multilinear symmetric form (Lemma 1) and prove that if a hypersurface X in $\mathbb{P}^{n+1}$ admits an additive action and the group $\mathrm{Aut}(X)^0$ is reductive, then X is either a hyperplane or a non-degenerate quadric (Proposition 5). Additive actions on non-degenerate quadrics and on quadrics of corank one are classified in Sect. 5 and Sect. 6, respectively.

2 Additive Actions on Projective Varieties

Let X be an irreducible algebraic variety of dimension n and $\mathbb{G}_a^n$ be the commutative unipotent group.

Definition 1. An *inner additive action* on X is an effective action $\mathbb{G}_a^n \times X \to X$ with an open orbit.

It is well known that for an action of a unipotent group on an affine variety all orbits are closed, see, e.g., [14, Sect. 1.3]. It implies that if an affine variety X admits an additive action, then X is isomorphic to the group $\mathbb{G}_a^n$ with the $\mathbb{G}_a^n$-action by left translations.

In general, the existence of an inner additive action implies that the variety X is rational. For X normal, the divisor class group $\mathrm{Cl}(X)$ is freely generated by prime divisors in the complement of the open $\mathbb{G}_a^n$-orbit. In particular, $\mathrm{Cl}(X)$ is a free finitely generated abelian group.

The most interesting case is the study of inner additive actions on complete varieties X. In this case an inner additive action determines a maximal commutative unipotent subgroup of the linear algebraic group $\mathrm{Aut}(X)^0$. Two inner additive actions are said to be *equivalent*, if the corresponding subgroups are conjugate in $\mathrm{Aut}(X)$.

Proposition 1. *Let X be a complete variety with an inner additive action. Assume that the group* $\mathrm{Aut}(X)^0$ *is reductive. Then X is a generalized flag variety G/P, where G is a linear semisimple group and P is a parabolic subgroup.*

Proof. Let X' be the normalization of X. Then the action of $\mathrm{Aut}(X)^0$ lifts to X'. By the assumption, some unipotent subgroup of $\mathrm{Aut}(X)^0$ acts on X' with an open orbit. Then a maximal unipotent subgroup of the reductive group $\mathrm{Aut}(X)^0$ acts on X' with an open orbit. It means that X' is a spherical variety of rank zero, see [16, Sect. 1.5.1] for details. It yields that X' is a generalized flag variety G/P, see [16, Proposition 10.1], and $\mathrm{Aut}(X)^0$ acts on X' transitively. The last condition implies that $X = X'$. □

A classification of generalized flag varieties admitting an inner additive action is obtained in [1]. In particular, the parabolic subgroup P is maximal in this case.

Definition 2. Let X be a closed subvariety of dimension n in the projective space $\mathbb{P}^m$. Then an *additive action* on X is an effective action $\mathbb{G}_a^n \times \mathbb{P}^m \to \mathbb{P}^m$ such that X is $\mathbb{G}_a^n$-invariant and the induced action $\mathbb{G}_a^n \times X \to X$ has an open orbit. Two additive actions on X are said to be *equivalent* if one is obtained from another via automorphism of $\mathbb{P}^m$ preserving X.

Clearly, any additive action on a projective subvariety X induces an inner additive action on X. The converse is not true, i.e., not any action $\mathbb{G}_a^n \times X \to X$ with an open orbit on a projective subvariety X can be extended to the ambient space $\mathbb{P}^m$.

Example 1. Consider a subvariety

$$X = V(x^2z - y^3) \subseteq \mathbb{P}^2$$

and a rational $\mathbb{G}_a$-action on X given by

$$\left(\frac{y}{x}, a\right) \mapsto \frac{y}{x} + a.$$

Using affine charts one can check that this action is regular. On the other hand, it cannot be extended to $\mathbb{P}^2$, because the closure of a $\mathbb{G}_a$-orbit on $\mathbb{P}^2$ cannot be a cubic, see Example 2.

At the same time, if the subvariety X is linearly normal in $\mathbb{P}^m$ and X is normal, then an extension of a $\mathbb{G}_a^n$-action to $\mathbb{P}^m$ exists. Indeed, the restriction $L =: \mathcal{O}(1)|_X$ of the line bundle $\mathcal{O}(1)$ on $\mathbb{P}^m$ can be linearized with respect to the action $\mathbb{G}_a^n \times X \to X$, see, e.g., [12]. The linearization defines a structure of a rational $\mathbb{G}_a^n$-module on the space of sections $H^0(X, L)$. Since X is linearly normal, the restriction $H^0(\mathbb{P}^m, \mathcal{O}(1)) \to H^0(X, L)$ is surjective. Consider a vector space decomposition $H^0(\mathbb{P}^m, \mathcal{O}(1)) = V_1 \oplus V_2$, where V_1 is the kernel of the restriction. The complementary subspace V_2 projects to $H^0(X, L)$ isomorphically. This isomorphism induces a structure of a rational $\mathbb{G}_a^n$-module on V_2. Further, we regard V_1 as the trivial $\mathbb{G}_a^n$-module. This gives a structure of a rational $\mathbb{G}_a^n$-module on

$H^0(\mathbb{P}^m, \mathcal{O}(1))$. Since $\mathbb{P}^m$ is the projectivization of the dual space to $H^0(\mathbb{P}^m, \mathcal{O}(1))$, we obtain a required extended action $\mathbb{G}_a^n \times \mathbb{P}^m \to \mathbb{P}^m$.

From now on we consider additive actions on projective subvarieties $X \subseteq \mathbb{P}^m$.

3 The Hassett–Tschinkel Correspondence

In [11] Brendan Hassett and Yuri Tschinkel established a remarkable correspondence between additive actions on the projective space $\mathbb{P}^n$ and local algebras of $\mathbb{K}$-dimension $n + 1$. Moreover, they described rational cyclic $\mathbb{G}_a^n$-modules in terms of local algebras. In this section we recall these results. The proofs given here are taken from [2]. By a local algebra we always mean a commutative associative local algebra with unit.

Let $\rho : \mathbb{G}_a^n \to \mathrm{GL}_{m+1}(\mathbb{K})$ be a faithful rational representation. The differential defines a representation $d\rho : \mathfrak{g} \to \mathfrak{gl}_{m+1}(\mathbb{K})$ of the tangent algebra $\mathfrak{g} = \mathrm{Lie}(\mathbb{G}_a^n)$ and the induced representation $\tau : U(\mathfrak{g}) \to \mathrm{Mat}_{m+1}(\mathbb{K})$ of the universal enveloping algebra $U(\mathfrak{g})$. Since the group $\mathbb{G}_a^n$ is commutative, the algebra $U(\mathfrak{g})$ is isomorphic to the polynomial algebra $\mathbb{K}[x_1, \dots, x_n]$, where $\mathfrak{g}$ is identified with the subspace $\langle x_1, \dots, x_n \rangle$. The algebra $R := \tau(U(\mathfrak{g}))$ is isomorphic to the factor algebra $U(\mathfrak{g})/\operatorname{Ker}\tau$. As $\tau(x_1), \dots, \tau(x_n)$ are commuting nilpotent operators, the algebra R is finite-dimensional and local. Let us denote by $X_1, \dots, X_n$ the images of the elements $x_1, \dots, x_n$ in R. Then the maximal ideal of R is $\mathfrak{m} := (X_1, \dots, X_n)$. The subspace $W := \tau(\mathfrak{g}) = \langle X_1, \dots, X_n \rangle$ generates R as an algebra with unit.

Assume that $\mathbb{K}^{m+1}$ is a cyclic $\mathbb{G}_a^n$-module with a cyclic vector v, i.e., $\langle \rho(\mathbb{G}_a^n)v \rangle = \mathbb{K}^{m+1}$. The subspace $\tau(U(\mathfrak{g}))v$ is $\mathfrak{g}$- and $\mathbb{G}_a^n$-invariant; it contains the vector v and therefore coincides with the space $\mathbb{K}^{m+1}$. Let $I = \{y \in U(\mathfrak{g}) : \tau(y)v = 0\}$. Since the vector v is cyclic, the ideal I coincides with $\operatorname{Ker}\tau$, and we obtain identifications

$$R \cong U(\mathfrak{g})/I \cong \tau(U(\mathfrak{g}))v = \mathbb{K}^{m+1}.$$

Under these identifications the action of an element $\tau(y)$ on $\mathbb{K}^{m+1}$ corresponds to the operator of multiplication by $\tau(y)$ on the factor algebra R, and the vector $v \in \mathbb{K}^{m+1}$ goes to the residue class of unit. Since $\mathbb{G}_a^n = \exp(\mathfrak{g})$, the $\mathbb{G}_a^n$-action on $\mathbb{K}^{m+1}$ corresponds to the multiplication by elements of $\exp(W)$ on R.

Conversely, let R be a local $(m + 1)$-dimensional algebra with a maximal ideal $\mathfrak{m}$, and $W \subseteq \mathfrak{m}$ be a subspace that generates R as an algebra with unit. Fix a basis $X_1, \dots, X_n$ in W. Then R admits a presentation $\mathbb{K}[x_1, \dots, x_n]/I$, where I is the kernel of the homomorphism

$$\mathbb{K}[x_1, \dots, x_n] \to R,\ x_i \mapsto X_i.$$

These data define a faithful representation ρ of the group $\mathbb{G}_a^n := \exp(W)$ on the space R: the operator $\rho((a_1, \dots, a_n))$ acts as multiplication by the element $\exp(a_1 X_1 + \dots + a_n X_n)$. Since W generates R as an algebra with unit, one checks that the representation is cyclic with unit in R as a cyclic vector.

Summarizing, we obtain the following result.

Theorem 1 ([11, Theorem 2.14]). *The correspondence described above establishes a bijection between*

(1) *equivalence classes of faithful cyclic rational representations*

$$\rho : \mathbb{G}_a^n \to \underset{m+1}{\mathrm{GL}}(\mathbb{K});$$

(2) *isomorphism classes of pairs (R, W), where R is a local $(m+1)$-dimensional algebra with the maximal ideal $\mathfrak{m}$ and W is an n-dimensional subspace of $\mathfrak{m}$ that generates R as an algebra with unit.*

Remark 1. Let $\rho : \mathbb{G}_a^n \to \mathrm{GL}_{m+1}(\mathbb{K})$ be a faithful cyclic rational representation. The set of cyclic vectors in $\mathbb{K}^{m+1}$ is an open orbit of a commutative algebraic group C with $\rho(\mathbb{G}_a^n) \subseteq C \subseteq \mathrm{GL}_{m+1}(\mathbb{K})$, and the complement of this set is a hyperplane. In our notation, the group C is the extension of the commutative unipotent group $\exp(\mathfrak{m}) \cong \mathbb{G}_a^m$ by scalar matrices.

A faithful linear representation $\rho : \mathbb{G}_a^n \to \mathrm{GL}_{m+1}(\mathbb{K})$ determines an effective action of the group $\mathbb{G}_a^n$ on the projectivization $\mathbb{P}^m$ of the space $\mathbb{K}^{m+1}$. Conversely, let G be a connected affine algebraic group with the trivial Picard group, and X be a normal G-variety. By [12, Sect. 2.4], every line bundle on X admits a G-linearization. Moreover, if G has no non-trivial characters, then a G-linearization is unique. This shows that every effective $\mathbb{G}_a^n$-action on $\mathbb{P}^m$ comes from a (unique) faithful rational $(m+1)$-dimensional $\mathbb{G}_a^n$-module.

An effective $\mathbb{G}_a^n$-action on $\mathbb{P}^m$ has an open orbit if and only if $n = m$. In this case the corresponding $\mathbb{G}_a^n$-module is cyclic. In terms of Theorem 1 the condition $n = m$ means $W = \mathfrak{m}$, and we obtain the following theorem.

Proposition 2 ([11, Proposition 2.15]). *There is a one-to-one correspondence between*

(1) *equivalence classes of additive actions on $\mathbb{P}^n$;*
(2) *isomorphism classes of local $(n+1)$-dimensional algebras.*

Remark 2. It follows from Remark 1 that if the group $\mathbb{G}_a^n$ acts on $\mathbb{P}^m$ and some orbit is not contained in a hyperplane, then the action can be extended to an additive action $\mathbb{G}_a^m \times \mathbb{P}^m \to \mathbb{P}^m$. It seems that such an extension exists without any extra assumption.

Given the projectivization $\mathbb{P}^m$ of a faithful rational $\mathbb{G}_a^n$-module and a point $x \in \mathbb{P}^m$ with the trivial stabilizer, the closure X of the orbit $\mathbb{G}_a^n \cdot x$ is a projective variety equipped with an additive action. Closures of generic orbits are hypersurfaces if and only if $n = m - 1$. If such a hypersurface is not a hyperplane, then $\mathbb{P}^m$ comes from the projectivization of a cyclic $\mathbb{G}_a^n$-module, it is given by a pair (R, W), and the condition $n = m - 1$ means that W is a hyperplane in $\mathfrak{m}$. We obtain the following result.

Proposition 3. *There is a one-to-one correspondence between*

(1) *equivalence classes of additive actions on hypersurfaces in $\mathbb{P}^{n+1}$ of degree at least two;*
(2) *isomorphism classes of pairs (R, W), where R is a local $(n+2)$-dimensional algebra with the maximal ideal $\mathfrak{m}$ and W is a hyperplane in $\mathfrak{m}$ that generates R as an algebra with unit.*

It is shown in [2, Theorem 5.1] that the degree of the hypersurface corresponding to a pair (R, W) is the maximal exponent d such that the subspace W does not contain the ideal $\mathfrak{m}^d$.

Example 2. There exist two 3-dimensional local algebras,

$$\mathbb{K}[x]/(x^3) \quad \text{and} \quad \mathbb{K}[x,y]/(x^2, xy, y^2).$$

In the first case $\mathfrak{m}^3 = 0$, and in the second one we have $\mathfrak{m}^2 = 0$. This shows that for every $\mathbb{G}_a$-action on $\mathbb{P}^2$ the orbit closures are either lines or quadrics.

4 Invariant Multilinear Forms on Local Algebras

Consider a pair (R, W) as in Proposition 3 and let $H \subseteq \mathbb{P}^{n+1}$ be the corresponding hypersurface. Let us fix a coordinate system on $R = \langle 1 \rangle \oplus \mathfrak{m}$ such that x_0 is the coordinate along $\langle 1 \rangle$ and $x_1, \ldots, x_{n+1}$ are coordinates on $\mathfrak{m}$.

Assume that H is defined by a homogeneous equation

$$f(x_0, x_1, \ldots, x_{n+1}) = 0$$

of degree d. Since H is invariant under the action of $\mathbb{G}_a^n$, the polynomial f is $\mathbb{G}_a^n$-semi-invariant [14, Theorem 3.1]. But the group $\mathbb{G}_a^n$ has no non-trivial characters, and the polynomial f is $\mathbb{G}_a^n$-invariant. Equivalently, f is annihilated by the Lie algebra $\mathfrak{g}$.

It is well known that for a given homogeneous polynomial f of degree d on a vector space R there exists a unique d-linear symmetric map

$$F\colon R \times R \times \cdots \times R \to \mathbb{K}$$

such that $f(v) = F(v, v, \ldots, v)$ for all $v \in R$, see, e.g., [14, Sect. 9.1]. The map F is called the *polarization* of the polynomial f.

Since the representation $d\rho$ of the Lie algebra $\mathfrak{g}$ on R is given by multiplication by elements of W, a homogeneous polynomial f on R is annihilated by $\mathfrak{g}$ if and only if

$$F(ab_1, b_2, \ldots, b_d) + F(b_1, ab_2, \ldots, b_d) + \cdots + F(b_1, b_2, \ldots, ab_d) = 0 \ \ \forall\, a \in W, \quad b_1, \ldots, b_d \in R. \tag{1}$$

Definition 3. Let R be a local algebra with the maximal ideal $\mathfrak{m}$. An *invariant d-linear form* on R is a d-linear symmetric map

$$F\colon R \times R \times \cdots \times R \to \mathbb{K}$$

such that $F(1,1,\dots,1) = 0$, the restriction of F to $\mathfrak{m}$ is nonzero, and there exists a hyperplane W in $\mathfrak{m}$ which generates R as an algebra with unit and such that condition (1) holds.

If F_1 (resp. F_2) are invariant d_1-linear (resp. d_2-linear) forms on R with respect to the same hyperplane W, then the product F_1F_2 defines an invariant $(d_1 + d_2)$-linear form. An invariant multilinear form is said to be *irreducible*, if it cannot be represented as such a product.

One can show that there is no invariant linear form. It implies that any invariant bilinear or 3-linear form is irreducible.

We are ready to formulate our main result.

Theorem 2. *Additive actions on hypersurfaces of degree $d \geq 2$ in $\mathbb{P}^{n+1}$ are in natural one-to-one correspondence with pairs (R, F), where R is a local algebra of dimension $n + 2$ and F is an irreducible invariant d-linear form on R up to a scalar.*

Proof. An additive action on a hypersurface H in $\mathbb{P}^{n+1}$ is given by a faithful rational representation $\rho : \mathbb{G}_a^n \to \mathrm{GL}_{n+2}(\mathbb{K})$ making $\mathbb{K}^{n+2}$ a cyclic $\mathbb{G}_a^n$-module. In our correspondence we identify $\mathbb{K}^{n+2}$ with the local algebra R. We choose coordinates $x_0, x_1, \dots, x_{n+1}$ compatible with the decomposition $R = \langle 1 \rangle \oplus \mathfrak{m}$. Let $f(x_0, x_1, \dots, x_{n+1}) = 0$ be the equation of the hypersurface H, where f is irreducible. Then the algebra of $\mathbb{G}_a^n$-invariants on $\mathbb{K}^{n+2}$ is freely generated by x_0 and f.

Every $\mathbb{G}_a^n$-invariant hypersurface in $\mathbb{P}^{n+1}$ is given by

$$\alpha f(x_0, x_1, \dots, x_{n+1}) + \beta x_0^d = 0, \quad (\alpha, \beta) \in \mathbb{K}^2 \setminus \{(0,0)\}.$$

So we may assume that f does not contain the term x_0^d. Let F be the polarization of f. Then condition (1) holds and $F(1,\dots,1) = 0$. If the restriction of F to $\mathfrak{m}$ is zero, then x_0 divides f, and f is not irreducible, a contradiction.

Conversely, let (R, F) be such a pair and W be a subspace from the definition of F. Then (R, W) gives rise to a structure of a rational $\mathbb{G}_a^n$-module on R. Consider the hypersurface $f = 0$ in $\mathbb{P}(R) \cong \mathbb{P}^{n+1}$, where $f(v) = F(v, v, \dots, v)$. It is invariant under the action of $\mathbb{G}_a^n$. By the assumptions, f is irreducible and thus the hypersurface $f = 0$ coincides with the closure of a generic $\mathbb{G}_a^n$-orbit. $\square$

Given a hyperplane W in the maximal ideal $\mathfrak{m}$ of a local algebra R that generates R as an algebra with unit, let d be the maximal exponent such that the subspace W does not contain the ideal $\mathfrak{m}^d$. By Theorem 2 and [2, Theorem 5.1], there exists a unique up to a scalar irreducible invariant (with respect to W) d-linear form F_W on R. Let us write down this form explicitly.

By linearity, we may assume that each argument of F_W is either the unit 1 or an element of $\mathfrak{m}$. Fix an isomorphism $\mathfrak{m}/W \cong \mathbb{K}$ and consider the projection $\pi\colon \mathfrak{m} \to \mathfrak{m}/W \cong \mathbb{K}$. We define the form

$$F_W(b_1,\ldots,b_d) := (-1)^k k!(d-k-1)!\,\pi(b_1\ldots b_d),$$

where k is the number of units among $b_1,\ldots,b_d$, and for $k = d$ we let $F_W(1,\ldots,1) = 0$.

Lemma 1. *The form F_W is an irreducible invariant d-linear form on R.*

Proof. We begin with condition (1). Since $ab_i \in \mathfrak{m}$ for all $a \in W$ and $b_i \in R$, with $0 < k < d$ this condition can be rewritten as

$$(k(-1)^{k-1}(k-1)!(d-k)! + (d-k)(-1)^k k!(d-k-1)!)\,\pi(ab_1\ldots b_d) = 0,$$

and it is obvious. For $k = 0$ we have $ab_1\ldots b_d \in \mathfrak{m}^{d+1} \subseteq W$, and thus $\pi(ab_1\ldots b_d) = 0$. For $k = d$ we have $-(d-2)!\pi(a) = 0$, because $a \in W$.

The restriction of F to $\mathfrak{m}$ is nonzero since $\mathfrak{m}^d$ is not contained in W. It follows from [2, Theorem 5.1] that the form F is irreducible. Finally, we have $F_W(1,\ldots,1) = 0$ by definition. □

The next proposition follows immediately from [13, Theorems 1 and 4]. Let us obtain this result using our technique.

Proposition 4. *Let H be a smooth hypersurface in $\mathbb{P}^{n+1}$ admitting an additive action. Then H is either a hyperplane or a non-degenerate quadric.*

Proof. Assume that an additive action on H is given by a triple (R, W, F). Let $e_0, e_1, \ldots, e_{n+1}$ be a basis of R compatible with the decomposition

$$R = \langle 1\rangle \oplus W \oplus \langle e_{n+1}\rangle.$$

Moreover, we may assume that $e_{n+1} \in \mathfrak{m}^d$, where $\mathfrak{m}^{d+1}$ is contained in W. Then in the notation of Lemma 1 we have $\pi(be_{n+1}) = 0$ for all $b \in \mathfrak{m}$. It means that the variable x_{n+1} can appear in the equation $f(x_0,\ldots,x_{n+1}) = 0$ of the hyperplane H only in the term $x_0^{d-1}x_{n+1}$. Thus the point $[0 : \ldots : 0 : 1]$ lies on H and it is singular provided $d \geq 3$. It remains to note that the only smooth quadric is a non-degenerate one. □

Proposition 5. *Let H be a hypersurface in $\mathbb{P}^{n+1}$ which admits an additive action and such that the group* $\mathrm{Aut}(H)^0$ *is reductive. Then H is either a hyperplane or a non-degenerate quadric.*

Proof. By Proposition 1, the variety H is smooth, and the assertion follows from Proposition 4. □

Remark 3. Take a triple (R, W, F) as in Definition 3 and consider the sum I of all ideals of the algebra R contained in W. It is the biggest ideal of R contained in W. Taking a compatible basis of R, we see that the equation of the corresponding hypersurface does not depend on the coordinates in I. Moreover, for the factor algebra $R' = R/I$ we have

$$R' = \langle 1' \rangle \oplus W' \oplus (\mathfrak{m}')^d$$

with $\mathfrak{m}' = \mathfrak{m}/I$, $W' = W/I$, and $\dim(\mathfrak{m}')^d = 1$. The invariant form F descents to R', the subspace W' contains no ideal of R', and the algebra R' is Gorenstein. Such a reduction is useful in classification problems.

5 Non-degenerate Quadrics

In this section we classify non-degenerate invariant bilinear symmetric forms on local algebras. These results are obtained in [15], but we give a short elementary proof.

Let R be a local algebra of dimension $n + 2$ with the maximal ideal $\mathfrak{m}$ and F a non-degenerate bilinear symmetric form on R such that $F(1, 1) = 0$. Assume that for some hyperplane W in $\mathfrak{m}$ generating R we have

$$F(ab_1, b_2) + F(b_1, ab_2) = 0 \quad \text{for all } b_1, b_2 \in R \text{ and } a \in W. \tag{2}$$

We choose a basis $e_0 = 1, e_1, \ldots, e_n, e_{n+1}$ of R such that $W = \langle e_1, \ldots, e_n \rangle$ and $\mathfrak{m} = \langle e_1, \ldots, e_{n+1} \rangle$. For any $b \in R$ let $b = b^{(0)} + b^{(1)} + \cdots + b^{(n+1)}$ be the decomposition corresponding to this basis.

Lemma 2. *(1) $F(1, a) = 0$ for all $a \in W$;*
(2) $F(1, b) = F(1, b^{(n+1)})$ for all $b \in R$;
(3) If $a, a' \in W$ and $aa' \in W$, then $aa' = 0$;
(4) The restriction of the form F to W is non-degenerate.

Proof. Assertion (1) follows from (2) with $b_1 = b_2 = 1$. For (2), note that

$$F(1, b) = F(1, b^{(0)}) + F(1, b^{(1)} + \cdots + b^{(n)}) + F(1, b^{(n+1)}).$$

The first term is 0 because of $F(1, 1) = 0$, and the second one is 0 by (1). If $a, a', aa' \in W$, then for any $b \in R$ we have

$$F(b, aa') = -F(ab, a') = F(aa'b, 1) = -F(b, aa'),$$

and $F(b, aa') = 0$. Since F is non-degenerate, we obtain (3).

For (4), assume that for some $0 \neq a \in W$ we have $F(a, a') = 0$ for all $a' \in W$. Since F is non-degenerate, it yields $F(a, e_{n+1}) = \lambda$ for some nonzero $\lambda \in \mathbb{K}$. If $F(1, e_{n+1}) = \mu$, then $F(\mu a - \lambda 1, e_{n+1}) = 0$, and the vector $\mu a - \lambda 1$ is in the kernel of the form F. □

Let us denote by $M(F)$ the matrix of a bilinear form F in a given basis.

Proposition 6. *In the notation as above, the triple (R, W, F) can be transformed into the form*

$$R = \mathbb{K}[e_1, \ldots, e_n]/(e_i^2 - e_j^2, e_i e_j;\ 1 \leq i < j \leq n), \qquad W = \langle e_1, \ldots, e_n \rangle,$$

$$M(F) = \begin{pmatrix} 0 & 0 & \ldots & 0 & 1 \\ 0 & -1 & \ldots & 0 & 0 \\ \vdots & \vdots & \ddots & \vdots & \vdots \\ 0 & 0 & \ldots & -1 & 0 \\ 1 & 0 & \ldots & 0 & 0 \end{pmatrix}$$

Proof. Since F is non-degenerate, we may assume that $F(1, e_{n+1}) = 1$. Using Lemma 2, (4), we may suppose that $F(e_i, e_j) = -\delta_{ij}$ for all $1 \leq i, j \leq n$. Now the matrix of the form F looks like

$$\begin{pmatrix} 0 & 0 & \ldots & 0 & 1 \\ 0 & -1 & \ldots & 0 & * \\ \vdots & \vdots & \ddots & \vdots & \vdots \\ 0 & 0 & \ldots & -1 & * \\ 1 & * & \ldots & * & * \end{pmatrix}$$

For all $1 \leq i < j \leq n$ we have $F(1, e_i e_j) = -F(e_i, e_j) = 0$. It follows from Lemma 2, (2) that $(e_i e_j)^{(n+1)} = 0$. We conclude that $e_i e_j \in W$ and thus $e_i e_j = 0$ by Lemma 2, (3).

Since $F(1, e_i^2) = -F(e_i, e_i) = 1$, we have $e_i^2 = e_{n+1} + f_i$, where $f_i \in W$. Then

$$(e_1 + e_i)(e_1 - e_i) = e_1^2 - e_i^2 = f_1 - f_i \in W.$$

By Lemma 2, (3) we obtain $e_1^2 = e_i^2$.

Without loss of generality it can be assumed that $e_{n+1} = e_1^2$. Let $n \geq 2$. Then $e_{n+1} e_i = e_j^2 e_i = 0$, where $1 \leq i \neq j \leq n$. If $n = 1$ then $e_{n+1} e_1 = e_1^3 \in \mathfrak{m}^3 = 0$.

Hence $e_{n+1} b = 0$ for any element $b \in \mathfrak{m}$, and R is isomorphic to $\mathbb{K}[e_1, \ldots, e_n]/(e_i^2 - e_j^2, e_i e_j)$.

It remains to prove that $e_{n+1} \perp_F \mathfrak{m}$. Indeed, we have

$$F(e_{n+1}, b) = F(e_1^2, b) = -F(e_1, e_1 b) = F(1, e_1^2 b) = F(1, 0) = 0 \quad \forall b \in \mathfrak{m}.$$

This completes the proof of the proposition. □

As a corollary we obtain the result of [15, Theorem 4].

Corollary 1. *A non-degenerate quadric $Q_n \subseteq \mathbb{P}^{n+1}$ admits a unique additive action up to equivalence.*

6 Quadrics of Corank One

Let us classify invariant bilinear symmetric forms of rank $n+1$ on local $(n+2)$-dimensional algebras. Geometrically these results can be interpreted as a classification of additive actions on quadrics of corank one in $\mathbb{P}^{n+1}$.

Let R be a local algebra of dimension $n+2$, $n \geq 2$, with the maximal ideal $\mathfrak{m}$ and F a bilinear symmetric form of rank $n+1$ on R such that $F(1,1) = 0$. Assume that for some hyperplane W in $\mathfrak{m}$ condition (2) holds. We choose a basis $e_0 = 1, e_1, \ldots, e_n, e_{n+1}$ of R such that $W = \langle e_1, \ldots, e_n \rangle$ and $\mathfrak{m} = \langle e_1, \ldots, e_{n+1} \rangle$.

Lemma 3. *The kernel* Ker F *is contained in* W.

Proof. Let Ker $F = \langle l \rangle$. Assume that l is not in W. Then we should consider four alternatives.

1. Let $\langle l \rangle = \langle 1 \rangle$. Then $F(a,b) = -F(1,ab) = 0$ for all $a \in W, b \in R$, and dim Ker $F \geq 2$, a contradiction.
2. Let $\langle l \rangle \subseteq \mathfrak{m} \setminus W$. Without loss of generality it can be assumed that $l = e_{n+1}$. As we have seen, $F(1,b) = F(1, b^{(n+1)}) = 0$ for all $b \in R$. Thus we have $1 \in$ Ker F, which leads to a contradiction.
3. Let $\langle l \rangle \subseteq R \setminus (\mathfrak{m} \cup \langle 1, W \rangle)$. Without loss of generality it can be assumed that $l = 1 + e_{n+1}$. We have $0 = F(1,l) = F(1,1) + F(1, e_{n+1}) = F(1, e_{n+1})$. It again follows that $1 \in$ Ker F.
4. Let $\langle l \rangle \subseteq \langle 1, W \rangle \setminus W$. We can assume that $l = 1 + f$, where $W \ni f \neq 0$. Then

$$F(1,b) = -F(f,b) = F(1,fb) = \cdots = F(1, f^{n+3} b) = 0 \quad \forall\, b \in R.$$

Thus we again have $1 \in$ Ker F.

This completes the proof of the lemma. □

Proposition 7. *In the notation as above, the triple (R, W, F) can be transformed into the form*

$$M(F) = \begin{pmatrix} 0 & 0 & \dots & 0 & 0 & 1 \\ 0 & -1 & \dots & 0 & 0 & 0 \\ \vdots & \vdots & \ddots & \vdots & \vdots & \vdots \\ 0 & 0 & \dots & -1 & 0 & 0 \\ 0 & 0 & \dots & 0 & 0 & 0 \\ 1 & 0 & \dots & 0 & 0 & 0 \end{pmatrix}, \quad W = \langle e_1, \dots, e_n \rangle,$$

and R is isomorphic to one of the following algebras:

1. $\mathbb{K}[e_1, \dots, e_n]/(e_i e_j - \lambda_{ij} e_n, e_i^2 - e_j^2 - (\lambda_{ii} - \lambda_{jj}) e_n, e_s e_n, 1 \le i < j \le n-1, 1 \le s \le n, n \ge 3)$ *where* λ_{ij} *are elements of a symmetric block diagonal* $(n-1) \times (n-1)$*-matrix* Λ *such that each block* Λ_k *is*

$$\lambda_k \begin{pmatrix} 1 & 0 & & 0 \\ 0 & \ddots & \ddots & \\ & \ddots & \ddots & 0 \\ 0 & & 0 & 1 \end{pmatrix} + \frac{1}{2} \begin{pmatrix} 0 & 1 & & 0 \\ 1 & \ddots & \ddots & \\ & \ddots & \ddots & 1 \\ 0 & & 1 & 0 \end{pmatrix} + \frac{i}{2} \begin{pmatrix} 0 & & 1 & 0 \\ & \ddots & \ddots & -1 \\ 1 & \ddots & \ddots & \\ 0 & -1 & & 0 \end{pmatrix}$$

with some $\lambda_k \in \mathbb{K}$;

2. $\mathbb{K}[e_1, e_2]/(e_1^3, e_1 e_2, e_2^2)$ *or* $\mathbb{K}[e_1]/(e_1^4)$ *with* $e_2 = e_1^3, e_3 = e_1^2$.

Remark 4. Blocks Λ_k of size 1 are (λ_k). Blocks Λ_k of size 2 are

$$\begin{pmatrix} \lambda_k + \frac{i}{2} & \frac{1}{2} \\ \frac{1}{2} & \lambda_k - \frac{i}{2} \end{pmatrix}$$

Proof of Proposition 7. By Lemma 3 we may assume that $\operatorname{Ker} F = \langle e_n \rangle$ and $F(1, e_{n+1}) = 1$, because of $F(1, a) = 0$ for all $a \in W$. Let V be the linear span $\langle e_1, \dots, e_{n-1} \rangle$. As in Lemma 2, (4) one can show that the restriction of F to V is non-degenerate. Thus we can assume that the matrix of F has the form

$$\begin{pmatrix} 0 & 0 & \dots & 0 & 0 & 1 \\ 0 & -1 & \dots & 0 & 0 & * \\ \vdots & \vdots & \ddots & \vdots & \vdots & \vdots \\ 0 & 0 & \dots & -1 & 0 & * \\ 0 & 0 & \dots & 0 & 0 & * \\ 1 & * & \dots & * & * & * \end{pmatrix}.$$

We have $1 = -F(e_i, e_i) = F(1, e_i^2) = F(1, (e_i^2)^{(n+1)}) \Rightarrow e_i^2 = e_{n+1} + f_i$ for some $f_i \in W$ and every $i = 1, \dots, n-1$. We may assume that $e_{n+1} = e_1^2$.

Then $f_i = e_i^2 - e_1^2 = (e_i + e_1)(e_i - e_1)$ and, as in Lemma 2, (3), we obtain $f_i = \lambda_{ii} e_n$.

Again as in Lemma 2, (3), we have $e_i e_j = \lambda_{ij} e_n$ for all $1 \leq i < j \leq n-1$.

Thus multiplication on the subspace V is given by the matrix $I_n e_{n+1} + \Lambda e_n$, where I_n is the identity matrix and a symmetric matrix Λ is defined up to adding a scalar matrix.

It is easy to check that the symmetric matrix $\Lambda = (\lambda_{ij})$ under orthogonal transformations on V transforms as the matrix of a bilinear symmetric form. It follows from [10, Chap. 11, Sect. 3] that Λ can be transformed into the canonical block diagonal form by orthogonal transformation. Here each block Λ_k has the form

$$\lambda_k \begin{pmatrix} 1 & 0 & & 0 \\ 0 & \ddots & \ddots & \\ & \ddots & \ddots & 0 \\ 0 & & 0 & 1 \end{pmatrix} + \frac{1}{2} \begin{pmatrix} 0 & 1 & & 0 \\ 1 & \ddots & \ddots & \\ & \ddots & \ddots & 1 \\ 0 & & 1 & 0 \end{pmatrix} + \frac{i}{2} \begin{pmatrix} 0 & & 1 & 0 \\ & \ddots & \ddots & -1 \\ 1 & \ddots & \ddots & \\ 0 & -1 & & 0 \end{pmatrix}, \quad \lambda_k \in \mathbb{K}.$$

We claim that $e_n \mathfrak{m} = 0$. Indeed, $F(ae_n, b) = -F(e_n, ab) = 0$ for all $a \in W$ and $b \in R$, and thus $ae_n = \alpha e_n$ for some $\alpha \in \mathbb{K}$. But a is nilpotent, and $\alpha = 0$. Finally, we have $e_n e_{n+1} = e_n e_1^2 = 0$.

Further,

$$\begin{aligned} F(e_{n+1}a, 1) &= -F(e_{n+1}, a) = -F(e_1^2, a) = -F(1, e_1^2 a) \\ &= -F(1, e_{n+1}a) \Rightarrow F(e_{n+1}, a) = 0 \end{aligned} \tag{3}$$

for all $a \in W$.

1. Let $n \geq 3$. We claim that $e_{n+1}\mathfrak{m} = 0$. Indeed, for $1 \leq i \neq j \leq n-1$ we have

$$e_{n+1} e_i = (e_j^2 - \lambda_{jj} e_n) e_i = \lambda_{ij} e_j e_n - \lambda_{jj} e_n e_i = 0.$$

In this case the algebra R is isomorphic to

$$\begin{aligned} &\mathbb{K}[e_1, \ldots, e_n]/(e_i e_j - \lambda_{ij} e_n, e_i^2 - e_j^2 - (\lambda_{ii} - \lambda_{jj}) e_n, e_s e_n, \\ &\quad 1 \leq i < j \leq n-1, 1 \leq s \leq n). \end{aligned}$$

2. Let $n = 2$. We have $e_3^2 = e_1^4 \in \mathfrak{m}^4 = 0 \Rightarrow e_3^2 = 0$. Since $F(e_1 e_3, 1) = -F(e_3, e_1) = 0$, it follows that $e_1 e_3 \in W$. Thus $e_1 e_3 = \alpha e_1 + \beta e_2$ and we have

$$0 = e_1^4 = (e_1 e_3) e_1 = \alpha e_1^2 + \beta e_1 e_2 = \alpha e_3 \Rightarrow \alpha = 0.$$

If $\beta = 0$, then $R \cong \mathbb{K}[e_1, e_2]/(e_1^3, e_1 e_2, e_2^2)$. If $\beta \neq 0$, then we may assume that $\beta = 1$, and $R \simeq \mathbb{K}[e_1]/(e_1^4)$ with $e_2 = e_1^3, e_3 = e_1^2$.

In all cases $e_{n+1}^2 = e_1^2 e_{n+1} = 0$, and it follows that $F(e_{n+1}, e_{n+1}) = F(1, e_{n+1}^2) = 0$. Combining this with (3), we obtain

$$M(F) = \begin{pmatrix} 0 & 0 & \dots & 0 & 0 & 1 \\ 0 & -1 & \dots & 0 & 0 & 0 \\ \vdots & \vdots & \ddots & \vdots & \vdots & \vdots \\ 0 & 0 & \dots & -1 & 0 & 0 \\ 0 & 0 & \dots & 0 & 0 & 0 \\ 1 & 0 & \dots & 0 & 0 & 0 \end{pmatrix}.$$

Proposition 7 is proved. □

Remark 5. The normal form of a symmetric matrix Λ is unique up to permutation of blocks. Indeed, we conjugate the matrix Λ by the symmetric block diagonal matrix T such that each block T_k is

$$\frac{1}{2}\begin{pmatrix} 1 & 0 & \dots & 0 & i \\ 0 & 1 & & i & 0 \\ \vdots & & \ddots & & \vdots \\ 0 & i & & 1 & 0 \\ i & 0 & \dots & 0 & 1 \end{pmatrix},$$

and obtain the Jordan normal form of Λ with the same block sizes and the same eigenvalues.

We claim that the matrix Λ defining a triple (R, W, F) is unique up to permutation of blocks, scalar multiplication, and adding a scalar matrix. To see this, let two matrices Λ, Λ' define the same triple (R, W, F). Notice that adding a scalar matrix to Λ we do not change the defining relations of R. Denote by ϕ an automorphism of R such that $W = \phi(W)$ and

$$F = \phi^{-1T} F \phi^{-1}.$$

It yields $\operatorname{Ker} F = \phi(\operatorname{Ker} F)$ and $\phi(e_n) = \alpha e_n$. Multiplying the matrix Λ' by α^{-1} we obtain $\phi(e_n) = e_n$. Moreover, ϕ induces on $W/\operatorname{Ker} F$ an orthogonal transformation, and thus two canonical forms of the matrix Λ can differ only by the order of blocks.

Example 3. Two cases in Proposition 7, (2), correspond to two non-equivalent actions of $\mathbb{G}_a^2$ on the quadric $2x_0x_3 - x_1^2 = 0$ in $\mathbb{P}^3$, namely,

$$(a_1, a_2) \cdot [x_0 : x_1 : x_2 : x_3] = \left[x_0 : x_1 + a_1x_0 : x_2 + a_2x_0 : x_3 + \frac{a_1^2}{2}x_0 + a_1x_1\right]$$

and

$$(a_1, a_2) \cdot [x_0 : x_1 : x_2 : x_3]$$
$$= \left[x_0 : x_1 + a_1 x_0 : x_2 + \left(a_2 + \frac{a_1^3}{6} \right) x_0 + \frac{a_1^2}{2} x_1 + a_1 x_3 : x_3 + \frac{a_1^2}{2} x_0 + a_1 x_1 \right].$$

For the first action there is a line of fixed points, while the second one has three orbits.

Example 4. Let $n = 3$. If the matrix Λ is diagonal, then up to scalar addition and multiplication we have $\Lambda = \begin{pmatrix} 0 & 0 \\ 0 & 0 \end{pmatrix}$ or $\Lambda = \begin{pmatrix} 0 & 0 \\ 0 & 1 \end{pmatrix}$. With non-diagonal Λ we have $\begin{pmatrix} i/2 & 1/2 \\ 1/2 & -i/2 \end{pmatrix}$. So there are three equivalence classes of additive actions in this case, and they can be easily written down explicitly.

Example 5. Consider the case $n = 4$. We have six types of the matrix Λ with one depending on a parameter. Namely, in the diagonal matrix $\Lambda = \mathrm{diag}(0, 1, t)$, where $t \in \mathbb{K} \setminus \{0, 1\}$, the parameter t is defined up to transformations $\{t, \frac{1}{t}, 1 - t, \frac{t-1}{t}, \frac{t}{t-1}, \frac{1}{1-t}\}$. Therefore, the parameters t and t' determine equivalent actions if and only if

$$\frac{(t^2 - t + 1)^3}{t^2(1 - t)^2} = \frac{(t'^2 - t' + 1)^3}{t'^2(1 - t')^2}.$$

The action of $\mathbb{G}_a^4$ on the quadric $2x_0x_5 - x_1^2 - x_2^2 - x_3^2 = 0$ in this case has the form

$$(a_1, a_2, a_3, a_4) \cdot [x_0 : x_1 : x_2 : x_3 : x_4 : x_5]$$
$$= \left[x_0 : x_1 + a_1 x_0 : x_2 + a_2 x_0 : x_3 + a_3 x_0 : x_4 + \frac{2a_4 + a_2^2 + t a_3^2}{2} x_0 \right.$$
$$\left. + a_2 x_2 + t a_3 x_3 : x_5 + \frac{a_1^2 + a_2^2 + a_3^2}{2} x_0 + a_1 x_1 + a_2 x_2 + a_3 x_3 \right].$$

This agrees with the results of [2, Sect. 4].

Acknowledgements The authors are grateful to Ernest B. Vinberg for useful consultations. Special thanks are due to the referees and the editors for careful reading, constructive criticism and important suggestions.

Both authors were partially supported by the Ministry of Education and Science of Russian Federation, project 8214. The first author was supported by the Dynasty Foundation, the Simons-IUM Grant, and the RFBR grant 12-01-00704.

References

1. I. Arzhantsev, Flag varieties as equivariant compactifications of $\mathbb{G}_a^n$. Proc. Am. Math. Soc. **139**(3), 783–786 (2011)
2. I. Arzhantsev, E. Sharoyko, Hassett-Tschinkel correspondence: Modality and projective hypersurfaces. J. Algebra **348**(1), 217–232 (2011)
3. I. Bazhov, Additive structures on cubic hypersurfaces, 8 pp. [arXiv:1307.6085]
4. A. Chambert-Loir, Y. Tschinkel, On the distribution of points of bounded height on equivariant compactifications of vector groups. Invent. Math. **148**(2), 421–452 (2002)
5. U. Derenthal, D. Loughran, Singular del Pezzo surfaces that are equivariant compactifications. J. Math. Sci. **171**(6), 714–724 (2010)
6. U. Derenthal, D. Loughran, Equivariant compactifications of two-dimensional algebraic groups, to appear in: Proceedings of the Edinburgh Mathematical Society 19 pp. [arXiv:1212.3518]
7. R. Devyatov, Unipotent commutative group actions on flag varieties and nilpotent multiplications, 25 pp. [arXiv:1309.3480]
8. E. Feigin, $\mathbb{G}_a^M$ degeneration of flag varieties. Selecta Math. New Ser. **18**(3), 513–537 (2012)
9. B. Fu, J.-M. Hwang, Uniqueness of equivariant compactifications of $\mathbb{C}^n$ by a Fano manifold of Picard number 1, 5 pp. [arXiv:1301.5486]
10. F. Gantmacher, *The Theory of Matrices*, Transl. from Russian by K.A. Hirsch (American Mathematical Society, Providence, 2000)
11. B. Hassett, Y. Tschinkel, Geometry of equivariant compactifications of $\mathbb{G}_a^n$. Int. Math. Res. Not. **20**, 1211–1230 (1999)
12. F. Knop, H. Kraft, D. Luna, T. Vust, Local properties of algebraic group actions, in *Algebraische Transformationsgruppen und Invariantentheorie*. DMV Seminar, vol. 13 (Birkhäuser, Basel, 1989), pp. 63–75
13. H. Matsumura, P. Monsky, On the automorphisms of hypersurfaces. J. Math. Kyoto Univ. **3**(3), 347–361 (1963)
14. V. Popov, E. Vinberg, Invariant theory, in *Algebraic Geometry IV*, ed. by A.N. Parshin, I.R. Shafarevich (Springer, Berlin, 1994)
15. E. Sharoiko, Hassett-Tschinkel correspondence and automorphisms of a quadric. Sb. Math. **200**(11), 1715–1729 (2009)
16. D. Timashev, in *Homogeneous Spaces and Equivariant Embeddings*. Encyclopaedia of Mathematical Sciences, vol. 138 (Springer, Berlin, 2011)

Cremona Groups of Real Surfaces

Jérémy Blanc and Frédéric Mangolte

Abstract We give an explicit set of generators for various natural subgroups of the real Cremona group $\mathrm{Bir}_{\mathbb{R}}(\mathbb{P}^2)$. This completes and unifies former results by several authors.

MSC 2000: 14E07, 14P25, 14J26

1 Introduction

1.1 On the Real Cremona Group $\mathrm{Bir}_{\mathbb{R}}(\mathbb{P}^2)$

The classical Noether–Castelnuovo Theorem [3] (see also [1, Chap. 8] for a modern exposition of the proof) gives generators of the group $\mathrm{Bir}_{\mathbb{C}}(\mathbb{P}^2)$ of birational transformations of the complex projective plane. The group is generated by the biregular automorphisms, which form the group $\mathrm{Aut}_{\mathbb{C}}(\mathbb{P}^2) = \mathrm{PGL}(3, \mathbb{C})$ of projectivities, and by the standard quadratic transformation

$$\sigma_0 \colon (x : y : z) \dashrightarrow (yz : xz : xy).$$

J. Blanc (✉)
Mathematisches Institut, Universität Basel, Rheinsprung 21, CH-4051 Basel, Schweiz
e-mail: Jeremy.Blanc@unibas.ch

F. Mangolte
LUNAM Université, LAREMA, Université d'Angers, Bd. Lavoisier, 49045 Angers Cedex 01, France
e-mail: frederic.mangolte@univ-angers.fr

I. Cheltsov et al. (eds.), *Automorphisms in Birational and Affine Geometry*, Springer Proceedings in Mathematics & Statistics 79,
DOI 10.1007/978-3-319-05681-4_3,

This result does not work over the real numbers. Indeed, recall that a *base point* of a birational transformation is a (possibly infinitely near) point of indeterminacy; and note that two of the base points of the quadratic involution

$$\sigma_1\colon (x:y:z) \dashrightarrow (y^2+z^2 : xy : xz)$$

are not real. Thus σ_1 cannot be generated by projectivities and σ_0. More generally, we cannot generate this way maps having non real base-points. Hence the group $\mathrm{Bir}_{\mathbb{R}}(\mathbb{P}^2)$ of birational transformations of the real projective plane is not generated by $\mathrm{Aut}_{\mathbb{R}}(\mathbb{P}^2) = \mathrm{PGL}(3,\mathbb{R})$ and σ_0.

The first result of this note is that $\mathrm{Bir}_{\mathbb{R}}(\mathbb{P}^2)$ is generated by $\mathrm{Aut}_{\mathbb{R}}(\mathbb{P}^2)$, σ_0, σ_1, and a family of birational maps of degree 5 having only non real base-points.

Theorem 1.1. *The group* $\mathrm{Bir}_{\mathbb{R}}(\mathbb{P}^2)$ *is generated by* $\mathrm{Aut}_{\mathbb{R}}(\mathbb{P}^2)$, σ_0, σ_1, *and the* standard quintic *transformations of* $\mathbb{P}^2$ *(defined in Example 3.1).*

The proof of this result follows the so-called Sarkisov program, which amounts to decompose a birational map between Mori fibre spaces as a sequence of simple maps, called *Sarkisov links*. The description of all possible links has been done in [9] for perfect fields, and in [14] for real surfaces. We recall it in Sect. 2 and show how to deduce Theorem 1.1 from the list of Sarkisov links.

Note that a family of generators of $\mathrm{Bir}_{\mathbb{K}}(\mathbb{P}^2)$ is given in [8], for any perfect field $\mathbb{K}$. When taking $\mathbb{K} = \mathbb{R}$, the list is however longer than the one given in Theorem 1.1.

Let X be an algebraic variety defined over $\mathbb{R}$ (always assumed to be geometrically irreducible), we denote as usual by $X(\mathbb{R})$ the set of real points endowed with the induced algebraic structure. The topological space $\mathbb{P}^2(\mathbb{R})$ is then the real projective plane, letting $\mathbb{F}_0 := \mathbb{P}^1 \times \mathbb{P}^1$, the space $\mathbb{F}_0(\mathbb{R})$ is the torus $\mathbb{S}^1 \times \mathbb{S}^1$ and letting $Q_{3,1} = \{(w:x:y:z) \in \mathbb{P}^3 \mid w^2 = x^2+y^2+z^2\}$, the real locus $Q_{3,1}(\mathbb{R})$ is the sphere $\mathbb{S}^2$.

Recall that an *isomorphism* $X(\mathbb{R}) \to Y(\mathbb{R})$ is a birational map $\varphi\colon X \dashrightarrow Y$ defined over $\mathbb{R}$ such that φ is defined at all real points of X and φ^{-1} at all real points of Y. The set of automorphisms of $X(\mathbb{R})$ form a group $\mathrm{Aut}(X(\mathbb{R}))$, and we have natural inclusions

$$\mathrm{Aut}_{\mathbb{R}}(X) \subset \mathrm{Aut}(X(\mathbb{R})) \subset \mathrm{Bir}_{\mathbb{R}}(X).$$

The strategy used to prove Theorem 1.1 allows us to treat similarly the case of natural subgroups of $\mathrm{Bir}_{\mathbb{R}}(\mathbb{P}^2)$, namely the groups $\mathrm{Aut}(\mathbb{P}^2(\mathbb{R}))$, $\mathrm{Aut}(Q_{3,1}(\mathbb{R}))$ and $\mathrm{Aut}(\mathbb{F}_0(\mathbb{R}))$ of three *minimal* real rational surfaces (see Corollary 2.10). This way, we give a unified treatment to prove three theorems on generators, the first two of them already proved in a different way in [11, 15].

Observe that $\mathrm{Aut}(Q_{3,1}(\mathbb{R}))$ and $\mathrm{Aut}(\mathbb{F}_0(\mathbb{R}))$ are not really subgroups of $\mathrm{Bir}_{\mathbb{R}}(\mathbb{P}^2)$, but each of them is isomorphic to a subgroup which is determined up to conjugation. In fact, for any choice of a birational map $\psi\colon \mathbb{P}^2 \dashrightarrow X$ ($X = Q_{3,1}$ or $\mathbb{F}_0$), $\psi^{-1}\,\mathrm{Aut}(X(\mathbb{R}))\psi \subset \mathrm{Bir}_{\mathbb{R}}(\mathbb{P}^2)$.

Theorem 1.2 ([15]). *The group* $\mathrm{Aut}(\mathbb{P}^2(\mathbb{R}))$ *is generated by*

$$\mathrm{Aut}_{\mathbb{R}}(\mathbb{P}^2) = \mathrm{PGL}(3, \mathbb{R})$$

and by standard quintic transformations.

Note that, up to the action of $\mathrm{PGL}(3, \mathbb{R})$, the standard quintic transformations form an algebraic variety of (real) dimension 4. This is in contrast with the complex case, where the set of standard quadratic transformations is $\{\sigma_0\}$, up to the action of $\mathrm{PGL}(3, \mathbb{C})$.

Theorem 1.3 ([11]). *The group* $\mathrm{Aut}(Q_{3,1}(\mathbb{R}))$ *is generated by*

$$\mathrm{Aut}_{\mathbb{R}}(Q_{3,1}) = \mathrm{PO}(3, 1)$$

and by standard cubic transformations.

Here the real dimension of the variety of standard cubic transformations, modulo $\mathrm{PO}(3, 1)$, is 2.

Theorem 1.4. *The group* $\mathrm{Aut}(\mathbb{F}_0(\mathbb{R}))$ *is generated by*

$$\mathrm{Aut}_{\mathbb{R}}(\mathbb{F}_0) \cong \mathrm{PGL}(2, \mathbb{R})^2 \rtimes \mathbb{Z}/2\mathbb{Z}$$

and by the involution

$$\tau_0\colon ((x_0 : x_1), (y_0 : y_1)) \dashrightarrow ((x_0 : x_1), (x_0 y_0 + x_1 y_1 : x_1 y_0 - x_0 y_1)).$$

In each case, we don't know any easy way of computing the relations between the given generators. (See [10] for a description in a more general setting, whose application to the real case does not fit our set of generators.)

The proof of Theorems 1.1, 1.2, 1.3, 1.4 is given in Sects. 4, 3, 5, 6, respectively. Section 7 is devoted to present some related recent results on birational geometry of real projective surfaces.

In the sequel, surfaces and maps are assumed to be real. In particular if we consider that a real surface is a complex surface endowed with a Galois-action of $G := \mathrm{Gal}(\mathbb{C}|\mathbb{R})$, a map is G-equivariant. On the contrary, points and curves are not assumed to be real a priori.

We would like to thank the referee whose remarks helped us to improve the exposition and Igor Dolgachev for indicating us references.

2 Mori Theory for Real Rational Surfaces and Sarkisov Program

We work with the tools of Mori theory. A good reference in dimension 2, over any perfect field, is [9]. The theory, applied to smooth projective real rational surfaces, becomes really simple. The description of Sarkisov links between real rational surfaces has been done in [14], together with a study of relations between these links. In order to state this classification, we first recall the following classical definitions (which can be found in [9]).

Definition 2.1. A smooth projective real rational surface X is said to be *minimal* if any birational morphism $X \to Y$, where Y is another smooth projective real surface, is an isomorphism.

Definition 2.2. A *Mori fibration* is a morphism $\pi: X \to W$ where X is a smooth projective real rational surface and one of the following occurs:

(1) $\rho(X) = 1$, W is a point (usually denoted $\{*\}$), and X is a del Pezzo surface;
(2) $\rho(X) = 2$, $W = \mathbb{P}^1$ and the map π is a conic bundle.

Note that for an arbitrary surface, the curve W in the second case should be any smooth curve, but we restrict ourselves to rational surfaces which implies that W is isomorphic to $\mathbb{P}^1$.

Proposition 2.3. *Let X be a smooth projective real rational surface. If X is minimal, then it admits a morphism $\pi: X \to W$ which is a Mori fibration.*

Proof. Follows from [7, Theorem 1]. See also [13]. □

Definition 2.4. A *Sarkisov link* between two Mori fibrations $\pi_1: X_1 \to W_1$ and $\pi_2: X_2 \to W_2$ is a birational map $\varphi: X_1 \dashrightarrow X_2$ of one of the following four types, where each of the diagrams is commutative:

(1) LINK OF TYPE I

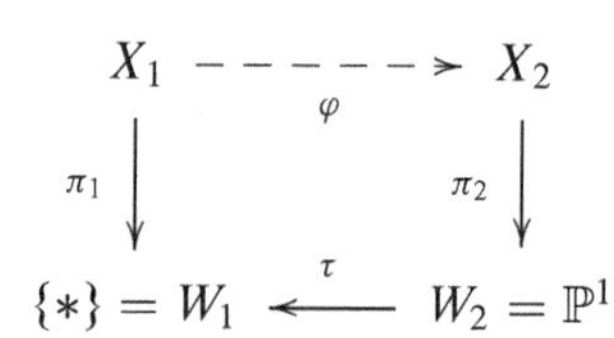

where $\varphi^{-1}: X_2 \to X_1$ is a birational morphism, which is the blow-up of either a real point or two non-real conjugate points of X_1, and where τ is the contraction of $W_2 = \mathbb{P}^1$ to the point W_1.

(2) LINK OF TYPE II

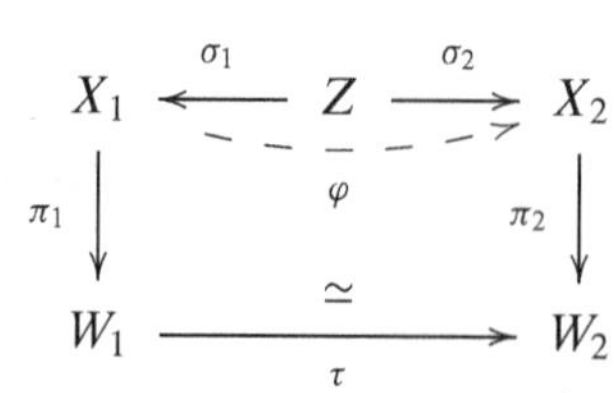

where $\sigma_i : Z \to X_i$ is a birational morphism, which is the blow-up of either a real point or two non-real conjugate points of X_i, and where τ is an isomorphism between W_1 and W_2.

(3) LINK OF TYPE III

$$\begin{array}{ccc} X_1 & \xrightarrow{\varphi} & X_2 \\ {\scriptstyle \pi_1}\downarrow & & \downarrow{\scriptstyle \pi_2} \\ \mathbb{P}^1 = W_1 & \xrightarrow{\tau} & W_2 = \{*\} \end{array}$$

where $\varphi: X_1 \to X_2$ is a birational morphism, which is the blow-up of either a real point or two non-real conjugate points of X_2, and where τ is the contraction of $W_1 = \mathbb{P}^1$ to the point W_2. (*It is the inverse of a link of type* I.)

(4) LINK OF TYPE IV

$$\begin{array}{ccc} X_1 & \xrightarrow[\varphi]{\simeq} & X_2 \\ {\scriptstyle \pi_1}\downarrow & & \downarrow{\scriptstyle \pi_2} \\ \mathbb{P}^1 = W_1 & & W_2 = \mathbb{P}^1 \end{array}$$

where $\varphi: X_1 \to X_2$ is an isomorphism and $\pi_1, \pi_2 \circ \varphi$ are conic bundles on X_1 with distinct fibres.

Remarks 2.5.

(1) The morphism τ is important only for links of type II, between two surfaces with a Picard group of rank 2 (in higher dimension τ is important also for other links).
(2) There is only one possible W_1 and one possible W_2 in cases I, III, IV but a priori several possibilities in case II.
(3) We shall see in Example 2.13(2) that indeed there exists links of type II where $W_1 = W_2 = \{*\}$. This is a feature of the real case that does not arise in the complex case.

Definition 2.6. If $\pi: X \to W$ and $\pi': X' \to W'$ are two (Mori) fibrations, an isomorphism $\psi: X \to X'$ is called an *isomorphism of fibrations* if there exists an isomorphism $\tau: W \to W'$ such that $\pi'\psi = \tau\pi$.

Note that the composition $\alpha\varphi\beta$ of a Sarkisov link φ with some automorphisms of fibrations α and β is again a Sarkisov link. We have the following fundamental result:

Proposition 2.7. *If $\pi: X \to W$ and $\pi': X' \to W'$ are two Mori fibrations, then any birational map $\psi: X \dashrightarrow X'$ is either an isomorphism of fibrations or admits a decomposition into Sarkisov links $\psi = \varphi_n \ldots \varphi_1$ such that*

(*i*) *for $i = 1, \ldots, n-1$, the birational map $\varphi_{i+1}\varphi_i$ is not biregular;*
(*ii*) *for $i = 1, \ldots, n$, every base-point of φ_i is a base-point of $\varphi_n \ldots \varphi_i$.*

Proof. Follows from [9, Theorem 2.5] (see also the appendix of [5]).

Let us give an idea of the strategy here, and refer to [9] for the details. If ψ is not an isomorphism of fibrations, then one can associate with it a Sarkisov degree, which is a triple of numbers (a, r, m) (see Definition at page 601 of [9]). The number $a \in \mathbb{Q}$ is given by the degree of the linear system $\mathcal{H}_X$ on X associated with ψ, the number $r \in \mathbb{N}$ is the maximal multiplicity of the base-points of this system and m is the number of base-points that realise this maximum. Then, we have the following dichotomy:

(*i*) If $r > a$, we denote by $\pi: \hat{X} \to X$ the blow-up of one real point or two conjugate non-real points that realise the multiplicity, then find that either $\hat{X}$ admits a structure of Mori fibration and $\varphi_1 = \pi^{-1}$ is a link of type I, or find a contraction $\pi': \hat{X} \to X_1$ such that $\varphi_1 = \pi'\pi^{-1}: X \dashrightarrow X_1$ is a link of type III.
(*ii*) If $r \leq a$, we either find a contraction $\varphi_1: X \to X_1$ which is a link of type III or find a link of type IV, which is an automorphism $\varphi_1: X \to X$.

In each case, it is shown that the Sarkisov degree of $\psi(\varphi_1)^{-1}: X_1 \dashrightarrow X'$ is smaller than the one of ψ, for the lexicographical ordering. The set of all possible Sarkisov degrees being discrete and bounded from below ([9, last paragraph of page 601]), the procedure ends at some point.

Moreover, the construction of the links implies that the two properties described above hold. □

Remark 2.8. In the above decomposition, if ψ has no real base-point (for instance when ψ induces an isomorphism $X(\mathbb{R}) \to X'(\mathbb{R})$), then φ_1 and φ_2 have no real base-point. However, the maps φ_i for $i \geq 3$ can have some real base-points, which have been artificially created, and correspond in fact to the base-points of $(\varphi_j)^{-1}$ for $j < i$.

This phenomenon happens for any $\psi \in \mathrm{Aut}(\mathbb{P}^2(\mathbb{R})) \setminus \mathrm{Aut}(\mathbb{P}^2)$, as the first link φ_1 will blow-up two non-real base-point and contract the line through these two, onto a real point, base-point of $(\varphi_1)^{-1}$ and of $\varphi_n \ldots \varphi_2$ (see Example 2.13(2) below).

Theorem 2.9 ([4] (see also [7])). *Let X be a real rational surface, if X is minimal, then it is isomorphic to one of the following:*

(1) $\mathbb{P}^2$,
(2) *the quadric* $Q_{3,1} = \{(w : x : y : z) \in \mathbb{P}^3 \mid w^2 = x^2 + y^2 + z^2\}$,
(3) *a Hirzebruch surface* $\mathbb{F}_n = \{((x : y : z), (u : v)) \in \mathbb{P}^2 \times \mathbb{P}^1 \mid yv^n = zu^n\}$ *with* $n \neq 1$.

By [12], if $n - n' \equiv 0 \mod 2$, $\mathbb{F}_n(\mathbb{R})$ is isomorphic to $\mathbb{F}_{n'}(\mathbb{R})$. We get:

Corollary 2.10. *Let $X(\mathbb{R})$ be the real locus of a real rational surface. If X is minimal, then $X(\mathbb{R})$ is isomorphic to one of the following:*

(1) $\mathbb{P}^2(\mathbb{R})$,
(2) $Q_{3,1}(\mathbb{R})$, *diffeomorphic to* $\mathbb{S}^2$,
(3) $\mathbb{F}_0(\mathbb{R})$, *diffeomorphic to* $\mathbb{S}^1 \times \mathbb{S}^1$,
(4) $\mathbb{F}_3(\mathbb{R})$, *diffeomorphic to the Klein bottle.*

Remark 2.11. Note that $\mathbb{F}_3(\mathbb{R})$ and $\mathbb{F}_1(\mathbb{R})$ are isomorphic. However, $\mathbb{F}_1$ is not minimal although $\mathbb{F}_3$ is.

In the same vein, there exists a birational morphism $\mathbb{P}^2(\mathbb{R}) \to Q_{3,1}(\mathbb{R})$, that contracts a real line (the map φ^{-1} in Example 2.13(2)).

We give a list of Mori fibrations on real rational surfaces and will show that, up to isomorphisms of fibrations, this list is exhaustive.

Example 2.12. The following morphisms $\pi: X \to W$ are Mori fibrations on the plane, the sphere, the Hirzebruch surfaces, and a particular Del Pezzo surface of degree 6.

(1) $\mathbb{P}^2 \to \{*\}$;
(2) $Q_{3,1} = \{(w : x : y : z) \in \mathbb{P}^3_{\mathbb{R}} \mid w^2 = x^2 + y^2 + z^2\} \to \{*\}$;
(3) $\mathbb{F}_n = \{((x : y : z), (u : v)) \in \mathbb{P}^2 \times \mathbb{P}^1 \mid yv^n = zu^n\} \to \mathbb{P}^1$ for $n \geq 0$ (the map is the projection on the second factor);
(4) $\mathcal{D}_6 = \{(w : x : y : z), (u : v) \in Q_{3,1} \times \mathbb{P}^1 \mid wv = xu\} \to \mathbb{P}^1$ (the map is the projection on the second factor).

Example 2.13. The following maps between the surfaces of Example 2.12 are Sarkisov links (in the list, fibres refer to those of the Mori fibrations introduced in Example 2.12):

(1) The contraction of the exceptional curve of $\mathbb{F}_1$ (or equivalently the blow-up of a real point of $\mathbb{P}^2$), is a link $\mathbb{F}_1 \to \mathbb{P}^2$ of type III. Note that the inverse of this link is of type I.
(2) The stereographic projection from the North pole $p_N = (1 : 0 : 0 : 1)$, $\varphi: Q_{3,1} \dashrightarrow \mathbb{P}^2$ given by

$$\varphi: (w : x : y : z) \dashrightarrow (x : y : w - z)$$

and its inverse $\varphi^{-1}: \mathbb{P}^2 \dashrightarrow Q_{3,1}$ given by

$$\varphi^{-1}: (x : y : z) \dashrightarrow (x^2 + y^2 + z^2 : 2xz : 2yz : x^2 + y^2 - z^2)$$

are both Sarkisov links of type II.

The map φ decomposes into the blow-up of p_N, followed by the contraction of the strict transform of the curve $z = w$ (intersection of $Q_{3,1}$ with the tangent plane at p_N), which is the union of two non-real conjugate lines. The map φ^{-1}

decomposes into the blow-up of the two non-real points $(1 : \pm\mathbf{i} : 0)$, followed by the contraction of the strict transform of the line $z = 0$.

(3) The projection on the first factor $\mathcal{D}_6 \to \mathcal{Q}_{3,1}$ which contracts the two disjoint conjugate non-real (-1)-curves $(0 : 0 : 1 : \pm\mathbf{i}) \times \mathbb{P}^1 \subset \mathcal{D}_6$ onto the two conjugate non-real points $(0 : 0 : 1 : \pm\mathbf{i}) \in \mathcal{Q}_{3,1}$ is a link of type III.

(4) The blow-up of a real point $q \in \mathbb{F}_n$, lying on the exceptional section if $n > 0$ (or any point if $n = 0$), followed by the contraction of the strict transform of the fibre passing through q onto a real point of $\mathbb{F}_{n+1}$ not lying on the exceptional section is a link $\mathbb{F}_n \dashrightarrow \mathbb{F}_{n+1}$ of type II.

(5) The blow-up of two conjugate non-real points $p, \bar{p} \in \mathbb{F}_n$ lying on the exceptional section if $n > 0$, or on the same section of self-intersection 0 if $n = 0$, followed by the contraction of the strict transform of the fibres passing through $p, \bar{p}$ onto two non-real conjugate points of $\mathbb{F}_{n+2}$ not lying on the exceptional section is a link $\mathbb{F}_n \dashrightarrow \mathbb{F}_{n+2}$ of type II.

(6) The blow-up of two conjugate non-real points $p, \bar{p} \in \mathbb{F}_n$, $n \in \{0, 1\}$ not lying on the same fibre (or equivalently not lying on a real fibre) and not on the same section of self-intersection $-n$ (or equivalently not lying on a real section of self-intersection $-n$), followed by the contraction of the fibres passing through $p, \bar{p}$ onto two non-real conjugate points of $\mathbb{F}_n$ having the same properties is a link $\mathbb{F}_n \dashrightarrow \mathbb{F}_n$ of type II.

(7) The exchange of the two components $\mathbb{P}^1 \times \mathbb{P}^1 \to \mathbb{P}^1 \times \mathbb{P}^1$ is a link $\mathbb{F}_0 \to \mathbb{F}_0$ of type IV.

(8) The blow-up of a real point $p \in \mathcal{D}_6$, not lying on a singular fibre (or equivalently $p \neq ((1 : 1 : 0 : 0), (1 : 1))$, $p \neq ((1 : -1 : 0 : 0), (1 : -1))$), followed by the contraction of the strict transform of the fibre passing through p onto a real point of $\mathcal{D}_6$, is a link $\mathcal{D}_6 \dashrightarrow \mathcal{D}_6$ of type II.

(9) The blow-up of two non-real conjugate points $p, \bar{p} \in \mathcal{D}_6$, not lying on the same fibre (or equivalently not lying on a real fibre), followed by the contraction of the strict transform of the fibres passing through $p, \bar{p}$ onto two non-real points of $\mathcal{D}_6$ is a link $\mathcal{D}_6 \dashrightarrow \mathcal{D}_6$ of type II.

Remark 2.14. Note that in the above list, the three links $\mathbb{F}_n \dashrightarrow \mathbb{F}_m$ of type II can be put in one family, and the same is true for the two links $\mathcal{D}_6 \dashrightarrow \mathcal{D}_6$. We distinguished here the possibilities for the base points to describe more precisely the geometry of each link. The two links $\mathcal{D}_6 \dashrightarrow \mathcal{D}_6$ could also be arranged into extra families, by looking if the base points belong to the two exceptional sections of self-intersection -1, but go in any case from $\mathcal{D}_6$ to $\mathcal{D}_6$ (see Definition 2.16 below).

Proposition 2.15. *Any Mori fibration $\pi: X \to W$, where X is a smooth projective real rational surface, belongs to the list of Example 2.12.*

Any Sarkisov link between two such Mori fibrations is equal to $\alpha\varphi\beta$, where φ or φ^{-1} belongs to the list described in Example 2.13 and where α and β are isomorphisms of fibrations.

Proof. Since any birational map between two surfaces with Mori fibrations decomposes into Sarkisov links and all links of Example 2.13 involve only the Mori fibrations of Example 2.12, it suffices to check that any link starting from one of the Mori fibrations of Example 2.12 belongs to the list 2.13. This is an easy case-by-case study; here are the steps.

Starting from a Mori fibration $\pi: X \to W$ where W is a point, the only links we can perform are links of type I or II centered at a real point or two conjugate non-real points. From Theorem 2.9, the surface X is either $\mathcal{Q}_{3,1}$ or $\mathbb{P}^2$, and both are homogeneous under the action of $\mathrm{Aut}(X)$, so the choice of the point is not relevant. Blowing-up a real point in $\mathbb{P}^2$ or two non-real points in $\mathcal{Q}_{3,1}$ gives rise to a link of type I to $\mathbb{F}_1$ or $\mathcal{D}_6$. The remaining cases correspond to the stereographic projection $\mathcal{Q}_{3,1} \dashrightarrow \mathbb{P}^2$ and its converse.

Starting from a Mori fibration $\pi: X \to W$ where $W = \mathbb{P}^1$, we have additional possibilities. If the link is of type IV, then X admits two conic bundle structures and by Theorem 2.9, the only possibility is $\mathbb{F}_0 = \mathbb{P}^1 \times \mathbb{P}^1$. If the link is of type III, then we contract a real (-1)-curve of X or two disjoint conjugate non-real (-1)-curves. The only possibilities for X are respectively $\mathbb{F}_1$ and $\mathcal{D}_6$, and the image is respectively $\mathbb{P}^2$ and $\mathcal{Q}_{3,1}$ (these are the inverses of the links described before). The last possibility is to perform a link a type II, by blowing up a real point or two conjugate non-real points, on respectively one or two smooth fibres, and to contract the strict transform. We go from $\mathcal{D}_6$ to $\mathcal{D}_6$ or from $\mathbb{F}_m$ to $\mathbb{F}_{m'}$ where $m' - m \in \{-2, -1, 0, 1, 2\}$. All possibilities are described in Example 2.13. □

We end this section by reducing the number of links of type II needed for the classification. For this, we introduce the notion of standard links.

Definition 2.16. The following links of type II are called *standard*:

(1) links $\mathbb{F}_m \dashrightarrow \mathbb{F}_n$, with $m, n \in \{0, 1\}$;
(2) links $\mathcal{D}_6 \dashrightarrow \mathcal{D}_6$ which do not blow-up any point on the two exceptional section of self-intersection -1.

The other links of type II will be called *special*.

The following result allows us to simplify the set of generators of our groups.

Lemma 2.17. *Any Sarkisov link of type* IV *decomposes into links of type* I, III, *and standard links of type* II.

Proof. Note that a link of type IV is, up to automorphisms preserving the fibrations, equal to the following automorphism of $\mathbb{P}^1 \times \mathbb{P}^1$

$$\tau: ((x_1 : x_2), (y_1 : y_2)) \mapsto ((y_1 : y_2), (x_1 : x_2)).$$

We denote by $\psi: \mathbb{P}^2 \dashrightarrow \mathbb{P}^1 \times \mathbb{P}^1$ the birational map $(x : y : z) \dashrightarrow ((x : y), (x : z))$ and observe that $\tau\psi = \psi\sigma$, where $\sigma \in \mathrm{Aut}_{\mathbb{R}}(\mathbb{P}^2)$. Hence, $\tau = \psi\tau\psi^{-1}$. Observing that ψ decomposes into the blow-up of the point $(0 : 0 : 1)$, which is a link of type III, followed by a standard link of type II, we get the result. □

Lemma 2.18. *Let $\pi: X \to \mathbb{P}^1$ and $\pi': X' \to \mathbb{P}^1$ be two Mori fibrations, where X, X' belong to the list $\mathbb{F}_0, \mathbb{F}_1, \mathcal{D}_6$. Let $\psi: X \dashrightarrow X'$ be a birational map, such that $\pi'\psi = \alpha\pi$ for some $\alpha \in \mathrm{Aut}_{\mathbb{R}}(\mathbb{P}^1)$. Then, ψ is either an automorphism or $\psi = \varphi_n \cdots \varphi_1$, where each φ_i is a* standard *link of type* II. *Moreover, if ψ is an isomorphism on the real points (i.e. is an isomorphism $X(\mathbb{R}) \to X'(\mathbb{R})$), the standard links φ_i can also be chosen to be isomorphisms on the real points.*

Proof. We first show that $\psi = \varphi_n \cdots \varphi_1$, where each φ_i is a link of type II, not necessarily standard. This is done by induction on the number of base-points of ψ (recall that we always count infinitely near points as base-points). If ψ has no base-point, it is an isomorphism. If q is a real proper base-point, or $q, \bar{q}$ are two proper non-real base-points (here proper means not infinitely near), we denote by φ_1 a Sarkisov link of type II centered at q (or $q, \bar{q}$). Then, $(\varphi_1)^{-1}\psi$ has less base-points than ψ. The result follows then by induction. Moreover, if ψ is an isomorphism on the real points, i.e. if ψ and ψ^{-1} have no real base-point, then so are all φ_i.

Let $\varphi: \mathcal{D}_6 \dashrightarrow \mathcal{D}_6$ be a special link of type II. Then, it is centered at two points $p_1, \bar{p}_1$ lying on the (-1)-curves $E_1, \bar{E}_1$. We choose then two general non-real conjugate points $q_1, \bar{q_1}$, and let $q_2 := \varphi(q_1)$ and $\bar{q_2} := \varphi(\bar{q_1})$. For $i = 1, 2$, we denote by $\varphi_i: \mathcal{D}_6 \dashrightarrow \mathcal{D}_6$ a standard link centered at $q_i, \bar{q_i}$. The image by φ_2 of E_1 is a curve of self-intersection 1. Hence, $\varphi_2\varphi(\varphi_1)^{-1}$ is a standard link of type II.

It remains to consider the case where each φ_i is a link $\mathbb{F}_{n_i} \dashrightarrow \mathbb{F}_{n_{i+1}}$. We denote by N the maximum of the integers n_i. If $N \leq 1$, we are done because all links of type II between $\mathbb{F}_j$ and $\mathbb{F}_{j'}$ with $j, j' \leq 1$ are standard. We can thus assume $N \geq 2$, which implies that there exists i such that $n_i = N$, $n_{i-1} < N, n_{i+1} \leq N$. We choose two general non-real points $q_{i-1}, \overline{q_{i-1}} \in \mathbb{F}_{n_{i-1}}$, and write $q_i = \varphi_{i-1}(q_{i-1})$, $q_{i+1} = \varphi_i(q_i)$. For $j \in \{i-1, i, i+1\}$, we denote by $\tau_j: \mathbb{F}_{n_j} \dashrightarrow \mathbb{F}_{n'_j}$ a Sarkisov link centered at $q_j, \overline{q_j}$. We obtain then the following commutative diagram

$$\begin{array}{ccccc}
\mathbb{F}_{n_{i-1}} & \overset{\varphi_{i-1}}{\dashrightarrow} & \mathbb{F}_{n_i} & \overset{\varphi_i}{\dashrightarrow} & \mathbb{F}_{n_{i+1}} \\
\downarrow \tau_{i-1} & & \downarrow \tau_i & & \downarrow \tau_{i+1} \\
\mathbb{F}_{n'_{i-1}} & \overset{\varphi'_{i-1}}{\dashrightarrow} & \mathbb{F}_{n'_i} & \overset{\varphi'_i}{\dashrightarrow} & \mathbb{F}_{n'_{i+1}},
\end{array}$$

where $\varphi'_{i-1}, \varphi'_i$ are Sarkisov links. By construction, $n'_{i-1}, n'_i, n'_{i+1} < N$, we can then replace $\varphi_i\varphi_{i-1}$ with $(\tau_{i+1})^{-1}\varphi'_i\varphi'_{i-1}\tau_{i-1}$ and "avoid" $\mathbb{F}_N$. Repeating this process if needed, we end up with a sequence of Sarkisov links passing only through $\mathbb{F}_1$ and $\mathbb{F}_0$. Moreover, since this process does not add any real base-point, it preserves the regularity at real points. □

Corollary 2.19. *Let $\pi: X \to W$ and $\pi': X' \to W'$ be two Mori fibrations, where X, X' are either $\mathbb{F}_0, \mathbb{F}_1, \mathcal{D}_6$ or $\mathbb{P}^2$. Any birational map $\psi: X \dashrightarrow X'$ is either an isomorphism preserving the fibrations or decomposes into links of type* I, III, *and standard links of type* II.

Proof. Follows from Proposition 2.7, Lemmas 2.17 and 2.18, and the description of Example 2.13. □

3 Generators of the Group $\mathrm{Aut}(\mathbb{P}^2(\mathbb{R}))$

We start this section by describing three kinds of elements of $\mathrm{Aut}(\mathbb{P}^2(\mathbb{R}))$, which are birational maps of $\mathbb{P}^2$ of degree 5. These maps are associated with three pairs of conjugate non-real points; the description is then analogue to the description of quadratic maps, which are associated with three points.

Example 3.1. Let $p_1, \bar{p}_1, p_2, \bar{p}_2, p_3, \bar{p}_3 \in \mathbb{P}^2$ be three pairs of non-real points of $\mathbb{P}^2$, not lying on the same conic. Denote by $\pi\colon X \to \mathbb{P}^2$ the blow-up of the six points, which induces an isomorphism $X(\mathbb{R}) \to \mathbb{P}^2(\mathbb{R})$. Note that X is isomorphic to a smooth cubic of $\mathbb{P}^3$. The set of strict transforms of the conics passing through five of the six points corresponds to three pairs of non-real (-1)-curves (or lines on the cubic), and the six curves are disjoint. The contraction of the six curves gives a birational morphism $\eta\colon X \to \mathbb{P}^2$, inducing an isomorphism $X(\mathbb{R}) \to \mathbb{P}^2(\mathbb{R})$, which contracts the curves onto three pairs of non-real points $q_1, \bar{q}_1, q_2, \bar{q}_2, q_3, \bar{q}_3 \in \mathbb{P}^2$; we choose the order so that q_i is the image of the conic not passing through p_i. The map $\psi = \eta\pi^{-1}$ is a birational map $\mathbb{P}^2 \dashrightarrow \mathbb{P}^2$ inducing an isomorphism

$$\mathbb{P}^2(\mathbb{R}) \to \mathbb{P}^2(\mathbb{R}).$$

Let $L \subset \mathbb{P}^2$ be a general line of $\mathbb{P}^2$. The strict transform of L on X by π^{-1} has self-intersection 1 and intersects the six curves contracted by η into two points (because these are conics). The image $\psi(L)$ has then six singular points of multiplicity 2 and self-intersection 25; it is thus a quintic passing through the q_i with multiplicity 2. The construction of ψ^{-1} being symmetric as the one of ψ, the linear system of ψ consists of quintics of $\mathbb{P}^2$ having multiplicity 2 at $p_1, \bar{p}_1, p_2, \bar{p}_2, p_3, \bar{p}_3$.

One can moreover check that ψ sends the pencil of conics through $p_1, \bar{p}_1, p_2, \bar{p}_2$ onto the pencil of conics through $q_1, \bar{q}_1, q_2, \bar{q}_2$ (and the same holds for the two other real pencil of conics, through $p_1, \bar{p}_1, p_3, \bar{p}_3$ and through $p_2, \bar{p}_2, p_3, \bar{p}_3$).

Example 3.2. Let $p_1, \bar{p}_1, p_2, \bar{p}_2 \in \mathbb{P}^2$ be two pairs of non-real points of $\mathbb{P}^2$, not on the same line. Denote by $\pi_1\colon X_1 \to \mathbb{P}^2$ the blow-up of the four points, and by $E_2, \bar{E}_2 \subset X_1$ the curves contracted onto $p_2, \bar{p}_2$ respectively. Let $p_3 \in E_2$ be a point, and $\bar{p}_3 \in \bar{E}_2$ its conjugate. We assume that there is no conic of $\mathbb{P}^2$ passing through $p_1, \bar{p}_1, p_2, \bar{p}_2, p_3, \bar{p}_3$ and let $\pi_2\colon X_2 \to X_1$ be the blow-up of $p_3, \bar{p}_3$.

On X, the strict transforms of the two conics $C, \bar{C}$ of $\mathbb{P}^2$, passing through $p_1, \bar{p}_1, p_2, \bar{p}_2, p_3$ and $p_1, \bar{p}_1, p_2, \bar{p}_2, \bar{p}_3$ respectively, are non-real conjugate disjoint (-1) curves. The contraction of these two curves gives a birational morphism $\eta_2\colon X_2 \to Y_1$, contracting $C, \bar{C}$ onto two points $q_3, \bar{q}_3$. On Y_1, we find two pairs of non-real (-1)-curves, all four curves being disjoint. These are the strict transforms of the exceptional curves associated with $p_2, \bar{p}_2$, and of the conics passing through $p_1, p_2, \bar{p}_2, p_3, \bar{p}_3$ and $\bar{p}_1, p_2, \bar{p}_2, p_3, \bar{p}_3$ respectively. The contraction of these curves gives a birational morphism $\eta_1\colon Y_1 \to \mathbb{P}^2$, and the images of the four curves are points $q_2, \bar{q_2}, q_1, \bar{q_1}$ respectively. Note that the four maps $\pi_1, \pi_2, \eta_1, \eta_2$ are blow-ups of non-real points, so the birational map $\psi = \eta_1\eta_2(\pi_1\pi_2)^{-1}\colon \mathbb{P}^2 \dashrightarrow \mathbb{P}^2$ induces an isomorphism $\mathbb{P}^2(\mathbb{R}) \to \mathbb{P}^2(\mathbb{R})$.

In the same way as in Example 3.1, we find that the linear system of ψ is of degree 5, with multiplicity 2 at the points $p_i, \bar{p}_i$. The situation is similar for ψ^{-1}, with the six points $q_i, \bar{q_i}$ in the same configuration: $q_1, \bar{q_1}, q_2, \bar{q_2}$ lie on the plane and $q_3, \bar{q_3}$ are infinitely near to $q_2, \bar{q_2}$ respectively.

One can moreover check that ψ sends the pencil of conics through $p_1, \bar{p}_1, p_2, \bar{p}_2$ onto the pencil of conics through $q_1, \bar{q_1}, q_2, \bar{q_2}$ and the pencil of conics through $p_2, \bar{p}_2, p_3, \bar{p}_3$ onto the pencil of conics through $q_2, \bar{q_2}, q_3, \bar{q_3}$. But, contrary to Example 3.1, there is no pencil of conics through $q_1, \bar{q_1}, q_3, \bar{q_3}$ (because $q_3, \bar{q_3}$ are infinitely near to $q_2, \bar{q_2}$).

Example 3.3. Let $p_1, \bar{p}_1$ be a pair of two conjugate non-real points of $\mathbb{P}^2$. We choose a point p_2 in the first neighbourhood of p_1, and a point p_3 in the first neighbourhood of p_2, not lying on the exceptional divisor corresponding to p_1. We denote by $\pi\colon X \to \mathbb{P}^2$ the blow-up of $p_1, \bar{p}_1, p_2, \bar{p}_2, p_3\bar{p}_3$. We denote by $E_i, \bar{E}_i \subset X$ the irreducible exceptional curves corresponding to the points $p_i, \bar{p}_i$, for $i = 1, 2, 3$. The strict transforms of the two conics through $p_1, \bar{p}_1, p_2, \bar{p}_2, p_3$ and $p_1, \bar{p}_1, p_2, \bar{p}_2, \bar{p}_3$ respectively are disjoint (-1)-curves on X, intersecting the exceptional curves $E_1, \bar{E}_1, E_2, \bar{E}_2$ similarly as $E_3, \bar{E}_3$. Hence, there exists a birational morphism $\eta\colon X \to \mathbb{P}^2$ contracting the strict transforms of the two conics and the curves $E_1, \bar{E}_1, E_2, \bar{E}_2$.

As in Examples 3.1 and 3.2, the linear system of $\psi = \eta\pi^{-1}$ consists of quintics with multiplicity two at the six points $p_1, \bar{p}_1, p_2, \bar{p}_2, p_3, \bar{p}_3$.

Definition 3.4. The birational maps of $\mathbb{P}^2$ of degree 5 obtained in Example 3.1 will be called *standard quintic transformations* and those of Examples 3.2 and 3.3 will be called *special quintic transformations* respectively.

Lemma 3.5. *Let* $\psi\colon \mathbb{P}^2 \dashrightarrow \mathbb{P}^2$ *be a birational map inducing an isomorphism* $\mathbb{P}^2(\mathbb{R}) \to \mathbb{P}^2(\mathbb{R})$. *The following hold:*

(1) *The degree of* ψ *is* $4k + 1$ *for some integer* $k \geq 0$.
(2) *Every multiplicity of the linear system of* ψ *is even.*
(3) *Every curve contracted by* ψ *is of even degree.*
(4) *If* ψ *has degree* 1, *it belongs to* $\mathrm{Aut}_{\mathbb{R}}(\mathbb{P}^2) = \mathrm{PGL}(3, \mathbb{R})$.
(5) *If* ψ *has degree 5, then it is a standard or special quintic transformation, described in Examples 3.1, 3.2 or 3.3, and has thus exactly* 6 *base-points.*
(6) *If* ψ *has at most* 6 *base-points, then* ψ *has degree* 1 *or* 5.

Remark 3.6. Part (1) is [15, Teorema 1].

Proof. Denote by d the degree of ψ and by $m_1, \ldots, m_k$ the multiplicities of the base-points of ψ. The Noether equalities yield $\sum_{i=1}^{k} m_i = 3(d-1)$ and $\sum_{i=1}^{k}(m_i)^2 = d^2 - 1$.

Let $C, \bar{C}$ be a pair of two curves contracted by ψ. Since $C \cap \bar{C}$ does not contain any real point, the degree of C and $\bar{C}$ is even. This yields (3), and implies that all multiplicities of the linear system of ψ^{-1} are even, giving (2).

In particular, $3(d-1)$ is a multiple of 4 (all multiplicities come by pairs of even integers), which implies that $d = 4k+1$ for some integer k. Hence (1) is proved.

If the number of base-points is at most $k = 6$, then by Cauchy–Schwartz we get

$$9(d-1)^2 = \left(\sum_{i=1}^{k} m_i\right)^2 \leq k \sum_{i=1}^{k} (m_i)^2 = k(d^2-1) = 6(d^2-1)$$

This yields $9(d-1) \leq 6(d+1)$, hence $d \leq 5$.

If $d = 5$, the Noether equalities yield $k = 6$ and $m_1 = m_2 = \cdots = m_6 = 2$. Hence, the base-points of ψ consist of three pairs of conjugate non-real points $p_1, \bar{p}_1, p_2, \bar{p}_2, p_3, \bar{p}_3$. Moreover, if a conic passes through 5 of the six points, its free intersection with the linear system is zero, so it is contracted by ψ, and there is no conic through the six points.

(*a*) If the six points belong to $\mathbb{P}^2$, the map is a standard quintic transformation, described in Example 3.1.
(*b*) If two points are infinitely near, the map is a special quintic transformation, described in Example 3.2.
(*c*) If four points are infinitely near, the map is a special quintic transformation, described in Example 3.3. □

Before proving Theorem 1.2, we will show that all quintic transformations are generated by linear automorphisms and standard quintic transformations:

Lemma 3.7. *Every quintic transformation* $\psi \in \mathrm{Aut}(\mathbb{P}^2(\mathbb{R}))$ *belongs to the group generated by* $\mathrm{Aut}_{\mathbb{R}}(\mathbb{P}^2)$ *and standard quintic transformations.*

Proof. By Lemma 3.5, we only need to show the result when ψ is a special quintic transformation as in Example 3.2 or Example 3.3.

We first assume that ψ is a special quintic transformation as in Example 3.2, with base-points $p_1, \bar{p}_1, p_2, \bar{p}_2, p_3, \bar{p}_3$, where $p_3, \bar{p}_3$ are infinitely near to $p_2, \bar{p}_2$. For $i = 1, 2$, we denote by $q_i \in \mathbb{P}^2$ the point which is the image by ψ of the conic passing through the five points of $\{p_1, \bar{p}_1, p_2, \bar{p}_2, p_3, \bar{p}_3\} \setminus \{p_i\}$. Then, the base-points of ψ^{-1} are $q_1, \bar{q}_1, q_2, \bar{q}_2, q_3, \bar{q}_3$, where $q_3, \bar{q}_3$ are points infinitely near to $q_2, \bar{q}_2$ respectively (see Example 3.2). We choose a general pair of conjugate non-real points $p_4, \bar{p}_4 \in \mathbb{P}^2$, and write $q_4 = \psi(p_4)$, $\bar{q}_4 = \psi(\bar{p}_4)$. We denote by φ_1 a standard quintic transformation having base-points at $p_1, \bar{p}_1, p_2, \bar{p}_2, p_4, \bar{p}_4$, and by φ_2 a standard quintic transformation having base-points at $q_1, \bar{q}_1, q_2, \bar{q}_2, q_4, \bar{q}_4$. We now prove that $\varphi_2 \psi (\varphi_1)^{-1}$ is a standard quintic transformation; this will yield the result. Denote by $p_i', \bar{p}_i{}'$ the base-points of $(\varphi_1)^{-1}$, with the order associated with the p_i, which means that p_i' is the image by φ_i of a conic not passing through p_i (see Example 3.1). Similarly, we denote by $q_i', \bar{q}_i{}'$ the base-points of $(\varphi_2)^{-1}$. We obtain the following commutative of birational maps, where the arrows indexed by points are blow-ups of these points:

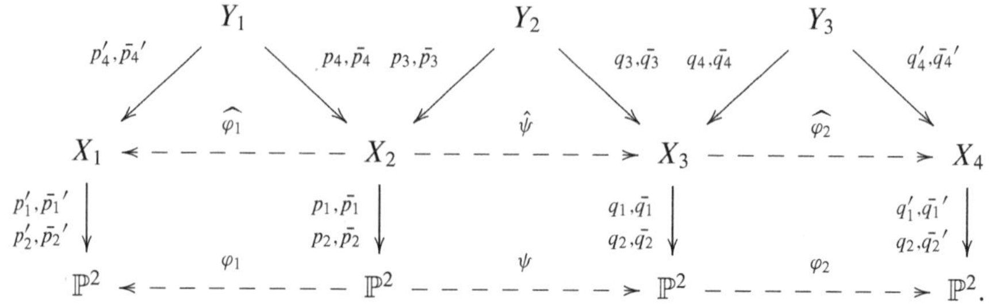

Each of the surfaces X_1, X_2, X_3, X_4 admits a conic bundle structure $\pi_i: X_i \to \mathbb{P}^1$, which fibres correspond to the conics passing through the four points blown-up on $\mathbb{P}^2$ to obtain X_i. Moreover, $\hat{\varphi}_1$, $\hat{\psi}$, $\hat{\varphi}_2$ preserve these conic bundle structures. The map $(\hat{\varphi}_1)^{-1}$ blows-up $p_4, \bar{p}_4'$ and contract the fibres associated with them, then $\hat{\psi}$ blows-up $p_3, \bar{p}_3$ and contract the fibres associated with them. The map $\hat{\varphi}_2$ blow-ups the points $q_4, \bar{q}_4$, which correspond to the image of the curves contracted by $(\hat{\varphi}_1)^{-1}$, and contracts their fibres, corresponding to the exceptional divisors corresponding to the points $p_4, \bar{p}_4'$. Hence, $\hat{\varphi}_2\hat{\psi}\hat{\varphi}_1$ is the blow-up of two conjugate non-real points $p_3', \bar{p}_3' \in X_1$, followed by the contraction of their fibres. We obtain the following commutative diagram:

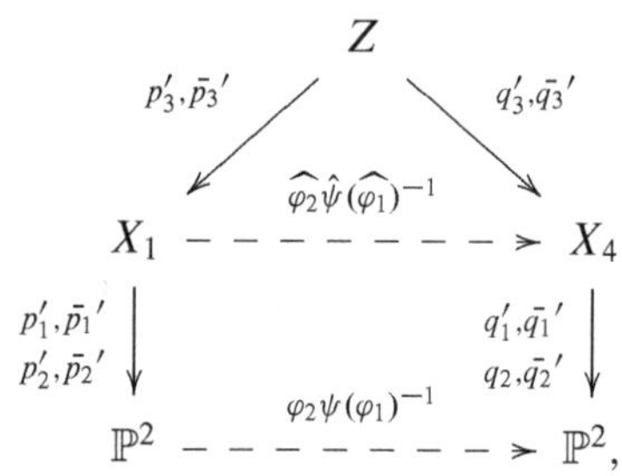

and the points $p_3', \bar{p}_3'$ correspond to the point of $\mathbb{P}^2$, hence $\varphi_2\psi(\varphi_1)^{-1}$ is a standard quintic transformation.

The remaining case is when ψ is a special quintic transformation as in Example 3.3, with base-points $p_1, \bar{p}_1, p_2, \bar{p}_2, p_3, \bar{p}_3$, where $p_3, \bar{p}_3$ are infinitely near to $p_2, \bar{p}_2$ and these latter are infinitely near to $p_1, \bar{p}_1$. The map ψ^{-1} has base-points $q_1, \bar{q}_1, q_2, \bar{q}_2, q_3, \bar{q}_3$, having the same configuration (see Example 3.3). We choose a general pair of conjugate non-real points $p_4, \bar{p}_4 \in \mathbb{P}^2$, and write $q_4 = \psi(p_4)$, $\bar{q}_4 = \psi(\bar{p}_4)$. We denote by φ_1 a special quintic transformation having base-points at $p_1, \bar{p}_1, p_2, \bar{p}_2, p_4, \bar{p}_4$, and by φ_2 a special quintic transformation having base-points at $q_1, \bar{q}_1, q_2, \bar{q}_2, q_4, \bar{q}_4$. The maps φ_1, φ_2 have four proper base-points, and are thus given in Example 3.2. The same proof as before implies that $\varphi_2\psi(\varphi_1)^{-1}$ is a special quintic transformation with four base-points. This gives the result. □

Lemma 3.8. *Let $\varphi\colon \mathbb{P}^2 \dashrightarrow \mathbb{P}^2$ be a birational map that decomposes as $\varphi = \varphi_5 \cdots \varphi_1$, where $\varphi_i\colon X_{i-1} \dashrightarrow X_i$ is a Sarkisov link for each i, where $X_0 = \mathbb{P}^2$, $X_1 = \mathcal{Q}_{3,1}$, $X_2 = X_3 = \mathcal{D}_6$, $X_4 = \mathcal{Q}_{3,1}$, $X_5 = \mathbb{P}^2$. If φ_2 is an automorphism of $\mathcal{D}_6(\mathbb{R})$ and $\varphi_4\varphi_3\varphi_2$ sends the base-point of $(\varphi_1)^{-1}$ onto the base-point of φ_5, then φ is an automorphism of $\mathbb{P}^2(\mathbb{R})$ of degree 5.*

Proof. We have the following commutative diagram, where each π_i is the blow-up of two conjugate non-real points and each η_i is the blow-up of one real point. The two maps $(\varphi_2)^{-1}$ and φ_4 are also blow-ups of non-real points.

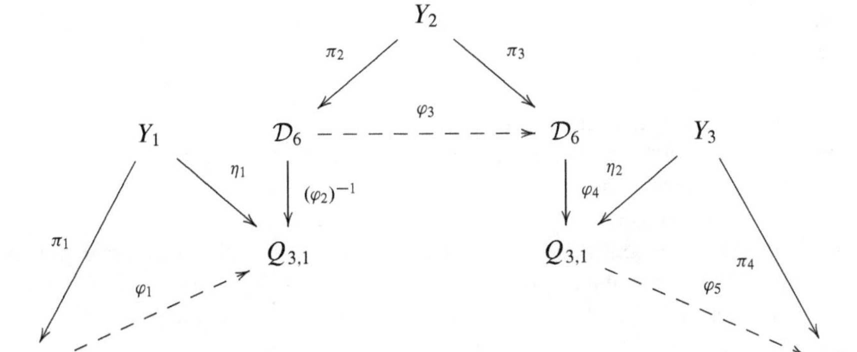

The only real base-points are those blown-up by η_1 and η_2. Since η_2 blows-up the image by $\varphi_4\varphi_3\varphi_2$ of the real point blown-up by η_1, the map φ has at most 6 base-points, all being non-real, and the same holds for φ^{-1}. Hence, φ is an automorphism of $\mathbb{P}^2(\mathbb{R})$ with at most 6 base-points. We can moreover see that $\varphi \notin \mathrm{Aut}_{\mathbb{R}}(\mathbb{P}^2)$, since the two curves of Y_2 contracted by π_2 are sent by $\varphi_4\pi_3$ onto conics of $\mathcal{Q}_{3,1}$, which are therefore not contracted by φ_5.

Lemma 3.5 implies that ψ has degree 5. □

Proposition 3.9. *The group* $\mathrm{Aut}(\mathbb{P}^2(\mathbb{R}))$ *is generated by* $\mathrm{Aut}_{\mathbb{R}}(\mathbb{P}^2)$ *and by elements of* $\mathrm{Aut}(\mathbb{P}^2(\mathbb{R}))$ *of degree 5.*

Proof. Let us prove that any $\varphi \in \mathrm{Aut}(\mathbb{P}^2(\mathbb{R}))$ is generated by $\mathrm{Aut}_{\mathbb{R}}(\mathbb{P}^2)$ and elements of $\mathrm{Aut}(\mathbb{P}^2(\mathbb{R}))$ of degree 5 . Applying Proposition 2.7, we decompose φ into Sarkisov links $\varphi = \varphi_r \cdots \varphi_1$ such that

(*i*) for $i = 1, \dots, r-1$, the map $\varphi_{i+1}\varphi_i$ is not biregular;
(*ii*) for $i = 1, \dots, r$, every real base-point of φ_i is a base-point of $\varphi_r \dots \varphi_i$.

In particular, φ_1 has no real base-point.

We proceed by induction on r, the case $r = 0$, corresponding to $\varphi \in \mathrm{Aut}_{\mathbb{R}}(\mathbb{P}^2)$, being obvious. We also observe that the decompositions of smaller length obtained by the induction process still satisfy property (ii) above. We can assume that (i) is also satisfied, because removing a biregular map $\varphi_{i+1}\varphi_i$ produces a decomposition of smaller length, still having the property (ii).

Since φ_1 has no real base-point, the first link φ_1 is then of type II from $\mathbb{P}^2$ to $Q_{3,1}$, and $\varphi_r \cdots \varphi_2$ has a unique real base-point $r \in Q_{3,1}$, which is the base-point of $(\varphi_1)^{-1}$. If φ_2 would blow-up this point, then $\varphi_2\varphi_1$ would be biregular, hence φ_2 is a link of type I from $Q_{3,1}$ to $\mathcal{D}_6$. We can write the map $\varphi_2\varphi_1$ as $\eta\pi^{-1}$, where $\pi\colon X \to \mathbb{P}^2$ is the blow-up of two pairs of non-real points, say $p_1, \bar{p_1}, p_2, \bar{p_2}$ and $\eta\colon X \to \mathcal{D}_6$ is the contraction of the strict transform of the real line passing through $p_1, \bar{p_1}$, onto a real point $q \in \mathcal{D}_6$. Note that $p_1, \bar{p_1}$ are proper points of $\mathbb{P}^2$, blown-up by φ_1 and $p_2, \bar{p_2}$ either are proper base-points or are infinitely near to $p_1, \bar{p_1}$.

The fibration $\mathcal{D}_6 \to \mathbb{P}^1$ corresponds to conics through $p_1, \bar{p_1}, p_2, \bar{p_2}$. If φ_3 was a link of type III, then $\varphi_3\varphi_2$ would be biregular, so φ_3 is of type II.

If q is a base-point of φ_3, then $\varphi_3 = \eta'\eta^{-1}$, where $\eta'\colon X \to \mathcal{D}_6$ is the contraction of the strict transform of the line through $p_2, \bar{p_2}$. We can then write $\varphi_3\varphi_2\varphi_1$ into only two links, exchanging p_1 with p_2 and $\bar{p_1}$ with $\bar{p_2}$, and this decreases r and preserves property (ii) on real base-points.

The remaining case is when φ_3 is the blow-up of two non-real points $p_3, \bar{p_3}$ of $\mathcal{D}_6$, followed by the contraction of the strict transforms of their fibres. We denote by $q' \in \mathcal{D}_6(\mathbb{R})$ the image of q by φ_3, consider $\psi_4 = (\varphi_2)^{-1}\colon \mathcal{D}_6 \to Q_{3,1}$, which is a link of type III, and write $\psi_5\colon Q_{3,1} \dashrightarrow \mathbb{P}^2$ the stereographic projection by $\psi_4(q')$, which is a link of type II centered at $\psi_4(q')$. By Lemma 3.8, the map $\chi = \psi_5\psi_4\varphi_3\varphi_2\varphi_1$ is an element of $\mathrm{Aut}(\mathbb{P}^2(\mathbb{R}))$ of degree 5. Since $\varphi\chi^{-1}$ decomposes into one link less than φ, with a decomposition having still property (ii), this concludes the proof by induction. □

Proof of Theorem 1.2. By Proposition 3.9, $\mathrm{Aut}(\mathbb{P}^2(\mathbb{R}))$ is generated by $\mathrm{Aut}_{\mathbb{R}}(\mathbb{P}^2)$ and by elements of $\mathrm{Aut}(\mathbb{P}^2(\mathbb{R}))$ of degree 5. Thanks to Lemma 3.7, $\mathrm{Aut}(\mathbb{P}^2(\mathbb{R}))$ is indeed generated by projectivities and standard quintic transformations. □

4 Generators of the Group $\mathrm{Bir}_{\mathbb{R}}(\mathbb{P}^2)$

Lemma 4.1. *Let $\varphi\colon Q_{3,1} \dashrightarrow Q_{3,1}$ be a birational map that decomposes as $\varphi = \varphi_3\varphi_2\varphi_1$, where $\varphi_i\colon X_{i-1} \dashrightarrow X_i$ is a Sarkisov link for each i, where $X_0 = Q_{3,1} = X_2$, $X_1 = \mathcal{D}_6$. If φ_2 has a real base-point, then φ can be written as $\varphi = \psi_2\psi_1$, where $\psi_1, (\psi_2)^{-1}$ are links of type* II *from $Q_{3,1}$ to $\mathbb{P}^2$.*

Proof. We have the following commutative diagram, where each of the maps η_1, η_2 blow-ups a real point, and each of the maps $(\varphi_1)^{-1}, \varphi_3$ is the blow-up of two conjugate non-real points.

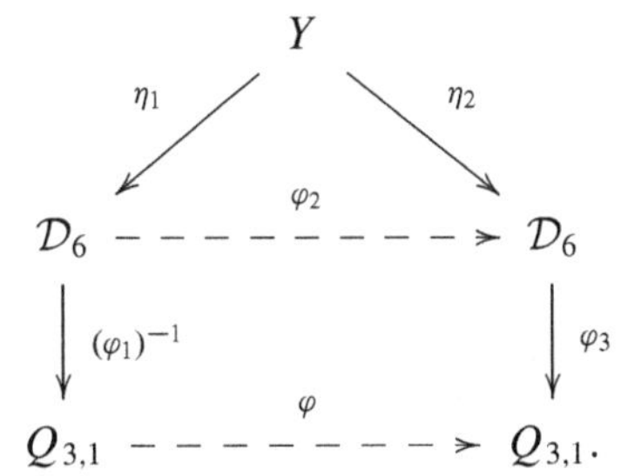

The map φ has thus exactly three base-points, two of them being non-real and one being real; we denote them by $p_1, \bar{p}_1, q$. The fibres of the Mori fibration $\mathcal{D}_6 \to \mathbb{P}^1$ correspond to conics of $\mathcal{Q}_{3,1}$ passing through the points $p_1, \bar{p}_1$. The real curve contracted by η_2 is thus the strict transform of the conic C of $\mathcal{Q}_{3,1}$ passing through $p_1, \bar{p}_1$ and q. The two curves contracted by φ_3 are the two non-real sections of self-intersection -1, which corresponds to the strict transforms of the two non-real lines L_1, L_2 of $\mathcal{Q}_{3,1}$ passing through q.

We can then decompose φ as the blow-up of p_1, p_2, q, followed by the contraction of the strict transforms of C, L_1, L_2. Denote by $\psi_1 \colon \mathcal{Q}_{3,1} \dashrightarrow \mathbb{P}^2$ the link of type II centered at q, which is the blow-up of q followed by the contraction of the strict transform of L_1, L_2, or equivalently the stereographic projection centered at q. The curve $\psi_1(C)$ is a real line of $\mathbb{P}^2$, which contains the two points $\psi_1(p_1)$, $\psi_1(\bar{p}_1)$. The map $\psi_2 = \varphi(\psi_1)^{-1} \colon \mathbb{P}^2 \dashrightarrow \mathcal{Q}_{3,1}$ is then the blow-up of these two points, followed by the contraction of the line passing through both of them. It is then a link of type II. □

Proof of Theorem 1.1. Let us prove that any $\varphi \in \mathrm{Bir}_{\mathbb{R}}(\mathbb{P}^2)$ is in the group generated by $\mathrm{Aut}_{\mathbb{R}}(\mathbb{P}^2)$, σ_0, σ_1, and standard quintic transformations of $\mathbb{P}^2$. We decompose φ into Sarkisov links: $\varphi = \varphi_r \cdots \varphi_1$. By Corollary 2.19, we can assume that all the φ_i are links of type I, III, or standard links of type II.

We proceed by induction on r, the case $r = 0$, corresponding to $\varphi \in \mathrm{Aut}_{\mathbb{R}}(\mathbb{P}^2)$, being obvious.

Note that φ_1 is either a link of type I from $\mathbb{P}^2$ to $\mathbb{F}_1$, or a link of type II from $\mathbb{P}^2$ to $\mathcal{Q}_{3,1}$. We now study the possibilities for the base-points of φ_1 and the next links:

(1) Suppose that $\varphi_1 \colon \mathbb{P}^2 \dashrightarrow \mathbb{F}_1$ is a link of type I, and that φ_2 is a link $\mathbb{F}_1 \dashrightarrow \mathbb{F}_1$. Then, φ_2 blows-up two non-real base-points of $\mathbb{F}_1$, not lying on the exceptional curve. Hence, $\psi = (\varphi_1)^{-1}\varphi_2\varphi_1$ is a quadratic transformation of $\mathbb{P}^2$ with three proper base-points, one real and two non-real. It is thus equal to $\alpha\sigma_1\beta$ for some $\alpha, \beta \in \mathrm{Aut}_{\mathbb{R}}(\mathbb{P}^2)$. Replacing φ with $\varphi\psi^{-1}$, we obtain a decomposition with less Sarkisov links, and conclude by induction.

(2) Suppose that $\varphi_1 \colon \mathbb{P}^2 \dashrightarrow \mathbb{F}_1$ is a link of type I, and that φ_2 is a link $\mathbb{F}_1 \dashrightarrow \mathbb{F}_0$. Then, $\varphi_2\varphi_1$ is the blow-up of two real points p_1, p_2 of $\mathbb{P}^2$ followed by the contraction of the line through p_1, p_2. The exceptional divisors corresponding to p_1, p_2 are two (0)-curves of $\mathbb{F}_0 = \mathbb{P}^1 \times \mathbb{P}^1$, intersecting at one real point.

($2a$) Suppose first that φ_3 has a base-point which is real, and not lying on E_1, E_2. Then, $\psi = (\varphi_1)^{-1}\varphi_3\varphi_2\varphi_1$ is a quadratic transformation of $\mathbb{P}^2$ with three proper base-points, all real. It is thus equal to $\alpha\sigma_0\beta$ for some $\alpha, \beta \in \mathrm{Aut}_{\mathbb{R}}(\mathbb{P}^2)$. Replacing φ with $\varphi\psi^{-1}$, we obtain a decomposition with less Sarkisov links and conclude by induction.

($2b$) Suppose now that φ_3 has non-real base-points, which are $q, \bar{q}$. Since φ_3 is a standard link of type II, it goes from $\mathbb{F}_0$ to $\mathbb{F}_0$, so q and $\bar{q}$ do not lie on a (0)-curve, and then do not belong to the curves E_1, E_2. We can then decompose $\varphi_2\varphi_3 \colon \mathbb{F}_1 \dashrightarrow \mathbb{F}_2$ into a Sarkisov link centered at two non-real points, followed by a Sarkisov link centered at a real point. This reduces to case (1), already treated before.

($2c$) The remaining case (for (2)) is when φ_3 has a base-point p_3 which is real, but lying on E_1 or E_2. We choose a general real point $p_4 \in \mathbb{F}_0$, and denote by $\theta \colon \mathbb{F}_0 \dashrightarrow \mathbb{F}_1$ a Sarkisov link centered at p_4. We observe that $\psi = (\varphi_1)^{-1}\theta\varphi_2\varphi_1$ is a quadratic map as in case ($2a$), and that $\varphi\psi^{-1} = \varphi_n \ldots \varphi_3\theta^{-1}\varphi_1$ admits now a decomposition of the same length, but which is in case ($2a$).

(3) Suppose now that $\varphi_1 \colon \mathbb{P}^2 \dashrightarrow Q_{3,1}$ is a link of type II and that φ_2 is a link of type II from $Q_{3,1}$ to $\mathbb{P}^2$. If φ_2 and $(\varphi_1)^{-1}$ have the same real base-point, the map $\varphi_2\varphi_1$ belongs to $\mathrm{Aut}_{\mathbb{R}}(\mathbb{P}^2)$. Otherwise, $\varphi_2\varphi_1$ is a quadratic map with one unique real base-point q and two non-real base-points. It is then equal to $\alpha\sigma_0\beta$ for some $\alpha, \beta \in \mathrm{Aut}_{\mathbb{R}}(\mathbb{P}^2)$. We conclude as before by induction hypothesis.

(4) Suppose that $\varphi_1 \colon \mathbb{P}^2 \dashrightarrow Q_{3,1}$ is a link of type II and φ_2 is a link of type I from $Q_{3,1}$ to $\mathcal{D}_6$. If φ_3 is a Sarkisov link of type III, then $\varphi_3\varphi_2$ is an automorphism of $Q_{3,1}$, so we can decrease the length. We only need to consider the case where φ_3 is a link of type II from $\mathcal{D}_6$ to $\mathcal{D}_6$. If φ_3 has a real base-point, we apply Lemma 4.1 to write $(\varphi_2)^{-1}\varphi_3\varphi_2 = \psi_2\psi_1$ where $\psi_1, (\psi_2)^{-1}$ are links $Q_{3,1} \dashrightarrow \mathbb{P}^2$. By (3), the map $\chi = \psi_1\varphi_1$ is generated by $\mathrm{Aut}_{\mathbb{R}}(\mathbb{P}^2)$ and σ_0. We can then replace φ with $\varphi\chi^{-1} = \varphi_r \cdots \varphi_3\varphi_2(\psi_1)^{-1} = \varphi_r \cdots \varphi_4\varphi_2\psi_2$, which has a shorter decomposition. The last case is when φ_3 has two non-real base-points. We denote by $q \in Q_{3,1}$ the real base-point of $(\varphi_1)^{-1}$, write $q' = (\varphi_2)^{-1}\varphi_3\varphi_2(q) \in Q_{3,1}$ and denote by $\psi \colon Q_{3,1} \dashrightarrow \mathbb{P}^2$ the stereographic projection centered at q'. By Lemma 3.8, the map $\chi = \psi(\varphi_2)^{-1}\varphi_3\varphi_2\varphi_1$ is an automorphism of $\mathbb{P}^2(\mathbb{R})$ of degree 5, which is generated by $\mathrm{Aut}_{\mathbb{R}}(\mathbb{P}^2)$ and standard automorphisms of $\mathbb{P}^2(\mathbb{R})$ of degree 5 (Lemma 3.7). We can thus replace φ with $\varphi\chi^{-1}$, which has a decomposition of shorter length. □

5 Generators of the Group $\mathrm{Aut}(Q_{3,1}(\mathbb{R}))$

Example 5.1. Let $p_1, \bar{p}_1, p_2, \bar{p}_2 \in Q_{3,1} \subset \mathbb{P}^3$ be two pairs of conjugate non-real points, not on the same plane of $\mathbb{P}^3$. Let $\pi \colon X \to Q_{3,1}$ be the blow-up of these points. The non-real plane of $\mathbb{P}^3$ passing through $p_1, \bar{p}_2, \bar{p}_2$ intersects $Q_{3,1}$ onto a conic, having self-intersection 2: two general different conics on $Q_{3,1}$ are the trace of hyperplanes, and intersect then into two points, being on the line of intersection of the two planes. The strict transform of this conic on X is thus a (-1)-curve.

Doing the same for the other conics passing through 3 of the points $p_1, \bar{p}_1, p_2, \bar{p}_2$, we obtain four disjoint (-1)-curves on X, that we can contract in order to obtain a birational morphism $\eta: X \to Q_{3,1}$; note that the target is $Q_{3,1}$ because it is a smooth projective rational surface of Picard rank 1. We obtain then a birational map $\psi = \eta\pi^{-1}: Q_{3,1} \dashrightarrow Q_{3,1}$ inducing an isomorphism $Q_{3,1}(\mathbb{R}) \to Q_{3,1}(\mathbb{R})$.

Denote by $H \subset Q_{3,1}$ a general hyperplane section. The strict transform of H on X by π^{-1} has self-intersection 2 and has intersection 2 with the 4 curves contracted. The image $\psi(H)$ has thus multiplicity 2 and self-intersection 18; it is then the trace of a cubic section. The construction of ψ and ψ^{-1} being similar, the linear system of ψ consists of cubic sections with multiplicity 2 at $p_1, \bar{p}_1, p_2, \bar{p}_2$.

Example 5.2. Let $p_1, \bar{p}_1 \in Q_{3,1} \subset \mathbb{P}^3$ be two conjugate non-real points and let $\pi_1: X_1 \to Q_{3,1}$ be the blow-up of the two points. Denote by $E_1, \bar{E}_1 \subset X_1$ the curves contracted onto $p_1, \bar{p}_1$ respectively. Let $p_2 \in E_1$ be a point, and $\bar{p}_2 \in \bar{E}_1$ its conjugate. We assume that there is no conic of $Q_{3,1} \subset \mathbb{P}^3$ passing through $p_1, \bar{p}_1, p_2, \bar{p}_2$ and let $\pi_2: X_2 \to X_1$ be the blow-up of $p_2, \bar{p}_2$.

On X, the strict transforms of the two conics $C, \bar{C}$ of $\mathbb{P}^2$, passing through $p_1, \bar{p}_1, p_2$ and $p_1, \bar{p}_1, \bar{p}_2$ respectively, are non-real conjugate disjoint (-1) curves. The contraction of these two curves gives a birational morphism $\eta_2: X_2 \to Y_1$. On this latter surface, we find two disjoint conjugate non-real (-1)-curves. These are the strict transforms of the exceptional curves associated with $p_1, \bar{p}_1$. The contraction of these curves gives a birational morphism $\eta_1: Y_1 \to Q_{3,1}$. The birational map $\psi = \eta_1\eta_2(\pi_1\pi_2)^{-1}: Q_{3,1} \dashrightarrow Q_{3,1}$ induces an isomorphism $Q_{3,1}(\mathbb{R}) \to Q_{3,1}(\mathbb{R})$.

Definition 5.3. The birational maps of $Q_{3,1}$ of degree 3 obtained in Example 5.1 will be called *standard cubic transformations* and those of Example 5.2 will be called *special cubic transformations*.

Note that since $\mathrm{Pic}(Q_{3,1}) = \mathbb{Z}H$, where H is an hyperplane section, we can associate with any birational map $Q_{3,1} \dashrightarrow Q_{3,1}$, an integer d, which is the *degree of the map*, such that $\psi^{-1}(H) = dH$.

Lemma 5.4. *Let $\psi: Q_{3,1} \dashrightarrow Q_{3,1}$ be a birational map inducing an isomorphism $Q_{3,1}(\mathbb{R}) \to Q_{3,1}(\mathbb{R})$. The following hold:*

(1) *The degree of ψ is $2k+1$ for some integer $k \geq 0$.*
(2) *If ψ has degree* 1, *it belongs to* $\mathrm{Aut}_{\mathbb{R}}(Q_{3,1}) = \mathrm{PO}(3,1)$.
(3) *If ψ has degree 3, then it is a standard or special cubic transformation, described in Examples 5.1 and 5.2, and has thus exactly* 4 *base-points.*
(4) *If ψ has at most* 4 *base-points, then ψ has degree* 1 *or* 3.

Proof. Denote by d the degree of ψ and by $a_1, \ldots, a_n$ the multiplicities of the base-points of ψ. Denote by $\pi: X \to Q_{3,1}$ the blow-up of the base-points, and by $E_1, \ldots, E_n \in \mathrm{Pic}(X)$ the divisors being the total pull-back of the exceptional (-1)-curves obtained after blowing-up the points. Writing $\eta: X \to Q_{3,1}$ the birational morphism $\psi\pi$, we obtain

$$\begin{aligned} \eta^*(H) &= d\pi^*(H) - \textstyle\sum_{i=1}^n a_i E_i \\ K_X &= \pi^*(-2H) + \textstyle\sum_{i=1}^n E_i. \end{aligned}$$

Since H corresponds to a smooth rational curve of self-intersection 2, we have $(\eta^*(H))^2$ and $\eta^*(H) \cdot K_X = -4$. We find then

$$\begin{aligned} 2 = \quad (\eta^*(H))^2 &= 2d^2 - \textstyle\sum_{i=1}^n (a_i)^2 \\ 4 = -K_X \cdot \eta^*(H) &= 4d - \textstyle\sum_{i=1}^n a_i. \end{aligned}$$

Since multiplicities come by pairs, $n = 2m$ for some integer m and we can order the a_i so that $a_i = a_{n+1-i}$ for $i = 1, \dots, m$. This yields

$$\begin{aligned} d^2 - 1 &= \textstyle\sum_{i=1}^m (a_i)^2 \\ 2(d-1) &= \textstyle\sum_{i=1}^m a_i \end{aligned}$$

Since $(a_i)^2 \equiv a_i \pmod 2$, we find $d^2 - 1 \equiv 2(d-1) \equiv 0 \pmod 2$, hence d is odd. This gives (1).

If the number of base-points is at most 4, we can choose $m = 2$, and obtain by Cauchy–Schwartz

$$4(d-1)^2 = \left(\sum_{i=1}^m a_i\right)^2 \leq m \sum_{i=1}^m (a_i)^2 = m(d^2-1) = 2(d^2-1).$$

This yields $2(d-1) \leq d+1$, hence $d \leq 3$.

If $d = 1$, all a_i are zero, and $\psi \in \mathrm{Aut}_{\mathbb{R}}(Q_{3,1})$.

If $d = 3$, we get $\sum_{i=1}^m (a_i)^2 = 8$, $\sum_{i=1}^m a_i = 4$, so $m = 2$ and $a_1 = a_2 = 2$. Hence, the base-points of ψ consist of two pairs of conjugate non-real points $p_1, \bar{p_1}, p_2, \bar{p_2}$. Moreover, if a conic passes through 3 of the points, its free intersection with the linear system is zero, so it is contracted by ψ, and there is no conic through the four points.

(*a*) If the four points belong to $Q_{3,1}$, the map is a standard cubic transformation, described in Example 5.1.

(*b*) If two points are infinitely near, the map is a special cubic transformation, described in Example 5.2. □

Lemma 5.5. *Let $\varphi: Q_{3,1} \dashrightarrow Q_{3,1}$ be a birational map that decomposes as $\varphi = \varphi_3\varphi_2\varphi_1$, where $\varphi_i: X_{i-1} \dashrightarrow X_i$ is a Sarkisov link for each i, where $X_0 = Q_{3,1} = X_2$, $X_1 = \mathcal{D}_6$. If φ_2 is an automorphism of $\mathcal{D}_6(\mathbb{R})$, then φ is a cubic automorphism of $Q_{3,1}(\mathbb{R})$ of degree* 3 *described in Examples 5.1 and 5.2. Moreover, φ is a standard cubic transformation if and only if the link φ_2 of type* II *is a standard link of type* II.

Proof. We have the following commutative diagram, where each of the maps π_1, π_2, $(\varphi_1)^{-1}$, φ_3 is the blow-up of two conjugate non-real points.

$$\begin{array}{ccccc} & & Y & & \\ & \swarrow^{\pi_1} & & \searrow^{\pi_2} & \\ \mathcal{D}_6 & & \overset{\varphi_2}{\dashrightarrow} & & \mathcal{D}_6 \\ \downarrow{\scriptstyle (\varphi_1)^{-1}} & & & & \downarrow{\scriptstyle \varphi_3} \\ \mathcal{Q}_{3,1} & & \overset{\varphi}{\dashrightarrow} & & \mathcal{Q}_{3,1}. \end{array}$$

Hence, φ is an automorphism of $\mathbb{P}^2(\mathbb{R})$ with at most 4 base-points. We can moreover see that $\varphi \notin \mathrm{Aut}_{\mathbb{R}}(\mathcal{Q}_{3,1})$, since the two curves of Y contracted by π_2 are sent by $\varphi_3\pi_2$ onto conics of $\mathcal{Q}_{3,1}$, contracted by φ^{-1}.

Lemma 3.5 implies that φ is cubic automorphism of $\mathcal{Q}_{3,1}(\mathbb{R})$ of degree 3 described in Examples 5.1 and 5.2. In particular, φ has exactly four base-points, blown-up by $(\varphi_1)^{-1}\pi_1$. Moreover, φ is a standard cubic transformation if and only these four points are proper base-points of $\mathcal{Q}_{3,1}$. This corresponds to saying that the two base-points of φ_2 do not belong to the exceptional curves contracted by $(\varphi_1)^{-1}$, and is thus the case exactly when φ_2 is a standard link of type II. □

Proof of Theorem 1.3. Let us prove that any $\varphi \in \mathrm{Aut}(\mathcal{Q}_{3,1}(\mathbb{R}))$ is generated by $\mathrm{Aut}_{\mathbb{R}}(\mathcal{Q}_{3,1})$ and standard cubic transformations of $\mathrm{Aut}(\mathcal{Q}_{3,1}(\mathbb{R}))$ of degree 3. Applying Proposition 2.7, we decompose φ into Sarkisov links: $\varphi = \varphi_r \cdots \varphi_1$, and assume that every real base-point of φ_i is a base-point of $\varphi_r \ldots \varphi_i$. This property implies that all links are either of type I, from $\mathcal{Q}_{3,1}$ to $\mathcal{D}_6$, of type II from $\mathcal{D}_6$ to $\mathcal{D}_6$ with non-real base-points, or of type III from $\mathcal{D}_6$ to $\mathcal{Q}_{3,1}$. In particular, all base-points of the φ_i and their inverses are non-real. (Note that here the situation is easier than in the case of $\mathbb{P}^2$, since no link produces "artificial" real base-points).

By Lemma 2.18, we can also assume that all links of type II are standard.

We proceed by induction on r. The first link φ_1 is of type I from $\mathcal{Q}_{3,1}$ to $\mathcal{D}_6$. If φ_2 is of type III, then $\varphi_2\varphi_1 \in \mathrm{Aut}_{\mathbb{R}}(\mathcal{Q}_{3,1})$. We replace these two links and conclude by induction. If φ_2 is a standard link of type II, then $\psi = (\varphi_1)^{-1}\varphi_2\varphi_1$ is a standard cubic transformation. Replacing φ with $\varphi\psi^{-1}$ decreases the number of links, so we conclude by induction. □

5.1 *Twisting Maps and Factorisation*

Choose a real line $L \subset \mathbb{P}^3$, which does not meet $\mathcal{Q}_{3,1}(\mathbb{R})$. The projection from L gives a morphism $\pi_L \colon \mathcal{Q}_{3,1}(\mathbb{R}) \to \mathbb{P}^1(\mathbb{R})$, which induces a conic bundle structure on the blow-up $\tau_L \colon \mathcal{D}_6 \to \mathcal{Q}_{3,1}$ of the two non-real points of $L \cap \mathcal{Q}_{3,1}$.

We denote by $T(\mathcal{Q}_{3,1}, \pi_L) \subset \mathrm{Aut}(\mathcal{Q}_{3,1}(\mathbb{R}))$ the group of elements $\varphi \in \mathrm{Aut}(\mathcal{Q}_{3,1}(\mathbb{R}))$ such that $\pi_L\varphi = \pi_L$ and such that the lift $(\tau_L)^{-1}\varphi\tau_L \in \mathrm{Aut}(\mathcal{D}_6(\mathbb{R}))$

preserves the set of two non-real (-1)-curves which are sections of the conic bundle $\pi_L \tau_L$.

Any element $\varphi \in T(Q_{3,1}, \pi_L)$ is called a *twisting map of* $Q_{3,1}$ *with axis* L.

Choosing the line $w = x = 0$ for L, we can get the more precise description given in [6, 11]: the twisting maps corresponds in local coordinates $(x, y, z) \mapsto (1 : x : y : z)$ to

$$\varphi_M\colon (x, y, z) \mapsto \big(x, (y, z) \cdot M(x)\big)$$

where $M\colon [-1, 1] \to O(2) \subset \mathrm{PGL}(2, \mathbb{R}) = \mathrm{Aut}(\mathbb{P}^1)$ is a real algebraic map.

Proposition 5.6. *Any twisting map with axis L is a composition of twisting maps with axis L, of degree* 1 *and* 3.

Proof. We can assume that L is the line $y = z = 0$.

The blow-up $\tau_L\colon \mathcal{D}_6 \to Q_{3,1}$ is a link of type III, described in Example 2.13(3), which blows-up two non-real points of $Q_{3,1}$. The fibres of the Mori Fibration $\pi\colon \mathcal{D}_6 \to \mathbb{P}^1$ correspond then, via τ_L, to the fibres of $\pi_L\colon Q_{3,1}(\mathbb{R}) \to \mathbb{P}^1(\mathbb{R})$. Hence, a twisting map of $Q_{3,1}$ corresponds to a map of the form $\tau\varphi\tau^{-1}$, where $\varphi\colon \mathcal{D}_6 \dashrightarrow \mathcal{D}_6$ is a birational map such that $\pi\varphi = \pi$, and which preserves the set of two (-1)-curves. This implies that φ has all its base-points on the two (-1)-curves. It remains to argue as in Lemma 2.18 and decompose φ into links that have only base-points on the set of two (-1)-curves. □

6 Generators of the Group Aut($\mathbb{F}_0(\mathbb{R})$)

Proof of Theorem 1.4. Let us prove that any $\varphi \in \mathrm{Aut}(\mathbb{F}_0(\mathbb{R}))$ is generated by $\mathrm{Aut}_{\mathbb{R}}(\mathbb{F}_0)$ and by the involution

$$\tau_0\colon ((x_0 : x_1), (y_0 : y_1)) \dashrightarrow ((x_0 : x_1), (x_0y_0 + x_1y_1 : x_1y_0 - x_0y_1)).$$

Observe that τ_0 is a Sarkisov link $\mathbb{F}_0 \dashrightarrow \mathbb{F}_0$ that is the blow-up of the two non-real points $p = ((\mathbf{i} : 1), (\mathbf{i} : 1))$, $\bar{p} = ((-\mathbf{i} : 1), (-\mathbf{i} : 1))$, followed by the contraction of the two fibres of the first projection $\mathbb{F}_0 \to \mathbb{P}^1$ passing through $p, \bar{p}$.

Applying Proposition 2.7, we decompose φ into Sarkisov links: $\varphi = \varphi_r \cdots \varphi_1$, and assume that every real base-point of φ_i is a base-point of $\varphi_r \ldots \varphi_i$. This property implies that all links are either of type IV from $\mathbb{F}_0$ to $\mathbb{F}_0$, or of type II, from $\mathbb{F}_{2d}$ to $\mathbb{F}_{2d'}$, with exactly two non-real base-points. In particular, as for the case of $Q_{3,1}$, there is no real base-point which is artificially created.

By Lemma 2.18, we can also assume that all links of type II are standard, so all go from $\mathbb{F}_0$ to $\mathbb{F}_0$.

Each link of type IV is an element of $\mathrm{Aut}_{\mathbb{R}}(\mathbb{F}_0)$.

Each link φ_i of type II consists of the blow-up of two non-real points $q, \bar{q}$, followed by the contraction of the fibres of the first projection $\mathbb{F}_0 \to \mathbb{P}^1$ passing through $q, \bar{q}$. Since the two points do not belong to the same fibre by any projection, we have $q = ((a + \mathbf{i}b : 1), (c + \mathbf{i}d : 1))$, for some $a, b, c, d \in \mathbb{R}$, $bd \neq 0$. There exists thus an element $\alpha \in \mathrm{Aut}_{\mathbb{R}}(\mathbb{F}_0)$ that sends q onto p and then $\bar{q}$ onto $\bar{p}$. In consequence, $\tau_0\alpha(\varphi_i)^{-1} \in \mathrm{Aut}_{\mathbb{R}}(\mathbb{P}^2)$. This yields the result. □

7 Other Results

7.1 *Infinite Transitivity on Surfaces*

The group of automorphisms of a complex projective algebraic variety is small: in most of the cases it is a finite dimensional algebraic group. Moreover, the group of automorphisms is 3-transitive only if the variety is $\mathbb{P}^1$. On the other hand, it was proved in [6] that for a real rational surface X, the group of automorphisms $\mathrm{Aut}(X(\mathbb{R}))$ acts n-transitively on $X(\mathbb{R})$ for any n. The next theorem determines all real algebraic surfaces X having a group of automorphisms which acts infinitely transitively on $X(\mathbb{R})$.

Definition 7.1. Let G be a topological group acting continuously on a topological space M. We say that two n-tuples of distinct points $(p_1, \dots, p_n)$ and $(q_1, \dots, q_n)$ are *compatible* if there exists an homeomorphism $\psi: M \to M$ such that $\psi(p_i) = q_i$ for each i. The action of G on M is then said to be *infinitely transitive* if for any pair of compatible n-tuples of points $(p_1, \dots, p_n)$ and $(q_1, \dots, q_n)$ of M, there exists an element $g \in G$ such that $g(p_i) = q_i$ for each i. More generally, the action of G is said to be infinitely transitive *on each connected component* if we require the above condition only in case, for each i, p_i and q_i belong to the same connected component of M.

Theorem 7.2 ([2]). *Let X be a nonsingular real projective surface. The group $\mathrm{Aut}\big(X(\mathbb{R})\big)$ is then infinitely transitive on each connected component if and only if X is geometrically rational and $\#X(\mathbb{R}) \leq 3$.*

7.2 *Density of Automorphisms in Diffeomorphisms*

In [11], it is proved that $\mathrm{Aut}\big(X(\mathbb{R})\big)$ is dense in $\mathrm{Diff}\big(X(\mathbb{R})\big)$ for the $\mathcal{C}^\infty$-topology when X is a geometrically rational surface with $\#X(\mathbb{R}) = 1$ (or equivalently when X is rational). In the cited paper, it is said that $\#X(\mathbb{R}) = 2$ is probably the only other case where the density holds. The following collect the known results in this direction.

Theorem 7.3 ([2, 11]).
Let X be a smooth real projective surface.

- *If X is not a geometrically rational surface, then* $\overline{\mathrm{Aut}(X(\mathbb{R}))} \neq \mathrm{Diff}(X(\mathbb{R}))$*;*
- *If X is a geometrically rational surface, then*
  - *If* $\#X(\mathbb{R}) \geq 5$*, then* $\overline{\mathrm{Aut}(X(\mathbb{R}))} \neq \mathrm{Diff}(X(\mathbb{R}))$*;*
  - *if* $\#X(\mathbb{R}) = 1$*, then* $\overline{\mathrm{Aut}(X(\mathbb{R}))} = \mathrm{Diff}(X(\mathbb{R}))$.

For $i = 3, 4$, there exists smooth real projective surfaces X with $\#X(\mathbb{R}) = i$ such that $\overline{\mathrm{Aut}(X(\mathbb{R}))} \neq \mathrm{Diff}(X(\mathbb{R}))$.

In the above statements, the closure is taken in the C^∞-topology.

Acknowledgements The first author was supported by the SNSF grant no. PP00P2_128422 /1. This research was partially supported by ANR Grant "BirPol" ANR-11-JS01-004-01.

References

1. M. Alberich-Carramiñana, in *Geometry of the Plane Cremona Maps*. Lecture Notes in Mathematics, vol. 1769 (Springer, Berlin, 2002)
2. J. Blanc, F. Mangolte, Geometrically rational real conic bundles and very transitive actions. Compos. Math. **147**(1), 161–187 (2011)
3. G. Castelnuovo, Le trasformazioni generatrici del gruppo cremoniano nel piano. Atti della R. Accad. delle Scienze di Torino **36**, 861–874 (1901)
4. A. Comessatti, Sulla connessione delle superficie razionali reali. Ann. Mat. Pura Appl. **23**, 215–283 (1914)
5. A. Corti, Factoring birational maps of threefolds after Sarkisov. J. Algebr. Geom. **4**(2), 223–254 (1995)
6. J. Huisman, F. Mangolte, The group of automorphisms of a real rational surface is n-transitive. Bull. Lond. Math. Soc. **41**, 563–568 (2009)
7. V.A. Iskovskikh, Minimal models of rational surfaces over arbitrary fields. Izv. Akad. Nauk SSSR Ser. Mat. **43**(1), 19–43, 237 (1979)
8. V.A. Iskovskikh, Generators of the two-dimensional Cremona group over a nonclosed field. Trudy Mat. Inst. Steklov. **200**, 157–170 (1991); translation in Proc. Steklov Inst. Math. **200**(2), 173–188 (1993)
9. V.A. Iskovskikh, Factorization of birational mappings of rational surfaces from the point of view of Mori theory. Uspekhi Mat. Nauk **51**(4(310)), 3–72 (1996)
10. V.A. Iskovskikh, F.K. Kabdykairov, S.L. Tregub, Relations in a two-dimensional Cremona group over a perfect field. Izv. Ross. Akad. Nauk Ser. Mat. **57**(3), 3–69 (1993)
11. J. Kollár, F. Mangolte, Cremona transformations and diffeomorphisms of surfaces. Adv. Math. **222**, 44–61 (2009)
12. F. Mangolte, Real algebraic morphisms on 2-dimensional conic bundles. Adv. Geom. **6**, 199–213 (2006)
13. S. Mori, Threefolds whose canonical bundles are not numerically effective. Ann. Math. **116**(2), 133–176 (1982)
14. Yu. Polyakova, Factorization of birational mappings of rational surfaces over the field of real numbers. Fundam. Prikl. Mat. **3**(2), 519–547 (1997)
15. F. Ronga, T. Vust, Diffeomorfismi birazionali del piano proiettivo reale. Comm. Math. Helv. **80**, 517–540 (2005)

On Automorphisms and Endomorphisms of Projective Varieties

Michel Brion

Abstract We first show that any connected algebraic group over a perfect field is the neutral component of the automorphism group scheme of some normal projective variety. Then we show that very few connected algebraic semigroups can be realized as endomorphisms of some projective variety X, by describing the structure of all connected subsemigroup schemes of $\mathrm{End}(X)$.

MSC classes: 14J50, 14L30, 20M20

1 Introduction and Statement of the Results

By a result of Winkelmann (see [22]), every connected real Lie group G can be realized as the automorphism group of some complex Stein manifold X, which may be chosen complete, and hyperbolic in the sense of Kobayashi. Subsequently, Kan showed in [14] that we may further assume $\dim_{\mathbb{C}}(X) = \dim_{\mathbb{R}}(G)$.

We shall obtain a somewhat similar result for connected algebraic groups. We first introduce some notation and conventions and recall general results on automorphism group schemes.

Throughout this chapter, we consider schemes and their morphisms over a fixed field k. Schemes are assumed to be separated; subschemes are locally closed unless mentioned otherwise. By a *point* of a scheme S, we mean a T-valued point $f : T \to S$ for some scheme T. A *variety* is a geometrically integral scheme of finite type.

M. Brion (✉)
Institut Fourier, Université de Grenoble, B.P. 74, 38402 Saint-Martin d'Hères Cedex, France
e-mail: Michel.Brion@ujf-grenoble.fr

I. Cheltsov et al. (eds.), *Automorphisms in Birational and Affine Geometry*, Springer Proceedings in Mathematics & Statistics 79,
DOI 10.1007/978-3-319-05681-4_4,

We shall use [18] as a general reference for group schemes. We denote by e_G the neutral element of a group scheme G, and by G^o the neutral component. An *algebraic group* is a smooth group scheme of finite type.

Given a projective scheme X, the functor of automorphisms,

$$T \longmapsto \mathrm{Aut}_T(X \times T),$$

is represented by a group scheme, locally of finite type, that we denote by $\mathrm{Aut}(X)$. The Lie algebra of $\mathrm{Aut}(X)$ is identified with the Lie algebra of global vector fields, $\mathrm{Der}(\mathcal{O}_X)$ (these results hold more generally for projective schemes over an arbitrary base, see [12, p. 268]; they also hold for proper schemes of finite type over a field, see [17, Theorem 3.7]). In particular, the neutral component, $\mathrm{Aut}^o(X)$, is a group scheme of finite type; when k is perfect, the reduced subscheme, $\mathrm{Aut}^o(X)_{\mathrm{red}}$, is a connected algebraic group. As a consequence, $\mathrm{Aut}^o(X)$ is a connected algebraic group if $\mathrm{char}(k) = 0$, since every group scheme of finite type is reduced under that assumption. Yet $\mathrm{Aut}^o(X)$ is not necessarily reduced in prime characteristics (see e.g. the examples in [17, Sect. 4]).

We may now state our first result:

Theorem 1. *Let G be a connected algebraic group, and n its dimension.*
If $\mathrm{char}(k) = 0$, *then there exists a smooth projective variety X such that* $\mathrm{Aut}^o(X) \cong G$ *and* $\dim(X) = 2n$.
If $\mathrm{char}(k) > 0$ *and k is perfect, then there exists a normal projective variety X such that* $\mathrm{Aut}^0_{\mathrm{red}}(X) \cong G$ *and* $\dim(X) = 2n$ *(resp.* $\mathrm{Aut}^o(X) \cong G$ *and* $\dim(X) = 2n + 2$).

This result is proved in Sect. 2, first in the case where $\mathrm{char}(k) = 0$; then we adapt the arguments to the case of prime characteristics, which is technically more involved due to group schemes issues. We rely on fundamental results about the structure and actions of algebraic groups over an algebraically closed field, for which we refer to the recent exposition [5].

Theorem 1 leaves open many basic questions about automorphism group schemes. For instance, can one realize every connected algebraic group over an arbitrary field (or more generally, every connected group scheme of finite type) as the full automorphism group scheme of a normal projective variety? Also, very little seems to be known about the group of components, $\mathrm{Aut}(X)/\mathrm{Aut}^o(X)$, where X is a projective variety. In particular, the question of the finite generation of this group is open, already when X is a complex projective manifold.

As a consequence of Theorem 1, we obtain the following characterization of Lie algebras of vector fields:

Corollary 1. *Let $\mathfrak{g}$ be a finite-dimensional Lie algebra over a field k of characteristic* 0. *Then the following conditions are equivalent:*

(i) $\mathfrak{g} \cong \mathrm{Der}(\mathcal{O}_X)$ *for some proper scheme X of finite type.*
(ii) $\mathfrak{g}$ *is the Lie algebra of a linear algebraic group.*

Under either condition, X may be chosen projective, smooth, and unirational of dimension $2n$, where $n := \dim(\mathfrak{g})$. If k is algebraically closed, then we may further choose X rational.

This result is proved in Sect. 2.3. The Lie algebras of linear algebraic groups over a field of characteristic 0 are called algebraic Lie algebras; they have been characterized by Chevalley in [6, 7]. More specifically, a finite-dimensional Lie algebra $\mathfrak{g}$ is algebraic if and only if its image under the adjoint representation is an algebraic Lie subalgebra of $\mathfrak{gl}(\mathfrak{g})$ (see [7, Chap. V, Sect. 5, Proposition 3]). Moreover, the algebraic Lie subalgebras of $\mathfrak{gl}(V)$, where V is a finite-dimensional vector space, are characterized in [6, Chap. II, Sect. 14]. Also, recall a result of Hochschild (see [13]): the isomorphism classes of algebraic Lie algebras are in bijective correspondence with the isomorphism classes of connected linear algebraic groups with unipotent center.

In characteristic $p > 0$, one should rather consider restricted Lie algebras, also known as p-Lie algebras. In this setting, characterizing Lie algebras of vector fields seems to be an open question. This is related to the question of characterizing automorphism group schemes, via the identification of restricted Lie algebras with infinitesimal group schemes of height ≤ 1 (see [18, Exp. VIIA, Theorem 7.4]).

Next, we turn to the monoid schemes of endomorphisms of projective varieties; we shall describe their connected subsemigroup schemes. For this, we recall basic results on schemes of morphisms.

Given two projective schemes X and Y, the functor of morphisms,

$$T \longmapsto \mathrm{Hom}_T(X \times T, Y \times T) \cong \mathrm{Hom}(X \times T, Y),$$

is represented by an open subscheme of the Hilbert scheme $\mathrm{Hilb}(X \times Y)$, by assigning to each morphism its graph (see [12, p. 268], and [15, Sect. 1.10], [19, Sect. 4.6.6] for more details). We denote that open subscheme by $\mathrm{Hom}(X, Y)$. The composition of morphisms yields a natural transformation of functors, and hence a morphism of schemes

$$\mathrm{Hom}(X, Y) \times \mathrm{Hom}(Y, Z) \longrightarrow \mathrm{Hom}(X, Z), \quad (f, g) \longmapsto gf$$

where Z is another projective scheme.

As a consequence of these results, the functor of endomorphisms of a projective scheme X is represented by a scheme, $\mathrm{End}(X)$; moreover, the composition of endomorphisms equips $\mathrm{End}(X)$ with a structure of monoid scheme with neutral element being of course the identity, id_X. Each connected component of $\mathrm{End}(X)$ is of finite type, and these components form a countable set. The automorphism group scheme $\mathrm{Aut}(X)$ is open in $\mathrm{End}(X)$ by [12, p. 267] (see also [15, Lemma I.1.10.1]). If X is a variety, then $\mathrm{Aut}(X)$ is also closed in $\mathrm{End}(X)$, as follows from [3, Lemma 4.4.4]; thus, $\mathrm{Aut}(X)$ is a union of connected components of $\mathrm{End}(X)$. In particular, $\mathrm{Aut}^o(X)$ is the connected component of id_X in $\mathrm{End}(X)$.

As another consequence, given a morphism $f : X \to Y$ of projective schemes, the functor of sections of f is represented by a scheme that we shall denote by $\mathrm{Sec}(f)$: the fiber at id_Y of the morphism

$$\lambda_f : \mathrm{Hom}(Y, X) \longrightarrow \mathrm{End}(Y), \quad g \longmapsto fg.$$

Every section of f is a closed immersion; moreover, $\mathrm{Sec}(f)$ is identified with an open subscheme of $\mathrm{Hilb}(X)$ by assigning to each section its image (see [12, p. 268] again; our notation differs from the one used there). Given a section $s \in \mathrm{Sec}(f)(k)$, we may identify Y with the closed subscheme $Z := s(Y)$; then f is identified with a *retraction* of X onto that subscheme, i.e., to a morphism $r : X \to Z$ such that $ri = \mathrm{id}_Z$, where $i : Z \to X$ denotes the inclusion. Moreover, the endomorphism $e := ir$ of X is *idempotent*, i.e., satisfies $e^2 = e$.

Conversely, every idempotent k-rational point of $\mathrm{End}(X)$ can be written uniquely as $e = ir$, where $i : Y \to X$ is the inclusion of the image of e (which coincides with its fixed point subscheme), and $r : X \to Y$ is a retraction. When X is a variety, Y is a projective variety as well. We now analyze the connected component of e in $\mathrm{End}(X)$:

Proposition 1. *Let X be a projective variety, $e \in \mathrm{End}(X)(k)$ an idempotent, and C the connected component of e in $\mathrm{End}(X)$. Write $e = ir$, where $i : Y \to X$ denotes the inclusion of a closed subvariety, and $r : X \to Y$ is a retraction.*

(i) *The morphism*

$$\rho_r : \mathrm{Hom}(Y, X) \longrightarrow \mathrm{End}(X), \quad f \longmapsto fr$$

restricts to an isomorphism from the connected component of i in $\mathrm{Hom}(Y, X)$, to C. Moreover, C is a subsemigroup scheme of $\mathrm{End}(X)$, and $f = fe$ for any $f \in C$.

(ii) *The morphism*

$$\lambda_i \rho_r : \mathrm{End}(Y) \longrightarrow \mathrm{End}(X), \quad f \longmapsto ifr$$

restricts to an isomorphism of semigroup schemes $\mathrm{Aut}^o(Y) \xrightarrow{\cong} eC$. In particular, eC is a group scheme with neutral element e.

(iii) *ρ_r restricts to an isomorphism from the connected component of i in $\mathrm{Sec}(r)$, to the subscheme $E(C)$ of idempotents in C. Moreover, $f_1 f_2 = f_1$ for all $f_1, f_2 \in E(C)$; in particular, $E(C)$ is a closed subsemigroup scheme of C.*

(iv) *The morphism*

$$\varphi : E(C) \times eC \longrightarrow C, \quad (f, g) \longmapsto fg$$

is an isomorphism of semigroup schemes, where the semigroup law on the left-hand side is given by $(f_1, g_1) \cdot (f_2, g_2) = (f_1, g_1 g_2)$.

This is proved in Sect. 3.1 by using a version of the rigidity lemma (see [3, Sect. 4.4]). As a straightforward consequence, the maximal connected subgroup schemes of $\mathrm{End}(X)$ are exactly the $\lambda_i \rho_r(\mathrm{Aut}^o(Y))$ with the above notation (this fact is easily be checked directly).

As another consequence of Proposition 1, the endomorphism scheme of a projective variety can have everywhere nonreduced connected components, even in characteristic 0. Consider for example a ruled surface

$$r : X = \mathbb{P}(\mathcal{E}) \longrightarrow Y,$$

where Y is an elliptic curve and $\mathcal{E}$ is a locally free sheaf on Y which belongs to a nonsplit exact sequence

$$0 \longrightarrow \mathcal{O}_Y \longrightarrow \mathcal{E} \longrightarrow \mathcal{O}_Y \longrightarrow 0$$

(such a sequence exists in view of the isomorphisms $\mathrm{Ext}^1(\mathcal{O}_Y, \mathcal{O}_Y) \cong H^1(Y, \mathcal{O}_Y) \cong k$). Let $i : Y \to X$ be the section associated with the projection $\mathcal{E} \to \mathcal{O}_Y$. Then the image of i yields an isolated point of $\mathrm{Hilb}(X)$ with Zariski tangent space of dimension 1 (see e.g. [19, Exercise 4.6.7]). Thus, the connected component of i in $\mathrm{Sec}(r)$ is a nonreduced fat point. By Proposition 1(iv), the connected component of $e := ir$ in $\mathrm{End}(X)$ is isomorphic to the product of that fat point with $\mathrm{Aut}^o(Y) \cong Y$, and hence is nonreduced everywhere. This explains a posteriori why we have to be so fussy with semigroup schemes.

A further consequence of Proposition 1 is the following:

Proposition 2. *Let X be a projective variety, S a connected subsemigroup scheme of $\mathrm{End}(X)$, and $E(S) \subset S$ the closed subscheme of idempotents. Assume that S has a k-rational point.*

(i) *$E(S)$ is a connected subsemigroup scheme of S, with semigroup law given by $f_1 f_2 = f_1$. Moreover, $E(S)$ has a k-rational point.*
(ii) *For any $e \in E(S)(k)$, the closed subsemigroup scheme $eS \subset S$ is a group scheme. Moreover, the morphism*

$$\varphi : E(S) \times eS \longrightarrow S, \quad (f, g) \longmapsto fg$$

is an isomorphism of semigroup schemes.
(iii) *Identifying S with $E(S) \times eS$ via φ, the projection $\pi : S \to E(S)$ is the unique retraction of semigroup schemes from S to $E(S)$. In particular, π is independent of the choice of the k-rational idempotent e.*

This structure result is proved in Sect. 3.2; a new ingredient is the fact that a subsemigroup scheme of a group scheme of finite type is a subgroup scheme (Lemma 10). The case where S has no k-rational point is discussed in Remark 5 at the end of Sect. 3.2.

Proposition 2 yields strong restrictions on the structure of connected subsemigroup schemes of $\mathrm{End}(X)$, where X is a projective variety. For example, if such a subsemigroup scheme is commutative or has a neutral element, then it is just a group scheme.

As another application of that proposition, we shall show that the dynamics of an endomorphism of X which belongs to some algebraic subsemigroup is very restricted. To formulate our result, we need the following:

Definition 1. Let X be a projective variety, and f a k-rational endomorphism of X.

We say that f is *bounded*, if f belongs to a subsemigroup of finite type of $\mathrm{End}(X)$. Equivalently, the powers f^n, where $n \geq 1$, are all contained in a finite union of subvarieties of $\mathrm{End}(X)$.

We say that a k-rational point $x \in X$ is *periodic*, if x is fixed by some f^n.

Proposition 3. *Let f be a bounded endomorphism of a projective variety X.*

(i) *There exists a smallest closed algebraic subgroup $G \subset \mathrm{End}(X)$ such that $f^n \in G$ for all $n \gg 0$. Moreover, G is commutative.*
(ii) *When k is algebraically closed, f has a periodic point if and only if G is linear. If X is normal, this is equivalent to the assertion that some positive power f^n acts on the Albanese variety of X via an idempotent.*

This result is proved in Sect. 3.3. As a direct consequence, every bounded endomorphism of a normal projective variety X has a periodic point, whenever the Albanese variety of X is trivial (e.g., when X is unirational); we do not know if any such endomorphism has a fixed point. In characteristic 0, it is known that every endomorphism (not necessarily bounded) of a smooth projective unirational variety X has a fixed point: this follows from the Woods Hole formula (see [1, Theorem 2], [8, Exp. 3, Corollary 6.12]) in view of the vanishing of $H^i(X, \mathcal{O}_X)$ for all $i \geq 1$, proved e.g. in [20, Lemma 1].

Also, it would be interesting to extend the above results to endomorphism schemes of complete varieties. In this setting, the rigidity lemma of [3, Sect. 4.4] still hold. Yet the representability of the functor of morphisms by a scheme is unclear: the Hilbert functor of a complete variety is generally not represented by a scheme (see e.g. [15, Exercise 5.5.1]), but no such example seems to be known for morphisms.

2 Proofs of Theorem 1 and of Corollary 1

2.1 *Proof of Theorem 1 in Characteristic* 0

We begin by setting notation and recalling a standard result of Galois descent, for any perfect field k.

We fix an algebraic closure $\bar{k}$ of k and denote by Γ the Galois group of $\bar{k}/k$. For any scheme X, we denote by $X_{\bar{k}}$ the $\bar{k}$-scheme obtained from X by the base change $\mathrm{Spec}(\bar{k}) \to \mathrm{Spec}(k)$. Then $X_{\bar{k}}$ is equipped with an action of Γ such that the structure map $X_{\bar{k}} \to \mathrm{Spec}(\bar{k})$ is equivariant; moreover, the natural morphism $X_{\bar{k}} \to X$ may be viewed as the quotient by this action. The assignment $Y \mapsto Y_{\bar{k}}$ defines a bijective correspondence between closed subschemes of X and Γ-stable closed subschemes of $X_{\bar{k}}$.

Next, recall Chevalley's structure theorem: every connected algebraic group G has a largest closed connected linear normal subgroup L, and the quotient G/L is an abelian variety (see e.g. [5, Theorem. 1.1.1] when $k = \bar{k}$; the general case follows by the above result of Galois descent).

We shall also need the existence of a normal projective equivariant compactification of G, in the following sense:

Lemma 1. *There exists a normal projective variety Y equipped with an action of $G \times G$ and containing an open orbit isomorphic to G, where $G \times G$ acts on G by left and right multiplication.*

Proof. When k is algebraically closed, this statement is [5, Proposition 3.1.1 (iv)]. For an arbitrary k, we adapt the argument of [loc. cit.].

If $G = L$ is linear, then we may identify it to a closed subgroup of some GL_n, and hence of PGL_{n+1}. The latter group has an equivariant compactification by the projectivization, $\mathbb{P}(M_{n+1})$, of the space of matrices of size $(n+1) \times (n+1)$, where $\mathrm{PGL}_{n+1} \times \mathrm{PGL}_{n+1}$ acts via the action of $\mathrm{GL}_{n+1} \times \mathrm{GL}_{n+1}$ on M_{n+1} by left and right matrix multiplication. Thus, we may take for Y the normalization of the closure of L in $\mathbb{P}(M_{n+1})$.

In the general case, choose a normal projective equivariant compactification Z of L and let

$$Y := G \times^L Z \longrightarrow G/L$$

be the fiber bundle associated with the principal L-bundle $G \to G/L$ and with the L-variety Z, where L acts on the left. Then Y is a normal projective variety, since so is L and hence L has an ample $L \times L$-linearized line bundle. Moreover, Y is equipped with a G-action having an open orbit, $G \times^L L \cong G$.

We now extend this G-action to an action of $G \times G$, where the open G-orbit is identified to the $G \times G$-homogeneous space $(G \times G)/\mathrm{diag}(G)$, and the original G-action, to the action of $G \times e_G$. For this, consider the scheme-theoretic center $Z(G)$ (resp. $Z(L)$) of G (resp. L). Then $Z(L) = Z(G) \cap L$, since $Z(L)_{\bar{k}} = Z(G)_{\bar{k}} \cap L_{\bar{k}}$ in view of [5, Proposition 3.1.1 (ii)]. Moreover, $G_{\bar{k}} = Z(G)_{\bar{k}} L_{\bar{k}}$ by [loc. cit.]; hence the natural map $Z(G)/Z(L) \to G/L$ is an isomorphism of group schemes. It follows that G/L is isomorphic to

$$(Z(G) \times Z(G))/(Z(L) \times Z(L))\mathrm{diag}(Z(G)) \cong (G \times G)/(L \times L)\mathrm{diag}(Z(G)).$$

Moreover, the $L \times L$-action on Z extends to an action of $(L \times L)\mathrm{diag}(Z(G))$, where $Z(G)$ acts trivially: indeed, $(L \times L)\mathrm{diag}(Z(G))$ is isomorphic to

$$(L \times L \times Z(G))/(L \times L) \cap \mathrm{diag}(Z(G)) = (L \times L \times Z(G))/\mathrm{diag}(Z(L)),$$

and the subgroup scheme $\mathrm{diag}(Z(L)) \subset L \times L$ acts trivially on Z by construction. This yields an isomorphism

$$G \times^{L} Z \cong (G \times G) \times^{(L\times L)\mathrm{diag}(Z(G))} Z,$$

which provides the desired action of $G \times G$. □

From now on in this subsection, we assume that $\mathrm{char}(k) = 0$. We shall construct the desired variety X from the equivariant compactification Y, by proving a succession of lemmas.

Denote by $\mathrm{Aut}^G(Y)$ the subgroup scheme of $\mathrm{Aut}(Y)$ consisting of automorphisms which commute with the left G-action. Then the right G-action on Y yields a homomorphism

$$\varphi : G \longrightarrow \mathrm{Aut}^G(Y).$$

Lemma 2. *With the above notation, φ is an isomorphism.*

Proof. Note that $\mathrm{Aut}^G(Y)$ stabilizes the open orbit for the left G-action, and this orbit is isomorphic to G. This yields a homomorphism $\mathrm{Aut}^G(Y) \to \mathrm{Aut}^G(G)$. Moreover, $\mathrm{Aut}^G(G) \cong G$ via the action of G on itself by right multiplication, and the resulting homomorphism $\psi : \mathrm{Aut}^G(Y) \to G$ is readily seen to be inverse of φ. □

Lemma 3. *There exists a finite subset $F \subset G(\bar{k})$ which generates a dense subgroup of $G_{\bar{k}}$.*

Proof. We may assume that $k = \bar{k}$. If the statement holds for some closed normal subgroup H of G and for the quotient group G/H, then it clearly holds for G. Thus, we may assume that G is simple, in the sense that it has no proper closed connected normal subgroup. Then, by Chevalley's structure theorem, G is either a linear algebraic group or an abelian variety. In the latter case, there exists $g \in G(k)$ of infinite order, and every such point generates a dense subgroup of G (actually, every abelian variety, not necessarily simple, is topologically generated by some k-rational point, see [10, Theorem 9]). In the former case, G is either the additive group $\mathbb{G}_a$, the multiplicative group $\mathbb{G}_m$, or a connected semisimple group. Therefore, G is generated by finitely many copies of $\mathbb{G}_a$ and $\mathbb{G}_m$, each of which is topologically generated by some k-rational point (specifically, by any nonzero $t \in k$ for $\mathbb{G}_a$, and by any $u \in k^*$ of infinite order for $\mathbb{G}_m$). □

Choose $F \subset G(\bar{k})$ as in Lemma 3. We may further assume that F contains id_Y and is stable under the action of the Galois group Γ; then $F = E_{\bar{k}}$ for a unique finite reduced subscheme $E \subset G$. We have

$$\mathrm{Aut}^F(Y_{\bar{k}}) = \mathrm{Aut}^{G(\bar{k})}(Y_{\bar{k}}) = \mathrm{Aut}^{G_{\bar{k}}}(Y_{\bar{k}})$$

and the latter is isomorphic to $G_{\bar{k}}$ via φ, in view of Lemma 2. Thus, φ yields an isomorphism $G \cong \mathrm{Aut}^E(Y)$.

Next, we identify G with a subgroup of $\mathrm{Aut}(Y \times Y)$ via the closed embedding of group schemes

$$\iota : \mathrm{Aut}(Y) \longrightarrow \mathrm{Aut}(Y \times Y), \quad \varphi \longmapsto \varphi \times \varphi.$$

For any $f \in F$, let $\Gamma_f \subset Y_{\bar{k}} \times Y_{\bar{k}}$ be the graph of f; in particular, Γ_{id_Y} is the diagonal, $\mathrm{diag}(Y_{\bar{k}})$. Then there exists a unique closed reduced subscheme $Z \subset Y \times Y$ such that $Z_{\bar{k}} = \bigcup_{f \in F} \Gamma_f$. We may now state the following key observation:

Lemma 4. *With the above notation, we have*

$$\iota(G) = \mathrm{Aut}^o(Y \times Y, Z),$$

where the right-hand side denotes the neutral component of the stabilizer of Z in $\mathrm{Aut}(Y \times Y)$.

Proof. We may assume that $k = \bar{k}$, so that $Z = \bigcup_{f \in F} \Gamma_f$. Moreover, by connectedness, $\mathrm{Aut}^o(Y \times Y, Z)$ is the neutral component of the intersection $\bigcap_{f \in F} \mathrm{Aut}(Y \times Y, \Gamma_f)$. On the other hand, $\mathrm{Aut}^o(Y \times Y, Z) \subset \mathrm{Aut}^o(Y \times Y)$, and the latter is isomorphic to $\mathrm{Aut}^o(Y) \times \mathrm{Aut}^o(Y)$ via the natural homomorphism

$$\mathrm{Aut}^o(Y) \times \mathrm{Aut}^o(Y) \longrightarrow \mathrm{Aut}^o(Y \times Y), \quad (\varphi, \psi) \longmapsto \varphi \times \psi$$

(see [5, Corollary 4.2.7]). Also, $\varphi \times \psi$ stabilizes a graph Γ_f if and only if $\psi f = f \varphi$. In particular, $\varphi \times \psi$ stabilizes $\mathrm{diag}(Y) = \Gamma_{\mathrm{id}_Y}$ iff $\psi = \varphi$, and $\varphi \times \varphi$ stabilizes Γ_f iff φ commutes with f. As a consequence, $\mathrm{Aut}^o(Y \times Y, Z)$ is the neutral component of $\iota(\mathrm{Aut}^F(Y))$. Since $\mathrm{Aut}^F(Y) = G$ is connected, this yields the assertion. □

Next, denote by X the normalization of the blow-up of $Y \times Y$ along Z. Then X is a normal projective variety equipped with a birational morphism

$$\pi : X \longrightarrow Y \times Y$$

which induces a homomorphism of group schemes

$$\pi^* : G \longrightarrow \mathrm{Aut}(X),$$

since Z is stable under the action of G on $Y \times Y$.

Lemma 5. *Keep the above notation and assume that $n \geq 2$. Then π^* yields an isomorphism of algebraic groups $G \xrightarrow{\cong} \mathrm{Aut}^o(X)$.*

Proof. It suffices to show the assertion after base change to $\mathrm{Spec}(\bar{k})$; thus, we may assume again that $k = \bar{k}$.

The morphism π is proper and birational, and $Y \times Y$ is normal (since normality is preserved under separable field extension). Thus, $\pi_*(\mathcal{O}_X) = \mathcal{O}_{Y \times Y}$ by Zariski's Main Theorem. It follows that π induces a homomorphism of algebraic groups

$$\pi_* : \mathrm{Aut}^o(X) \to \mathrm{Aut}^o(Y \times Y)$$

(see e.g. [5, Corollary 4.2.6]). In particular, $\mathrm{Aut}^o(X)$ preserves the fibers of π, and hence stabilizes the exceptional divisor E of that morphism; as a consequence, the image of π_* stabilizes $\pi(E)$. By connectedness, this image stabilizes every irreducible component of $\pi(E)$; but these components are exactly the graphs Γ_f, where $f \in F$ (since the codimension of any such graph in $Y \times Y$ is $\dim(Y) = n \geq 2$). Thus, the image of π_* is contained in $\iota(G)$: we may view π_* as a homomorphism $\mathrm{Aut}^o(X) \to G$. Since π is birational, both maps π^*, π_* are injective and the composition $\pi_* \pi^*$ is the identity. It follows that π_* is inverse to π^*. □

We may now complete the proof of Theorem 1 when $n \geq 2$. Let X be as in Lemma 5; then X admits an equivariant desingularization, i.e., there exists a smooth projective variety X' equipped with an action of G and with a G-equivariant birational morphism

$$f : X' \longrightarrow X$$

(see [16, Proposition 3.9.1, Theorem 3.36]). We check that the resulting homomorphism of algebraic groups

$$f^* : G \longrightarrow \mathrm{Aut}^o(X')$$

is an isomorphism. For this, we may assume that $k = \bar{k}$; then again, [5, Corollary 4.2.6] yields a homomorphism of algebraic groups

$$f_* : \mathrm{Aut}^o(X') \longrightarrow \mathrm{Aut}^o(X) = G$$

which is easily seen to be inverse of f^*.

On the other hand, if $n = 1$, then G is either an elliptic curve, or $\mathbb{G}_a$, or a k-form of $\mathbb{G}_m$. We now construct a smooth projective surface X such that $\mathrm{Aut}^o(X) \cong G$, via case-by-case elementary arguments.

When G is an elliptic curve, we have $G \cong \mathrm{Aut}^o(G)$ via the action of G by translations on itself. It follows that $G \cong \mathrm{Aut}^o(G \times C)$, where C is any smooth projective curve of genus ≥ 2.

When $G = \mathbb{G}_a$, we view G as the group of automorphisms of the projective line $\mathbb{P}^1$ that fix the point ∞ and the tangent line at that point, $T_\infty(\mathbb{P}^1)$. Choose $x \in \mathbb{P}^1(k)$ such that $0, x, \infty$ are all distinct, and let X be the smooth projective surface obtained by blowing up $\mathbb{P}^1 \times \mathbb{P}^1$ at the three points $(\infty, 0)$, (∞, x), and (∞, ∞). Arguing as in the proof of Lemma 5, one checks that $\mathrm{Aut}^o(X)$ is isomorphic to the neutral component of the stabilizer of these three points, in $\mathrm{Aut}^o(\mathbb{P}^1 \times \mathbb{P}^1) \cong \mathrm{PGL}_2 \times \mathrm{PGL}_2$. This identifies $\mathrm{Aut}^o(X)$ with the stabilizer of ∞ in PGL_2, i.e, with the automorphism group Aff_1 of the affine line, acting on the first copy of $\mathbb{P}^1$. Thus, $\mathrm{Aut}^o(X)$ acts on each exceptional line via the natural action of Aff_1 on $\mathbb{P}(T_\infty(\mathbb{P}^1) \oplus k)$, with an obvious notation; this action factors through an action of $\mathbb{G}_m = \mathrm{Aff}_1/\mathbb{G}_a$, isomorphic to the $\mathbb{G}_m$-action on $\mathbb{P}^1$ by multiplication. Let X' be the smooth projective surface obtained by blowing up X at a k-rational point of some exceptional line, distinct from 0 and ∞; then $\mathrm{Aut}^o(X') \cong \mathbb{G}_a$.

Finally, when G is a k-form of $\mathbb{G}_m$, we consider the smooth projective curve C that contains G as a dense open subset; then C is a k-form of the projective line $\mathbb{P}^1$ on which $\mathbb{G}_m$ acts by multiplication. Thus, the complement $P := C \setminus G$ is a point of degree 2 on C (a k-form of $\{0, \infty\}$); moreover, G is identified with the stabilizer of P in $\mathrm{Aut}(C)$. Let X be the smooth projective surface obtained by blowing up $C \times C$ at $(P \times P) \cup (P \times e_G)$, where the neutral element e_G is viewed as a k-point of C. Arguing as in the proof of Lemma 5 again, one checks that $\mathrm{Aut}^o(X)$ is isomorphic to the neutral component of the stabilizer of $(P \times P) \cup (P \times e_G)$ in $\mathrm{Aut}^o(C \times C) \cong \mathrm{Aut}^0(C) \times \mathrm{Aut}^0(C)$, i.e., to G acting on the first copy of C. □

Remark 1. One may ask for analogues of Theorem 1 for automorphism groups of compact complex spaces. Given any such space X, the group of biholomorphisms, $\mathrm{Aut}(X)$, has the structure of a complex Lie group acting biholomorphically on X (see [9]). If X is Kähler, or more generally in Fujiki's class $\mathcal{C}$, then the neutral component $\mathrm{Aut}^0(X) =: G$ has a meromorphic structure, i.e., a compactification G^* such that the multiplication $G \times G \to G$ extends to a meromorphic map $G^* \times G^* \to G^*$ which is holomorphic on $(G \times G^*) \cup (G^* \times G)$; moreover, G is Kähler and acts biholomorphically and meromorphically on X (see [11, Theorem 5.5, Corollary 5.7]).

Conversely, every connected meromorphic Kähler group of dimension n is the connected automorphism group of some compact Kähler manifold of dimension $2n$; indeed, the above arguments adapt readily to that setting. But it seems to be unknown whether any connected complex Lie group can be realized as the connected automorphism group of some compact complex manifold.

2.2 *Proof of Theorem 1 in Prime Characteristic*

In this subsection, the base field k is assumed to be perfect, of characteristic $p > 0$. Let Y be an equivariant compactification of G as in Lemma 1. Consider the closed subgroup scheme $\mathrm{Aut}^G(Y) \subset \mathrm{Aut}(Y)$, defined as the centralizer of G acting on the left; then the G-action on the right still yields a homomorphism of group schemes $\varphi : G \to \mathrm{Aut}^G(Y)$.

As in Lemma 2, *φ is an isomorphism.* To check this claim, note that φ induces an isomorphism $G(\bar{k}) \xrightarrow{\cong} \mathrm{Aut}^G(Y)(\bar{k})$ by the argument of that lemma. Moreover, the induced homomorphism of Lie algebras

$$\mathrm{Lie}(\varphi) : \mathrm{Lie}(G) \longrightarrow \mathrm{Lie}\,\mathrm{Aut}^G(Y)$$

is an isomorphism as well: indeed, $\mathrm{Lie}(\varphi)$ is identified with the natural map

$$\psi : \mathrm{Lie}(G) \longrightarrow \mathrm{Der}^G(\mathcal{O}_Y),$$

where $\mathrm{Der}^G(\mathcal{O}_Y)$ denotes the Lie algebra of left G-invariant derivations of $\mathcal{O}_Y$. Furthermore, the restriction to the open dense subset G of Y yields an injective map

$$\eta : \mathrm{Der}^G(\mathcal{O}_Y) \longrightarrow \mathrm{Der}^G(\mathcal{O}_G) \cong \mathrm{Lie}(G)$$

such that $\eta\psi = \mathrm{id}$; thus, η is the inverse of ψ. It follows that $\mathrm{Aut}^G(Y)$ is reduced; this completes the proof of the claim.

Next, *Lemma 3 fails in positive characteristics*, already for $\mathbb{G}_a$ since every finite subset of $\bar{k}$ generates a finite additive group; that lemma also fails for $\mathbb{G}_m$ when $\bar{k}$ is the algebraic closure of a finite field. Yet we have the following replacement:

Lemma 6. *With the above notation, there exists a finite subset F of $G(\bar{k})$ such that $\mathrm{Aut}^{G_{\bar{k}}}(Y_{\bar{k}}) = \mathrm{Aut}^{F,o}(Y_{\bar{k}})$, where the right-hand side denotes the neutral component of $\mathrm{Aut}^F(Y_{\bar{k}})$.*

Proof. We may assume that $k = \bar{k}$; then $\mathrm{Aut}^G(Y) = \mathrm{Aut}^{G(k)}(Y)$. The subgroup schemes $\mathrm{Aut}^{E,o}(Y)$, where E runs over the finite subsets of $G(k)$, form a family of closed subschemes of $\mathrm{Aut}^o(Y)$. Thus, there exists a minimal such subgroup scheme, say, $\mathrm{Aut}^{F,o}(Y)$. For any $g \in G(k)$, the subgroup scheme $\mathrm{Aut}^{F\cup\{g\},o}(Y)$ is contained in $\mathrm{Aut}^{F,o}(Y)$; thus, equality holds by minimality. In other words, $\mathrm{Aut}^{F,o}(Y)$ centralizes g; hence F satisfies the assertion. □

It follows from Lemmas 2 and 6 that $\mathrm{Aut}^{F,o}(Y_{\bar{k}}) \cong G_{\bar{k}}$ for some finite subset $F \subset G(\bar{k})$; we may assume again that F contains id_Y and is stable under the action of the Galois group Γ. Thus, $G \cong \mathrm{Aut}^E(Y)$, where $E \subset G$ denotes the finite reduced subscheme such that $E_{\bar{k}} = F$.

Next, *Lemma 4 still holds* with the same proof, in view of [5, Corollary 4.2.7]. In other words, we may again identify G with the connected stabilizer in $\mathrm{Aut}(Y \times Y)$ of the unique closed reduced subscheme $Z \subset Y \times Y$ such that $Z_{\bar{k}} = \bigcup_{f\in F} \Gamma_f$.

Consider again the morphism $\pi : X \to Y \times Y$ obtained as the normalization of the blow-up of Z. Then X is a normal projective variety, and π induces a homomorphism of group schemes

$$\pi^* : G \longrightarrow \mathrm{Aut}^o(X).$$

Now the statement of Lemma 5 adapts as follows:

Lemma 7. *Keep the above notation and assume that $n \geq 2$. Then π^* yields an isomorphism of algebraic groups $G \xrightarrow{\cong} \mathrm{Aut}^o(X)_{\mathrm{red}}$.*

Proof. Using the fact that normality is preserved under separable field extension, we may assume that $k = \bar{k}$. By [5, Corollary 4.2.6] again, we have a homomorphism of group schemes

$$\pi_* : \mathrm{Aut}^o(X) \longrightarrow \mathrm{Aut}^o(Y \times Y)$$

and hence a homomorphism of algebraic groups

$$\pi_{*,\mathrm{red}} : \mathrm{Aut}^o(X)_{\mathrm{red}} \longrightarrow \mathrm{Aut}^o(Y \times Y)_{\mathrm{red}}.$$

Arguing as in the proof of Lemma 5, one checks that $\pi_{*,\mathrm{red}}$ maps $\mathrm{Aut}^o(X)_{\mathrm{red}}$ onto $\iota(G)$, and is injective on k-rational points. Also, the homomorphism of Lie algebras $\mathrm{Lie}(\pi_{*,\mathrm{red}})$ is injective, as it extends to a homomorphism

$$\mathrm{Lie}(\pi_*) : \mathrm{Lie}\,\mathrm{Aut}^o(X) = \mathrm{Der}(\mathcal{O}_X) \longrightarrow \mathrm{Der}(\mathcal{O}_{Y\times Y}) = \mathrm{Lie}\,\mathrm{Aut}^o(Y \times Y)$$

which is injective, since π is birational. Thus, we obtain an isomorphism $\pi_{*,\mathrm{red}} : \mathrm{Aut}^o(X)_{\mathrm{red}} \xrightarrow{\cong} \iota(G)$ which is the inverse of π^*. □

To realize G as a connected automorphism group scheme, we now prove:

Lemma 8. *With the above notation, the homomorphism of Lie algebras*

$$\mathrm{Lie}(\pi^*) : \mathrm{Lie}(G) \longrightarrow \mathrm{Der}(\mathcal{O}_X)$$

is an isomorphism if $n \geq 2$ and $n - 1$ is not a multiple of p.

Proof. We may assume again that $k = \bar{k}$. Since π is birational, both maps $\mathrm{Lie}(\pi^*)$ and $\mathrm{Lie}(\pi_*)$ are injective and the composition $\mathrm{Lie}(\pi_*)\,\mathrm{Lie}(\pi^*)$ is the identity of $\mathrm{Lie}(G)$. Thus, it suffices to show that the image of $\mathrm{Lie}(\pi_*)$ is contained in $\mathrm{Lie}(G)$. For this, we use the natural action of $\mathrm{Der}(\mathcal{O}_X)$ on the jacobian ideal of π, defined as follows. Consider the sheaf Ω^1_X of Kähler differentials on X. Recall that $\Omega^1_X \cong \mathcal{I}_{\mathrm{diag}(X)}/\mathcal{I}^2_{\mathrm{diag}(X)}$ with an obvious notation; thus, Ω^1_X is equipped with an $\mathrm{Aut}(X)$-linearization (see [18, Exp. I, Sect. 6] for background on linearized sheaves, also called equivariant). Likewise, $\Omega^1_{Y\times Y}$ is equipped with an $\mathrm{Aut}(Y \times Y)$-linearization, and hence with an $\mathrm{Aut}^o(X)$-linearization via the homomorphism π_*. Moreover, the natural map $\pi^*(\Omega^1_{Y\times Y}) \to \Omega^1_X$ is a morphism of $\mathrm{Aut}^0(X)$-linearized sheaves, since it arises from the inclusion $\pi^{-1}(\mathcal{I}_{\mathrm{diag}(Y\times Y)}) \subset \mathcal{I}_{\mathrm{diag}(X)}$. This yields a morphism of $\mathrm{Aut}^o(X)$-linearized sheaves

$$\pi^*(\Omega^{2n}_{Y\times Y}) \longrightarrow \Omega^{2n}_X.$$

Since the composition

$$\Omega^{2n}_X \times \mathrm{Hom}(\Omega^{2n}_X, \mathcal{O}_X) \longrightarrow \mathcal{O}_X$$

is also a morphism of linearized sheaves, we obtain a morphism of linearized sheaves

$$\mathrm{Hom}(\Omega_X^{2n}, \pi^*(\Omega_{Y\times Y}^{2n})) \longrightarrow \mathcal{O}_X$$

with image the jacobian ideal $\mathcal{I}_\pi$. Thus, $\mathcal{I}_\pi$ is equipped with an $\mathrm{Aut}^o(X)$-linearization. In particular, for any open subset U of X, the Lie algebra $\mathrm{Der}(\mathcal{O}_X)$ acts on $\mathcal{O}(U)$ by derivations that stabilize $\Gamma(U, \mathcal{I}_\pi)$.

We now take $U = \pi^{-1}(V)$, where V denotes the open subset of $Y \times Y$ consisting of those smooth points that belong to at most one of the graphs Γ_f. Then the restriction

$$\pi_U : U \longrightarrow V$$

is the blow-up of the smooth variety V along a closed subscheme W, the disjoint union of smooth subvarieties of codimension n. Thus, $\mathcal{I}_{\pi_U} = \mathcal{O}_U(-(n-1)E)$, where E denotes the exceptional divisor of π_U. Hence we obtain an injective map

$$\mathrm{Der}(\mathcal{O}_X) = \mathrm{Der}(\mathcal{O}_X, \mathcal{I}_\pi) \longrightarrow \mathrm{Der}(\mathcal{O}_U, \mathcal{O}_U(-(n-1)E)),$$

with an obvious notation. Since $n-1$ is not a multiple of p, we have

$$\mathrm{Der}(\mathcal{O}_U, \mathcal{O}_U(-(n-1)E)) = \mathrm{Der}(\mathcal{O}_U, \mathcal{O}_U(-E)).$$

(Indeed, if $D \in \mathrm{Der}(\mathcal{O}_U, \mathcal{O}_U(-(n-1)E))$ and z is a local generator of $\mathcal{O}_U(-E)$ at $x \in X$, then $z^{n-1}\mathcal{O}_{X,x}$ contains $D(z^{n-1}) = (n-1)z^{n-2}D(z)$, and hence $D(z) \in z\mathcal{O}_{X,x}$). Also, the natural map

$$\mathrm{Der}(\mathcal{O}_U) \longrightarrow \mathrm{Der}(\pi_{U,*}(\mathcal{O}_U)) = \mathrm{Der}(\mathcal{O}_V)$$

is injective and sends $\mathrm{Der}(\mathcal{O}_U, \mathcal{O}_U(-E))$ to $\mathrm{Der}(\mathcal{O}_V, \pi_{U,*}(\mathcal{O}_U(-E)))$. Moreover, $\pi_{U,*}(\mathcal{O}_U(-E))$ is the ideal sheaf of W, and hence is stable under $\mathrm{Der}(\mathcal{O}_X)$ acting via the composition

$$\mathrm{Der}(\mathcal{O}_X) \longrightarrow \mathrm{Der}(\pi_*(\mathcal{O}_X)) = \mathrm{Der}(\mathcal{O}_{Y\times Y}) \longrightarrow \mathrm{Der}(\mathcal{O}_V).$$

It follows that the image of $\mathrm{Lie}(\pi_*)$ stabilizes the ideal sheaf of the closure of W in $Y \times Y$, i.e., of the union of the graphs Γ_f. In view of Lemma 4, we conclude that $\mathrm{Lie}(\pi_*)$ sends $\mathrm{Der}(\mathcal{O}_X)$ to $\mathrm{Lie}(G)$. □

Lemmas 7 and 8 yield an isomorphism $G \cong \mathrm{Aut}^o(X)$ when $n \geq 2$ and p does not divide $n-1$. Next, when $n \geq 2$ and p divides $n-1$, we choose a smooth projective curve C of genus $g \geq 2$, and consider $Y' := Y \times C$. This is a normal projective variety of dimension $n+1$, equipped with an action of $G \times G$. Moreover, we have isomorphisms

$$\mathrm{Aut}^o(Y) \xrightarrow{\cong} \mathrm{Aut}^o(Y) \times \mathrm{Aut}^o(C) \xrightarrow{\cong} \mathrm{Aut}^o(Y'), \quad \varphi \longmapsto \varphi \times \mathrm{id}_C$$

(where the second isomorphism follows again from [5, Corollary 4.2.6]); this identifies $G \cong \mathrm{Aut}^{G,o}(Y)$ with $\mathrm{Aut}^{G,o}(Y')$. We may thus replace everywhere Y with Y' in the above arguments, to obtain a normal projective variety X' of dimension $2n+2$ such that $\mathrm{Aut}^0(X') \cong G$.

Finally, if $n = 1$ then G is again an elliptic curve, or $\mathbb{G}_a$, or a k-form of $\mathbb{G}_m$ (since every form of $\mathbb{G}_a$ over a perfect field is trivial). It follows that $G \cong \mathrm{Aut}^o(X)$ for some smooth projective surface X, constructed as at the end of Sect. 2.1. □

Remark 2. If G is linear, then there exists a normal projective *unirational* variety X such that $\mathrm{Aut}^o(X)_{\mathrm{red}} \cong G$ and $\dim(X) = 2n$. Indeed, G itself is unirational (see [18, Exp. XIV, Corollary 6.10], and hence so is the variety X considered in the above proof when $n \geq 2$; on the other hand, when $n = 1$, the above proof yields a smooth projective rational surface X such that $\mathrm{Aut}^o(X) \cong G$. If in addition k is algebraically closed, then G is rational; hence we may further choose X rational.

Conversely, if X is a normal projective variety having a trivial Albanese variety (e.g., X is unirational), then $\mathrm{Aut}^o(X)$ is linear. Indeed, the Albanese variety of $\mathrm{Aut}^o(X)_{\mathrm{red}}$ is trivial in view of [2, Theorem 2]. Thus, $\mathrm{Aut}^o(X)_{\mathrm{red}}$ is affine by Chevalley's structure theorem. It follows that $\mathrm{Aut}^o(X)$ is affine, or equivalently linear.

Returning to a connected linear algebraic group G, the above proof adapts to show that there exists a normal projective unirational variety X such that $\mathrm{Aut}^o(X) \cong G$: in the argument after Lemma 8, it suffices to replace the curve C with a normal projective rational variety Z such that $\mathrm{Aut}^o(Z)$ is trivial and $\dim(Z) \geq 2$ is not a multiple of p. Such a variety may be obtained by blowing up $\mathbb{P}^2$ at 4 points in general position when $p \geq 3$; if $p = 2$, then we blow up $\mathbb{P}^3$ along a smooth curve which is neither rational nor contained in a plane.

Remark 3. It is tempting to generalize the above proof to the setting of an arbitrary base field k. Yet this raises many technical difficulties; for instance, Chevalley's structure theorem fails over any imperfect field (see [18, Exp. XVII, Appendix III, Proposition 5.1], and [21] for a remedy). Also, normal varieties need not be geometrically normal, and hence the differential argument of Lemma 8 also fails in that setting.

2.3 *Proof of Corollary 1*

(i)⇒(ii) Let $G := \mathrm{Aut}^o(X)$. Recall from [17, Lemma 3.4, Theorem 3.7] that G is a connected algebraic group with Lie algebra $\mathfrak{g}$. Also, recall that

$$G_{\bar{k}} = Z(G)_{\bar{k}}\, L_{\bar{k}},$$

where $Z(G)$ denotes the center of G, and L the largest closed connected normal linear subgroup of G. As a consequence, $\mathfrak{g}_{\bar{k}} = \mathrm{Lie}(Z(G))_{\bar{k}} + \mathrm{Lie}(L)_{\bar{k}}$. It follows that $\mathfrak{g} = \mathrm{Lie}(Z(G)) + \mathrm{Lie}(L)$, and hence we may choose a subspace $V \subset \mathrm{Lie}(Z(G))$ such that

$$\mathfrak{g} = V \oplus \mathrm{Lie}(L)$$

as vector spaces. This decomposition also holds as Lie algebras, since $[V, V] = 0 = [V, \mathrm{Lie}(L)]$. Hence $\mathfrak{g} = \mathrm{Lie}(U \times L)$, where U is the (commutative, connected) unipotent algebraic group with Lie algebra V.

(ii)⇒(i) Let G be a connected linear algebraic group such that $\mathfrak{g} = \mathrm{Lie}(G)$. By Theorem 1 and Remark 2, there exists a smooth projective unirational variety X of dimension $2n$ such that $G \cong \mathrm{Aut}^o(X)$; when k is algebraically closed, we may further choose X rational. Then of course $\mathfrak{g} \cong \mathrm{Der}(\mathcal{O}_X)$. □

3 Proofs of the Statements About Endomorphisms

3.1 Proof of Proposition 1

(i) Since C is connected and has a k-rational point, it is geometrically connected in view of [18, Exp. VIB, Lemma 2.1.2]. Likewise, the connected component of i in $\mathrm{Hom}(Y, X)$ is geometrically connected. To show the first assertion, we may thus assume that k is algebraically closed. But then that assertion follows from [3, Proposition 4.4.2, Remark 4.4.3].

The scheme-theoretic image of $C \times C$ under the morphism

$$\mathrm{End}(X) \times \mathrm{End}(X) \longrightarrow \mathrm{End}(X), \quad (f, g) \longmapsto gf$$

is connected and contains $e^2 = e$; thus, this image is contained in C. Therefore, C is a subsemigroup scheme of $\mathrm{End}(X)$. Also, every $g \in \mathrm{Hom}(Y, X)$ satisfies $gre = grir = gr$. Thus, $f = fe$ for any $f \in C$.

(ii) Since $(if_1r)(if_2r) = if_1f_2r$ for all $f_1, f_2 \in \mathrm{End}(Y)$, we see that $\lambda_i\rho_r$ is a homomorphism of semigroup schemes which sends id_Y to e. Also, $eifr = irifr = ifr$ for all $f \in \mathrm{End}(Y)$, so that $\lambda_i\rho_r$ sends $\mathrm{End}(Y)$ to $e\mathrm{End}(X)$. Since Y is a projective variety, $\mathrm{Aut}^o(Y)$ is the connected component of i in $\mathrm{End}(Y)$, and hence is sent by $\lambda_i\rho_r$ to $C \cap e\mathrm{End}(X) = eC$.

To show that $\lambda_i\rho_r$ is an isomorphism, note that $eC = eCe = irCir$ by (i). Moreover, the morphism

$$\lambda_r\rho_i : \mathrm{End}(X) \longrightarrow \mathrm{End}(Y), \quad f \longmapsto rfi$$

sends e to id_Y, and hence C to $\mathrm{Aut}^o(Y)$. Finally, $\lambda_r\rho_i(\lambda_i\rho_r(f)) = r(ifr)i = f$ for all $f \in \mathrm{End}(Y)$, and $\lambda_i\rho_r(\lambda_r\rho_i(f)) = i(rfi)r = efe = f$ for all $f \in eC$. Thus, $\lambda_r\rho_i$ is the desired inverse.

(iii) Let $f \in C$ such that $f^2 = f$. Then $fef = f$ by (i), and hence $ef \in eC$ is idempotent. But eC is a group scheme by (ii); thus, $ef = e$. Write $f = gr$, where g is a point of the connected component of i in $\mathrm{Hom}(Y, X)$. Then $egr = e$ and hence $rgr = r$, so that $rg = \mathrm{id}_Y$. Conversely, if $g \in \mathrm{Sec}(r)$, then gr is idempotent as already noted. This shows the first assertion. For the second assertion, just note that $(g_1r)(g_2r) = g_1r$ for all $g_1, g_2 \in \mathrm{Sec}(r)$.

(iv) We have with an obvious notation $\varphi(f_1, g_1)\varphi(f_2, g_2) = f_1g_1f_2g_2 = f_1g_1ef_2g_2$ by (i). Since $ef_2 = e$ by (iv), it follows that $\varphi(f_1, g_1)\varphi(f_2, g_2) = f_1g_1eg_2 = f_1g_1g_2$. Thus, φ is a homomorphism of semigroup schemes.

We now construct the inverse of φ. Let $f \in C$; then $ef \in eC$ has a unique inverse, $(ef)^{-1}$, in eC. Moreover, $f = fe = f(ef)^{-1}ef$ and $f(ef)^{-1}$ is idempotent, since

$$f(ef)^{-1}f(ef)^{-1} = f(ef)^{-1}ef(ef)^{-1} = f(ef)^{-1}e = f(ef)^{-1}.$$

We may thus define a morphism

$$\psi : C \longrightarrow E(C) \times eC, \quad f \longmapsto (f(ef)^{-1}, ef).$$

Then $\varphi\psi(f) = f(ef)^{-1}ef = fe = f$ for all $f \in C$, and $\psi\varphi(f, g) = (fg(efg)^{-1}, efg) = (fgg^{-1}, eg) = (f, g)$ for all $f \in E(C)$ and $g \in eC$. Thus, ψ is the desired inverse. □

3.2 *Proof of Proposition 2*

(i) Consider the connected component C of $\mathrm{End}(X)$ that contains S. Then C is of finite type, and hence so is S. Choose a k-rational point f of S and denote by $\langle f \rangle$ the smallest closed subscheme of S containing all the powers f^n, where $n \geq 1$. Then $\langle f \rangle$ is a reduced commutative subsemigroup scheme of S. By the main result of [4], it follows that $\langle f \rangle$ has an idempotent k-rational point. In particular, $E(S)$ has a k-rational point.

Since $E(S) \subset E(C)$, we have $f_1f_2 = f_1$ for any $f_1, f_2 \in E(S)$, by Proposition 1. It remains to show that $E(S)$ is connected; this will follow from (ii) in view of the connectedness of S.

(ii) By Proposition 1 again, φ yields an isomorphism $E(C) \times eC \xrightarrow{\cong} C$. Moreover, $fe = f$ for all $f \in C$, and $eC = eCe$ is a group scheme. Thus, $eS = eSe$ is a submonoid scheme of eC, and hence a closed subgroup scheme by Lemma 9 below. In other words, ef is invertible in eS for any $f \in S$. One may now check as in the proof of Proposition 1(iv) that the morphism

$$\psi : S \longrightarrow E(S) \times eS, \quad f \longmapsto (f(ef)^{-1}, ef)$$

yields an isomorphism of semigroup schemes, with inverse φ.

(iii) Since φ is an isomorphism of semigroup schemes, π is a homomorphism of such schemes. Moreover, $\pi(f) = f(ef)^{-1}$ for all $f \in S$, since ψ is the inverse of φ. If $f \in E(S)$, then $f = f^2 = fef$ and hence $\pi(f) = fef(ef)^{-1} = fe = f$. Thus, π is a retraction.

Let $\rho : S \to E(S)$ be a retraction of semigroup schemes. For any $f \in S$, we have $\rho(f) = \rho(f(ef)^{-1}ef) = \rho(f(ef)^{-1})\rho(ef)$. Moreover, $\rho(f(ef)^{-1}) = f(ef)^{-1}$, since $f(ef)^{-1} \in E(S)$; also, $\rho(ef) = \rho(ef)\rho((ef)^{-1}) = \rho(e) = e$. Hence $\rho(f) = f(ef)^{-1}e = f(ef)^{-1} = \pi(f)$. □

Lemma 9. *Let G be a group scheme of finite type, and $S \subset G$ a subsemigroup scheme. Then S is a closed subgroup scheme.*

Proof. We have to prove that S is closed and stable under the automorphism $g \mapsto g^{-1}$ of G. It suffices to check these assertions after base extension to any larger field; hence we may assume that k is algebraically closed.

Arguing as at the beginning of the proof of Proposition 2(i), we see that S has an idempotent k-rational point; hence S contains the neutral element, e_G. In other words, S is a submonoid scheme of G. By Lemma 10 below, there exists an open subgroup scheme $G(S) \subset S$ which represents the invertibles in S. In particular, $G(S)_{\mathrm{red}}$ is the unit group of the algebraic monoid S_{red}. Since that monoid has a unique idempotent, it is an algebraic group by [3, Proposition 2.2.5]. In other words, we have $G(S)_{\mathrm{red}} = S_{\mathrm{red}}$. As $G(S)$ is open in S, it follows that $G(S) = S$. Thus, S is a subgroup scheme of G, and hence is closed by [18, Exp. VIA, Corollary 0.5.2]. □

To complete the proof, it remains to show the following result of independent interest:

Lemma 10. *Let M be a monoid scheme of finite type. Then the group functor of invertibles of M is represented by a group scheme $G(M)$, open in M.*

Proof. We first adapt the proof of the corresponding statement for (reduced) algebraic monoids (see [3, Theorem 2.2.4]). Denote for simplicity the composition law of M by $(x, y) \mapsto xy$, and the neutral element by 1. Consider the closed subscheme $G \subset M \times M$ defined in set-theoretic notation by

$$G = \{(x, y) \in M \times M \mid xy = yx = 1\}.$$

Then G is a subgroup scheme of the monoid scheme $M \times M^{\mathrm{op}}$, where M^{op} denotes the opposite monoid, that is, the scheme M equipped with the composition law $(x, y) \mapsto yx$. Moreover, the first projection

$$p : G \longrightarrow M$$

is a homomorphism of monoid schemes, which sends the T-valued points of G isomorphically to the T-valued invertible points of M for any scheme T. It follows that the group scheme G represents the group functor of invertibles in M.

To complete the proof, it suffices to check that p is an open immersion; for this, we may again assume that k is algebraically closed. Clearly, p is universally injective; we now show that it is étale. Since that condition defines an open subscheme of G, stable under the action of $G(k)$ by left multiplication, we only need to check that p is étale at the neutral element 1 of G. For this, the argument of [loc. cit.] does not adapt readily, and we shall rather consider the formal completion of M at 1,

$$N := \mathrm{Spf}(\hat{\mathcal{O}}_{M,1}).$$

Then N is a formal scheme having a unique point; moreover, N has a structure of formal monoid scheme, defined as follows. The composition law $\mu : M \times M \to M$ sends $(1,1)$ to 1, and hence yields a homomorphism of local rings $\mu^{\#} : \mathcal{O}_{M,1} \to \mathcal{O}_{M\times M,(1,1)}$. In turn, $\mu^{\#}$ yields a homomorphism of completed local rings

$$\Delta : \hat{\mathcal{O}}_{M,1} \longrightarrow \hat{\mathcal{O}}_{M\times M,(1,1)} = \hat{\mathcal{O}}_{M,1} \hat{\otimes} \hat{\mathcal{O}}_{M,1}.$$

We also have the homomorphism

$$\varepsilon : \hat{\mathcal{O}}_{M,1} \longrightarrow k$$

associated with 1. One readily checks that Δ and ε satisfy conditions (i) (co-associativity) and (ii) (co-unit) of [18, Exp. VIIB, 2.1]; hence they define a formal monoid scheme structure on N. In view of [loc. cit., 2.7. Proposition], it follows that N is in fact a group scheme. As a consequence, p is an isomorphism after localization and completion at 1; in other words, p is étale at 1. □

Remark 4. Proposition 2 gives back part of the description of all algebraic semigroup structures on a projective variety X, obtained in [3, Theorem 4.3.1].

Specifically, every such structure $\mu : X \times X \to X$, $(x,y) \mapsto xy$ yields a homomorphism of semigroup schemes $\lambda : X \longrightarrow \mathrm{End}(X)$, $x \longmapsto (y \mapsto xy)$ (the "left regular representation"). Thus, $S := \lambda(X)$ is a closed subsemigroup scheme of End(X). Choose an idempotent $e \in X(k)$. In view of Proposition 2, we have $\lambda(x)\lambda(e) = \lambda(x)$ for all $x \in X$; moreover, $\lambda(e)\lambda(x)$ is invertible in $\lambda(e)S$. It follows that $xey = xy$ for all $x, y \in X$. Moreover, for any $x \in X$, there exists $y \in eX$ such that $yexz = exyz = ez$ for all $z \in X$. In particular, $(exe)(eye) = (eye)(exe) = e$, and hence eXe is an algebraic group.

These results are the main ingredients in the proof of [3, Theorem 4.3.1]. They are deduced there from the classical rigidity lemma, while the proof of Proposition 2 relies on a generalization of that lemma.

Remark 5. If k is not algebraically closed, then connected semigroup schemes of endomorphisms may well have no k-rational point. For example, let X be a projective variety having no k-rational point; then the subsemigroup scheme $S \subset \mathrm{End}(X)$ consisting of constant endomorphisms (i.e., of those endomorphisms that factor through the inclusion of a closed point in X) is isomorphic to X itself, equipped with the composition law $(x, y) \mapsto y$. Thus, S has no k-rational point either.

Yet Proposition 2 can be extended to any geometrically connected subsemigroup scheme $S \subset \mathrm{End}(X)$, not necessarily having a k-rational point. Specifically, $E(S)$ is a nonempty, geometrically connected subsemigroup scheme, with semigroup law given by $f_1 f_2 = f_1$. Moreover, there exists a unique retraction of semigroup schemes

$$\pi : S \longrightarrow E(S);$$

it assigns to any point $f \in S$, the unique idempotent $e \in E(S)$ such that $ef = f$. Finally, the above morphism π defines a structure of $E(S)$-monoid scheme on S, with composition law induced by that of S, and with neutral section the inclusion

$$\iota : E(S) \longrightarrow S.$$

In fact, this monoid scheme is a group scheme: consider indeed the closed subscheme $T \subset S \times S$ defined in set-theoretic notation by

$$T = \{(x, y) \in S \times S \mid xy = yx,\ x^2 y = x,\ xy^2 = y\},$$

and the morphism

$$\rho : T \longrightarrow S, \quad (x, y) \longmapsto xy.$$

Then one may check that ρ is a retraction from T to $E(S)$, with section

$$\varepsilon : E(S) \longrightarrow T, \quad x \longmapsto (x, x).$$

Moreover, T is a group scheme over $E(S)$ via ρ, with composition law given by $(x, y)(x', y') := (xx', y'y)$, neutral section ε, and inverse given by $(x, y)^{-1} := (y, x)$. Also, the first projection

$$p_1 : T \longrightarrow S, \quad (x, y) \longrightarrow x$$

is an isomorphism which identifies ρ with π; furthermore, p_1 is an isomorphism of monoid schemes. This yields the desired group scheme structure.

When k is algebraically closed, all these assertions are easily deduced from the structure of S obtained in Proposition 2; the case of an arbitrary field follows by descent.

3.3 *Proof of Proposition 3*

(i) can be deduced from the results of [4]; we provide a self-contained proof by adapting some of the arguments from [loc. cit.].

As in the beginning of the proof of Proposition 2, we denote by $\langle f \rangle$ the smallest closed subscheme of $\mathrm{End}(X)$ containing all the powers f^n, where $n \geq 1$. In view of the boundedness assumption, $\langle f \rangle$ is an algebraic subsemigroup; clearly, it is also commutative. The subsemigroups $\langle f^m \rangle$, where $m \geq 1$, form a family of closed subschemes of $\langle f \rangle$; hence there exists a minimal such subsemigroup, $\langle f^{n_0} \rangle$. Since $\langle f^m \rangle \cap \langle f^n \rangle \supset \langle f^{mn} \rangle$, we see that $\langle f^{n_0} \rangle$ is the smallest such subsemigroup.

The connected components of $\langle f^{n_0} \rangle$ form a finite set F, equipped with a semigroup structure such that the natural map $\varphi : \langle f^{n_0} \rangle \to F$ is a homomorphism of semigroups. In particular, the finite semigroup F is generated by $\varphi(f^{n_0})$. It follows readily that F has a unique idempotent, say $\varphi(f^{n_0 n})$. Then the fiber $\varphi^{-1}\varphi(f^{n_0 n})$ is a closed connected subsemigroup of $\langle f^{n_0} \rangle$, and contains $\langle f^{n_0 n} \rangle$. By the minimality assumption, we must have $\langle f^{n_0 n} \rangle = \varphi^{-1}\varphi(f^{n_0 n}) = \langle f^{n_0} \rangle$. As a consequence, $\langle f^{n_0} \rangle$ is connected.

Also, recall that $\langle f^{n_0} \rangle$ is commutative. In view of Proposition 2, it follows that this algebraic semigroup is in fact a group. In particular, $\langle f^{n_0} \rangle$ contains a unique idempotent, say e. Therefore, e is also the unique idempotent of $\langle f \rangle$: indeed, if $g \in \langle f \rangle$ is idempotent, then $g = g^{n_0} \in \langle f^{n_0} \rangle$, and hence $g = e$.

Thus, $e\langle f \rangle = \langle ef \rangle$ is a closed submonoid of $\langle f \rangle$ with neutral element e and no other idempotent. In view of [3, Proposition 2.2.5], it follows that $e\langle f \rangle$ is a group, say, G. Moreover, $f^{n_0} = ef^{n_0} \in e\langle f \rangle$, and hence $f^n \in G$ for all $n \geq n_0$. On the other hand, if H is a closed subgroup of $\mathrm{End}(X)$ and n_1 is a positive integer such that $f^n \in H$ for all $n \geq n_1$, then H contains $\langle f^{n_1} \rangle$, and hence $\langle f^{n_0} \rangle$ by minimality. In particular, the neutral element of H is e. Let g denote the inverse of f^{n_1} in H; then H contains $f^{n_1+1}g = ef$, and hence $G \subset H$. Thus, G satisfies the assertion.

(ii) Assume that $f^n(x) = x$ for some $n \geq 1$ and some $x \in X(k)$. Replacing n with a large multiple, we may assume that $f^n \in G$. Let $Y := e(X)$, where e is the neutral element of G as above, and let $y := e(x)$. Then Y is a closed subvariety of X, stable by f and hence by G; moreover, G acts on Y by automorphisms. Also, $y \in Y$ is fixed by f^n. Since $f^n = (ef)^n$, it follows that the $(ef)^m(y)$, where $m \geq 1$, form a finite set. As the positive powers of ef are dense in G, the G-orbit of y must be finite. Thus, y is fixed by the neutral component G^o. In view of [5, Proposition 2.1.6], it follows that G^o is linear; hence so is G.

Conversely, if G is linear, then G^o is a connected linear commutative algebraic group, and hence fixes some point $y \in Y(k)$ by Borel's fixed point theorem. Then y is periodic for f.

Next, assume that X is normal; then so is Y by Lemma 11 below. In view of [2, Theorem 2], it follows that G^o acts on the Albanese variety $A(Y)$ via a finite quotient of its own Albanese variety, $A(G^o)$. In particular, G is linear if and only if G^o acts trivially on $A(Y)$. Also, note that $A(Y)$ is isomorphic to a summand of the abelian variety $A(X)$: the image of the idempotent $A(e)$ induced by e. If G^o acts trivially on $A(Y)$, then it acts on $A(X)$ via $A(e)$, since $G^o = G^o e$. Thus, some positive power f^n acts on $A(X)$ via $A(e)$ as well. Conversely, if f^n acts on $A(X)$ via some idempotent g, then we may assume that $f^n \in G^o$ by taking a large power. Thus, $f^n = ef^n = f^n e$ and hence $g = A(e)g = gA(e)$; in other words, g acts on $A(X)$ as an idempotent of the summand $A(Y)$. On the other hand, $g = A(f^n)$ yields an automorphism of $A(Y)$; it follows that $g = A(e)$.

Lemma 11. *Let X be a normal variety, and $r : X \to Y$ a retraction. Then Y is a normal variety as well.*

Proof. Consider the normalization map, $\nu : \tilde{Y} \to Y$. By the universal property of ν, there exists a unique morphism $\tilde{r} : X \to \tilde{Y}$ such that $r = \nu\tilde{r}$. Since r has a section, so has ν. As $\tilde{Y}$ is a variety and ν is finite, it follows that ν is an isomorphism. □

Remark 6. With the notation of the proof of (i), the group G is the closure of the subgroup generated by ef. Hence G is *monothetic* in the sense of [10], which obtains a complete description of this class of algebraic groups. Examples of monothetic algebraic groups include all the semiabelian varieties, except when k is the algebraic closure of a finite field (then the monothetic algebraic groups are exactly the finite cyclic groups).

Acknowledgements This work was began during a staying at Tsinghua University in March 2013. I warmly thank the Mathematical Sciences Center for its hospitality, and Jérémy Blanc, Corrado De Concini, Stéphane Druel, Baohua Fu, Eduard Looijenga, Jörg Winkelmann, and De-Qi Zhang for helpful discussions or e-mail exchanges.

References

1. M. Atiyah, R. Bott, A Lefschetz fixed point formula for elliptic differential operators. Bull. Am. Math. Soc. **72**(2), 245–250 (1966)
2. M. Brion, Some basic results on actions of nonaffine algebraic groups, in *Symmetry and Spaces*. Progress in Mathematics, vol. 278 (Birkhäuser, Boston, 2010), pp. 1–20
3. M. Brion, On algebraic semigroups and monoids, preprint. arXiv:1208.0675
4. M. Brion, L. Renner, Algebraic Semigroups are Strongly π-regular, preprint. arXiv:1209.2042
5. M. Brion, P. Samuel, V. Uma, *Lectures on the Structure of Algebraic Groups and Geometric Applications*. CMI Lecture Series in Mathematics, vol. 1 (Hindustan Book Agency, New Dehli, 2013)
6. C. Chevalley, *Théorie des groupes de Lie. Tome II. Groupes algébriques*. Actualités Sci. Ind., vol. 1152 (Hermann & Cie., Paris, 1951)
7. C. Chevalley, *Théorie des groupes de Lie. Tome III. Théorèmes généraux sur les algèbres de Lie*. Actualités Sci. Ind., vol. 1226 (Hermann & Cie., Paris, 1955)
8. Cohomologie l-adique et fonctions L (SGA5), *Séminaire de Géométrie Algébrique du Bois Marie 1965–66*. Lecture Notes in Mathematics, vol. 589 (Springer, Berlin, 1977)

9. A. Douady, Le problème des modules pour les sous-espaces analytiques compacts d'un espace analytique donné. Ann. Inst. Fourier (Grenoble) **16**(1), 1–95 (1966)
10. G. Falcone, P. Plaumann, K. Strambach, Monothetic algebraic groups. J. Aust. Math. Soc. **82**(3), 315–324 (2007)
11. A. Fujiki, On automorphism groups of compact Kähler manifolds. Invent. Math. **44**(3), 225–258 (1978)
12. A. Grothendieck, Techniques de construction et théorèmes d'existence en géométrie algébrique IV: les schémas de Hilbert. Séminaire Bourbaki, vol. 5 (1960–1961), Exposé No. 221, 28 pp.
13. G. Hochschild, Note on algebraic Lie algebras. Proc. Am. Math. Soc. **29**(1), 10–16 (1971)
14. S.-J. Kan, Complete hyperbolic Stein manifolds with prescribed automorphism groups. Comment. Math. Helv. **82**(2), 371–383 (2007)
15. J. Kollár, *Rational Curves on Algebraic Varieties*. Ergebnisse der Mathematik und ihrer Grenzgebiete, vol. 32 (Springer, Berlin, 1996)
16. J. Kollár, *Lectures on Resolution of Singularities*. Annals of Mathematics Studies, vol. 166 (Princeton University Press, Princeton, 2007)
17. H. Matsumura, F. Oort, Representability of group functors, and automorphisms of algebraic schemes. Invent. Math. **4**(1), 1–25 (1967)
18. Schémas en groupes I, II, III (SGA3), *Séminaire de Géométrie Algébrique du Bois Marie 1962–1964*. Lecture Notes in Mathematics, vols. 151, 152, 153 (Springer, Berlin, 1970). Revised version ed. by P. Polo, P. Gille, vols. I, III (Soc. Math. France, 2011)
19. E. Sernesi, *Deformations of Algebraic Schemes*. Grundlehren Math. Wiss., vol. 334 (Springer, Berlin, 2006)
20. J.-P. Serre, On the fundamental group of a unirational variety. J. Lond. Math. Soc. **34**(4), 481–484 (1959)
21. B. Totaro, Pseudo-abelian varieties. Ann. Sci. Éc. Norm. Sup. **46**(5), 693–721 (2013)
22. J. Winkelmann, Realizing connected Lie groups as automorphism groups of complex manifolds. Comment. Math. Helv. **79**(2), 285–299 (2004)

Del Pezzo Surfaces and Local Inequalities

Ivan Cheltsov

Abstract I prove new local inequality for divisors on smooth surfaces, describe its applications, and compare it to a similar local inequality that is already known by experts.

2010 *Mathematics Subject Classification.* 14E05, 14J45, 14J26, 14Q10, 32Q20

Let X be a Fano variety of dimension $n \geqslant 1$ with at most Kawamata log terminal singularities (see [6, Definition 6.16]). In many applications, it is useful to *measure* how singular effective $\mathbb{Q}$-divisors D on X can be provided that $D \sim_{\mathbb{Q}} -K_X$. Of course, this can be done in many ways depending on what I mean by *measure*. A possible *measurement* can be given by the so-called α-invariant of the Fano variety X that can be defined as

$$\alpha(X) = \sup\left\{\lambda \in \mathbb{Q} \;\middle|\; \begin{array}{l}\text{the pair } (X, \lambda D) \text{ is Kawamata log terminal}\\ \text{for every effective } \mathbb{Q}\text{-divisor } D \sim_{\mathbb{Q}} -K_X.\end{array}\right\} \in \mathbb{R}.$$

The invariant $\alpha(X)$ has been studied intensively by many people who used different notation for $\alpha(X)$. The notation $\alpha(X)$ is due to Tian who defined $\alpha(X)$ in a different way. However, his definition coincides with the one I just gave by [4, Theorem A.3]. The α-invariants play a very important role in Kähler geometry due to

Throughout this chapter, I assume that most of the considered varieties are algebraic, normal, and defined over complex numbers.

I. Cheltsov (✉)
Room 5618, James Clerk Maxwell Building, School of Mathematics, University of Edinburgh, Kings Buildings, Mayfield Road, Edinburgh EH9 3JZ, Scotland
e-mail: I.Cheltsov@ed.ac.uk

I. Cheltsov et al. (eds.), *Automorphisms in Birational and Affine Geometry*, Springer Proceedings in Mathematics & Statistics 79,
DOI 10.1007/978-3-319-05681-4_5,

Theorem 1 ([13], [7, Criterion 6.4]). *Let X be a Fano variety of dimension n that has at most quotient singularities. If $\alpha(X) > \frac{n}{n+1}$, then X admits an orbifold Kähler–Einstein metric.*

The α-invariants are usually very tricky to compute. But they are computed in many cases. For example, the α-invariants of smooth del Pezzo surfaces have been computed as follows:

Theorem 2 ([1, Theorem 1.7]). *Let S_d be a smooth del Pezzo surface of degree d. Then*

$$\alpha(S_d) = \begin{cases} \dfrac{1}{3} \text{ if } d = 9, 7 \text{ or } S_d = \mathbb{F}_1, \\ \dfrac{1}{2} \text{ if } d = 5, 6 \text{ or } S_d = \mathbb{P}^1 \times \mathbb{P}^1, \\ \dfrac{2}{3} \text{ if } d = 4, \end{cases}$$

$$\alpha(S_3) = \begin{cases} \dfrac{2}{3} \text{ if } S_3 \text{ is a cubic surface in } \mathbb{P}^3 \text{ with an Eckardt point}, \\ \dfrac{3}{4} \text{ if } S_3 \text{ is a cubic surface in } \mathbb{P}^3 \text{ without Eckardt points}, \end{cases}$$

$$\alpha(S_2) = \begin{cases} \dfrac{3}{4} \text{ if } |-K_{S_2}| \text{ has a tacnodal curve}, \\ \dfrac{5}{6} \text{ if } |-K_{S_2}| \text{ has no tacnodal curves}, \end{cases}$$

$$\alpha(S_1) = \begin{cases} \dfrac{5}{6} \text{ if } |-K_{S_1}| \text{ has a cuspidal curve}, \\ 1 \text{ if } |-K_{S_1}| \text{ has no cuspidal curves}. \end{cases}$$

Note that $\alpha(X) < 1$ if and only if there exists an effective $\mathbb{Q}$-divisor D on X such that $D \sim_{\mathbb{Q}} -K_X$ and the pair (X, D) is not log canonical. Such divisors (if they exist) are called non-log canonical *special tigers* by Keel and McKernan (see [9, Definition 1.13]). They play an important role in birational geometry of X. How does one describe non-log canonical special tigers? Note that if D is a non-log canonical special tiger on X, then

$$(1-\mu)D + \mu D'$$

is also a non-log canonical special tiger on X for *any* effective $\mathbb{Q}$-divisor D' on X such that $D' \sim_{\mathbb{Q}} -K_X$ and *any* sufficiently small $\mu \geqslant 0$. Thus, to describe non-log canonical special tigers on X, I only need to consider those of them whose supports do not contain supports of other non-log canonical special tigers. Let me call such non-log canonical special tigers *Siberian tigers*. Unfortunately, Siberian tigers are not easy to describe in general. But sometimes it is possible. For example, Kosta proved

Lemma 3 ([11, Lemma 3.1]). *Let S be a hypersurface of degree 6 in $\mathbb{P}(1,1,2,3)$ that has exactly one singular point O. Suppose that O is a Du Val singular point of type $\mathbb{A}_3$. Then all Siberian tigers on X are cuspidal curves in $|-K_S|$, which implies, in particular, that*

$$\alpha(S) = \begin{cases} \frac{5}{6} \text{ if there is a cuspidal curve in } |-K_S|, \\ 1 \text{ otherwise.} \end{cases}$$

The original proof of Lemma 3 is *global* and lengthy. In [11], Kosta applied the very same *global* method to compute the α-invariants of all del Pezzo surfaces of degree 1 that has at most Du Val singularities (in most of cases her computations do not give description of Siberian tigers). Later I noticed that the nature of her *global* method is, in fact, purely *local*. Implicitly, Kosta proved

Theorem 4 ([3, Corollary 1.29]). *Let S be a surface, let P be a smooth point in S, let Δ_1 and Δ_2 be two irreducible curves on S that are both smooth at P and intersect transversally at P, and let a_1 and a_2 be non-negative rational numbers. Suppose that $\frac{2n-2}{n+1}a_1 + \frac{2}{n+1}a_2 \leqslant 1$ for some positive integer $n \geqslant 3$. Let Ω be an effective $\mathbb{Q}$-divisor on the surface S whose support does not contain the curves Δ_1 and Δ_2. Suppose that the log pair $(S, a_1\Delta_1 + a_2\Delta_2 + \Omega)$ is not log canonical at P. Then* $\mathrm{mult}_P(\Omega \cdot \Delta_1) > 2a_1 - a_2$ *or* $\mathrm{mult}_P(\Omega \cdot \Delta_2) > \frac{n}{n-1}a_2 - a_1$.

Unfortunately, Theorem 4 has a very limited application scope. Together with Kosta, I generalized Theorem 4 as

Theorem 5 ([3, Theorem 1.28]). *Let S be a surface, let P be a smooth point in S, let Δ_1 and Δ_2 be two irreducible curves on S that both are smooth at P and intersect transversally at P, let a_1 and a_2 be non-negative rational numbers, and let Ω be an effective $\mathbb{Q}$-divisor on the surface S whose support does not contain the curves Δ_1 and Δ_2. Suppose that the log pair $(S, a_1\Delta_1 + a_2\Delta_2 + \Omega)$ is not log canonical at P. Suppose that there are non-negative rational numbers α, β, A, B, M, and N such that $\alpha a_1 + \beta a_2 \leqslant 1$, $A(B-1) \geqslant 1$, $M \leqslant 1$, $N \leqslant 1$, $\alpha(A + M - 1) \geqslant A^2(B + N - 1)\beta$, $\alpha(1-M) + A\beta \geqslant A$. Suppose, in addition, that $2M + AN \leqslant 2$ or $\alpha(B + 1 - MB - N) + \beta(A + 1 - AN - M) \geqslant AB - 1$. Then* $\mathrm{mult}_P(\Omega \cdot \Delta_1) > M + Aa_1 - a_2$ *or* $\mathrm{mult}_P(\Omega \cdot \Delta_2) > N + Ba_2 - a_1$.

Despite the fact that Theorem 5 looks very ugly, it is much more flexible and much more applicable than Theorem 4. By [6, Excercise 6.26], an analogue of Theorem 5 holds for surfaces with at most quotient singularities. This helped me to find in [2] many new applications of Theorem 5 that do not follow from Theorem 4.

Remark 6. How does one apply Theorem 5? Let me say few words about this. Let S be a smooth surface, and let D be an effective $\mathbb{Q}$-divisor on S. The purpose of

Theorem 5 is to prove that (S, D) is log canonical provided that D satisfies some *global* numerical conditions. To do so, I assume that (S, D) is not log canonical at P and seek for a contradiction. First, I look for some nice curves that pass through P that has very small intersection with D. Suppose I found two such curves, say Δ_1 and Δ_2, that are both irreducible and both pass through P. If Δ_1 or Δ_2 are not contained in the support of the divisor D, I can bound $\mathrm{mult}_P(D)$ by $D \cdot \Delta_1$ or $D \cdot \Delta_2$ and, hopefully, get a contradiction with $\mathrm{mult}_P(D) > 1$, which follows from the fact (S, D) is not log canonical at P. This shows that I should look for the curves Δ_1 and Δ_2 among the curves which are close enough to the boundary of the Mori cone $\overline{\mathbb{NE}}(S)$. Suppose that both curves Δ_1 and Δ_2 lie in the boundary of the Mori cone $\overline{\mathbb{NE}}(S)$. Then $\Delta_1^2 \leqslant 0$ and $\Delta_2^2 \leqslant 0$. Keeping in mind, that the curves Δ_1 and Δ_2 can, a priori, be contained in the support of the divisor D, I must put $D = a_1\Delta_1 + a_2\Delta_2 + \Omega$ for some non-negative rational numbers a_1 and a_2, where Ω is an effective $\mathbb{Q}$-divisor on S whose support does not contain the curves Δ_1 and Δ_2. Then I try to bound a_1 and a_2 using some *global* methods. Usually, I end up with two non-negative rational numbers α and β such that $\alpha a_1 + \beta a_2 \leqslant 1$. Put $M = D \cdot \Delta_1$, $N = C \cdot \Delta_2$, $A = -\Delta_1^2$, and $B = -\Delta_1^2$. Suppose that Δ_1 and Δ_2 are both smooth at P and intersect transversally at P (otherwise I need to blow up the surface S and replace the pair (S, D) by its *log pull back*). If I am lucky, then $A(B-1) \geqslant 1$, $M \leqslant 1$, $N \leqslant 1$, $\alpha(A + M - 1) \geqslant A^2(B + N - 1)\beta$, $\alpha(1 - M) + A\beta \geqslant A$, and either $2M + AN \leqslant 2$ or $\alpha(B + 1 - MB - N) + \beta(A + 1 - AN - M) \geqslant AB - 1$ (or both), which implies that

$$M + Aa_1 - a_2 \geqslant M + Aa_1 - a_2\Delta_1 \cdot \Delta_2 = \Omega \cdot \Delta_1 \geqslant \mathrm{mult}_P\Big(\Omega \cdot \Delta_1\Big) > M + Aa_1 - a_2$$

or

$$N + Ba_2 - a_1 \geqslant N + Ba_2 - a_1\Delta_1 \cdot \Delta_2 = \Omega \cdot \Delta_2 \geqslant \mathrm{mult}_P\Big(\Omega \cdot \Delta_2\Big) > N + Ba_2 - a_1$$

by Theorem 5. This is the contradiction I was looking for.

Unfortunately, the hypotheses of Theorem 5 are not easy to verify in general. Moreover, the proof of Theorem 5 is very lengthy. It seems likely that Theorem 5 is a special case or, perhaps, a corollary of a more general statement that looks better and has a shorter proof. Ideally, the proof of such generalization, if it exists, should be inductive like the proof of

Theorem 7 ([6, Excercise 6.31]). *Let S be a surface, let P be a smooth point in S, let Δ be an irreducible curve on S that is smooth at P, let a be a non-negative rational number such that $a \leqslant 1$, and let Ω be an effective $\mathbb{Q}$-divisor on the surface S whose support does not contain the curve Δ. Suppose that the log pair $(S, a\Delta + \Omega)$ is not log canonical at P. Then* $\mathrm{mult}_P(\Omega \cdot \Delta) > 1$.

Proof. Put $m = \mathrm{mult}(\Omega)$. If $m > 1$, then I am done, since $\mathrm{mult}_P(\Omega \cdot \Delta) \geqslant m$. In particular, I may assume that the log pair $(S, a\Delta + \Omega)$ is log canonical in a punctured neighborhood of the point P. Since the log pair $(S, a\Delta + \Omega)$ is not log canonical at P, there exists a birational morphism $h: \hat{S} \to S$ that is a composition of $r \geqslant 1$ blow ups of smooth points dominating P, and there exists an h-exceptional divisor, say E_r, such that $e_r > 1$, where e_r is a rational number determined by

$$K_{\hat{S}} + a\hat{\Delta} + \hat{\Omega} + \sum_{i=1}^{r} e_i E_i \sim_{\mathbb{Q}} h^*\big(K_S + a\Delta + \Omega\big),$$

where each e_i is a rational number, each E_i is an h-exceptional divisor, $\hat{\Omega}$ is a proper transform on $\hat{S}$ of the divisor Ω, and $\hat{\Delta}$ is a proper transform on $\hat{S}$ of the curve Δ.

Let $f: \tilde{S} \to S$ be the blow up of the point P, let $\tilde{\Omega}$ be the proper transform of the divisor Ω on the surface $\tilde{S}$, let E be the f-exceptional curve, and let $\tilde{\Delta}$ be the proper transform of the curve Δ on the surface $\tilde{S}$. Then the log pair $(\tilde{S}, a\tilde{\Delta} + (a + m - 1)E + \tilde{\Omega})$ is not log canonical at some point $Q \in E$.

Let me prove the inequality $\mathrm{mult}_P(\Omega \cdot \Delta) > 1$ by induction on r. If $r = 1$, then $a+m-1 > 1$, which implies that $m > 2-a \geqslant 1$. This implies that $\mathrm{mult}_P(\Omega \cdot \Delta) > 1$ if $r = 1$. Thus, I may assume that $r \geqslant 2$. Since

$$\mathrm{mult}_P\Big(\Omega \cdot \Delta\Big) \geqslant m + \mathrm{mult}_Q\Big(\tilde{\Omega} \cdot \tilde{\Delta}\Big),$$

it is enough to prove that $m + \mathrm{mult}_Q(\tilde{\Omega} \cdot \tilde{\Delta}) > 1$. Moreover, I may assume that $m \leqslant 1$, since $\mathrm{mult}_P(\Omega \cdot \Delta) \geqslant m$. Then the log pair $(\tilde{S}, a\tilde{\Delta} + (a+m-1)E + \tilde{\Omega})$ is log canonical at a punctured neighborhood of the point $Q \in E$, since $a+m-1 \leqslant 2$.

If $Q \notin \tilde{\Delta}$, then the log pair $(\tilde{S}, (a + m - 1)E + \tilde{\Omega})$ is not log canonical at the point Q, which implies that

$$m = \tilde{\Omega} \cdot E \geqslant \mathrm{mult}_Q\Big(\tilde{\Omega} \cdot E\Big) > 1$$

by induction. The latter implies that $Q = \tilde{\Delta} \cap E$, since $m \leqslant 1$. Then

$$a + m - 1 + \mathrm{mult}_Q\Big(\tilde{\Omega} \cdot \tilde{\Delta}\Big) = \mathrm{mult}_Q\Bigg(\Big((a + m - 1)E + \tilde{\Omega}\Big) \cdot \tilde{\Delta}\Bigg) > 1$$

by induction. This implies that $\mathrm{mult}_Q(\tilde{\Omega} \cdot \tilde{\Delta}) > 2-a-m$. Then $m + \mathrm{mult}_Q(\tilde{\Omega} \cdot \tilde{\Delta}) > 2 - a \geqslant 1$ as required. □

Recently, I jointly with Park and Won proved that all Siberian tigers on smooth cubic surfaces are just anticanonical curves that have non-log canonical singularities (see [5, Theorem 1.12]). This follows from

Theorem 8 ([5, Corollary 1.13]). *Let S be a smooth cubic surface in $\mathbb{P}^3$, let P be a point in S, let T_P be the unique hyperplane section of the surface S that is singular at P, let D be any effective $\mathbb{Q}$-divisor on the surface S such that $D \sim_{\mathbb{Q}} -K_S$. Then*

(S, D) is log canonical at P provided that $\mathrm{Supp}(D)$ *does not contain at least one irreducible component of* $\mathrm{Supp}(T_P)$.

Siberian tigers on smooth del Pezzo surfaces of degree 1 and 2 are also just anticanonical curves that have non-log canonical singularities (see [5, Theorem 1.12]). This follows easily from the proofs of [1, Lemmas 3.1 and 3.5]. Surprisingly, smooth del Pezzo surfaces of degree 4 contains much more Siberian tigers.

Example 9. Let S be a smooth complete intersection of two quadric hypersurfaces in $\mathbb{P}^4$, let L be a line on S, and let P_0 be a point in L such that L is the only line in S that passes though P_0. Then there exists exactly five conics in S that pass through P_0. Let me denote them by C_1^0, C_2^0, C_3^0, C_4^0, and C_5^0. Then

$$\frac{\sum_{i=1}^{5} C_i^0}{3} + \frac{2}{3}L \sim_{\mathbb{Q}} -K_S,$$

is a Siberian tiger. Let Z be a general smooth rational cubic curve in S such that $Z + L$ is cut out by a hyperplane section and $P \in Z$. Then $Z \cap L$ consists of a point P and another point which I denote by Q. Let $f\colon \tilde{S} \to S$ be a blow up of the point Q, and let E be its exceptional curve. Denote by $\tilde{L}$ and $\tilde{Z}$ the proper transforms of the curves L and Z on the surface $\tilde{S}$, respectively. Then $\tilde{Z} \cap \tilde{L} = \varnothing$. Let $g\colon \hat{S} \to \tilde{S}$ be the blow up of the point $\tilde{Z} \cap E$, and let F be its exceptional curve. Denote by $\hat{E}$, $\hat{L}$ and $\hat{Z}$ the proper transforms of the curves E, $\tilde{L}$ and $\tilde{Z}$ on the surface $\hat{S}$, respectively. Then $\hat{S}$ is a minimal resolution of a singular del Pezzo surface of degree 2, and $|-K_{\hat{S}}|$ gives a morphism $\hat{S} \to \mathbb{P}^2$ that is a double cover away from the curves $\hat{E}$ and $\hat{L}$. This double cover induces an involution $\tau \in \mathrm{Bir}(S)$. Put $C_i^1 = \tau(C_i^0)$ for every i. Then C_1^1, C_2^1, C_3^1, C_4^1 and C_5^1 are curves of degree 5 that all intersect exactly in one point in L. Denote this point by P_1. Iterate this constriction k times. This gives me five irreducible curves C_1^k, C_2^k, C_3^k, C_4^k and C_5^k that intersect exactly in one point P_k. Then

$$\frac{\sum_{i=1}^{5} C_i^k}{a_{2k+1} + a_{2k+3}} + \frac{4a_{2k+1} - a_{2k+3}}{a_{2k+1} + a_{2k+3}} L \sim_{\mathbb{Q}} -K_S, \tag{1}$$

where a_i is the i-th Fibonacci number. Moreover, each curve C_i^k is a curve of degree a_{2k+3}. Furthermore, the log canonical threshold of the divisor (1) is

$$\frac{a_{2k+3}(a_{2k+1} + a_{2k+3})}{1 + a_{2k+3}(a_{2k+1} + a_{2k+3})} < 1,$$

which easily implies that the divisor (1) is a Siberian tiger.

Quite surprisingly, Theorem 8 has other applications as well. For example, it follows from [10, Corollary 2.12], [5, Lemma 1.10] and Theorem 8 that every cubic cone in $\mathbb{A}^4$ having unique singular point does not admit non-trivial regular $\mathbb{G}_a$-actions (cf. [8, Question 2.22]).

The crucial part in the proof of Theorem 8 is played by two sibling lemmas. The first one is

Lemma 11 ([5, Lemma 4.8]). *Let S be a smooth cubic surface in $\mathbb{P}^3$, let P be a point in S, let T_P be the unique hyperplane section of the surface S that is singular at P, let D be any effective $\mathbb{Q}$-divisor on the surface S such that $D \sim_{\mathbb{Q}} -K_S$. Suppose that T_P consists of three lines such that one of them does not pass through P. Then (S, D) is log canonical at P.*

Its younger sister is

Lemma 12 ([5, Lemma 4.9]). *Let S be a smooth cubic surface in $\mathbb{P}^3$, let P be a point in S, let T_P be the unique hyperplane section of the surface S that is singular at P, let D be any effective $\mathbb{Q}$-divisor on the surface S such that $D \sim_{\mathbb{Q}} -K_S$. Suppose that T_P consists of a line and a conic intersecting transversally. Then (S, D) is log canonical at P.*

The proofs of Lemmas 11 and 12 we found in [5] are *global*. In fact, they resemble the proofs of classical results by Segre and Manin on cubic surfaces (see [6, Theorems 2.1 and 2.2]). Once the paper [5] has been written, I asked myself a question: can I prove Lemmas 11 and 12 using just *local* technique? To answer this question, let me sketch their *global* proofs first.

Global proof of Lemma 11. Let me use the notation and assumptions of Lemma 11. I write $T_P = L + M + N$, where L, M, and N are lines on the cubic surface S. Without loss of generality, I may assume that the line N does not pass through the point P. Let D be any effective $\mathbb{Q}$-divisor on the surface S such that $D \sim_{\mathbb{Q}} -K_S$. I must show that (S, D) is log canonical at P. Suppose that the log pair (S, D) is not log canonical at the point P. Let me seek for a contradiction.

Put $D = aL + bM + cN + \Omega$, where a, b, and c are non-negative rational numbers and Ω is an effective $\mathbb{Q}$-divisor on S whose support contains none of the lines L, M and N. Put $m = \mathrm{mult}_P(\Omega)$. Then $a \leqslant 1$, $b \leqslant 1$ and $c \leqslant 1$. Moreover, the pair (S, D) is log canonical outside finitely many points. This follows from [6, Lemma 5.3.6] and is very easy to prove (see, for example, [5, Lemma 4.1] or the proof of [1, Lemma 3.4]).

Since (S, D) is not log canonical at the point P, I have

$$m + a + b = \mathrm{mult}_P(D) > 1$$

by [6, Excercise 6.18] (this also follows from Theorem 7). In particular, the rational number a must be positive, since otherwise I would have

$$1 = L \cdot D \geqslant \mathrm{mult}_P(D) > 1.$$

Similarly, the rational number b must be positive as well.

The inequality $m + a + b > 1$ is very handy. However, a stronger inequality $m + a + b > c + 1$ holds. Indeed, there exists a non-negative rational number μ

such that the divisor $(1+\mu)D-\mu T_P$ is effective and its support does not contain at least one components of T_P. Now to obtain $m+a+b>c+1$, it is enough to apply [6, Excercise 6.18] to the divisor $(1+\mu)D-\mu T_P$, since $(S,(1+\mu)D-\mu T_P)$ is not log canonical at P.

Since a, b, c do not exceed 1 and $(S, L+M+N)$ is log canonical, $\Omega \neq 0$. Let me write $\Omega=\sum_{i=1}^r e_iC_i$, where every e_i is a positive rational number, and every C_i is an irreducible reduced curve of degree $d_i>0$ on the surface S. Then

$$a+b+c+\sum_{i=1}^r e_id_i=3,$$

since $-K_S\cdot D=3$.

Let $f\colon \tilde{S}\to S$ be a blow up of the point P, and let E be the exceptional divisor of f. Denote by $\tilde{L}$, $\tilde{M}$ and $\tilde{N}$ the proper transforms on $\tilde{S}$ of the lines L, M and N, respectively. For each i, denote by $\tilde{C}_i$ the proper transform of the curve C_i on the surface $\tilde{S}$. Then

$$K_{\tilde{S}}+a\tilde{L}+b\tilde{M}+c\tilde{N}+(a+b+m-1)E+\sum_{i=1}^r e_i\tilde{C}_i\sim_{\mathbb{Q}} f^*(K_S+D),$$

which implies that the log pair $(\tilde{S}, a\tilde{L}+b\tilde{M}+c\tilde{N}+(a+b+m-1)E+\sum_{i=1}^r e_i\tilde{C}_i)$ is not log canonical at some point $Q\in E$.

I claim that either $Q\in\tilde{L}\cap E$ or $Q\in\tilde{M}\cap E$. Indeed, it follows from

$$\begin{cases} 1=D\cdot L=\Big(aL+bM+cN+\Omega\Big)\cdot L=-a+b+c+\Omega\cdot L\geqslant -a+b+c+m,\\ 1=D\cdot M=\Big(aL+bM+cN+\Omega\Big)\cdot M=a-b+c+\Omega\cdot M\geqslant a-b+c+m,\\ 1=D\cdot N=\Big(aL+bM+cN+\Omega\Big)\cdot N=a+b-c+\Omega\cdot N\geqslant a+b-c,\end{cases}$$

that $m\leqslant 1-c$ and $a+b+m-1\leqslant 1$, because $a\leqslant 1$ and $b\leqslant 1$. On the other hand, if $Q\notin\tilde{L}\cup\tilde{M}$, then the log pair $(\tilde{S},(a+b+m-1)E+\sum_{i=1}^r e_i\tilde{C}_i)$ is not log canonical at Q, which implies that

$$m=\Big(\sum_{i=1}^r e_i\tilde{C}_i\Big)\cdot E>1$$

by Theorem 7. This shows that either $Q\in\tilde{L}\cap E$ or $Q\in\tilde{M}\cap E$, since $m\leqslant 1-c\leqslant 1$. Without loss of generality, I may assume that $Q=\tilde{L}\cap E$.

Let $\rho\colon S\dashrightarrow\mathbb{P}^2$ be the linear projection from the point P. Then ρ is a generically two-to-one rational map. Thus the map ρ induces an involution $\tau\in\mathrm{Bir}(S)$ known as the Geiser involution (see [6, Sect. 2.14]). The involution τ is biregular outside $P\cup N$, $\tau(L)=L$ and $\tau(M)=M$.

For each i, denote by $\hat{d}_i$ the degree of the curve $\tau(C_i)$. Put $\hat{\Omega} = \sum_{i=1}^{r} e_i \tau(C_i)$. Then

$$aL + bM + (a + b + m - 1)N + \hat{\Omega} \sim_{\mathbb{Q}} -K_S,$$

and $(S, aL + bM + (a + b + m - 1)M + \hat{\Omega})$ is not log canonical at the point $L \cap N$. Thus, I can replace the original effective $\mathbb{Q}$-divisor D by the divisor

$$aL + bM + (a + b + m - 1)N + \hat{\Omega} \sim_{\mathbb{Q}} -K_S$$

that has the same properties as D. Moreover, I have

$$\sum_{i=1}^{r} e_i \hat{d}_i < \sum_{i=1}^{r} e_i d_i,$$

since $m + a + b > c + 1$. Iterating this process, I obtain a contradiction after finitely many steps. □

Global proof of Lemma 12. Let me use the notations and assumptions of Lemma 12. I write $T_P = L + C$, where L is a line, and C is a conic. Let D be any effective $\mathbb{Q}$-divisor on the surface S such that $D \sim_{\mathbb{Q}} -K_S$. I must show that the log pair (S, D) is log canonical at P. Suppose that (S, D) is not log canonical at the point P. Let me seek for a contradiction.

Let me write $D = nL + kC + \Omega$, where n and k are non-negative rational numbers and Ω is an effective $\mathbb{Q}$-divisor on S whose support contains none of the curves L and C. Put $m = \mathrm{mult}_P(\Omega)$. Then $2n + m \leqslant 2$ and $2k + m \leqslant 1 + n$, since

$$\begin{cases} 1 = D \cdot L = \Big(nL + kC + \Omega\Big) \cdot L = -n + 2k + \Omega \cdot L \geqslant -n + 2k + m, \\ 2 = D \cdot C = \Big(nL + kC + \Omega\Big) \cdot C = 2n + \Omega \cdot C \geqslant 2n + m. \end{cases}$$

Arguing as in the proof [1, Lemma 3.4], I see that the log pair (S, D) is log canonical outside finitely many points (this follows, for example, from [6, Lemma 5.3.6]). In particular, both rational numbers n and k do not exceed 1. On the other hand, it follows from [6, Excercise 6.18] that

$$m + n + k = \mathrm{mult}_P(D) > 1,$$

because the log pair (S, D) is not log canonical at the point P. The later implies that $n > 0$, since $1 = L \cdot D \geqslant \mathrm{mult}_P(D)$ if $n = 0$.

I claim that $n > k$ and $m + n > 1$. Indeed, there exists a non-negative rational number μ such that the divisor $(1 + \mu)D - \mu T_P$ is effective and its support does not contain at least one components of T_P. Then $(S, (1 + \mu)D - \mu T_P)$ is not log canonical at P. If $n \leqslant k$, then the support of $(1+\mu)D-\mu T_P$ does not contain L, which is impossible, since

$$\mathrm{mult}_P\Big((1+\mu)D-\mu T_P\Big)>1$$

and $1=L\cdot((1+\mu)D-\mu T_P)$. Thus, I proved that $n>k$. Now I can apply [6, Excercise 6.18] to the divisor $(1+\mu)D-\mu T_P$ and obtain $m+n>1$.

Let $f\colon \tilde{S}\to S$ be the blow up of the point P, let $\tilde{\Omega}$ be the proper transform of the divisor Ω on the surface $\tilde{S}$, let $\tilde{L}$ be the proper transform of the line L on the surface $\tilde{S}$, let $\tilde{C}$ be the proper transform of the conic C on the surface $\tilde{S}$, and let E be the f-exceptional curve. Then

$$K_{\tilde{S}}+n\tilde{L}+k\tilde{C}+\tilde{\Omega}+\big(n+k+m-1\big)E\sim_{\mathbb{Q}} f^{*}\big(K_S+D\big)\sim_{\mathbb{Q}}0,$$

which implies that the log pair $(\tilde{S},n\tilde{L}+k\tilde{C}+(n+k+m-1)E+\tilde{\Omega})$ is not log canonical at some point $Q\in E$. On the other hand, I must have $n+k+m-1\leqslant 1$, because $2n+m\leqslant 2$, $2k+m\leqslant 1+n$ and $n\leqslant 1$.

I claim that $Q\in\tilde{L}$. Indeed, if $Q\in\tilde{C}$, then the log pair $(\tilde{S},k\tilde{C}+(n+k+m-1)E+\tilde{\Omega})$ is not log canonical at Q, which implies that $k>n$, since

$$1-n+k=\Big(\tilde{\Omega}+\big(n+k+m-1\big)E\Big)\cdot\tilde{C}>1,$$

by Theorem 7. Since I proved already that $n>k$, the curve $\tilde{C}$ does not contain Q. Thus, if $Q\notin\tilde{L}$, then $Q\notin\tilde{L}\cup\tilde{C}$, which contradicts [5, Lemma 3.2], since

$$n\tilde{L}+k\tilde{C}+\tilde{\Omega}+(n+k+m-1)E\sim_{\mathbb{Q}}-K_{\tilde{S}}.$$

Since n and k do not exceed 1 and the log pair $(S,L+C)$ is log canonical, the effective $\mathbb{Q}$-divisor Ω cannot be the zero-divisor. Let r be the number of the irreducible components of the support of the $\mathbb{Q}$-divisor Ω. Let me write $\Omega=\sum_{i=1}^{r}e_iC_i$, where every e_i is a positive rational number, and every C_i is an irreducible reduced curve of degree $d_i>0$ on the surface S. Then

$$n+2k+\sum_{i=1}^{r}a_id_i=3,$$

since $-K_S\cdot D=3$.

Let $\rho\colon S\dashrightarrow\mathbb{P}^2$ be the linear projection from the point P. Then ρ is a generically 2-to-1 rational map. Thus the map ρ induces a birational involution τ of the cubic surface S. This involution is also known as the Geiser involution (cf. the proof of Lemma 11). The involution τ is biregular outside of the conic C, and $\tau(L)=L$.

For every i, put $\hat{C}_i=\tau(C_i)$, and denote by $\hat{d}_i$ the degree of the curve $\hat{C}_i$. Put $\hat{\Omega}=\sum_{i=1}^{r}e_i\hat{C}_i$. Then

$$nL + (n + k + m - 1)C + \hat{\Omega} \sim_{\mathbb{Q}} -K_S,$$

and $(S, nL + (n + k + m - 1)C + \hat{\Omega})$ is not log canonical at the point $L \cap C$ that is different from P. Thus, I can replace the original effective $\mathbb{Q}$-divisor D by $nL + (n + k + m - 1)C + \hat{\Omega}$ that has the same properties as D. Moreover, since $m + n > 1$, the inequality

$$\sum_{i=1}^{r} e_i \hat{d}_i < \sum_{i=1}^{r} e_i d_i$$

holds. Iterating this process, I obtain a contradiction in a finite number of steps as in the proof of Lemma 11. □

It came as a surprise that Theorem 5 can be used to replace the *global* proof of Lemma 12 by its *local* counterpart. Let me show how to do this.

Local proof of Lemma 12. Let me use the assumptions and notation of Lemma 12. I write $T_P = L + C$, where L is a line, and C is a conic. Let D be any effective $\mathbb{Q}$-divisor on the surface S such that $D \sim_{\mathbb{Q}} -K_S$. I must show that the log pair (S, D) is log canonical at P. Suppose that (S, D) is not log canonical at the point P. Let me seek for a contradiction.

Put $D = nL + kC + \Omega$, where n and k are non-negative rational numbers and Ω is an effective $\mathbb{Q}$-divisor on S whose support contains none of the curves L and C. Put $m = \mathrm{mult}_P(\Omega)$. Then

$$m + n + k = \mathrm{mult}_P(D) > 1,$$

since (S, D) is not log canonical at P. The later implies that $n > 0$, since $1 = L \cdot D \geqslant \mathrm{mult}_P(D)$ if $n = 0$.

Replacing D by an effective $\mathbb{Q}$-divisor $(1+\mu)D - \mu T_P$ for an appropriate $\mu \geqslant 0$, I may assume that $k = 0$. Then $2 = C \cdot D = 2n + \Omega \cdot C \geqslant 2n + m$. Moreover, the log pair (S, D) is log canonical outside finitely many points. The latter follows, for example, from [6, Lemma 5.3.6] and is very easy to prove (cf. the proof of [1, Lemma 3.4]).

Let $f: \tilde{S} \to S$ be the blow up of the point P, let $\tilde{\Omega}$ be the proper transform of the divisor Ω on the surface $\tilde{S}$, let $\tilde{L}$ be the proper transform of the line L on the surface $\tilde{S}$, and let E be the f-exceptional curve. Then

$$K_{\tilde{S}} + n\tilde{L} + \tilde{\Omega} + \big(n + m - 1\big)E \sim_{\mathbb{Q}} f^*\big(K_S + D\big) \sim_{\mathbb{Q}} 0,$$

which implies that $(\tilde{S}, n\tilde{L} + (n + m - 1)E + \tilde{\Omega})$ is not log canonical at some point $Q \in E$. Arguing as in the proof of [1, Lemma 3.5], I get $Q = \tilde{L} \cap E$. Now I can apply Theorem 5 to the log pair $(\tilde{S}, n\tilde{L} + (n + m - 1)E + \tilde{\Omega})$ at the point Q.

Put $\Delta_1 = E$, $\Delta_2 = \tilde{L}$, $M = 1$, $A = 1$, $N = 0$, $B = 2$, and $\alpha = \beta = 1$. Check that all hypotheses of Theorem 5 are satisfied. By Theorem 5, I have

$$m = \mathrm{mult}_Q(\tilde{\Omega} \cdot E) > 1 + (n + m - 1) - n = m$$

or $1+n-m = \mathrm{mult}_Q(\tilde{\Omega}\cdot\tilde{L}) > 2n-(n+m-1) = 1+n-m$, which is absurd. □

I tried to apply Theorem 5 to find a *local* proof of Lemma 11 as well. But I failed. This is not surprising. Let me explain why. The proof of Theorem 5 is *asymmetric* with respect to the curves Δ_1 and Δ_2. The *global* proof of Lemma 12 is also *asymmetric* with respect to the curves L and C. The proof of Theorem 5 is based on uniquely determined iterations of blow ups: I must keep blowing up the point of the proper transform of the curve Δ_2 that dominates the point P. The *global* proof of Lemma 12 is based on uniquely determined composition of Geiser involutions. So, Lemma 12 can be considered as a *global* wrap up of a purely *local* special case of Theorem 5, where the line L plays the role of the curve Δ_2 in Theorem 5. On the other hand, Lemma 11 is *symmetric* with respect to the lines L and M. Moreover, its proof is not deterministic at all, since the composition of Geiser involutions in the proof of Lemma 11 is not uniquely determined by the initial data, i.e., every time I apply Geiser involution, I have exactly two possible candidates for the next one: either I can use the Geiser involution induced by the projection from $L \cap N$ or I can use the Geiser involution induced by the projection from $M \cap N$. So, there is a little hope that Theorem 5 can be used to replace the usage of Geiser involutions in the proof of Lemma 11. Of course, there is a chance that the proof of Lemma 11 cannot be *localized* like the proof of Lemma 12. Fortunately, this is not the case. Indeed, instead of using Geiser involutions in the *global* proof of Lemma 11, I can use

Theorem 13. *Let S be a surface, let P be a smooth point in S, let Δ_1 and Δ_2 be two irreducible curves on S that both are smooth at P and intersect transversally at P, let a_1 and a_2 be non-negative rational numbers, and let Ω be an effective $\mathbb{Q}$-divisor on the surface S whose support does not contain the curves Δ_1 and Δ_2. Suppose that the log pair $(S, a_1\Delta_1 + a_2\Delta_2 + \Omega)$ is not log canonical at P. Put $m = \mathrm{mult}_P(\Omega)$. Suppose that $m \leqslant 1$. Then $\mathrm{mult}_P(\Omega\cdot\Delta_1) > 2(1-a_2)$ or $\mathrm{mult}_P(\Omega\cdot\Delta_2) > 2(1-a_1)$.*

Proof. I may assume that $a_1 \leqslant 1$ and $a_2 \leqslant 1$. Then the log pair $(S, a_1\Delta_1 + a_2\Delta_2 + \Omega)$ is log canonical in a punctured neighborhood of the point P, because $m \leqslant 1$.

Since the log pair $(S, a_1\Delta_1 + a_2\Delta_2 + \Omega)$ is not log canonical at P, there exists a birational morphism $h\colon \hat{S} \to S$ that is a composition of $r \geqslant 1$ blow ups of smooth points dominating P, and there exists an h-exceptional divisor, say E_r, such that $e_r > 1$, where e_r is a rational number determined by

$$K_{\hat{S}} + a_1\hat{\Delta}_1 + a_2\hat{\Delta}_2 + \hat{\Omega} + \sum_{i=1}^{r} e_i E_i \sim_{\mathbb{Q}} h^*\big(K_S + a_1\Delta_1 + a_2\Delta_2 + \Omega\big),$$

where e_i is a rational number, each E_i is an h-exceptional divisor, $\hat{\Omega}$ is a proper transform on $\hat{S}$ of the divisor Ω, $\hat{\Delta}_1$ and $\hat{\Delta}_2$, are proper transforms on $\hat{S}$ of the curves Δ_1 and Δ_2, respectively.

Let $f\colon \tilde{S} \to S$ be the blow up of the point P, let $\tilde{\Omega}$ be the proper transform of the divisor Ω on the surface $\tilde{S}$, let E be the f-exceptional curve, let $\tilde{\Delta}_1$ and $\tilde{\Delta}_2$ be the proper transforms of the curves Δ_1 and Δ_2 on the surface $\tilde{S}$, respectively. Then

$$K_{\tilde{S}} + a_1\tilde{\Delta}_1 + a_2\tilde{\Delta}_2 + \big(a_1 + a_2 + m - 1\big)E + \tilde{\Omega} \sim_{\mathbb{Q}} f^*\big(K_S + a_1\Delta_1 + a_2\Delta_2 + \Omega\big).$$

which implies that the log pair $(\tilde{S}, a_1\tilde{\Delta}_1 + a_2\tilde{\Delta}_2 + \big(a_1 + a_2 + m - 1\big)E + \tilde{\Omega})$ is not log canonical at some point $Q \in E$.

If $r = 1$, then $a_1 + a_2 + m - 1 > 1$, which implies that $m > 2 - a_1 - a_2$. On the other hand, if $m > 2 - a_1 - a_2$, then either $m > 2(1 - a_1)$ or $m > 2(1 - a_2)$, because otherwise I would have $2m \leqslant 4 - 2(a_1 + a_2)$, which contradicts to $m > 2 - a_1 - a_2$. Thus, if $r = 1$, them $\mathrm{mult}_P(\Omega \cdot \Delta_1) > 2(1 - a_2)$ or $\mathrm{mult}_P(\Omega \cdot \Delta_2) > 2(1 - a_1)$.

Let me prove the required assertion by induction on r. The case $r = 1$ is done. Thus, I may assume that $r \geqslant 2$. If $Q \neq E \cap \tilde{\Delta}_1$ and $Q \neq E \cap \tilde{\Delta}_2$, then it follows from Theorem 7 that $m = \tilde{\Omega} \cdot E > 1$, which is impossible, since $m \leqslant 1$ by assumption. Thus, either $Q = E \cap \tilde{\Delta}_1$ or $Q = E \cap \tilde{\Delta}_2$. Without loss of generality, I may assume that $Q = E \cap \tilde{\Delta}_1$.

By induction, I can apply the required assertion to $(\tilde{S}, a_1\tilde{\Delta}_1 + (a_1 + a_2 + m - 1)E + \tilde{\Omega})$ at the point Q. This implies that either

$$\mathrm{mult}_Q\Big(\tilde{\Omega} \cdot \tilde{\Delta}_1\Big) > 2\Big(1 - (a_1 + a_2 + m - 1)\Big) = 4 - 2a_1 - 2a_2 - 2m$$

or $\mathrm{mult}_Q(\tilde{\Omega} \cdot E) > 2(1 - a_1)$. In the latter case, I have

$$\mathrm{mult}_P\Big(\Omega \cdot \Delta_2\Big) \geqslant m > 2(1 - a_1),$$

since $m = \mathrm{mult}_Q(\tilde{\Omega} \cdot E) > 2(1 - a_1)$, which is exactly what I want. Thus, to complete the proof, I may assume that $\mathrm{mult}_Q(\tilde{\Omega} \cdot \tilde{\Delta}_1) > 4 - 2a_1 - 2a_2 - 2m$.

If $\mathrm{mult}_P(\Omega \cdot \Delta_2) > 2(1 - a_1)$, then I am done. Thus, to complete the proof, I may assume that $\mathrm{mult}_P(\Omega \cdot \Delta_2) \leqslant 2(1 - a_1)$. This gives me $m \leqslant 2(1 - a_1)$, since $\mathrm{mult}_P(\Omega \cdot \Delta_2) \geqslant m$. Then

$$\mathrm{mult}_P\Big(\Omega \cdot \Delta_1\Big) \geqslant m + \mathrm{mult}_Q\Big(\tilde{\Omega} \cdot \tilde{\Delta}_1\Big) > m + 4 - 2a_1 - 2a_2 - 2m$$
$$= 4 - 2a_1 - 2a_2 - m > 2(1 - a_2),$$

because $m \leqslant 2(1 - a_1)$. This completes the proof. □

Let me show how to prove Lemma 11 using Theorem 13. This is very easy.

Local proof of Lemma 11. Let me use the assumptions and notation of Lemma 11. I write $T_P = L + M + N$, where L, M, and N are lines on the cubic surface S.

Without loss of generality, I may assume that the line N does not pass through the point P. Let D be any effective $\mathbb{Q}$-divisor on the surface S such that $D \sim_{\mathbb{Q}} -K_S$. I must show that the log pair (S, D) is log canonical at P. Suppose that the log pair (S, D) is not log canonical at P. Let me seek for a contradiction.

The log pair (S, D) is log canonical in a punctured neighborhood of the point P (use [6, Lemma 5.3.6] or the proof of [1, Lemma 3.4]). Put $D = aL+bM+cN+\Omega$, where a, b, and c are non-negative rational numbers and Ω is an effective $\mathbb{Q}$-divisor on S whose support contains none of the lines L, M, and N. Put $m = \mathrm{mult}_P(\Omega)$.

Since $(S, L + M + N)$ is log canonical, $D \neq L + M + N$. Then there exists a non-negative rational number μ such that the divisor $(1 + \mu)D - \mu T_P$ is effective and its support does not contain at least one components of $T_P = L + M + N$. Thus, replacing D by $(1+\mu)D-\mu T_P$, I can assume that at least one number among a, b, and c is zero. On the other hand, I know that

$$\mathrm{mult}_P(D) = m + a + b > 1,$$

because the log pair (S, D) is not log canonical at P. Thus, if $a = 0$, then

$$1 = L \cdot D \geqslant \mathrm{mult}_P\big(L\big)\mathrm{mult}_P\big(D\big) = \mathrm{mult}_P\big(D\big) = m + b > 1,$$

which is absurd. This shows that $a > 0$. Similarly, $b > 0$. Therefore, $c = 0$. Then

$$1 = N \cdot D = N \cdot (aL + bM + \Omega) = a + b + N \cdot \Omega \geqslant a + b,$$

which implies that $a + b \leqslant 1$. On the other hand, I know that

$$\begin{cases} 1 = L \cdot \Big(aL + bM + \Omega\Big) = -a + b + L \cdot \Omega \geqslant -a + b + m, \\ 1 = M \cdot \Big(aL + bM + \Omega\Big) = a - b + M \cdot \Omega \geqslant a - b + m, \end{cases}$$

which implies that $m \leqslant 1$. Thus, I can apply Theorem 13 to $(S, aL+bM+\Omega)$. This gives either

$$1 + a - b = \mathrm{mult}_P(\Omega \cdot L) > 2(1 - b)$$

or $1 - a + b = \mathrm{mult}_P(\Omega \cdot M) > 2(1 - a)$. Then either $1 + a - b > 2 - 2b$ or $1 - a + b > 2 - 2a$. In both cases, $a + b > 1$, which is not the case (I proved this earlier). □

I was very surprised to find out that Theorem 13 has many other applications as well. Let me show how to use Theorem 13 to give a short proof of Lemma 3.

Proof of Lemma 3. Let me use the assumptions and notation of Lemma 3. Every cuspidal curve in $| - K_S|$ is a Siberian tigers, since all curves in $| - K_S|$ are irreducible. Let D be a Siberian tiger. I must prove that D is a cuspidal curve in $|-K_S|$.

The pair (S, D) is not log canonical at some point $P \in S$. Let C be a curve in $|-K_S|$ that contains P. If P is the base locus of the pencil $|-K_S|$, then (S, C) is log canonical at P, because every curve in the pencil $|-K_S|$ is smooth at its unique base point. Moreover, if $P = O$, then (S, C) is also log canonical at P by [12, Theorem 3.3]. In the latter case, the curve C has an ordinary double point at P by [12, Theorem 3.3], which also follows from Kodaira's table of singular fibers of elliptic fibration. Furthermore, if C is singular at P and (S, C) is not log canonical at P, then C has an ordinary cusp at P.

If $D = C$ and C is a cuspidal curve, then I am done. Thus, I may assume that this is not the case. Let me seek for a contradiction.

I claim that $C \not\subseteq \mathrm{Supp}(D)$. Indeed, if C is cuspidal curve, then $C \not\subseteq \mathrm{Supp}(D)$, since D is a Siberian tiger. If (S, C) is log canonical, put $D = aC + \Omega$, where a is a non-negative rational number, and Ω is an effective $\mathbb{Q}$-divisor on S whose support does not contain the curve C. Then $a < 1$, since $D \sim_{\mathbb{Q}} C$ and $D \neq C$. Then

$$\frac{1}{1-a}D - \frac{a}{1-a}C = \frac{1}{1-a}(aC + \Omega) - \frac{a}{1-a}C = \frac{1}{1-a}\Omega \sim_{\mathbb{Q}} -K_S$$

and the log pair $(S, \frac{1}{1-a}\Omega)$ is not log canonical at P, because (S, C) is log canonical at P, and (S, D) is not log canonical at P. Since D is a Siberian tiger, I see that $a = 0$, i.e., $C \not\subset \mathrm{Supp}(D)$.

If $P \neq O$, then

$$1 = C \cdot D \geqslant \mathrm{mult}_P(D),$$

which is impossible by [6, Excercise 6.18], since the log pair (S, D) is not log canonical at the point P. Thus, I see that $P = O$.

Let $f\colon \tilde{S} \to S$ be a minimal resolution of singularities of the surface S. Then there are three f-exceptional curves, say E_1, E_2, and E_3, such that $E_1^2 = E_2^2 = E_3^2 = -2$. I may assume that $E_1 \cdot E_3 = 0$ and $E_1 \cdot E_2 = E_2 \cdot E_3 = 1$. Let $\tilde{C}$ be the proper transform of the curve C on the surface $\tilde{S}$. Then $\tilde{C} \sim_{\mathbb{Q}} f^*(C) - E_1 - E_2 - E_3$.

Let $\tilde{D}$ be the proper transform of the $\mathbb{Q}$-divisor D on the surface $\tilde{S}$. Then

$$\tilde{D} \sim_{\mathbb{Q}} f^*(D) - a_1E_1 - a_2E_2 - a_3E_3$$

for some non-negative rational numbers a_1, a_2 and a_3. Then

$$\begin{cases} 1 - a_1 - a_3 = \tilde{D} \cdot \tilde{C} \geqslant 0, \\ 2a_1 - a_2 = \tilde{D} \cdot E_1 \geqslant 0, \\ 2a_2 - a_1 - a_3 = \tilde{D} \cdot E_2 \geqslant 0, \\ 2a_3 - a_2 = \tilde{D} \cdot E_3 \geqslant 0, \end{cases}$$

which gives $1 \geqslant a_1 + a_3$, $2a_1 \geqslant a_2$, $3a_2 \geqslant 2a_3$, $2a_3 \geqslant a_2$, $3a_2 \geqslant 2a_1$, $a_1 \leqslant \frac{3}{4}$, $a_2 \leqslant 1$, $a_3 \leqslant \frac{3}{4}$. On the other hand, I have

$$K_{\tilde{S}} + \tilde{D} + \sum_{i=1}^{3} a_i E_i \sim_{\mathbb{Q}} f^*(K_S + D) \sim_{\mathbb{Q}} 0,$$

which implies that $(\tilde{S}, \tilde{D} + a_1E_1 + a_2E_2 + a_3E_3)$ is not log canonical at some point $Q \in E_1 \cup E_2 \cup E_3$.

Suppose that $Q \in E_1$ and $Q \notin E_2$. Then $(\tilde{S}, \tilde{D} + a_1E_1)$ is not log canonical at Q. Then $2a_1 - a_2 = \tilde{D} \cdot E_1 > 1$ by Theorem 7. Therefore, I have

$$1 \geqslant \frac{4}{3}a_1 \geqslant 2a_1 - \frac{2}{3}a_1 \geqslant 2a_1 - a_2 > 1,$$

which is absurd. Thus, if $Q \in E_1$, then $Q = E_1 \cap E_2$. Similarly, I see that if $Q \in E_3$, then $Q = E_3 \cap E_2$.

Suppose that $Q \in E_2$ and $Q \notin E_1 \cup E_3$. Then $(\tilde{S}, \tilde{D} + a_2E_2)$ is not log canonical at Q. Then $2a_2 - a_1 - a_3 = \tilde{D} \cdot E_2 > 1$ by Theorem 7. Therefore, I have

$$1 \geqslant a_2 = 2a_2 - \frac{a_2}{2} - \frac{a_2}{2} \geqslant 2a_2 - a_1 - a_3 > 1,$$

which is absurd. Thus, I proved that either $Q = E_1 \cap E_2$ or $Q = E_3 \cap E_2$. Without loss of generality, I may assume that $Q = E_1 \cap E_2$.

The log pair $(\tilde{S}, \tilde{D} + a_1E_1 + a_2E_2)$ is not log canonical at Q. Put $m = \mathrm{mult}_Q(\tilde{D})$. Then

$$\begin{cases} 2a_1 - a_2 = \tilde{D} \cdot E_1 \geqslant m, \\ 2a_2 - a_1 - a_3 = \tilde{D} \cdot E_2 \geqslant m, \\ 2a_3 - a_2 = \tilde{D} \cdot E_3 \geqslant 0, \end{cases}$$

which implies that $a_1 + a_3 \geqslant 2m$. Since I already proved that $a_1 + a_3 \leqslant 1$, $m \leqslant \frac{1}{2}$. Applying Theorem 13 to the log pair $(\tilde{S}, \tilde{D} + a_1E_1 + a_2E_2)$ at the point Q, I see that $\tilde{D} \cdot E_1 > 2(1 - a_2)$ or $\tilde{D} \cdot E_2 > 2(1 - a_1)$. In the former case, one has

$$2a_1 - a_2 = \tilde{D} \cdot E_1 > 2(1 - a_2),$$

which implies that $2 \geqslant 2a_1 + 2a_3 \geqslant 2a_1 + a_2 > 2$, since $1 \geqslant a_1 + a_3$ and $2a_3 \geqslant a_2$. Thus, I proved that

$$2a_2 - a_1 - a_3 = \tilde{D} \cdot E_2 > 2(1 - a_1),$$

which implies that $2a_2 + a_1 > 2 + a_3$. Then $2a_2 + 1 - a_3 > 2a_2 + a_1 > 2 + a_3$, since $a_1 + a_3 \leqslant 1$. The last inequality implies that $2a_2 > 1 + 2a_3$. Since I already proved that $2a_3 \geqslant a_2$, I conclude that $2a_2 > 1 + a_2$, which is impossible, since $a_1 \leqslant 1$. The obtained contradiction completes the proof. □

Similarly, I can use Theorem 13 instead of Theorem 5 in the *local* proof of Lemma 12 (I leave the details to the reader). Theorem 13 has a nice and clean inductive proof like Theorem 7 has. So, what if Theorem 13 is the desired generalization of Theorem 5? This may seem unlikely keeping in mind how both theorems look like. However, Theorem 13 does generalize Theorem 4, which is the ancestor and a special case of Theorem 5. The latter follows from

Remark 14. Let S be a surface, let Δ_1 and Δ_2 be two irreducible curves on S that are both smooth at P and intersect transversally at P. Take an effective $\mathbb{Q}$-divisor $a_1\Delta_1 + a_2\Delta_2 + \Omega$, where a_1 and a_2 are non-negative rational numbers, and Ω is an effective $\mathbb{Q}$-divisor on the surface S whose support does not contain the curves Δ_1 and Δ_2. Put $m = \mathrm{mult}_P(\Omega)$. Let n be a positive integer such that $n \geqslant 3$. Theorem 4 asserts that $\mathrm{mult}_P(\Omega \cdot \Delta_1) > 2a_1 - a_2$ or

$$\mathrm{mult}_P\Big(\Omega \cdot \Delta_2\Big) > \frac{n}{n-1}a_2 - a_1$$

provided that $\frac{2n-2}{n+1}a_1 + \frac{2}{n+1}a_2 \leqslant 1$ and the log pair $(S, a_1\Delta_1 + a_2\Delta_2 + \Omega)$ is not log canonical at P. On the other hand, $\mathrm{mult}_P(\Omega\cdot\Delta_1) \geqslant m$ and $\mathrm{mult}_P(\Omega\cdot\Delta_2) \geqslant m$. Thus, Theorem 4 asserts something non-obvious only if

$$\begin{cases} 2a_1 - a_2 \geqslant m, \\ \dfrac{n}{n-1}a_2 - a_1 \geqslant m, \\ \dfrac{2n-2}{n+1}a_1 + \dfrac{2}{n+1}a_2 \leqslant 1. \end{cases} \tag{15}$$

Note that (15) implies that $a_1 \leqslant \frac{1}{2}$, $a_2 \leqslant 1$, and $m \leqslant 1$. Thus, if (15) holds, then I can apply Theorem 13 to the log pair $(S, a_1\Delta_1 + a_2\Delta_2 + \Omega)$ to get $\mathrm{mult}_P(\Omega\cdot\Delta_1) > 2(1 - a_2)$ or $\mathrm{mult}_P(\Omega \cdot \Delta_2) > 2(1 - a_1)$. On the other hand, if (15) holds, then $2(1 - a_2) \geqslant 2a_1 - a_2$ and

$$2(1 - a_1) \geqslant \frac{2n-2}{n+1}a_1 + \frac{2}{n+1}a_2.$$

Nevertheless, Theorem 13 is not a generalization of Theorem 5, i.e., I cannot use Theorem 13 instead of Theorem 5 in general. I checked this in many cases considered in [2]. To convince the reader, let me give

Example 16. Put $S = \mathbb{P}^2$. Take some integers $m \geqslant 2$ and $k \geqslant 2$. Put $r = km(m-1)$. Let C be a curve in S that is given by $z^{r-1}y = x^r$, where $[x : y : z]$ are projective coordinates on S. Put $\Omega = \lambda C$ for some positive rational number λ. Let Δ_1 be a line in S that is given by $x = 0$, and let Δ_2 be a line in S that is given by $y = 0$. Put $a_1 = \frac{1}{m}$ and $a_2 = 1 - \frac{1}{m}$. Let P be the intersection point $\Delta_1 \cap \Delta_2$. Then $(S, a_1\Delta_1 + a_2\Delta_2 + \Omega)$ is log canonical P if and only if $\lambda \leqslant \frac{1}{m} + \frac{1}{km^2}$. Take any $\lambda > \frac{1}{m} + \frac{1}{km^2}$ such that $\lambda < \frac{k}{km-1}$. Then $\mathrm{mult}_P(\Omega) = \lambda < \frac{2}{m} \leqslant 1$ and

$$\mathrm{mult}_P\Big(\Omega \cdot \Delta_1\Big) = \lambda < \frac{k}{km-1} < \frac{2}{m} = 2(1 - a_2),$$

which implies that

$$k(m-1) + \frac{m-1}{m} > km(m-1)\lambda = \mathrm{mult}_P\Big(\Omega \cdot \Delta_2\Big) > 2(1 - a_1) = \frac{2m-2}{m}$$

by Theorem 13. Taking λ close enough to $\frac{1}{m} + \frac{1}{km^2}$, I can get $\mathrm{mult}_P(\Omega \cdot \Delta_2)$ as close to $k(m-1) + \frac{m-1}{m}$ as I want. Thus, the inequality $\mathrm{mult}_P(\Omega \cdot \Delta_2) > \frac{2m-2}{m}$ provided by Theorem 13 is not very good when $k \gg 0$. Now let me apply Theorem 5 to the log pair $(S, a_1\Delta_1 + a_2\Delta_2 + \Omega)$ to get much better estimate for $\mathrm{mult}_P(\Omega \cdot \Delta_2)$. Put $\alpha = 1$, $\beta = 1$, $M = 1$, $B = km$, $A = \frac{1}{km-1}$, and $N = 0$. Then

$$\begin{cases} 1 = \alpha a_1 + \beta a_2 \leqslant 1, \\ 1 = A(B-1) \geqslant 1, \\ 1 = M \leqslant 1, \\ 0 = N \leqslant 1, \\ \dfrac{1}{km-1} = \alpha(A + M - 1) \geqslant A^2(B + N - 1)\beta = \dfrac{1}{km-1}, \\ \dfrac{1}{km-1} = \alpha(1 - M) + A\beta \geqslant A = \dfrac{1}{km-1}, \\ 2 = 2M + AN \leqslant 2. \end{cases}$$

By Theorem 5, $\mathrm{mult}_P(\Omega \cdot \Delta_1) > M + Aa_1 - a_2$ or $\mathrm{mult}_P(\Omega \cdot \Delta_2) > N + Ba_2 - a_1$. Since $\mathrm{mult}_P\big(\Omega \cdot \Delta_1\big) = \lambda < \frac{k}{km-1} = M + Aa_1 - a_2$, it follows from Theorem 5 that

$$\mathrm{mult}_P(\Omega \cdot \Delta_2) > N + Ba_2 - a_1 = k(m-1) - \frac{1}{m}.$$

For $k \gg 0$, the latter inequality is much stronger than $\mathrm{mult}_P(\Omega \cdot \Delta_2) > \frac{2m-2}{m}$ given by Theorem 13. Moreover, I can always choose λ close enough to $\frac{1}{m} + \frac{1}{km^2}$ so that the multiplicity $\mathrm{mult}_P(\Omega \cdot \Delta_2) = km(m-1)\lambda$ is as close to $k(m-1) + \frac{m-1}{m}$ as I want. This shows that the inequality $\mathrm{mult}_P(\Omega \cdot \Delta_2) > k(m-1) - \frac{1}{m}$ provided by Theorem 5 is almost sharp.

I have a strong feeling that Theorems 5 and 13 are special cases of some more general result that is not yet found. Perhaps, it can be found by analyzing the proofs of Theorems 5 and 13.

References

1. I. Cheltsov, Log canonical thresholds of del Pezzo surfaces. Geom. Funct. Anal. **11**, 1118–1144 (2008)
2. I. Cheltsov, Two local inequalties. Izv. Math. **78**, 375–426 (2014)
3. I. Cheltsov, D. Kosta, Computing α-invariants of singular del Pezzo surfaces. J. Geom. Anal. **24**, 798–842 (2014) doi:10.1007/s12220-012-9357-6
4. I. Cheltsov, C. Shramov, Log canonical thresholds of smooth Fano threefolds (with an appendix by Jean-Pierre Demailly). Russ. Math. Surv. **63**, 73–180 (2008)
5. I. Cheltsov, J. Park, J. Won, Affine cones over smooth cubic surfaces (2013) [arXiv:1303.2648]
6. A. Corti, J. Kollár, K. Smith, *Rational and Nearly Rational Varieties*. Cambridge Studies in Advanced Mathematics, vol. 92 (Cambridge University Press, Cambridge, 2004)
7. J.-P. Demailly, J. Kollár, Semi-continuity of complex singularity exponents and Kähler-Einstein metrics on Fano orbifolds. Ann. Sci. École Norm. Sup. **34**, 525–556 (2001)
8. H. Flenner, M. Zaidenberg, Rational curves and rational singularities. Math. Z. **244**, 549–575 (2003)
9. S. Keel, J. McKernan, Rational curves on quasi-projective surfaces (English summary). Mem. Am. Math. Soc. **140**(669), viii+153 pp. (1999)
10. T. Kishimoto, Yu. Prokhorov, M. Zaidenberg, *$\mathbb{G}_a$-actions on affine cones*. Transf. Groups **18**(4), 1137–1153 (2013)
11. D. Kosta, Del Pezzo surfaces with Du Val singularities. Ph.D. Thesis, University of Edinburgh, 2009
12. J. Park, A note on del Pezzo fibrations of degree 1. Commun. Algebra **31**, 5755–5768 (2003)
13. G. Tian, On Kähler–Einstein metrics on certain Kähler manifolds with $c_1(M) > 0$. Invent. Math. **89**, 225–246 (1987)

Fano Hypersurfaces and their Birational Geometry

Tommaso de Fernex

Abstract We survey some results on the nonrationality and birational rigidity of certain hypersurfaces of Fano type. The focus is on hypersurfaces of Fano index one, but hypersurfaces of higher index are also discussed.

2010 *Mathematics Subject Classification.* Primary: 14E08; Secondary: 14J45, 14E05, 14B05, 14N30.

1 Introduction

This paper gives an account of the main result of [9], which states that every smooth complex hypersurface of degree N in $\mathbb{P}^N$, for $N \geq 4$, is birationally superrigid. The result is contextualized within the framework of smooth Fano hypersurfaces in projective spaces and the problem of rationality. The paper overviews the history of the problem and the main ideas that come into play in its solution, from the method of maximal singularities to the use of arc spaces and multiplier ideals.

Working over fields that are not necessarily algebraically closed, we also discuss an extension of a theorem of Segre and Manin stating that every smooth projective cubic surface of Picard number one over a perfect field is birationally rigid. The proof, which is an adaptation of the arguments of Segre and Manin, is a simple manifestation of the method of maximal singularities.

T. de Fernex (✉)
Department of Mathematics, University of Utah, 155 South 1400 East, Salt Lake City, UT 48112-0090, USA
e-mail: defernex@math.utah.edu

I. Cheltsov et al. (eds.), *Automorphisms in Birational and Affine Geometry*, Springer Proceedings in Mathematics & Statistics 79,
DOI 10.1007/978-3-319-05681-4_6,

The last section of the paper explores hypersurfaces in $\mathbb{P}^N$ of degree $d < N$. We suspect that the result of [9] is an extreme case of a more general phenomenon, and propose two problems which suggest that the birational geometry of Fano hypersurfaces should progressively become more rigid as their degree d approaches N. A theorem of [20] brings some evidence to this phenomenon.

Unless stated otherwise, we work over the field of complex numbers $\mathbb{C}$. Some familiarity with the basic notions of singularities of pairs and multiplier ideals will be assumed; basic references on the subject are [21, 24].

2 Mori Fiber Spaces and Birational Rigidity

Projective hypersurfaces form a rich class of varieties from the point of view of rationality problems and related questions. We focus on smooth hypersurfaces, and let

$$X = X_d \subset \mathbb{P}^N$$

denote a smooth complex projective hypersurface of dimension $N-1$ and degree d. By adjunction, X is Fano (i.e., its anticanonical class $-K_X$ is ample) if and only if $d \le N$.

If $d \le 2$ then X is clearly rational with trivial moduli, and there is no much else to say. However, already in degree $d = 3$ the situation becomes rather delicate. Cubic surfaces are rational, but cubic threefolds are nonrational by a theorem of Clemens and Griffiths [4]. Moving up in dimension, we find several examples of families of rational cubics fourfolds [1, 15, 17, 18, 26, 36, 37], with those due to Hassett filling up a countable union of irreducible families of codimension 2 in the moduli space of cubic hypersurfaces in $\mathbb{P}^5$. By contrast, a conjecture of Kuznetsov [23] predicts that the very general cubic fourfold should be nonrational. Apart from simple considerations (e.g., rationality of cubic hypersurfaces of even dimension containing disjoint linear subspaces of half the dimension) no much is known in higher dimensions, and there is no clear speculation on what the picture should be. In degree $d = 4$, we only have Iskovskikh and Manin's theorem on the nonrationality of $X_4 \subset \mathbb{P}^4$ [19].

The situation starts to show a more uniform behavior if one bounds the degree from below in terms of the dimension. A result in this direction is due to Kollár [20].

Theorem 1. *Let $X = X_d \subset \mathbb{P}^N$ be a very general hypersurface.*

(a) *If $2\lceil (N+2)/3 \rceil \le d \le N$, then X is not ruled (hence is nonrational).*
(b) *If $3\lceil (N+2)/4 \rceil \le d \le N$, then X is not birationally equivalent to any conic bundle.*

This result suggests a certain trend: as the degree approaches (asymptotically) the dimension, the birational geometry of the hypersurface tends to "rigidify."

This principle can be formulated precisely in the extreme case $d = N$, where the geometry becomes as "rigid" as it can be.

A *Mori fiber space* is a normal $\mathbb{Q}$-factorial projective variety with terminal singularities, equipped with a morphism of relative Picard number one with connected fibers of positive dimension such that the anticanonical class is relatively ample. Examples of Mori fiber spaces are conic bundles and Del Pezzo fibrations. A Fano manifold with Picard number one can be regarded as a Mori fiber space over $\operatorname{Spec}\mathbb{C}$.

Theorem 2. *Let $X = X_N \subset \mathbb{P}^N$ be any (smooth) hypersurface. If $N \geq 4$, then every birational map from X to a Mori fiber space X'/S' is an isomorphism (and in fact a projective equivalence). In particular, X is nonrational.*

We say that $X_N \subset \mathbb{P}^N$, for $N \geq 4$, is *birationally superrigid*. In general, a Fano manifold X of Picard number one is said to be *birationally rigid* if every birational map ϕ from X to a Mori fiber space X'/S' is, up to an isomorphism, a birational automorphism of X; it is said to be *birationally superrigid* if any such ϕ is an isomorphism.[1]

Theorem 2 has a long history, tracing back to the work of Fano on quartic threefolds [13, 14]. Let $X = X_4 \subset \mathbb{P}^4$. Fano claimed that $\operatorname{Bir}(X) = \operatorname{Aut}(X)$, a fact that alone suffices to show that X is nonrational as $\operatorname{Aut}(X)$ if finite and $\operatorname{Bir}(\mathbb{P}^3)$ is not. Fano's method is inspired to Noether's factorization of planar Cremona maps. The idea is to look at the indeterminacy locus of a given birational self-map $\phi\colon X \dashrightarrow X$. If ϕ is not an isomorphism, then the base scheme $B \subset X$ of a linear system defining ϕ must be "too singular" with respect to the equations cutting out B in X. As this is impossible, one concludes that ϕ is a regular automorphism.

Fano's proof is incomplete. The difficulty that Fano had to face is that, differently from the surface case where the multiplicities of the base scheme are a strong enough invariant to quantify how "badly singular" the map is, in higher dimension one needs to dig further into a resolution of singularities to extract the relevant information.

The argument was eventually corrected and completed in [19]. In their paper, Iskovskikh and Manin only look at the birational group $\operatorname{Bir}(X)$, but it soon became clear that the proof itself leads to the stronger conclusion that X is birationally superrigid. In fact, the very definition of birational superrigidity was originally motivated by their work.[2]

Following [19], significant work has been done throughout the years to extend this result to higher dimensions, starting from Pukhlikov who proved it first for

---

[1]This definition can be generalized to all Mori fiber spaces, see [5].

[2]Mori fiber spaces are the output of the minimal model program for projective manifolds of negative Kodaira dimension. It is natural to motivate the notion of birational rigidity also from this point of view: a Mori fiber space is birationally rigid (resp., superrigid) if, within its own birational class, it is the unique answer of the program up to birational (resp., biregular) automorphisms preserving the fibration.

$X_5 \subset \mathbb{P}^5$ [29], and then in all dimensions under a suitable condition of "local regularity" on the equation defining the hypersurface [30]. Some low dimensional cases were established in [3, 10], and the complete proof of Theorem 2 was finally given in [9].

While in this paper we focus on smooth projective hypersurfaces, the birational rigidity problem has been extensively studied for many other Fano varieties and Mori fiber spaces, especially in dimension 3. There is a large literature on the subject that is too vast to be included here. For further reading, a good place to start is [7].

The study of birational rigidity has also ties with other birational properties of algebraic varieties such as unirationality and rational connectedness. The work of Iskovskikh and Manin was originally motivated by the Lüroth problem, which asked whether unirational varieties are necessarily rational. It was known by work of Segre [35] that there are smooth quartic threefolds $X_4 \subset \mathbb{P}^4$ that are unirational, and it is easy to see that all smooth cubic threefolds $X_3 \subset \mathbb{P}^4$ are unirational. The results of [4, 19] gave the first counter-examples to the Lüroth problem.

Birational rigidity also relates to stability properties. A recent theorem of Odaka and Okada [28] proves that any birationally superrigid Fano manifold of index 1 is slope stable in the sense of Ross and Thomas [33].

3 Cubic Surfaces of Picard Number One

Before Fano's idea could be made work in dimension three, Segre found a clever way to apply Noether's method once more to dimension two. Cubic surfaces are certainly rational over the complex numbers, but they may fail to be rational when the ground field is not algebraically closed. The method of Noether works perfectly well, in fact, to prove that every smooth projective cubic surface of Picard number one over a field κ is nonrational [34]. Later, Manin observed that if the field is perfect then the proof can be adapted to show that if two such cubic surfaces are birational equivalent, then they are projectively equivalent [25].[3] For a thorough discussion of these results, see also [22].

The theorems of Segre and Manin extend rather straightforwardly to the following result, which implies that cubic surfaces of Picard number one are *birationally rigid* (over their ground field).

Theorem 3. *Let $X_\kappa \subset \mathbb{P}^3_\kappa$ be a smooth cubic surface of Picard number one over a perfect field κ. Suppose that there is a birational map $\phi_\kappa \colon X_\kappa \dashrightarrow X'_\kappa$ where X'_κ is either a Del Pezzo surface of Picard number one, or a conic bundle over a curve S'_κ. Then X'_κ is a smooth cubic surface of Picard number one, and there is a birational automorphism $\beta_\kappa \in \mathrm{Bir}(X_\kappa)$ such that $\phi_\kappa \circ \beta_\kappa \colon X_\kappa \to X'_\kappa$ is a projective equivalence. In particular, X_κ is nonrational.*

[3]The hypothesis in Manin's theorem that κ be perfect can be removed, cf. [22].

Proof. Fix an integer $r' \geq 1$ and a divisor A'_κ on X'_κ, given by the pullback of a very ample divisor on S'_κ, such that $-r'K_{X'_\kappa} + A'_\kappa$ is very ample. Here we set $S'_\kappa = \operatorname{Spec} \kappa$ and $A'_\kappa = 0$ if X'_κ is a Del Pezzo surface of Picard number one.

Since X_κ has Picard number one, its Picard group is generated by the hyperplane class, which is linearly equivalent to $-K_{X_\kappa}$. Then there is a positive integer r such that

$$(\phi_\kappa)_*^{-1}(-r'K_{X'_\kappa} + A'_\kappa) \sim -rK_{X_\kappa}.$$

Let $\overline{\kappa}$ be the algebraic closure of κ, and denote $X = X_{\overline{\kappa}}$, $X' = X'_{\overline{\kappa}}$, $S' = S'_{\overline{\kappa}}$, $A' = A'_{\overline{\kappa}}$ and $\phi = \phi_{\overline{\kappa}}$. Note that A' is zero if $\dim S' = 0$, and is the pullback of a very ample divisor on S' if $\dim S' = 1$. Let $D' \in |-r'K_{X'} + A'|$ be a general element, and let

$$D = \phi_*^{-1} D' \in |-rK_X|.$$

We split the proof in two cases.

Case 1. Assume that $\operatorname{mult}_x(D) > r$ for some $x \in X$.

The idea is to use these points of high multiplicity to construct a suitable birational involution of X (defined over κ) that, pre-composed to ϕ, untwists the map. This part of the proof is the same as in the proof of Manin's theorem, and we only sketch it. The construction is also explained in [22], to which we refer for more details.

The Galois group of $\overline{\kappa}$ over κ acts on the base points of ϕ and preserves the multiplicities of D at these points. Since D belongs to a linear system with zero-dimensional base locus and $\deg D = 3r$ (as a cycle in $\mathbb{P}^3$), there are at most two points at which D has multiplicity larger than r, and the union of these points is preserved by the Galois action. If there is only one point $x \in X$ (not counting infinitely near ones), then x is defined over κ. Otherwise, we have two distinct points x, y on X whose union $\{x, y\} \subset X$ is defined over κ.

In the first case, consider the rational map $X \dashrightarrow \mathbb{P}^2$ given by the linear system $|\mathcal{O}_X(1) \otimes \mathfrak{m}_x|$ (i.e., induced by the linear projection $\mathbb{P}^3 \dashrightarrow \mathbb{P}^2$ with center x). The blow-up $g\colon \tilde{X} \to X$ of X at x resolves the indeterminacy of the map, and we get a double cover $h\colon \tilde{X} \to \mathbb{P}^2$. The Galois group of this cover is generated by an involution $\tilde{\alpha}_1$ of $\tilde{X}$, which descends to a birational involution α_1 of X. In the second case, consider the map $X \dashrightarrow \mathbb{P}^3$ given by the linear system $|\mathcal{O}_X(2) \otimes \mathfrak{m}_x^2 \otimes \mathfrak{m}_y^2|$. In this case, we obtain a double cover $h\colon \tilde{X} \to Q \subset \mathbb{P}^3$ where now $g\colon \tilde{X} \to X$ is the blow-up of X at $\{x, y\}$ and Q is a smooth quadric surface. As before, we denote by $\tilde{\alpha}_1$ the Galois involution of the cover and by α_1 the birational involution induced on X. In both cases, α_1 is defined over κ. Therefore the composition

$$\phi_1 = \phi \circ \alpha_1\colon X \dashrightarrow X'$$

is defined over κ and hence is given by a linear system in $|-r_1K_X|$ for some r_1. A point x with $\mathrm{mult}_x(D) > r$ cannot be an Eckardt point, and thus is a center of indeterminacy for ϕ_1.

In either case, we have $r_1 < r$. To see this, let E be the exceptional divisor of $g\colon \tilde{X} \to X$, and let L be the pullback to $\tilde{X}$ of the hyperplane class of $\mathbb{P}^2$ (resp., of $Q \subset \mathbb{P}^3$) by h. Note that $L \sim g^*(-K_X) - E$ by construction, and $g_*\tilde{\alpha}_{1*}E \sim -sK_X$ for some $s \geq 1$ since it is supported on a nonempty curve (by Zariski's Main Theorem) that is defined over κ. If m is the multiplicity of D at x (and hence at y in the second case) and $\tilde{D}$ is the proper transform of D on $\tilde{X}$, then $\tilde{D}+(m-r)E \sim rL$. Applying $(\tilde{\alpha}_1)_*$ to this divisor and pushing down to X, we obtain $\alpha_{1*}D \sim -r_1K_X$ where $r_1 = r-(m-r)s < r$ since $m > r$. Therefore, this operation lowers the degree of the equations defining the map.

Let $D_1 = \phi_{1*}^{-1}D' \in |-r_1K_X|$. If $\mathrm{mult}_x(D_1) > r_1$ for some $x \in X$, then we proceed as before to construct a new involution α_2, and proceed from there. Since the degree decreases each time, this process stops after finitely many steps. It stops precisely when, letting

$$\phi_i = \phi \circ \alpha_1 \circ \cdots \circ \alpha_i \colon X \dashrightarrow X'$$

and $D_i = \phi_{i*}^{-1}D' \in |-r_iK_X|$, we have $\mathrm{mult}_x(D_i) \leq r_i$ for every $x \in X$. Note that ϕ_i is defined over κ. Then, replacing ϕ by ϕ_i, we reduce to the next case.

Case 2. Assume that $\mathrm{mult}_x(D) \leq r$ for every $x \in X$.

Taking a sequence of blow-ups, we obtain a resolution of indeterminacy

$$\begin{array}{ccc} & Y & \\ {\scriptstyle p}\swarrow & & \searrow{\scriptstyle q} \\ X & \overset{\phi}{\dashrightarrow} & X' \end{array}$$

with Y smooth. Write

$$\begin{aligned} K_Y + \tfrac{1}{r'}D_Y &= p^*(K_X + \tfrac{1}{r'}D) + E' \\ &= q^*(K_{X'} + \tfrac{1}{r'}D') + F' \end{aligned}$$

where E' is p-exceptional, F' is q-exceptional, and $D_Y = p_*^{-1}D = q_*^{-1}D'$. Since X' is smooth and D' is a general hyperplane section, we have $F' \geq 0$ and $\mathrm{Supp}(F') = \mathrm{Ex}(q)$. Note that $K_{X'} + \frac{1}{r'}D'$ is nef. Intersecting with the image in Y of a general complete intersection curve $C \subset X$ we see that $(K_X + \frac{1}{r'}D)\cdot C \geq 0$, and this implies that $r \geq r'$.

Next, we write

$$\begin{aligned} K_Y + \tfrac{1}{r}D_Y &= p^*(K_X + \tfrac{1}{r}D) + E \\ &= q^*(K_{X'} + \tfrac{1}{r}D') + F \end{aligned}$$

where, again, E is p-exceptional and F is q-exceptional. The fact that $\mathrm{mult}_x(D) \leq r$ for all $x \in X$ implies that $E \geq 0$. Intersecting this time with the image in Y of a general complete intersection curve C' in a general fiber of $X' \to S'$, we get $(K_{X'} + \frac{1}{r}D') \cdot C' \geq 0$, and therefore $r = r'$. Note also that $E = E'$ and $F = F'$.

The difference $E - F$ is numerically equivalent to the pullback of A'. In particular, $E - F$ is nef over X and is numerically trivial over X'. Since $p_*(E - F) \leq 0$, the Negativity Lemma, applied to p, implies that $E \leq F$. Similarly, since $q_*(E - F) \geq 0$, the Negativity Lemma, applied to q, implies that $E \geq F$. Therefore $E = F$. This means that A' is numerically trivial, and hence $S' = \mathrm{Spec}\,\overline{\kappa}$. Furthermore, we have $\mathrm{Ex}(q) \subset \mathrm{Ex}(p)$, and therefore Zariski's Main Theorem implies that the inverse map

$$\sigma = \phi^{-1}: X' \dashrightarrow X$$

is a morphism.

To conclude, just observe that if $S'_\kappa = \mathrm{Spec}\,\kappa$, then X'_κ must have Picard number one. But σ, being the inverse of ϕ, is defined over κ. It follows that σ is an isomorphism, as otherwise it would increase the Picard number. Therefore X'_κ is a smooth cubic surface of Picard number one. Since we can assume without loss of generality to have picked $r' = 1$ to start with, we conclude that, after the reduction step performed in Case 1, ϕ is a projective equivalence defined over κ. The second assertion of the theorem follows by taking β_κ given by $\alpha_1 \circ \cdots \circ \alpha_i$ over κ. □

4 The Method of Maximal Singularities

The proof of Theorem 3 already shows the main features of the *method of maximal singularities.*

The reduction performed in Case 1 of the proof is clearly inspired by Noether's untwisting process used to factorize planar Cremona maps into quadratic transformations [2, 27]. This procedure has been generalized in higher dimensions to build the *Sarkisov's program*, which provides a way of factorize birational maps between Mori fiber spaces into *elementary links*, see [5, 16].

The discussion of Case 2 of the proof generalizes to the following property, due to [5, 19].[4]

Proposition 4 (Noether–Fano Inequality). *Let $\phi: X \dashrightarrow X'$ be a birational map from a Fano manifold X of Picard number one to a Mori fiber space X'/S'. Fix a sufficiently divisible integer r' and a sufficiently ample divisor on S' such that if A' is the pullback of this divisor to X' then $-r'K_{X'} + A'$ is a very ample divisor*

[4]For a comparison, one should notice how similar the arguments are. We decided to use the same exact wording when the argument is the same so that the differences will stand out.

(if $S' = \operatorname{Spec}\mathbb{C}$ *then take* $A' = 0$*). Let* r *be the positive rational number such that* $\phi_*^{-1}(-r'K_{X'} + A) \sim_{\mathbb{Q}} -rK_X$*, and let* $B \subset X$ *be the base scheme of the linear system* $\phi_*^{-1}|-r'K_{X'} + A| \subset |-rK_X|$*. If the pair* $(X, \frac{1}{r}B)$ *is canonical, then* $r = r'$ *and* ϕ *is an isomorphism.*

Proof. Let

$$\begin{array}{ccc} & Y & \\ {}^{p}\swarrow & & \searrow^{q} \\ X & \overset{\phi}{\dashrightarrow} & X' \end{array}$$

be a resolution of singularities. Note that the exceptional loci $\operatorname{Ex}(p)$ and $\operatorname{Ex}(q)$ have pure codimension 1. Fix a general element $D' \in |-r'K_{X'} + A|$ and let $D_Y = q_*^{-1}D$ (which is the same as q^*D) and $D = p_*D_Y$. Note that $D_Y = p_*^{-1}D$ and $D = \phi_*^{-1}D' \in |-rK_X|$.

Write

$$\begin{aligned} K_Y + \tfrac{1}{r'}D_Y &= p^*(K_X + \tfrac{1}{r'}D) + E' \\ &= q^*(K_{X'} + \tfrac{1}{r'}D') + F' \end{aligned}$$

where E' is p-exceptional and F' is q-exceptional. Since X' has terminal singularities and D' is a general hyperplane section, we have $F' \ge 0$ and $\operatorname{Supp}(F') = \operatorname{Ex}(q)$. Note that $K_{X'} + \frac{1}{r'}D'$ is numerically equivalent to the pullback of A', which is nef. Intersecting with the image in Y of a general complete intersection curve $C \subset X$ we see that $(K_X + \frac{1}{r'}D) \cdot C \ge 0$, and this implies that $r \ge r'$.

Next, we write

$$\begin{aligned} K_Y + \tfrac{1}{r}D_Y &= p^*(K_X + \tfrac{1}{r}D) + E \\ &= q^*(K_{X'} + \tfrac{1}{r}D') + F \end{aligned}$$

where, again, E is p-exceptional and F is q-exceptional. Assume that the pair $(X, \frac{1}{r}B)$ is canonical. Since D is defined by a general element of the linear system of divisors cutting out B, and $r \ge 1$, it follows that $(X, \frac{1}{r}D)$ is canonical. This means that $E \ge 0$. Intersecting this time with the image in Y of a general complete intersection curve C' in a general fiber of $X' \to S'$, we get $(K_{X'} + \frac{1}{r}D') \cdot C' \ge 0$, and therefore $r = r'$. Note also that $E = E'$ and $F = F'$.

The difference $E - F$ is numerically equivalent to the pullback of A'. In particular, $E - F$ is nef over X and is numerically trivial over X'. Since $p_*(E - F) \le 0$, the Negativity Lemma, applied to p, implies that $E \le F$. Similarly, since $q_*(E - F) \ge 0$, the Negativity Lemma, applied to q, implies that $E \ge F$. Therefore $E = F$. This means that A' is numerically trivial, and hence

$S' = \operatorname{Spec}\mathbb{C}$ and X' is a Fano variety of Picard number one. Furthermore, we have $\operatorname{Ex}(q) \subset \operatorname{Ex}(p)$,

By computing the Picard number of Y in two ways (from X and from X'), we conclude that $\operatorname{Ex}(p) = \operatorname{Ex}(q)$, and thus the difference $p^*D - q^*D'$ is q-exceptional. Since D is ample, this implies that ϕ is a morphism. Since X and X' have the same Picard number and X' is normal, it follows that ϕ is an isomorphism.

Remark 5. A more general version of this property gives a criterion for a birational map $\phi: X \dashrightarrow X'$ between two Mori fiber spaces X/S and X'/S' to be an isomorphism preserving the fibration. Given the correct statement, the proof easily adapts to this setting. For more details, see [5, 8] (the proof in [5] uses, towards the end, some results from the minimal model program; this is replaced in [8] by an easy computation of Picard numbers similar to the one done at the end of the proof of the proposition).

The idea at this point is to relate this condition on the singularities of the pair $(X, \frac{1}{r}B)$ to intersection theoretic invariants such as multiplicities, which can be easily related to the degrees of the equations involved when, say, X is a hypersurface in a projective space.

If X is a smooth surface and D is an effective divisor, then $(X, \frac{1}{r}D)$ is canonical if and only if $\operatorname{mult}_x(D) \le r$ for every $x \in X$. In higher dimension, however, being canonical cannot be characterized by a simple condition on multiplicities.

The way [19] deals with this problem is by carefully keeping track of all valuations and discrepancies along the exceptional divisors appearing on a resolution of singularities. The combinatorics of the whole resolution, encoded in a suitable graph which remembers all centers of blow-up, becomes an essential ingredient of the computation. This approach has been used to study birational rigidity problems for several years until Corti proposed in [6] an alternative approach based on the Shokurov–Kollár Connectedness Theorem. Corti's approach has led to a significant simplification of the proof of Iskovskikh–Manin's theorem, and has provided a starting point for setting up the proof of Theorem 2.

5 Cutting Down the Base Locus

Let $X = X_4 \subset \mathbb{P}^4$, and suppose that $\phi: X \dashrightarrow X'$ is a birational map to a Mori fiber space X'/S' which is not an isomorphism. Using the same notation as in Proposition 4, it follows that the pair $(X, \frac{1}{r}B)$ is not canonical. The following property, due to [30], implies that the pair is canonical away from a finite set.

Lemma 6. *Let $X \subset \mathbb{P}^N$ be a smooth hypersurface, and let $D \in |\mathcal{O}_X(r)|$. Then $\operatorname{mult}_C(D) \le r$ for every irreducible curve $C \subset X$.*

Proof. Let $g: X \to \mathbb{P}^{N-1}$ be the morphism induced by projecting from a general point of $\mathbb{P}^N$, and write $g^{-1}(g(C)) = C \cup C'$. The residual component C' has degree $(d-1)\deg(C)$ where $d = \deg(X)$. Taking a sufficiently general projection, the ramification divisor intersects C transversely at $(d-1)\deg(C)$ distinct points x_i, which are exactly the points of intersection $C \cap C'$. If $\mathrm{mult}_C(D) > r$, then we get

$$\deg(D|_{C'}) \geq \sum_i \mathrm{mult}_{x_i}(D|_{C'}) > r(d-1)\deg(C) = \deg(D|_{C'}),$$

a contradiction. □

Therefore there is a prime exceptional divisor E on some resolution $f\colon \tilde{X} \to X$, lying over a point $x \in X$, such that

$$\tfrac{1}{r} \cdot \mathrm{ord}_E(B) > \mathrm{ord}_E(K_{\tilde{X}/X}).$$

where $K_{\tilde{X}/X}$ is the relative canonical divisor. In the left hand side we regard ord_E as a valuation on the function field of X, and $\mathrm{ord}_E(B) = \mathrm{ord}_E(\mathcal{I}_B)$ denotes the smallest valuation of an element of the stalk of the ideal sheaf $\mathcal{I}_B \subset \mathcal{O}_X$ of B at the center of valuation x.

Corti's idea, at this point, is to take a general hyperplane section $Y \subset X$ through x. This has two effects:

(a) the restriction $B|_Y$ of the base scheme B is a zero-dimensional scheme, and
(b) the pair $(Y, \frac{1}{r}B|_Y)$ is not log canonical.

The first assertion is clear. Let us discuss why (b) is true. Suppose for a moment that the proper transform $\tilde{Y} \subset \tilde{X}$ of Y intersects (transversely) E, and let F be an irreducible component of $E|_{\tilde{Y}}$. By adjunction, we have

$$K_{\tilde{Y}/Y} = (K_{\tilde{X}/X} + \tilde{Y} - f^*Y)|_{\tilde{Y}}.$$

Since $\mathrm{ord}_E(Y) \geq 1$ and $\mathrm{ord}_F(B|_Y) \geq \mathrm{ord}_E(B)$, we have

$$\tfrac{1}{r} \cdot \mathrm{ord}_F(B|_Y) > \mathrm{ord}_F(K_{\tilde{Y}/Y}) + 1,$$

and this implies (b). In general, we cannot expect that $\tilde{Y}$ intersects E. Nevertheless, the Connectedness Theorem tells us that, after possibly passing to a higher resolution, $\tilde{Y}$ will intersect some other prime divisor E' over X, with center x, such that

$$\tfrac{1}{r} \cdot \mathrm{ord}_{E'}(B) + \mathrm{ord}_{E'}(Y) > \mathrm{ord}_{E'}(K_{\tilde{X}/X}) + 1.$$

Then the same computation using the adjunction formula produces a divisor F over Y satisfying the previous inequality.

The property that $(Y, \frac{1}{r}B|_Y)$ is not log canonical can be equivalently formulated in terms of log canonical thresholds. It says that the log canonical threshold $c = \mathrm{lct}(Y, B|_Y)$ of the pair $(Y, B|_Y)$ satisfies the inequality

$$c < 1/r.$$

The advantage now is that we know how to compare log canonical thresholds to multiplicities. The following result is due to [6, 11].

Theorem 7. *Let V be a smooth variety of dimension n, let $Z \subset V$ be a scheme supported at a closed point $x \in V$, and let $c = \operatorname{lct}(V, Z)$. Then Z has multiplicity*

$$e_x(Z) \geq (n/c)^n.$$

For the purpose of establishing birational rigidity, one only needs the case $n = 2$ of this theorem, which is the case first proved by Corti. The case $n = 2$ can be deduced from a more general formula which can be easily proven by induction on the number of blow-ups needed to produce a log resolution. Here we sketch the proof in all dimension which, although perhaps less direct, has the advantage of explaining the nature of the result as a manifestation of the classical inequality between arithmetic mean and geometric mean.

Sketch of the proof of Theorem 7. The proof uses a flat degeneration to monomial ideals. It is easy to prove the theorem in this case. If $\mathfrak{a} \subset \mathbb{C}[u_1, \dots, u_n]$ is a $(u_1, \dots, u_n)$-primary monomial ideal then the log canonical threshold c can be computed directly from the Newton polyhedron. This allows to reduce to the case in which $\mathfrak{a} = (u_1^{a_1}, \dots, u_n^{a_n})$, where the log canonical threshold is equal to $\sum 1/a_i$ and the Samuel multiplicity is equal to $\prod a_i$. In this special case, the stated inequality is just the usual inequality between arithmetic mean and geometric mean. □

Applying the case $n = 2$ of this theorem to our setting, we get

$$e_x(B|_Y) \geq (2/c)^2 > 4r^2,$$

which is impossible because $B|_Y$, being cut out on Y by equations of degree r, is contained in a zero-dimensional complete intersection scheme of degree $4r^2$. This finishes the proof of Iskovskikh–Manin's theorem.

6 Beyond Connectedness: First Considerations

Consider now the general case $X = X_N \subset \mathbb{P}^N$. We would like to apply Corti's strategy for all $N \geq 4$. Again, we use the notation of Proposition 4 and assume the existence of a non-regular birational map $\phi\colon X \dashrightarrow X'$. Then there is a prime divisor E on a resolution $f\colon \tilde{X} \to X$ such that

$$\tfrac{1}{r} \cdot \operatorname{ord}_E(B) > \operatorname{ord}_E(K_{\tilde{X}/X}).$$

and E maps to a closed point $x \in X$ by Lemma 6.

In order to cut down the base scheme B to a zero-dimensional scheme, we need to restrict to a surface. Let $Y \subset X$ be the surface cut out by $N - 3$ general hyperplane sections through x. Then $B|_Y$ is zero dimensional. As before, the Connectedness Theorem implies that $(Y, \frac{1}{r}B|_Y)$ is not log canonical at x, and we get the inequality $e_x(B|_Y) > 4r^2$.

If X is sufficiently *general* in moduli, then it contains certain cycles of low degree and high multiplicity at x, and the inequality is still sufficient to conclude that X is birationally superrigid, as shown in [30]. However, if X is *arbitrary* in moduli, then for $N \geq 5$ the inequality is not strong enough to give a contradiction as now B is cut out by equations of degree r on a surface of degree N (rather than 4).

The issue is that we are cutting down several times, but we are not keeping track of this. Morally, we should expect that as we keep cutting down, the singularities of the pair get "worse" at each step. One can try to measure this by looking at the multiplier ideal of the pair. If we cut down to a general hyperplane section $H \subset X$ through x, then the pair $(H, \frac{1}{r}B|_H)$ is not log canonical, and this implies that its multiplier ideal $\mathcal{J}(H, \frac{1}{r}B|_H)$ is nontrivial at x. In fact, we can do better: if we set $c = \operatorname{lct}(H, B|_H)$ then $\mathcal{J}(H, cB|_H)$ is nontrivial at x, which is a stronger condition since $c < 1/r$. The question is: What happens when we cut further down? Optimally, the multiplier ideal will "get deeper" at each step and we can use this information to get a better bound on the multiplicity of $B|_Y$.

Suppose for instance that the proper transform $\tilde{Y}$ intersects (transversely) E, and let F be a component of $E|_{\tilde{Y}}$. Since Y has codimension $N - 3$, the adjunction formula gives, this time,

$$c \cdot \operatorname{ord}_F(B|_Y) - (N-4) \cdot \operatorname{ord}_F(\mathfrak{m}_{Y,x}) \geq \operatorname{ord}_F(K_{\tilde{Y}/Y}) + 1.$$

This condition can be interpreted in the language of multiplier ideals by saying that, locally at x,

$$(\mathfrak{m}_{Y,x})^{N-4} \not\subset \mathcal{J}(Y, cB|_Y).$$

Applying Theorem 8 below, we get

$$e_x(B|_Y) \geq 4(N-3)/c^2 > 4(N-3)r^2.$$

For $N \geq 4$, this contradicts the fact that $B|_Y$ is contained in a zero-dimensional complete intersection scheme of degree Nr^2.

The result we have applied is the following reformulation of Theorem 2.1 of [10], which in turn is a small variant of Theorem 7.

Theorem 8. *Let V be a smooth variety of dimension n, let $Z \subset V$ be a scheme supported at a closed point $x \in V$, and let $c > 0$. Assume that $(\mathfrak{m}_{V,x})^k \not\subset \mathcal{J}(V, cZ)$ locally near x. Then Z has multiplicity*

$$e_x(Z) \geq (k+1)(n/c)^n.$$

Unfortunately, this computation breaks down if $\tilde{Y}$ is disjoint from E, and there is no stronger version of the Connectedness Theorem to fix it. In fact, in general the multiplier ideal simply fails to "get deeper." This is already the case in the following simple example.

Example 9. Let $D = (y^2 = x^3) \subset \mathbb{A}^2$ and $c = 5/6$. Then $\mathcal{J}(\mathbb{A}^2, cD) = \mathfrak{m}_{\mathbb{A}^2,0}$.

(a) If $L \subset \mathbb{A}^2$ is a general line through the origin, then $\mathcal{J}(L, cD|_L) = \mathfrak{m}_{L,0}$.
(b) If $L = (y = 0) \subset \mathbb{A}^2$, then $\mathcal{J}(L, cD|_L) = (\mathfrak{m}_{L,0})^2$.

7 The Role of the Space of Arcs

Let us discuss the example a little further. The multiplier ideal of $(\mathbb{A}^2, cD)$ can be computed by taking the well-known log resolution $f: \tilde{X} \to \mathbb{A}^2$ of the cusp given by a sequence of three blow-ups. The exceptional divisor E extracted by the third blow-up computes the log canonical threshold of the pair (which is $c = 5/6$), and is responsible for the nontrivial multiplier ideal. That is, we have $c \cdot \operatorname{ord}_E(D) = \operatorname{ord}_E(K_{\tilde{X}/\mathbb{A}^2}) + 1$, and

$$\mathcal{J}(\mathbb{A}^2, cD) = f_*\mathcal{O}_{\tilde{X}}(\lceil K_{\tilde{X}/\mathbb{A}^2} - cf^*D \rceil) = f_*\mathcal{O}_{\tilde{X}}(-E) = \mathfrak{m}_{\mathbb{A}^2,0}.$$

No matter how we choose L, the proper transform $\tilde{L}$ will always be disjoint from E, so we cannot rely on the computation done in the previous section. What makes the choice of L in case (b) more special is that in this case the proper transform of L on the first blow-up $\mathrm{Bl}_0 \mathbb{A}^2$ contains the center of ord_E on the blow-up. The intuition is that this choice brings $\tilde{L}$ "closer" to E, at least "to the first order."

In order to understand what is really happening, we work with formal arcs. Given a variety X, the *arc space* is given, set theoretically, by

$$X_\infty = \{\alpha \colon \operatorname{Spec} \mathbb{C}[[t]] \to X\}.$$

This space inherits a scheme structure from his description as the inverse limit of the jet schemes which parametrize maps $\operatorname{Spec} \mathbb{C}[t]/(t^{m+1}) \to X$, and comes with a morphism $\pi\colon X_\infty \to X$ mapping an arc $\alpha(t)$ to $\alpha(0) \in X$. Note that X_∞ is not Noetherian, is not of finite type, and does not have finite topological dimension.

Given a resolution $f\colon \tilde{X} \to X$ and a smooth prime divisor E on it, we consider the diagram

$$\begin{array}{ccccccc} \tilde{\pi}^{-1}(E) & \subset & \tilde{X}_\infty & \xrightarrow{f_\infty} & X_\infty & \supset & C_X(E) \\ \downarrow & & {\scriptstyle\tilde{\pi}}\downarrow & & \downarrow{\scriptstyle\pi} & & \downarrow \\ E & \subset & \tilde{X} & \xrightarrow{f} & X & \ni & x \end{array}$$

where f_∞ is the map on arc spaces given by composition and $C_X(E)$ is the *maximal divisorial set* of E, defined by

$$C_X(E) = \overline{f_\infty(\tilde{\pi}^{-1}(E))}.$$

This set is irreducible and only depends on the valuation ord_E. The following theorem due to [12] is the key to relate this construction to discrepancies and multiplier ideals.

Theorem 10. *Suppose that X is a smooth variety.*

(a) *The generic point α of $C_X(E)$ defines a valuation $\mathrm{ord}_\alpha \colon \mathbb{C}(X)^* \to \mathbb{Z}$, and this valuation coincides with ord_E.*
(b) *The set $C_X(E)$ has finite topological codimension in X_∞, and this codimension is equal to $\mathrm{ord}_E(K_{\tilde{X}/X}) + 1$.*

It is easy to guess how the valuation is defined: the generic point of $C_X(E)$ is a K-valued arc $\alpha \colon \mathrm{Spec}\, K[[t]] \to X$, the pullback map $\alpha^* \colon \mathcal{O}_{X,\pi(\alpha)} \to K[[t]]$ extends to an inclusion of fields $\alpha^* \colon \mathbb{C}(X) \hookrightarrow K((t))$, and the valuation is obtained by simply composing with the valuation $\mathrm{ord}_t \colon K((t))^* \to \mathbb{Z}$. The computation of the codimension of $C_X(E)$ uses the description of the fibers of the maps at the jet levels $f_m \colon \tilde{X}_m \to X_m$ and is essentially equivalent to the change-of-variable formula in motivic integration.

For our purposes, the advantage of working with divisorial sets in arc spaces is that, given a subvariety $Y \subset X$ containing the center of ord_E, even if the proper transform $\tilde{Y}$ is disjoint from E, the arc space Y_∞ (which is naturally embedded in X_∞) always intersects $C_X(E)$. We can then pick an irreducible component

$$C \subset (Y_\infty \cap C_X(E)).$$

In general, C itself may not be a maximal divisorial set. However, it is not too far from it. In the language of [12], one says that C is a *cylinder* in Y_∞, which essentially means that C is cut out by finitely many equations (maximal divisorial sets in arc spaces of smooth varieties are examples of cylinders). The upshot is that the generic point β of C defines a valuation val_β of $\mathbb{C}(Y)$, and we can find a prime divisor F over Y and a positive integer q such that

$$\mathrm{val}_\beta = q \cdot \mathrm{val}_F \quad \text{and} \quad \mathrm{codim}(C, Y_\infty) \geq q \cdot (\mathrm{ord}_F(K_{\tilde{Y}/Y}) + 1).$$

Now we can start relating multiplier ideals, since we can easily compare $q \cdot \mathrm{ord}_F$ to val_E, and we control the equations cutting out Y_∞ inside X_∞ and hence how the codimension of C compares to that of $C_X(E)$.

In order to control how the multiplier ideal behaves under restriction, and to show that it gets deeper, we need to ensure that if E has center $x \in X$ then $q \cdot \mathrm{ord}_F(\mathfrak{m}_{Y,x}) = \mathrm{val}_E(\mathfrak{m}_{X,x})$. In general, there is only an inequality. A tangency condition on Y is the first step to achieve this. This condition alone is not enough in general, but it suffices in the homogeneous setting, when $X = \mathbb{A}^n$ and the valuation ord_E is invariant under the homogeneous $\mathbb{C}^*$-action on a system of coordinates centered at x. The following result is proved in [9].

Theorem 11. *Let $X = \mathbb{A}^n$, let $Z \subset \mathbb{A}^n$ be a closed subscheme, and let $c > 0$. Assume that there is a prime divisor E on some resolution $\tilde{X} \to \mathbb{A}^n$ with center a point $x \in X$ such that*

(a) $c \cdot \mathrm{ord}_E(Z) \geq \mathrm{ord}_E(K_{\tilde{X}/\mathbb{A}^n}) + 1$, *and*
(b) *the valuation* ord_E *is invariant under the homogeneous $\mathbb{C}^*$-action on a system of affine coordinates centered at x.*

Let $Y = \mathbb{A}^{n-k} \subset \mathbb{A}^n$ be a linear subspace of codimension k through x that is tangent to the direction determined by a general point of the center of E in $\mathrm{Bl}_x \mathbb{A}^n$. Then

$$(\mathfrak{m}_{Y,x})^k \not\subset \mathcal{J}(Y, cZ|_Y).$$

Note that the conclusion of the theorem is precisely the condition assumed in Theorem 8. Unfortunately we cannot apply this theorem directly to where we left in the proof of Theorem 2 because we first need to reduce to a homogeneous setting. This reduction is probably the most delicate part of the proof of Theorem 2. To make the reduction, we use linear projections to linear spaces and flat degenerations to homogeneous ideals. The idea of using linear projections first appeared in [31], where a proof of birational rigidity was proposed but turned out to contain a gap. Theorem 8 is hidden behind the proof of a certain inequality on log canonical thresholds under generic projection, just like it was in the proof of the main theorem of [10]. Nadel's vanishing theorem is used in the end to draw the desired contradiction. This part of the proof is technical and goes beyond the purpose of this note; for more details, we refer the interested reader to [9].

8 What to Expect for Fano Hypersurfaces of Higher Index

Birational rigidity fails for hypersurfaces $X = X_d \subset \mathbb{P}^N$ of degree $d < N$ for a very simple reason: each linear projection $\mathbb{P}^N \dashrightarrow \mathbb{P}^k$, for $0 \leq k \leq N-d$, induces a Mori fiber structure on the hypersurface. If the center of projection is not contained in X, then the general fiber of the Mori fiber space is given by a Fano hypersurface of index $N+1-d-k$. Note that, by the Lefschetz Hyperplane Theorem, if $k \geq N/2$ (and $d \geq 2$) then the center of projection cannot be contained in X.

In low dimensions, other Mori fiber spaces may appear. Consider for instance, the case $X = X_3 \subset \mathbb{P}^4$. As explained above, X admits fibrations in cubic surfaces over $\mathbb{P}^1$, each induced by a linear projection $\mathbb{P}^4 \dashrightarrow \mathbb{P}^1$. Additionally, for every line $L \subset X$ the linear projection $\mathbb{P}^4 \dashrightarrow \mathbb{P}^2$ centered at L induces, birationally, a conic bundle structure of X onto $\mathbb{P}^2$. Furthermore, the projection from any point $x \in X$

induces a birational involution of X which swaps the two sheets of the rational cover $X \dashrightarrow \mathbb{P}^3$. However, these two last constructions are more specific of the low dimension and low degree of the hypersurface, and do not generalize when the degree and dimension get larger.

With this in mind, it is natural to consider the following problem.

Problem 12. Find a (meaningful) function $g(N)$ such that for every $X_d \subset \mathbb{P}^N$, with $g(N) \leq d \leq N$, the only Mori fiber spaces birational to X are those induced by the linear projections $\mathbb{P}^N \dashrightarrow \mathbb{P}^k$ for $0 \leq k \leq N - d$.

Remark 13. Taking $g(N) = N$ will of course work for $N \geq 4$ by Theorem 2. The problem is to determine, if it exists, a better function which includes Fano hypersurfaces of higher index. Already proving that $g(N) = N - 1$ works for $N \gg 1$ would be very interesting.[5]

A similar problem is the following.

Problem 14. Find a (meaningful) function $h(m, N)$ such that there is no $X_d \subset \mathbb{P}^N$, with $h(m, N) \leq d \leq N$, birational to a Mori fiber space of fiber dimension $\leq m$ (other than $X \to \operatorname{Spec} \mathbb{C}$ if $m = N - 1$).

As a first step, one can try to find solutions to these problems that work for general (or, even, very general) hypersurfaces. Part (b) of Theorem 1 gives a great solution to the case $m = 1$ of Problem 14 for very general hypersurfaces.

Acknowledgements The research was partially supported by NSF CAREER grant DMS-0847059 and a Simons Fellowship.

The author would like to thank the referee for useful comments and suggestions.

References

1. A. Beauville, R. Donagi, La variété des droites d'une hypersurface cubique de dimension 4. C. R. Acad. Sci. Paris Sér. I Math. **301**(14), 703–706 (1985)
2. G. Castelnuovo, Le trasformazioni generatrici del gruppo cremoniano del piano. Atti Accad. Torino **36**, 861–874 (1901)
3. I. Cheltsov, On a smooth four-dimensional quintic. Mat. Sb. **191**(9), 139–160 (2000) [English translation, Sb. Math. **191**(9–10), 1399–1419 (2000)]
4. C.H. Clemens, Phillip A. Griffiths, The intermediate Jacobian of the cubic threefold. Ann. Math. **95**(2), 281–356 (1972)
5. A. Corti, Factoring birational maps of threefolds after Sarkisov. J. Algebr. Geom. **4**(2), 223–254 (1995)
6. A. Corti, Singularities of linear systems and 3-fold birational geometry, in *Explicit Birational Geometry of 3-Folds*. London Mathematical Society Lecture Note Series, vol. 281 (Cambridge University Press, Cambridge, 2000), pp. 259–312

[5]A partial solution to Problem 12 in the special case $g(N) = N - 1$ has been recently announced in [32].

7. A. Corti, M. Reid (eds.), *Explicit Birational Geometry of 3-Folds*. London Mathematical Society Lecture Note Series, vol. 281 (Cambridge University Press, Cambridge, 2000)
8. T. de Fernex, *Birational Transformations of Varieties*. Thesis (Ph.D.)-University of Illinois at Chicago, 2002
9. T. de Fernex, Birationally rigid hypersurfaces. Invent. Math. **192**, 533–566 (2013)
10. T. de Fernex, L. Ein, M. Mustaţă, Bounds for log canonical thresholds with applications to birational rigidity. Math. Res. Lett. **10**(2–3), 219–236 (2003)
11. T. de Fernex, L. Ein, M. Mustaţă, Multiplicities and log canonical threshold. J. Algebr. Geom. **13**(3), 603–615 (2004)
12. L. Ein, R. Lazarsfeld, M. Mustaţă, Contact loci in arc spaces. Compos. Math. **140**(5), 1229–1244 (2004)
13. G. Fano, Sopra alcune varietà algebriche a tre dimensioni aventi tutti i generi nulli. Atti Accad. Torino **43**, 973–377 (1907/1908)
14. G. Fano, Osservazioni sopra alcune varietà non razionali aventi tutti i generi nulli. Atti Accad. Torino **47**, 1067–1071 (1915)
15. G. Fano, Alcune questioni sulla forma cubica dello spazio a cinque dimensioni. Comment. Math. Helv. **16**, 274–283 (1944)
16. C.D. Hacon, J. McKernan, The Sarkisov program. J. Algebr. Geom. **22**(2), 389–405 (2013)
17. B. Hassett, Some rational cubic fourfolds. J. Algebr. Geom. **8**(1), 103–114 (1999)
18. B. Hassett, Special cubic fourfolds. Compos. Math. **120**(1), 1–23 (2000)
19. V.A. Iskovskih, Ju.I. Manin, Three-dimensional quartics and counterexamples to the Lüroth problem. Mat. Sb. (N.S.) **86**(128), 140–166 (1971)
20. J. Kollár, Nonrational hypersurfaces. J. Am. Math. Soc. **8**(1), 241–249 (1995)
21. J. Kollár, S. Mori, in *Birational Geometry of Algebraic Varieties*. Cambridge Tracts in Mathematics, vol. 134 (Cambridge University Press, Cambridge, 1998) [With the collaboration of C.H. Clemens, A. Corti; Translated from the 1998 Japanese original]
22. J. Kollár, K.E. Smith, A. Corti, in *Rational and Nearly Rational Varieties*. Cambridge Studies in Advanced Mathematics, vol. 92 (Cambridge University Press, Cambridge, 2004)
23. A. Kuznetsov, Derived categories of cubic fourfolds, in *Cohomological and Geometric Approaches to Rationality Problems*. Progress in Mathematics, vol. 282 (Birkhäuser, Boston, 2010), pp. 219–243
24. R. Lazarsfeld, *Positivity in algebraic geometry. II: Positivity for Vector Bundles, and Multiplier Ideals*. Ergebnisse der Mathematik und ihrer Grenzgebiete, 3. Folge. A Series of Modern Surveys in Mathematics [Results in Mathematics and Related Areas, 3rd Series. A Series of Modern Surveys in Mathematics], vol. 49 (Springer, Berlin, 2004)
25. Ju.I. Manin, Rational surfaces over perfect fields. Inst. Hautes Études Sci. Publ. Math. **30**, 55–113 (1966)
26. U. Morin, Sulla razionalità dell'ipersuperficie cubica generale dello spazio lineare S_5. Rend. Sem. Mat. Univ. Padova **11**, 108–112 (1940)
27. M. Noether, Zur Theorie der eindentigen Ebenentrasformationen. Math. Ann. **5**,(4), 635–639 (1872)
28. Y. Odaka, T. Okada, *Birational Superrigidity and Slope Stability of Fano Manifolds* [arXiv:1107.3910] (2011)
29. A.V. Pukhlikov, Birational isomorphisms of four-dimensional quintics. Invent. Math. **87**(2), 303–329 (1987)
30. A.V. Pukhlikov, Birational automorphisms of Fano hypersurfaces. Invent. Math. **134**(2), 401–426 (1998)
31. A.V. Pukhlikov, Birationally rigid Fano hypersurfaces. Izv. Ross. Akad. Nauk Ser. Mat. **66**(6), 159–186 (2002) [English translation, Izv. Math. **66**(6), 1243–1269 (2002)]
32. A.V. Pukhlikov, *Birational Geometry of Fano Hypersurfaces of Index Two* [arXiv:1310.2572] (2013)
33. J. Ross, R. Thomas, An obstruction to the existence of constant scalar curvature Kähler metrics. J. Differ. Geom. **72**(3), 429–466 (2006)

34. B. Segre, On the rational solutions of homogeneous cubic equations in four variables. Math. Notae **11**, 1–68 (1951)
35. B. Segre, Variazione continua ed omotopia in geometria algebrica. Ann. Mat. Pura Appl. **50**(4), 149–186 (1960)
36. S.L. Tregub, Three constructions of rationality of a cubic fourfold. Vestnik Moskov. Univ. Ser. I Mat. Mekh. **3**, 8–14 (1984)
37. Y.G. Zarhin, Algebraic cycles over cubic fourfolds. Boll. Un. Mat. Ital. B (7) **4**(4), 833–847 (1990)

On the Locus of Nonrigid Hypersurfaces

Thomas Eckl and Aleksandr Pukhlikov

Abstract We show that the Zariski closure of the set of hypersurfaces of degree M in $\mathbb{P}^M$, where $M \geq 5$, which are either not factorial or not birationally superrigid, is of codimension at least $\binom{M-3}{2} + 1$ in the parameter space.

Mathematics Subject Classification: 14E05, 14E07, 14J45

1 Formulation of the Main Result and Scheme of the Proof

Let $\mathbb{P}^M$, where $M \geq 5$, be the complex projective space, $\mathcal{F} = \mathbb{P}(H^0(\mathbb{P}^M, \mathcal{O}_{\mathbb{P}^M}(M)))$ the space parameterizing hypersurfaces of degree M. There are Zariski open subsets $\mathcal{F}_{\text{reg}} \subset \mathcal{F}_{\text{sm}} \subset \mathcal{F}$, consisting of hypersurfaces, regular and smooth, respectively (the regularity condition, first introduced in [14], is now well known; for the convenience of the reader we reproduce it below in Sect. 2). The well-known theorem proven in [14] claims that every regular hypersurface $V \in \mathcal{F}_{\text{reg}}$ is birationally superrigid. Let $\mathcal{F}_{\text{srigid}} \subset \mathcal{F}$ be the set of (possibly singular) hypersurfaces that are factorial and birationally superrigid. The aim of this note is to show the following claim.

Theorem 1. *The Zariski closure* $\overline{\mathcal{F} \backslash \mathcal{F}_{\text{srigid}}}$ *of the complement is of codimension at least* $\binom{M-3}{2} + 1$ *in* $\mathcal{F}$.

Note that we do not discuss the question of whether $\mathcal{F}_{\text{srigid}}$ is open or not.

T. Eckl • A. Pukhlikov (✉)
Department of Mathematical Sciences, The University of Liverpool, M &O Building, Peach Street, Liverpool, L69 7ZL, UK
e-mail: pukh@liv.ac.uk

I. Cheltsov et al. (eds.), *Automorphisms in Birational and Affine Geometry*, Springer Proceedings in Mathematics & Statistics 79,
DOI 10.1007/978-3-319-05681-4_7,

Remark 1. The concept of birational (super)rigidity is now very well known; there are quite a few papers containing basic definitions and examples of birationally rigid varieties and their properties. The most recent (and most detailed) reference is the book [21], but all necessary information on birational rigidity can be also found in [4, 14, 15, 19]. Here we just remind the reader that a factorial birationally superrigid Fano variety V of index 1 with Picard number 1 has the following most spectacular geometric properties:

- every birational map to a $\mathbb{Q}$-factorial Fano variety of the same dimension with Picard number 1 is a biregular isomorphism (in particular, the groups of birational and biregular automorphisms of V coincide),
- V admits no rational dominant maps onto a variety of positive dimension, the general fiber of which is rationally connected (in a different wording, V admits no structures of a rationally connected fiber space).

We prove Theorem 1, directly constructing a set in $\mathcal{F}$, every point of which corresponds to a factorial and birationally superrigid hypersurface, with the Zariski closure of its complement of codimension at least $\binom{M-3}{2} + 1$. More precisely, let $\mathcal{F}_{\mathrm{qsing}\geq r}$ be the set of hypersurfaces, every point of which is either smooth or a quadratic singularity of rank at least r. We do *not* assume that singularities are isolated, but it is obvious that for $V \in \mathcal{F}_{\mathrm{qsing}\geq r}$ the following estimate holds:

$$\operatorname{codim} \operatorname{Sing} V \geq r - 1.$$

In particular, by the famous Grothendieck theorem ([7, XI. Cor. 3.14], [1]) any $V \in \mathcal{F}_{\mathrm{qsing}\geq 5}$ is a factorial variety, therefore a Fano variety of index 1: the group of classes of divisors modulo linear equivalence is the same as the Picard group and since $V \subset \mathbb{P}$ is a hypersurface of degree M, we have

$$\operatorname{Pic} V = \mathbb{Z} K_V, \quad K_V = -H,$$

where $H \in \operatorname{Pic} V$ is the class of a hyperplane section.

It is easy to see (Proposition 2) that $\operatorname{codim}(\mathcal{F} \backslash \mathcal{F}_{\mathrm{qsing}\geq 5}) \geq \binom{M-3}{2} + 1$.

Denote by $\mathcal{F}_{\mathrm{reg,\,qsing}\geq 5} \subset \mathcal{F}_{\mathrm{qsing}\geq 5}$ the subset, consisting of such Fano hypersurfaces $V \in \mathcal{F}$ that:

(1) at every smooth point the regularity condition of [14] is satisfied;
(2) through every singular point there are only finitely many lines on V.

We obtain Theorem 1 from the following two facts.

Theorem 2. *The codimension of the complement of* $\mathcal{F}_{\mathrm{reg,\,qsing}\geq 5}$ *in* $\mathcal{F}$ *is at least* $\frac{(M-3)(M-4)}{2} + 1$ *if* $M \geq 5$.

Theorem 3. *Every hypersurface* $V \in \mathcal{F}_{\mathrm{reg,\,qsing}\geq 5}$ *is birationally superrigid.*

Proof of Theorem 2 is straightforward and follows the arguments of [14, 16]; it is given in Sect. 2.

Proof of Theorem 3 starts in the usual way [14, 16, 19]: take a mobile linear system $\Sigma \subset |nH|$ on a hypersurface $V \in \mathcal{F}_{\mathrm{reg,qsing}\geq 5}$. Assume that for a generic $D \in \Sigma$ the pair $(V, \frac{1}{n}D)$ is not canonical, that is, the system Σ has a *maximal singularity* $E \subset V^+$, where $\varphi\colon V^+ \to V$ is a birational morphism, V^+ a smooth projective variety, E a φ-exceptional divisor and the *Noether–Fano inequality*

$$\mathrm{ord}_E\, \varphi^*\Sigma > na(E)$$

is satisfied (see [19] for definitions and details). We need to get a contradiction, which would immediately imply birational superrigidity and complete the proof of Theorem 3.

We proceed in the standard way.

Let $D_1, D_2 \in \Sigma$ be generic divisors and $Z = (D_1 \circ D_2)$ the effective cycle of their scheme-theoretic intersection, the *self-intersection* of the system Σ. Further, let $B = \varphi(E)$ be the center of the maximal singularity E. If $\mathrm{codim}_V B = 2$, then

$$\mathrm{codim}_B(B \cap \mathrm{Sing}\, V) \geq 2,$$

so we can take a curve $C \subset B$, $C \cap \mathrm{Sing}\, V = \emptyset$, and applying [14, Sect. 3], conclude that

$$\mathrm{mult}_C\, \Sigma \leq n.$$

As $\mathrm{mult}_B\, \Sigma > n$, we get a contradiction. So we may assume that $\mathrm{codim}_V B \geq 3$.

Proposition 1 (The $4n^2$-Inequality). *The following estimate holds:*

$$\mathrm{mult}_B\, Z > 4n^2.$$

If $B \not\subset \mathrm{Sing}\, V$, then the $4n^2$-inequality is a well-known fact going back to the paper on the quartic threefold [10], so in this case no proof is needed. The details can be found in any of the above-mentioned references for the method of maximal singularities: [4, 14, 15, 19, 21]. The treatment in [4] is based on somewhat different approach, the remaining four texts making use of the same technique as in Sect. 3 below.

Therefore we assume that $B \subset \mathrm{Sing}\, V$. In that case Proposition 1 is a nontrivial new result, proved below in Sect. 3. The proof makes use of the fact that the condition of having at most quadratic singularities of rank $\geq r$ is stable with respect to blow ups, in some a bit subtle way. That fact is shown in Sect. 4.

Now we complete the proof of Theorem 3, repeating word for word the arguments of [14]. Namely, we choose an irreducible component Y of the effective cycle Z, satisfying the inequality

$$\frac{\mathrm{mult}_o\, Y}{\deg Y} > \frac{4}{M},$$

where $o \in B$ is a point of general position. Applying the technique of hypertangent divisors in precisely the same way as it is done in [14] (see also [19, Chap. 3]), we construct a curve $C \subset Y$, satisfying the inequality $\mathrm{mult}_o C > \deg C$, which is impossible. It is here that we need the regularity conditions. This contradiction completes the proof of Theorem 3.

Remark 2. (i) $4n^2$-inequality is not true for a quadratic singularity of rank ≤ 4: the non-degenerate quadratic point of a threefold shows that $2n^2$ is the best we can achieve.

(ii) Birational superrigidity of Fano hypersurfaces with non-degenerate quadratic singularities was shown in [18]. Birational (super)rigidity of Fano hypersurfaces with isolated singular points of higher multiplicities $3 \leq m \leq M - 2$ was proved in [17], but the argument is really hard. These two results show that the estimate for the codimension of the nonrigid locus could most probably be considerably sharpened.

(iii) There are a few other papers where various classes of singular Fano varieties were studied from the viewpoint of their birational rigidity. The most popular object was three-dimensional quartics [5, 11, 13, 22]. Other families were investigated in [2, 3]. A family of Fano varieties (Fano double spaces of index one) with a higher dimensional singular locus was recently proven to be birationally superrigid in [12].

(iv) A recent preprint of de Fernex [6] proves birational superrigidity of a class of Fano hypersurfaces of degree M in $\mathbb{P}^M$ with not necessarily isolated singularities without assuming regularity. But the dimension of the singularity locus is bounded by $\frac{1}{2}M - 4$, and no estimate of the codimension of the complement of this class is given.

2 The Estimates for the Codimension

Let us prove Theorem 2.

First we discuss the regularity conditions in more details. Let x be a smooth point on a hypersurface V of degree M in $\mathbb{P}^M$. Choose homogeneous coordinates $(X_0 : \ldots : X_M)$ on $\mathbb{P}^M$ such that $x = (1 : 0 : \ldots : 0)$. Then $V \cap \{X_0 \neq 0\}$ is the vanishing locus of a polynomial

$$q_1 + \cdots + q_M$$

where each q_i is a homogeneous polynomial of degree i in M variables $X_1, \ldots, X_M$. According to [14] the hypersurface V is called regular in x if $q_1, \ldots, q_{M-1}$ is a regular sequence in $\mathbb{C}[x_1, \ldots, x_M]$. This is equivalent to

$$\mathrm{codim}_{\mathbb{A}^M}(\{q_1 = \cdots = q_{M-1} = 0\}) = M - 1. \qquad (1)$$

Since all the vanishing loci $\{q_i = 0\}$ are cones with vertex in x, the set $\{q_1 = \cdots = q_{M-1} = 0\}$ must consist of a finite number of lines through x. Vice versa, by homogeneity every line on X through x must be contained in each of the loci $\{q_i = 0\}$. Hence there is at most a finite number of lines on V through x.

If x is a singular point on V, then $q_1 \equiv 0$, and V is called regular in x if

$$\operatorname{codim}_{\mathbb{A}^M}(\{q_2 = \cdots = q_M = 0\}) = M - 1. \tag{2}$$

The set $\{q_2 = \cdots = q_M = 0\} \subset V$ consists of all the lines on V through x, since homogeneity implies that every line through x on V also lies in $\{q_i = 0\}$.

Finally, since regularity is a Zariski-open condition on sequences of polynomials, it does not depend on the choice of homogeneous coordinates on $\mathbb{P}^M$ such that $x = (1 : 0 : \ldots : 0)$ whether x satisfies conditions (1) and (2).

The set $\mathcal{F}_{\text{reg}} \subset \mathcal{F}$ consists of all hypersurfaces $V \subset \mathbb{P}^M$ of degree M such that V is regular in all points $x \in V$. It is not known whether the set $\mathcal{F}_{\text{reg}}$ is Zariski-open in $\mathcal{F}$, but it certainly contains a Zariski-open subset of $\mathcal{F}$. The codimension in $\mathcal{F}$ of its complement $\mathcal{F} \setminus \mathcal{F}_{\text{reg}}$ is defined as the codimension of the Zariski closure of the complement. On the other hand, $\mathcal{F}_{\text{qsing}\geq 5}$ is certainly Zariski-open, hence $\mathcal{F} \setminus \mathcal{F}_{\text{qsing}\geq 5}$ is Zariski-closed. We have

$$\operatorname{codim}_{\mathcal{F}}(\mathcal{F} \setminus \mathcal{F}_{\text{reg, qsing}\geq 5}) = \min(\operatorname{codim}_{\mathcal{F}}(\mathcal{F} \setminus \mathcal{F}_{\text{reg}}), \operatorname{codim}_{\mathcal{F}}(\mathcal{F} \setminus \mathcal{F}_{\text{qsing}\geq 5})).$$

Hence the estimate of Theorem 2 follows from the following two propositions:

Proposition 2. *The codimension of the complement $\mathcal{F} \setminus \mathcal{F}_{\text{qsing}\geq 5}$ in $\mathcal{F}$ is at least $\binom{M-3}{2} + 1$ if $M \geq 5$.*

Proposition 3. *The codimension of the (Zariski closure of the) complement $\mathcal{F}\setminus\mathcal{F}_{\text{reg}}$ in $\mathcal{F}$ is at least $\frac{M(M-5)}{2} + 4$ if $M \geq 5$.*

Proof of Proposition 2. Let $S_M := \mathbb{P}^{\binom{M+1}{2}-1}$ be the projectivized space of all symmetric $M \times M$-matrices with complex entries. Let $S_{M,r}$ be the projectivized algebraic subset of $M \times M$ symmetric matrices of rank $\leq r$. The locus $Q_r(P)$ of hypersurfaces $H \in \mathcal{F}$ with $P \in H$ a singularity that is at least a quadratic point of rank at most r has codimension in $\mathcal{F}$ equal to

$$\begin{aligned}\operatorname{codim}_{\mathcal{F}} Q_r(P) &= 1 + M + \operatorname{codim}_{S_M} S_{M,r} = 1 + M + \dim S_M - \dim S_{M,r} = \\ &= M + \binom{M+1}{2} - \dim S_{M,r}.\end{aligned}$$

Here, we use that a point $P \in H$ is quadratic of rank at most r if the Hessian of H in P has rank $\leq r$.

Let $G(M-r, M)$ be the Grassmann variety parameterizing $(M-r)$-dimensional subspaces of $\mathbb{C}^M$. To calculate $\dim S_{M,r}$ we consider the incidence correspondence (see [8, Ex. 12.4])

$$\Phi := \{(A, \Lambda) : \Lambda^T \cdot A = A \cdot \Lambda = 0\} \subset S_M \times G(M - r, M).$$

Since the fibers of the natural projection $\pi_2 : \Phi \to G(M - r, M)$ is given by a linear subspace of S_M of dimension $\binom{r+1}{2} - 1$, the variety Φ is irreducible of

$$\dim \Phi = \binom{r+1}{2} - 1 + r(M - r).$$

Since on the other hand the natural projection $\pi_1 : \Phi \to S_M$ is generically $1 : 1$ onto $S_{M,r}$, $\dim \Phi = \dim S_{M,r}$.

Consequently, since the $Q_r(P)$ cover Q_r and P varies in $\mathbb{P}^M$,

$$\mathrm{codim}_{\mathcal{F}} Q_r \geq \mathrm{codim}_{\mathcal{F}} Q_r(P) - M = \binom{M - r + 1}{2} + 1.$$

This completes the proof of Proposition 2.

Proof of Proposition 3. As a first step we compare the codimension of $\overline{\mathcal{F} \setminus \mathcal{F}_{\mathrm{reg}}}$ in $\mathcal{F}$ with that of related algebraic sets, leading to the estimate (3).

Let $\Phi = \{(x, H) : x \in H\} \subset \mathbb{P}^M \times \mathcal{F}$ be the incidence variety of hypersurfaces of degree M in $\mathbb{P}^M$. Let Φ_{reg} be the subset of pairs (x, H) satisfying the regularity conditions (1) and (2). Note that the Zariski closure $\overline{\Phi \setminus \Phi_{\mathrm{reg}}}$ in Φ maps onto the Zariski closure $\overline{\mathcal{F} \setminus \mathcal{F}_{\mathrm{reg}}}$ in $\mathcal{F}$. Denote the fibers of Φ and Φ_{reg} over a point $x \in \mathbb{P}^M$ under the natural projection $\pi_1 : \mathbb{P}^M \times \mathcal{F} \to \mathbb{P}^M$ by $\Phi(x)$ and $\Phi_{\mathrm{reg}}(x)$.

Claim. For any two points $x, x' \in \mathbb{P}^M$ there is a projective-linear automorphism $\alpha_{\mathcal{F}}$ on $\mathcal{F} = \mathbb{P}H^0(\mathbb{P}^M, \mathcal{O}_{\mathbb{P}^M}(M))$ such that

$$\alpha_{\mathcal{F}}(\Phi(x)) = \Phi(x') \text{ and } \alpha_{\mathcal{F}}(\Phi_{\mathrm{reg}}(x)) = \Phi_{\mathrm{reg}}(x').$$

Proof of Claim. It is possible to choose two sets of homogeneous coordinates $(X_0 : \ldots : X_M)$ and $(X'_0 : \ldots : X'_M)$ on $\mathbb{P}^M$ such that $x = (1:0:\ldots:0)$ in terms of the X-coordinates, $x' = (1:0:\ldots:0)$ in terms of the X'-coordinates, and we have $X'_0 = X_0$. These coordinates allow to decompose the vector space $H^0(\mathbb{P}^M, \mathcal{O}_{\mathbb{P}^M}(M))$ into the direct sums

$$\bigoplus_{i=0}^{M} \mathcal{P}_{i,M} \cdot X_0^{M-i} \text{ resp. } \bigoplus_{i=0}^{M} \mathcal{P}'_{i,M} \cdot (X'_0)^{M-i},$$

where $\mathcal{P}_{i,M}$ resp. $\mathcal{P}'_{i,M}$ denote the vector spaces of homogeneous polynomials of degree i in $X_1, \ldots, X_M$ resp. $X'_1, \ldots, X'_M$. Then a pair $(x, H) \in \Phi(x)$ if $H \in \bigoplus_{i=1}^{M} \mathcal{P}_{i,M} \cdot X_0^{M-i}$ and $(x, H) \in \Phi_{\mathrm{reg}}(x)$ if $H = \sum_{i=1}^{M} Q_i \cdot X_0^{M-i}$ and the

Q_i satisfy the regularity conditions (1) resp. (2). Analogous statements hold for pairs (x', H').

Let α be the projective-linear automorphism on $\mathbb{P}^M$ describing the coordinate change from the X- to the X'-coordinates. In particular, we have $\alpha(x) = x'$. The induced linear automorphism on $H^0(\mathbb{P}^M, \mathcal{O}_{\mathbb{P}^M}(M))$ is denoted by $\tilde{\alpha}_{\mathcal{F}}$, and its projectivization on $\mathcal{F}$ by $\alpha_{\mathcal{F}}$. The choice of coordinates shows that $\tilde{\alpha}_{\mathcal{F}}$ maps $\mathcal{P}_{i,M} \cdot X_0^{M-i}$ onto $\mathcal{P}'_{i,M} \cdot (X'_0)^{M-i}$, hence $\alpha_{\mathcal{F}}(\Phi(x)) = \Phi(x')$.

Furthermore, α maps the hyperplane $\{X_0 = 0\}$ onto the hyperplane $\{X'_0 = 0\}$, hence α is an affine isomorphism of the complements $\cong \mathbb{A}^M$. Consequently, if $H = \sum_{i=1}^M Q_i \cdot X_0^{M-i}$ and $H' = \tilde{\alpha}_{\mathcal{F}}(H) = \sum_{i=1}^M Q'_i \cdot (X'_0)^{M-i}$, then the vanishing loci $\{Q_i = \cdots = Q_j = 0\} \subset \mathbb{A}^M$ are mapped bijectively onto $\{Q'_i = \cdots = Q'_j = 0\} \subset \mathbb{A}^M$, for all possible $1 \leq i \leq j \leq M$. This implies $\alpha_{\mathcal{F}}(\Phi_{\text{reg}}(x)) = \Phi_{\text{reg}}(x')$. □

From now on we fix $x \in \mathbb{P}^M$, and we also fix a homogeneous coordinate system $(X_0 : \ldots : X_M)$ such that $x = (1 : 0 : \ldots : 0)$. The claim shows that whatever the choice of x is,

$$\dim \overline{\Phi \setminus \Phi_{\text{reg}}} = \dim \overline{\Phi(x) \setminus \Phi_{\text{reg}}(x)} + M.$$

Let $\tilde{\Phi}(x) = \prod_{i=1}^M \mathcal{P}_{i,M}$ and $\tilde{\Phi}_{\text{reg}}(x)$ be the preimages of $\Phi(x)$, $\Phi_{\text{reg}}(x)$ in the affine cone $H^0(\mathbb{P}^M, \mathcal{O}_{\mathbb{P}^M}(M)) = \prod_{i=0}^M \mathcal{P}_{i,M}$ over $\mathcal{F}$. Obviously,

$$\operatorname{codim}_{\Phi(x)} \overline{\Phi(x) \setminus \Phi_{\text{reg}}(x)} = \operatorname{codim}_{\tilde{\Phi}(x)} \overline{\tilde{\Phi}(x) \setminus \tilde{\Phi}_{\text{reg}}(x)}.$$

Since $\dim \overline{\mathcal{F} \setminus \mathcal{F}_{\text{reg}}} \leq \dim \overline{\Phi \setminus \Phi_{\text{reg}}}$ we finally calculate

$$\begin{aligned}
\operatorname{codim}_{\mathcal{F}} \overline{\mathcal{F} \setminus \mathcal{F}_{\text{reg}}} &\geq \dim \mathcal{F} - \dim \overline{\Phi \setminus \Phi_{\text{reg}}} = \dim \Phi - (M-1) - \dim \overline{\Phi \setminus \Phi_{\text{reg}}} = \\
&= \dim \Phi(x) + M - (M-1) - \dim \overline{\Phi(x) \setminus \Phi_{\text{reg}}(x)} - M = \\
&= \operatorname{codim}_{\Phi(x)} \overline{\Phi(x) \setminus \Phi_{\text{reg}}(x)} - (M-1),
\end{aligned}$$

hence

$$\operatorname{codim}_{\mathcal{F}} \overline{\mathcal{F} \setminus \mathcal{F}_{\text{reg}}} \geq \operatorname{codim}_{\tilde{\Phi}(x)} \overline{\tilde{\Phi}(x) \setminus \tilde{\Phi}_{\text{reg}}(x)} - (M-1). \tag{3}$$

In a next step we stratify $\tilde{\Phi}(x) \setminus \tilde{\Phi}_{\text{reg}}(x)$ according to the place where the sequence $q_1, \ldots, q_{M-1}$ resp. $q_2, \ldots, q_M$ first fails to be regular.

Setting $\mathcal{P}^*_{1,M} := \mathcal{P}_{1,M} \setminus \{0\}$ we first split up $\tilde{\Phi}(x) \setminus \tilde{\Phi}_{\text{reg}}(x) \subset \prod_{i=1}^M \mathcal{P}_{i,M}$ as the union of the intersections

$$S_1 := \tilde{\Phi}(x) \setminus \tilde{\Phi}_{\text{reg}}(x) \cap \left(\mathcal{P}^*_{1,M} \times \prod_{i=2}^M \mathcal{P}_{i,M} \right)$$

and

$$S_2 := \tilde{\Phi}(x) \setminus \tilde{\Phi}_{\mathrm{reg}}(x) \cap \left(\{0\} \times \prod_{i=2}^{M} \mathcal{P}_{i,M} \right).$$

S_1 stratifies further into strata

$$S_{1,k} := \left\{ (q_1, \dots, q_M) \in \mathcal{P}^*_{1,M} \right.$$
$$\left. \times \prod_{i=2}^{M} \mathcal{P}_{i,M} : q_1, \dots, q_{k-1} \text{ regular}, q_1, \dots, q_k \text{ not regular} \right\},$$

where $k = 2, \dots, M-1$, and S_2 stratifies further into strata

$$S_{2,l} := \left\{ (0, q_2, \dots, q_M) \in \{0\} \right.$$
$$\left. \times \prod_{i=2}^{M} \mathcal{P}_{i,M} : q_2, \dots, q_{l-1} \text{ regular}, q_2, \dots, q_l \text{ not regular} \right\},$$

where $l = 2, \dots, M$. Obviously,

$$\operatorname{codim}_{\tilde{\Phi}(x)} \overline{\tilde{\Phi}(x) \setminus \tilde{\Phi}_{\mathrm{reg}}(x)} = \min_{2 \le k \le M-1, 2 \le l \le M} (\operatorname{codim}_{\tilde{\Phi}(x)} \overline{S_{1,k}}, \operatorname{codim}_{\tilde{\Phi}(x)} \overline{S_{2,l}}). \tag{4}$$

So we need to bound the codimension of the strata $S_{1,k}$, $S_{2,l}$ in $\tilde{\Phi}(x)$ from below.

Starting with the strata $S_{1,k}$ we follow the notation of [14] and set for $k = 1, \dots, M-2$

$$R_k := \{(q_1, \dots, q_k) : q_1, \dots, q_k \text{ regular}\} \subset \mathcal{P}^*_{1,M} \times \prod_{i=2}^{k} \mathcal{P}_{i,M}.$$

Note that the natural projection $\pi_{k-1} : \mathcal{P}^*_{1,M} \times \prod_{i=2}^{M} \mathcal{P}_{i,M} \to \mathcal{P}^*_{1,M} \times \prod_{i=2}^{k-1} \mathcal{P}_{i,M}$ maps $S_{1,k}$ onto R_{k-1}. Hence for $k = 2, \dots, M-1$ we have

$$\operatorname{codim}_{\tilde{\Phi}(x)} \overline{S_{1,k}} \ge \min_{(q_1, \dots, q_{k-1}) \in R_{k-1}} \operatorname{codim}_{\pi_{k-1}^{-1}(q_1, \dots, q_{k-1})} S_{1,k} \cap \pi_{k-1}^{-1}(q_1, \dots, q_{k-1}). \tag{5}$$

Using the technique of [14], we obtain for $k = 2, \dots, M-1$

$$\min_{(q_1,\dots,q_{k-1})\in R_{k-1}} \operatorname{codim}_{\pi_{k-1}^{-1}(q_1,\dots,q_{k-1})} S_{1,k} \cap \pi_{k-1}^{-1}(q_1,\dots,q_{k-1}) \geq \binom{M}{k}. \tag{6}$$

Unfortunately these estimates are too weak for our purposes if $k = M - 1$. The technique of [16] yields the better estimate

$$\operatorname{codim}_{\mathcal{P}_{1,M}^* \times \prod_{i=2}^{k-1} \mathcal{P}_{i,M}} S_{1,M-1} \geq \frac{M(M-3)}{2} + 3. \tag{7}$$

To prove this estimate we first observe that for each $(q_1, \dots, q_M) \in S_{1,M-1}$ there exist $2 \leq b \leq M$, $1 \leq i_1 < \cdots < i_{b-2} \leq M-2$ and a b-dimensional linear subspace $L_b \subset \mathbb{A}^M$ (that is, $0 \in L_b$) such that

1. L_b is linearly generated by one of the two-dimensional irreducible components B of $\{q_1 = \cdots = q_{M-2} = 0\}$. Note that all the irreducible components of $\{q_1 = \cdots = q_{M-2} = 0\}$ are two-dimensional because $S_{1,M-1} \subset \pi_{M-2}^{-1}(R_{M-2})$.
2. B is an irreducible component of $\{q_{i_1} = \cdots = q_{i_{b-2}} = 0\} \cap L_b$. In the terminology of [16] $q_{i_1}, \dots, q_{i_{b-2}}$ is called a *good sequence* for $B \subset L_b$. Note further that this condition is empty if $b = 2$.

Vice versa, fix $2 \leq b \leq M$ and $1 \leq i_1 < \cdots < i_{b-2} \leq M-2$ and let $G(b, M)$ be the Grassmann variety of b-dimensional linear subspaces in $\mathbb{A}^M$. Let

$$Z(b; i_1, \dots, i_{b-2}) \subset G(b, M) \times \mathcal{P}_{1,M}^* \times \prod_{i=2}^{M} \mathcal{P}_{i,M}$$

be the set of tuples $(L_b, q_1, \dots, q_M)$ such that there exists a two-dimensional irreducible component B of $\{q_{i_1} = \cdots = q_{i_{b-2}} = 0\} \cap L_b$ satisfying the conditions

(1)′ B linearly spans L_b,
(2)′ $q_{i|B} \equiv 0$ for all $1 \leq i \leq M-1$.

Note that for $b = 2$ only the second condition is relevant.

Then $Z(b; i_1, \dots, i_{b-2})$ is locally Zariski-closed in $G(b, M) \times \mathcal{P}_{1,M}^* \times \prod_{i=2}^{M} \mathcal{P}_{i,M}$ since the existence of a B satisfying (1)′ is an open condition, and (2)′ is a closed condition. Furthermore, the first observation shows that $S_{1,M-1}$ is contained in the image of $\bigcup_{b;i_1,\dots,i_{b-2}} Z(b; i_1, \dots, i_{b-2})$ under the projection to $\mathcal{P}_{1,M}^* \times \prod_{i=2}^{M} \mathcal{P}_{i,M}$. Since this image is covered by the fibers $Z(b; i_1, \dots, i_{b-2}; L_b)$ of $Z(b; i_1, \dots, i_{b-2})$ over $L_b \in G(b, M)$ we obtain the estimate

$$\begin{aligned} &\operatorname{codim}_{\mathcal{P}_{1,M}^* \times \prod_{i=2}^{k-1} \mathcal{P}_{i,M}} S_{1,M-1} \\ &\geq \min_{b;i_1,\dots,i_{b-2};L_b} \left(\operatorname{codim}_{\mathcal{P}_{1,M}^* \times \prod_{i=2}^{k-1} \mathcal{P}_{i,M}} Z(b; i_1, \dots, i_{b-2}; L_b) - b(M-b) \right), \end{aligned} \tag{8}$$

where $b(M-b)$ is the dimension of the Grassmann variety $G(b, M)$.

Denote a fiber of $Z(b; i_1, \dots, i_{b-2}; L_b)$ over $(q_{i_1}, \dots, q_{i_{b-2}}) \subset \prod_{j=1}^{b-2} \mathcal{P}_{i_j,M}$ by

$$Z(b; q_{i_1}, \dots, q_{i_{b-2}}; L_b) \subset \prod_{i \notin \{i_1,\dots,i_{b-2}\}} \mathcal{P}_{i,M}.$$

Since these fibers are determined by condition $(2)'$ they are algebraic subsets, and

$$\begin{aligned} &\operatorname{codim}_{\mathcal{P}^*_{1,M} \times \prod_{i=2}^{k-1} \mathcal{P}_{i,M}} Z(b; i_1, \dots, i_{b-2}; L_b) \\ &\geq \min_{q_{i_1},\dots,q_{i_{b-2}}} \operatorname{codim}_{\prod_{i \notin \{i_1,\dots,i_{b-2}\}} \mathcal{P}_{i,M}} Z(b; q_{i_1}, \dots, q_{i_{b-2}}; L_b). \end{aligned} \tag{9}$$

Obviously, $(q_i)_{i \notin \{i_1,\dots,i_{b-2}\}} \in Z(b; q_{i_1}, \dots, q_{i_{b-2}}; L_b)$ implies

$$(\lambda_i q_i)_{i \notin \{i_1,\dots,i_{b-2}\}} \in Z(b; q_{i_1}, \dots, q_{i_{b-2}}; L_b), \ \lambda_i \in \mathbb{C}.$$

Hence $Z(b; q_{i_1}, \dots, q_{i_{b-2}}; L_b)$ maps to a closed algebraic set $Y \subset \prod_{i \notin \{i_1,\dots,i_{b-2}\}} \mathbb{P}\mathcal{P}_{i,M}$ (the projectivization of $\prod_{i \notin \{i_1,\dots,i_{b-2}\}} \mathcal{P}_{i,M}$ in each factor $\mathcal{P}_{i,M}$ separately) that has the same codimension as $Z(b; q_{i_1}, \dots, q_{i_{b-2}}; L_b)$. Note that Y is a product of algebraic subsets in $\mathbb{P}\mathcal{P}_{i,M}$. Choosing linear coordinates $x_1, \dots, x_n$ such that $L_b = \{x_{b+1} = \dots = x_M = 0\}$ such a factor of Y cannot intersect the projectivized algebraic subset of polynomials p_i of the form $\prod_{k=1}^{i}(a_{k,1}x_1 + \dots + a_{k,b}x_b)$. This is the case because such a p_i cannot vanish on an irreducible algebraic subset of L_b that locally spans L_b. Hence the factor of Y in the projective space $\mathbb{P}\mathcal{P}_{i,M}$ must have codimension at least $(b-1) \cdot i + 1$. Since furthermore we always have $M - 1 \notin \{q_{i_1}, \dots, q_{i_{b-2}}\}$, we obtain the estimate

$$\begin{aligned} &\operatorname{codim}_{\prod_{i \notin \{i_1,\dots,i_{b-2}\}} \mathcal{P}_{i,M}} Z(b; q_{i_1}, \dots, q_{i_{b-2}}; L_b) \\ &\geq \sum_{\substack{1 \leq i \leq M-1 \\ i \notin \{i_1,\dots,i_{b-2}\}}} ((b-1) \cdot i + 1) \\ &\geq (b-1)(1 + \dots + (M-b) + (M-1)) + (M-b+1) \\ &= (b-1)\frac{(M-b+1)(M-b)}{2} + (M-2)b + 2. \end{aligned} \tag{10}$$

Note that this estimate also holds when $b = 2$ since the codimension of the set of polynomials q_i vanishing on a two-dimensional linear subspace is $i + 1$.

The estimates (8)–(10) show that the codimension of $S_{1,M-1}$ in $\mathcal{P}^*_{1,M} \times \prod_{i=2}^{M} \mathcal{P}_{i,M}$ is at least the minimum of the numbers

$$F(b) := (b-1)\frac{(M-b+1)(M-b)}{2} + (M-2)b + 2 - b(M-b), \ 2 \leq b \leq M.$$

An analysis of the derivative of $F(b)$ shows that $F(b)$ has no local minimum for $2 < b < M$. Hence on the interval $[2, M]$ the function $F(b)$ has its minimum in $F(2) = \frac{M(M-3)}{2} + 3$ or in $F(M) = (M-2)M + 2$, and estimate (7) follows.

For the strata $S_{2,l}$ of S_2 we again follow the notation of [14] and set for $l = 2, \ldots, M-1$

$$Q_l := \left\{ (q_2, \ldots, q_l) \in \prod_{j=2}^{l} \mathcal{P}_{j,M} : q_2, \ldots, q_l \text{ regular} \right\}.$$

Arguing as before we have for $l = 2, \ldots, M$

$$\operatorname{codim}_{\tilde{\Phi}(x)} \overline{S_{2,l}} \geq \min_{(q_2,\ldots,q_{l-1}) \in Q_{l-1}} \operatorname{codim}_{\pi_{l-1}^{-1}(0,q_2,\ldots,q_{l-1})} S_{2,l} \cap \pi_{l-1}^{-1}(0, q_2, \ldots, q_{l-1}) + M, \tag{11}$$

where the additional summand M comes from the fact that always $q_1 = 0$ for sequences in $S_{2,l}$. The technique of [14] shows that for $l = 2, \ldots, M$

$$\min_{(q_2,\ldots,q_{l-1}) \in Q_{l-1}} \operatorname{codim}_{\pi_{l-1}^{-1}(0,q_2,\ldots,q_{l-1})} S_{2,l} \cap \pi_{l-1}^{-1}(0, q_2, \ldots, q_{l-1}) \geq \binom{M+1}{l}. \tag{12}$$

For $l = M$ we deduce completely analogous to (7) that

$$\operatorname{codim}_{\{0\} \times \prod_{i=2}^{M} \mathcal{P}_{i,M}} S_{2,M} \geq \frac{(M+1)(M-2)}{2} + 3. \tag{13}$$

Finally, the estimates (3)–(13) imply that $\operatorname{codim}_{\mathcal{F}} \overline{\mathcal{F} \setminus \mathcal{F}_{\text{reg}}}$ is bounded from below by the minimum of the numbers

$$\binom{M}{k} - (M-1), \; 2 \leq k \leq M-2,$$

$$\frac{M(M-3)}{2} + 3 - (M-1),$$

$$\binom{M+1}{l} + M - (M-1), \; 2 \leq l \leq M-1,$$

$$\frac{(M+1)(M-2)}{2} + 3 + M - (M-1),$$

that is

$$\frac{M(M-3)}{2} + 3 - (M-1) = \frac{M(M-5)}{2} + 4.$$

3 The $4n^2$-Inequality

Let us prove Proposition 1. We fix a mobile linear system Σ on V and a maximal singularity $E \subset V^+$ satisfying the Noether–Fano inequality $\mathrm{ord}_E\, \varphi^*\Sigma > na(E)$. We assume the centre $B = \varphi(E)$ of E on V to be maximal, that is, B is not contained in the center of another maximal singularity of the system Σ. In other words, the pair $(V, \frac{1}{n}\Sigma)$ is canonical outside B in a neighborhood of the generic point of B.

Further, we assume that $B \subset \mathrm{Sing}\, V$ (otherwise the claim is well known), so that $\mathrm{codim}(B \subset V) \geq 4$. Let

$$\begin{array}{rccc} \varphi_{i,i-1}\colon & V_i & \to & V_{i-1} \\ & \cup & & \cup \\ & E_i & \to & B_{i-1} \end{array}$$

$i = 1, \ldots, K$, be the *resolution* of E, that is, $V_0 = V$, $B_0 = B$, $\varphi_{i,i-1}$ blows up $B_{i-1} = \mathrm{centre}(E, V_{i-1})$, $E_i = \varphi_{i,i-1}^{-1}(B_{i-1})$ the exceptional divisor, and, finally, the divisorial valuations, determined by E and E_K, coincide.

As explained in Sect. 4 below, for every $i = 0, \ldots, K-1$ there is a Zariski open subset $U_i \subset V_i$ such that $U_i \cap B_i \neq \emptyset$ is smooth and either V_i is smooth along $U_i \cap B_i$, or every point $p \in U_i \cap B_i$ is a quadratic singularity of V_i of rank at least 5. In particular, the quasi-projective varieties $\varphi_{i,i-1}^{-1}(U_{i-1})$, $i = 1, \ldots, K$, are factorial and the exceptional divisor

$$E_i^* = E_i \cap \varphi_{i,i-1}^{-1}(U_{i-1})$$

is either a projective bundle over $U_{i-1} \cap B_{i-1}$ (in the non-singular case) or a fibration into quadrics of rank ≥ 5 over $U_{i-1} \cap B_{i-1}$ (in the singular case). We may assume that $U_i \subset \varphi_{i,i-1}^{-1}(U_{i-1})$ for $i = 1, \ldots, K-1$. The exceptional divisors E_i^* are all irreducible.

As usual, we break the sequence of blow ups into the *lower* $(1 \leq i \leq L)$ and *upper* $(L+1 \leq i \leq K)$ parts: $\mathrm{codim}\, B_{i-1} \geq 3$ if and only if $1 \leq i \leq L$. It may occur that $L = K$ and the upper part is empty (see [14, 15, 19]). Set

$$L_* = \max\{i = 1, \ldots, K \mid \mathrm{mult}_{B_{i-1}} V_{i-1} = 2\}.$$

Obviously, $L_* \leq L$. Set also

$$\delta_i = \mathrm{codim}\, B_{i-1} - 2 \quad \text{for} \quad 1 \leq i \leq L_*$$

and

$$\delta_i = \mathrm{codim}\, B_{i-1} - 1 \quad \text{for} \quad L_* + 1 \leq i \leq K.$$

We denote strict transforms on V_i by adding the upper index i: say, Σ^i means the strict transform of the system Σ on V_i. Let $D \in \Sigma$ be a generic divisor. Obviously,

$$D^i|_{U_i} = \varphi^*_{i,i-1}(D^{i-1}|_{U_{i-1}}) - \nu_i E^*_i,$$

where the integer coefficients $\nu_i = \frac{1}{2}\operatorname{mult}_{B_{i-1}} \Sigma^{i-1}$ for $i = 1, \dots, L_*$ and $\nu_i = \operatorname{mult}_{B_{i-1}} \Sigma^{i-1}$ for $i = L_* + 1, \dots, K$.

Now the Noether–Fano inequality takes the traditional form

$$\sum_{i=1}^{K} p_i \nu_i > n \left(\sum_{i=1}^{K} p_i \delta_i \right), \tag{14}$$

where p_i is the number of paths from the top vertex E_K to the vertex E_i in the oriented graph Γ of the sequence of blow ups $\varphi_{i,i-1}$, see [14, 15, 19, 21] for details.

We may assume that $\nu_1 < \sqrt{2}n$, otherwise for generic divisors $D_1, D_2 \in \Sigma$ we have

$$\operatorname{mult}_B(D_1 \circ D_2) \geq 2\nu_1^2 > 4n^2$$

and the $4n^2$-inequality is shown.

Remark 3. It is worth mentioning, although we do not use it in the proof, that the inequality $\nu_1 > n$ holds. To show this inequality, one needs to take a point $p \in B$ of general position and a generic complete intersection 3-germ $Y \ni p$. This operation reduces the claim to the case of a non log canonical singularity centered at a non-degenerate quadratic point, when the claim is well known, see [4, 20].

□

Obviously, the multiplicities ν_i satisfy the inequalities

$$\nu_1 \geq \cdots \geq \nu_{L_*} \tag{15}$$

and, if $K \geq L_* + 1$, then

$$2\nu_{L_*} \geq \nu_{L_*+1} \geq \cdots \geq \nu_K. \tag{16}$$

Now let $Z = (D_1 \circ D_2)$ be the self-intersection of the mobile system Σ and set $m_i = \operatorname{mult}_{B_{i-1}} Z^{i-1}$ for $1 \leq i \leq L$. Applying the technique of counting multiplicities in word for word the same way as in [14,15,19], we obtain the estimate

$$\sum_{i=1}^{L} p_i m_i \geq 2 \sum_{i=1}^{L_*} p_i \nu_i^2 + \sum_{i=L_*+1}^{K} p_i \nu_i^2.$$

Denote the right-hand side of this inequality by $q(\nu_1, \ldots, \nu_K)$. We see that

$$\sum_{i=1}^{L} p_i m_i > \mu,$$

where μ is the minimum of the positive definite quadratic form $q(\nu_*)$ on the compact convex polytope Δ defined on the hyperplane

$$\Pi = \left\{ \sum_{i=1}^{K} p_i \nu_i = n \left(\sum_{i=1}^{K} p_i \delta_i \right) \right\}$$

by the inequalities (15,16). Let us estimate μ.

We use the standard optimization technique in two steps. First, we minimize $q|_\Pi$ separately for the two groups of variables

$$\nu_1, \ldots, \nu_{L_*} \quad \text{and} \quad \nu_{L_*+1}, \ldots, \nu_K.$$

Easy computations show that the minimum is attained for

$$\nu_1 = \cdots = \nu_{L_*} = \theta_1 \quad \text{and} \quad \nu_{L_*+1} = \cdots = \nu_K = \theta_2,$$

satisfying the inequality $2\theta_1 \geq \theta_2$. Putting

$$\Sigma_* = \sum_{i=1}^{L_*} p_i \quad \text{and} \quad \Sigma^* = \sum_{i=L_*+1}^{K} p_i,$$

we get the extremal problem: to find the minimum of the positive definite quadratic form

$$\bar{q}(\theta_1, \theta_2) = 2\Sigma_* \theta_1^2 + \Sigma^* \theta_2^2$$

on the ray, cut out on the line

$$\Lambda = \left\{ \Sigma_* \theta_1 + \Sigma^* \theta_2 = n \sum_{i=1}^{K} p_i \delta_i \right\}$$

by the inequality $2\theta_1 \geq \theta_2$.

First we minimize $\bar{q}|_\Lambda$ on the whole line Λ. The minimum is attained for $\theta_1 = \theta$, $\theta_2 = 2\theta$, where θ is obtained from the equation of the line Λ:

$$\theta = \frac{n}{\Sigma_* + 2\Sigma^*} \sum_{i=1}^{K} p_i \delta_i.$$

We see that the condition $2\theta_1 \geq \theta_2$ is satisfied and for that reason can be ignored.

Now set

$$\Sigma_l = \sum_{i=1}^{L} p_i, \quad \Sigma_l^* = \sum_{i=L_*+1}^{L} p_i, \quad \Sigma_u = \sum_{i=L+1}^{K} p_i$$

(if $L \geq L_* + 1$; otherwise set $\Sigma_l^* = 0$). Obviously, the relations

$$\Sigma_l = \Sigma_* + \Sigma_l^* \quad \text{and} \quad \Sigma^* = \Sigma_l^* + \Sigma_u \tag{17}$$

hold. Recall that, due to our assumptions on the singularities of V_i we have $\delta_i \geq 2$ for $i \leq L$. Therefore,

$$\theta \geq \frac{2\Sigma_l + \Sigma_u}{\Sigma_* + 2\Sigma^*} n$$

and so

$$\mu \geq 2\frac{(2\Sigma_l + \Sigma_u)^2}{\Sigma_* + 2\Sigma^*} n^2.$$

Since

$$\Sigma_l \operatorname{mult}_B Z \geq \sum_{i=1}^{L} p_i m_i,$$

we finally obtain the estimate

$$\operatorname{mult}_B Z > 2\frac{(2\Sigma_l + \Sigma_u)^2}{\Sigma_l(\Sigma_* + 2\Sigma^*)} n^2.$$

Therefore, the $4n^2$-inequality follows from the estimate

$$(2\Sigma_l + \Sigma_u)^2 \geq 2\Sigma_l(\Sigma_* + 2\Sigma^*).$$

Replacing in the right-hand side $\Sigma_* + 2\Sigma^*$ by

$$\Sigma_* + 2(\Sigma_l^* + \Sigma_u) = \Sigma_l + \Sigma_l^* + 2\Sigma_u,$$

we bring the required estimate to the following form:

$$2\Sigma_l^2 + \Sigma_u^2 \geq 2\Sigma_l \Sigma_l^*,$$

which is an obvious inequality. Proof of Proposition 1 is now complete. □

4 Stability of the Quadratic Singularities Under Blow Ups

We start with the following essential

Definition 1. Let $X \subset Y$ be a subvariety of codimension 1 in a smooth quasi-projective complex variety Y of dimension n. A point $P \in X$ is called a quadratic point of rank r if there are analytic coordinates $z = (z_1, \ldots, z_n)$ of Y around P and a quadratic form $q_2(z)$ of rank r such that the germ of X in P is given by

$$(P \in X) \cong \{q_2(z) + \text{terms of higher degree} = 0\} \subset Y.$$

Theorem 4. *Let $X \subset Y$ be a subvariety of codimension 1 in a smooth quasi-projective complex variety Y of dimension n, with at most quadratic points of rank $\geq r$ as singularities. Let $B \subset X$ be an irreducible subvariety. Then there exists an open set $U \subset Y$ such that*

(i) $B \cap U$ is smooth, and
(ii) the blow up $\tilde{X}_U$ of $X \cap U$ along $B \cap U$ has at most quadratic points of rank $\geq r$ as singularities.

Proof. The statement is obvious if $B \not\subset \mathrm{Sing}(X)$ because then we can pick an open subset $U \subset Y$ such that $U \cap \mathrm{Sing}(X) = \emptyset$. So we assume from now on that $B \subset \mathrm{Sing}(X)$.

By restricting to a Zariski-open subset of Y we may assume that $B \subset \mathrm{Sing}(X)$ is a smooth subvariety. By assumption there exist analytic coordinates $z = (z_1, \ldots, z_n)$ around each point $P \in B \subset Y$ such that the germ

$$(P \in X) \cong \left\{ f(z) = z_1^2 + \cdots + z_r^2 + \text{terms of higher degree} = 0 \right\} \subset Y.$$

Then the singular locus $\mathrm{Sing}(X)$ is contained in the vanishing locus of the partial derivatives of this equation, hence in

$$\left\{ \frac{\partial f}{\partial z_1} = \cdots = \frac{\partial f}{\partial z_r} = 0 \right\}.$$

Since

$$\frac{\partial f}{\partial z_i} = 2z_i + \text{terms of higher degree}, 1 \leq i \leq r,$$

setting $z_1' := \frac{1}{2}\frac{\partial f}{\partial z_1}, \ldots, z_r' := \frac{1}{2}\frac{\partial f}{\partial z_r}$ yields new analytic coordinates

$$z_1', \ldots, z_r', z_{r+1}, \ldots, z_n$$

of Y around P. In these new coordinates the defining equation of X still is of the form

$$z_1'^2 + \cdots + z_r'^2 + \text{terms of higher degree} = 0,$$

and $B \subset \{z_1' = \cdots = z_r' = 0\}$. Since B is smooth around P on the linear subspace $\{z_1' = \cdots = z_r' = 0\}$ we can find further coordinates $z_{r+1}' \ldots, z_n'$ such that around P,

$$B = \{z_1' = \cdots = z_k' = 0\}, k \geq r.$$

Claim. $(P \in X) \cong \left\{z_1'^2 + \cdots + z_r'^2 + f_{\geq 3} = 0\right\}$ where $f_{\geq 3}$ consists of terms of degree ≥ 3 and is an element of $(z_1', \ldots, z_k')^2$.

Proof of Claim. $B \subset \mathrm{Sing}(X)$ must be contained in $\left\{\frac{\partial f_{\geq 3}}{\partial z_j'} = 0\right\}$, hence $\frac{\partial f_{\geq 3}}{\partial z_j'} \in (z_1', \ldots, z_k')$ for all $k+1 \leq j \leq n$. This is only possible if $f_{\geq 3} \in (z_1', \ldots, z_k')$. Write $f_{\geq 3} = z_1' f_1' + \cdots + z_k' f_k'$. Then as before $\frac{\partial f_{\geq 3}}{\partial z_i'} = f_i' + \sum_{1 \leq j \leq k, j \neq i} z_j' \frac{\partial f_j'}{\partial z_i'} \in (z_1', \ldots, z_k')$ for all $1 \leq i \leq k$. But this is only possible if $f_i' \in (z_1', \ldots, z_k')$ for all $1 \leq i \leq k$. □

Using the coordinates $z_1', \ldots, z_n'$ we can cover the blow up of Y along B over $P \in Y$ by k charts with coordinates

$$t_1^{(i)}, \ldots, z_i, \ldots, t_k^{(i)}, z_{k+1}, \ldots, z_n, 1 \leq i \leq k,$$

where $z_j' = t_j^{(i)} z_i$ for $1 \leq j \leq k$, $j \neq i$, $z_i' = z_i$ and $z_l' = z_l$ for $k+1 \leq l \leq n$. To prove the theorem we only need to check in each chart that along the fiber of the exceptional divisor over $P \in B$ there are at most quadratic points of rank $\geq r$ as singularities. We distinguish several cases:

Case 1. $1 \leq i \leq r$, say $i = 1$.
Then the strict transform of X is given by the equation

$$1 + (t_2^{(1)})^2 + \cdots + (t_r^{(1)})^2 + z_1 \cdot F + G = 0,$$

where G is a polynomial in $(t_2^{(1)}, \ldots, t_k^{(1)})^2 \cap (z_{k+1}, \ldots, z_n)$. On the fiber of the exceptional divisor over P given by $\{z_1 = z_{k+1} = \cdots = z_n = 0\}$, the gradient of this function can only vanish when $t_2^{(1)} = \cdots = t_r^{(1)} = 0$. But this locus does not intersect the strict transform, hence in this chart the strict transform is smooth along the fiber of the exceptional divisor over P.

Case 2. $r+1 \leq i \leq k$, say $i = k$.

Then the strict transform of X is given by the equation

$$(t_1^{(k)})^2 + \cdots + (t_r^{(k)})^2 + z_k \cdot F + G = 0,$$

where G is a polynomial in $(t_1^{(1)}, \dots, t_k^{(1)})^2 \cap (z_{k+1}, \dots, z_n)$. On the fiber of the exceptional divisor over P which is given by $\{z_k = z_{k+1} = \dots = z_n = 0\}$, the gradient of this function can only vanish when $t_1^{(k)} = \dots = t_r^{(k)} = 0$. Consequently, we have to discuss all points P of the form

$$(0, \dots, 0, a_{r+1}, \dots, a_{k-1}, 0, 0, \dots, 0) \in \{t_1^{(k)} = \dots = t_r^{(k)} = z_k = z_{k+1} = \dots = z_n = 0\}$$

on the strict transform of X. If we change to coordinates $t_1^{(k)}, \dots, t_r^{(k)}, t'_{r+1} := t_{r+1}^{(k)} - a_{r+1}, \dots, t'_{k-1} := t_{k-1}^{(k)} - a_{k-1}, z_k, z_{k+1}, \dots, z_n$ centered in P, then we obtain the equation defining the strict transform of X by substituting $t_{r+1}^{(k)}$ with $t'_{r+1} + a_{r+1}$, $\dots$, $t_{k-1}^{(k)}$ with $t'_{k-1} + a_{k-1}$. Three cases may occur:

First, the defining equation may contain linear terms in the ideal

$$(t'_{r+1}, \dots, t'_{k-1}, z_k, z_{k+1}, \dots, z_n).$$

Then the strict transform of X is smooth in P.

Second, the defining equation may contain no linear terms, but quadratic terms in $(t'_{r+1}, \dots, t'_{k-1}, z_k, z_{k+1}, \dots, z_n)$. Since these quadratic terms are added to $(t_1^{(k)})^2 + \dots + (t_r^{(k)})^2$, the quadratic term of the equation has still rank $\geq r$.

Finally, all terms of the defining equation besides those in $(t_1^{(k)})^2 + \dots + (t_r^{(k)})^2$ may be at least of degree ≥ 3. Then the quadratic term of the defining equation is $(t_1^{(k)})^2 + \dots + (t_r^{(k)})^2$, hence of rank r. □

Remark 2. Note that $\tilde{X}_U$ is again a subvariety of codimension 1 in the smooth quasi-projective blow up of U along $B \cap U$. The universal property of blow ups [9, Prop. II.7.14] and the calculations in the proof above tell us that the exceptional locus $E_U \subset \tilde{X}_U$ is a Cartier divisor on $\tilde{X}_U$ such that the morphism $E_U \to B \cap U$ is a fibration into quadrics of rank $\geq r$ in a $\mathbb{P}^{\operatorname{codim}_Y B}$-bundle.

References

1. F. Call, G. Lyubeznik, A simple proof of Grothendieck's theorem on the parafactoriality of local rings. Contemp. Math. **159**, 15–18 (1994)
2. I.A. Cheltsov, A double space with a double line. Sb. Math. **195**(9–10), 1503–1544 (2004)
3. I.A. Cheltsov, On nodal sextic fivefold. Math. Nachr. **280**(12), 1344–1353 (2007)
4. A. Corti, Singularities of linear systems and 3-fold birational geometry, in *Explicit Birational Geometry of 3-Folds* (Cambridge University Press, Cambridge, 2000), pp. 259–312
5. A. Corti, M. Mella, Birational geometry of terminal quartic 3-folds. I. Am. J. Math. **126**(4), 739–761 (2004)
6. T. De Fernex, Birational geometry of singular Fano hypersurfaces (2012, preprint) [arXiv:1208.6073]

7. A. Grothendieck, Cohomologie locale des faisceaux cohérents et théorèmes de Lefschetz locaux et globaux (SGA 2). Documents Mathématiques (Paris) 4, Société Mathématique de France (2005)
8. J. Harris, in *Algebraic Geometry*. Graduate Texts in Mathematics, vol. 133 (Springer, New York, 1992)
9. R. Hartshorne, in *Algebraic Geometry*. Graduate Texts in Mathematics, vol. 52 (Springer, New York, 1977)
10. V.A. Iskovskikh, Yu.I. Manin, Three-dimensional quartics and counterexamples to the Lüroth problem. Math. USSR Sb. **86**(1), 140–166 (1971)
11. M. Mella, Birational geometry of quartic 3-folds. II. Math. Ann. **330**(1), 107–126 (2004)
12. R. Mullany, Fano double spaces with a big singular locus. Math. Notes **87**(3), 444–448 (2010)
13. A.V. Pukhlikov, Birational automorphisms of a three-dimensional quartic with an elementary singularity. Math. USSR Sb. **63**, 457–482 (1989)
14. A.V. Pukhlikov, Birational automorphisms of Fano hypersurfaces. Invent. Math. **134**(2), 401–426 (1998)
15. A.V. Pukhlikov, Essentials of the method of maximal singularities, in *Explicit Birational Geometry of 3-Folds* (Cambridge University Press, Cambridge, 2000), pp. 73–100
16. A.V. Pukhlikov, Birationally rigid Fano complete intersections. Crelle J. für die reine und angew. Math. **541**, 55–79 (2001)
17. A.V. Pukhlikov, Birationally rigid Fano hypersurfaces with isolated singularities. Sb. Math. **193**(3), 445–471 (2002)
18. A.V. Pukhlikov, Birationally rigid singular Fano hypersurfaces. J. Math. Sci. **115**(3), 2428–2436 (2003)
19. A.V. Pukhlikov, Birationally rigid varieties, I. Fano varieties. Russ. Math. Surv. **62**(5), 857–942 (2007)
20. A.V. Pukhlikov, Birational geometry of singular Fano varieties. Proc. Steklov Inst. Math. **264**, 159–177 (2009)
21. A.V. Pukhlikov, *Birationally Rigid Varieties*. Mathematical Surveys and Monographs, vol. 190 (American Mathematical Society, Providence, 2013)
22. K. Shramov, Birational automorphisms of nodal quartic threefolds (2008) [arXiv:0803.4348]

On the Genus of Birational Maps Between Threefolds

Stéphane Lamy

Abstract In this note we present two equivalent definitions for the genus of a birational map $\varphi\colon X \dashrightarrow Y$ between smooth complex projective threefolds. The first one is the definition introduced by Frumkin [Mat. Sb. (N.S.) 90(132):196–213, 325, 1973], and the second one was recently suggested to me by S. Cantat. By focusing first on proving that these two definitions are equivalent, one can obtain all the results in M.A. Frumkin [Mat. Sb. (N.S.) 90(132):196–213, 325, 1973] in a much shorter way. In particular, the genus of an automorphism of $\mathbb{C}^3$, view as a birational self-map of $\mathbb{P}^3$, will easily be proved to be 0.

2010 *Mathematics Subject Classification.* 14E07.

1 Preliminaries

By a n-**fold** we always mean a smooth projective variety of dimension n over $\mathbb{C}$.

Let $\varphi\colon X \dashrightarrow Y$ be a birational map between n-folds. We assume that a projective embedding of Y is fixed once and for all, hence φ corresponds to the linear system on X given by preimages by φ of hyperplane sections on Y.

We call **base locus** of φ the base locus of the linear system associated with φ: this is a subvariety of codimension at least 2 of X, which corresponds to the indeterminacy set of the map.

S. Lamy (✉)
Institut de Mathématiques de Toulouse, Université Paul Sabatier, 118 route de Narbonne, 31062 Toulouse Cedex 9, France
e-mail: slamy@math.univ-toulouse.fr

I. Cheltsov et al. (eds.), *Automorphisms in Birational and Affine Geometry*, Springer Proceedings in Mathematics & Statistics 79,
DOI 10.1007/978-3-319-05681-4_8,

Another subvariety of X associated with φ is the **exceptional set**, which is defined as the complement of the maximal open subset where φ is a local isomorphism. If $X = Y = \mathbb{P}^n$ the exceptional set (given by the single equation Jacobian $= 0$) has pure codimension 1, but this is not the case in general: consider for instance the case of a flop, or more generally of any isomorphism in codimension 1, where the exceptional set coincides with the base locus.

A **regular resolution** of φ is a morphism $\sigma: Z \to X$ which is a sequence of blow-ups $\sigma = \sigma_1 \circ \cdots \circ \sigma_r$ along smooth irreducible centers, such that $Z \to Y$ is a birational morphism, and such that each center B_i of the blow-up $\sigma_i: Z_i \to Z_{i-1}$ is contained in the base locus of the induced map $Z_{i-1} \dashrightarrow Y$. Recall that as a standard consequence of resolution of singularities, a regular resolution always exists.

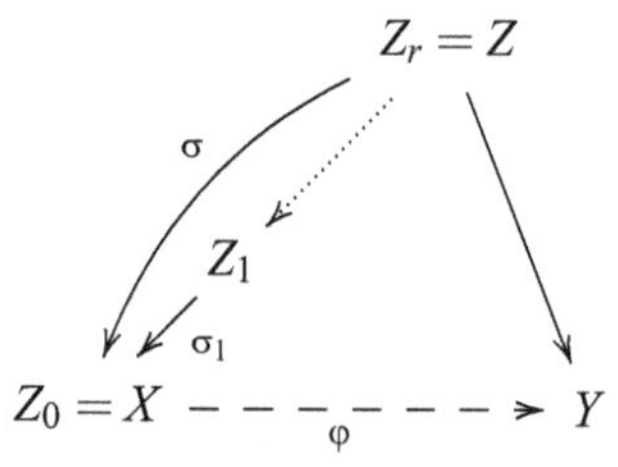

We shall use the following basic observations about the exceptional set and the base locus of a birational map.

Lemma 1. *(1) Let $\tau: X \to Y$ be a birational morphism between threefolds. Then through a general point of any component of the exceptional set of τ, there exists a rational curve contracted by τ.*

(2) Let $\varphi: X \dashrightarrow Y$ be a birational map between threefolds, and let $E \subset X$ be an irreducible divisor contracted by φ. Then E is birational to $\mathbb{P}^1 \times C$ for a smooth curve C.

Proof. For the first assertion (which is in fact true in any dimension), see for instance [2, Proposition 1.43]. When φ is a morphism, the second assertion is in fact what is first proved in [2]. Finally, when φ is not a morphism, we reduce to the previous case by considering a resolution of φ. □

Lemma 2. *Let $\varphi: X \dashrightarrow Y$ be a birational map between n-folds, and consider*

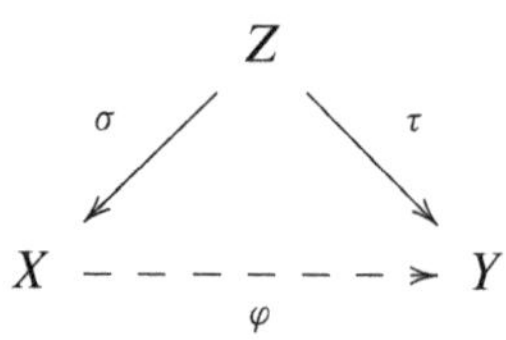

a regular resolution of φ. Then a point $p \in X$ is in the base locus of φ if and only if the set $\tau(\sigma^{-1}(p))$ has dimension at least 1.

Proof. If p is not in the base locus of φ then by regularity of the resolution $\sigma^{-1}(p)$ is a single point, and thus $\tau(\sigma^{-1}(p))$ as well.

Now suppose that p is in the base locus of φ, and consider H_Y a general hyperplane section of Y. Denote by H_X, H_Z the strict transform of H_Y on X and Z respectively. By definition of the base locus, we have $p \in H_X$, hence

$$\sigma^{-1}(p) \cap H_Z \neq \emptyset \quad \text{and} \quad \tau(\sigma^{-1}(p)) \cap H_Y \neq \emptyset.$$

This implies that $\tau(\sigma^{-1}(p))$ has positive dimension. □

We will consider blow-ups of smooth irreducible centers in threefolds. If B is such a center, B is either a point or a smooth curve. We define the **genus** $g(B)$ to be 0 if B is a point, and the usual genus if B is a curve. Similarly, if E is an irreducible divisor contracted by a birational map between threefolds, then by Lemma 1 E is birational to a product $\mathbb{P}^1 \times C$ where C is a smooth curve, and we define the **genus** $g(E)$ of the contracted divisor to be the genus of C.

2 The Two Definitions

Consider now a birational map $\varphi\colon X \dashrightarrow Y$ between threefolds, and let $\sigma\colon Z \to Y$ be a regular resolution of φ^{-1}.

Frumkin [3] defines the genus $g(\varphi)$ of φ to be the maximum of the genus among the centers of the blow-ups in the resolution σ. Remark that this definition depends a priori from a choice of regular resolution, and Frumkin spends a few pages in order to show that in fact it does not.

During the social dinner of the conference *Groups of Automorphisms in Birational and Affine Geometry*, S. Cantat suggested to me another definition, which is certainly easier to handle in practice: define the genus of φ to be the maximum of the genus among the irreducible divisors in X contracted by φ.

Denote by $F_1, \dots, F_r$ the exceptional divisors of the sequence of blow-ups $\sigma = \sigma_1 \circ \cdots \circ \sigma_r$, or more precisely their strict transforms on Z. On the other hand, denote by $E_1, \dots, E_s$ the strict transforms on Z of the irreducible divisors contracted by φ.

Note that if φ^{-1} is a morphism, then both collections $\{F_i\}$ and $\{E_i\}$ are empty. In this case, by convention we say that the genus of φ is 0. In this section we prove:

Proposition 3. *Assume φ^{-1} is not a morphism. Then*

$$\max_{i=1,\dots,s} g(E_i) = \max_{i=1,\dots,r} g(F_i).$$

In other words the definition of the genus by Frumkin coincides with the one suggested by Cantat, and in particular it does not depend on a choice of a regular resolution.

Denote by B_i the center of the blow-up σ_i producing F_i. We define a **partial order** on the divisors F_i by saying that $F_j \succcurlyeq F_k$ if one of the following conditions is verified:

(i) $j = k$;
(ii) $j > k$, B_k is a point, and B_j is contained in the strict transform of F_k;
(iii) $j > k$, B_k is a curve, and B_j intersects the general fiber of the strict transform of the ruled surface F_k.

We say that F_i is **essential** if F_i is a maximal element for the order $\succcurlyeq$.

Lemma 4. *The maximum* $\max_i g(F_i)$ *is realized by an essential divisor.*

Proof. We can assume that the maximum is not 0, otherwise there is nothing to prove. Consider F_k realizing the maximum, and $F_j \succcurlyeq F_k$ with $j > k$. Then the centers B_j, B_k of σ_j and σ_k are curves, and B_j dominates B_k by a morphism. By the Riemann–Hurwitz formula, we get $g(F_j) \geq g(F_k)$, and the claim follows. □

Lemma 5. *The subset of the essential divisors* F_i *with* $g(F_i) \geq 1$ *is contained in the set of the contracted divisors* E_j.

Proof. Let $B_i \subset Z_{i-1}$ be the center of a blow-up producing a non-rational essential divisor F_i, and consider the diagram:

$$\begin{array}{ccc} & Z & \\ {\scriptstyle\tau}\swarrow & & \searrow{\scriptstyle\tilde{\sigma}=\sigma_i\circ\cdots\circ\sigma_r} \\ X & \overset{\psi_{i-1}}{\dashleftarrow} & Z_{i-1} \end{array}$$

By applying Lemma 2 to $\psi_{i-1}\colon Z_{i-1} \dashrightarrow X$, we get $\dim \tau(\tilde{\sigma}^{-1}(p)) \geq 1$ for any point $p \in B_i$. Since F_i is essential, $l_p := \tilde{\sigma}^{-1}(p)$ is a smooth rational curve contained in F_i for all except finitely many $p \in B_i$. So $\tau(l_p)$ is also a curve. If $\tau(l_p)$ varies with p, then $\tau(F_i)$ is a divisor, which is one of the E_i. Now suppose $\tau(l_p)$ is a curve independent of p, that means that F_i is contracted to this curve by τ. Consider q a general point of F_i. By Lemma 1 there is a rational curve $C \subset F_i$ passing through q and contracted by τ, but this curve should dominate the curve B_i of genus ≥ 1: contradiction. □

Proof of Proposition 3. Observe that the strict transform of a divisor contracted by φ must be contracted by σ, hence we have the inclusion $\{E_i\} \subset \{F_i\}$. This implies $\max_i g(E_i) \leq \max_i g(F_i)$. If all F_i are rational, then the equality is obvious.

Suppose at least one of the F_i is non-rational. By Lemma 5 we have the inclusions

$$\{F_i; F_i \text{ is non-rational and essential}\} \subset \{E_i\} \subset \{F_i\}.$$

Taking maximums, this yields the inequalities

$$\max_i\{g(F_i); F_i \text{ is non-rational and essential}\} \leq \max_i g(E_i) \leq \max_i g(F_i).$$

By Lemma 4 we conclude that these three maximums are equal. □

3 Some Consequences

The initial motivation for a reworking of the paper of Frumkin was to get a simple proof of the fact that a birational self-map of $\mathbb{P}^3$ coming from an automorphism of $\mathbb{C}^3$ admits a resolution by blowing-up points and rational curves:

Corollary 6. *The genus of φ is zero in the following two situations:*

(1) $\varphi \in \mathrm{Bir}(\mathbb{P}^3)$ is the completion of an automorphism of $\mathbb{C}^3$;
(2) $\varphi: X \dashrightarrow Y$ is a pseudo-isomorphism (i.e. an isomorphism in codimension 1).

In particular for such a map φ any regular resolution only involves blow-ups of points and of smooth rational curves.

Proof. Both results are obvious using the definition via contracted divisors! □

I mention the following result for the sake of completeness, even if I essentially follow the proof of Frumkin (with some slight simplifications).

Proposition 7 (Compare with [3, Proposition 2.2]). *Let $\varphi: X \dashrightarrow Y$ be a birational map between threefolds, and let $\sigma: Z \to X$, $\sigma': Z' \to Y$ be resolutions of φ, φ^{-1} respectively. Assume that σ is a regular resolution and denote by $h: Z \dashrightarrow Z'$ the induced birational map. Then $g(h) = 0$.*

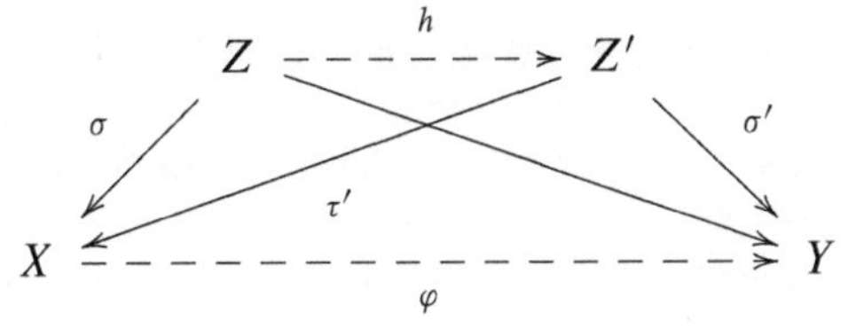

Proof. We write $\sigma = \sigma_1 \circ \cdots \circ \sigma_r$, where $\sigma_i: Z_i \to Z_{i-1}$ is the blow-up of a smooth center B_i. Note that $Z_0 = X$, and $Z_r = Z$. Assume that the induced map $h_i: Z_i \dashrightarrow Z'$ has genus 0 (this is clearly the case for $h_0 = \tau'^{-1}$), and let us prove the same for $h_{i+1} = h_i \circ \sigma_{i+1}$. We can assume that σ_{i+1} is the blow-up of a non-rational smooth curve B_{i+1}, otherwise there is nothing to prove. Consider

$$\begin{array}{ccc} & W_i & \\ {}^{p_i}\swarrow & & \searrow^{q_i} \\ Z_i & \underset{h_i}{\dashrightarrow} & Z' \end{array}$$

a regular resolution of h_i^{-1}. Since W_i dominates Y via the morphism $\sigma' \circ q_i$, the curve B_{i+1} is in the base locus of p_i^{-1}: otherwise B_{i+1} would not be in the base locus of $Z_i \dashrightarrow Y$, contradicting the regularity of the resolution σ. Thus for any point $x \in B_{i+1}$, $p_i^{-1}(x)$ is a curve, and there exists an open set $U \subset Z_i$ such that $p_i^{-1}(U \cap B_{i+1})$ is a nonempty divisor. By applying over U the universal property of blow-up (see [4, Proposition II.7.14]), we get that there exists an irreducible divisor on W_i whose strict transform on Z_{i+1} is the exceptional divisor E_{i+1} of σ_{i+1}. Hence the birational map $p_i^{-1} \circ \sigma_{i+1}: Z_{i+1} \dashrightarrow W_i$ does not contract any divisor and so has genus 0. Composing by q_i which also has genus 0 by hypothesis we obtain $g(h_{i+1}) = 0$. By induction we obtain $g(h_r) = 0$, hence the result since $h_r = h$. □

Remark 8. In the setting of Proposition 7, even if σ' is also a regular resolution, there is no reason for $h: Z \dashrightarrow Z'$ to be a pseudo-isomorphism. For instance, if φ admits a curve C in its base locus, one could construct regular resolutions of φ starting blowing-up arbitrary many points on C, and so the Picard number of Z can be arbitrary large (thanks to the referee who pointed this fact out).

On the other hand, it is not clear if we could restrict the definition of a regular resolution (for instance allowing the blow-up of a point only if it is a singular point of the base locus), such that the regular resolutions σ and σ' would lead to threefolds Z and Z' with the same Picard numbers, and with h a pseudo-isomorphism. In such a case, Proposition 7 would follow from Corollary 6.

The next result is less elementary.

Proposition 9. *Let X be a threefold with Hodge numbers $h^{0,1} = h^{0,3} = 0$, and let $\varphi: X \dashrightarrow X$ be a birational self-map. Then $g(\varphi) = g(\varphi^{-1})$.*

For the proof, which relies on the use of intermediate Jacobians, I refer to the original paper of Frumkin [3, Proposition 2.6], or to [5] where it is proved that the exceptional loci of φ and φ^{-1} are birational (and even more piecewise isomorphic). Note that Frumkin does not mention any restriction on the Hodge numbers of X, but it seems implicit since the proof uses the fact, through the reference to [1, 3.23], that the complex torus $\mathcal{J}(X)$ is a principally polarized abelian variety.

Corollary 10. *Let $g \geq 0$ be an integer. The set of birational self-maps of $\mathbb{P}^3$ of genus at most g is a subgroup of* $\mathrm{Bir}(\mathbb{P}^3)$.

Proof. Stability under taking inverse is Proposition 9, and stability under composition comes from the fact that any divisor contracted by $\varphi \circ \varphi'$ is contracted either by φ or by φ'. □

Question 11. The last corollary could be stated for any threefold satisfying the assumptions of Proposition 9, but I am not aware of any relevant example. For instance, if $X \subset \mathbb{P}^4$ is a smooth cubic threefold, is there any birational self-map of X with genus $g \geq 1$?

Acknowledgements I thank J. Blanc and A. Dubouloz who pointed out and helped me fix some silly mistakes in an early version of this work. This research was supported by ANR Grant "BirPol" ANR-11-JS01-004-01.

References

1. C.H. Clemens, P.A. Griffiths, The intermediate Jacobian of the cubic threefold. Ann. Math. **95**(2), 281–356 (1972)
2. O. Debarre, in *Higher-Dimensional Algebraic Geometry*. Universitext (Springer, New York, 2001)
3. M.A. Frumkin, A filtration in the three-dimensional Cremona group. Mat. Sb. (N.S.) **90**(132), 196–213, 325 (1973)
4. R. Hartshorne, in *Algebraic Geometry*. Graduate Texts in Mathematics, vol. 52 (Springer, New York, 1977)
5. S. Lamy, J. Sebag, Birational self-maps and piecewise algebraic geometry. J. Math. Sci. Univ. Tokyo **19**(3), 325–357 (2012)

On the Automorphisms of Moduli Spaces of Curves

Alex Massarenti and Massimiliano Mella

Abstract In the last years the biregular automorphisms of Deligne–Mumford's and Hassett's compactifications of the moduli space of n-pointed genus g smooth curves have been extensively studied by A. Bruno and the authors. In this paper we give a survey of these recent results and extend our techniques to some moduli spaces appearing as intermediate steps of Kapranov's and Keel's realizations of $\overline{M}_{0,n}$, and to the degenerations of Hassett's spaces obtained by allowing zero weights.

2010 *Mathematics Subject Classification.* Primary 14H10; Secondary 14D22, 14D23, 14D06

1 Introduction and Survey on the Automorphisms of Moduli Spaces of Curves

The moduli space of n-pointed genus g curves is a central object in algebraic geometry. The scheme $M_{g,n}$ parametrizing genus g smooth curves with n marked points satisfying the inequality $2g-2+n > 0$ has been compactified by Deligne and

Partially supported by Progetto PRIN 2010 "*Geometria sulle varietà algebriche*" MIUR and GRIFGA

A. Massarenti (✉)
SISSA, Via Bonomea 265, 34136 Trieste, Italy
e-mail: alex.massarenti@sissa.it

M. Mella
Dipartimento di Matematica e Informatica, Università di Ferrara, Via Machiavelli 35, 44100 Ferrara, Italy
e-mail: mll@unife.it

I. Cheltsov et al. (eds.), *Automorphisms in Birational and Affine Geometry*, Springer Proceedings in Mathematics & Statistics 79,
DOI 10.1007/978-3-319-05681-4_9,

Mumford in [3] by adding Deligne–Mumford stable curves as boundary points. In [7] Hassett introduced alternative compactifications of $M_{g,n}$ by allowing the marked points to have rational weights $0 < a_i \leqslant 1$. In the last years A. Bruno and the authors focused on the problem of determining the biregular automorphisms of all these compactifications, see [1, 2, 11, 12].

In what follows we will summarize and contextualize these results. Furthermore, in Sect. 2 we will extend our techniques to other moduli spaces of curves, namely Hassett's spaces appearing as intermediate steps of Kapranov's Construction 2.1 and of Keel's Construction 2.2. Finally, in Sect. 3 we will compute the automorphism groups of the degenerations of Hassett's spaces obtained by allowing some of the weights a_i to be zero.

1.1 The Automorphism Groups of $\overline{M}_{g,n}$

The first fundamental result about the automorphisms of moduli spaces of curves is due to Royden [15] and dates back to 1971.

Theorem 1.1. *Let $M_{g,n}^{\mathrm{un}}$ be the moduli space of genus g smooth curves marked by n unordered points. If $2g-2+n \geqslant 3$ then $M_{g,n}^{\mathrm{un}}$ has no nontrivial automorphisms.*

For a contextualization of this result in the Teichmüller-theoretic literature we refer to [13]. If $2g-2+n \geqslant 3$ the general genus g smooth curve with n unordered marked points does not have nontrivial automorphisms. Furthermore, $M_{g,n}^{\mathrm{un}}$ has at most finite quotient singularities and the stack $\mathcal{M}_{g,n}^{\mathrm{un}}$ is a smooth Deligne–Mumford stack. Therefore, by the argument used in [11, Sect. 4], we have that $\mathrm{Aut}(\mathcal{M}_{g,n}^{\mathrm{un}})$ is trivial if $2g-2+n \geqslant 3$.

The symmetric group on n elements S_n acts naturally on the moduli spaces $M_{g,n}$ and on its Deligne–Mumford compactification $\overline{M}_{g,n}$. Therefore $S_n \subseteq \mathrm{Aut}(\overline{M}_{g,n})$ and Theorem 1.1 gave a strong evidence for the equality to hold, see also [10] for the genus zero case. In particular Farkas [5, Question 4.6] asked if it is true that $\mathrm{Aut}(\overline{M}_{0,n}) \cong S_n$ for any $n \geqslant 5$, and it seems that also W. Fulton pointed to this question. Farkas himself brought Kapranov's paper [8] to the attention of the second author.

In [1, 2], Bruno and the second author, thanks to Kapranov's works [8], managed to translate issues on the moduli space $\overline{M}_{0,n}$ in terms of classical projective geometry of $\mathbb{P}^{n-3}$. Studying linear systems on $\mathbb{P}^{n-3}$ with particular base loci they derived a theorem on the fibrations of $\overline{M}_{0,n}$.

Theorem 1.2 ([2, Theorem 2]). *Let $f : \overline{M}_{0,n} \to \overline{M}_{0,r_1} \times \cdots \times \overline{M}_{0,r_h}$ be a dominant morphism with connected fibers. Then f factors with a forgetful map.*

In its original form [2, Theorem 2] states that f is a forgetful map. This is because in [2] a forgetful map is defined as the composition of a forgetful morphism $\varphi_I : \overline{M}_{0,n} \to \overline{M}_{0,r}$ with an automorphism of $\overline{M}_{0,r}$ [2, Definition 1.1].

Furthermore they realized that via this theorem on fibrations they could construct a morphism of groups between $\mathrm{Aut}(\overline{M}_{0,n})$ and S_n. Indeed if $\varphi : \overline{M}_{0,n} \to \overline{M}_{0,n}$ is an automorphism and $\pi_i : \overline{M}_{0,n} \to \overline{M}_{0,n-1}$ is a forgetful morphism by Theorem 1.2 we have the following diagram:

$$\begin{array}{ccc} \overline{M}_{0,n} & \xrightarrow{\varphi^{-1}} & \overline{M}_{0,n} \\ {\scriptstyle \pi_{j_i}}\downarrow & & \downarrow{\scriptstyle \pi_i} \\ \overline{M}_{0,n-1} & \xrightarrow{\tilde{\varphi}} & \overline{M}_{0,n-1} \end{array}$$

where π_{j_i} is a forgetful map. This allows us to define a surjective morphism of groups

$$\begin{array}{rcl} \chi : \mathrm{Aut}(\overline{M}_{0,n}) & \longrightarrow & S_n \\ \varphi & \longmapsto & \sigma_\varphi \end{array} \tag{1}$$

where

$$\begin{array}{rcl} \sigma_\varphi : \{1,\dots,n\} & \longrightarrow & \{1,\dots,n\} \\ i & \longmapsto & j_i \end{array}$$

Note that in order to have a morphism of groups we have to consider φ^{-1} instead of φ. Furthermore, the factorization of $\pi_i \circ \varphi^{-1}$ is unique. Indeed if $\pi_i \circ \varphi^{-1}$ admits two factorizations $\tilde{\varphi}_1 \circ \pi_j$ and $\tilde{\varphi}_2 \circ \pi_h$, then the equality $\tilde{\varphi}_1 \circ \pi_j([C, x_1, \dots, x_n]) = \tilde{\varphi}_2 \circ \pi_h([C, x_1, \dots, x_n])$ for any $[C, x_1, \dots, x_n] \in \overline{M}_{0,n}$ implies $\tilde{\varphi}_1([C, y_1, \dots, y_{n-1}]) = \tilde{\varphi}_2([C, y_1, \dots, y_{n-1}])$ for any $[C, y_1, \dots, y_{n-1}] \in \overline{M}_{0,n-1}$. Now $\tilde{\varphi}_1 = \tilde{\varphi}_2$ implies $\tilde{\varphi}_1 \circ \pi_j = \tilde{\varphi}_1 \circ \pi_h$ and since $\tilde{\varphi}_1$ is an isomorphism we have $\pi_j = \pi_h$.

Once again, via the projective geometry inherited by Kapranov's construction, the kernel of χ consists of automorphisms inducing on $\mathbb{P}^{n-3}$ a birational self-map that stabilizes lines and rational normal curves through $(n-1)$ fixed points. This proves that the kernel is trivial, see the proof of Theorem 2.6 for the details, and gives the following positive answer to [5, Question 4.6].

Theorem 1.3 ([2, Theorem 3]). *The automorphism group of $\overline{M}_{0,n}$ is isomorphic to S_n for any $n \geqslant 5$.*

Although a similar statement in higher genus was expected for many years, the problem of computing $\mathrm{Aut}(\overline{M}_{g,n})$ for $g \geqslant 1$ was not explicitly settled. However, A. Gibney, S. Keel, and I. Morrison gave an explicit description of the fibrations $\overline{M}_{g,n} \to X$ of $\overline{M}_{g,n}$ on a projective variety X in the case $g \geqslant 1$, providing an analogue of Theorem 1.2. Let N be the set $\{1,\dots,n\}$ of the markings. If $S \subset N$, then S^c denotes its complement.

Theorem 1.4 ([6, Theorem 0.9]). *Let $D \in \mathrm{Pic}(\overline{M}_{g,n})$ be a nef divisor.*

- *If $g \geqslant 2$ either D is the pull-back of a nef divisor on $\overline{M}_{g,n-1}$ via one of the forgetful morphisms or D is big and the exceptional locus of D is contained in $\partial\overline{M}_{g,n}$.*

- *If* $g = 1$ *either* D *is the tensor product of pull-backs of nef divisors on* $\overline{M}_{1,S}$ *and* $\overline{M}_{1,S^c}$ *via the tautological projection for some subset* $S \subseteq N$ *or* D *is big and the exceptional locus of* D *is contained in* $\partial\overline{M}_{g,n}$.

An immediate consequence of Theorem 1.4 is that for $g \geqslant 2$ any fibration of $\overline{M}_{g,n}$ to a projective variety factors through a projection to some $\overline{M}_{g,i}$ with $i < n$, while $\overline{M}_g$ has no nontrivial fibrations. Such a clear description of the fibrations of $\overline{M}_{g,n}$ is no longer true for $g = 1$. An explicit counterexample was given by R. Pandharipande [2, Example A.2] who also observed that Theorem 1.4 could be the starting point to compute the automorphism groups of $\overline{M}_{g,n}$. In order to compute $\mathrm{Aut}(\overline{M}_{1,n})$ the first author provided a factorization result for a particular type of fibration.

Lemma 1.5 ([11, Lemma 1.3]). *Let* φ *be an automorphism of* $\overline{M}_{1,n}$. *Any fibration of the type* $\pi_i \circ \varphi$ *factorizes through a forgetful morphism* $\pi_j : \overline{M}_{1,n} \to \overline{M}_{1,n-1}$.

Thanks to Theorem 1.4 and Lemma 1.5 in [11] the first author constructed the analogue of the morphism (1) for $g \geqslant 1$ and proved the following theorem.

Theorem 1.6 ([11, Theorem 3.9]). *Let* $\overline{\mathcal{M}}_{g,n}$ *be the moduli stack parametrizing Deligne–Mumford stable* n*-pointed genus* g *curves, and let* $\overline{M}_{g,n}$ *be its coarse moduli space. If* $2g - 2 + n \geqslant 3$ *then*

$$\mathrm{Aut}(\overline{\mathcal{M}}_{g,n}) \cong \mathrm{Aut}(\overline{M}_{g,n}) \cong S_n.$$

For $2g - 2 + n < 3$ *we have the following special behavior:*

- $\mathrm{Aut}(\overline{M}_{1,2}) \cong (\mathbb{C}^*)^2$ *while* $\mathrm{Aut}(\overline{\mathcal{M}}_{1,2})$ *is trivial,*
- $\mathrm{Aut}(\overline{M}_{0,4}) \cong \mathrm{Aut}(\overline{\mathcal{M}}_{0,4}) \cong \mathrm{Aut}(\overline{M}_{1,1}) \cong PGL(2)$ *while* $\mathrm{Aut}(\overline{\mathcal{M}}_{1,1}) \cong \mathbb{C}^*$,
- $\mathrm{Aut}(\overline{M}_{0,3})$ *and* $\mathrm{Aut}(\overline{\mathcal{M}}_{0,3})$ *are trivial,*
- $\mathrm{Aut}(\overline{M}_2)$ *and* $\mathrm{Aut}(\overline{\mathcal{M}}_2)$ *are trivial [6, Corollary 0.12].*

The proof of Theorem 1.6 is divided into two parts: the cases $2g-2+n \geqslant 3$ and $2g-2+n < 3$. When $2g-2+n \geqslant 3$ the proof uses extensively Theorem 1.4. This result, combined with the triviality of the automorphism group of the generic curve of genus $g \geqslant 3$, leads the first author to prove that the automorphism group of $\overline{M}_{g,1}$ is trivial for any $g \geqslant 3$. However, any genus two curve is hyperelliptic and has a nontrivial automorphism: the hyperelliptic involution. Therefore the argument used in the case $g \geqslant 3$ completely fails and a different strategy is needed: the first author proved that any automorphism of $\overline{M}_{2,1}$ preserves the boundary and then applied Theorem 1.1 to conclude that $\mathrm{Aut}(\overline{M}_{2,1})$ is trivial. Finally he tackled the general case by induction on n.

The case $g = 1, n = 2$ requires an explicit description of the moduli space $\overline{M}_{1,2}$. In [11, Theorem 2.3] the first author proved that $\overline{M}_{1,2}$ is isomorphic to a weighted blow-up of $\mathbb{P}(1,2,3)$ in the point $[1:0:0]$. In particular $\overline{M}_{1,2}$ is toric. From this he derived that $\mathrm{Aut}(\overline{M}_{1,2})$ is isomorphic to $(\mathbb{C}^*)^2$ [11, Proposition 3.8].

For the stack $\mathcal{M}_{g,n}$ and the coarse moduli space $M_{g,n}$ the automorphism groups are not known. However, it is reasonable that they both are isomorphic to S_n for $2g-2+n \geqslant 3$. A possible strategy to prove this would be to show that any automorphism of $\overline{\mathcal{M}}_{g,n}$ and $\overline{M}_{g,n}$ preserves the boundary and to apply Theorem 1.6.

1.2 Automorphisms of Hassett's Moduli Spaces

Many of the techniques used to deal with the automorphisms of $\overline{M}_{g,n}$ apply also to moduli spaces of weighted pointed curves. These are the compactifications of $M_{g,n}$ introduced by Hassett in [7]. Hassett constructed new compactifications $\overline{\mathcal{M}}_{g,A[n]}$ of the moduli stack $\mathcal{M}_{g,n}$ and $\overline{M}_{g,A[n]}$ for the coarse moduli space by assigning rational weights $A=(a_1,\dots,a_n)$, $0<a_i\leqslant 1$ to the markings.

We would like to stress that there is a dichotomy in our understanding of the automorphism groups of Hassett's moduli spaces: the case $g\geqslant 1$ where everything is known and the case $g=0$ where we have a limited understanding. Indeed in genus zero we manage to compute just the automorphism groups of some of the Hassett's spaces satisfying Definition 1.8. Namely the spaces appearing in Constructions 1.9 and 2.1, and some of the spaces in Construction 2.2.

In [7, Sect. 2.1.1] Hassett considers a natural variation on the moduli problem of weighted pointed stable curves by allowing some of the marked points to have weight zero. We introduce these more general spaces because we will compute their automorphisms in Sect. 3. Consider the data $(g,\tilde{A}) := (g,a_1,\dots,a_n)$ such that $a_i\in\mathbb{Q}$, $0\leqslant a_i\leqslant 1$ for any $i=1,\dots,n$, and

$$2g-2+\sum_{i=1}^{n} a_i > 0.$$

Definition 1.7. A family of nodal curves with marked points $\pi:(C,s_1,\dots,s_n)\to S$ is stable of type $(g,\tilde{A})$ if

- the sections $s_1,\dots,s_n$ with positive weights lie in the smooth locus of π, and for any subset $\{s_{i_1},\dots,s_{i_r}\}$ with nonempty intersection we have $a_{i_1}+\dots+a_{i_r}\leqslant 1$,
- $K_\pi+\sum_{i=1}^n a_i s_i$ is π-relatively ample.

There exists a connected Deligne–Mumford stack $\overline{\mathcal{M}}_{g,\tilde{A}[n]}$ representing the moduli problem of pointed stable curves of type $(g,\tilde{A})$. The corresponding coarse moduli scheme $\overline{M}_{g,\tilde{A}[n]}$ is projective over $\mathbb{Z}$.

If $a_i>0$ for any $i=1,\dots,n$, by [7, Theorem 3.8] a weighted pointed stable curve admits no infinitesimal automorphisms and its infinitesimal deformation space is unobstructed of dimension $3g-3+n$. Then $\overline{\mathcal{M}}_{g,A[n]}$ is a smooth Deligne–Mumford stack of dimension $3g-3+n$. However, when some of the marked points are allowed to have weight zero even the stack $\overline{\mathcal{M}}_{g,\tilde{A}[n]}$ may be singular, see [7, Sect. 2.1.1].

Following Hassett we denote by A the subset of $\tilde{A}$ containing all the positive weights so that $|\tilde{A}| = |A| + N$ where N is the number of zero weights. Note that an $\tilde{A}$-stable curve is an A-stable curve with N additional arbitrary marked points. Furthermore, the points with weight zero are allowed to be singular points. So at the level of stacks

$$\overline{\mathcal{M}}_{g,\tilde{A}[n]} \cong \underbrace{\mathcal{C}_{g,A[n]} \times_{\overline{\mathcal{M}}_{g,A[n]}} \cdots \times_{\overline{\mathcal{M}}_{g,A[n]}} \mathcal{C}_{g,A[n]}}_{N \text{ times}}$$

where $\mathcal{C}_{g,A[n]}$ is the universal curve over $\overline{\mathcal{M}}_{g,A[n]}$.

In the more general setting of zero weights we still have natural morphisms between Hassett's spaces. For fixed g, n, consider two collections of weight data $\tilde{A}[n], \tilde{B}[n]$ such that $a_i \geqslant b_i \geqslant 0$ for any $i = 1, \ldots, n$. Then there exists a birational *reduction morphism*

$$\rho_{\tilde{B}[n],\tilde{A}[n]} : \overline{M}_{g,\tilde{A}[n]} \to \overline{M}_{g,\tilde{B}[n]}$$

associating to a curve $[C, s_1, \ldots, s_n] \in \overline{M}_{g,\tilde{A}[n]}$ the curve $\rho_{\tilde{B}[n],\tilde{A}[n]}([C, s_1, \ldots, s_n])$ obtained by collapsing components of C along which $K_C + b_1 s_1 + \cdots + b_n s_n$ fails to be ample.

Furthermore, for any g consider a collection of weight data $\tilde{A}[n] = (a_1, \ldots, a_n)$ and a subset $\tilde{A}[r] := (a_{i_1}, \ldots, a_{i_r}) \subset \tilde{A}[n]$ such that $2g - 2 + a_{i_1} + \cdots + a_{i_r} > 0$. Then there exists a *forgetful morphism*

$$\pi_{\tilde{A}[n],\tilde{A}[r]} : \overline{M}_{g,\tilde{A}[n]} \to \overline{M}_{g,\tilde{A}[r]}$$

associating to a curve $[C, s_1, \ldots, s_n] \in \overline{M}_{g,\tilde{A}[n]}$ the curve $\pi_{\tilde{A}[n],\tilde{A}[r]}([C, s_1, \ldots, s_n])$ obtained by collapsing components of C along which $K_C + a_{i_1} s_{i_1} + \cdots + a_{i_r} s_{i_r}$ fails to be ample. For the details see [7, Sect. 4].

Some of the spaces $\overline{M}_{0,A[n]}$ appear as intermediate steps of Kapranov's blow-up construction of $\overline{M}_{0,n}$ [7, Sect. 6.1]. In higher genus $\overline{M}_{g,A[n]}$ may be related to the log minimal model program on $\overline{M}_{g,n}$, see for instance [14].

In the more general setting of Hassett's spaces not all forgetful maps are well defined as morphisms. However, in [12, Theorem 2.6, Proposition 2.7] the authors manage to derive a weighted version of Theorems 1.2 and 1.4 and thanks to these they construct a morphism of groups

$$\begin{aligned} \chi : \operatorname{Aut}(\overline{M}_{g,A[n]}) &\longrightarrow S_r \\ \varphi \quad &\longmapsto \sigma_\varphi \end{aligned} \tag{2}$$

where

$$\begin{aligned} \sigma_\varphi : \{1, \ldots, r\} &\longrightarrow \{1, \ldots, r\} \\ i \quad &\longmapsto \quad j_i \end{aligned}$$

and r is the number of well-defined forgetful maps of relative dimension one on $\overline{M}_{g,A[n]}$. We would like to stress that the morphism (2) makes sense for any space $\overline{M}_{g,A[n]}$ when $g \geqslant 1$ and in the genus zero case the morphism (2) is well defined only for Hassett's spaces factorizing Kapranov in the sense of the following definition.

Definition 1.8 ([12, Definition 2.1]). We say that Hassett's moduli space $\overline{M}_{0,A[n]}$ *factors Kapranov* if there exists a morphism ρ_2 that makes the following diagram commutative:

$$\begin{array}{ccc} \overline{M}_{0,n} & & \\ \Big\downarrow{\scriptstyle \rho_1} & \searrow^{f_i} & \\ \overline{M}_{0,A[n]} & \xrightarrow{\ \rho_2\ } & \mathbb{P}^{n-3} \end{array}$$

where f_i is Kapranov's map and ρ_1 is a reduction. We call such a ρ_2 *Kapranov's factorization*.

Let us recall Kapranov's construction.

Construction 1.9 ([8]). Fixed $(n-1)$-points $p_1, \ldots, p_{n-1} \in \mathbb{P}^{n-3}$ in linear general position:

(1) Blow-up the points $p_1, \ldots, p_{n-2}$, then the lines $\langle p_i, p_j \rangle$ for $i, j = 1, \ldots, n-2, \ldots$, the $(n-5)$-planes spanned by $n-4$ of these points.
(2) Blow-up p_{n-1}, the lines spanned by pairs of points including p_{n-1} but not $p_{n-2}, \ldots$, the $(n-5)$-planes spanned by $n-4$ of these points including p_{n-1} but not p_{n-2}.

⋮

(r) Blow-up the linear spaces spanned by subsets $\{p_{n-1}, p_{n-2}, \ldots, p_{n-r+1}\}$ so that the order of the blow-ups is compatible with the partial order on the subsets given by inclusion, the $(r-1)$-planes spanned by r of these points including $p_{n-1}, p_{n-2}, \ldots, p_{n-r+1}$ but not $p_{n-r}, \ldots$, the $(n-5)$-planes spanned by $n-4$ of these points including $p_{n-1}, p_{n-2}, \ldots, p_{n-r+1}$ but not p_{n-r}.

⋮

($n-3$) Blow-up the linear spaces spanned by subsets $\{p_{n-1}, p_{n-2}, \ldots, p_4\}$.

The composition of these blow-ups is the morphism $f_n : \overline{M}_{0,n} \to \mathbb{P}^{n-3}$ induced by the psi-class Ψ_n.

In [7, Sect. 6.1] Hassett interprets the intermediate steps of Construction 1.9 as moduli spaces of weighted rational curves. Consider the weight data

$$A_{r,s}[n] := (\underbrace{1/(n-r-1), \ldots, 1/(n-r-1)}_{(n-r-1) \text{ times}}, s/(n-r-1), \underbrace{1, \ldots, 1}_{r \text{ times}})$$

for $r = 1, \ldots, n-3$ and $s = 1, \ldots, n-r-2$. Then the variety obtained at the r-th step once we finish blowing-up the subspaces spanned by subsets S with $|S| \leqslant s+r-2$ is isomorphic to $\overline{M}_{0,A_{r,s}[n]}$.

To help the reader getting acquainted with Construction 1.9 we develop in detail the simplest case.

Example 1.10. Let $n = 5$, and fix $p_1, \ldots, p_4 \in \mathbb{P}^2$ points in general position. Kapranov's map f_5 is as follows: blow-up p_1, p_2, p_3 and then blow-up p_4.

At the step $r = 1, s = 1$ we get $\overline{M}_{0,A_{1,1}[n]} = \mathbb{P}^2$ and the weights are

$$A_{1,1}[5] := (1/3, 1/3, 1/3, 1/3, 1).$$

While for $r = 2, s = 1$ we get $\overline{M}_{0,A_{2,1}[n]} \cong \overline{M}_{0,5}$, indeed in this case the weight data are

$$A_{2,1}[5] := (1/2, 1/2, 1/2, 1, 1).$$

Note that as long as all the weights are strictly greater than $1/3$, Hassett's space is isomorphic to $\overline{M}_{0,n}$ because at most two points can collide, so the only components that get contracted are rational tail components with exactly two marked points. Since these have exactly three special points they have no moduli and contracting them does not affect the coarse moduli space even though it does change the universal curve, see also [7, Corollary 4.7]. In our case $\overline{M}_{0,A_{2,1}[5]} \cong \overline{M}_{0,5}$.

We have only one intermediate step, namely $r = 1, s = 2$. The moduli space $\overline{M}_{0,A_{1,2}[5]}$ parametrizes weighted pointed curves with weight data

$$A_{1,2}[5] := (1/3, 1/3, 1/3, 2/3, 1).$$

Since $a_4 + a_i = 1$ for $i = 1, 2, 3$ and $a_4 + a_5 > 1$ the point p_4 is allowed to collide with p_1, p_2, p_3 but not with p_5 which has not yet been blown-up. Kapranov's map $f_5 : \overline{M}_{0,5} \to \mathbb{P}^2$ factorizes as

$$\begin{array}{ccc}
\overline{M}_{0,5} \cong \overline{M}_{0,A_{2,1}[5]} & & \\
\Big\downarrow f_5 & \searrow^{\rho_1} & \\
 & & \overline{M}_{0,A_{1,2}[5]} \\
 & \swarrow_{\rho_2} & \\
\mathbb{P}^2 \cong \overline{M}_{0,A_{1,1}[5]} & &
\end{array}$$

where ρ_1, ρ_2 are the corresponding reduction morphisms. Let us analyze these two morphisms.

- Given $(C, s_1, \ldots, s_5) \in \overline{M}_{0,A_{2,1}[5]}$ the curve $\rho_1(C, s_1, \ldots, s_5)$ is obtained by collapsing components of C along which $K_C + \frac{1}{3}s_1 + \frac{1}{3}s_2 + \frac{1}{3}s_3 + \frac{2}{3}s_4 + s_5$

fails to be ample. So it contracts the two-pointed components of the following curves:

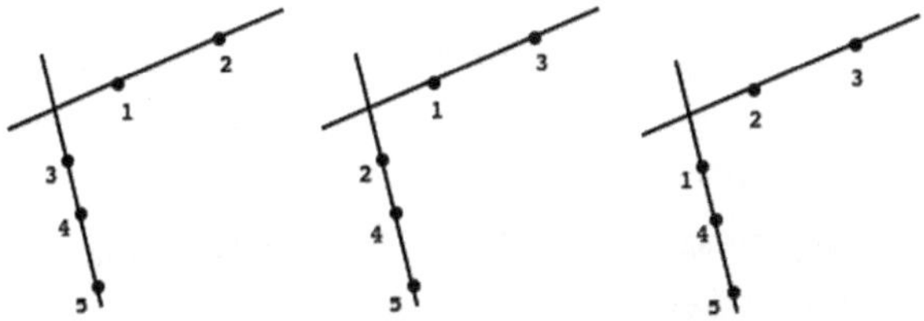

along which $K_C + \frac{1}{3}s_1 + \frac{1}{3}s_2 + \frac{1}{3}s_3 + \frac{2}{3}s_4 + s_5$ is anti-ample, and the two-pointed components of the following curves:

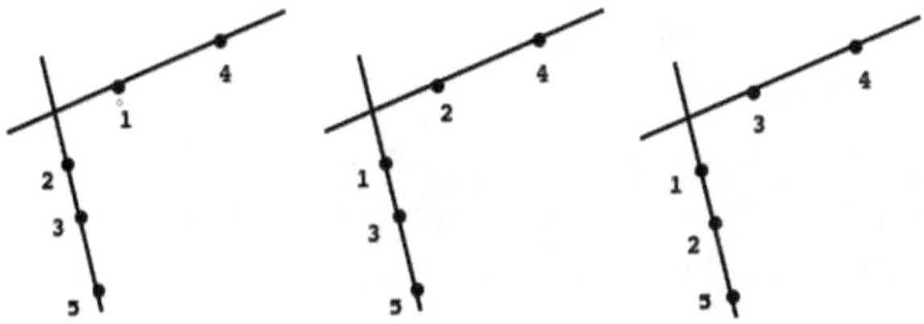

along which $K_C + \frac{1}{3}s_1 + \frac{1}{3}s_2 + \frac{1}{3}s_3 + \frac{2}{3}s_4 + s_5$ is nef but not ample. However, all the contracted components have exactly three special points, and therefore they do not have moduli. This affects only the universal curve but not the coarse moduli space.

Finally $K_C + \frac{1}{3}s_1 + \frac{1}{3}s_2 + \frac{1}{3}s_3 + \frac{2}{3}s_4 + s_5$ is nef but not ample on the three-pointed component of the curve

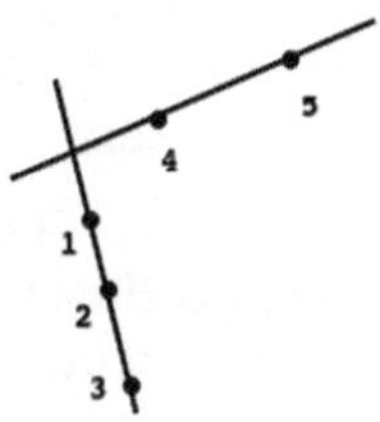

In fact this corresponds to the contraction of the divisor $E_{5,4} = f_5^{-1}(p_4)$.

– The morphism ρ_2 contracts the three-pointed components of the curves

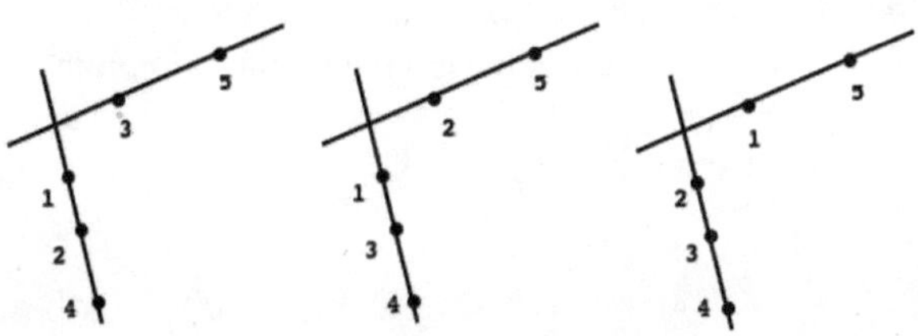

along which $K_C + \frac{1}{3}s_1 + \frac{1}{3}s_2 + \frac{1}{3}s_3 + \frac{1}{3}s_4 + s_5$ has degree zero. This corresponds to the contractions of the divisors $E_{5,3} = f_5^{-1}(p_3)$, $E_{5,2} = f_5^{-1}(p_2)$ and $E_{5,1} = f_5^{-1}(p_1)$.

We do not have a complete classification in terms of the weights A of Hassett's spaces $\overline{M}_{0,A[n]}$ that factor Kapranov. However, this is enough to compute the automorphisms of all intermediate steps of Construction 1.9.

Theorem 1.11 ([12, Theorem 3.3]). *The automorphism groups of Hassett's spaces appearing in Construction 1.9 are given by*

- $\mathrm{Aut}(\overline{M}_{0,A_{r,s}[n]}) \cong (\mathbb{C}^*)^{n-3} \times S_{n-2}$, *if* $r = 1$, $1 < s < n-3$,
- $\mathrm{Aut}(\overline{M}_{0,A_{r,s}[n]}) \cong (\mathbb{C}^*)^{n-3} \times S_{n-2} \times S_2$, *if* $r = 1$, $s = n-3$,
- $\mathrm{Aut}(\overline{M}_{0,A_{r,s}[n]}) \cong S_n$, *if* $r \geqslant 2$.

The automorphisms appearing in Theorem 1.11 are induced explicitly by the morphism (2). More explicitly:

- $\mathrm{Ker}(\chi) \cong (\mathbb{C}^*)^{n-3}$ and $\mathrm{Im}(\chi) \cong S_{n-2}$, if $r = 1$ and $1 < s < n-3$,
- $\mathrm{Ker}(\chi) \cong (\mathbb{C}^*)^{n-3} \times S_2$ and $\mathrm{Im}(\chi) \cong S_{n-2}$, if $r = 1$ and $s = n-3$,
- $\mathrm{Ker}(\chi)$ is trivial and $\mathrm{Im}(\chi) \cong S_n$, if $r \geqslant 2$.

Furthermore, we have $\overline{M}_{0,A_{1,1}[n]} \cong \mathbb{P}^{n-3}$ and $\mathrm{Aut}(\overline{M}_{0,A_{1,1}[n]}) \cong PGL(n-2)$.

Remark 1.12. Hassett's space $\overline{M}_{0,A_{1,2}[5]}$ is the blow-up of $\mathbb{P}^2$ in three points in general position, that is Del Pezzo surface $\mathcal{S}_6$ of degree 6. By Theorem 1.11 we recover the classical result on its automorphism group $\mathrm{Aut}(\mathcal{S}_6) \cong (\mathbb{C}^*)^2 \times S_3 \times S_2$. For a proof not using the theory of moduli of curves see [4, Sect. 6].

Furthermore, note that we are allowed to permute the points labeled by $1, 2, 3$ and to exchange the marked points $4, 5$. However, any permutation mapping $1, 2$ or 3 to 4 or 5 contracts a boundary divisor isomorphic to $\mathbb{P}^1$ to the point $\rho_1(E_{5,4})$, so it does not induce an automorphism. Furthermore, Cremona transformation lifts to the automorphism of $\overline{M}_{0,A_{1,2}[5]}$ corresponding to the transposition $4 \leftrightarrow 5$.

Remark 1.13. The step $r = 1, s = n-3$ of Construction 1.9 is Losev–Manin's space $\overline{L}_{n-2}$, see [7, Sect. 6.4]. This space is a toric variety of dimension $n-3$. By Theorem 1.11 we recover $(\mathbb{C}^*)^{n-3} \subset \mathrm{Aut}(\overline{L}_{n-2})$. The automorphisms in $S_{n-2} \times S_2$ reflect on the toric setting as automorphisms of the fan of $\overline{L}_{n-2}$.

For example consider Del Pezzo surface of degree six $\overline{M}_{0,A_{1,2}[5]} \cong \overline{L}_3 \cong \mathcal{S}_6$. Let us say that $\mathcal{S}_6$ is the blow-up of $\mathbb{P}^2$ at the coordinate points p_1, p_2, p_3 with exceptional divisors e_1, e_2, e_3 and let us denote by $l_i = \langle p_j, p_k \rangle$, $i \neq j, k$, $i = 1, 2, 3$ the three lines generated by p_1, p_2, p_3.

Such surface can be realized as the complete intersection in $\mathbb{P}^2 \times \mathbb{P}^2$ cut out by the equations $x_0y_0 = x_1y_1 = x_2y_2$. The six lines are given by $e_i = \{x_j = x_k = 0\}$, $l_i = \{y_j = y_k = 0\}$ for $i \neq j, k$, $i = 1, 2, 3$. The torus $T = (\mathbb{C}^*)^3/\mathbb{C}^*$ acts on $\mathbb{P}^2 \times \mathbb{P}^2$ by

$$(\lambda_0, \lambda_1, \lambda_2) \cdot ([x_0 : x_1 : x_2], [y_0 : y_1 : y_2]) = ([\lambda_0 x_0 : \lambda_1 x_1 : \lambda_2 x_2], [\lambda_0^{-1} y_0 : \lambda_1^{-1} y_1 : \lambda_2^{-1} y_2]).$$

This torus action stabilizes $\mathcal{S}_6$. Furthermore, S_2 acts on $\mathcal{S}_6$ by the transpositions $x_i \leftrightarrow y_i$, and S_3 acts on $\mathcal{S}_6$ by permuting the two sets of homogeneous coordinates separately. The action of S_3 corresponds to the permutations of the three points of $\mathbb{P}^2$ we are blowing-up, while the S_2-action is the switch of roles of exceptional divisors between the sets of lines $\{e_1, e_2, e_3\}$ and $\{l_1, l_2, l_3\}$. These six lines are arranged in a hexagon inside $\mathcal{S}_6$

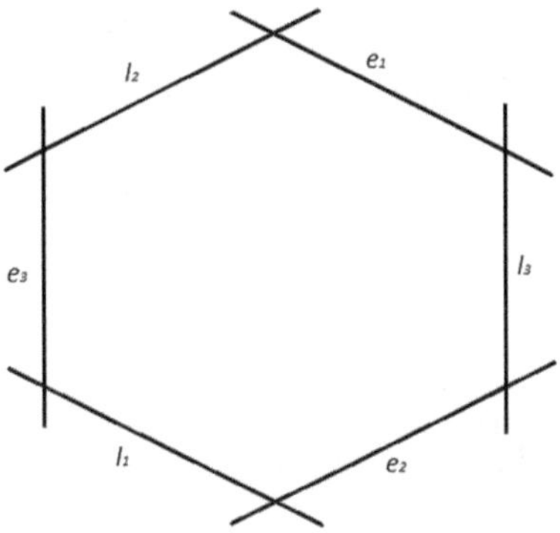

which is stabilized by the action of $S_3 \times S_2$. The fan of $\mathcal{S}_6$ is the following

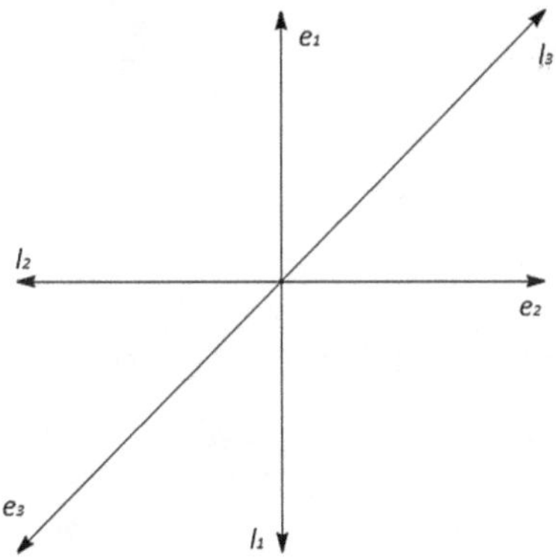

where the six 1-dimensional cones correspond to the toric divisors e_1, l_3, e_2, l_1, e_3 and l_2. It is clear from the picture that the fan has many symmetries given by permuting $\{e_1, e_2, e_3\}$, $\{l_1, l_2, l_3\}$ and switching e_i with l_i for $i = 1, 2, 3$.

In higher genus all the forgetful maps are well defined as morphisms [12, Lemma 3.8]. However, a transposition $i \leftrightarrow j$ of the marked points in order to induce an automorphism of $\overline{M}_{g,A[n]}$ has to preserve the weight data in a suitable sense.

Example 1.14. In $\overline{M}_{3,A[6]}$ with weights $(1/4, 1/4, 1/2, 3/4, 1, 1)$ consider the divisor parametrizing reducible curves $C_1 \cup C_2$, where C_1 has genus zero and weights $(1/4, 1/4, 3/4)$, and C_2 has genus three and weights $(1/2, 1, 1)$.

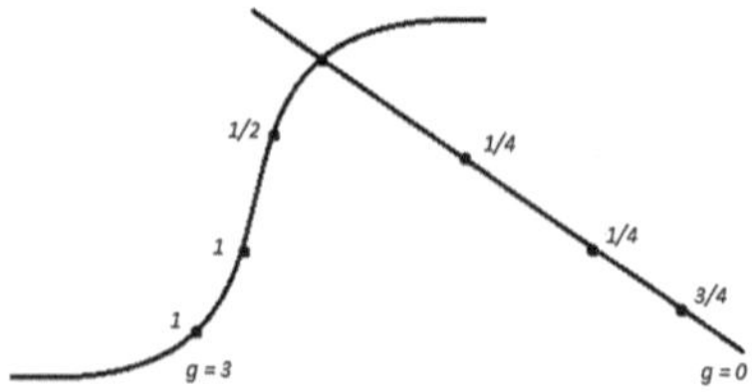

After the transposition $3 \leftrightarrow 4$ the genus zero component has markings $(1/4, 1/4, 1/2)$, so it is contracted. This means that the transposition induces a birational map

$$\overline{M}_{3,A[6]} \overset{3\leftrightarrow 4}{\dashrightarrow} \overline{M}_{3,A[6]}$$

contracting a divisor on a codimension two subscheme of $\overline{M}_{3,A[6]}$.

We see that troubles come from rational tails with at least three marked points. To avoid this, the authors introduced the following definition.

Definition 1.15 ([12, Definition 3.10]). A transposition $i \leftrightarrow j$ of two marked points with positive weights in A is *admissible* if and only if for any $h_1, \dots, h_r \in \{1, \dots, n\}$, with $r \geqslant 2$,

$$a_i + \sum_{k=1}^{r} a_{h_k} \leqslant 1 \iff a_j + \sum_{k=1}^{r} a_{h_k} \leqslant 1.$$

In [12, Lemma 3.13] the authors proved that, if $g \geqslant 1$ then the image of the morphism (2) is the subgroup $\mathcal{A}_{A[n]}$ of S_n generated by the admissible transpositions. Finally in [12, Theorems 3.15, 3.18] they managed to control the kernel of the morphism (2) and proved the following generalization of Theorem 1.6.

Theorem 1.16. *Let $\overline{\mathcal{M}}_{g,A[n]}$ be the Hassett's moduli stack parametrizing weighted n-pointed genus g stable curves, and let $\overline{M}_{g,A[n]}$ be its coarse moduli space. If $g \geqslant 1$ and $2g - 2 + n \geqslant 3$, then*

$$\mathrm{Aut}(\overline{\mathcal{M}}_{g,A[n]}) \cong \mathrm{Aut}(\overline{M}_{g,A[n]}) \cong \mathcal{A}_{A[n]}.$$

Furthermore

- $\mathrm{Aut}(\overline{M}_{1,A[2]}) \cong (\mathbb{C}^*)^2$ *while* $\mathrm{Aut}(\overline{\mathcal{M}}_{1,A[2]})$ *is trivial,*
- $\mathrm{Aut}(\overline{M}_{1,A[1]}) \cong PGL(2)$ *while* $\mathrm{Aut}(\overline{\mathcal{M}}_{1,A[1]}) \cong \mathbb{C}^*$,
- $\mathrm{Aut}(\overline{\mathcal{M}}_2)$ *and* $\mathrm{Aut}(\overline{M}_2)$ *are trivial.*

2 Kapranov's and Keel's Spaces

Construction 1.9 provides a factorization of Kapranov's blow-up construction of $\overline{M}_{0,n}$. There are many other factorizations of the morphisms $f_i : \overline{M}_{0,n} \to \mathbb{P}^{n-3}$ as compositions of reduction morphisms. We consider other two factorizations. The first is due to Kapranov [7, Sect. 6.2].

Construction 2.1. Fixed $(n-1)$-points $p_1, \ldots, p_{n-1} \in \mathbb{P}^{n-3}$ in linear general position:

(1) Blow-up the points $p_1, \ldots, p_{n-1}$,
(2) Blow-up the strict transforms of the lines $\langle p_{i_1}, p_{i_2}\rangle$, $i_1, i_2 = 1, \ldots, n-1$,

⋮

(k) Blow-up the strict transforms of the $(k-1)$-planes $\langle p_{i_1}, \ldots, p_{i_k}\rangle$, $i_1, \ldots, i_k = 1, \ldots, n-1$,

⋮

$(n-4)$ Blow-up the strict transforms of the $(n-5)$-planes $\langle p_{i_1}, \ldots, p_{i_{n-4}}\rangle$, $i_1, \ldots, i_{n-4} = 1, \ldots, n-1$.

Now, consider Hassett's spaces $X_k[n] := \overline{M}_{0,A[n]}$ for $k = 1, \ldots, n-4$, such that

- $a_i + a_n > 1$ for $i = 1, \ldots, n-1$,
- $a_{i_1} + \cdots + a_{i_r} \leqslant 1$ for each $\{i_1, \ldots, i_r\} \subset \{1, \ldots, n-1\}$ with $r \leqslant n-k-2$,
- $a_{i_1} + \cdots + a_{i_r} > 1$ for each $\{i_1, \ldots, i_r\} \subset \{1, \ldots, n-1\}$ with $r > n-k-2$.

Then $X_k[n]$ is isomorphic to the variety obtained at the step k of the blow-up construction. Clearly the spaces $X_k[n]$ satisfy Definition 1.8.

Hassett's spaces appearing in [7, Sect. 6.3] are strictly related to the construction of $\overline{M}_{0,n}$ provided by Keel in [9]. These spaces give another factorization of Kapranov's map $f_i : \overline{M}_{0,n} \to \mathbb{P}^{n-3}$.

Construction 2.2 ([7, Sect. 6.3]). We start with the variety $Y_0[n] := (\mathbb{P}^1)^{n-3}$ which can be realized as Hassett's space $\overline{M}_{0,A[n]}$ where $A[n] = (a_1, \ldots, a_n)$ satisfy the following conditions:

- $a_i + a_j > 1$ where $\{i, j\} \subset \{1, 2, 3\}$,
- $a_i + a_{j_1} + \cdots + a_{j_r} \leqslant 1$ for $i = 1, 2, 3$, $\{j_1, \ldots, j_r\} \subseteq \{4, \ldots, n\}$, with $r \geqslant 2$.

Let Δ_d be the locus in $(\mathbb{P}^1)^{n-3}$ where at least $n-2-d$ of the points coincide, that is the d-dimensional diagonal. Let $\pi_i : (\mathbb{P}^1)^{n-3} \to \mathbb{P}^1$ for $i = 1, \ldots, n-3$ be the projections, and let

$$F_0 := \pi_1^{-1}(0) \cup \ldots \cup \pi_{n-3}^{-1}(0).$$

We define F_1 and F_∞ similarly and use the same notation for proper transforms. Consider the following sequence of blow-ups

(1) Blow-up $\Delta_1 \cap (F_0 \cup F_1 \cup F_\infty)$.

$\vdots$

(h) Blow-up $\Delta_k \cap (F_0 \cup F_1 \cup F_\infty)$.

$\vdots$

$(n-4)$ Blow-up $\Delta_{n-4} \cap (F_0 \cup F_1 \cup F_\infty)$.

The variety $Y_h[n]$ obtained at the step h can be realized as Hassett's space $\overline{M}_{0,A[n]}$ where the weights satisfy the following conditions:

- $a_i + a_j > 1$ if $\{i, j\} \subset \{1, 2, 3\}$,
- $a_i + a_{j_1} + \cdots + a_{j_r} \leqslant 1$ if $i \in \{1, 2, 3\}$ and $\{j_1, \ldots, j_r\} \subset \{4, \ldots, n\}$ with $0 < r \leqslant n - h - 3$,
- $a_i + a_{j_1} + \cdots + a_{j_r} > 1$ if $i \in \{1, 2, 3\}$ and $\{j_1, \ldots, j_r\} \subset \{4, \ldots, n\}$ with $r > n - h - 3$.

Now, we consider another sequence of blow-ups starting from $Y_{n-4}[n]$.

$(n-3)$ Blow-up Δ_1.
$(n-2)$ Blow-up Δ_2.

$\vdots$

$(2n-9)$ Blow-up Δ_{n-5}.

The variety $Y_h[n]$ obtained at the step h can be realized as Hassett's space $\overline{M}_{0,A[n]}$ where the weights satisfy the following conditions:

- $a_i + a_j > 1$ if $\{i, j\} \subset \{1, 2, 3\}$,
- $a_{j_1} + \cdots + a_{j_r} \leqslant 1$ if $\{j_1, \ldots, j_r\} \subset \{4, \ldots, n\}$ with $0 < r \leqslant 2n - h - 7$,
- $a_{j_1} + \cdots + a_{j_r} > 1$ if $\{j_1, \ldots, j_r\} \subset \{4, \ldots, n\}$ with $r > 2n - h - 7$.

Remark 2.3. For instance taking

$$A = (1 - (n-3)\epsilon, 1 - (n-3)\epsilon, 1 - (n-3)\epsilon, \epsilon, \ldots, \epsilon)$$

where ϵ is an arbitrarily small positive rational number, we have $\overline{M}_{0,A[n]} \cong (\mathbb{P}^1)^{n-3}$. Note that $(\mathbb{P}^1)^2$ does not admit any birational morphism to $\mathbb{P}^2$. However, at the first step of Construction 2.2 we get $(\mathbb{P}^1)^2$ blown-up at three points on the diagonal. Such blow-up is isomorphic to the blow-up of $\mathbb{P}^2$ at four general points that is $\overline{M}_{0,5}$. In the following we will prove that this fact holds also in higher dimension. More precisely the spaces $Y_h[n]$ factor Kapranov, in the sense of Definition 1.8, for any $n - 4 \leqslant h \leqslant 2n - 9$.

Recall that Hassett's spaces $\overline{M}_{0,A_{r,s}[n]}$ are the intermediate steps of Construction 1.9.

Lemma 2.4. *Hassett's spaces $Y_h[n]$ admit a reduction morphism to $\overline{M}_{0,A_{2,1}[n]}$ for any $n - 4 \leqslant h \leqslant 2n - 9$. Furthermore, $Y_{n-3}[n] \cong \overline{M}_{0,A_{2,2}[n]}$.*

Proof. By construction $Y_h[n]$ has a reduction morphism to $Y_{n-4}[n]$ for any $h \geqslant n-3$. So it is enough to prove the statement for $Y_{n-4}[n]$. This variety is isomorphic to $\overline{M}_{0,A[n]}$ with

$$A[n] = (1-\epsilon, 1-\epsilon, 1-\epsilon, \epsilon, \dots, \epsilon).$$

By [7, Corollary 4.7], the reduction morphism

$$A^{'}[n] := (1, 1, 1-\epsilon, \epsilon, \dots, \epsilon) \mapsto A[n]$$

is an isomorphism. So we may proceed with $A^{'}[n]$ instead of $A[n]$. Now, take $\epsilon = \frac{1}{n-3}$ and consider the reduction morphism

$$A^{'}[n] := (1, 1, 1-1/(n-3), 1/(n-3), \dots, 1/(n-3)) \mapsto A_{2,1}[n].$$

The space $Y_{n-3}[n]$ can be realized as a Hassett's space with weight data

$$A[n] = (1, 1, 1/(n-3), 1/(n-3), \dots, 1/(n-3), 2/(n-3)).$$

This is the weight data of Hassett's spaces produced at the step $r = s = 2$ of Construction 1.9. □

Proposition 2.5. *Hassett's spaces $Y_h[n]$ factor Kapranov for any $n-4 \leqslant h \leqslant 2n-9$.*

Proof. By Lemma 2.4 we have a reduction morphism $\rho : Y_h[n] \to \overline{M}_{0,A_{2,1}[n]}$. Since $\overline{M}_{0,A_{2,1}[n]}$ factors Kapranov we get the following commutative diagram:

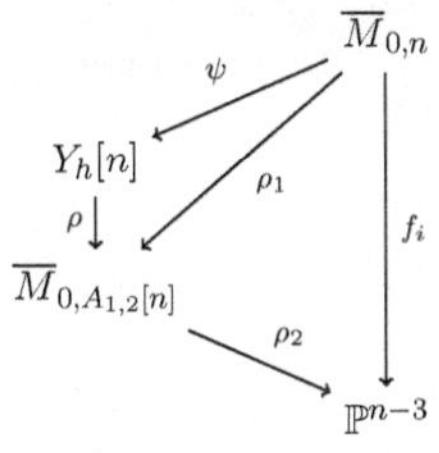

where ψ is a reduction morphism. Since ρ, ρ_1, and ψ are reduction morphisms $\rho_1 = \rho \circ \psi$ and $f_i = \rho_2 \circ (\rho \circ \psi)$. □

Note that, by Remark 2.3, in general the spaces $Y_h[n]$ do not factor Kapranov. However, we do not know if the bound $h \geqslant n-4$ in Proposition 2.5 is sharp.

Theorem 2.6. *The automorphism groups of Hassett's spaces $X_k[n]$ for $1 \leqslant k \leqslant n-4$ and $Y_h[n]$ for $n-4 \leqslant h \leqslant 2n-9$ are given by*

$$\mathrm{Aut}(X_k[n]) \cong \mathrm{Aut}(Y_h[n]) \cong S_n.$$

Proof. Since step $k = 1$ of Construction 2.1 we have blown-up $n-1$ points in $\mathbb{P}^{n-3}$. Furthermore, the same is true for $\overline{M}_{0,A_{2,1}[n]}$, which is the step $r = 2, s = 1$ of Construction 1.9, and therefore, by Lemma 2.4, for $Y_h[n]$ with $n-4 \leqslant h \leqslant 2n-9$.

We may proceed by considering the spaces $X_k[n]$ because our proof works exactly in the same way for the spaces $Y_h[n]$. The key fact is that in both these classes of spaces we have blown-up $n-1$ points in linear general position in $\mathbb{P}^{n-3}$. By Construction 2.1 for $k = 1$ we see that with weights

$$A[n] = (1/(n-3), \dots, 1/(n-3), 1)$$

we have $\overline{M}_{0,A[n]} \cong X_1[n]$. Clearly any transposition of the first $n-1$ marked points gives an automorphism of $X_1[n]$. Let $C_1 \cup C_2$ be a stable curve with C_1 not necessarily irreducible and $C_2 \cong \mathbb{P}^1$ with two marked points x_i, x_n. A transposition $x_n \leftrightarrow x_j$ induces the contraction of C_2. On the other hand C_2 is a smooth rational curve with three special points. Therefore it does not have moduli and $x_n \leftrightarrow x_j$ induces an automorphism of $X_1[n]$. We conclude that any permutation induces an automorphism of $X_1[n]$. Furthermore, for any $k \geqslant 1$ there is a birational morphism $X_k[n] \to X_1[n]$. Then any permutation induces an automorphism of $X_k[n]$ for $k > 1$ as well. By Construction 2.1 any $X_k[n]$ factors Kapranov and the morphism (2)

$$\chi : \mathrm{Aut}(X_k[n]) \to S_n$$

is surjective. Let $\varphi \in \mathrm{Ker}(\chi)$ be an automorphism inducing the trivial permutation. Then φ preserves the fibers, say F_i, of all forgetful maps. Let $f_n : X_k[n] \to \mathbb{P}^{n-3}$ be the Kapranov's map corresponding to the marked point p_n. Then $f_n(F_i)$ is a line through p_i for $i = 1, \dots, n-1$ and $f_n(F_n)$ is a rational normal curve through $p_1, \dots, p_{n-1}$. Therefore φ induces a birational map $\varphi_{\mathcal{H}} : \mathbb{P}^{n-3} \dashrightarrow \mathbb{P}^{n-3}$ preserving the lines L_i through p_i and the rational normal curves C through $p_1, \dots, p_{n-1}$. Let $|\mathcal{H}| \subseteq |\mathcal{O}_{\mathbb{P}^{n-3}}(d)|$ be the linear system associated with $\varphi_{\mathcal{H}}$. The equalities

$$\deg(\varphi_{\mathcal{H}}(L_i)) = d - \mathrm{mult}_{p_i} \mathcal{H} = 1,$$
$$\deg(\varphi_{\mathcal{H}}(C)) = (n-3)d - \textstyle\sum_{i=1}^{n-1} \mathrm{mult}_{p_i} \mathcal{H} = n-3.$$

yield $d = 1$. So $\varphi_{\mathcal{H}}$ is an automorphism of $\mathbb{P}^{n-3}$ fixing $n-1$ points in general position, this forces $\varphi_{\mathcal{H}} = Id$. Then χ is injective and $\mathrm{Aut}(\overline{M}_{0,A_{r,s}[n]}) \cong S_n$. □

3 Hassett's Spaces with Zero Weights

In this section we compute the automorphisms of the moduli spaces of weighted pointed curves $\overline{M}_{g,\tilde{A}[n]}$ of Definition 1.7. Recall that we denote by $A \subseteq \tilde{A}$ the set of positive weights and by N the number of zero weights.

Lemma 3.1. *If $g \geqslant 2$, then all the forgetful morphisms $\overline{M}_{g,\tilde{A}[n]} \to \overline{M}_{g,\tilde{A}[r]}$ are well defined morphisms.*

Proof. If $g \geqslant 2$ we have $2g - 2 + a_{i_1} + \cdots + a_{i_r} \geqslant 2 + a_{i_1} + \cdots + a_{i_r} > 0$. To conclude it is enough to apply [7, Theorem 4.3]. □

Remark 3.2. Lemma 3.1 does not hold for $g = 1$. For instance consider $\overline{M}_{1,\tilde{A}[2]}$ with $\tilde{A} := (1/3, 0)$. The second forgetful morphism is well defined but the first is not, being $2g - 2 + a_2 = 0$. To avoid this problem when $g = 1$ we will consider Hassett's spaces $\overline{M}_{1,\tilde{A}[n]}$ such that at least two of the weights are different from zero.

Lemma 3.3. *If $\overline{M}_{1,\tilde{A}[n]}$ is a Hassett's space with at least two weights a_{i_1}, a_{i_2} different from zero, then all the forgetful morphisms $\overline{M}_{g,\tilde{A}[n]} \to \overline{M}_{g,\tilde{A}[n-1]}$ are well-defined morphisms.*

Proof. In any case we have $2g - 2 + a_{j_1} + \cdots + a_{j_{n-1}} \geqslant 2g - 2 + a_{i_1} > 0$ or $2g - 2 + a_{j_1} + \cdots + a_{j_{n-1}} \geqslant 2g - 2 + a_{i_2} > 0$. Again it is enough to apply [7, Theorem 4.3]. □

The following proposition describes the fibrations of Hassett's spaces $\overline{M}_{g,\tilde{A}[n]}$. Its proof derives easily from the suitable variations on the proofs of [12, Proposition 2.7, Lemma 2.8].

Proposition 3.4. *Let $f : \overline{M}_{g,\tilde{A}[n]} \to X$ be a dominant morphism with connected fibers.*

- *If $g \geq 2$ either f is of fiber type and factorizes through a forgetful morphism $\pi_I : \overline{M}_{g,\tilde{A}[n]} \to \overline{M}_{g,\tilde{A}[r]}$, or f is birational and* $\mathrm{Exc}(f) \subseteq \partial\overline{M}_{g,\tilde{A}[n]}$.
- *If $g = 1$, φ is an automorphism of $\overline{M}_{1,\tilde{A}[n]}$ and $\tilde{A}[n]$ has at least two nonzero weights then any fibration of the type $\pi_i \circ \varphi$ factorizes through a forgetful morphism $\pi_j : \overline{M}_{1,\tilde{A}[n]} \to \overline{M}_{1,\tilde{A}[n-1]}$.*

From now on we consider the case $g \geqslant 2$ and when $g = 1$ we restrict two Hassett's spaces having at least two nonzero weights. Let $\varphi : \overline{M}_{g,\tilde{A}[n]} \to \overline{M}_{g,\tilde{A}[n]}$ be an automorphism and $\pi_i : \overline{M}_{g,\tilde{A}[n]} \to \overline{M}_{g,\tilde{A}[n-1]}$ a forgetful morphism. By Proposition 3.4 we have the following diagram:

$$\begin{array}{ccc} \overline{M}_{g,\tilde{A}[n]} & \xrightarrow{\varphi^{-1}} & \overline{M}_{g,\tilde{A}[n]} \\ {\scriptstyle \pi_{j_i}}\downarrow & & \downarrow{\scriptstyle \pi_i} \\ \overline{M}_{g,\tilde{A}[n-1]} & \xrightarrow{\tilde{\varphi}} & \overline{M}_{g,\tilde{A}[n-1]} \end{array}$$

where π_{j_i} is a forgetful map. This allows us to associate with an automorphism a permutation in S_r, where r is the number of well-defined forgetful maps. By Lemmas 3.1 and 3.3 we have $r = n$. Therefore we get a morphism of groups

$$\begin{aligned} \chi : \mathrm{Aut}(\overline{M}_{g,\tilde{A}[n]}) &\longrightarrow S_n \\ \varphi &\longmapsto \sigma_\varphi \end{aligned}$$

where

$$\begin{array}{rccc} \sigma_\varphi : \{1,\ldots,n\} & \longrightarrow & \{1,\ldots,n\} \\ i & \longmapsto & j_i \end{array}$$

Note that in order to have a morphism of groups we have to consider φ^{-1} instead of φ. As for Hassett's spaces with nonzero weights the image of χ depends on the weight data. Recall that A is the set of positive weights in $\tilde{A}$. By [12, Proposition 3.12] a transposition $i \leftrightarrow j$ of two marked points with weights in A induces an automorphism of $\overline{M}_{g,A[n]}$ and therefore of $\overline{M}_{g,\tilde{A}[n]}$ if and only if $i \leftrightarrow j$ is admissible.

Furthermore, the symmetric group S_N permuting the marked points with zero weights acts on $\overline{M}_{g,\tilde{A}[n]}$. Note that if $i \leftrightarrow j$ is a transposition switching a marked point with positive weight a_j and a marked point with weight zero a_i then $i \leftrightarrow j$ induces just a birational automorphism

$$\overline{M}_{g,\tilde{A}[n]} \dashrightarrow \overline{M}_{g,\tilde{A}[n]}$$

because it is not defined on the loci parametrizing curves $[C, x_1, \ldots, x_n]$ where the marked point with weight zero x_i lies in the singular locus of C.

Example 3.5. In $\overline{M}_{1,\tilde{A}[6]}$ with weights $(0, 0, 1/3, 1/3, 1/3, 2/3)$ consider the divisor parametrizing reducible curves $C_1 \cup C_2$, where C_1 has genus zero and markings $(1/3, 1/3, 2/3)$, and C_2 has genus one and markings $(0, 0, 1/3)$.

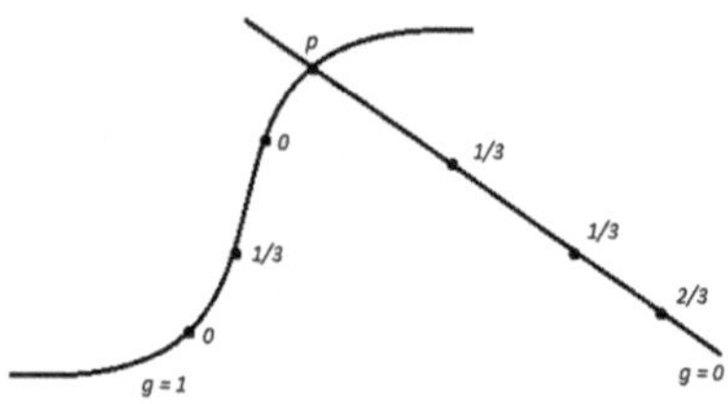

After the transposition $3 \leftrightarrow 6$ the genus zero component has markings $(1/3, 1/3, 1/3)$, so it is contracted. This means that the transposition induces a birational map

$$\overline{M}_{1,\tilde{A}[6]} \overset{3\leftrightarrow 6}{\dashrightarrow} \overline{M}_{1,\tilde{A}[6]}$$

contracting a divisor on a codimension two subscheme of $\overline{M}_{1,A[6]}$. Similarly the transposition $1 \leftrightarrow 6$ does not define an automorphism because it is not defined on the locus where the first marked point x_1 coincides with the node $p = C_1 \cap C_2$.

Let us consider the subgroup $\mathcal{A}_{A[n-N]} \subseteq S_{n-N}$ generated by admissible transpositions of points with positive weights and the symmetric group S_N permuting the marked points with zero weights. The actions of $\mathcal{A}_{A[n-N]}$ and S_N on $\overline{M}_{g,\tilde{A}[n]}$ are independent and $\mathcal{A}_{A[n-N]} \times S_N \subseteq \mathrm{Im}(\chi)$. Furthermore, by [12, Lemma 3.13] we have

$$\mathrm{Im}(\chi) = \mathcal{A}_{A[n-N]} \times S_N.$$

Finally, with the suitable variations in the proofs of [12, Proposition 3.14] and [12, Theorems 3.15, 3.18] we have the following theorem.

Theorem 3.6. *If* $g \geqslant 2$ *and if* $g = 1$, $n \geqslant 3$, $|A| \geqslant 2$ *then*

$$\mathrm{Aut}(\overline{\mathcal{M}}_{g,\tilde{A}[n]}) \cong \mathrm{Aut}(\overline{M}_{g,\tilde{A}[n]}) \cong \mathcal{A}_{A[n-N]} \times S_N.$$

References

1. A. Bruno, M. Mella, On some fibrations of $\overline{M}_{0,n}$. (2011) http://arxiv.org/abs/1105.3293v1 [arXiv:1105.3293v1]
2. A. Bruno, M. Mella, The automorphism group of $\overline{M}_{0,n}$. J. Eur. Math. Soc. **15**(3), 949–968 (2013)
3. P. Deligne, D. Mumford, The irreducibility of the space of curves of given genus. Inst. Hautes Études Sci. Publ. Math. **36**, 75109 (1969)
4. I.V. Dolgachev, V.A. Iskovskikh, Finite subgroups of the plane Cremona group, in *Algebra, Arithmetic, and Geometry: In Honor of Y.I. Manin, Vol. I.* Progress in Mathematics, vol. 269 (Birkhäuser, Boston, 2009), pp. 443–548
5. G. Farkas, The global geometry of the moduli space of curves. Proc. Symp. Pure Math. **80**, 125–147 (2009)
6. A. Gibney, S. Keel, I. Morrison, Towards the ample cone of $\overline{M}_{g,n}$. J. Am. Math. Soc. **15**, 273–294 (2002)
7. B. Hassett, Moduli spaces of weighted pointed stable curves. Adv. Math. **173**(2), 316–352 (2003)
8. M. Kapranov, Veronese curves and Grothendieck-Knudsen moduli spaces $\overline{M}_{0,n}$. J. Algebr. Geom. **2**, 239–262 (1993)
9. S. Keel, Intersection theory of moduli space of stable N-pointed curves of genus zero. Trans. Am. Math. Soc. **330**(2), 545–574 (1992)
10. S. Keel, J. McKernan, Contractible extremal rays of $\overline{M}_{0,n}$. (1997) http://arxiv.org/abs/alg-geom/9607009v1 [arXiv:alg-geom/9607009v1]
11. A. Massarenti, The automorphism group of $\overline{M}_{g,n}$. J. Lond. Math. Soc. (2013). doi: 10.1112/jlms/jdt057
12. A. Massarenti, M. Mella, On the automorphisms of Hassett's moduli spaces. http://arxiv.org/abs/1307.6828v1 [arXiv:1307.6828v1]
13. S. Mochizuki, Correspondences on hyperbolic curves. J. Pure Appl. Algebr. **131**, 227–244 (1998)
14. H. Moon, Log canonical models for $\overline{M}_{g,n}$. http://arxiv.org/abs/1111.5354 [arXiv:1111.5354]
15. H.L. Royden, Automorphisms and isometries of Teichmüller spaces, in *Advances in the Theory of Riemann Surfaces*, ed. by L.V. Ahlfors, L. Bers, H.M. Farkas, R.C. Gunning, I. Kra, H.E. Rauch. Annals of Mathematics Studies, vol. 66 (Princeton University Press, Princeton, 1971), pp. 369–383

Normal Analytic Compactifications of $\mathbb{C}^2$

Pinaki Mondal

Abstract This is a survey of some results on the structure and classification of normal analytic compactifications of $\mathbb{C}^2$. Mirroring the existing literature, we especially emphasize the compactifications for which the curve at infinity is irreducible.

2010 Mathematics Subject Classification: 32C20, 14M27, 14J26.

1 Introduction

A compact normal analytic surface $\bar{X}$ is called a compactification of $\mathbb{C}^2$ if there is a subvariety C (the *curve at infinity*) such that $\bar{X} \setminus C$ is isomorphic to $\mathbb{C}^2$. *Non-singular* compactifications of $\mathbb{C}^2$ have been studied at least since 1954 when Hirzebruch included the problem of finding all such compactifications in his list of problems on differentiable and complex manifolds [11]. Remmert and Van de Ven [23] proved that $\mathbb{P}^2$ is the only non-singular analytic compactification of $\mathbb{C}^2$ for which the curve at infinity is irreducible. Kodaira as part of his classification of surfaces, and independently Morrow [20] showed that every non-singular compactification of $\mathbb{C}^2$ is *rational* (i.e., bimeromorphic to $\mathbb{P}^2$) and can be obtained from $\mathbb{P}^2$ or some Hirzebruch surface via a sequence of blow-ups. Moreover, Morrow [20] gave the complete classification (modulo extraneous blow-ups of points at infinity) of non-singular compactifications of $\mathbb{C}^2$ for which the curve at infinity has normal crossing singularities.

P. Mondal (✉)
Weizmann Institute of Sciences, Rehovot, Israel

I. Cheltsov et al. (eds.), *Automorphisms in Birational and Affine Geometry*, Springer Proceedings in Mathematics & Statistics 79,
DOI 10.1007/978-3-319-05681-4_10,

The main topic of this chapter is therefore *singular* normal analytic compactifications of $\mathbb{C}^2$. The studies on singular normal analytic compactifications so far have concentrated mostly on the (simplest possible) case of compactifications for which the curve at infinity is irreducible; following [21], we call these *primitive* compactifications (of $\mathbb{C}^2$). These were studied from different perspectives in [2–4, 8, 12, 13, 16, 21], and more recently in [17–19]. The primary motive of this chapter is to describe these results. For relatively more technical of the results, however, we omit the precise statements and prefer to give only a "flavour". The only new results of this chapter are Proposition 3.2 and parts of Proposition 4.1.

Notation 1.1. Unless otherwise stated, by a "compactification" we mean throughout a normal analytic compactification of $\mathbb{C}^2$.

2 Analytic vs. Algebraic Compactifications

As mentioned in the introduction, non-singular compactifications of $\mathbb{C}^2$ are projective, and therefore, *algebraic* (i.e., analytifications of proper schemes). In particular this implies that every compactification $\bar{X}$ of $\mathbb{C}^2$ is necessarily *Moishezon*, or equivalently, analytification of a proper *algebraic space*. Moreover, if $\pi : \bar{X}' \to X$ is a resolution of singularities of $\bar{X}$, then the intersection matrix of the curves contracted by π is negative definite. On the other hand, by the contractibility criterion of Grauert [10], for every non-singular compactification $\bar{X}'$ of $\mathbb{C}^2$ and a (possibly reducible) curve $C \subseteq \bar{X}' \setminus \mathbb{C}^2$ with negative definite intersection matrix, there is a compactification $\bar{X}$ of $\mathbb{C}^2$ and a birational holomorphic map $\pi : \bar{X}' \to \bar{X}$ such that π contracts only C (and no other curve). The preceding observation, combined with the classification of n on-singular compactifications of $\mathbb{C}^2$ due to Kodaira and Morrow, forms the basis of our understanding of (normal) compactifications of $\mathbb{C}^2$. However, it is an open question how to determine if a (singular) compactification of $\mathbb{C}^2$ constructed via contraction of a given (possibly reducible) negative definite curve (from a non-singular compactification) is algebraic. [19] solves this question in the special case of *primitive* compactifications of $\mathbb{C}^2$ (for which, in particular, algebraicity is equivalent to projectivity—see Theorem 5.4).

More precisely, let $X := \mathbb{C}^2$ and $\bar{X}^0 := \mathbb{P}^2 \supseteq X$. Let $\bar{X}$ be a primitive compactification of X which is not isomorphic to $\mathbb{P}^2$ and $\sigma : \bar{X}^0 \dashrightarrow \bar{X}$ be the bimeromorphic map induced by identification of X. Then σ maps the *line at infinity* $L_\infty := \mathbb{P}^2 \setminus X$ (minus the points of indeterminacy) to a point $P_\infty \in C_\infty := \bar{X} \setminus X$.

Theorem 2.1 ([19, Corollary 1.6]). *$\bar{X}$ is algebraic iff there is an algebraic curve $C \subseteq X$ with one place at infinity*[1] *such that P_∞ does* not *belong to the closure of C in $\bar{X}$.*

[1]Recall that C has *one place at infinity* iff C meets the line at infinity at only one point Q and C is unibranch at Q.

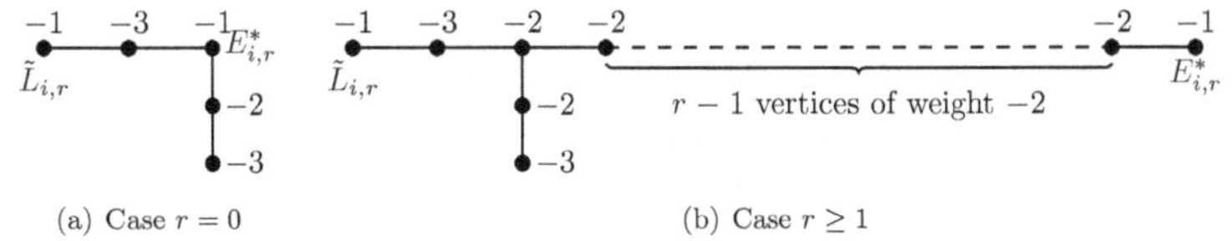

Fig. 1 Dual graph of $\tilde{E}^{(i,r)} \cup E^*_{i,r}$

Remark 2.2. Theorem 2.1 can be viewed as the *effective* version of (a special case of) some other algebraicity criteria (e.g., those of [22, 25]). More precisely, in the situation of Theorem 2.1, both [25, Theorem 3.3] and [22, Lemma 2.4] imply that $\bar{X}$ is algebraic iff there is an algebraic curve $C \subseteq X$ which satisfies the following (weaker) condition:

$$P_\infty \text{ does not belong to the closure of } C \text{ in } \bar{X}. \qquad (*)$$

Theorem 2.1 implies that in the algebraic case it is possible to choose C with an additional property, namely that it has one place at infinity. A possible way to construct such curves is via the *key forms* of the *divisorial valuation* on $\mathbb{C}(X)$ associated with C_∞ (see Remarks 5.2 and 5.16). The key forms are in general not polynomials, but if they are indeed polynomials, then the *last* key form defines a curve C with one place at infinity which satisfies $(*)$. On the other hand, if the last key form is not a polynomial, then it turns out that there are no curve $C \subseteq X$ which satisfies $(*)$ [19, Proposition 4.2], so that $\bar{X}$ is not algebraic.

Example 2.3 ([19, Examples 1.3 and 2.5]). Let (u, v) be a system of "affine" coordinates near a point $O \in \mathbb{P}^2$ ("affine" means that both $u = 0$ and $v = 0$ are lines on $\mathbb{P}^2$) and L be the line $\{u = 0\}$. Let C_1 and C_2 be curve-germs at O defined respectively by $f_1 := v^5 - u^3$ and $f_2 := (v - u^2)^5 - u^3$. For each i, r, $1 \leq i \leq 2$ and $r \geq 0$, let $\tilde{X}_{i,r}$ be the surface constructed by resolving the singularity of C_i at O and then blowing up r more times the point of intersection of the (successive) strict transform of C_i with the exceptional divisor. Let $\tilde{E}^{(i,r)}$ be the union of the strict transform $\tilde{L}_{i,r}$ (on $\tilde{X}_{i,r}$) of L and (the strict transforms of) all exceptional curves except the exceptional curve $E^*_{i,r}$ for the *last* blow up. It is straightforward to compute that for $r \leq 9$ the intersection matrix of $\tilde{E}^{(i,r)}$ is negative definite, so that $\tilde{E}^{(i,r)}$ can be analytically contracted to the unique singular point $P_{i,r}$ on a normal surface $\bar{X}_{i,r}$ which is a primitive compactification of $\mathbb{C}^2$. Note that the weighted dual graphs of $\tilde{E}^{(i,r)} \cup E^*_{i,r}$ are *identical* (see Fig. 1).

Choose coordinates $(x, y) := (1/u, v/u)$ on $X := \mathbb{P}^2 \setminus L$. Then $C_1 \cap X = V(y^5 - x^2)$ and $C_2 \cap X = V((xy - 1)^5 - x^7)$. Let $\tilde{C}_{i,r}$ (resp. $C_{i,r}$) be the strict transform of C_i on $\tilde{X}_{i,r}$ (resp. $\bar{X}_{i,r}$). Note that each $C_{1,r}$ satisfies $(*)$, so that all $\bar{X}_{1,r}$ are algebraic by the criteria of Schröer and Palka. On the other hand, if $L'_{2,r}$ is the pullback on $\tilde{X}_{2,r}$ of a general line in $\mathbb{P}^2$, then $\tilde{C}_{2,r} - 5L'_{2,r}$ intersects components of

$\tilde{E}^{(2,r)}$ trivially and $E^*_{2,r}$ positively, so its positive multiples are the only candidates for total transforms of curves on $\bar{X}_{2,r}$ satisfying $(*)$, provided the latter surface is algebraic. In other words, Schröer and Palka's criteria imply that $\bar{X}_{2,r}$ is algebraic if and only if some positive multiple of $\tilde{C}_{2,r} - 5L'_{2,r}$ is numerically equivalent to an effective divisor. Theorem 2.1 implies that such a divisor does not exist for $r = 8, 9$. Indeed, the sequence of key forms associated to (the divisorial valuation on $\mathbb{C}(x, y)$ corresponding to) $E^*_{i,r}$ for $0 \leq r \leq 9$ are as follows [19, Example 3.22]:

$$\begin{matrix}\text{key forms} \\ \text{for } E^*_{1,r}\end{matrix} = \begin{cases} x, y & \text{if } r = 0, \\ x, y, y^5 - x^2 & \text{if } r \geq 1. \end{cases}$$

$$\begin{matrix}\text{key forms} \\ \text{for } E^*_{2,r}\end{matrix} = \begin{cases} x, y & \text{if } r = 0, \\ x, y, y^5 - x^2 & \text{if } 1 \leq r \leq 7, \\ x, y, y^5 - x^2, y^5 - x^2 - 5y^4x^{-1} & \text{if } 8 \leq r \leq 9. \end{cases}$$

In particular, for $8 \leq r \leq 9$, the last key form for $E^*_{2,r}$ is *not* a polynomial. It follows (from Remark 2.2 and Theorem 2.1) that $\bar{X}_{2,r}$ are algebraic for $r \leq 7$, but $\bar{X}_{2,8}$ and $\bar{X}_{2,9}$ are *not* algebraic. On the other hand, the key forms for $E^*_{1,r}$ are polynomials for each r, $0 \leq r \leq 9$, which implies via the same arguments that $\bar{X}_{1,r}$ are algebraic, as we have already seen via Schröer and Palka's criteria.

Remark 2.4. It can be shown (by explicitly computing the geometric genus and multiplicity) that the singularities at $P_{i,8}$ (of Example 2.3) are in fact *hypersurface singularities* which are *Gorenstein* and *minimally elliptic* (in the sense of [14]). Minimally elliptic Gorenstein singularities have been extensively studied, and in a sense they form the simplest class of non-rational singularities. Since having only rational singularities implies algebraicity of the surface (via a result of Artin), it follows that the non-algebraic surface $\bar{X}_{2,8}$ of Example 2.3 is a normal non-algebraic Moishezon surface with the "simplest possible" singularity.

We do not know to what extent the properties of C and $\bar{X}$ of Theorem 2.1 influence one another (in the case that $\bar{X}$ is algebraic). It is not hard to see that $\bar{X} \setminus \{P_\infty\}$ has at most one singular point, and the singularity, if exists, is a cyclic quotient singularity (Proposition 4.1). The following question was suggested by Tommaso de Fernex.

Question 2.5. Let $\bar{X}$ be a primitive algebraic compactification of $\mathbb{C}^2$ formed by (minimally) resolving the singularities of a curve-germ at a point on the line L_∞ at infinity on $\mathbb{P}^2$ and then contracting the strict transform of L_∞ and all exceptional curves other than the last one. Let P_∞ be as in Theorem 2.1 and g be the smallest integer such that there exists a curve C on $\bar{X}$ with geometric genus g which does not pass through P_∞. What is the relation between g and the singularity of $\bar{X}$ at P_∞?

Some computed examples suggest the following conjectural answer to the first case of Question 2.5:

Conjecture 2.6. In the situation of Question 2.5, $g = 0$ iff the singularity at P_∞ is rational.

A motivation behind Conjecture 2.6 is to understand the relation between rational singularity at a point and existence of rational curves that do not pass through the singularity, as discovered e.g. in [6, Theorem 0.3]. Another motivation is Abhyankar's question about the relation between the genus and *semigroup of poles* of plane curves with one place at infinity [24, Question 3]. More precisely, if $\bar{X}$ and C are as in Question 2.5, then the condition that $\bar{X}$ has a rational singularity induces (via assertion (1) of Corollary 5.7) a restriction on the semigroup of poles of C (cf. Remark 5.2). In particular, if Conjecture 2.6 is true, then it (together with Corollary 5.7) will answer the genus zero case of Abhyankar's question.

3 Curve at Infinity

Let $\bar{X}$ be a normal compactification of $X := \mathbb{C}^2$ and $C_\infty := \bar{X} \setminus X$ be the curve at infinity. An application of the classification results of non-singular compactifications of $\mathbb{C}^2$ to the desingularization of $\bar{X}$ immediately yields that C_∞ is a connected *tree* of (possibly singular) rational curves. In this section we take a deeper look at the structure of C_∞ and describe a somewhat stronger version of a result of Brenton [2].

Let $\Gamma_1, \ldots, \Gamma_k$ be the irreducible components of C_∞. Choose a copy $\bar{X}^0$ of $\mathbb{P}^2$ such that the center (i.e., image under the natural bimeromorphic map $\bar{X} \dashrightarrow \bar{X}^0$ induced by identification of X) of each Γ_j on $\bar{X}^0$ is a point $O_j \in L_\infty$, where $L_\infty := \bar{X}^0 \setminus X$ is the "line at infinity" on $\bar{X}^0$. Fix a Γ_j, $1 \le j \le k$. For each pair of (distinct) points P_1, P_2 on Γ_j, define a positive integer $m_j(P_1, P_2)$ as follows:

$$\begin{aligned} m_j(P_1, P_2) := \min\{i_{O_j}(C_1, C_2) : &\text{for each } i,\ 1 \le i \le 2,\ C_i \text{ is an analytic curve germ} \\ &\text{at } O_j \text{ distinct from (the germ of) } L_\infty \text{ and the closure of the strict} \\ &\text{transform of } C_i \text{ on } \bar{X} \text{ passes through } P_i\}, \text{ where} \\ i_{O_j}(C_1, C_2) := &\text{ intersection multiplicity of } C_1 \text{ and } C_2 \text{ at } O_j. \end{aligned}$$

It is not hard to see (e.g., using [17, Proposition 4.2]) that there exists an integer $\tilde{m}_j$ and a unique point $\tilde{P}_j \in \Gamma_j$ such that

(1) $m_j(P_1, P_2) = \tilde{m}_j$ for all $P_1, P_2 \in \Gamma_j \setminus \{\tilde{P}_j\}$, and
(2) $m_j(\tilde{P}_j, P') < \tilde{m}_j$ for all $P' \in \Gamma_j \setminus \{\tilde{P}_j\}$.

Remark 3.1. $\tilde{P}_j$ has the following interpretation in the language of the *valuative tree* [5]: the valuative tree $\mathcal{V}_j$ at O_j is the space of all valuations centered at

O_j (which has a natural tree-structure rooted at ord_{O_j}). The order of vanishing ord_{Γ_j} along Γ_j is an element of $\mathcal{V}_j$ and the points on Γ_j are in a one-to-one correspondence with the *tangent vectors* at ord_{Γ_j} [5, Theorem B.1]. Then $\tilde{P}_j$ is the point on Γ_j which corresponds to the (unique) tangent vector at ord_j which is represented by ord_{O_j}.

The result below follows from a combination of [17, Proposition 4.2] and [18, Proposition 3.1].

Proposition 3.2 (cf. the Proposition in [2]).

(1) $\Gamma_j \setminus \tilde{P}_j \cong \mathbb{C}$.
(2) Either $\tilde{P}_j$ *is a singular point of* $\bar{X}$ *or* $\tilde{P}_j \in \Gamma_i$ *for some* $i \neq j$.

Remark 3.3. Assertion 1 of Proposition 3.2 implies that for every proper birational map $\tilde{\Gamma}_i \to \Gamma_i$, the pre-image of $\tilde{P}_i$ consists of only one point and $\tilde{\Gamma}_i$ is uni-branched at that point. In particular, in the language of [2], $\tilde{\Gamma}_i$ has a *totally extraordinary singularity* at $\tilde{P}_i$. Consequently, Proposition 3.2 strengthens the main result of [2].

Remark 3.4. Assertion 2 implies in particular that if $\bar{X}$ is non-singular and C_∞ is irreducible, then C_∞ is non-singular as well. More precisely, a theorem of Remmert and Van de Ven in [23] states that in this scenario $\bar{X}$ is isomorphic to $\mathbb{P}^2$. On the other hand, it was shown in [2] that Proposition 3.2 together with Morrow's classification [20] of "minimal normal compactifications[2]" of $\mathbb{C}^2$ implies the theorem of Remmert and Van de Ven.

Remark 3.5. If C_∞ is not irreducible, then it is possible that some Γ_i is singular, even if $\bar{X}$ is non-singular. One such example was constructed in [2] for which C_∞ has two irreducible components.

For special types of compactifications one can say more about the curve at infinity. We say that a compactification $\bar{X}$ of $\mathbb{C}^2$ is *minimal* if $\bar{X}$ does not dominate any other (normal analytic) compactification of $\mathbb{C}^2$, or equivalently (by Grauert's theorem), if the self-intersection number of every irreducible component of C_∞ is non-negative.

Proposition 3.6 ([17, Proposition 3.7], [18, Corollary 3.6]).

(1) If $\bar{X}$ *is minimal, then there is a unique point* $P_\infty \in C_\infty$ *such that* $\Gamma_i \cap \Gamma_j = \{P_\infty\}$ *for all* $i \neq j$. *In particular,* $\tilde{P}_i = P_\infty$ *for all* i.
(2) If $\bar{X}$ *is primitive algebraic, then* $\Gamma_1 = C_\infty$ *is non-singular off* $\tilde{P}_1$, *and it has at worst a (non-normal)* toric singularity *at* $\tilde{P}_1$.

[2] $\bar{X}$ is a "minimal normal compactification" (in the sense of Morrow), or in modern terminology, a minimal SNC-compactification of $X := \mathbb{C}^2$ iff (i) $\bar{X}$ is non-singular, (ii) each Γ_i is non-singular, (iii) C_∞ has at most normal-crossing singularities, and (iv) for all Γ_i with self-intersection -1, contracting Γ_i destroys some of the preceding properties.

4 Singular Points

As in the preceding section, let $\bar{X}$ be a normal compactification of $X := \mathbb{C}^2$ and C_∞ be the curve at infinity. In Proposition 4.1 below we give upper bounds for $|\operatorname{Sing}(\bar{X})|$ in the general case and in the case that $\bar{X}$ is a minimal compactification. Note that both of these upper bounds are sharp [17, Examples 3.9 and 4.8]. Moreover, it is not hard to see that the lower bound for $|\operatorname{Sing}(\bar{X})|$ in both cases is zero, i.e., for each $k \geq 1$, there are non-singular minimal compactifications of $\mathbb{C}^2$ with k irreducible curves at infinity.

Proposition 4.1. *Assume that C_∞ has k irreducible components. Let* $\operatorname{Sing}(\bar{X})$ *be the set of singular points of $\bar{X}$.*

(1) (a) $|\operatorname{Sing}(\bar{X})| \leq 2k$.
(b) $\bar{X}$ has at most one singular point which is not sandwiched.[3]
(2) Assume $\bar{X}$ is a minimal compactification. Then
(a) $|\operatorname{Sing}(\bar{X})| \leq k + 1$.
(b) Let P_∞ be as in assertion 1 of Proposition 3.6. Then $\left|\operatorname{Sing}(\bar{X}) \setminus \{P_\infty\}\right| \leq k$. *Moreover, every point in* $\operatorname{Sing}(\bar{X}) \setminus \{P_\infty\}$ *is a cyclic quotient singularity.*

Proof. Assertions 1a and 2a and the first statement of assertion 2b follows from [17, Proposition 3.7]. We now prove assertion 1b. If $\bar{X}$ dominates $\mathbb{P}^2$, then every singularity of $\bar{X}$ is sandwiched, as required. So assume that $\bar{X}^0$ does not dominate $\mathbb{P}^2$. Let $\bar{X}^1$ be the normalization of the closure of the image of $\mathbb{C}^2$ in $\bar{X} \times \mathbb{P}^2$ defined via identification of X with a copy of $\mathbb{C}^2$ in $\mathbb{P}^2$. Then all singularities of $\bar{X}^1$ are sandwiched. Assertion 1b now follows from the observation that the natural projection $\bar{X}^1 \to \bar{X}$ is an isomorphism over the complement of the strict transform on $\bar{X}^1$ of the line at infinity on $\mathbb{P}^2$. The last statement of assertion 2b follows from similar reasoning and an application of [18, Proposition 3.1]. □

5 Classification Results for Primitive Compactifications

5.1 Primitive Algebraic Compactifications

Using the correspondence with plane curves with one place at infinity (Theorem 2.1), it is possible to explicitly describe the defining equations of all primitive algebraic compactifications of $\mathbb{C}^2$. In particular, it turns out that every primitive

[3]Recall that an isolated singular point P on a surface Y is *sandwiched* if there exists a birational map $Y \dashrightarrow Y'$ such that the image of P is non-singular. Sandwiched singularities are *rational* [15, Proposition 1.2].

algebraic compactification is a "weighted complete intersection" (embedded in a weighted projective variety). We now describe this result.

Definition 5.1 ([18, Definition 3.2]). A sequence $\vec{\omega} := (\omega_0, \ldots, \omega_{n+1})$, $n \in \mathbb{Z}_{\geq 0}$, of positive integers is called a *key sequence* if it has the following properties: let $d_k := \gcd(\omega_0, \ldots, \omega_k)$, $0 \leq k \leq n+1$ and $p_k := d_{k-1}/d_k$, $1 \leq k \leq n+1$. Then

(1) $d_{n+1} = 1$, and
(2) $\omega_{k+1} < p_k \omega_k$, $1 \leq k \leq n$.

A key sequence $(\omega_0, \ldots, \omega_{n+1})$ is called *algebraic* if in addition

(3) $p_k \omega_k \in \mathbb{Z}_{\geq 0}\langle \omega_0, \ldots, \omega_{k-1} \rangle$, $1 \leq k \leq n$.

Finally, a key sequence $(\omega_0, \ldots, \omega_{n+1})$ is called *essential* if $p_k \geq 2$ for $1 \leq k \leq n$. Given an arbitrary key sequence $(\omega_0, \ldots, \omega_{n+1})$, it has an associated *essential subsequence* $(\omega_0, \omega_{i_1}, \ldots, \omega_{i_l}, \omega_{n+1})$ where $\{i_j\}$ is the collection of all k, $1 \leq k \leq n$, such that $p_k \geq 2$.

Remark 5.2. Let $\bar{X}$ be a primitive algebraic compactification of $\mathbb{C}^2$. Theorem 5.4 below states that $\bar{X}$ has an associated *algebraic key sequence* $\vec{\omega}$. On the other hand, Theorem 2.1 attaches to $\bar{X}$ a curve C with one place at infinity. It turns out that the essential subsequence $\vec{\omega}_e$ of $\vec{\omega}$ is "almost the same as" the *δ-sequence* of C (defined e.g. in [26, Sect. 3])—see [19, Remark 2.10] for the precise relation. Moreover, recall (from Remark 2.2) that the *last key form* g of the divisorial valuation associated to the curve at infinity on $\bar{X}$ is a polynomial and defines a curve C as in the preceding sentence. Then it can be shown that the polynomials $G_1, \ldots, G_n$ (which induces an embedding of $\bar{X}$ into a weighted projective space) defined in Theorem 5.4 below contains a subsequence $G_{i_1}, \ldots, G_{i_l}$ such that $G_{i_j}|_{\mathbb{C}^2}$ are precisely the *approximate roots* (introduced by Abhyankar and Moh [1]) of g.

Remark 5.3. Let $\vec{\omega} := (\omega_0, \ldots, \omega_{n+1})$ be a key sequence. It is straightforward to see that property 2 implies the following: for each k, $1 \leq k \leq n$, $p_k \omega_k$ can be *uniquely* expressed in the form $p_k \omega_k = \beta_{k,0}\omega_0 + \beta_{k,1}\omega_1 + \cdots + \beta_{k,k-1}\omega_{k-1}$, where $\beta_{k,j}$'s are integers such that $0 \leq \beta_{k,j} < p_j$ for all $j \geq 1$. $\beta_{k,0} \geq 0$. If $\vec{\omega}$ is in additional *algebraic*, then $\beta_{k,0}$'s of the preceding sentence are *non-negative*.

Theorem 5.4 ([18, Proposition 3.5]). *Let $\vec{\omega} := (\omega_0, \ldots, \omega_{n+1})$ be an algebraic key sequence. Let $w, y_0, \ldots, y_{n+1}$ be indeterminates. Pick $\theta_1, \ldots, \theta_n \in \mathbb{C}^*$ and define polynomials $G_1, \ldots, G_n \in \mathbb{C}[w, y_0, \ldots, y_{n+1}]$ as follows:*

$$G_k := w^{p_k \omega_k - \omega_{k+1}} y_{k+1} - \left(y_k^{p_k} - \theta_k \prod_{j=0}^{k-1} y_j^{\beta_{k,j}} \right) \tag{1}$$

where p_k's and $\beta_{k,j}$'s are as in Remark 5.3. Let $\bar{X}_{\vec{\omega},\vec{\theta}}$ be the subvariety of the weighted projective space $\mathbf{WP} := \mathbb{P}^{n+2}(1, \omega_0, \ldots, \omega_{n+1})$ *(with weighted homogeneous coordinates $[w : y_0 : y_1 : \cdots : y_{n+1}]$) defined by $G_1, \ldots, G_n$. Then $\bar{X}_{\vec{\omega},\vec{\theta}}$ is a primitive compactification of $\mathbb{C}^2 \cong \bar{X}_{\vec{\omega},\vec{\theta}} \setminus V(w)$. Conversely, every primitive algebraic compactification of $\mathbb{C}^2$ is of the form $\bar{X}_{\vec{\omega},\vec{\theta}}$ for some $\vec{\omega}, \vec{\theta}$.*

A more or less straightforward corollary is:

Corollary 5.5 ([18, Proposition 3.1, Corollary 3.6]). *Let $\bar{X}$ be a primitive algebraic compactification of $\mathbb{C}^2$. Consider the equations of $\bar{X}$ from Proposition 5.4. Let $C_\infty := \bar{X} \setminus X = \bar{X} \setminus V(w)$ and P_∞ (resp. P_0) be the point on C_∞ with coordinates $[0 : \cdots : 0 : 1]$ (resp. $[0 : 1 : \bar{\theta}_1 : \cdots : \bar{\theta}_n : 0]$), where $\bar{\theta}_k$ is an p_k-th root of θ_k, $1 \le k \le n$). Then*

(1) $\bar{X} \setminus \{P_0, P_\infty\}$ is non-singular.
(2) If $\bar{X}$ is not a weighted projective space, then P_∞ is a singular point of $\bar{X}$.
(3) Let $\tilde{\omega} := \gcd(\omega_0, \ldots, \omega_n)$. Then P_0 is a cyclic quotient singularity of type $\frac{1}{\tilde{\omega}}(1, \omega_{n+1})$.
(4) $C_\infty \setminus P_\infty \cong \mathbb{C}$. In particular, C_∞ is non-singular off P_∞.
(5) Let S be the subsemigroup of $\mathbb{Z}^2$ generated by $\{(\omega_k, 0) : 0 \le k \le n\} \cup \{(0, \omega_{n+1})\}$. Then $C_\infty \cong \operatorname{Proj} \mathbb{C}[S]$, where $\mathbb{C}[S]$ is the semigroup algebra generated by S, and the grading in $\mathbb{C}[S]$ is induced by the sum of coordinates of elements in S.
(6) Let $\tilde{S} := \mathbb{Z}_{\ge 0}\langle p_{n+1}\omega_{n+1}\rangle \cap \mathbb{Z}_{\ge 0}\langle \omega_0, \ldots, \omega_n\rangle$. Then $\mathbb{C}[C_\infty \setminus P_0] \cong \mathbb{C}[\tilde{S}]$. In particular, C_∞ has at worst a (non-normal) toric singularity *at P_∞.*

Let $\bar{X}_{\vec{\omega},\vec{\theta}}$ be an algebraic primitive compactification of $\mathbb{C}^2$. We can compute the canonical divisor of $\bar{X}_{\vec{\omega},\vec{\theta}}$ in terms of $\vec{\omega}$:

Theorem 5.6 ([18, Theorem 4.1]). *Let $p_1, \ldots, p_{n+1}$ be as in the definition of algebraic key sequences. Then the canonical divisor of $\bar{X}_{\vec{\omega},\vec{\theta}}$ is*

$$K_{\bar{X}_{\vec{\omega},\vec{\theta}}} = -\left(\omega_0 + \omega_{n+1} + 1 - \sum_{k=1}^{n}(p_k - 1)\omega_k\right)[C_\infty], \tag{2}$$

where $[C_\infty]$ is the Weil divisor corresponding to C_∞. Moreover, the index *of $\bar{X}_{\vec{\omega},\vec{\theta}}$ (i.e., the smallest positive integer m such that $mK_{\bar{X}_{\vec{\omega},\vec{\theta}}}$ is Cartier) is*

$$\operatorname{ind}(K_{\bar{X}_{\vec{\omega},\vec{\theta}}}) = \min\left\{m \in \mathbb{Z}_{\ge 0} : m\left(\omega_0 + \omega_{n+1} + 1 - \sum_{k=1}^{n}(p_k - 1)\omega_k\right) \in \mathbb{Z}p_{n+1} \cap \mathbb{Z}\omega_{n+1}\right\}. \tag{3}$$

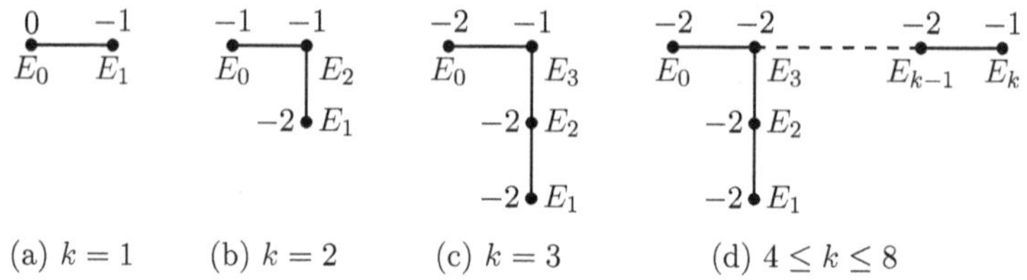

Fig. 2 Dual graph of curves at infinity on Y_k (from Definition 5.9)

5.2 *Special Types of Primitive Algebraic Compactifications*

Straightforward applications of Theorem 5.6 yield the following characterizations of primitive algebraic compactifications of $\mathbb{C}^2$ which have only rational or elliptic singularities, and those which are Gorenstein. For these results, let $\bar{X}_{\vec{\omega},\vec{\theta}}$ be, as in Theorem 5.4, the primitive algebraic compactification corresponding to an algebraic key sequence $\vec{\omega} := (\omega_0, \ldots, \omega_{n+1})$ and $\vec{\theta} \in (\mathbb{C}^*)^n$.

Corollary 5.7 (Simple singularities, [18, Corollary 4.4]).

(1) $\bar{X}_{\vec{\omega},\vec{\theta}}$ *has only rational singularities iff* $\omega_0 + \omega_{n+1} + 1 - \sum_{k=1}^{n}(p_k - 1)\omega_k > 0$.
(2) $\bar{X}_{\vec{\omega},\vec{\theta}}$ *has only elliptic singularities iff* $0 \geq \omega_0 + \omega_{n+1} + 1 - \sum_{k=1}^{n}(p_k - 1)\omega_k > -\omega_{\min}$, *where* $\omega_{\min} := \min\{\omega_0, \ldots, \omega_{l+1}\}$.

Corollary 5.8 (Gorenstein, [18, Proposition 4.5]). $\bar{X}_{\vec{\omega},\vec{\theta}}$ *is Gorenstein iff the following properties hold:*

(1) p_{n+1} *divides* $\omega_{n+1} + 1$, *and*
(2) ω_{n+1} *divides* $\omega_0 + \omega_{n+1} + 1 - \sum_{k=1}^{n}(p_k - 1)\omega_k$.

In the case that the anti-canonical divisor of $\bar{X}_{\vec{\omega},\vec{\theta}}$ is ample, a deeper examination of conditions 1 and 2 of Corollary 5.8 yields the following result which is originally due to [3, 4]. We will use the following construction:

Definition 5.9. For $1 \leq k \leq 8$, we now describe a procedure to construct a compactification Y_k of $\mathbb{C}^2$ via n successive blow ups from $\mathbb{P}^2$. We will denote by E_k, $1 \leq k \leq 8$, the k-th exceptional divisor on Y_k. Let E_0 be the line at infinity in $\mathbb{P}^2$ and pick a point $O \in E_0$. Let Y_1 be the blow up of $\mathbb{P}^2$ at O, and for $2 \leq k \leq 3$, let Y_k be the blow up of Y_{k-1} at the intersection of the strict transform of E_0 and E_{k-1}. Finally, for $3 \leq k \leq 7$, pick a point O_k on E_k which is not on the strict transform of any E_j, $0 \leq j \leq k-1$, and define Y_{k+1} to be the blow up of Y_k at O_k (Fig. 2).

Corollary 5.10 (Gorenstein plus vanishing geometric genus, [3, Proposition 2], [4, Theorem 6]). *Let* $\bar{X}$ *be a primitive Gorenstein compactification of* $\mathbb{C}^2$. *Then the following are equivalent :*

(i) *the geometric genus* $p_g(\bar{X})$ *of* $\bar{X}$ *is zero,*
(ii) *each singular point of* $\bar{X}$ *is a rational double point, and*
(iii) *the canonical bundle* $K_{\bar{X}}$ *of* $\bar{X}$ *is anti-ample.*

If any of these holds, then one of the following holds:

(1) $\bar{X} \cong \mathbb{P}^2$,
(2) $\bar{X}$ *is the singular quadric hypersurface* $x^2 + y^2 + z^2 = 0$ *in* $\mathbb{P}^3$, *or*
(3) $\bar{X}$ *is obtained from some* Y_k *(from Definition 5.9),* $3 \leq k \leq 8$, *by contracting the strict transforms of all* E_j *for* $0 \leq j < k$.

In particular, if $\bar{X}$ *is singular, then the dual curve for the resolution of singularities of* $\bar{X}$ *is one of the Dynkin diagrams* A_1, $A_1 + A_2$, A_4, E_5, E_6, E_7 *or* E_8 *(with the weight of each vertex being* -2*).*

Miyanishi and Zhang in [16] proved a converse to Corollary 5.10. Recall that a surface S is called *log del Pezzo* if S has only quotient singularities and the anticanonical divisor $-K_S$ is ample.

Theorem 5.11 ([16, Theorem 1]). *Let* S *be a Gorenstein log del Pezzo surface of rank one. Then* S *is a compactification of* $\mathbb{C}^2$ *iff the dual curve for the resolution of singularities of* $\bar{X}$ *is one of the Dynkin diagrams* A_1, $A_1 + A_2$, A_4, E_5, E_6, E_7 *or* E_8.

In the same article Miyanishi and Zhang give a topological characterization of primitive Gorenstein compactification of $\mathbb{C}^2$ with vanishing geometric genus:

Theorem 5.12 ([16, Theorem 2]). *Let* S *be a Gorenstein log del Pezzo surface. Suppose that either* S *is singular or that there are no* (-1)*-curves contained in the smooth locus of* S. *Then* S *is a compactification of* $\mathbb{C}^2$ *iff the smooth locus of* S *is simply connected.*

From Theorem 5.6 and the classification of dual graph of resolution of singularities of primitive compactifications discussed in Sect. 5.3, it is possible to obtain classifications of primitive compactifications with ample anti-canonical divisors and log terminal and log canonical singularities obtained originally by Kojima [12] and Kojima and Takahashi [13]. Both of these classifications consist of explicit lists of dual graphs of resolution of singularities, and we omit their statements. However, they also prove converse results in the spirit of Theorems 5.11 and 5.12.

Theorem 5.13 ([12, Theorem 0.1]). *Let* S *be a log del Pezzo surface of rank one. Assume that the singularity type of* S *is one of the possible choices (listed in* [12, Appendix C]*) for the singularity type of primitive compactifications of* $\mathbb{C}^2$ *with at most quotient singularities. If* $\mathrm{ind}(S) \leq 3$, *then* S *is a primitive compactification of* $\mathbb{C}^2$.

Theorem 5.14 ([13, Theorem 1.2]). *Let* S *be a numerical del Pezzo surface (i.e., the intersection of the anti-canonical divisor of* S *with itself and every irreducible curve on* S *is positive) with at most rational singularities. Assume the singularity type of* S *is one of the possible choices (listed in* [13]*) for the singularity type of primitive numerical del Pezzo compactifications of* $\mathbb{C}^2$ *with rational singularities. Then* S *is a primitive compactification of* $\mathbb{C}^2$.

From a slightly different perspective, Furushima [8] and Ohta [21] studied primitive compactifications of $\mathbb{C}^2$ which are hypersurfaces in $\mathbb{P}^3$. The following was conjectured and proved for $d \leq 4$ by Furushima, and then proved in the general case by Ohta:

Theorem 5.15 ([8, 9, 21]). *Let $\bar{X}_d$ be a minimal compactification of $\mathbb{C}^2$ which is a hypersurface of degree $d \geq 2$ in $\mathbb{P}^3$ and $C_d := \bar{X}_d \setminus \mathbb{C}^2$ be the curve at infinity. Assume $\bar{X}_d$ has a singular point P of multiplicity $d-1$. Then*

(1) P is the unique singular point of $\bar{X}_d$ and the geometric genus of P is $p_g(P) = (d-1)(d-2)(d-3)/6$.
(2) C_d is a line on $\mathbb{P}^3$.
(3) $(\bar{X}_d, C_d) \cong (V_d, L_d)$ (up to a linear change of coordinates), where

$$V_d := \{[z_0 : z_1 : z_2 : z_3] \in \mathbb{P}^3 : z_0^d = z_1^{d-1} z_2 + z_2^{d-1} z_3\}$$
$$L_d := \{z_0 = z_2 = 0\}.$$

5.3 Dual Graphs for the Resolution of Singularities

Let $\vec{\omega} := (\omega_0, \dots, \omega_{n+1})$ be a key sequence. Then to every $\vec{\theta} := (\theta_1, \dots, \theta_n) \in (\mathbb{C}^*)^n$, we can associate a primitive compactification $\bar{X}_{\vec{\omega},\vec{\theta}}$ of $\mathbb{C}^2$. Moreover, $\bar{X}_{\vec{\omega},\vec{\theta}}$ is algebraic iff $\vec{\omega}$ is an *algebraic* key sequence, and the correspondence $(\vec{\omega}, \vec{\theta}) \mapsto \bar{X}_{\vec{\omega},\vec{\theta}}$ is given by Theorem 5.4. The correspondence in the general case is treated in [17]; in our notation it can be described as follows: define $G_1, \dots, G_n \in A := \mathbb{C}[w, y_0, y_0^{-1}, y_1, \dots, y_{n+1}]$ as in Theorem 5.4 (if $\vec{\omega}$ is not algebraic, then at least one of the G_k's will not be a polynomial). Let I be the ideal in A generated by $w-1, G_1, \dots, G_n$. Then $A/I \cong \mathbb{C}[x, x^{-1}, y]$ via the map $y_0 \mapsto x, y_1 \mapsto y$. Let $f_k \in \mathbb{C}[x, x^{-1}, y]$ be the image of G_k, $1 \leq k \leq n$. Consider the family of curves $C_\xi \subseteq \mathbb{C}^2 \setminus V(x)$, $\xi \in \mathbb{C}$, defined by $f_n^{\omega_0} = \xi x^{\omega_{n+1}}$. Then $\bar{X}_{\vec{\omega},\vec{\theta}}$ is the unique primitive compactification of $\mathbb{C}^2 = \operatorname{Spec} \mathbb{C}[x, y]$ which *separates (some branches of) the curves C_ξ at infinity*, i.e., for generic ξ, the closure of the curve C_ξ in $\bar{X}_{\vec{\omega},\vec{\theta}}$ intersects generic points of the curve at infinity. It follows from the results of [17, Corollary 4.11] that every primitive compactification of $\mathbb{C}^2$ is of the form $\bar{X}_{\vec{\omega},\vec{\theta}}$ for some appropriate $\vec{\omega}$ and $\vec{\theta}$.

Remark 5.16 (A valuation theoretic characterization of $\bar{X}_{\vec{\omega},\vec{\theta}}$). Let $f_1, \dots, f_n$ be as in the preceding paragraph. Then $\bar{X}_{\vec{\omega},\vec{\theta}}$ is the unique primitive compactification of $\mathbb{C}^2 = \operatorname{Spec} \mathbb{C}[x, y]$ such that the *key forms* (see Remark 2.2) of the valuation on $\mathbb{C}[x, y]$ corresponding to the curve at infinity on $\bar{X}_{\vec{\omega},\vec{\theta}}$ are $x, y, f_1, \dots, f_n$.

The dual graph of the minimal resolution of singularities of $\bar{X}_{\vec{\omega},\vec{\theta}}$ depends only on the *essential subsequence* (Definition 5.1) $\vec{\omega}_e$ of $\vec{\omega}$. The precise description of

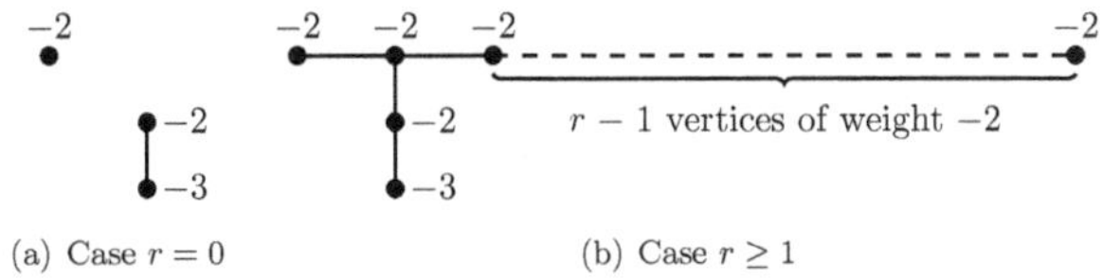

Fig. 3 Dual graph of minimal resolution of singularities of $\bar{X}_{i,r}$ from Example 2.3

Table 1 Some key sequences $\vec{\omega}$ and corresponding $\vec{d}$, $\vec{p}$

$(\omega_0, \ldots, \omega_{n+1})$	$(d_0, \ldots, d_{n+1})$	$(p_1, \ldots, p_{n+1})$	$(p_1\omega_1, \ldots, p_n\omega_n)$
(2,5)	(2,1)	(2)	∅
(2,5,10-r)	(2,1,1)	(2,1)	(10)
(4,10,3,2)	(4,2,1,1)	(2,2,1)	(20,6)

the dual graph in terms of $\vec{\omega}_e$ is a bit technical and it essentially corresponds to the resolution of singularities of a point at infinity on (the closure of) the curve C_ξ from the preceding paragraph for generic ξ—we refer to [17, Appendix] for details. Rather we now state the characterization from [19] of those dual graphs which appear only for algebraic, only for non-algebraic, and for both algebraic and non-algebraic compactifications.

Theorem 5.17 ([19, Theorem 2.8]). *Let $\vec{\omega} := (\omega_0, \ldots, \omega_{n+1})$ be an* essential *key sequence and let $\Gamma_{\vec{\omega}}$ be the dual graph for the minimal resolution of singularities for some (and therefore, every) primitive compactification $\bar{X}_{\vec{\omega}',\vec{\theta}}$ of $\mathbb{C}^2$ where $\vec{\omega}'$ is a key sequence with essential subsequence $\vec{\omega}$. Then*

(1) There exists a primitive algebraic *compactification $\bar{X}$ of $\mathbb{C}^2$ such that the dual graph for the minimal resolution of singularities of $\bar{X}$ is $\Gamma_{\vec{\omega}}$ iff $\vec{\omega}$ is an* algebraic *key sequence.*
(2) There exists a primitive non-algebraic *compactification $\bar{X}$ of $\mathbb{C}^2$ such that the dual graph for the minimal resolution of singularities of $\bar{X}$ is $\Gamma_{\vec{\omega}}$ iff*

(a) either $\vec{\omega}$ is not algebraic, *or*
(b) $\bigcup_{1 \le k \le n} \{\alpha \in \mathbb{Z}\langle\omega_0, \ldots, \omega_k\rangle \setminus \mathbb{Z}_{\ge 0}\langle\omega_0, \ldots, \omega_k\rangle : \omega_{k+1} < \alpha < p_k\omega_k\} \neq \emptyset$.

Example 5.18 ([19, Corollary 2.13, Example 2.15]). The dual graph of the minimal resolution of singularities of $\bar{X}_{i,r}$ from Example 2.3 (see Fig. 3) corresponds to the essential key sequence $(2, 5)$ for $r = 0$ and $(2, 5, 10-r)$ for $1 \le r \le 9$. A glance at Table 1 shows that $(2, 5)$ and $(2, 5, 10-r)$, $1 \le r \le 9$, are algebraic key sequences, so that Theorem 5.17 implies that each of these sequences corresponds to some algebraic primitive compactifications. Now note that for $\vec{\omega} := (2, 5, 10-r)$,

$$\mathbb{Z}\langle\omega_0, \omega_1\rangle \setminus \mathbb{Z}_{\ge 0}\langle\omega_0, \omega_1\rangle = \mathbb{Z}\langle 2, 5\rangle \setminus \mathbb{Z}_{\ge 0}\langle 2, 5\rangle = \mathbb{Z} \setminus \mathbb{Z}_{\ge 0}\langle 2, 5\rangle = \{1, 3\}.$$

Since for $r = 8, 9$, we have $\omega_2 = 10 - r < 3$, Theorem 5.17 implies that in this case $\vec{\omega}$ also corresponds to some non-algebraic primitive compactifications. In summary, $(2, 5)$ and $(2, 5, 10-r)$, $1 \le r \le 7$, correspond to *only algebraic*

Fig. 4 $\Gamma_{(4,10,3,2)}$

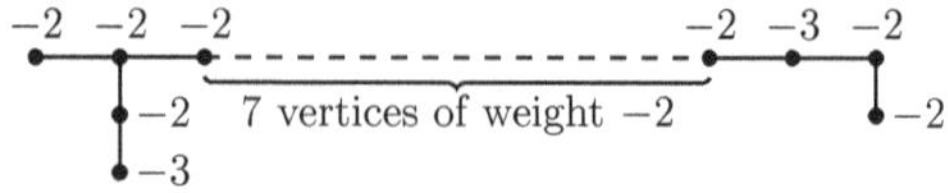

primitive compactifications, and $(2, 5, 10 - r)$, $8 \leq r \leq 9$, corresponds to *both* algebraic and non-algebraic compactifications, as it was shown in Example 2.3.

On the other hand, for $\vec{\omega} = (4, 10, 3, 2)$, Table 1 shows that $p_2\omega_2 = 6 \notin \mathbb{Z}_{\geq 0}\langle 4, 10\rangle = \mathbb{Z}_{\geq 0}\langle \omega_0, \omega_1\rangle$, so that $\vec{\omega}$ is *not* an algebraic key sequence. Consequently Theorem 5.17 implies that $\Gamma_{(4,10,3,2)}$ corresponds to *only non-algebraic* primitive compactifications (see Fig. 4).

6 Groups of Automorphism and Moduli Spaces of Primitive Compactifications

In [18, Sect. 5] the groups of automorphisms and moduli spaces of primitive compactifications have been precisely worked out. Here we omit the precise statements and content ourselves with the description of some special cases.

Definition 6.1. A key sequence $\vec{\omega} = (\omega_0, \ldots, \omega_{n+1})$ is in the *normal form* iff

(1) either $n = 0$, or
(2) ω_0 does not divide ω_1 and $\omega_1/\omega_0 > 1$.

Theorem 6.2 (cf. [18, Corollary 5.4]). *Let $\bar{X}$ be a primitive compactification of $\mathbb{C}^2$. Then $\bar{X} \cong \bar{X}_{\vec{\omega},\vec{\theta}}$ for some key sequence $\vec{\omega} := (\omega_0, \ldots, \omega_{n+1})$ in the normal form (and some appropriate $\vec{\theta}$). Moreover,*

(1) $n = 0$ iff $\bar{X}$ is isomorphic to some weighted projective space $\mathbb{P}^2(1, p, q)$.
(2) If $\bar{X} \ncong \mathbb{P}^2(1, 1, q)$ for any $q \geq 1$, then there are coordinates (x, y) on $\mathbb{C}^2$ such that for every automorphism F of $\bar{X}$, $F|_{\mathbb{C}^2}$ is of the form $(x, y) \mapsto (ax + b, a'y + f(x))$ for some $a, a', b \in \mathbb{C}$ and $f \in \mathbb{C}[x]$. Moreover, if $n > 1$ then a and a' are some roots of unity and $b = f = 0$.

Theorem 6.3 (cf. [18, Corollary 5.8]). *Let $\vec{\omega} := (\omega_0, \ldots, \omega_{n+1})$ be an essential key sequence in the normal form and $\mathcal{X}_{\vec{\omega}}$ (resp. $\mathcal{X}^{alg}_{\vec{\omega}}$) be the space of normal analytic (resp. algebraic) surfaces which are isomorphic to $\bar{X}_{\vec{\omega}',\vec{\theta}}$ for some key sequence $\vec{\omega}'$ with essential subsequence $\vec{\omega}$ and some $\vec{\theta}$. Then*

(1) $\mathcal{X}_{\vec{\omega}}$ is of the form $\left((\mathbb{C}^*)^k \times \mathbb{C}^l\right)/G$ for some subgroup G of $\mathbb{C}^*$.
(2) $\mathcal{X}^{alg}_{\vec{\omega}}$ is either empty (in the case that $\vec{\omega}$ is not algebraic), or a closed subset of $\mathcal{X}_{\vec{\omega}}$ of the form $\left((\mathbb{C}^*)^k \times \mathbb{C}^{l'}\right)/G$ for some $l' \leq l$.

Remark 6.4. The correspondence of Theorem 2.1 between primitive algebraic compactifications with $\mathbb{C}^2$ and planar curves with one place at infinity extends to their moduli spaces. The moduli spaces of curves with one place at infinity are of the form $(\mathbb{C}^*)^k \times \mathbb{C}^l$ for some $k, l \geq 0$ [7, Corollary 1]. The extra complexity (i.e., action by the group G from Theorem 6.3) in the structure of the moduli spaces of primitive algebraic compactifications comes from the action of their groups of automorphisms.

(The precise version in [18, Corollary 5.8] of) Theorem 6.2 yields a classification of all normal analytic surfaces of Picard rank 1 which admits a $\mathbb{G}_a^2$ action [18, Corollary 6.2].

References

1. S.S. Abhyankar, T.T. Moh, Newton-Puiseux expansion and generalized Tschirnhausen transformation, I, II. J. Reine Angew. Math. **260**, 47–83 , (1973); J. Reine Angew. Math. **261**, 29–54 (1973)
2. L. Brenton, A note on compactifications of $\mathbf{C}^2$. Math. Ann. **206**, 303–310 (1973)
3. L. Brenton, On singular complex surfaces with negative canonical bundle, with applications to singular compactifications of $\mathbf{C}^2$ and to 3-dimensional rational singularities. Math. Ann. **248**(2), 117–124 (1980)
4. L. Brenton, D. Drucker, G.C.E. Prins, Graph theoretic techniques in algebraic geometry, I, in *The Extended Dynkin Diagram* $\bar{E}_8$ *and Minimal Singular Compactifications of* $\mathbf{C}^2$. Annals of Mathematics Studies, vol. 100 (Princeton University Press, Princeton, 1981), pp. 47–63
5. C. Favre, M. Jonsson, in *The Valuative Tree*. Lecture Notes in Mathematics, vol. 1853 (Springer, Berlin, 2004)
6. H. Flenner, M. Zaidenberg, Rational curves and rational singularities. Math. Z. **244**(3), 549–575 (2003)
7. M. Fujimoto, M. Suzuki, Construction of affine plane curves with one place at infinity. Osaka J. Math. **39**(4), 1005–1027 (2002)
8. M. Furushima, On minimal compactifications of $\mathbf{C}^2$. Math. Nachr. **186**, 115–129 (1997)
9. M. Furushima, Errata to the paper: On the minimal compactifications of $\mathbf{C}^2$ [Math. Nachr. **186**, 115–129 (1997); MR1461216 (98e:32049)]. Math. Nachr. **193**, 255 (1998)
10. H. Grauert, Über Modifikationen und exzeptionelle analytische Mengen. Math. Ann. **146**, 331–368 (1962)
11. F. Hirzebruch, Some problems on differentiable and complex manifolds. Ann. Math. **60**(2), 213–236 (1954)
12. H. Kojima, Minimal singular compactifications of the affine plane. Nihonkai Math. J. **12**(2), 165–195 (2001)
13. H. Kojima, T. Takahashi, Notes on minimal compactifications of the affine plane. Ann. Mat. Pura Appl. (4) **188**(1), 153–169 (2009)
14. H.B. Laufer, On minimally elliptic singularities. Am. J. Math. **99**(6), 1257–1295 (1977)
15. J. Lipman, Rational singularities, with applications to algebraic surfaces and unique factorization. Inst. Hautes Études Sci. Publ. Math. **36**, 195–279 (1969)
16. M. Miyanishi, D.-Q. Zhang, Gorenstein log del Pezzo surfaces of rank one. J. Algebra **118**(1), 63–84 (1988)
17. P. Mondal, Analytic compactifications of $\mathbb{C}^2$. Part I: Curvettes at infinity (2011), http://arxiv.org/abs/1110.6905

18. P. Mondal, Analytic compactifications of $\mathbb{C}^2$. II: One irreducible curve at infinity (2013), http://arxiv.org/abs/1307.5577
19. P. Mondal, An effective criterion for algebraic contractibility of rational curves (2013), http://arxiv.org/abs/1301.0126
20. J.A. Morrow, Compactifications of $\mathbf{C}^2$. Bull. Am. Math. Soc. **78**, 813–816 (1972)
21. T. Ohta, Normal hypersurfaces as a compactification of $\mathbb{C}^2$. Kyushu J. Math. **55**(1), 165–181 (2001)
22. K. Palka, Classification of singular $\mathbb{Q}$-homology planes. I: Structures and singularities. Israel J. Math. **195**, 37–69 (2012)
23. R. Remmert, T. van de Ven, Zwei Sätze über die komplex-projektive Ebene. Nieuw. Arch. Wisk. **8**(3), 147–157 (1960)
24. A. Sathaye, On planar curves. Am. J. Math. **99**(5), 1105–1135 (1977)
25. S. Schröer, On contractible curves on normal surfaces. J. Reine Angew. Math. **524**, 1–15 (2000)
26. M. Suzuki, Affine plane curves with one place at infinity. Ann. Inst. Fourier (Grenoble) **49**(2), 375–404 (1999)

Jordan Groups and Automorphism Groups of Algebraic Varieties

Vladimir L. Popov

Abstract The first section of this paper is focused on Jordan groups in abstract setting, the second on that in the settings of automorphisms groups and groups of birational self-maps of algebraic varieties. The appendix contains formulations of some open problems and the relevant comments.

MSC 2010: 20E07, 14E07

This is the expanded version of my talk, based on [43, Sect. 2], at the workshop *Groups of Automorphisms in Birational and Affine Geometry*, October 29 to November 3, 2012, Levico Terme, Italy. The appendix is the expanded version of my notes on open problems posted on the site of this workshop [46].

Below k is an algebraically closed field of characteristic zero. Variety means algebraic variety over k in the sense of Serre (so algebraic group means algebraic group over k). We use without explanation standard notation and conventions of [5,48,61]. In particular, $k(X)$ denotes the field of rational functions of an irreducible variety X. $\mathrm{Bir}(X)$ denotes the group of birational self-maps of an irreducible variety X. Recall that if X is the affine n-dimensional space $\mathbf{A}^n$, then $\mathrm{Bir}(X)$ is called the *Cremona group over k of rank n*; we denote it by Cr_n (cf. [44,45]).

I am indebted to the referee for careful reading and remarks.

V.L. Popov (✉)
Steklov Mathematical Institute, Russian Academy of Sciences, Gubkina 8, Moscow 119991, Russia

National Research University Higher School of Economics, Myasnitskaya 20, Moscow 101000, Russia
e-mail: popovvl@mi.ras.ru

I. Cheltsov et al. (eds.), *Automorphisms in Birational and Affine Geometry*, Springer Proceedings in Mathematics & Statistics 79,
DOI 10.1007/978-3-319-05681-4_11,

1 Jordan Groups

1.1 Main Definition

The notion of Jordan group was introduced in [43]:

Definition 1 ([43, Def. 2.1]). A group G is called a *Jordan group* if there exists a positive integer d, depending on G only, such that every finite subgroup K of G contains a normal abelian subgroup whose index in K is at most d. The minimal such d is called the *Jordan constant of* G and is denoted by J_G.

Informally, this means that all finite subgroups of G are "almost" abelian in the sense that they are extensions of abelian groups by finite groups taken from a finite list.

Actually, one obtains the same class of groups if the assumption of normality in Definition 1 is dropped. Indeed, for any group P containing a subgroup Q of finite index, there is a normal subgroup N of P such that $[P : N] \leqslant [P : Q]!$ and $N \subseteq Q$ (see, e.g., [26, Exer. 12 to Chap. I]).

1.2 Examples

1.2.1 Jordan's Theorem

The first example that led to Definition 1 justifies the coined name. It is given by the classical Jordan's theorem [23] (see, e.g., [13, Sect. 36] for a modern exposition). In terms of Definition 1 the latter can be reformulated as follows:

Theorem 1 ([23]). *The group* $\mathbf{GL}_n(k)$ *is Jordan for every* n.

Since the symmetric group Sym_{n+1} admits a faithful n-dimensional representation and the alternating group Alt_{n+1} is the only non-identity proper normal subgroup of Sym_{n+1} for $n \geqslant 2$, $n \neq 3$, Definition 1 yields the lower bound

$$(n+1)! \leqslant J_{\mathbf{GL}_n(k)} \quad \text{for } n \geqslant 4. \tag{1}$$

Frobenius, Schur, and Blichfeldt initiated exploration of the upper bounds for $J_{\mathbf{GL}_n(k)}$. In 2007, using the classification of finite simple groups, M. J. Collins [12] gave optimal upper bounds and thereby found the precise values of $J_{\mathbf{GL}_n(k)}$ for all n. In particular, in [12] is proved that

(a) the equality in (1) holds for all $n \geqslant 71$ and $n = 63, 65, 67, 69$;
(b) $J_{\mathbf{GL}_n(k)} = 60^r r!$ if $n = 2r$ or $2r + 1$ and either $20 \leqslant n \leqslant 62$ or $n = 64, 66, 68, 70$;
(c) $J_{\mathbf{GL}_n(k)} = 60,\ 360,\ 25{,}920,\ 25{,}920,\ 6{,}531{,}840$ resp., for $n = 2, 3, 4, 5, 6$.

The values of $J_{\mathbf{GL}_n(k)}$ for $7 \leqslant n \leqslant 19$ see in [12].

1.2.2 Affine Algebraic Groups

Since any subgroup of a Jordan groups is Jordan, Theorem 1 yields

Corollary 1. *Every linear group is Jordan.*

Since every affine algebraic group is linear [61, 2.3.7], this, in turn, yields the following generalization of Theorem 1:

Theorem 2. *Every affine algebraic group is Jordan.*

1.2.3 Nonlinear Jordan Groups

Are there nonlinear Jordan groups? The next example, together with Theorem 1, convinced me that Definition 1 singles out an interesting class of groups and therefore deserves to be introduced.

Example 1. By Serre [59, Thm. 5.3], [58, Thm. 3.1], the planar Cremona group Cr_2 is Jordan. On the other hand, by [9, Prop. 5.1] (see also [10, Prop. 2.2]), Cr_2 is not linear. Note that in [59, Thm. 5.3] one also finds a "multiplicative" upper bound for J_{Cr_2}: as is specified there, a crude computation shows that every finite subgroup G of Cr_2 contains a normal abelian subgroup A of rank $\leqslant 2$ with $[G : A]$ dividing $2^{10} \cdot 3^4 \cdot 5^2 \cdot 7$ (it is also mentioned that the exponents of 2 and 3 can be somewhat lowered, but those of 5 and 7 cannot). □

Example 2. Let F_d be a free group with d free generators and let F_d^n be its normal subgroup generated by the nth powers of all elements. As is known (see, e.g., [1, Thm. 2]), the group $B(d,n) := F_d/F_d^n$ is infinite for $d \geqslant 2$ and odd $n \geqslant 665$ (recently S. Adian announced in [2] that 665 may be replaced by 100). On the other hand, by I. Schur, finitely generated linear torsion groups are finite (see, e.g., [13, Thm. 36.2]). Hence infinite $B(d,n)$ is nonlinear. On the other hand, for $d \geqslant 2$ and odd $n \geqslant 665$, every finite subgroup in $B(d,n)$ is cyclic (see [1, Thm. 8]); hence $B(d,n)$ is Jordan and $J_{B(d,n)} = 1$. □

Example 3. Let p be a positive prime integer and let $T(p)$ be a Tarski monster group, i.e., an infinite group, such that every its proper subgroup is a cyclic group of order p. By Olshanskii [40], for big p (e.g., $\geqslant 10^{75}$), such a group exists. $T(p)$ is necessarily simple and finitely generated (and, in fact, generated by every two non-commuting elements). By the same reason as in Example 2, $T(p)$ is not linear. The definitions imply that T_p is Jordan and $J_{T(p)} = 1$. (I thank A. Yu. Ol'shanskiĭ who drew in [41] my attention to this example.) □

1.2.4 Diffeomorphism Groups of Smooth Topological Manifolds

Let M be a compact connected n-dimensional smooth topological manifold. Assume that M admits an unramified covering $\tilde{M} \to M$ such that $H^1(\tilde{M}, \mathbf{Z})$ contains the cohomology classes $\alpha_1, \dots, \alpha_n$ satisfying $\alpha_1 \cup \dots \cup \alpha_n \neq 0$. Then, by Mundet i Riera [35, Thm. 1.4(1)], the group Diff(M) is Jordan. This result is applicable to $\mathbf{T}^n$, the product of n circles, and, more generally, to the connected sum $N \sharp \mathbf{T}^n$, where N is any compact connected orientable smooth topological manifold. (I thank I. Mundet i Riera who drew in [36] my attention to [17, 35, 51].)

Recently, in [38] has been proven that if N is a compact smooth topological manifold whose integral cohomology is torsion free and supported in even degrees, then the group Diff(N) is Jordan.[1]

1.2.5 Non-Jordan Groups

Are there non-Jordan groups?

Example 4. The group Sym_∞ of all permutations of $\mathbf{Z}$ contains the alternating group Alt_n for every n. Hence Sym_∞ is non-Jordan because Alt_n is simple for $n \geqslant 5$ and $|\mathrm{Alt}_n| = n!/2 \xrightarrow[n\to\infty]{} \infty$. □

Using Example 4 one obtains a finitely generated non-Jordan group:

Example 5. Let $\mathcal{N}$ be the subgroup of Sym_∞ generated by the transposition $\sigma := (1, 2)$ and the "translation" δ defined by the condition

$$\delta(i) = i + 1 \quad \text{for every } i \in \mathbf{Z}.$$

Then $\delta^m \sigma \delta^{-m}$ is the transposition $(m + 1, m + 2)$ for every m. Since the set of transpositions $(1, 2), (2, 3), \ldots, (n - 1, n)$ generates the symmetric group Sym_n, this shows that $\mathcal{N}$ contains Alt_n for every n; whence $\mathcal{N}$ is non-Jordan. □

One can show that $\mathcal{N}$ is not finitely presented. Here is an example of a finitely presented non-Jordan group which is also simple.

Example 6. Consider Richard J. Thompson's group V, see [6, Sect. 6]. It is finitely presented, simple and contains a subgroup isomorphic with Sym_n for every $n \geqslant 2$. The latter implies, as in Example 4, that V is non-Jordan. (I thank Vic. Kulikov who drew my attention to this example.) □

1.3 General Properties

1.3.1 Subgroups, Quotient Groups, and Products

Exploring whether a group is Jordan or not leads to the questions on the connections between Jordaness of a group, its subgroup, and its quotient group.

---

[1] It is proved in the recent preprint I. Mundet i Riera, Finite group actions on spheres, Euclidean spaces, and compact manifolds with $\chi \neq 0$ (March 2014) [arXiv:1403.0383] that if M is a sphere, an Euclidean space $\mathbb{R}^n$, or a compact manifold (possibly with boundary) with nonzero Euler characteristic, then Diff(M) is Jordan.

Theorem 3 ([43, Lemmas 2.6, 2.7, 2.8]).

(1) *Let H be a subgroup of a group G.*

  (i) *If G is Jordan, then H is Jordan and $J_H \leqslant J_G$.*
  (ii) *If G is Jordan and H is normal in G, then G/H is Jordan and $J_{G/H} \leqslant J_G$ in either of the cases:*

    (a) *H is finite;*
    (b) *the extension $1 \to H \to G \to G/H \to 1$ splits.*

  (iii) *If H is torsion-free, normal in G, and G/H is Jordan, then G is Jordan and $J_G \leqslant J_{G/H}$.*

(2) *Let G_1 and G_2 be two groups. Then $G_1 \times G_2$ is Jordan if and only if G_1 and G_2 are. In this case, $J_{G_i} \leqslant J_{G_1 \times G_2} \leqslant J_{G_1} J_{G_2}$ for every i.*

Proof. (1)(i). This follows from Definition 1.

If H is normal in G, let $\pi\colon G \to G/H$ be the natural projection.

(1)(ii)(a). Let F be a finite subgroup of G/H. Since H is finite, $\pi^{-1}(F)$ is finite. Since G is Jordan, $\pi^{-1}(F)$ contains a normal abelian subgroup A whose index is at most J_G. Hence $\pi(A)$ is a normal abelian subgroup of F whose index in F is at most J_G.

(1)(ii)(b). By the condition, there is a subgroup S in G such that $\pi|_S : S \to G/H$ is an isomorphism; whence the claim by (1)(i).

(1)(iii). Let F be a finite subgroup of G. Since H is torsion free, $F \cap H = \{1\}$; whence $\pi|_F\colon S \to \pi(F)$ is an isomorphism. Therefore, as G/H is Jordan, F contains a normal abelian subgroup whose index in F is at most $J_{G/H}$.

(2) If $G := G_1 \times G_2$ is Jordan, then (1)(i) implies that G_1 and G_2 are Jordan and $J_{G_i} \leqslant J_G$ for every i. Conversely, let G_1 and G_2 be Jordan. Let $\pi_i\colon G \to G_i$ be the natural projection. Take a finite subgroup F of G. Then $F_i := \pi_i(F)$ contains an abelian normal subgroup A_i such that

$$[F_i : A_i] \leqslant J_{G_i}. \tag{2}$$

The subgroup $\tilde{A}_i := \pi_i^{-1}(A_i) \cap F$ is normal in F and $F/\tilde{A}_i$ is isomorphic to F_i/A_i. From (2) we then conclude that

$$[F : \tilde{A}_i] \leqslant J_{G_i}. \tag{3}$$

Since $A := \tilde{A}_1 \cap \tilde{A}_2$ is the kernel of the diagonal homomorphism

$$F \longrightarrow F/\tilde{A}_1 \times F/\tilde{A}_2$$

determined by the canonical projection $F \to F/\tilde{A}_i$, we infer from (3) that

$$[F : A] = |F/A| \leqslant |F/\tilde{A}_1 \times F/\tilde{A}_2| = |F_1/A_1||F_2/A_2| \leqslant J_{G_1} J_{G_2}. \tag{4}$$

By construction, $A \subseteq A_1 \times A_2$ and A_i is abelian. Hence A is abelian as well. Since A is normal in F, the claim then follows from (4). □

Theorem 4. *Let H be a normal subgroup of a group G. If H and G/H are Jordan, then any set of pairwise nonisomorphic simple nonabelian finite subgroups of G is finite.*

Proof. Since up to isomorphism there are only finitely many finite groups of a fixed order, Definition 1 implies that any set of pairwise nonisomorphic simple nonabelian finite subgroups of a given Jordan group is finite. This implies the claim because simplicity of a finite subgroup S of G yields that either $S \subseteq H$ or the canonical projection $G \to G/H$ embeds S in G/H. □

1.3.2 Counterexample

For a normal subgroup H of G, it is not true, in general, that G is Jordan if H and G/H are.

Example 7. For every integer $n > 0$ fix a finite group G_n with the properties:

(i) G_n has an abelian normal subgroup H_n such that G_n/H_n is abelian;
(ii) there is a subgroup Q_n of G_n such that the index in Q_n of every abelian subgroup of Q_n is greater or equal than n.

Such a G_n exists, see below. Now take $G := \prod_n G_n$ and $H := \prod_n H_n$. Then H and G/H are abelian by (i), hence Jordan, but G is not Jordan by (ii).

The following construction from [62, Sect. 3] proves the existence of such a G_n. Let K be a finite commutative group of order n written additively and let $\hat{K} := \mathrm{Hom}(K, k^*)$ be the group of characters of K written multiplicatively. The formula

$$(\alpha, g, \ell)(\alpha', g', \ell') := (\alpha\alpha'\ell'(g), g + g', \ell\ell') \tag{5}$$

endows the set $k^* \times K \times \hat{K}$ with the group structure. Denote by G_K the obtained group. It is embedded in the exact sequence of groups

$$\{1\} \to k^* \xrightarrow{\iota} G_K \xrightarrow{\pi} K \times \hat{K} \to \{(0,1)\},$$
$$\text{where}\quad \iota(\alpha) := (\alpha, 0, 1) \quad\text{and}\quad \pi((\alpha, g, \ell)) := (g, \ell).$$

Thus, if one takes $G_n := G_K$ and $H_n := \iota(k^*)$, then property (i) holds. Let μ_n be the subgroup of all nth roots of unity in k^*. From (5) and $|K| = n$ we infer that the subset $Q_K := \mu_n \times K \times \hat{K}$ is a subgroup of G_K. In [62, Sect. 3] is proved that for $Q_n = Q_K$ property (ii) holds. □

1.3.3 Bounded Groups

However, under certain conditions, G is Jordan if and only if H and G/H are. An example of such a condition is given in Theorem 5 below; it is based on Definition 2 below introduced in [43].

Given a group G, put

$$b_G := \sup_F |F|,$$

where F runs over all finite subgroups of G.

Definition 2 ([43, Def. 2.9]). A group G is called *bounded* if $b_G \neq \infty$.

Example 8. Finite groups and torsion free groups are bounded. □

Example 9. It is immediate from Definition 2 that every extension of a bounded group by bounded is bounded. □

Example 10. By the classical Minkowski's theorem $\mathbf{GL}_n(\mathbf{Z})$ is bounded (see, e.g., [19, Thm. 39.4]). Since every finite subgroup of $\mathbf{GL}_n(\mathbf{Q})$ is conjugate to a subgroup of $\mathbf{GL}_n(\mathbf{Z})$ (see, e.g., [13, Thm. 73.5]), this implies that $\mathbf{GL}_n(\mathbf{Q})$ is bounded and $b_{\mathbf{GL}_n(\mathbf{Q})} = b_{\mathbf{GL}_n(\mathbf{Z})}$. H. Minkowski and I. Schur obtained the following upper bound for $b_{\mathbf{GL}_n(\mathbf{Z})}$, see, e.g., [19, Sect. 39]. Let $\mathscr{P}(n)$ be the set of all primes $p \in \mathbf{N}$ such that $[n/(p-1)] > 0$. Then

$$b_{\mathbf{GL}_n(\mathbf{Z})} \leqslant \prod_{p \in \mathscr{P}(n)} p^{d_p}, \quad \text{where} \quad d_p = \sum_{i=0}^{\infty} \left[\frac{n}{p^i(p-1)} \right]. \tag{6}$$

In particular, the right-hand side of the inequality in (6) is

$$2,\ 24,\ 48,\ 5{,}760,\ 11{,}520,\ 2{,}903{,}040 \quad \text{resp., for } n = 1, 2, 3, 4, 5, 6. \quad \square$$

Example 11. Maintain the notation and assumption of Sect. 1.2.4. If $\chi(M) \neq 0$, then by Mundet i Riera [35, Thm. 1.4(2)], the group Diff(M) is bounded. Further information on smooth manifolds with bounded diffeomorphism groups is contained in [51]. □

Example 12. Every bounded group G is Jordan with $J_G \leqslant b_G$, and there are non-bounded Jordan groups (e.g., $\mathbf{GL}_n(k)$). □

Theorem 5 ([43, Lemma 2.11]). *Let H be a normal subgroup of a group G such that G/H is bounded. Then G is Jordan if and only if H is Jordan, and in this case*

$$J_G \leqslant b_{G/H} J_H^{b_{G/H}}.$$

Proof. A proof is needed only for the sufficiency. So let H be Jordan and let F be a finite subgroup of G. By Definition 1

$$L := F \cap H \tag{7}$$

contains a normal abelian subgroup A such that

$$[L : A] \leqslant J_H. \tag{8}$$

Let g be an element of F. Since L is a normal subgroup of F, we infer that gAg^{-1} is a normal abelian subgroup of L and

$$[L : A] = [L : gAg^{-1}]. \tag{9}$$

The abelian subgroup

$$M := \bigcap_{g \in F} gAg^{-1}. \tag{10}$$

is normal in F. We intend to prove that $[F : M]$ is upper bounded by a constant not depending on F. To this end, fix the representatives $g_1, \dots, g_{|F/L|}$ of all cosets of L in F. Then (10) and normality of A in L imply that

$$M = \bigcap_{i=1}^{|F/L|} g_i A g_i^{-1}. \tag{11}$$

From (11) we deduce that M is the kernel of the diagonal homomorphism

$$L \longrightarrow \prod_{i=1}^{|F/L|} L/g_i A g_i^{-1}$$

determined by the canonical projections $L \to L/g_i A g_i^{-1}$. This, (9), and (8) yield

$$[L : M] \leqslant [L : A]^{|F/L|} \leqslant J_H^{|F/L|}. \tag{12}$$

Let $\pi\colon G \to G/H$ be the canonical projection. By (7) the finite subgroup $\pi(F)$ of G/H is isomorphic to F/L. Since G/H is bounded, this yields $|F/L| \leqslant b_{G/H}$. We then deduce from (12) and $[F : M] = [F : L][L : M]$ that

$$[F : M] \leqslant b_{G/H} J_H^{b_{G/H}};$$

whence the claim. □

The following corollary should be compared with statement (1)(ii)(a) of Theorem 3:

Corollary 2. *Let H be a finite normal subgroup of a group G such that the center of H is trivial. If G/H is Jordan, then G is Jordan and*

$$J_G \leqslant |\mathrm{Aut}(H)| J_{G/H}^{|\mathrm{Aut}(H)|}.$$

Proof. Let $\varphi\colon G \to \mathrm{Aut}(H)$ be the homomorphism determined by the conjugating action of G on H. Triviality of the center of H yields $H \cap \ker\varphi = \{1\}$. Hence the restriction of the natural projection $G \to G/H$ to $\ker\varphi$ is an embedding $\ker\varphi \hookrightarrow G/H$. Therefore, $\ker\varphi$ is Jordan since G/H is. But $G/\ker\varphi$ is finite since it is isomorphic to a subgroup of $\mathrm{Aut}(H)$ for the finite group H. By Theorem 5 this implies the claim. □

2 When Are Aut(X) and Bir(X) Jordan?

2.1 *Problems A and B*

In [43, Sect. 2] were posed the following two problems:

Problem A. Describe algebraic varieties X for which $\mathrm{Aut}(X)$ is Jordan.

Problem B. The same with $\mathrm{Aut}(X)$ replaced by $\mathrm{Bir}(X)$.

Note that for rational varieties X Problem B means finding n such that the Cremona group Cr_n is Jordan; in this case, it was essentially posed in [59, 6.1].

Describing finite subgroups of the groups $\mathrm{Aut}(X)$ and $\mathrm{Bir}(X)$ for various varieties X is a classical research direction, currently flourishing. Understanding which of these groups are Jordan sheds a light on the structure of these subgroups. Varieties X with non-Jordan group $\mathrm{Bir}(X)$ or $\mathrm{Aut}(X)$ are, in a sense, more "symmetric" and, therefore, more remarkable than those with Jordan group. The discussion below supports the conclusion that they occur "rarely" and their finding is a challenge.

2.2 *Groups* Aut(X)

In this subsection we shall consider Problem A.

Lemma 1. *Let $X_1, \ldots, X_n$ be all the irreducible components of a variety X. If every* $\mathrm{Aut}(X_i)$ *is Jordan, then* $\mathrm{Aut}(X)$ *is Jordan.*

Proof. Define the homomorphism $\pi\colon \mathrm{Aut}(X) \to \mathrm{Sym}_n$ by $g \cdot X_i = X_{\pi(g)}$ for $g \in \mathrm{Aut}(X)$. Then $g \cdot X_i = X_i$ for every $g \in \mathrm{Ker}(\pi)$ and i, so the homomorphism $\pi_i\colon \mathrm{Ker}(\pi) \to \mathrm{Aut}(X_i)$, $g \mapsto g|_{X_i}$, arises. The definition implies that $\pi_1 \times \cdots \times \pi_n\colon \mathrm{Ker}(\pi) \to \prod_{i=1}^n \mathrm{Aut}(X_i)$ is an injection; whence $\mathrm{Ker}(\pi)$ is Jordan by Theorem 3(2). Hence $\mathrm{Aut}(X)$ is Jordan by Theorem 5. □

At this writing (October 2013), not a single variety X with non-Jordan $\mathrm{Aut}(X)$ is known (to me).

Question 1 ([43, Quest. 2.30 and 2.14]). Is there an irreducible variety X such that $\mathrm{Aut}(X)$ is non-Jordan? Is there an irreducible affine variety X with this property?

Since every automorphism of X uniquely lifts to the normalization X^ν of X, the group $\mathrm{Aut}(X)$ is isomorphic to a subgroup of $\mathrm{Aut}(X^\nu)$. Therefore, the existence of an irreducible variety with non-Jordan automorphism group implies the existence of a normal such variety.

Remark 1. One may consider the counterpart of the first question replacing X by a connected smooth topological manifold M, and $\mathrm{Aut}(X)$ by $\mathrm{Diff}(M)$. The following yields the affirmative answer:

Theorem 6 ([47]). *There is a simply connected noncompact smooth oriented four-dimensional manifold M such that* $\mathrm{Diff}(M)$ *contains an isomorphic copy of every finitely presented* (*in particular, of every finite*) *group. This copy acts on M properly discontinuously.*

Clearly, $\mathrm{Diff}(M)$ is non-Jordan. By Popov [47, Thm. 2], "noncompact" in Theorem 6 cannot be replaced by "compact". The following question (I reformulate it using Definition 1) was posed by É. Ghys (see [17, Quest. 13.1]): Is the diffeomorphism group of any compact smooth manifold Jordan? In fact, according to [36], É. Ghys conjectured the affirmative answer.

On the other hand, in many cases it can be proven that $\mathrm{Aut}(X)$ is Jordan. Below are described several extensive classes of X with this property.

2.2.1 Toral Varieties

First, consider the wide class of affine varieties singled out by the following

Definition 3 ([43, Def. 1.13]). A variety is called *toral* if it is isomorphic to a closed subvariety of some $\mathbf{A}^n \setminus \bigcup_{i=1}^n H_i$, where H_i is the set of zeros of the ith standard coordinate function x_i on $\mathbf{A}^n$.

Remark 2. $\mathbf{A}^n \setminus \bigcup_{i=1}^n H_i$ is the group variety of the n-dimensional affine torus; whence the terminology.[2] Warning: "toral" does not imply "affine toric" in the sense of [18].

[2] Recently I found that in some papers toral varieties are called very affine varieties.

The class of toral varieties is closed with respect to taking products and closed subvarieties.

Lemma 2 ([43, Lemma 1.14(a)]). *The following properties of an affine variety X are equivalent:*

(i) *X is toral;*
(ii) *$k[X]$ is generated by $k[X]^*$, the group of units of $k[X]$.*

Proof. If X is closed in $\mathbf{A}^n \setminus \bigcup_{i=1}^n H_i$, then the restriction of functions is an epimorphism $k\big[\mathbf{A}^n \setminus \bigcup_{i=1}^n H_i\big] \to k[X]$. Since $k\big[\mathbf{A}^n \setminus \bigcup_{i=1}^n H_i\big] = k[x_1, \ldots, x_n, 1/x_1, \ldots 1/x_n]$, this proves (i) $\Rightarrow$ (ii).

Conversely, assume that (ii) holds and let

$$k[X] = k[f_1, \ldots, f_n] \tag{13}$$

for some $f_1, \ldots, f_n \in k[X]^*$. Since X is affine, (13) implies that $\iota\colon X \to \mathbf{A}^n$, $x \mapsto (f_1(x), \ldots, f_n(x))$, is a closed embedding. The standard coordinate functions on $\mathbf{A}^n$ do not vanish on $\iota(X)$ since every f_i does not vanish on X. Hence $\iota(X) \subseteq \mathbf{A}^n \setminus \bigcup_{i=1}^n H_i$. This proves (ii)$\Rightarrow$(i). □

Lemma 3. *Any quasiprojective variety X endowed with a finite automorphism group G is covered by G-stable toral open subsets.*

Proof. First, any point $x \in X$ is contained in a G-stable affine open subset of X. Indeed, since the orbit $G \cdot x$ is finite and X is quasiprojective, there is an affine open subset U of X containing $G \cdot x$. Hence $V := \bigcap_{g \in G} g \cdot U$ is a G-stable open subset containing x, and, since every $g \cdot U$ is affine, V is affine as well, see, e.g., [61, Prop. 1.6.12(i)].

Thus, the problem is reduced to the case where X is affine. Assume then that X is affine, and let $k[X] = k[h_1, \ldots, h_s]$. Replacing h_i by $h_i + \alpha_i$ for an appropriate $\alpha_i \in k$, we may (and shall) assume that every h_i vanishes nowhere on the $G \cdot x$. Expanding the set $\{h_1, \ldots, h_s\}$ by including $g \cdot h_i$ for every i and $g \in G$, we may (and shall) assume that $\{h_1, \ldots, h_s\}$ is G-stable. Then $h := h_1 \cdots h_s \in k[X]^G$. Hence the affine open set $X_h := \{z \in X \mid h(z) \neq 0\}$ is G-stable and contains $G \cdot x$. Since $k[X_h] = k[h_1, \ldots, h_s, 1/h]$ and $h_1, \ldots, h_s, 1/h \in k[X_h]^*$, the variety X_h is toral by Lemma 2. □

Remark 3. Lemma 3 and its proof remain true for any variety X such that every G-orbit is contained in an affine open subset; whence the following

Corollary 3. *Every variety is covered by open toral subsets.*

For irreducible toral varieties the following was proved in [43, Thm. 2.16].

Theorem 7. *The automorphism group of every toral variety is Jordan.*

Proof. By Theorem 1 it suffices to prove this for irreducible toral varieties.

By Rosenlicht [56], for any irreducible variety X,

$$\Gamma := k[X]^*/k^*$$

is a free abelian group of finite rank. Let X be toral and let H be the kernel of the natural action of $\mathrm{Aut}(X)$ on Γ. We claim that H is abelian. Indeed, for every function $f \in k[X]^*$, the line in $k[X]$ spanned over k by f is H-stable. Since $\mathbf{GL}_1$ is abelian, this yields that

$$h_1h_2 \cdot f = h_2h_1 \cdot f \quad \text{for any elements} \quad h_1, h_2 \in H. \tag{14}$$

As X is toral, $k[X]^*$ generates the k-algebra $k[X]$ by Lemma 2. Hence (14) holds for every $f \in k[X]$. Since X is affine, the automorphisms of X coincide if and only if they induce the same automorphisms of $k[X]$. Therefore, H is abelian, as claimed.

Let n be the rank of Γ. Then $\mathrm{Aut}(\Gamma)$ is isomorphic to $\mathbf{GL}_n(\mathbf{Z})$. By the definition of H, the natural action of $\mathrm{Aut}(X)$ on Γ induces an embedding of $\mathrm{Aut}(X)/H$ into $\mathrm{Aut}(\Gamma)$. Hence $\mathrm{Aut}(X)/H$ is isomorphic to a subgroup of $\mathbf{GL}_n(\mathbf{Z})$ and therefore is bounded by Example 8(2). Thus, $\mathrm{Aut}(X)$ is an extension of a bounded group by an abelian group, hence Jordan by Theorem 5. This completes the proof. □

Remark 4. Maintain the notation of the proof of Theorem 7 and assume that X is irreducible. Let $f_1, \dots, f_n$ be a basis of Γ. There are the homomorphisms $\lambda_i \colon H \to k^*$, $i = 1, \dots, n$, such that $h \cdot f_i = \lambda(h) f_i$ for every $h \in H$ and i. Since $k[X]^*$ generates $k[X]$, the diagonal map $H \to (k^*)^n$, $h \mapsto (\lambda_1(h), \dots, \lambda_n(h))$, is injective. This and the proof of Theorem 7 show that for any irreducible toral variety X with $\mathrm{rk}\, k[X]^*/k^* = n$, there is an exact sequence

$$\{1\} \to D \to \mathrm{Aut}(X) \to B \to \{1\},$$

where D is a subgroup of the torus $(k^*)^n$ and B is a subgroup of $\mathbf{GL}_n(\mathbf{Z})$.

Combining Theorem 7 with Corollary of Lemma 3, we get the following:

Theorem 8. *Any point of any variety has an open neighborhood U such that* $\mathrm{Aut}(U)$ *is Jordan.*

2.2.2 Toric Varieties

By [14] (see also [39, Sect. 3.4]), if X is a smooth complete toric variety, then $\mathrm{Aut}(X)$ is an affine algebraic group. Whence $\mathrm{Aut}(X)$ is Jordan by Theorem 2.

2.2.3 Affine Spaces

Next, consider the fundamental objects of algebraic geometry, the affine spaces $\mathbf{A}^n$. The group $\mathrm{Aut}(\mathbf{A}^n)$ is the "affine Cremona group of rank n".

Since $\mathrm{Aut}(\mathbf{A}^1)$ is the affine algebraic group Aff_1, it is Jordan by Theorem 2.

Since $\mathrm{Aut}(\mathbf{A}^2)$ is the subgroup of Cr_2, it is Jordan by Example 1. Another proof: By Igarashi [20] every finite subgroup of $\mathrm{Aut}(\mathbf{A}^2)$ is conjugate to a subgroup of $\mathbf{GL}_2(k)$, so the claim follows from Theorem 1.

The group $\mathrm{Aut}(\mathbf{A}^3)$ is Jordan being the subgroup of Cr_3 that is Jordan by Corollary 13 below.

At this writing (October 2013) is unknown whether $\mathrm{Aut}(\mathbf{A}^n)$ is Jordan for $n \geqslant 4$ or not. By Theorem 16, if the so-called BAB Conjecture (see Sect. 2.3.5 below) holds true in dimension n, then Cr_n is Jordan, hence $\mathrm{Aut}(\mathbf{A}^n)$ is Jordan as well.

We note here that $\mathrm{Diff}(\mathbb{R}^n)$ is Jordan for $n = 1, 2$ [36], $n = 3$ [32], and $n = 4$ [25] (I thank I. Mundet i Riera who drew in [37] my attention to the last two references). It would be interesting to understand whether $\mathrm{Diff}(\mathbb{R}^n)$ is Jordan for every n or not (cf. Theorem 6).[3]

2.2.4 Fixed Points and Jordaness

The following method of proving Jordaness of $\mathrm{Aut}(X)$ was suggested in [43, Sect. 2] and provides extensive classes of X with Jordan $\mathrm{Aut}(X)$. It is based on the use of the following fact:

Lemma 4. *Let X be an irreducible variety, let G be a finite subgroup of* $\mathrm{Aut}(X)$, *and let $x \in X$ be a fixed point of G. Then the natural action of G on $\mathrm{T}_{x,X}$, the tangent space of X at x, is faithful.*

Proof. Let $\mathfrak{m}_{x,X}$ be the maximal ideal of $\mathcal{O}_{x,X}$, the local ring of X at x. Being finite, G is reductive. Since $\mathrm{char}\, k = 0$, this implies that $\mathfrak{m}_{x,X} = L \oplus \mathfrak{m}_{x,X}^2$ for some submodule L of the G-module $\mathfrak{m}_{x,X}$. Let K be the kernel of the action of G on L and let L^d be the k-linear span in $\mathfrak{m}_{x,X}$ of the dth powers of all the elements of L. By the Nakayama's Lemma, the restriction to L^d of the natural projection $\mathfrak{m}_{x,X} \to \mathfrak{m}_{x,X}/\mathfrak{m}_{x,X}^{d+1}$ is surjective. Hence K acts trivially on $\mathfrak{m}_{x,X}/\mathfrak{m}_{x,X}^{d+1}$ for every d.

Take an element $f \in \mathfrak{m}_{x,X}$. Since G is finite, the k-linear span $\langle K \cdot f \rangle$ of the K-orbit of f in $\mathfrak{m}_{x,X}$ is finite-dimensional. This and $\bigcap_s \mathfrak{m}_{x,X}^s = \{0\}$ (see, e.g., [3, Cor. 10.18]) implies that $\langle K \cdot f \rangle \cap \mathfrak{m}_{x,X}^{d+1} = \{0\}$ for some d. Since $f - g \cdot f \in \mathfrak{m}_{x,X}^{d+1}$

[3] The answer is obtained in the recent preprint I. Mundet i Riera, Finite group actions on spheres, Euclidean spaces, and compact manifolds with $\chi \neq 0$ (March 2014) [arXiv:1403.0383] (see also the footnote at the end of Sect. 1.2.4): $\mathrm{Diff}(\mathbb{R}^n)$ is Jordan for every n. Given Theorem 3(1)(i), this yields the following

Theorem. $\mathrm{Aut}(\mathbf{A}^n)$ *is Jordan for every n.*

for every element $g \in K$, we conclude that $f = g \cdot f$, i.e., f is K-invariant. Thus, K acts trivially on $\mathfrak{m}_{x,X}$, hence on $\mathcal{O}_{x,X}$ as well. Since $k(X)$ is the field of fractions of $\mathcal{O}_{x,X}$, K acts trivially on $k(X)$, and therefore, on X. But K acts on X faithfully because $K \subseteq \mathrm{Aut}(X)$. This proves that K is trivial. Since L is the dual of the G-module $\mathrm{T}_{x,X}$, this completes the proof. □

The idea of the method is to use the fact that if a finite subgroup G of $\mathrm{Aut}(X)$ has a fixed point $x \in X$, then, by Lemma 4 and Theorem 1, there is a normal abelian subgroup of G whose index in G is at most $J_{\mathbf{GL}_n(k)}$ for $n = \dim \mathrm{T}_{x,X}$.

This yields the following:

Theorem 9. *Let X be an irreducible variety and let G be a finite subgroup of* $\mathrm{Aut}(X)$. *If G has a fixed point in X, then there is a normal abelian subgroup of G whose index in G is at most $J_{\mathbf{GL}_m(k)}$, where*

$$m = \max_x \dim \mathrm{T}_{x,X}. \tag{15}$$

Corollary 4. *If every finite automorphism group of an irreducible variety X has a fixed point in X, then* $\mathrm{Aut}(X)$ *is Jordan and*

$$J_{\mathrm{Aut}(X)} \leqslant J_{\mathbf{GL}_m(k)},$$

where m is defined by (15).

Corollary 5. *Let p be a prime number. Then every finite p-subgroup G of* $\mathrm{Aut}(\mathbf{A}^n)$ *contains an abelian normal subgroup whose index in G is at most $J_{\mathbf{GL}_n(k)}$.*[4]

Proof. This follows from Theorem 9 since in this case $(\mathbf{A}^n)^G \neq \varnothing$, see [60, Thm. 1.2]. □

Remark 5. At this writing (October 2013), it is unknown whether or not $(\mathbf{A}^n)^G \neq \varnothing$ for every finite subgroup G of $\mathrm{Aut}(\mathbf{A}^n)$. By Theorem 9 the affirmative answer would imply that $\mathrm{Aut}(\mathbf{A}^n)$ is Jordan (cf. Sect. 2.2.3).

Remark 6. The statement of Corollary 5 remains true if $\mathbf{A}^n$ is replaced by any p-acyclic variety X, and n in $J_{\mathbf{GL}_n(k)}$ is replaced by m (see (15)). This is because in this case $X^G \neq \varnothing$ for every finite p-subgroup G of $\mathrm{Aut}(X)$, see [60, Sect. 7–8].

The following applications are obtained by combining the above idea with Theorem 5.

Theorem 10. *Let X be an irreducible variety. Consider an* $\mathrm{Aut}(X)$*-stable equivalence relation $\sim$ on the set its points. If there is a finite equivalence class C of $\sim$, then* $\mathrm{Aut}(X)$ *is Jordan and*

[4] See the footnote at the end of Sect. 2.2.3

$$J_{\mathrm{Aut}(X)} \leqslant |C|!J_{\mathbf{GL}_m(k)}^{|C|!},$$

where m is defined by (15).

Proof. By the assumption, every equivalence class of $\sim$ is $\mathrm{Aut}(X)$-stable. The kernel K of the action of $\mathrm{Aut}(X)$ on C is a normal subgroup of $\mathrm{Aut}(X)$ and, since the elements of $\mathrm{Aut}(X)$ induce permutations of C,

$$[\mathrm{Aut}(X) : K] \leqslant |C|!. \tag{16}$$

By Theorem 5, Jordaness of $\mathrm{Aut}(X)$ follows from that of K. To prove that the latter holds, take a point of $x \in C$. Since x is fixed by every finite subgroup of K, Theorem 9 implies that K is Jordan and $J_K \leqslant J_{\mathbf{GL}_m(k)}$. By Theorem 5, this and (16) imply the claim. □

Example 13. Below are several examples of $\mathrm{Aut}(X)$-stable equivalence relations on an irreducible variety X:

(i) $x \sim y \Longleftrightarrow \mathcal{O}_{x,X}$ and $\mathcal{O}_{y,X}$ are k-isomorphic;
(ii) $x \sim y \Longleftrightarrow \dim \mathrm{T}_{x,X} = \dim \mathrm{T}_{y,X}$;
(iii) $x \sim y \Longleftrightarrow$ the tangent cones of X at x and y are isomorphic. □

Corollary 6. *If an irreducible variety X has a point x such that the set*

$$\{y \in X \mid \mathcal{O}_{x,X} \text{ and } \mathcal{O}_{y,X} \text{ are } k\text{-isomorphic}\}$$

is finite, then $\mathrm{Aut}(X)$ *is Jordan.*

Call a point $x \in X$ a *vertex* of X if

$$\dim \mathrm{T}_{x,X} \geqslant \dim \mathrm{T}_{y,X} \quad \text{for every point } y \in X.$$

Thus every point of X is a vertex of X if and only if X is smooth.

Corollary 7. *The automorphism group of every irreducible variety with only finitely many vertices is Jordan.*

Corollary 8. *The automorphism group of every nonsmooth irreducible variety with only finitely many singular points is Jordan.*

Corollary 9. *Let $X \subset \mathbf{A}^n$ be the affine cone of a smooth closed proper irreducible subvariety Z of $\mathbf{P}^{n-1}$ that does not lie in any hyperplane. Then* $\mathrm{Aut}(X)$ *is Jordan.*

Proof. The assumptions imply that the singular locus of X consists of a single point, the origin; whence the claim by Corollary 8. □

Corollary 10. *If an irreducible variety X has a point x such that there are only finitely many points $y \in X$ for which the tangent cones of X at x and at y are isomorphic, then* $\mathrm{Aut}(X)$ *is Jordan.*

Remark 7. Smoothness in Corollary 9 may be replaced by the assumption that Z is not a cone. Indeed, in this case the origin constitutes a single equivalence class of equivalence relation (iii) in Example 13; whence the claim by Corollary 10.

2.2.5 The Koras–Russell Threefolds

Let $X = X_{d,s,l}$ be the so-called Koras–Russell threefold of the first kind [33], i.e., the smooth hypersurface in $\mathbf{A}^4$ defined by the equation

$$x_1^d x_2 + x_3^s + x_4^l + x_1 = 0,$$

where $d \geqslant 2$ and $2 \leqslant s \leqslant l$ with s and l relatively prime; the case $d = s = 2$ and $l = 3$ is the famous Koras–Russell cubic. According to [33, Cor. 6.1], every element of $\mathrm{Aut}(X)$ fixes the origin $(0,0,0,0) \in X$. By Corollary 4 and item (iii) of Sect. 1.2.1 this implies that $\mathrm{Aut}(X)$ is Jordan and

$$J_{\mathrm{Aut}(X)} \leqslant 360.$$

Actually, during the conference I learned from L. Moser-Jauslin that X contains a line ℓ passing through the origin, stable with respect to $\mathrm{Aut}(X)$, and such that every element of $\mathrm{Aut}(X)$ fixing ℓ pointwise has infinite order. This implies that every finite subgroup of $\mathrm{Aut}(X)$ is cyclic and hence

$$J_{\mathrm{Aut}(X)} = 1.$$

2.2.6 Small Dimensions

Since $\mathrm{Aut}(X)$ is a subgroup of $\mathrm{Bir}(X)$, Jordaness of $\mathrm{Bir}(X)$ implies that of $\mathrm{Aut}(X)$. This and Theorem 14 below yield the following

Theorem 11. *Let X be an irreducible variety of dimension $\leqslant 2$ not birationally isomorphic to $\mathbf{P}^1 \times E$, where E is an elliptic curve. Then* $\mathrm{Aut}(X)$ *is Jordan.*

Note that if E is an elliptic curve and $X = \mathbf{P}^1 \times E$, then $\mathrm{Aut}(X) = \mathbf{PGL}_2(k) \times \mathrm{Aut}(E)$, see [29, pp. 98–99]. Fixing a point of E, endow E with a structure of abelian variety E_{ab}. Since $\mathrm{Aut}(E)$ is an extension of the finite group $\mathrm{Aut}(E_{\mathrm{ab}})$ by the abelian group E_{ab}, Theorems 2, 5, and 3(2) imply that $\mathrm{Aut}(\mathbf{P}^1 \times E)$ is Jordan.[5]

[5] It is proved in the recent preprint T.I. Bandman, Y.G. Zarhin, Jordan groups and algebraic surfaces (April 2014) [arXiv:1404.1581] that if X is birationally isomophic $\mathbf{P}^1 \times E$, where E is an elliptic curve, then $\mathrm{Aut}(X)$ is Jordan. The proof is based on the other recent preprint Y.G. Zarhin, Jordan groups and elliptic ruled surfaces (January 2014) [arXiv:1401.7596], where this is proved for projective X. Given Theorem 11, we then obtain

Theorem. *If X is a variety of dimension $\leqslant 2$, then* $\mathrm{Aut}(X)$ *is Jordan.*

Note also that all irreducible curves (not necessarily smooth and projective) whose automorphism group is infinite are classified in [42].

2.2.7 Non-uniruled Varieties

Again, using that Jordaness of $\mathrm{Bir}(X)$ implies that of $\mathrm{Aut}(X)$, we deduce from recent Theorem 17(i)(a) below the following

Theorem 12. $\mathrm{Aut}(X)$ *is Jordan for any irreducible non-uniruled variety* X.

2.3 *Groups* $\mathrm{Bir}(X)$

Now we shall consider Problem B (see Sect. 2.1). Exploring $\mathrm{Bir}(X)$, one may, maintaining this group, replace X by any variety birationally isomorphic to X. Note that by Theorem 8 one can always attain that after such a replacement $\mathrm{Aut}(X)$ becomes Jordan.

The counterpart of Question 1 is

Question 2 ([43, Quest. 2.31]). Is there an irreducible variety X such that $\mathrm{Bir}(X)$ is non-Jordan?

In contrast to the case of Question 1, at present we know the answer to Question 2: motivated by my question, Yu. Zarhin proved in [62] the following

Theorem 13 ([62, Cor. 1.3]). *Let* X *be an abelian variety of positive dimension and let* Z *be a rational variety of positive dimension. Then* $\mathrm{Bir}(X \times Z)$ *is non-Jordan.*

Sketch of proof. By Theorem 3(1)(i), it suffices to prove that $\mathrm{Bir}(X \times \mathbf{A}^1)$ is non-Jordan. Consider an ample divisor D on X and the sheaf $L := \mathcal{O}_X(D)$. For a positive integer n, consider the following group $\Theta(L^n)$. Its elements are all pairs $(x, [f])$ where $x \in X$ is such that $L^n \cong T_x^*(L^n)$ for the translation $T_x\colon X \to X$, $z \mapsto z + x$, and $[f]$ is the automorphism of the additive group of $k(X)$ induced by the multiplication by $f \in k(X)^*$. The group structure of $\Theta(L^n)$ is defined by $(x, [f])(y, [h]) = (x + y, [T_x^* h \cdot f])$. One proves that $\Theta(L^n)$ enjoys the properties: (i) $\varphi\colon \Theta(L^n) \to \mathrm{Bir}(X \times \mathbf{A}^1)$, $\varphi(x, [f])(y, t) = (x + y, f(y)t)$, is a group embedding; (ii) $\Theta(L^n)$ is isomorphic to a group G_K from Example 7 with $|K| \geqslant n$. This implies the claim (see Example 7). □

Below Problem B is solved for varieties of small dimensions ($\leqslant 2$).

2.3.1 Curves

If X is a curve, then the answer to Question 2 is negative.

Proving this, we may assume that X is smooth and projective; whence $\mathrm{Bir}(X) = \mathrm{Aut}(X)$.

If $g(X)$, the genus of X, is 0, then $X = \mathbf{P}^1$, so $\mathrm{Aut}(X) = \mathbf{PGL}_2(k)$. Hence $\mathrm{Aut}(X)$ is Jordan by Theorem 2.

If $g(X) = 1$, then X is an elliptic curve, hence $\mathrm{Aut}(X)$ is Jordan (see the penultimate paragraph in Sect. 2.2.6).

If $g(X) \geqslant 2$, then, being finite, $\mathrm{Aut}(X)$ is Jordan.

2.3.2 Surfaces

Answering Question 2 for surfaces X, we may assume that X is a smooth projective minimal model.

If X is of general type, then by Matsumura's theorem $\mathrm{Bir}(X)$ is finite, hence Jordan.

If X is rational, then $\mathrm{Bir}(X)$ is Cr_2, hence Jordan, see Example 1.

If X is a nonrational ruled surface, it is birationally isomorphic to $\mathbf{P}^1 \times B$ where B is a smooth projective curve such that $g(B) > 0$; we may then take $X = \mathbf{P}^1 \times B$. Since $g(B) > 0$, there are no dominant rational maps $\mathbf{P}^1 \dashrightarrow B$; whence the elements of $\mathrm{Bir}(X)$ permute fibers of the natural projection $\mathbf{P}^1 \times B \to B$. The set of elements inducing trivial permutation is a normal subgroup $\mathrm{Bir}_B(X)$ of $\mathrm{Bir}(X)$. The definition implies that $\mathrm{Bir}_B(X) = \mathbf{PGL}_2(k(B))$, hence $\mathrm{Bir}_B(X)$ is Jordan by Theorem 2. Identifying $\mathrm{Aut}(B)$ with the subgroup of $\mathrm{Bir}(X)$ in the natural way, we get the decomposition

$$\mathrm{Bir}(X) = \mathrm{Bir}_B(X) \rtimes \mathrm{Aut}(B). \tag{17}$$

If $g(B) \geqslant 2$, then $\mathrm{Aut}(B)$ is finite; whence $\mathrm{Bir}(X)$ is Jordan by virtue of (17) and Theorem 5. If $g(B) = 1$, then $\mathrm{Bir}(X)$ is non-Jordan by Theorem 13.

The canonical class of all the other surfaces X is numerically effective, so, for them, $\mathrm{Bir}(X) = \mathrm{Aut}(X)$, cf. [22, Sect. 7.1, Thm. 1 and Sect. 7.3, Thm. 2].

Let X be such a surface. The group $\mathrm{Aut}(X)$ has a structure of a locally algebraic group with finite or countably many components, see [31], i.e., there is a normal subgroup $\mathrm{Aut}(X)^0$ in $\mathrm{Aut}(X)$ such that

(i) $\mathrm{Aut}(X)^0$ is a connected algebraic group,
(ii) $\mathrm{Aut}(X)/\mathrm{Aut}(X)^0$ is either a finite or a countable group,

By (i) and the structure theorem on algebraic groups [4, 55] there is a normal connected affine algebraic subgroup L of $\mathrm{Aut}(X)^0$ such that $\mathrm{Aut}(X)^0/L$ is an abelian variety. By Matsumura [30, Cor. 1] nontriviality of L would imply that X is ruled. Since we assumed that X is not ruled, this means that $\mathrm{Aut}(X)^0$ is an abelian variety. Hence $\mathrm{Aut}(X)^0$ is abelian and, a fortiori, Jordan.

By (i) the group $\mathrm{Aut}(X)^0$ is contained in the kernel of the natural action of $\mathrm{Aut}(X)$ on $H^2(X, \mathbf{Q})$ (we may assume that $k = \mathbf{C}$). Therefore, this action defines a homomorphism $\mathrm{Aut}(X)/\mathrm{Aut}(X)^0 \to \mathbf{GL}(H^2(X, \mathbf{Q}))$. The kernel of this homomorphism is finite by Dolgachev [15, Prop. 1], and the image is bounded by Example 10. By Examples 8, 9 this yields that $\mathrm{Aut}(X)/\mathrm{Aut}(X)^0$ is bounded. In turn, since $\mathrm{Aut}(X)^0$ is Jordan, by Theorem 5 this implies that $\mathrm{Aut}(X)$ is Jordan.

2.3.3 The Upshot

The upshot of the last two subsections is

Theorem 14 ([43, Thm. 2.32]). *Let X be an irreducible variety of dimension $\leqslant 2$. Then the following two properties are equivalent:*

(a) *the group* $\mathrm{Bir}(X)$ *is Jordan;*
(b) *the variety X is not birationally isomorphic to $\mathbf{P}^1 \times B$, where B is an elliptic curve.*

2.3.4 Finite and Connected Algebraic Subgroups of Bir(*X*) and Aut(*X*)

Recall that the notions of algebraic subgroup of $\mathrm{Bir}(X)$ and $\mathrm{Aut}(X)$ make sense, and every algebraic subgroup of $\mathrm{Aut}(X)$ is that of $\mathrm{Bir}(X)$, see, e.g., [44, Sect. 1]. Namely, a map $\psi\colon S \to \mathrm{Bir}(X)$ of a variety S is called an *algebraic family* if the domain of definition of the partially defined map $\alpha\colon S \times X \to X$, $(s, x) \mapsto \psi(s)(x)$ contains a dense open subset of $S \times X$ and α coincides on it with a rational map $\varrho\colon S \times X \dashrightarrow X$. The group $\mathrm{Bir}(X)$ is endowed with the *Zariski topology* [58, Sect. 1.6], in which a subset Z of $\mathrm{Bir}(X)$ is closed if and only if $\psi^{-1}(Z)$ is closed in S for every family ψ. If S is an algebraic group and ψ is an algebraic family which is a homomorphism of abstract groups, then $\psi(S)$ is called an *algebraic subgroup of* $\mathrm{Bir}(X)$. In this case, $\ker(\psi)$ is closed in S and the restriction to $\psi(S)$ of the Zariski topology of $\mathrm{Bir}(X)$ coincides with the topology determined by the natural identification of $\psi(S)$ with the algebraic group $S/\ker(\psi)$. If $\psi(S) \subset \mathrm{Aut}(X)$ and ϱ is a morphism, then $\psi(S)$ is called an *algebraic subgroup of* $\mathrm{Aut}(X)$.

The following reveals a relation between embeddability of finite subgroups of $\mathrm{Bir}(X)$ in connected affine algebraic subgroups of $\mathrm{Bir}(X)$ and Jordaness of $\mathrm{Bir}(X)$ (and the same holds for $\mathrm{Aut}(X)$).

For every integer $n > 0$, consider the set of all isomorphism classes of connected reductive algebraic groups of rank $\leqslant n$, and fix a group in every class. The obtained set of groups $\mathscr{R}_n$ is finite. Therefore,

$$J_{\leqslant n} := \sup_{R \in \mathscr{R}_n} J_R \tag{18}$$

is a positive integer.

Theorem 15. *Let X be an irreducible variety of dimension n. Then every finite subgroup G of every connected affine algebraic subgroup of* $\mathrm{Bir}(X)$ *has a normal abelian subgroup whose index in G is at most $J_{\leqslant n}$.*

Proof. Let L be a connected affine algebraic subgroup of $\mathrm{Bir}(X)$ containing G. Being finite, G is reductive. Let R be a maximal reductive subgroup of L containing G. Then L is a semidirect product of R and the unipotent radical of L, see [34, Thm. 7.1]. Therefore, R is connected because L is. Faithfulness of the action R acts on X yields that $\mathrm{rk}\, R \leqslant \dim X$, see, e.g., [44, Lemma 2.4]. The claim then follows from (18), Theorem 2, and Definition 1. □

Theorem 15 and Definition 1 imply

Corollary 11. *Let X be an irreducible variety of dimension n such that* $\mathrm{Bir}(X)$ *(resp.* $\mathrm{Aut}(X)$*) is non-Jordan. Then for every integer $d > J_{\leqslant n}$, there is a finite subgroup G of* $\mathrm{Bir}(X)$ *(resp.* $\mathrm{Aut}(X)$*) with the properties:*

(i) *G does not lie in any connected affine algebraic subgroup of* $\mathrm{Bir}(X)$ *(resp.* $\mathrm{Aut}(X)$*);*
(ii) *for any abelian normal subgroup of G, its index in G is $\geqslant d$.*

Corollary 12. *If* Cr_n *(resp.* $\mathrm{Aut}(\mathbf{A}^n)$*) is non-Jordan,*[6] *then for every integer $d > J_{\leqslant n}$, there is a finite subgroup G of* Cr_n *(resp.* $\mathrm{Aut}(\mathbf{A}^n)$*) with the properties:*

(i) *the action of G on $\mathbf{A}^n$ is nonlinearizable;*
(ii) *for any abelian normal subgroup of G, its index in G is $\geqslant d$.*

Proof. This follows from Corollary 11 because $\mathbf{GL}_n(k)$ is a connected affine algebraic subgroup of $\mathrm{Aut}(\mathbf{A}^n)$ and nonlinearizability of the action of G on $\mathbf{A}^n$ means that G is not contained in a subgroup of Cr_n (resp. $\mathrm{Aut}(\mathbf{A}^n)$) conjugate to $\mathbf{GL}_n(k)$. □

2.3.5 Recent Developments

The initiated in [43] line of research of Jordaness of $\mathrm{Aut}(X)$ and $\mathrm{Bir}(X)$ for algebraic varieties X has generated interest of algebraic geometers in Moscow among whom I have promoted it, and led to a further progress in Problem B (hence A as well) in papers [49, 50, 62]. In [62], the earliest of them, the examples of non-Jordan groups $\mathrm{Bir}(X)$ only known to date (October 2013) have been constructed (see Theorem 13 above). Below are formulated the results obtained in [49, 50]. Some of them are conditional, valid under the assumption that the following general conjecture by A. Borisov, V. Alexeev, and L. Borisov holds true:

BAB Conjecture. All Fano varieties of a fixed dimension n and with terminal singularities are contained in a finite number of algebraic families.

[6] See the footnote at the end of Sect. 2.2.3

Theorem 16 ([49, Thm. 1.8]). *If the BAB Conjecture holds true in dimension n, then, for every rationally connected n-dimensional variety X, the group* $\mathrm{Bir}(X)$ *is Jordan and, moreover,* $J_{\mathrm{Bir}(X)} \leqslant u_n$ *for a number u_n depending only on n.*

Since for $n = 3$ the BAB Conjecture is proved [24], this yields

Corollary 13 ([49, Cor. 1.9]). *The space Cremona group* Cr_3 *is Jordan.*

Proposition 1 ([49, Prop. 1.11]). $u_3 \leqslant (25920 \cdot 20736)^{20736}$.

The pivotal idea of the proof of Theorem 16 is to use a technically refined form of the "fixed-point method" described in Sect. 2.2.4.

Theorem 17 ([50, Thm. 1.8]). *Let X be an irreducible smooth proper n-dimensional variety.*

(i) *The group* $\mathrm{Bir}(X)$ *is Jordan in either of the cases:*

  (a) *X is non-uniruled;*
  (b) *the BAB Conjecture holds true in dimension n and the irregularity of X (i.e., the dimension of its Picard variety) is* 0.

(ii) *If X is non-uniruled and its irregularity is* 0, *then the group* $\mathrm{Bir}(X)$ *is bounded (see Definition* 2).

3 Appendix: Problems

Below I add afew additional problems to those which have already been formulated above (Problems A and B in Sect. 2.1, and Questions 1, 2).

3.1 Cr_n*-Conjugacy of Finite Subgroups of* $\mathbf{GL}_n(k)$

Below $\mathbf{GL}_n(k)$ is identified in the standard way with the subgroup of Cr_n, which, in turn, is identified with the subgroup of Cr_m for any $m = n+1, n+2, \ldots, \infty$ (cf. [44, Sect. 1] or [45, Sect. 1]).

Question 3. Consider the following properties of two finite subgroups A and B of $\mathbf{GL}_n(k)$:

(i) A and B are isomorphic,
(ii) A and B are conjugate in Cr_n.

Does (i) imply (ii)?

Comments. 1. Direct verification based on the classification in [16] shows that the answer is affirmative for $n \leqslant 2$.

2. By Popov [45, Cor. 5], if A and B are abelian, then the answer is affirmative for every n.
3. If A and B are isomorphic, then they are conjugate in Cr_{2n}. This is the corollary of the following stronger statement:

Proposition 2. *For any finite group G and any injective homomorphisms*

$$G \underset{\alpha_2}{\overset{\alpha_1}{\rightrightarrows}} \mathbf{GL}_n(k), \tag{19}$$

there exists an element $\varphi \in \mathrm{Cr}_{2n}$ such that $\alpha_1 = \mathrm{Int}(\varphi) \circ \alpha_2$.

Proof. Every element $g \in \mathbf{GL}_n(k)$ is a linear transformation $x \mapsto g \cdot x$ of $\mathbf{A}^n$ (with respect to the standard structure of k-linear space on $\mathbf{A}^n$). The injections α_1 and α_2 determine two faithful linear actions of G on $\mathbf{A}^n$: the ith action ($i = 1, 2$) maps $(g, x) \in G \times \mathbf{A}^n$ to $\alpha_i(g) \cdot x$. Consider the product of these actions, i.e., the action of G on $\mathbf{A}^n \times \mathbf{A}^n$ defined by

$$G \times \mathbf{A}^n \times \mathbf{A}^n \to \mathbf{A}^n \times \mathbf{A}^n, \quad (g, x, y) \mapsto (\alpha_1(g) \cdot x, \alpha_2(g) \cdot y). \tag{20}$$

The natural projection of $\mathbf{A}^n \times \mathbf{A}^n \to \mathbf{A}^n$ to the ith factor is G-equivariant. By classical Speiser's Lemma (see [27, Lemma 2.12] and references therein), this implies that $\mathbf{A}^n \times \mathbf{A}^n$ endowed with G-action (20) is G-equivariantly birationally isomorphic to $\mathbf{A}^n \times \mathbf{A}^n$ endowed with the G-action via the ith factor by means of α_i. Therefore, $\mathbf{A}^n \times \mathbf{A}^n$ endowed with the G-action via the first factor by means of α_1 is G-equivariantly birationally isomorphic to $\mathbf{A}^n \times \mathbf{A}^n$ endowed with the G-action via the second factor by means of α_2; whence the claim. □

Remark 8. In general, it is impossible to replace Cr_{2n} by Cr_n in Proposition 8. Indeed, in [54] one finds the examples of finite abelian groups G and embeddings (19) such that $\alpha_1 \notin \mathrm{Int}(\mathrm{Cr}_n) \circ \alpha_2$. However, since the images of these embeddings are isomorphic finite abelian subgroups of $\mathbf{GL}_n(k)$, by Popov [45, Cor. 5] these images are conjugate in Cr_n.

3.2 *Torsion Primes*

Let X be an irreducible variety. The following definition is based on the fact that the notion of algebraic torus in $\mathrm{Bir}(X)$ makes sense.

Definition 4 ([45, Sect. 8]). Let G be a subgroup of $\operatorname{Bir}(X)$. A prime integer p is called a *torsion number* of G if there exists a finite abelian p-subgroup of G that does not lie in any torus of G.

Let $\operatorname{Tors}(G)$ be the set of all torsion primes of G. If G is a connected reductive algebraic subgroup of $\operatorname{Bir}(X)$, this set coincides with that of the torsion primes of algebraic group G in the sense of classical definition, cf., e.g., [57, 1.3].

Question 4 ([45, Quest. 3]). What are, explicitly,

$$\operatorname{Tors}(\mathrm{Cr}_n),\quad \operatorname{Tors}(\operatorname{Aut}\mathbf{A}^n),\quad \operatorname{Tors}(\operatorname{Aut}^*\mathbf{A}^n),\qquad n = 3, 4, \ldots, \infty$$

where $\operatorname{Aut}^*\mathbf{A}^n$ is the group of those automorphisms of $\mathbf{A}^n$ that preserve the volume form $dx_1 \wedge \cdots \wedge dx_n$ on $\mathbf{A}^n$ (here $x_1, \ldots, x_n$ are the standard coordinate functions on $\mathbf{A}^n$), cf. [45, Sect. 2]?

Comments. By Popov [45, Sect. 8],

$$\begin{aligned}
\operatorname{Tors}(\mathrm{Cr}_1) &= \{2\},\\
\operatorname{Tors}(\mathrm{Cr}_2) &= \{2,3,5\} \quad \text{(this coincides with } \operatorname{Tors}(E_8)),\\
\operatorname{Tors}(\mathrm{Cr}_n) &\supseteq \{2,3\} \quad \text{for any } n \geqslant 3,\\
\operatorname{Tors}(\operatorname{Aut}\mathbf{A}^n) = \operatorname{Tors}(\operatorname{Aut}^*\mathbf{A}^n) &= \varnothing \quad \text{for } n \leqslant 2.
\end{aligned}$$

Question 5 ([45, Quest. 4]). What is the minimal n such that seven lies in one of the sets $\operatorname{Tors}(\mathrm{Cr}_n)$, $\operatorname{Tors}(\operatorname{Aut}\mathbf{A}^n)$, $\operatorname{Tors}(\operatorname{Aut}^*\mathbf{A}^n)$?

Question 6 ([45, Quest. 5]). Are these sets finite?

Question 7. Are the sets

$$\textstyle\bigcup_{n\geqslant 1}\operatorname{Tors}(\mathrm{Cr}_n),\quad \bigcup_{n\geqslant 1}\operatorname{Tors}(\operatorname{Aut}\mathbf{A}^n),\quad \bigcup_{n\geqslant 1}\operatorname{Tors}(\operatorname{Aut}^*\mathbf{A}^n)$$

finite?

3.3 *Embeddability in* $\operatorname{Bir}(X)$

Not every group G is embeddable in $\operatorname{Bir}(X)$ for some X. For instance, by Cornulier [11, Thm. 1.2], if G is finitely generated, its embeddability in $\operatorname{Bir}(X)$ implies that G has a solvable word problem. Another example: by Cantat [7], $\mathbf{PGL}_\infty(k)$ is not embeddable in $\operatorname{Bir}(X)$ for $k = \mathbf{C}$ (I thank S. Cantat who informed me in [8] about these examples).

If $\operatorname{Bir}(X)$ is Jordan, then by Example 5 and Theorem 3(1)(i), $\mathscr{N}$ is not embeddable in $\operatorname{Bir}(X)$. Hence, by Theorem 17(i)(a), $\mathscr{N}$ is not embeddable in $\operatorname{Bir}(X)$ for any non-uniruled X.

Conjecture 1. The finitely generated group $\mathscr{N}$ defined in Example 5 is not embeddable in $\mathrm{Bir}(X)$ for every irreducible variety X.

Since $\mathscr{N}$ contains Sym_n for every n, and every finite group can be embedded in Sym_n for an appropriate n, the existence of an irreducible variety X for which $\mathrm{Bir}(X)$ contains an isomorphic copy of $\mathscr{N}$ implies that $\mathrm{Bir}(X)$ contains an isomorphic copy of every finite group and, in particular, every simple finite group. Therefore, Conjecture 1 follows from the affirmative answer to

Question 8. Let X be an irreducible variety. Is any set of pairwise nonisomorphic simple nonabelian finite subgroups of $\mathrm{Bir}(X)$ finite?

The affirmative answer looks likely. At this writing (October 2013) about this question I know the following:

Proposition 3. *If* $\dim X \leqslant 2$, *then the answer to Question 8 is affirmative.*

Proof. The claim immediately follows from Theorems 14 and 4 if X is not birationally isomorphic to $\mathbf{P}^1 \times B$, where B is an elliptic curve. For $X = \mathbf{P}^1 \times B$ the claim follows from (17) and Theorem 4 because $\mathrm{Aut}(B)$ and $\mathrm{Bir}_B(X)$ are Jordan groups. □

By Theorems 16, 17 and by Kollár et al. [24], the answer to Question 8 is affirmative also in each of the following cases:

1. X is non-uniruled;
2. X is rationally connected or smooth proper with irregularity 0, and
   (a) either $\dim X = 3$ or
   (b) $\dim X > 3$ and the BAB Conjecture holds true in dimension $\dim X$.

Note that if X and $\mathrm{Bir}(X)$ in Question 8 are replaced, respectively, by a connected smooth topological manifold M and $\mathrm{Diff}(M)$, then by Theorem 6, for a noncompact M, the answer, in general, is negative. But for a compact M a finiteness theorem [47, Thm. 2] holds.

3.4 Contractions

Developing the classical line of research, in recent years were growing activities aimed at description of finite subgroups of $\mathrm{Bir}(X)$ for various X; the case of rational X (i.e., that of the Cremona group $\mathrm{Bir}(X)$) was, probably, most actively explored with culmination in the classification of finite subgroups of Cr_2, [16]. In these studies, all the classified groups appear in the corresponding lists on equal footing. However, in fact, some of them are "more basic" than the others because the latter may be obtained from the former by a certain standard construction. Given this, it is natural to pose the problem of describing these "basic" groups.

Namely, let X_1 and X_2 be the irreducible varieties and let $G_i \subset \mathrm{Bir}(X_i)$, $i = 1, 2$, be the subgroups isomorphic to a finite group G. Assume that fixing the isomorphisms $G \to G_i$, $i = 1, 2$, we obtain the rational actions of G on X_1 and X_2 such that there is a G-equivariant rational dominant map $\varphi\colon X_1 \dashrightarrow X_2$. Let $\pi_{X_i}\colon X_i \dashrightarrow X_i \dot{/} G$, $i = 1, 2$ be the rational quotients (see, e.g., [44, Sect. 1]) and let $\varphi_G\colon X_1 \dot{/} G \dashrightarrow X_2 \dot{/} G$ be the dominant rational map induced by φ. Then the following holds (see, e.g., [52, Sect. 2.6]):

1. The appearing commutative diagram

$$\begin{array}{ccc} X_1 & \overset{\varphi}{\dashrightarrow} & X_2 \\ {\scriptstyle \pi_{X_1}}\downarrow & & \downarrow{\scriptstyle \pi_{X_2}} \\ X_1 \dot{/} G & \overset{\varphi_G}{\dashrightarrow} & X_2 \dot{/} G \end{array} \qquad (21)$$

   is, in fact, cartesian, i.e., $\pi_{X_1}\colon X_1 \dashrightarrow X_1 \dot{/} G$ is obtained from $\pi_{X_2}\colon X_2 \dashrightarrow X_2 \dot{/} G$ by the base change φ_G. In particular, X_1 is birationally G-isomorphic to

$$X_2 \times_{X_2 \dot{/} G} (X_1 \dot{/} G) := \overline{\{(x, y) \in X_2 \times (X_1 \dot{/} G) \mid \pi_{X_2}(x) = \varphi_G(y)\}}.$$

2. For every irreducible variety Z and every dominant rational map $\beta\colon Z \to X_2 \dot{/} G$ such that $X_2 \times_{X_2 \dot{/} G} Z$ is irreducible, the variety $X_2 \times_{X_2 \dot{/} G} Z$ inherits via X_2 a faithful rational action of G such that one obtains commutative diagram (21) with $X_1 = X_2 \times_{X_2 \dot{/} G} Z$, $\varphi_G = \beta$, and $\varphi = \mathrm{pr}_1$.

If such a φ exists, we say that G_1 is *induced* from G_2 *by a base change*. The latter is called *trivial* if φ is a birational isomorphism. If a finite subgroup G of $\mathrm{Bir}(X)$ is not induced by a nontrivial base change, we say that G is *incompressible*.

Example 14. The standard embedding $\mathrm{Cr}_n \hookrightarrow \mathrm{Cr}_{n+1}$ permits to consider the finite subgroups of Cr_n as that of Cr_{n+1}. Every finite subgroup of Cr_{n+1} obtained this way is induced by the nontrivial base change determined by the projection $\mathbf{A}^{n+1} \to \mathbf{A}^n$, $(a_1, \ldots, a_n, a_{n+1}) \mapsto (a_1, \ldots, a_n, a_{n+1})$. □

Example 15. This is Example 6 in [53]. Let G be a finite group that does not embed in $\mathrm{Bir}(Z)$ for any curve Z of genus $\leqslant 1$ (for instance, $G = \mathrm{Sym}_5$) and let X be a smooth projective curve of minimal possible genus such that G is isomorphic to a subgroup of $\mathrm{Aut}(X)$. Then this subgroup of $\mathrm{Bir}(X)$ is incompressible.[7] □

---

[7]The proof in [53] should be corrected as follows. Assume that there is a faithful action of G of a smooth projective curve Y and a dominant G-equivariant morphism $\varphi\colon X \to Y$ of degree $n > 1$. By the construction, X and Y have the same genus $g > 1$, and the Hurwitz formula yields that the

Example 16. By Example 5 in [53], a finite cyclic subgroup of order $\geqslant 2$ in $\mathrm{Bir}(X)$ is never incompressible. □

Example 17. Consider two rational actions of $G := \mathrm{Sym}_3 \times \mathbb{Z}/2\mathbb{Z}$ on $\mathbf{A}^3$. The subgroup Sym_3 acts by natural permuting the coordinates in both cases. The nontrivial element of $Z/2\mathbb{Z}$ acts by $(a_1, a_2, a_3) \mapsto (-a_1, -a_2, -a_3)$ in the first case and by $(a_1, a_2, a_3) \mapsto (a_1^{-1}, a_2^{-1}, a_3^{-1})$ in the second. The surfaces

$$P := \{(a_1, a_2, a_3) \in \mathbf{A}^3 \mid a_1 + a_2 + a_3 = 0\},$$
$$T := \{(a_1, a_2, a_3) \in \mathbf{A}^3 \mid a_1 a_2 a_3 = 1\}$$

are G-stable in, resp., the first and the second case. Since P and T are rational, these actions of G on P and T determine, up to conjugacy, resp., the subgroups G_P and G_T of Cr_2, both isomorphic to G. By Iskovskikh [21] (see also [27, 28]), these subgroups are not conjugate in Cr_2. However, by Lemire et al. [28, Sect. 5], G_T is induced from G_P by a nontrivial base change (of degree 2). □

In fact, Example 17 is a special case (related to the simple algebraic group $\mathbf{G}_2$) of the following.

Example 18. Let G be a connected reductive algebraic group. Recall [27, Def. 1.5] that G is called a Cayley group if there is a birational isomorphism of $\lambda\colon G \dashrightarrow \mathrm{Lie}(G)$, where $\mathrm{Lie}(G)$ is the Lie algebra of G, equivariant with respect to the conjugating and adjoint actions of G on the underlying varieties of G and $\mathrm{Lie}(G)$, respectively, i.e., such that

$$\lambda(gXg^{-1}) = \mathrm{Ad}_G g(\lambda(X)) \tag{22}$$

if g and $X \in G$ and both sides of (22) are defined.

Fix a maximal torus T of G and consider the natural actions of the Weyl group $W = N_G(T)/T$ on T and on $\mathfrak{t} := \mathrm{Lie}(T)$. Since these actions are faithful and T and $\mathfrak{t}$ are rational varieties, this determines, up to conjugacy, two embeddings of W in Cr_r, where $r = \dim T$. Let W_T and $W_{\mathfrak{t}}$ be the images of these embeddings. By Lemire et al. [27, Lemma 3.5(a) and Sect. 1.5], if G is not Cayley and W has no outer autormorphisms, then W_T and $W_{\mathfrak{t}}$ are not conjugate in Cr_r. On the other hand, by Lemire et al. [27, Lemma 10.3], W_T is induced from $W_{\mathfrak{t}}$ by a (nontrivial) base change (see also Lemma 5 below).

This yields, for arbitrary n, the examples of isomorphic nonconjugate finite subgroups of Cr_n one of which is induced from the other by a nontrivial base change. For instance, if $G = \mathbf{SL}_{n+1}$, then $r = n$ and $W = \mathrm{Sym}_n$. Since, by Lemire et al. [27, Thm. 1.31], $\mathbf{SL}_{n+1}$ is not Cayley for $n \geqslant 3$ and Sym_n has no

number of branch points of φ (counted with positive multiplicities) is the integer $(n-1)(2-2g)$. But the latter is negative—a contradiction.

outer automorphisms for $n \neq 6$, the above construction yields for these n two nonconjugate subgroups of Cr_n isomorphic to Sym_n, one of which is induced from the other by a nontrivial base change. □

The following gives a general way of constructing two finite subgroups of Cr_n one of which is induced from the other by a base change.

Consider an n-dimensional irreducible nonsingular variety X and a finite subgroup G of $\mathrm{Aut}(X)$. Suppose that $x \in X$ is a fixed point of G. By Lemma 4, the induced action of G on the tangent space of X at x is faithful. Therefore this action determines, up to conjugacy, a subgroup G_1 of Cr_n isomorphic to G. On the other hand, if X is rational, the action of G on X determines, up to conjugacy, another subgroup G_2 of Cr_n isomorphic to G.

Lemma 5. *G_2 is induced from G_1 by a base change.*

Proof. By Lemma 3 we may assume that X is affine, in which case the claim follows from [27, Lemma 10.3]. □

Corollary 14. *Let X be a nonrational irreducible variety and let G be an incompressible finite subgroup of* $\mathrm{Aut}(X)$. *Then $X^G = \varnothing$.*

Question 9. Which finite subgroups of Cr_2 are incompressible?

Acknowledgements Supported by grants РФФИ 14-01-00160, НШ–2998.2014.1, and the programme *Contemporary Problems of Theoretical Mathematics* of the Branch of Mathematics in the Russian Academy of Sciences.

References

1. S.I. Adian, TheBurnside problem and related topics. Russ. Math. Surv. **65**(5), 805–855 (2011)
2. S.I. Adian, Newbounds of odd periods for which we can prove the infinity of free Burnside groups, in *International Conference "Contemporary Problems of Mathematics, Mechanics, and Mathematical Physics"*. Steklov Math. Inst. RAS, Moscow, 13 May 2013. http://www.mathnet.ru/php/presentation.phtml?&presentid=6786&option_lang=eng
3. M.F. Atiyah, I.G. MacDonald, *Introduction to Commutative Algebra* (Addison-Wesley, Reading, 1969)
4. I. Barsotti, Structure theorems for group varieties. Ann. Mat. Pura Appl. **38**(4), 77–119 (1955)
5. A. Borel, *Linear Algebraic Groups*, 2nd edn. (Springer, New York, 1991)
6. J.W. Cannon, W.J. Floyd, W.R. Parry, Introductory notes on Richard Thompson's groups. L.'Enseignement Math. Revue Internat. IIe Sér. **42**(3), 215–256 (1996)
7. S. Cantat, Morphisms between Cremona groups and a characterization of rational varieties. Preprint (2012), http://perso.univ-rennes1.fr/serge.cantat/publications.html
8. S. Cantat,*Letter of May* 31, 2013 *to V.L. Popov*.
9. D. Cerveau, J. Deserti, Transformations birationnelles de petit degreé (April 2009) [arXiv:0811.2325]
10. Y. Cornulier, Nonlinearity of some subgroups of the planar Cremona group. Preprint (February 2013), http://www.normalesup.org/~cornulier/crelin.pdf
11. Y. Cornulier, Sofic profile and computability of Cremona groups (May 2013) [arXiv:1305.0993]

12. M.J. Collins, On Jordan's theorem for complex linear groups. J. Group Theory **10**, 411–423 (2007)
13. C.W. Curtis, I. Reiner, *Representation Theory of Finite Groups and Associative Algebras* (Wiley, New York, 1962)
14. M. Demazure, Sous-groupes algébriques de rang maximum du groupe de Cremona. Ann. Sci. École Norm. Sup. **3**(4), 507–588 (1970)
15. I. Dolgachev, Infinite Coxeter groups and automorphisms of algebraic surfaces. Contemp. Math. **58**(Part 1), 91–106 (1986)
16. I. Dolgachev, V. Iskovskikh, Finite subgroups of the plane Cremona group, in *Algebra, Arithmetic, and Geometry In Honor of Yu.I. Manin*. Progress in Mathematics, vol. 269 (Birkhäuser, Boston, 2009), pp. 443–548
17. D. Fisher, Groups acting on manifolds: around the Zimmer program, in *Geometry, Rigidity, and Group Actions*. Chicago Lectures in Mathematics (University of Chicago Press, Chicago, 2011), pp. 72–157
18. W. Fulton, *Introduction to Toric Varieties*. Annals of Mathematics Studies, vol. 131 (Princeton University Press, Princeton, 1993)
19. B. Huppert, *Character Theory of Finite Groups*. De Gruyter Expositions in Mathematics, vol. 25 (Walter de Gruyter, Berlin, 1998)
20. T. Igarashi, *Finite Subgroups of the Automorphism Group of the Affine Plane*. M.A. thesis, Osaka University, 1977
21. V.A. Iskovskikh, Two non-conjugate embeddings of $\mathrm{Sym}_3 \times \mathbb{Z}_2$ into the Cremona group. Proc. Steklov Inst. Math. **241**, 93–97 (2003)
22. V.A. Iskovskikh, I.R. Shafarevich, Algebraic surfaces, in *Algebraic Geometry, II*. Encyclopaedia of Mathematical Sciences, vol. 35 (Springer, Berlin, 1996), pp. 127–262
23. C. Jordan, Mémoire sur les equations différentielle linéaire à intégrale algébrique. J. Reine Angew. Math. **84**, 89–215 (1878)
24. J. Kollár, Y. Miyaoka, S. Mori, H. Takagi, Boundedness of canonical $\mathbb{Q}$-Fano 3-folds. Proc. Jpn. Acad. Ser. A Math. Sci. **76**, 73–77 (2000)
25. S. Kwasik, R. Schultz, *Finite symmetries of* $\mathbb{R}^4$ *and* S^4. Preprint (2012)
26. S. Lang, *Algebra* (Addison-Wesley, Reading, 1965)
27. N. Lemire, V. Popov, Z. Reichstein, Cayley groups. J. Am. Math. Soc. **19**(4), 921–967 (2006)
28. N. Lemire, V. Popov, Z. Reichstein, On the Cayley degree of an algebraic group, in *Proceedings of the XVIth Latin American Algebra Colloquium*. Bibl. Rev. Mat. Iberoamericana, Rev. Mat. (Iberoamericana, Madrid, 2007), pp. 87–97
29. M. Maruyama, On automorphisms of ruled surfaces. J. Math. Kyoto Univ. **11-1**, 89–112 (1971)
30. H. Matsumura, On algebraic groups of birational tansformations. Rend. Accad. Naz. Lincei Ser. VIII **34**, 151–155 (1963)
31. T. Matsusaka, Polarized varieties, fields of moduli and generalized Kummer varieties of polarized varieties. Am. J. Math. **80**, 45–82 (1958)
32. W.H. Meeks, Sh.-T. Yau, Group actions on $\mathbb{R}^3$, in *The Smith Conjecture* (New York, 1979). Pure Appl. Math., vol. 112 (Academic, Orlando, 1984), pp. 167–179
33. L. Moser-Jauslin, Automorphism groups of Koras–Russell threefolds of the first kind, in *Affine Algebraic Geometry*. CRM Proceedings and Lecture Notes, vol. 54 (American Mathematical Society, Providence, 2011), pp. 261–270
34. G. Mostow, Fully reducible subgroups of algebraic groups. Am. J. Math. **78**, 200–221 (1956)
35. I. Mundet i Riera, Jordan's theorem for the diffeomorphism group of some manifolds. Proc. Am. Math. Soc. **138**(6), 2253–2262 (2010)
36. I. Mundet i Riera, *Letter of July* 30, 2013 *to V.L. Popov*
37. I. Mundet i Riera, *Letters of October* 30 *and* 31, 2013 *to V.L. Popov*
38. I. Mundet i Riera, *Finite Group Actions on Manifolds Without Odd Cohomology* (November 2013) [arXiv:1310.6565v2]
39. T. Oda, *Covex Bodies and Algebraic Geometry*. Ergebnisse der Mathematik und ihrer Grenzgebiete 3. Folge, Bd. 15 (Springer, Berlin, 1988)

40. A.Yu. Olshanskii, Groups of bounded period with subgroups of prime order. Algebra Logic **21**, 369–418 (1983) [Translation of Algebra i Logika **21**, 553–618 (1982)]
41. A.Yu. Ol'shanskiĭ, *Letters of August* 15 *and* 23, 2013 *to V.L. Popov*
42. V.L. Popov, Algebraic curves with an infinite automorphism group. Math. Notes **23**, 102–108 (1978)
43. V.L. Popov, On the Makar-Limanov, Derksen invariants, and finite automorphism groups of algebraic varieties, in *Affine Algebraic Geometry: The Russell Festschrift*. CRM Proceedings and Lecture Notes, vol. 54 (American Mathematical Society, Providence, 2011), pp. 289–311 (January 2010) [arXiv:1001.1311]
44. V.L. Popov, Some subgroups of the Cremona groups, in *Affine Algebraic Geometry*. Proceedings (Osaka, Japan, 3–6 March 2011) (World Scientific, Singapore, 2013), pp. 213–242 (October 2011) [arXiv:1110.2410]
45. V.L. Popov, Tori in the Cremona groups. Izvestiya Math. **77**(4), 742–771 (2013) (July 2012) [arXiv:1207.5205]
46. V.L. Popov, *Problems for the Problem Session* (CIRM, Trento) (November 2012), http://www.science.unitn.it/cirm/Trento_postersession.html
47. V.L. Popov, *Finite Subgroups of Diffeomorphism Groups* (October 2013) [arXiv:1310.6548]
48. V.L. Popov, E.B. Vinberg, Invariant theory, in *Algebraic Geometry IV*. Encyclopaedia of Mathematical Sciences, vol. 55 (Springer, Berlin, 1994), pp. 123–284
49. Y. Prokhorov, C. Shramov, *Jordan Property for Cremona Groups* (June 2013) [arXiv:1211.3563]
50. Y. Prokhorov, C. Shramov, *Jordan Property for Groups of Birational Selfmaps* (July 2013) [arXiv:1307.1784]
51. V. Puppe, Do manifolds have little symmetry? J. Fixed Point Theory Appl. **2**(1), 85–96 (2007)
52. Z. Reichstein, On the notion of essential dimension for algebraic groups. Transform. Gr. **5**(3), 265–304 (2000)
53. Z. Reichstein, Compression of group actions, in *Invariant Theory in All Characteristics*. CRM Proceedings and Lecture Notes, vol. 35 (Amererican Mathematical Society, Providence, 2004), pp. 199–202
54. Z. Reichstein, B. Youssin, A birational invariant for algebraic group actions. Pac. J. Math. **204**, 223–246 (2002)
55. M. Rosenlicht, Some basic theorems on algebraic groups. Am. J. Math. **78**, 401–443 (1956)
56. M. Rosenlicht, Some rationality questions on algebraic groups. Ann. Mat. Pura Appl. **43**(4), 25–50 (1957)
57. J.-P. Serre, Sous-groupes finis des groupes de Lie, in *Séminaire N. Bourbaki* 1998/–99, Exp. no. 864. Astérisque, vol. 266 (Société Mathématique de France, Paris, 2000), pp. 415–430
58. J.-P. Serre, *Le groupe de Cremona et ses sous-groupes finis*. Séminaire Bourbaki, no. 1000, Novembre 2008, 24 pp.
59. J.-P. Serre, A Minkowski-style bound for the orders of the finite subgroups of the Cremona group of rank 2 over an arbitrary field. Moscow Math. J. **9**(1), 183–198 (2009)
60. J.-P. Serre, How to use finite fields for problems concerning infinite fields. Contemp. Math. **487**, 183–193 (2009)
61. T.A. Springer, *Linear Algebraic Groups*, 2nd edn. Progress in Mathematics, vol. 9 (Birkhäuser, Boston, 1998)
62. Y.G. Zarhin, Theta groups and products of abelian and rational varieties. Proc. Edinb. Math. Soc. (2) **57**(1), 299–304 (2014) [arXiv:1006.1112]

2-Elementary Subgroups of the Space Cremona Group

Yuri Prokhorov

Abstract We give a sharp bound for orders of elementary abelian two-groups of birational automorphisms of rationally connected threefolds.

Subject Classification: 14E07, 14E09, 14E30, 14E35, 14E08

1 Introduction

Throughout this paper we work over $\Bbbk$, an algebraically closed field of characteristic 0. Recall that the *Cremona group* $\mathrm{Cr}_n(\Bbbk)$ is the group of birational transformations of the projective space $\mathbb{P}^n_{\Bbbk}$. We are interested in finite subgroups of $\mathrm{Cr}_n(\Bbbk)$. For $n = 2$ these subgroups are classified basically (see [5] and references therein) but for $n \geq 3$ the situation becomes much more complicated. There are only a few, very specific classification results (see e.g. [14, 15, 18]).

Let p be a prime number. A group G is said to be *p-elementary abelian* of rank r if $G \simeq (\mathbb{Z}/p\mathbb{Z})^r$. In this case we denote $\mathrm{r}(G) := r$. A. Beauville [3] obtained a sharp bound for the rank of p-elementary abelian subgroups of $\mathrm{Cr}_2(\Bbbk)$.

Y. Prokhorov (✉)
Steklov Mathematical Institute, 8 Gubkina str., Moscow 119991, Russia

Laboratory of Algebraic Geometry, SU-HSE, 7 Vavilova str., Moscow 117312, Russia

Faculty of Mathematics, Department of Algebra, Moscow State University, Vorobievy Gory, Moscow 119991, Russia
e-mail: prokhoro@gmail.com

I. Cheltsov et al. (eds.), *Automorphisms in Birational and Affine Geometry*, Springer Proceedings in Mathematics & Statistics 79,
DOI 10.1007/978-3-319-05681-4_12,

Theorem 1.1 ([3]). *Let* $G \subset \mathrm{Cr}_2(\Bbbk)$ *be a 2-elementary abelian subgroup. Then* $\mathrm{r}(G) \leq 4$. *Moreover, this bound is sharp and such groups* G *with* $\mathrm{r}(G) = 4$ *are classified up to conjugacy in* $\mathrm{Cr}_2(\Bbbk)$.

The author [14] was able to get a similar bound for p-elementary abelian subgroups of $\mathrm{Cr}_3(\Bbbk)$ which is sharp for $p \geq 17$.

In this paper we improve this result in the case $p = 2$. We study 2-elementary abelian subgroups acting on rationally connected threefolds. In particular, we obtain a *sharp* bound for the rank of such subgroups in $\mathrm{Cr}_3(\Bbbk)$. Our main result is the following.

Theorem 1.2. *Let* Y *be a rationally connected three-dimensional algebraic variety over* $\Bbbk$ *and let* $G \subset \mathrm{Bir}_{\Bbbk}(Y)$ *be a 2-elementary abelian group. Then* $\mathrm{r}(G) \leq 6$.

Corollary 1.3. *Let* $G \subset \mathrm{Cr}_3(\Bbbk)$ *be a 2-elementary abelian group. Then* $\mathrm{r}(G) \leq 6$ *and the bound is sharp (see Example* 3.4).

Unfortunately we are not able to classify all the birational actions $G \hookrightarrow \mathrm{Bir}_{\Bbbk}(Y)$ as above attaining the bound $\mathrm{r}(G) \leq 6$ (cf. [3]). However, in some cases we get a description of these "extremal" actions.

The structure of the paper is as follows. In Sect. 3 we reduce the problem to the study of biregular actions of 2-elementary abelian groups on Fano-Mori fiber spaces and investigate the case of nontrivial base. A few facts about actions of 2-elementary abelian groups on Fano threefolds are discussed in Sect. 4. In Sect. 5 (resp. Sect. 6) we study actions on non-Gorenstein (resp. Gorenstein) Fano threefolds. Our main theorem is a direct consequence of Propositions 3.2, 5.1, and 6.1.

2 Preliminaries

Notation.

- For a group G, $\mathrm{r}(G)$ denotes the minimal number of generators. In particular, if G is an elementary abelian p-group, then $G \simeq (\mathbb{Z}/p\mathbb{Z})^{r(G)}$.
- $\mathrm{Fix}(G, X)$ (or simply $\mathrm{Fix}(G)$ if no confusion is likely) denotes the fixed point locus of an action of G on X.

Terminal Singularities. Recall that the *index* of a terminal singularity $(X \ni P)$ is a minimal positive integer r such that K_X is a Cartier divisor at P.

Lemma 2.1. *Let* $(X \ni P)$ *be a germ of a threefold terminal singularity and let* $G \subset \mathrm{Aut}(X \ni P)$ *be a 2-elementary abelian subgroup. Then* $\mathrm{r}(G) \leq 4$. *Moreover, if* $\mathrm{r}(G) = 4$, *then* $(X \ni P)$ *is not a cyclic quotient singularity.*

Proof. Let m be the index of $(X \ni P)$. Consider the index-one cover $\pi\colon (X^\sharp \ni P^\sharp) \to (X \ni P)$ (see [19]). Here $(X^\sharp \ni P^\sharp)$ is a terminal point of index 1 (or smooth) and π is a cyclic cover of degree m which is étale outside of P.

Thus $X \ni P$ is the quotient of $X^\sharp \ni P$ by a cyclic group of order m. If $m = 1$, we take π to be the identity map. We may assume that $\Bbbk = \mathbb{C}$ and then the map $X^\sharp \setminus \{P^\sharp\} \to X \setminus \{P\}$ can be regarded as the topological universal cover. Hence there exists a natural lifting $G^\sharp \subset \mathrm{Aut}(X^\sharp \ni P^\sharp)$ fitting in the following exact sequence

$$1 \longrightarrow C_m \longrightarrow G^\sharp \longrightarrow G \longrightarrow 1, \qquad (*)$$

where $C_m \simeq \mathbb{Z}/m\mathbb{Z}$. We claim that $G^\sharp$ is abelian. Assume the converse, then $m \geq 2$. The group $G^\sharp$ permutes the eigenspaces of C_m in the Zariski tangent space $T_{P^\sharp,X^\sharp}$. Let $n := \dim T_{P^\sharp,X^\sharp}$ be the embedded dimension. By the classification of three-dimensional terminal singularities [10, 19] we have one of the following:

$$\begin{array}{llll} (1)\ \tfrac{1}{m}(a,-a,b), & n=3, & \gcd(a,m)=\gcd(b,m)=1; & \\ (2)\ \tfrac{1}{m}(a,-a,b,0), & n=4, & \gcd(a,m)=\gcd(b,m)=1; & \qquad (**)\\ (3)\ \tfrac{1}{4}(a,-a,b,2), & n=4, & \gcd(a,2)=\gcd(b,2)=1, \quad m=4, & \end{array}$$

where $\frac{1}{m}(a_1,\dots,a_n)$ denotes the diagonal action

$$x_k \longmapsto \exp(2\pi i a_k/m)\cdot x_k, \quad k=1,\dots,n.$$

Put $T = T_{P^\sharp,X^\sharp}$ in the first case and denote by $T \subset T_{P^\sharp,X^\sharp}$ the three-dimensional subspace $x_4 = 0$ in the second and the third cases. Then C_m acts on T freely outside of the origin and T is $G^\sharp$-invariant. By (*) we see that the derived subgroup $[G^\sharp, G^\sharp]$ is contained in C_m. In particular, $[G^\sharp, G^\sharp]$ is abelian and also acts on T freely outside of the origin. Assume that $[G^\sharp, G^\sharp] \neq \{1\}$. Since $\dim T = 3$, this implies that the representation of $G^\sharp$ on T is irreducible (otherwise T has a one-dimensional invariant subspace, say T_1, and the kernel of the map $G^\sharp \to GL(T_1) \simeq \Bbbk^*$ must contain $[G^\sharp, G^\sharp]$). In particular, the eigenspaces of C_m on T have the same dimension. Since T is irreducible, the order of $G^\sharp$ is divisible by $3 = \dim T$ and so $m > 2$. In this case, by the above description of the action of C_m on $T_{P^\sharp,X^\sharp}$ we get that there are exactly three distinct eigenspaces $T_i \subset T$. The action of $G^\sharp$ on the set $\{T_i\}$ induces a transitive homomorphism $G^\sharp \to \mathfrak{S}_3$ whose kernel contains C_m. Hence we have a transitive homomorphism $G \to \mathfrak{S}_3$. Since G is a two-group, this is impossible.

Thus $G^\sharp$ is abelian. Then

$$\mathrm{r}(G) \leq \mathrm{r}(G^\sharp) \leq \dim T_{P^\sharp,X^\sharp}.$$

This proves our statement. □

Remark 2.2. If in the above notation the action of G on X is free in codimension one, then $\mathrm{r}(G) \leq \dim T_{P^\sharp,X^\sharp} - 1$.

For convenience of references, we formulate the following easy result.

Lemma 2.3. *Let G be a 2-elementary abelian group and let X be a G-threefold with isolated singularities.*

(i) *If* $\dim \mathrm{Fix}(G) > 0$, *then* $\dim \mathrm{Fix}(G) + \mathrm{r}(G) \le 3$.
(ii) *Let* $\delta \in G\backslash\{1\}$ *and let* $S \subset \mathrm{Fix}(\delta)$ *be the union of two-dimensional components. Then S is G-invariant and smooth in codimension* 1.

Sketch of the proof. Consider the action of G on the tangent space to X at a general point of a component of $\mathrm{Fix}(G)$ (resp. at a general point of $\mathrm{Sing}(S)$). □

3 G-Equivariant Minimal Model Program

Definition 3.1. Let G be a finite group. A *G-variety* is a variety X provided with a biregular faithful action of G. We say that a normal G-variety X is *$G\mathbb{Q}$-factorial* if any G-invariant Weil divisor on X is $\mathbb{Q}$-Cartier.

The following construction is standard (see e.g. [15]).

Let Y be a rationally connected three-dimensional algebraic variety and let $G \subset \mathrm{Bir}(Y)$ be a finite subgroup. Taking an equivariant compactification and running an equivariant minimal model program we get a G-variety X and a G-equivariant birational map $Y \dashrightarrow X$, where X has a structure a *G-Fano-Mori fibration* $f : X \to B$. This means that X has at worst terminal $G\mathbb{Q}$-factorial singularities, f is a G-equivariant morphism with connected fibers, B is normal, $\dim B < \dim X$, the anticanonical Weil divisor $-K_X$ is ample over B, and the relative G-invariant Picard number $\rho(X)^G$ equals to one. Obviously, in the case $\dim X = 3$ we have the following possibilities:

(C) B is a rational surface and a general fiber $f^{-1}(b)$ is a conic;
(D) $B \simeq \mathbb{P}^1$ and a general fiber $f^{-1}(b)$ is a smooth del Pezzo surface;
(F) B is a point and X is a *$G\mathbb{Q}$-Fano threefold*, that is, X is a Fano threefold with at worst terminal $G\mathbb{Q}$-factorial singularities and such that $\mathrm{Pic}(X)^G \simeq \mathbb{Z}$. In this situation we say that X is *G-Fano threefold* if X is Gorenstein, that is, K_X is a Cartier divisor.

Proposition 3.2. *Let G be a 2-elementary abelian group and let $f : X \to B$ be a G-Fano-Mori fibration with* $\dim X = 3$ *and* $\dim B > 0$. *Then* $\mathrm{r}(G) \le 6$. *Moreover, if* $\mathrm{r}(G) = 6$ *and* $B \simeq \mathbb{P}^1$, *then a general fiber $f^{-1}(b)$ is a del Pezzo surface of degree* 4 *or* 8.

Proof. Let $G_f \subset G$ (resp. $G_B \subset \mathrm{Aut}(B)$) be the kernel (resp. the image) of the homomorphism $G \to \mathrm{Aut}(B)$. Thus G_B acts faithfully on B and G_f acts faithfully on the generic fiber $F \subset X$ of f. Clearly, G_f and G_B are 2-elementary groups with $\mathrm{r}(G_f) + \mathrm{r}(G_B) = \mathrm{r}(G)$. Assume that $B \simeq \mathbb{P}^1$. Then $\mathrm{r}(G_B) \le 2$ by the classification of finite subgroups of $PGL_2(\Bbbk)$. By Theorem 1.1 we have $\mathrm{r}(G_f) \le 4$. If furthermore $\mathrm{r}(G) = 6$, then $\mathrm{r}(G_f) = 4$ and the assertion about F follows by Lemma 3.3 below. This proves our assertions in the case $B \simeq \mathbb{P}^1$. The case $\dim B = 2$ is treated similarly. □

Lemma 3.3 (cf. [3]). *Let F be a del Pezzo surface and let $G \subset \mathrm{Aut}(F)$ be a 2-elementary abelian group with $\mathrm{r}(F) \ge 4$. Then $\mathrm{r}(F) = 4$ and one of the following holds:*

(i) $K_F^2 = 4$, $\rho(F)^G = 1$;
(ii) $K_F^2 = 8$, $\rho(F)^G = 2$.

Proof. Similar to [3, §3]. □

Example 3.4. Let $F \subset \mathbb{P}^4$ be the quartic del Pezzo surface given by $\sum x_i^2 = \sum \lambda_i x_i^2 = 0$ with $\lambda_i \ne \lambda_j$ for $i \ne j$ and let $G_f \subset \mathrm{Aut}(F)$ be the 2-elementary abelian subgroup generated by involutions $x_i \mapsto -x_i$. Consider also a 2-elementary abelian subgroup $G_B \subset \mathrm{Aut}(\mathbb{P}^1)$ induced by a faithful representation $\mathrm{Q}_8 \to GL_2(\Bbbk)$ of the quaternion group Q_8. Then $\mathrm{r}(G_f) = 4$, $\mathrm{r}(G_B) = 2$, and $G := G_f \times G_B$ naturally acts on $X := F \times \mathbb{P}^1$. Two projections give us two structures of G-Fano-Mori fibrations of types (D) and (C). This shows that the bound $\mathrm{r}(G) \le 6$ in Proposition 3.2 is sharp. Moreover, X is rational and so we have an embedding $G \subset \mathrm{Cr}_3(\Bbbk)$.

4 Actions on Fano Threefolds

Main Assumption. From now on we assume that we are in the case (F), that is, X is a $G\mathbb{Q}$-Fano threefold.

Remark 4.1. The group G acts naturally on the space of anticanonical sections $H^0(X, -K_X)$. Assume that $H^0(X, -K_X) \ne 0$. Since G is an abelian group, there exists a decomposition if $H^0(X, -K_X)$ into eigenspaces. Then any eigensection $s \in H^0(X, -K_X)$ defines an invariant member $S \in |-K_X|$.

Lemma 4.2. *Let X be a $G\mathbb{Q}$-Fano threefold, where G is a 2-elementary abelian group with $\mathrm{r}(G) \ge 5$. Let S be an invariant effective Weil divisor such that $-(K_X + S)$ is nef. Then the pair (X, S) is log canonical (lc). In particular, S is reduced. If $-(K_X + S)$ is ample, then the pair (X, S) is purely log terminal (plt).*

Proof. Assume that the pair (X, S) is not lc. Since S is G-invariant and $\rho(X)^G = 1$, we see that S is numerically proportional to K_X. Hence S is ample. We apply quite standard connectedness arguments of Shokurov [22] (see, e.g., [11, Prop. 2.6]): for a suitable G-invariant boundary D, the pair (X, D) is lc, the divisor $-(K_X + D)$ is ample, and the minimal locus V of log canonical singularities is also G-invariant. Moreover, V is either a point or a smooth rational curve.

By Lemma 2.1 we may assume that G has no fixed points. Hence, $V \simeq \mathbb{P}^1$ and we have a map $\varsigma : G \to \mathrm{Aut}(\mathbb{P}^1)$. By Lemma 2.3 $\mathrm{r}(\ker \varsigma) \le 2$. Therefore, $\mathrm{r}(\varsigma(G)) \ge 3$. This contradicts the classification of finite subgroups of $PGL_2(\Bbbk)$.

If $-(K_X + S)$ is ample and (X, S) has a log canonical center of dimension ≤ 1, then by considering $(X, S' = S + \epsilon B)$, where B is a suitable invariant divisor and $0 < \epsilon \ll 1$, we get a non-lc pair (X, S'). This contradicts the above considered case. □

Corollary 4.3. *Let X be a $G\mathbb{Q}$-Fano threefold, where G is a 2-elementary abelian group with $\mathrm{r}(G) \ge 6$ and let S be an invariant Weil divisor. Then $-(K_X + S)$ is not ample.*

Proof. If $-(K_X + S)$ is ample, then by Lemma 4.2 the pair (X, S) is plt. By the adjunction principle [22] the surface S is irreducible, normal and has only quotient singularities. Moreover, $-K_S$ is ample. Hence S is rational. We get a contradiction by Theorem 1.1 and Lemma 2.3(i). □

Lemma 4.4. *Let S be a K3 surface with at worst Du Val singularities and let $\Gamma \subset \mathrm{Aut}(S)$ be a 2-elementary abelian group. Then $\mathrm{r}(\Gamma) \le 5$.*

Proof. Let $\tilde{S} \to S$ be the minimal resolution. Here $\tilde{S}$ is a smooth K3 surface and the action of Γ lists to $\tilde{S}$. Let $\Gamma_s \subset \Gamma$ be the largest subgroup that acts trivially on $H^{2,0}(\tilde{S}) \simeq \mathbb{C}$. The group Γ/Γ_s is cyclic. Hence, $\mathrm{r}(\Gamma/\Gamma_s) \le 1$. According to [13, Th. 4.5] we have $\mathrm{r}(\Gamma_s) \le 4$. Thus $\mathrm{r}(\Gamma) \le 5$. □

Corollary 4.5. *Let X be a $G\mathbb{Q}$-Fano threefold, where G is a 2-elementary abelian group. Let $S \in |-K_X|$ be a G-invariant member. If $\mathrm{r}(G) \ge 7$, then the singularities of S are worse than Du Val.*

Proposition 4.6. *Let X be a $G\mathbb{Q}$-Fano threefold, where G is a 2-elementary abelian group with $\mathrm{r}(G) \ge 6$. Let $S \in |-K_X|$ be a G-invariant member and let $G_\bullet \subset G$ be the largest subgroup that acts trivially on the set of components of S. One of the following holds:*

(i) *S is a K3 surface with at worst Du Val singularities, then $S \subset \mathrm{Fix}(\delta)$ for some $\delta \in G \setminus \{1\}$ and $G/\langle\delta\rangle$ faithfully acts on S. In this case $\mathrm{r}(G) = 6$.*
(ii) *The surface S is reducible (and reduced). The group G acts transitively on the components of S and $G_\bullet$ acts faithfully on each component $S_i \subset S$. There are two subcases:*
   (a) *any component $S_i \subset S$ is rational and $\mathrm{r}(G_\bullet) \le 4$.*
   (b) *any component $S_i \subset S$ is birationally ruled over an elliptic curve and $\mathrm{r}(G_\bullet) \le 5$.*

Proof. By Lemma 4.2 the pair (X, S) is lc. Assume that S is normal (and irreducible). By the adjunction formula $K_S \sim 0$. We claim that S has at worst Du Val singularities. Indeed, otherwise by the Connectedness Principle [22, Th. 6.9] S

has at most two non-Du Val points. These points are fixed by an index two subgroup $G' \subset G$. This contradicts Lemma 2.1. Taking Lemma 4.4 into account we get the case (i).

Now assume that S is not normal. Let $S_i \subset S$ be an irreducible component (the case $S_i = S$ is not excluded). If the action on components $S_i \subset S$ is not transitive, there is an invariant divisor $S' < S$. Since X is $G\mathbb{Q}$-factorial and $\rho(X)^G = 1$, the divisor $-(K_X + S')$ is ample. This contradicts Corollary 4.3.

By Lemma 2.3(ii) the action of $G_\bullet$ on each component S_i is faithful.

If S_i is a rational surface, then $\mathrm{r}(G_\bullet) \le 4$ by Theorem 1.1. Assume that S_i is not rational. Let $\nu\colon S' \to S_i$ be the normalization. Write $0 \sim \nu^*(K_X + S) = K_{S'} + D'$, where D' is the *different*, see [22, §3]. Here D' is an effective reduced divisor and the pair is lc [22, 3.2]. Since S is not normal, $D' \neq 0$. Consider the minimal resolution $\mu\colon \tilde{S} \to S'$ and let $\tilde{D}$ be the crepant pull-back of D', that is, $\mu_* \tilde{D} = D'$ and

$$K_{\tilde{S}} + \tilde{D} = \mu^*(K_{S'} + D') \sim 0.$$

Here $\tilde{D}$ is again an effective reduced divisor. Hence $\tilde{S}$ is a ruled surface. Consider the Albanese map $\alpha : \tilde{S} \to C$. Let $\tilde{D}_1 \subset \tilde{D}$ be an α-horizontal component. By the adjunction formula $\tilde{D}_1$ is an elliptic curve and so C is. Let Γ be the image of $G_\bullet$ in $\mathrm{Aut}(C)$. Then $\mathrm{r}(\Gamma) \le 3$ and so $\mathrm{r}(G_\bullet) \le 5$. So, the last assertion is proved. □

5 Non-Gorenstein Fano Threefolds

Let G be a 2-elementary abelian group and let X be $G\mathbb{Q}$-Fano threefold. In this section we consider the case where X is non-Gorenstein, i.e., it has at least one terminal point of index > 1. We denote by $\mathrm{Sing}'(X) = \{P_1, \dots, P_M\}$ the set of non-Gorenstein points.

Recall that any (analytic) threefold terminal singularity $U \ni P$ has a small deformation U_t, where $t \in$ (unit disk) $\subset \mathbb{C}$, such that for $0 < |t| \ll 1$ the threefold $U_t \ni P_{i,t}$ has only cyclic quotient singularities $U_t \ni P_{i,t}$ of the form $\frac{1}{m_i}(1, -1, a_i)$ with $\gcd(m_i, a_i) = 1$ [19]. The collection $\mathbf{B}(U, P) := \left\{\frac{1}{m_i}(1, -1, a_i)\right\}$ does not depend on the choice of deformation and called the *basket* of $U \ni P$. For a threefold X with terminal singularities we denote by $\mathbf{B} = \mathbf{B}(X)$ its *global basket*, the union of baskets of all singular points.

Proposition 5.1. *Let X be a non-Gorenstein Fano threefold with terminal singularities. Assume that X admits a faithful action of a 2-elementary abelian group G with $\mathrm{r}(G) \ge 6$. Then $\mathrm{r}(G) = 6$, G transitively acts on $\mathrm{Sing}'(X)$, $|-K_X| \neq \emptyset$, and the configuration of non-Gorenstein singularities is described below.*

(1) $M = 8$, $\mathbf{B}(X, P_i) = \left\{\frac{1}{2}(1, 1, 1)\right\}$;
(2) $M = 8$, $\mathbf{B}(X, P_i) = \left\{\frac{1}{3}(1, 1, 2)\right\}$;
(3) $M = 4$, $\mathbf{B}(X, P_i) = \left\{2 \times \frac{1}{2}(1, 1, 1)\right\}$;

(4) $M = 4$, $\mathbf{B}(X, P_i) = \left\{2 \times \frac{1}{3}(1,1,2)\right\}$;
(5) $M = 4$, $\mathbf{B}(X, P_i) = \left\{3 \times \frac{1}{2}(1,1,1)\right\}$;
(6) $M = 4$, $\mathbf{B}(X, P_i) = \left\{\frac{1}{4}(1,-1,1), \frac{1}{2}(1,1,1)\right\}$.

Proof. Let $P^{(1)}, \dots, P^{(n)} \in \operatorname{Sing}'(X)$ be representatives of distinct G-orbits and let G_i be the stabilizer of $P^{(i)}$. Let $r := \mathrm{r}(G)$, $r_i := \mathrm{r}(G_i)$, and let $m_{i,1}, \dots, m_{i,\nu_i}$ be the indices of points in the basket of $P^{(i)}$. We may assume that $m_{i,1} \ge \dots \ge m_{i,\nu_i}$ By the orbifold Riemann–Roch formula [19] and a form of Bogomolov–Miyaoka inequality [8, 9] we have

$$\sum_{i=1}^{n} 2^{r-r_i} \sum_{j=1}^{\nu_i} \left(m_{i,j} - \frac{1}{m_{i,j}}\right) < 24. \tag{***}$$

If P is a cyclic quotient singularity, then $\nu_i = 1$ and by Lemma 2.1 $r_i \le 3$. If P is not a cyclic quotient singularity, then $\nu_i \ge 2$ and again by Lemma 2.1 $r_i \le 4$. Since $m_{i,j} - 1/m_{i,j} \ge 3/2$, in both cases we have

$$2^{r-r_i} \sum_{j=1}^{\nu_i} \left(m_{i,j} - \frac{1}{m_{i,j}}\right) \ge 3 \cdot 2^{r-4} \ge 12.$$

Therefore, $n = 1$, i.e., G transitively acts on $\operatorname{Sing}'(X)$, and $r = 6$.

If P is not a point of type $cAx/4$ (i.e., it is not as in (3) of (**)), then by the classification of terminal singularities [19] $m_{1,1} = \dots = m_{1,\nu_i}$ and (***) has the form

$$24 > 2^{6-r_1} \nu_1 \left(m_{1,1} - \frac{1}{m_{1,1}}\right) \ge 8 \left(m_{1,1} - \frac{1}{m_{1,1}}\right).$$

Hence $r_1 \ge 3$, $\nu_1 \le 3$, $m_{1,1} \le 3$, and $3 \cdot 2^{r_1-3} \ge \nu_1 m_{1,1}$. If $r_1 = 3$, then $\nu_1 = 1$. If $r_1 = 4$, then $\nu_1 \ge 2$ and $\nu_1 m_{1,1} \le 6$. This gives us the possibilities (1)–(5).

Assume that P is a point of type $cAx/4$. Then $m_{1,1} = 4$, $\nu_1 > 1$, and $m_{1,j} = 2$ for $1 < j \le \nu_1$. Thus (***) has the form

$$24 > 2^{6-r_1} \left(\frac{15}{4} + \frac{3}{2}(\nu_1 - 1)\right) = 2^{4-r_1} (9 + 6\nu_1).$$

We get $\nu_1 = 2$, $r_1 = 4$, i.e., the case (6).

Finally, the computation of $\dim |-K_X|$ follows by the orbifold Riemann–Roch formula [19]

$$\dim |-K_X| = -\frac{1}{2} K_X^3 + 2 - \sum_{P \in \mathbf{B}(X)} \frac{b_P (m_P - b_P)}{2 m_P}.$$

6 Gorenstein Fano Threefolds

The main result of this section is the following:

Proposition 6.1. *Let G be a 2-elementary abelian group and let X be a (Gorenstein) G-Fano threefold. Then* $\mathrm{r}(G) \le 6$. *Moreover, if* $\mathrm{r}(G) = 6$, *then* $\mathrm{Pic}(X) = \mathbb{Z} \cdot K_X$ *and* $-K_X^3 \ge 8$.

Let X be a Fano threefold with at worst Gorenstein terminal singularities. Recall that the number

$$\iota(X) := \max\{i \in \mathbb{Z} \mid -K_X \sim iA,\ A \in \mathrm{Pic}(X)\}$$

is called the *Fano index* of X. The integer $g = \mathrm{g}(X)$ such that $-K_X^3 = 2g - 2$ is called the *genus* of X. It is easy to see that $\dim |-K_X| = g+1$ [7, Corollary 2.1.14]. In particular, $|-K_X| \ne \emptyset$.

Notation. Throughout this section G denotes a 2-elementary abelian group and X denotes a Gorenstein G-Fano threefold. There exists an invariant member $S \in |-K_X|$ (see 4.1). We write $S = \sum_{i=1}^{N} S_i$, where the S_i are irreducible components. Let $G_\bullet \subset G$ be the kernel of the homomorphism $G \to \mathfrak{S}_N$ induced by the action of G on $\{S_1, \dots, S_N\}$. Since G is abelian and the action of G on $\{S_1, \dots, S_N\}$ is transitive, the group $G_\bullet$ coincides with the stabilizer of any S_i. Clearly, $N = 2^{\mathrm{r}(G)-\mathrm{r}(G_\bullet)}$. If $\mathrm{r}(G) \ge 6$, then by Proposition 4.6 we have $\mathrm{r}(G_\bullet) \le 5$ and so $N \ge 2^{\mathrm{r}(G)-5}$.

Lemma 6.2. *Let $G \subset \mathrm{Aut}(\mathbb{P}^n)$ be a 2-elementary subgroup and n is even. Then G is conjugate to a diagonal subgroup. In particular,* $\mathrm{r}(G) \le n$.

Proof. Let $G^\sharp \subset SL_{n+1}(\Bbbk)$ be the lifting of G and let $G' \subset G^\sharp$ be a Sylow two-subgroup. Then $G' \simeq G$. Since G' is abelian, the representation $G' \hookrightarrow SL_{n+1}(\Bbbk)$ is diagonalizable. □

Corollary 6.3. *Let $Q \subset \mathbb{P}^4$ be a quadric and let $G \subset \mathrm{Aut}(Q)$ be a 2-elementary subgroup. Then* $\mathrm{r}(G) \le 4$.

Lemma 6.4. *Let $G \subset \mathrm{Aut}(\mathbb{P}^3)$ be a 2-elementary subgroup. Then* $\mathrm{r}(G) \le 4$.

Certainly, the fact follows by Blichfeldt's theorem which asserts that the lifting $G^\sharp \subset SL_4(\Bbbk)$ is a monomial representation (see e.g. [20, §3]). Here we give a short independent proof.

Proof. Assume that $\mathrm{r}(G) \ge 5$. Take any element $\delta \in G \setminus \{1\}$. By Lemma 2.1 the group G has no fixed points. Since the set $\mathrm{Fix}(\delta)$ is G-invariant, $\mathrm{Fix}(\delta) = L_1 \cup L_2$, where L_1, $L_2 \subset \mathbb{P}^3$ are skew lines.

Let $G_1 \subset G$ be the stabilizer of L_1. There is a subgroup $G_2 \subset G_1$ of index 2 having a fixed point $P \in L_1$. Thus $\mathrm{r}(G_2) \ge 3$ and the "orthogonal" plane Π is G_2-invariant. By Lemma 6.2 there exists an element $\delta' \in G_2$ that acts trivially on Π, i.e., $\Pi \subset \mathrm{Fix}(\delta')$. But then δ' has a fixed point, a contradiction. □

Lemma 6.5. *If* $\mathrm{Bs}|-K_X| \neq \emptyset$, *then* $\mathrm{r}(G) \leq 4$.

Proof. By Shin [21] the base locus $\mathrm{Bs}|-K_X|$ is either a single point or a rational curve. In both cases $\mathrm{r}(G) \leq 4$ by Lemmas 2.1 and 2.3. □

Lemma 6.6. *If* $-K_X$ *is not very ample, then* $\mathrm{r}(G) \leq 5$.

Proof. Assume that $\mathrm{r}(G) \geq 6$. By Lemma 6.5 the linear system $|-K_X|$ is base point free. It is easy to show that $|-K_X|$ defines a double cover $\phi : X \to Y \subset \mathbb{P}^{g+1}$ (cf. [6, Chap. 1, Prop. 4.9]). Here Y is a variety of degree $g-1$ in $\mathbb{P}^{g+1}$, a variety of minimal degree. Let $\bar{G}$ be the image of G in $\mathrm{Aut}(Y)$. Then $\mathrm{r}(\bar{G}) \geq \mathrm{r}(G) - 1$. If $g = 2$ (resp. $g = 3$), then $Y = \mathbb{P}^3$ (resp. $Y \subset \mathbb{P}^4$ is a quadric) and $\mathrm{r}(G) \leq 5$ by Lemma 6.4 (resp. by Corollary 6.3). Thus we may assume that $g \geq 4$. If Y is smooth, then according to the Enriques theorem (see, e.g., [6, Th. 3.11]) Y is a rational scroll $\mathbb{P}_{\mathbb{P}^1}(\mathscr{E})$, where $\mathscr{E}$ is a rank 3 vector bundle on $\mathbb{P}^1$. Then X has a G-equivariant projection to a curve. This contradicts $\rho(X)^G = 1$. Hence Y is singular. In this case, Y is a projective cone (again by the Enriques theorem). If its vertex $O \in Y$ is zero-dimensional, then $\dim T_{O,Y} \geq 5$. On the other hand, X has only hypersurface singularities. Therefore the double cover $X \to Y$ is not étale over O and so G has a fixed point on X. This contradicts Lemma 2.1. Thus Y is a cone over a curve with vertex along a line L. As above, L must be contained in the branch divisor and so $L' := \phi^{-1}(L)$ is a G-invariant rational curve. Since the image of G in $\mathrm{Aut}(L')$ is a 2-elementary abelian group of rank ≤ 2, by Lemma 2.3 we have $\mathrm{r}(G) \leq 4$. □

Remark 6.7. Recall that for a Fano threefold X with at worst Gorenstein terminal singularities one has $\iota(X) \leq 4$. Moreover, $\iota(X) = 4$ if and only if $X \simeq \mathbb{P}^3$ and $\iota(X) = 3$ if and only if X is a quadric in $\mathbb{P}^4$ [7]. In these cases we have $\mathrm{r}(G) \leq 4$ by Lemma 6.4 and Corollary 6.3, respectively. If $\iota(X) = 2$, then X is so-called *del Pezzo threefold*. The number $d := (-\frac{1}{2}K_X)^3$ is called the *degree* of X.

Lemma 6.8. *Assume that the divisor* $-K_X$ *is very ample,* $\mathrm{r}(G) \geq 6$, *and the action of* G *on* X *is not free in codimension* 1. *Let* $\delta \in G$ *be an element such that* $\dim \mathrm{Fix}(\delta) = 2$ *and let* $D \subset \mathrm{Fix}(\delta)$ *be the union of all two-dimensional components. Then* $\mathrm{r}(G) = 6$ *and* D *is a Du Val member of* $|-K_X|$. *Moreover,* $\iota(X) = 1$ *except, possibly, for the case where* $\iota(X) = 2$ *and* $-\frac{1}{2}K_X$ *is not very ample.*

Proof. Since G is abelian, $\mathrm{Fix}(\delta)$ and D are G-invariant and so $-K_X \sim_{\mathbb{Q}} \lambda D$ for some $\lambda \in \mathbb{Q}$. In particular, D is a $\mathbb{Q}$-Cartier divisor. Since X has only terminal Gorenstein singularities, D must be Cartier. Clearly, D is smooth outside of $\mathrm{Sing}(X)$. Further, D is ample and so it must be connected. Since D is a reduced Cohen–Macaulay scheme with $\dim \mathrm{Sing}(D) \leq 0$, it is irreducible and normal.

Let $X \hookrightarrow \mathbb{P}^{g+1}$ be the anticanonical embedding. The action of δ on X is induced by an action of a linear involution of $\mathbb{P}^{g+1}$. There are two disjointed linear subspaces $V_+, V_- \subset \mathbb{P}^{g+1}$ of δ-fixed points and the divisor D is contained in one of them. This means that D is a component of a hyperplane section $S \in |-K_X|$ and so $\lambda \geq 1$.

Since $\mathrm{r}(G) \geq 6$, by Corollary 4.3 we have $\lambda = 1$ and $-K_X \sim D$ (because $\mathrm{Pic}(X)$ is a torsion free group). Since D is irreducible, the case (i) of Proposition 4.6 holds.

Finally, if $\iota(X) > 1$, then by Remark 6.7 we have $\iota(X) = 2$. If furthermore the divisor A is very ample, then it defines an embedding $X \hookrightarrow \mathbb{P}^N$ so that D spans $\mathbb{P}^N$. In this case the action of δ must be trivial, a contradiction. □

Lemma 6.9. *If* $\rho(X) > 1$, *then* $\mathrm{r}(G) \leq 5$.

Proof. We use the classification of G-Fano threefolds with $\rho(X) > 1$ [17]. By this classification $\rho(X) \leq 4$. Let G_0 be the kernel of the action of G on $\mathrm{Pic}(X)$.

Consider the case $\rho(X) = 2$. Then $[G : G_0] = 2$. In the cases (1.2.1) and (1.2.4) of [17] the variety X has a structure of G_0-equivariant conic bundle over $\mathbb{P}^2$. As in Proposition 3.2 we have $\mathrm{r}(G_0) \leq 4$ and $\mathrm{r}(G) \leq 5$ in these cases. In the cases (1.2.2) and (1.2.3) of [17] the variety X has two birational contractions to $\mathbb{P}^3$ and a quadric $Q \subset \mathbb{P}^4$, respectively. As above we get $\mathrm{r}(G) \leq 5$ by Lemma 6.4 and Corollary 6.3.

Consider the case $\rho(X) = 3$. We show that in this case $\mathrm{Pic}(X)^G \not\simeq \mathbb{Z}$ (and so this case does not occur). Since G is a 2-elementary abelian group, its action on $\mathrm{Pic}(X) \otimes \mathbb{Q}$ is diagonalizable. Since, $\mathrm{Pic}(X)^G = \mathbb{Z} \cdot K_X$, the group G contains an element τ that acts on $\mathrm{Pic}(X) \simeq \mathbb{Z}^3$ as the reflection with respect to the orthogonal complement to K_X. Since the group G preserves the natural bilinear form $\langle \mathbf{x}_1, \mathbf{x}_2 \rangle := \mathbf{x}_1 \cdot \mathbf{x}_2 \cdot K_X$, the action must be as follows:

$$\tau : \mathbf{x} \longmapsto \mathbf{x} - \lambda K_X, \qquad \lambda = \frac{2\mathbf{x} \cdot K_X^2}{K_X^3}.$$

Hence λK_X is an integral element for any $\mathbf{x} \in \mathrm{Pic}(X)$. This gives a contradiction in all cases (1.2.5)–(1.2.7) of [17, Th. 1.2]. For example, in the case (1.2.5) of [17, Th. 1.2] our variety X has a structure (non-minimal) del Pezzo fibration of degree 4 and $-K_X^3 = 12$. For the fiber F we have $F \cdot K_X^2 = K_F^2 = 4$ and λK_X is not integral, a contradiction.

Finally, consider the case $\rho(X) = 4$. Then according to [17] X is a divisor of multidegree $(1,1,1,1)$ in $(\mathbb{P}^1)^4$. All the projections $\varphi_i : X \to \mathbb{P}^1$, $i = 1, \ldots, 4$ are G_0-equivariant. We claim that natural maps $\varphi_{i*} : G_0 \to \mathrm{Aut}(\mathbb{P}^1)$ are injective. Indeed, assume that $\varphi_{1*}(\vartheta)$ is the identity map in $\mathrm{Aut}(\mathbb{P}^1)$ for some $\vartheta \in G$. This means that $\vartheta \circ \varphi_1 = \varphi_1$. Since $\mathrm{Pic}(X)^G = \mathbb{Z}$, the group G permutes the classes $\varphi_i^* \mathscr{O}_{\mathbb{P}^1}(1) \in \mathrm{Pic}(X)$. Hence, for any $i = 1, \ldots, 4$, there exists $\sigma_i \in G$ such that $\varphi_i = \varphi_1 \circ \sigma_i$. Then

$$\vartheta \circ \varphi_i = \vartheta \circ \varphi_1 \circ \sigma_i = \varphi_1 \circ \sigma_i = \varphi_i.$$

Hence, $\varphi_{i*}(\vartheta)$ is the identity for any i. Since $\varphi_1 \times \cdots \times \varphi_4$ is an embedding, ϑ must be the identity as well. This proves our clam. Therefore, $\mathrm{r}(G_0) \leq 2$. The group G/G_0 acts on $\mathrm{Pic}(X)$ faithfully. By the same reason as above, an element of G/G_0 cannot act as the reflection with respect to K_X. Therefore, $\mathrm{r}(G/G_0) \leq 2$ and $\mathrm{r}(G) \leq 4$. □

Now we consider the case of del Pezzo threefolds.

Lemma 6.10. *If* $\iota(X) = 2$, *then* $\mathrm{r}(G) \le 5$.

Proof. By Lemma 6.9 we may assume that $\rho(X) = 1$. Let $A := -\frac{1}{2}K_X$ and let $d := A^3$ be the degree of X. Since $\rho(X) = 1$, we have $d \le 5$ (see e.g. [16]). Consider the possibilities for d case by case. We use the classification (see [21] and [16]).

If $d = 1$, then the linear system $|A|$ has a unique base point. This point is smooth and must be G-invariant. By Lemma 2.1 $\mathrm{r}(G) \le 3$. If $d = 2$, then the linear system $|A|$ defines a double cover $\varphi : X \to \mathbb{P}^3$. Then the image of G in $\mathrm{Aut}(\mathbb{P}^3)$ is a 2-elementary group $\bar{G}$ with $\mathrm{r}(\bar{G}) \ge \mathrm{r}(G) - 1$, where $\mathrm{r}(\bar{G}) \le 4$ by Lemma 6.4. If $d = 3$, then $X = X_3 \subset \mathbb{P}^4$ is a cubic hypersurface. By Lemma 6.2 $\mathrm{r}(G) \le 4$. If $d = 5$, then X is smooth, unique up to isomorphism, and $\mathrm{Aut}(X) \simeq PGL_2(\Bbbk)$ (see [7]).

Finally, consider the case $d = 4$. Then $X = Q_1 \cap Q_2 \subset \mathbb{P}^5$ is an intersection of two quadrics (see e.g. [21]). Let $\mathcal{Q}$ be the pencil generated by Q_1 and Q_2. Since X has a isolated singularities and it is not a cone, a general member of $\mathcal{Q}$ is smooth by Bertini's theorem and for any member $Q \in \mathcal{Q}$ we have $\dim \mathrm{Sing}(Q) \le 1$. Let D be the divisor of degree 6 on $\mathcal{Q} \simeq \mathbb{P}^1$ given by the vanishing of the determinant. The elements of $\mathrm{Supp}(D)$ are exactly degenerate quadrics. Clearly, for any point $P \in \mathrm{Sing}(X)$ there exists a unique quadric $Q \in \mathcal{Q}$ which is singular at P. This defines a map $\pi : \mathrm{Sing}(X) \to \mathrm{Supp}(D)$. Let $Q \in \mathrm{Supp}(D)$. Then $\pi^{-1}(Q) = \mathrm{Sing}(Q) \cap X = \mathrm{Sing}(Q) \cap Q'$, where $Q' \in \mathcal{Q}$, $Q' \ne Q$. In particular, $\pi^{-1}(Q)$ consists of at most two points. Hence the cardinality of $\mathrm{Sing}(X)$ is at most 12.

Assume that $\mathrm{r}(G) \ge 6$. Let $S \in |-K_X|$ be an invariant member. We claim that $S \supset \mathrm{Sing}(X)$ and $\mathrm{Sing}(X) \ne \emptyset$. Indeed, otherwise $S \cap \mathrm{Sing}(X) = \emptyset$. By Proposition 4.6 S is reducible: $S = S_1 + \cdots + S_N$, $N \ge 2$. Since $\iota(X) = 2$, we get $N = 2$ and $S_1 \sim S_2$, i.e., S_i is a hyperplane section of $X \subset \mathbb{P}^5$. As in the proof of Corollary 4.3 we see that S_i is rational. This contradicts Proposition 4.6 (ii). Thus $\emptyset \ne \mathrm{Sing}(X) \subset S$. By Lemma 6.8 the action of G on X is free in codimension 1. By Remark 2.2 for the stabilizer G_P of a point $P \in \mathrm{Sing}(X)$ we have $\mathrm{r}(G_P) \le 3$. Then by the above estimate the variety X has exactly 8 singular points and G acts on $\mathrm{Sing}(X)$ transitively.

Note that our choice of S is not unique: there is a basis $s^{(1)}, \dots, s^{(g+2)} \in H^0(X, -K_X)$ consisting of eigensections. This basis gives us G-invariant divisors $S^{(1)}, \dots, S^{(g+2)}$ generating $|-K_X|$. By the above $\mathrm{Sing}(X) \subset S^{(i)}$ for all i. Thus $\mathrm{Sing}(X) \subset \cap S^{(i)} = \mathrm{Bs}|-K_X|$. This contradicts the fact that $-K_X$ is very ample. □

The following two examples show that the inequality $\mathrm{r}(G) \le 5$ in the above lemma is sharp.

Example 6.11. Let $X = X_{2\cdot 2} \subset \mathbb{P}^5$ be the variety given by $\sum x_i^2 = \sum \lambda_i x_i^2 = 0$ with $\lambda_i \ne \lambda_j$ for $i \ne j$ and let $G \subset \mathrm{Aut}(X)$ be the 2-elementary abelian subgroup generated by involutions $x_i \mapsto -x_i$. Then X is a *rational* del Pezzo threefold of degree 4 and $\mathrm{r}(G) = 5$.

Example 6.12 (suggested by the referee). Let A be the Jacobian of a curve of genus 2 and let Θ be its theta-divisor. The linear system $|2\Theta|$ defines a finite morphism $\alpha : A \to B \subset \mathbb{P}^3$ of degree 2 whose image $B = \alpha(A)$ is a quartic with 16 nodes [2, Chap. VIII, Exercises]. Let $\varphi : X \to \mathbb{P}^3$ be the double cover branched along B. Then X is a del Pezzo threefold of degree 2 whose singular locus consists of 16 nodes. In this situation, the rank of the Weil divisor class group $\mathrm{Cl}(X)$ equals to 7 (see [16, Th. 7.1]) and X has a small resolution which can be obtained by blowing up of six points in general position on $\mathbb{P}^3$ (see e.g. [4, 23, Chap. 3] or [16, Th. 7.1]). In particular, X is rational. The translation by a two-torsion point $a \in A$ induces a projective involution τ_a of $B \subset \mathbb{P}^3$. These involutions lift to X and generate a 2-elementary subgroup $H \subset \mathrm{Aut}(X)$ with $\mathrm{r}(H) = 4$. The Galois involution γ of the double cover φ is contained in the center of $\mathrm{Aut}(X)$. Hence γ and H generate a 2-elementary subgroup $G \subset \mathrm{Aut}(X)$ of rank 5.

Note that the fixed point locus of γ on X is a Kummer surface isomorphic to B. On the other hand, the fixed point loci of involutions acting on $X_{2\cdot 2}$ are either rational surfaces or subvarieties of dimension ≤ 1. Hence the groups constructed in Examples 6.11 and 6.12 are not conjugate to each other in the Cremona group.

From now on we assume that $\mathrm{Pic}(X) = \mathbb{Z} \cdot K_X$. Let $g := \mathrm{g}(X)$.

Lemma 6.13. *If* $g \leq 4$, *then* $\mathrm{r}(G) \leq 5$. *If* $g = 5$, *then* $\mathrm{r}(G) \leq 6$.

Proof. We may assume that $-K_X$ is very ample. Automorphisms of X are induced by projective transformations of $\mathbb{P}^{g+1}$ that preserve $X \subset \mathbb{P}^{g+1}$. On the other hand, there is a natural representation of G on $H^0(X, -K_X)$ which is faithful. Thus the composition

$$\mathrm{Aut}(X) \hookrightarrow GL(H^0(X, -K_X)) = GL_{g+2}(\Bbbk) \to PGL_{g+2}(\Bbbk)$$

is injective. Since G is abelian, its image $\bar{G} \subset GL_{g+2}(\Bbbk)$ is contained in a maximal torus and by the above $\bar{G}$ contains no scalar matrices. Hence, $\mathrm{r}(G) \leq g + 1$. □

Example 6.14. Let G be the two-torsion subgroup of the diagonal torus of $PGL_7(\Bbbk)$. Then X faithfully acts on the Fano threefold in $\mathbb{P}^6$ given by the equations $\sum x_i^2 = \sum \lambda_i x_i^2 = \sum \mu_i x_i^2 = 0$. This shows that the bound $\mathrm{r}(G) \leq 6$ in the above lemma is sharp. Note however that X is not rational if it is smooth [1]. Hence in this case our construction does not give any embedding of G to $\mathrm{Cr}_3(\Bbbk)$.

Lemma 6.15. *If in the above assumptions* $\mathrm{g}(X) \geq 6$, *then* X *has at most* 29 *singular points.*

Proof. According to [12] the variety X has a *smoothing*. This means that there exists a flat family $\mathfrak{X} \to \mathfrak{T}$ over a smooth one-dimensional base $\mathfrak{T}$ with special fiber $X = \mathfrak{X}_0$ and smooth general fiber $X_t = \mathfrak{X}_t$. Using the classification of Fano threefolds [6] (see also [7]) we obtain $h^{1,2}(X_t) \leq 10$. Then by Namikawa [12] we have

$$\#\mathrm{Sing}(X) \leq 21 - \frac{1}{2}\mathrm{Eu}(X_t) = 20 - \rho(X_t) + h^{1,2}(X_t) \leq 29.$$

Proof of Proposition 6.1. Assume that $\mathrm{r}(G) \geq 7$. Let $S \in |-K_X|$ be an invariant member. By Corollary 4.5 the singularities of S are worse than Du Val. So S satisfies the conditions (ii) of Proposition 4.6. Write $S = \sum_{i=1}^{N} S_i$. By Proposition 4.6 the group $G_\bullet$ acts on S_i faithfully and

$$N = 2^{\mathrm{r}(G)-\mathrm{r}(G_\bullet)} \geq 4.$$

First we consider the case where X is smooth near S. Since $\rho(X) = 1$, the divisors S_i's are linear equivalent to each other and so $\iota(X) \geq 4$. This contradicts Lemma 6.10.

Therefore, $S \cap \mathrm{Sing}(X) \neq \emptyset$. By Lemma 6.8 the action of G on X is free in codimension 1 and by Remark 2.2 we see that $\mathrm{r}(G_P) \leq 3$, where G_P is the stabilizer of a point $P \in \mathrm{Sing}(X)$. Then by Lemma 6.15 the variety X has exactly 16 singular points and G acts on $\mathrm{Sing}(X)$ transitively. Since $S \cap \mathrm{Sing}(X) \neq \emptyset$, we have $\mathrm{Sing}(X) \subset S$. On the other hand, our choice of S is not unique: there is a basis $s^{(1)}, \ldots, s^{(g+2)} \in H^0(X, -K_X)$ consisting of eigensections. This basis gives us G-invariant divisors $S^{(1)}, \ldots, S^{(g+2)}$ generating $|-K_X|$. By the above $\mathrm{Sing}(X) \subset S^{(i)}$ for all i. Thus $\mathrm{Sing}(X) \subset \cap S^{(i)} = \mathrm{Bs}|-K_X|$. This contradicts Lemma 6.6. □

Acknowledgements The work was completed during the author's stay at the International Centre for Theoretical Physics, Trieste. The author would like to thank ICTP for hospitality and support.

I acknowledge partial supports by RFBR grants No. 11-01-00336-a, the grant of Leading Scientific Schools No. 4713.2010.1, Simons-IUM fellowship, and AG Laboratory SU-HSE, RF government grant ag. 11.G34.31.0023.

References

1. A. Beauville, Variétés de Prym et jacobiennes intermédiaires. Ann. Sci. École Norm. Sup. (4) **10**(3), 309–391 (1977)
2. A. Beauville, in *Surfaces algébriques complexes*. Astérisque, vol. 54 (Société Mathématique de France, Paris, 1978)
3. A. Beauville, *p*-elementary subgroups of the Cremona group. J. Algebra **314**(2), 553–564 (2007)
4. A.B. Coble, in *Algebraic Geometry and Theta Functions*. American Mathematical Society Colloquium Publications, vol. 10 (American Mathematical Society, New York, 1929)
5. I. Dolgachev, V. Iskovskikh, Finite subgroups of the plane Cremona group, in *Algebra, Arithmetic, and Geometry: In Honor of Yu.I. Manin. Vol. I*. Progress in Mathematics, vol. 269 (Birkhäuser, Boston, 2009), pp. 443–548
6. V.A. Iskovskikh, Anticanonical models of three-dimensional algebraic varieties. J. Sov. Math. **13**, 745–814 (1980)
7. V. Iskovskikh, Y. Prokhorov, in *Fano Varieties. Algebraic Geometry V.*. Encyclopaedia of Mathematical Sciences, vol. 47 (Springer, Berlin, 1999)

8. Y. Kawamata, Boundedness of **Q**-Fano threefolds, in *Proceedings of the International Conference on Algebra, Part 3 (Novosibirsk, 1989)*. Contemporary Mathematics, vol. 131 (American Mathematical Society, Providence, 1992), pp. 439–445
9. J. Kollár, Y. Miyaoka, S. Mori, H. Takagi, Boundedness of canonical **Q**-Fano 3-folds. Proc. Jpn. Acad. Ser. A Math. Sci. **76**(5), 73–77 (2000)
10. S. Mori, On 3-dimensional terminal singularities. Nagoya Math. J. **98**, 43–66 (1985)
11. S. Mori, Y. Prokhorov, Multiple fibers of del Pezzo fibrations. Proc. Steklov Inst. Math. **264**(1), 131–145 (2009)
12. Y. Namikawa, Smoothing Fano 3-folds. J. Algebr. Geom. **6**(2), 307–324 (1997)
13. V. Nikulin, Finite automorphism groups of Kähler K3 surfaces. Trans. Mosc. Math. Soc. **2**, 71–135 (1980)
14. Y. Prokhorov, p-elementary subgroups of the Cremona group of rank 3, in *Classification of Algebraic Varieties*. EMS Series of Congress Reports (European Mathematical Society, Zürich, 2011), pp. 327–338
15. Y. Prokhorov, Simple finite subgroups of the Cremona group of rank 3. J. Algebr. Geom. **21**, 563–600 (2012)
16. Y. Prokhorov, G-Fano threefolds, I. Adv. Geom. **13**(3), 389–418 (2013)
17. Y. Prokhorov, G-Fano threefolds, II. Adv. Geom. **13**(3), 419–434 (2013)
18. Y. Prokhorov, On birational involutions of $\mathbf{P}^3$. Izvestiya Math. Russ. Acad. Sci. **77**(3), 627–648 (2013)
19. M. Reid, Young person's guide to canonical singularities, in *Algebraic Geometry, Bowdoin, 1985 (Brunswick, Maine, 1985)*. Proceedings of Symposia in Pure Mathematics, vol. 46 (American Mathematical Society, Providence, 1987), pp. 345–414
20. J.-P. Serre, Bounds for the orders of the finite subgroups of $G(k)$, in *Group Representation Theory* (EPFL Press, Lausanne, 2007), pp. 405–450
21. K.-H. Shin, 3-dimensional Fano varieties with canonical singularities. Tokyo J. Math. **12**(2), 375–385 (1989)
22. V. Shokurov, 3-fold log flips. Russ. Acad. Sci. Izv. Math. **40**(1), 95–202 (1993)
23. R. Varley, Weddle's surfaces, Humbert's curves, and a certain 4-dimensional abelian variety. Am. J. Math. **108**(4), 931–952 (1986)

Birational Automorphism Groups of Projective Varieties of Picard Number Two

De-Qi Zhang

Abstract We slightly extend a result of Oguiso on birational automorphism groups (resp. of Lazić–Peternell on Morrison–Kawamata cone conjecture) from Calabi–Yau manifolds of Picard number 2 to arbitrary singular varieties X (resp. to klt Calabi–Yau pairs in broad sense) of Picard number 2. When X has only klt singularities and is not a complex torus, we show that either $\mathrm{Aut}(X)$ is almost infinite cyclic, or it has only finitely many connected components.

2000 Mathematics Subject Classification: 14J50, 14E07, 32H50.

1 Introduction

This note is inspired by Oguiso [8] and Lazić–Peternell [6].

Let X be a normal projective variety defined over the field $\mathbb{C}$ of complex numbers. The following subgroup (of the birational group $\mathrm{Bir}(X)$)

$$\mathrm{Bir}_2(X) := \{g : X \dashrightarrow X \mid g \text{ is an isomorphism outside codimension two subsets}\}$$

is also called the group of pseudo-automorphisms of X.

Let $\mathrm{NS}(X) = \{\text{Cartier divisors}\}/(\text{algebraic equivalence})$ be the Neron–Severi group, which is finitely generated. Let $\mathrm{NS}_{\mathbb{R}}(X) := \mathrm{NS}(X) \otimes \mathbb{R}$ with $\rho(X) := \dim_{\mathbb{R}} \mathrm{NS}_{\mathbb{R}}(X)$ the *Picard number*. Let $\mathrm{Eff}(X) \subset \mathrm{NS}_{\mathbb{R}}(X)$ be the *cone*

D.-Q. Zhang (✉)
Department of Mathematics, National University of Singapore, 10 Lower Kent Ridge Road, Singapore 119076
e-mail: matzdq@nus.edu.sg

I. Cheltsov et al. (eds.), *Automorphisms in Birational and Affine Geometry*, Springer Proceedings in Mathematics & Statistics 79,
DOI 10.1007/978-3-319-05681-4_13,

of effective $\mathbb{R}$*-divisor*; its closure $\overline{\mathrm{Eff}}(X)$ is called the *cone of pseudo-effective divisors*. The *ample cone* $\mathrm{Amp}(X) \subset \mathrm{NS}_{\mathbb{R}}(X)$ consists of classes of ample $\mathbb{R}$-Cartier divisors; its closure $\mathrm{Nef}(X)$ is called the *nef cone*. A divisor D is *movable* if $|mD|$ has no fixed component for some $m > 0$. The *closed movable cone* $\overline{\mathrm{Mov}}(X) \subset \mathrm{NS}_{\mathbb{R}}(X)$ is the closure of the convex hull of movable divisors. $\mathrm{Mov}(X)$ is the interior part of $\overline{\mathrm{Mov}}(X)$.

A pair (X, Δ) of a normal projective variety X and an effective Weil $\mathbb{R}$-divisor Δ is a *klt Calabi–Yau pair in broad sense* if it has at worst Kawamata log terminal (klt) singularities (cf. [5, Definition 2.34] or [1, §3.1]) and $K_X + \Delta \equiv 0$ (numerically equivalent to zero); in this case, if $K_X + \Delta$ is $\mathbb{Q}$-Cartier, then $K_X + \Delta \sim_{\mathbb{Q}} 0$, i.e., $r(K_X + \Delta) \sim 0$ (linear equivalence) for some $r > 0$, by Nakayama's abundance theorem in the case of zero numerical dimension. (X, Δ) is a *klt Calabi–Yau pair in narrow sense* if it is a klt Calabi–Yau pair in broad sense and if we assume further that the irregularity $q(X) := h^1(X, \mathcal{O}_X) = 0$. When $\Delta = 0$, a klt Calabi–Yau pair in broad/narrow sense is called a *klt Calabi–Yau variety in broad/narrow sense.*

On a terminal minimal variety (like a terminal Calabi–Yau variety) X, we have $\mathrm{Bir}(X) = \mathrm{Bir}_2(X)$. Totaro [9] formulated the following generalization of the Morrison–Kawamata cone conjecture (cf. [4]) and proved it in dimension two.

Conjecture 1.1. Let (X, Δ) be a *klt* Calabi–Yau pair in broad sense.

(1) There exists a rational polyhedral cone Π which is a fundamental domain for the action of $\mathrm{Aut}(X)$ on the effective nef cone $\mathrm{Nef}(X) \cap \mathrm{Eff}(X)$, i.e.,

$$\mathrm{Nef}(X) \cap \mathrm{Eff}(X) = \bigcup_{g \in \mathrm{Aut}(X)} g^*\Pi,$$

and $\mathrm{int}(\Pi) \cap \mathrm{int}(g^*\Pi) = \emptyset$ unless $g^*_{|\mathrm{NS}_{\mathbb{R}}(X)} = \mathrm{id}$.

(2) There exists a rational polyhedral cone Π' which is a fundamental domain for the action of $\mathrm{Bir}_2(X)$ on the effective movable cone $\overline{\mathrm{Mov}}(X) \cap \mathrm{Eff}(X)$.

If X has Picard number 1, then $\mathrm{Aut}(X)/\mathrm{Aut}_0(X)$ is finite; here $\mathrm{Aut}_0(X)$ is the *connected component of identity* in $\mathrm{Aut}(X)$; see [7, Prop. 2.2].

Now suppose that X has Picard number 2. Then $\dim_{\mathbb{R}} \mathrm{NS}_{\mathbb{R}}(X) = 2$. So the (strictly convex) cone $\overline{\mathrm{Eff}}(X)$ has exactly two extremal rays. Set

$$A := \mathrm{Aut}(X),\ A^- := A \backslash A^+,\ B_2 := \mathrm{Bir}_2(X),\ B_2^- := B_2 \backslash B_2^+, \text{ where}$$

$$B_2^+ = \mathrm{Bir}_2^+(X) := \{g \in B_2 \mid g^* \text{ preserves each of the two extremal rays of } \overline{\mathrm{Eff}}(X)\},$$

$$A^+ = \mathrm{Aut}^+(X) := \{g \in A \mid g^* \text{ preserves each of the two extremal rays of } \overline{\mathrm{Eff}}(X)\},$$

$$B_2^0 = \mathrm{Bir}_2^0(X) := \{g \in B_2 \mid g^*_{|\mathrm{NS}_{\mathbb{R}}(X)} = \mathrm{id}\}.$$

When X is a Calabi–Yau manifold, Theorem 1.2 is more or less contained in [8] or [6]. Our argument here for general X is slightly streamlined and direct.

Theorem 1.2. *Let X be a normal projective variety of Picard number 2. Then:*

(1) $|\operatorname{Aut}(X) : \operatorname{Aut}^+(X)| \le 2$; $|\operatorname{Bir}_2(X) : \operatorname{Bir}_2^+(X)| \le 2$.
(2) $\operatorname{Bir}_2^0(X)$ *coincides with both* $\operatorname{Ker}(\operatorname{Bir}_2(X) \to \operatorname{GL}(\operatorname{NS}_{\mathbb{R}}(X)))$ *and* $\operatorname{Ker}(\operatorname{Aut}(X) \to \operatorname{GL}(\operatorname{NS}_{\mathbb{R}}(X)))$. *Hence we have inclusions:*

$$\operatorname{Aut}_0(X) \subseteq \operatorname{Bir}_2^0(X) \subseteq \operatorname{Aut}^+(X) \subseteq \operatorname{Bir}_2^+(X) \subseteq \operatorname{Bir}_2(X).$$

(3) $|\operatorname{Bir}_2^0(X) : \operatorname{Aut}_0(X)|$ *is finite.*
(4) $\operatorname{Bir}_2^+(X)/\operatorname{Bir}_2^0(X)$ *is isomorphic to either* $\{\mathrm{id}\}$ *or* $\mathbb{Z}$. *In the former case,* $|\operatorname{Aut}(X) : \operatorname{Aut}_0(X)| \le |\operatorname{Bir}_2(X) : \operatorname{Aut}_0(X)| < \infty$.
(5) If one of the extremal rays of $\overline{\operatorname{Eff}}(X)$ *or of the movable cone of* X *is generated by a rational divisor class, then* $\operatorname{Bir}_2^+(X) = \operatorname{Bir}_2^0(X)$ *and* $|\operatorname{Bir}_2(X) : \operatorname{Aut}_0(X)| < \infty$.
(6) If one of the extremal rays of the nef cone of X *is generated by a rational divisor class, then* $\operatorname{Aut}^+(X) = \operatorname{Bir}_2^0(X)$ *and* $|\operatorname{Aut}(X) : \operatorname{Aut}_0(X)| < \infty$.

Theorem 1.2 and the proof of [6, Theorem 1.4] imply the following, and also a weak cone theorem as in [6, Theorem 1.4(1)] when $|\operatorname{Bir}_2(X) : \operatorname{Aut}_0(X)|$ is finite.

Theorem 1.3. *Let (X, Δ) be a klt Calabi–Yau pair in broad sense of Picard number 2. Suppose that* $|\operatorname{Bir}_2(X) : \operatorname{Aut}_0(X)|$ *(or equivalently* $|\operatorname{Bir}_2^+(X) : \operatorname{Bir}_2^0(X)|$*) is infinite. Then Conjecture 1.1 holds true.*

A group G is *almost infinite cyclic*, if there exists an infinite cyclic subgroup H such that the index $|G : H|$ is finite. If a group G_1 has a finite normal subgroup N_1 such that G_1/N_1 is almost infinite cyclic, then G_1 is also almost infinite cyclic (cf. [11, Lemma 2.6]).

Theorem 1.4. *Let X be a normal projective variety of Picard number 2. Then:*

(1) Either $\operatorname{Aut}(X)/\operatorname{Aut}_0(X)$ *is finite, or it is almost infinite cyclic and* $\dim X$ *is even.*
(2) Suppose that X has at worst Kawamata log terminal singularities. Then one of the following is true.
(2a) X is a complex torus.
(2b) $|\operatorname{Aut}(X) : \operatorname{Aut}_0(X)| \le |\operatorname{Bir}_2(X) : \operatorname{Aut}_0(X)| < \infty$.
(2c) X is a klt Calabi–Yau variety in narrow sense and $\operatorname{Aut}_0(X) = (1)$, *so both* $\operatorname{Aut}(X)$ *and* $\operatorname{Bir}_2(X)$ *are almost infinite cyclic.*

Below is a consequence of Theorem 1.4 and generalizes Oguiso [8, Theorem 1.2(1)].

Corollary 1.5. *Let X be an odd-dimensional projective variety of Picard number 2. Suppose that* $\operatorname{Aut}_0(X) = (1)$ *(e.g., X is non-ruled and* $q(X) = h^1(X, \mathcal{O}_X) = 0$*). Then* $\operatorname{Aut}(X)$ *is finite.*

For a linear transformation $T : V \to V$ of a vector space V over $\mathbb{R}$ or $\mathbb{C}$, the *spectral radius* $\rho(T)$ is defined as

$$\rho(T) := \max\{|\lambda| \,;\, \lambda \text{ is a real or complex eigenvalue of } T\}.$$

Corollary 1.6. *Let X be a normal projective variety of Picard number 2. Then:*

(1) Every $g \in \mathrm{Bir}_2^+(X) \setminus \mathrm{Bir}_2^0(X)$ acts on $\mathrm{NS}_{\mathbb{R}}(X)$ with spectral radius > 1.
(2) A class $g\,\mathrm{Aut}_0(X)$ in $\mathrm{Aut}(X)/\,\mathrm{Aut}_0(X)$ is of infinite order if and only if the spectral radius of $g^*_{|\mathrm{NS}_{\mathbb{R}}(X)}$ is > 1.

(1) above follows from the proof of Theorem 1.2, while (2) follows from (1) and again Theorem 1.2.

Remark 1.7. (1) The second alternative in Theorem 1.4(1) and (2c) in Theorem 1.4(2) do occur. Indeed, the complete intersection X of two general hypersurfaces of type $(1, 1)$ and $(2, 2)$ in $\mathbb{P}^2 \times \mathbb{P}^2$ is called Wehler's K3 surface (hence $\mathrm{Aut}_0(X) = (1)$) of Picard number 2 such that $\mathrm{Aut}(X) = \mathbb{Z}/(2) * \mathbb{Z}/(2)$ (a free product of two copies of $\mathbb{Z}/(2)$) which contains $\mathbb{Z}$ as a subgroup of index two; see [10].
(2) We cannot remove the possibility (2a) in Theorem 1.4(2). It is possible that $\mathrm{Aut}_0(X)$ has positive dimension and $\mathrm{Aut}(X)/\,\mathrm{Aut}_0(X)$ is almost infinite cyclic at the same time. Indeed, as suggested by Oguiso, using the Torelli theorem and the surjectivity of the period map for abelian surfaces, one should be able to construct an abelian surface X of Picard number 2 with irrational extremal rays of the nef cone of X and an automorphism g with $g^*_{|\mathrm{NS}(X)}$ of infinite order. Hence $g\,\mathrm{Aut}_0(X)$ is of infinite order in $\mathrm{Aut}(X)/\,\mathrm{Aut}_0(X)$ and g^* has spectral radius > 1 (cf. Corollary 1.5).
(3) See Oguiso [8] for more examples of Calabi–Yau threefolds and hyperkähler fourfolds with infinite $\mathrm{Bir}_2(X)$ or $\mathrm{Aut}(X)$.

2 Proof of Theorems

We use the notation and terminology in the book of Hartshorne and the book [5].

Proof of Theorem 1.2. Since X has Picard number 2, we can write the pseudo-effective closed cone as

$$\overline{\mathrm{Eff}}(X) = \mathbb{R}_{\ge 0}[f_1] + \mathbb{R}_{\ge 0}[f_2]$$

(1) is proved in [6, 8]. For reader's convenience, we reproduce here. Let $g \in B_2^-$ or A^-. Since g permutes extremal rays of $\overline{\mathrm{Eff}}(X)$, we can write $g^* f_1 = a f_2$, $g^* f_2 = b f_1$ with $a > 0$, $b > 0$. Since g^* is defined on the integral lattice $\mathrm{NS}(X)/(\text{torsion})$, $\deg(g^*) = \pm 1$. Hence $ab = 1$. Thus $\mathrm{ord}(g^*) = 2$ and $g^2 \in B_2^0$. Now (1) follows from the observation that $B_2^- = gB_2^+$ or $A^- = gA^+$.

(2) The first equality is by the definition of B_2^0. For the second equality, we just need to show that every $g \in B_2^0$ is in $\mathrm{Aut}(X)$. Take an ample divisor H on X. Then $g^*H = H$ as elements in $\mathrm{NS}_{\mathbb{R}}(X)$ over which B_2^0 acts trivially. Thus $\mathrm{Amp}(X) \cap g(\mathrm{Amp}(X)) \neq \emptyset$, where $\mathrm{Amp}(X)$ is the ample cone of X. Hence $g \in \mathrm{Aut}(X)$, g being isomorphic in codimension one (cf. e.g. [4, Proof of Lemma 1.5]).
(3) Applying Lieberman [7, Proof of Proposition 2.2] to an equivariant resolution, $\mathrm{Aut}_{[H]}(X) := \{g \in \mathrm{Aut}(X) \mid g^*[H] = [H]\}$ is a finite extension of $\mathrm{Aut}_0(X)$ for the divisor class $[H]$ of every ample (or even nef and big) divisor H on X. Since $B_2^0 \subseteq \mathrm{Aut}_{[H]}(X)$ (cf. (2)), (3) follows. See [8, Proposition 2.4] for a related argument.
(4) For $g \in B_2^+$, write $g^*f_1 = \chi(g)f_1$ for some $\chi(g) > 0$. Then $g^*f_2 = (1/\chi(g))f_2$ since $\deg(g^*) = \pm 1$. In fact, the spectral radius $\rho(g^*_{|\mathrm{NS}_{\mathbb{R}}(X)}) = \max\{\chi(g), 1/\chi(g)\}$. Consider the homomorphism

$$\varphi : B_2^+ \to (\mathbb{R}, +), \ \ g \mapsto \log \chi(g).$$

Then $\mathrm{Ker}(\varphi) = B_2^0$. We claim that $\mathrm{Im}(\varphi) \subset (\mathbb{R}, +)$ is discrete at the origin (and hence everywhere). Indeed, since g^* acts on $\mathrm{NS}(X)/(\text{torsion}) \cong \mathbb{Z}^{\oplus 2}$, its only eigenvalues $\chi(g)^{\pm}$ are quadratic algebraic numbers, the coefficients of whose minimal polynomial over $\mathbb{Q}$ are bounded by a function in $|\log \chi(g)|$. The claim follows. (Alternatively, as the referee suggested, B_2^+/B_2^0 sits in $\mathrm{GL}(\mathbb{Z}, \mathrm{NS}(X)/(\text{torsion})) \cap \mathrm{Diag}(f_1, f_2)$ which is a discrete group; here $\mathrm{Diag}(f_1, f_2)$ is the group of diagonal matrices with respect to the basis of $\mathrm{NS}_{\mathbb{R}}(X)$ given by f_1, f_2.) The claim implies that $\mathrm{Im}(\varphi) \cong \mathbb{Z}^{\oplus r}$ for some $r \leq 1$. (4) is proved. See [6, Theorem 3.9] for slightly different reasoning.
(5) is proved in Lemma 2.2 below while (6) is similar (cf. [8]).

This proves Theorem 1.2.

Lemma 2.2. *Let X be a normal projective variety of Picard number 2. Then* $\mathrm{Bir}_2^+(X) = \mathrm{Bir}_2^0(X)$ *and hence* $|\mathrm{Aut}(X) : \mathrm{Aut}_0(X)| \leq |\mathrm{Bir}_2(X) : \mathrm{Aut}_0(X)| < \infty$, *if one of the following conditions is satisfied.*

(1) There is an $\mathbb{R}$-Cartier divisor D such that, as elements in $\mathrm{NS}_{\mathbb{R}}(X)$, *$D \neq 0$ and $g^*D = D$ for all $g \in \mathrm{Bir}_2^+(X)$.*
(2) The canonical divisor K_X is $\mathbb{Q}$-Cartier, and $K_X \neq 0$ as element in $\mathrm{NS}_{\mathbb{R}}(X)$.
(3) At least one extremal ray of $\overline{\mathrm{Eff}}(X)$, *or of the movable cone of X is generated by a rational divisor class.*

Proof. We consider Case(1) (which implies Case(2)). In notation of proof of Theorem 1.2, for $g \in B_2^+ \setminus B_2^0$, we have $g^*f_1 = \chi(g)f_1$ with $\chi(g) \neq 1$, and further $\chi(g)^{\pm 1}$ are two eigenvalues of the action g^* on $\mathrm{NS}_{\mathbb{R}}(X) \cong \mathbb{R}^{\oplus 2}$ corresponding to the eigenvectors f_1, f_2. Since $g^*D = D$ as elements in $\mathrm{NS}_{\mathbb{R}}(X)$, g^* has three distinct eigenvalues: $1, \chi(g)^{\pm 1}$, contradicting the fact: $\dim \mathrm{NS}_{\mathbb{R}}(X) = 2$.

Consider Case(3). Since every $g \in \mathrm{Bir}_2(X)$ acts on both of the cones, g^2 preserves each of the two extremal rays of both cones, one of which is rational, by the assumption. Thus at least one of the eigenvalues of $(g^2)^*_{|\mathrm{NS}_{\mathbb{R}}(X)}$ is a rational number (and also an algebraic integer), so it is 1. Now the proof for Case(1) implies $g^2 \in B_2^0$. So B_2^+/B_2^0 is trivial (otherwise, it is isomorphic to $\mathbb{Z}$ and torsion free by Theorem 1.2). □

Proof of Theorem 1.4.

(1) follows from Theorem 1.2, and [8, Proposition 3.1] or [6, Lemma 3.1] for the observation that $\dim X$ is even when $\mathrm{Aut}^+(X)$ strictly contains B_2^0.

For (2), by Lemma 2.2, we may assume that $K_X = 0$ as element in $\mathrm{NS}_{\mathbb{R}}(X)$. Since X is klt, $rK_X \sim 0$ for some (minimal) $r > 0$, by Nakayama's abundance theorem in the case of zero numerical dimension.

Lemma 2.4. *Suppose that $q(X) = h^1(X, \mathcal{O}_X) > 0$. Then Theorem 1.4(2) is true.*

Proof. Since X is klt (and hence has only rational singularities) and a complex torus contains no rational curve, the albanese map $a = \mathrm{alb}_X : X \to A(X) := \mathrm{Alb}(X)$ is a well-defined morphism, where $\dim A(X) = q(X) > 0$; see [3, Lemma 8.1]. Further, $\mathrm{Bir}(X)$ descends to a regular action on $A(X)$, so that a is $\mathrm{Bir}(X)$-equivariant, by the universal property of alb_X. Let $X \to Y \to A(X)$ be the Stein factorization of $a : X \to A(X)$. Then $\mathrm{Bir}(X)$ descends to a regular action on Y.

If $X \to Y$ is not an isomorphism, then one has that $2 = \rho(X) > \rho(Y)$, so $\rho(Y) = 1$ and the generator of $\mathrm{NS}_{\mathbb{R}}(Y)$ gives a $\mathrm{Bir}(X)$-invariant class in $\mathrm{NS}_{\mathbb{R}}(X)$. Thus Lemma 2.2 applies, and Theorem 1.4(2b) occurs.

If $X \to Y$ is an isomorphism, then the Kodaira dimensions satisfy $\kappa(X) = \kappa(Y) \geq \kappa(a(X)) \geq 0$, by the well-known fact that every subvariety of a complex torus has nonnegative Kodaira dimension. Hence $\kappa(X) = \kappa(a(X)) = 0$, since $rK_X \sim 0$. Thus a is surjective and has connected fibers, so it is birational (cf. [2, Theorem 1]). Hence $X \cong Y = a(X) = A(X)$, and X is a complex torus. □

We continue the proof of Theorem 1.4(2). By Lemma 2.4, we may assume that $q(X) = 0$. This together with $rK_X \sim 0$ imply that X is a klt Calabi–Yau variety in narrow sense. $G_0 := \mathrm{Aut}_0(X)$ is a linear algebraic group, by applying [7, Theorem 3.12] to an equivariant resolution X' of X with $q(X') = q(X) = 0$, X having only rational singularities. The relation $rK_X \sim 0$ gives rise to the global index-one cover:

$$\hat{X} := \mathit{Spec}\ \oplus_{i=0}^{r-1} \mathcal{O}_X(-iK_X) \to X$$

which is étale in codimension one, where $K_{\hat{X}} \sim 0$. Every singularity of $\hat{X}$ is klt (cf. [5, Proposition 5.20]) and also Gorenstein, and hence canonical. Thus $\kappa(\hat{X}) = 0$, so $\hat{X}$ is non-uniruled. Hence $G_0 = (1)$, otherwise, since the class of K_X is G_0-stable, the linear algebraic group G_0 lifts to an action on $\hat{X}$, so $\hat{X}$ is ruled, a contradiction. Thus Theorem 1.4 (2b) or (2c) occurs (cf. Theorem 1.2).

This proves Theorem 1.4.

Proof of Theorem 1.3. It follows from the arguments in [6, Theorem 1.4], Theorem 1.2 and the following (replacing [6, Theorem 2.5]):

Lemma 2.6. *Let* (X, Δ) *be a* klt *Calabi–Yau variety in broad sense. Then both the cones* $\mathrm{Nef}(X)$ *and* $\overline{\mathrm{Mov}}(X)$ *are locally rational polyhedral inside the cone* $\mathrm{Big}(X)$ *of big divisors.*

Proof. Let $D \in \overline{\mathrm{Mov}}(X) \cap \mathrm{Big}(X)$. Since (X, Δ) is klt and klt is an open condition, replacing D by a small multiple, we may assume that $(X, \Delta + D)$ is klt. By Birkar [1, Theorem 1.2], there is a composition $\sigma : X \dashrightarrow X_1$ of divisorial and flip birational contractions such that $(X_1, \Delta_1 + D_1)$ is klt and $K_{X_1} + \Delta_1 + D_1$ is nef; here $\Delta_1 := \sigma_*\Delta_1$, $D_1 := \sigma_* D$, and $K_{X_1} + \Delta_1 = \sigma_*(K_X + \Delta) \equiv 0$. Since $K_X + \Delta + D \equiv D \in \overline{\mathrm{Mov}}(X)$, σ consists only of flips, so $D = \sigma^* D_1$. By Birkar [1, Theorem 3.9.1], $(K_{X_1} + \Delta_1) + D_1$ ($\equiv D_1$) is semi-ample (and big), so it equals $\tau^* D_2$, where $\tau : X_1 \to X_2$ is a birational morphism and D_2 is an $\mathbb{R}$-Cartier ample divisor. Write $D_2 = \sum_{i=1}^{s} c_i H_i$ with $c_i > 0$ and H_i ample and Cartier. Then $D \equiv \sum_{i=1}^{s} c_i \sigma^* \tau^* H_i$ with $\sigma^* \tau^* H_i$ movable and Cartier. We are done (letting $\sigma = \mathrm{id}$ when $D \in \mathrm{Nef}(X) \cap \mathrm{Big}(X)$). Alternatively, as the referee suggested, in the case when D_2 lies on the boundary of the movable cone, fix a rational effective divisor E close to D_2 outside the movable cone—but still inside the big cone. Then, for $\varepsilon \in \mathbb{Q}_{\geq 0}$ small enough, $\varepsilon E \equiv K_X + \Delta + \varepsilon E$ is klt and a rational divisor. Taking H an ample divisor, the rationality theorem in [5, Theorem 3.5, and Complement 3.6] shows that the ray spanned by D_2 is rational. □

Acknowledgements The author would like to thank the referee for very careful reading and valuable suggestions. He is partly supported by an ARF of NUS.

References

1. C. Birkar, P. Cascini, C.D. Hacon, J. McKernan, Existence of minimal models for varieties of log general type. J. Am. Math. Soc. **23**, 405–468 (2010)
2. Y. Kawamata, Characterization of abelian varieties. Compos. Math. **43**, 253–276 (1981)
3. Y. Kawamata, Minimal models and the Kodaira dimension of algebraic fiber spaces. J. Reine Angew. Math. **363**, 1–46 (1985)
4. Y. Kawamata, On the cone of divisors of Calabi-Yau fiber spaces. Internat. J. Math. **8**(5), 665–687 (1997)
5. J. Kollár, S. Mori, in *Birational Geometry of Algebraic Varieties*. Cambridge Tracts in Mathematics, vol. 134 (Cambridge University Press, Cambridge, 1998)
6. V. Lazić, T. Peternell, On the cone conjecture for Calabi-Yau manifolds with Picard number two (2012) [arXiv:1207.3653]
7. D.I. Lieberman, *Compactness of the Chow Scheme: Applications to Automorphisms and Deformations of Kähler Manifolds*. Lecture Notes in Mathematics, vol. 670 (Springer, Berlin, 1978), pp. 140–186
8. K. Oguiso, Automorphism groups of Calabi-Yau manifolds of Picard number two (2012) [arXiv:1206.1649]

9. B. Totaro, The cone conjecture for Calabi-Yau pairs in dimension 2. Duke Math. J. **154**(2), 241–263 (2010)
10. J. Wehler, $K3$-surfaces with Picard number 2. Arch. Math. (Basel) **50**(1), 73–82 (1988)
11. D.-Q. Zhang, Compact Kähler manifolds with automorphism groups of maximal rank. Trans. Am. Math. Soc. **366**, 3675–3692 (2014). doi:http://dx.doi.org/10.1090/S0002-9947-2014-06227-2

Part II
Automorphisms of Affine Varieties

Rational Curves with One Place at Infinity

Abdallah Assi

Abstract Let $\mathbb{K}$ be an algebraically closed field of characteristic zero. Given a polynomial $f(x, y) \in \mathbb{K}[x, y]$ with one place at infinity, we prove that either f is equivalent to a coordinate, or the family $(f_\lambda)_{\lambda \in \mathbb{K}}$ has at most two rational elements. When $(f_\lambda)_{\lambda \in \mathbb{K}}$ has two rational elements, we give a description of the singularities of these two elements.

2000 Mathematics Subject Classification: 14H20

1 Introduction and Notations

Let $\mathbb{K}$ be an algebraically closed field of characteristic zero, and let $f = y^n + a_1(x)y^{n-1} + \cdots + a_n(x)$ be a monic reduced polynomial of $\mathbb{K}[x][y]$. For all $\lambda \in \mathbb{K}$, we set $f_\lambda = f - \lambda$. Hence we get a family of polynomials $(f_\lambda)_{\lambda \in \mathbb{K}}$. We shall suppose that f_λ is a reduced polynomial for all $\lambda \in \mathbb{K}$. Let g be a nonzero polynomial of $\mathbb{K}[x][y]$. We define the intersection multiplicity of f with g, denoted $\mathrm{int}(f, g)$, to be the rank of the $\mathbb{K}$-vector space $\dfrac{\mathbb{K}[x][y]}{(f, g)}$. Note that $\mathrm{int}(f, g)$ is also the x-degree of the y-resultant of f and g. Let $p = (a, b) \in V(f) \cap V(g)$, where V denotes the set of zeros in $\mathbb{K}^2$. By setting $\bar{x} = x - a, \bar{y} = y - b$, we may assume that $p = (0, 0)$. We define the intersection multiplicity of f with g at p, denoted

A. Assi (✉)
Mathématiques, Université d'Angers, 49045 Angers ceded 01, Angers, France

Department of Mathematics, American University of Beirut, Beirut 1107 2020, Lebanon
e-mail: assi@univ-angers.fr

I. Cheltsov et al. (eds.), *Automorphisms in Birational and Affine Geometry*, Springer Proceedings in Mathematics & Statistics 79,
DOI 10.1007/978-3-319-05681-4_14,

$\text{int}_p(f,g)$, to be the rank of the $\mathbb{K}$-vector space $\dfrac{\mathbb{K}[[x,y]]}{(f,g)}$. Note that $\text{int}(f,g) = \sum_{p\in V(f)\cap V(g)} \text{int}_p(f,g)$. We define the local Milnor number of f at p, denoted $\mu_p(f)$, to be the intersection multiplicity $\text{int}_p(f_x, f_y)$, where f_x (resp. f_y) denotes the x-derivative (resp. the y-derivative) of f. We set $\mu(f) = \sum_{p\in V(f)} \mu_p(f)$ and $\mu = \text{int}(f_x, f_y)$ and we recall that $\mu = \sum_{\lambda\in\mathbb{K}} \mu(f_\lambda) = \sum_{\lambda\in\mathbb{K}} \sum_{p\in V(f_\lambda)} \mu_p(f_\lambda)$. Let q be a point in $V(f)$ and assume after possibly a change of variables that $q = (0,0)$. The number of places of f at q, denoted r_q, is defined to be the number of irreducible components of f in $\mathbb{K}[[x,y]]$.

Assume, after possibly a change of variables, that $\deg_x a_i(x) < i$ for all $i = 1,\ldots,n$ (where $\deg_x$ denotes the x-degree). In particular f has one point at infinity defined by $y = 0$. Let $h_f(x,y,u) = u^n f(\frac{x}{u}, \frac{y}{u})$. The local equation of f at infinity is nothing but $F(y,u) = h_f(1,y,u) \in \mathbb{K}[[u]][y]$. We define the Milnor number of f at infinity, denoted μ_∞, to be the rank of the $\mathbb{K}$-vector space $\dfrac{\mathbb{K}[[u]][y]}{(F_u, F_y)}$. We define the number of places at infinity of f, denoted r_∞, to be the number of irreducible components of $F(y,u)$ in $\mathbb{K}[[u]][y]$.

2 Curves with One Place at Infinity

Let the notations be as in Sect. 1, in particular $f = y^n + a_1(x)y^{n-1} + \cdots + a_n(x)$ is a monic reduced polynomial of $\mathbb{K}[x,y]$. Let $R(x,\lambda) = P_0(\lambda)x^i + \cdots + P_i(\lambda)$ be the y-resultant of f_λ, f_y. We say that $(f_\lambda)_{\lambda\in\mathbb{K}}$ is d-regular (discriminant-regular) if $P_0(\lambda) \in \mathbb{K}^*$. Note that $(f_\lambda)_{\lambda\in\mathbb{K}}$ is d-regular if and only if $\text{int}(f_\lambda, f_y) = i$ for all $\lambda \in \mathbb{K}$. Suppose that $(f_\lambda)_{\lambda\in\mathbb{K}}$ is not d-regular, and let $\lambda_1,\ldots,\lambda_s$ be the set of roots of $P_0(\lambda)$. We set $I(f) = \{\lambda_1,\ldots,\lambda_s\}$, and we call $I(f)$ the set of d-irregular values of $(f_\lambda)_{\lambda\in\mathbb{K}}$. Let $A_f = \sum_{k=1}^{s}(i - \text{int}(f-\lambda_k, f_y))$. For all $\lambda \in \mathbb{K} - I(f)$, we have $\text{int}(f_\lambda, f_y) = \mu + n - 1 + A_f$, where $\mu = \text{int}(f_x, f_y)$ (see [4, 5]).

Note that $A_f = \sum_{\lambda\in\mathbb{K}}(i - \text{int}(f_\lambda, f_y))$, in particular $(f_\lambda)_{\lambda\in\mathbb{K}}$, is d-regular if and only if $A_f = 0$. On the other hand, given $a \in \mathbb{K}$, if $\text{int}(f_a, f_y) = \mu + n - 1$, then either $(f_\lambda)_{\lambda\in\mathbb{K}}$ is d-regular or $I(f) = \{a\}$.

Assume that $\deg_x a_k(x) < k$ for all $k = 1,\ldots,n$, in such a way that $y = 0$ is the only point at infinity of f.

Proposition 2.1 (see [1–3]). *Let the notations be as above and assume that f has one place at infinity, i.e., the projective curve defined by the homogeneous equation $h_f(x,y,u) = f(\frac{x}{u}, \frac{y}{u})u^n$ is analytically irreducible at the point at infinity $(1:0:0)$. We have the following*

- *For all $\lambda \in \mathbb{K}$, $f - \lambda$ has one place at infinity.*
- *The family $(f_\lambda)_{\lambda\in\mathbb{K}}$ is d-regular. In particular, $\text{int}(f_\lambda, f_y) = \mu + n - 1$ for all $\lambda \in \mathbb{K}$.*
- *If $\mu = 0$, then $\deg_x a_n(x)$ divides n and there exists an automorphism σ of $\mathbb{K}^2$ such that $\sigma(f)$ is a coordinate of $\mathbb{K}^2$.*

Let the notations be as above. If δ_p (resp. δ_∞) denotes the order of the conductor of f at $p \in V(f)$ (resp. at the point at infinity), then $2\delta_p = \mu_p + r_p - 1$ (resp. $2\delta_\infty = \mu_\infty + r_\infty - 1$) (see [7]). Assume that f is an irreducible polynomial, and let $g(f)$ be the genus of the normalized curve of $V(f)$. By the genus formula we have:

$$2g(f) + \left(\sum_{p \in V(f)} 2\delta_p \right) + 2\delta_\infty = (n-1)(n-2).$$

Now $\text{int}(f, f_y) = \mu + n - 1 + A(f)$, where $A(f)$ is a nonnegative integer and $A(f) = 0$ if and only if $(f_\lambda)_{\lambda \in \mathbb{K}}$ has at most one d-irregular value at infinity. On the other hand, the local intersection multiplicity of f with f_y at the point at infinity is $\mu_\infty + n - 1$. In particular $\mu + \mu_\infty = (n-1)(n-2)$, consequently, if $\mu(f) = \sum_{p \in V(f)} \mu_p$, and $\overline{\mu}(f) = \mu - \mu(f)$, then

$$2g(f) + \left(\sum_{p \in V(f)} 2\delta_p \right) + 2\delta_\infty = \mu(f) + \overline{\mu}(f) + \mu_\infty + A(f).$$

We finally get:

$$2g(f) + \sum_{p \in V(f)} (r_p - 1) + r_\infty - 1 = \overline{\mu}(f) + A(f) \qquad (**)$$

in particular $g(f) = \sum_{p \in V(f)} (r_p - 1) + r_\infty - 1 = 0$ if and only if $A(f) = \overline{\mu}(f) = 0$. Roughly speaking, f is a rational unibranch curve (at infinity as well as at finite distance) if and only if the pencil $(f_\lambda)_{\lambda \in \mathbb{K}}$ has at most one d-irregular value at infinity and for all $\lambda \neq 0$, f_λ is a smooth curve. Under these hypotheses, Lin–Zaidenberg Theorem implies that f is equivalent to a quasihomogeneous curve $Y^a - X^b$ with $\gcd(a, b) = 1$ (see [6]). Note that these hypotheses are satisfied when $r_\infty - 1 = 0 = \mu$. Hence we get the third assertion of Proposition 2.1. since in this case, $\min(a, b) = 1$ and f is equivalent to a coordinate

3 Rational One Place Curves

Let $f = y^n + a_1(x)y^{n-1} + \cdots + a_n(x)$ be a polynomial of $\mathbb{K}[x, y]$ and let the notations be as in Sects. 1 and 2. Assume that f has one place at infinity, i.e., $r_\infty = 1$. If f is rational, then it follows from the equality (**) of Sect. 2 that $\sum_{p \in V(f)} (r_p - 1) = \overline{\mu}(f)$. We shall prove the following:

Theorem 3.1. *Assume that f has one place at infinity and let $(f_\lambda)_{\lambda \in \mathbb{K}}$ be the pencil of curves defined by f. If f is rational, then exactly one of the following holds:*

i) *For all* $\lambda \in \mathbb{K}$, f_λ *is rational, and* $\sigma(f)$ *is a coordinate of* $\mathbb{K}^2$ *for some automorphism* σ *of* $\mathbb{K}^2$.
ii) *The polynomial* $f-\lambda$ *is rational for at most one* $\lambda_1 \neq 0$, *i.e., the pencil* $(f_\lambda)_{\lambda\in\mathbb{K}}$ *has at most two rational elements.*

We shall prove first the following Lemma:

Lemma 3.2. *Let* $H = y^N + a_1(x)y^{N-1} + \cdots + a_N(x)$ *be a nonzero reduced polynomial of* $\mathbb{K}[[x]][y]$, *and let* $H = H_1 \ldots H_r$ *be the decomposition of* H *into irreducible components of* $\mathbb{K}[[x]][y]$. *Let* $\mu_{(0,0)}$ *denotes the Milnor number of* H *at* $(0,0)$ *(i.e.,* $\mu_{(0,0)}$ *is the rank of the* $\mathbb{K}$*-vector space* $\frac{\mathbb{K}[[x]][y]}{(H_x, H_y)}$*). We have the following:*

i) $\mu_{(0,0)} \geq r-1$.
ii) *If* $r \geq 3$, *then* $\mu_{(0,0)} > r-1$.
iii) *If* $r=2$ *and* $\mu_{(0,0)} = r-1 = 1$, *then* (H_1, H_2) *is a local system of coordinates at* $(0,0)$.

Proof. We have $\text{int}_{(0,0)}(H, H_y) = \mu_{(0,0)} + N - 1$, but

$$\text{int}_{(0,0)}(H, H_y) = \sum_{i=1}^{r} \text{int}(H_i, H_{i_y}) + 2\sum_{i\neq j} \text{int}_{(0,0)}(H_i, H_j)$$

$$= \sum_{i=1}^{r} \text{int}[(H_{i_x}, H_{i_y}) + \deg_y H_i - 1] + 2\sum_{i\neq j} \text{int}_{(0,0)}(H_i, H_j)$$

hence

$$\mu_{(0,0)} + N - 1 = \left(\sum_{i=1}^{r} \text{int}(H_{i_x}, H_{i_y})\right) + N - r + 2\sum_{i\neq j} \text{int}_{(0,0)}(H_i, H_j).$$

Finally we have $\mu_{(0,0)} = (\sum_{i=1}^{r} \text{int}(H_{i_x}, H_{i_y})) - r + 1 + 2\sum_{i\neq j} \text{int}_{(0,0)}(H_i, H_j)$. Now for all $1 \leq i \leq r$, $\text{int}_{(0,0)}(H_{i_x}, H_{i_y}) \geq 0$ and $\sum_{i\neq j} \text{int}_{(0,0)}(H_i, H_j) \geq C_2^r = \frac{r(r-1)}{2}$, hence $\mu_{(0,0)} \geq r(r-1)-(r-1) = (r-1)^2$ and (i), (ii) follow immediately. Assume that $r=2$. If $\mu_{(0,0)} = r-1$, then $\text{int}_{(0,0)}(H_{1_x}, H_{1_y}) = \text{int}_{(0,0)}(H_{2_x}, H_{2_y}) = 0$ and $\text{int}_{(0,0)}(H_1, H_2) = 1$. This implies (iii) □

Proof of Theorem 3.1. If $\mu(f) = 0$, then $\mu = 0$ and by Proposition 2.1., $\sigma(f)$ is a coordinate of $\mathbb{K}^2$ for some automorphism σ of $\mathbb{K}^2$. Assume that $\mu(f) > 0$ and let $p_1, \ldots, p_s$ be the set of singular points of $V(f)$. Let r_i denotes the number of places of f at p_i for all $1 \leq i \leq s$. By Lemma 3.2., for all $1 \leq i \leq s$, $\mu_{p_i} \geq r_i - 1$, on the other hand, equality (**) of Sect. 2 implies that $\sum_{i=1}^{s}(\mu_{p_i} + r_i - 1) = \mu$, in particular $\mu \leq \sum_{i=1}^{s} 2\mu_{p_i} = 2\mu(f)$, hence $\mu(f) \geq \frac{\mu}{2}$. If f_{λ_1} is rational for some

$\lambda_1 \neq 0$, then the same argument as above implies that $\mu(f_{\lambda_1}) \geq \frac{\mu}{2}$. This is possible only for at most one $\lambda_1 \neq 0$, hence (ii) follows immediately. □

The following proposition characterizes the case where the pencil $(f_\lambda)_{\lambda\in\mathbb{K}}$ has exactly two rational elements.

Proposition 3.3. *Let the notations be as in Theorem 3.1 and assume that the pencil $(f_\lambda)_{\lambda\in\mathbb{K}}$ has exactly two rational elements f and f_{λ_1}. We have $\mu(f) = \mu(f_{\lambda_1}) = \frac{\mu}{2}$, furthermore, given a singular point p of $V(f)$ (resp. $V(f_{\lambda_1})$), f (resp. f_{λ_1}) has two places at p and $\mu_p(f) = 1$ (resp. $\mu_p(f_{\lambda_1}) = 1$). In particular, f (resp. f_{λ_1}) has exactly $\frac{\mu}{2}$ singular points.*

Proof. It follows from the proof of Theorem 3.1. that $\mu(f) \geq \frac{\mu}{2}$ and that $\mu(f_{\lambda_1}) \geq \frac{\mu}{2}$. Clearly this holds only if $\mu(f) = \mu(f_{\lambda_1}) = \frac{\mu}{2}$. Let p be a singular point of $V(f)$. We have $\mu_p = r_p - 1$, hence, by Lemma 3.2(ii), $r_p \leq 2$. But $\mu_p > 0$, hence $r_p = 2$ and $\mu_p = 1$. This implies that f has $\frac{\mu}{2}$ singular points. Clearly the same holds for f_{λ_1}. □

The results above imply the following:

Proposition 3.4. *Assume that f has one place at infinity and let $(f_\lambda)_{\lambda\in\mathbb{K}}$ be the pencil of polynomials defined by f. Assume that f is a rational polynomial and that $\mu(f) > 0$. Let $p_1, \ldots, p_s$ be the set of singular points of f. We have the following*

- *i) If $r_{p_i} = 1$ (resp. $r_{p_i} \geq 3$) for some $1 \leq i \leq s$, then f is the only rational point of the pencil $(f_\lambda)_\lambda$.*
- *ii) If $r_{p_i} = 2$ for all $1 \leq i \leq s$ but $s \neq \frac{\mu}{2}$, then f is the only rational element of the pencil $(f_\lambda)_\lambda$.*

Proof. This is an immediate application of Theorem 3.1. and Proposition 3.3. □

Proposition 3.5. *Let $f \neq g$ be two monic polynomials of $\mathbb{K}[x][y]$ and assume that f, g are parameterized by polynomials of $\mathbb{K}[t]$. Under these hypotheses, exactly one of the following conditions holds:*

- *i) $f = g + \lambda_1$ for some $\lambda_1 \in \mathbb{K}^*$, and f is equivalent to a coordinate, i.e., $\sigma(f)$ is a coordinate of $\mathbb{K}^2$ for some automorphism σ of $\mathbb{K}^2$.*
- *ii) $f = g + \lambda_1$ for some $\lambda_1 \in \mathbb{K}^*$, $\mu(f) = \mu(g) = \frac{\mathrm{int}(f_x, f_y)}{2} > 0$, and f (resp. g) has $\frac{\mathrm{int}(f_x, f_y)}{2}$ singular points with two places at each of them.*
- *iii) $\mathrm{int}(f, g) > 0$, i.e., f, g meet in a least one point of $\mathbb{K}^2$.*

Proof. The polynomial f (resp. g) has one place at infinity. If $\text{int}(f, g) = 0$, then $f = ag + \lambda_1, a, \lambda_1 \in \mathbb{K}^*$. Since f and g are monic, then $a = 1$. Hence g and $g + \lambda_1$ are two rational elements of the pencil $(f_\lambda)_{\lambda \in \mathbb{K}}$. Now apply Theorem 3.1. and Proposition 3.3. □

Remark 3.6. Let $(x(t), y(t)) = (t^3 - 3t, t^2 - 2)$ and $(X(s), Y(s)) = (s^3 + 3s, s^2 + 2)$, and let $f(x, y) = \text{res}_t(x - x(t), y - y(t))$ (resp. $g(x, y) = \text{res}_s(x - X(s), y - Y(s))$). We have $(x(t) - X(s), y(t) - Y(s)) = \mathbb{K}[t, s]$, hence $\text{int}(f, g) = 0$. In fact,

$$f(x, y) = y^3 - x^2 - 3y + 2 = -x^2 + (y + 2)(y - 1)^2$$

and

$$g(x, y) = y^3 - x^2 - 3y - 2 = -x^2 + (y - 2)(y + 1)^2,$$

hence $f = g + 4$. The genus of a generic element of the family $(f_\lambda)_\lambda$ is 1, and $f, f - 4$ are the two rational elements of this family. Note that $\mu = 2$ and $\mu(f) = \mu(f - 4) = 1$. This example shows that the bound of Theorem 3.1. is sharp.

Remark 3.7. Let $(f_\lambda)_{\lambda \in \mathbb{K}}$ be a pencil of polynomials of $\mathbb{K}[x, y]$ and assume that $f - \lambda$ is irreducible for all $\lambda \in \mathbb{K}$. If the generic element of the pencil is rational, then for all $\lambda \in \mathbb{K}$, $f - \lambda$ is rational and irreducible. In this case, by Neumann and Norbury [8], f has one place at infinity and $\sigma(f)$ is a coordinate of $\mathbb{K}^2$ for some automorphism σ of $\mathbb{K}$. Assume that the genus of the generic element of the pencil $(f_\lambda)_{\lambda \in \mathbb{K}}$ is greater than or equal to one. Similarly to the case of curves with one place at infinity, it is natural to address the following question:

Question: Is there an integer $c \in \mathbb{N}$ such that, given a pencil of irreducible polynomials $(f_\lambda)_{\lambda \in \mathbb{K}}$, if $\mu + A_f > 0$, then the number of rational elements in the pencil is bounded by c?

Acknowledgements The author would like to thank the referee for the valuable and helpful comments.

References

1. S.S. Abhyankar, On the semigroup of a meromorphic curve, Part 1, in *Proceedings of the International Symposium on Algebraic Geometry*, Kyoto, 1977, pp. 240–414
2. S.S. Abhyankar, T.T. Moh, Newton-Puiseux expansion and generalized Tschirnhausen transformation. Crelle J. **260**, 47–83 (1973)
3. S.S. Abhyankar, T.T. Moh, Newton-Puiseux expansion, and generalized Tschirnhausen transformation II. Crelle J. **261**, 29–54 (1973)
4. A. Assi, Sur l'intersection des courbes méromorphes. C. R. Acad. Sci. Paris Sér. I Math. **329**(7), 625–628 (1999)
5. A. Assi, Meromorphic plane curves. Math. Z. **230**(1), 16–183 (1999)

6. V. Lin, M. Zaidenberg, An irreducible simply connected algebraic curve in $\mathbb{C}^2$ is equivalent to a quasihomogeneous curve. Dokl. Akad. Nauk SSSR **271**(5), 1048–1052 (1983)
7. J. Milnor, *Singular Points of Complex Hypersurfaces*. Annals of Mathematics Studies, vol. 61 (Princeton University Press, Princeton, 1968)
8. W. Neumann, P. Norbury, Nontrivial rational polynomials in two variables have reducible fibres. Bull. Austr. Math. Soc. **58**, 501–503 (1998)

The Jacobian Conjecture, Together with Specht and Burnside-Type Problems

Alexei Belov, Leonid Bokut, Louis Rowen, and Jie-Tai Yu

Dedicated to the memory of A.V. Yagzhev

Abstract We explore an approach to the celebrated Jacobian Conjecture by means of identities of algebras, initiated by the brilliant deceased mathematician, Alexander Vladimirovich Yagzhev (1951–2001), whose works have only been partially published. This approach also indicates some very close connections between mathematical physics, universal algebra, and automorphisms of polynomial algebras.

2010 Mathematics Subject Classification: Primary 13F20, 14E08, 14R15, 17A30, 17A40, 17A50; Secondary 13F25, 17A01, 17A05, 17A15, 17A65.

1 Introduction

This paper explores an approach to polynomial mappings and the Jacobian Conjecture and related questions, initiated by A.V. Yagzhev [95–105] whereby these questions are translated to identities of algebras, leading to a solution in [103]

A. Belov
Department of Mathematics, Bar-Ilan University, Ramat Gan 52900, Israel
e-mail: beloval@math.biu.ac.il

L. Bokut
Sobolev Institute of Mathematics, Novosibirsk, Russia

South China Normal University, Guangzhou, China
e-mail: bokut@math.nsc.ru

L. Rowen (✉)
Department of Mathematics, Bar-Ilan University, Ramat Gan 52900, Israel
e-mail: rowen@math.biu.ac.il

J.-T. Yu
Department of Mathematics, The University of Hong Kong, Hong Kong SAR, China
e-mail: yujt@hku.hk; yujietai@yahoo.com

I. Cheltsov et al. (eds.), *Automorphisms in Birational and Affine Geometry*, Springer Proceedings in Mathematics & Statistics 79,
DOI 10.1007/978-3-319-05681-4_15,

of the version of the Jacobian Conjecture for free associative algebras. (The first version, for two generators, was obtained by Dicks and Levin [27, 28], and the full version by Schofield [69].) We start by laying out the basic framework in this introduction. Next, we set up Yagzhev's correspondence to algebras in Sect. 2, leading to the basic notions of weak nilpotence and Engel type. In Sect. 3 we discuss the Jacobian Conjecture in the context of various varieties, including the free associative algebra.

Given any polynomial endomorphism ϕ of the n-dimensional affine space $\mathbb{A}_{\mathbf{k}}^n = \operatorname{Spec} \mathbf{k}[x_1, \dots, x_n]$ over a field $\mathbf{k}$, we define its *Jacobian matrix* to be the matrix

$$\left(\partial \phi^*(x_i)/\partial x_j\right)_{1 \le i,j \le n}.$$

The determinant of the Jacobian matrix is called the *Jacobian* of ϕ. The celebrated **Jacobian Conjecture** JC_n in dimension $n \geq 1$ asserts that *for any field* $\mathbf{k}$ *of characteristic zero, any polynomial endomorphism* ϕ *of* $\mathbb{A}_{\mathbf{k}}^n$ *having Jacobian* 1 *is an automorphism.* Equivalently, one can say that ϕ preserves the standard top-degree differential form $dx_1 \wedge \cdots \wedge dx_n \in \Omega^n(\mathbb{A}_{\mathbf{k}}^n)$. References to this well-known problem and related questions can be found in [5,35,48]. By the Lefschetz principle it is sufficient to consider the case $\mathbf{k} = \mathbb{C}$; obviously, JC_n implies JC_m if $n > m$. The conjecture JC_n is obviously true in the case $n = 1$, and it is open for $n \geq 2$.

The **Jacobian Conjecture**, denoted as JC, is the conjunction of the conjectures JC_n for all finite n. The Jacobian Conjecture has many reformulations (such as the Kernel Conjecture and the Image Conjecture, cf. [35, 36, 39, 108, 109] for details) and is closely related to questions concerning quantization. It is stably equivalent to the following conjecture of Dixmier, concerning automorphisms of the Weyl algebra W_n, otherwise known as the *quantum affine algebra*.

Dixmier Conjecture $\mathbf{DC}_n$: *Does* $\operatorname{End}(W_n) = \operatorname{Aut}(W_n)$?

The implication $\mathrm{DC}_n \to \mathrm{JC}_n$ is well known, and the inverse implication $\mathrm{JC}_{2n} \to \mathrm{DC}_n$ was recently obtained independently by Tsuchimoto [80] (using p-curvature) and Belov and Kontsevich [14, 15] (using Poisson brackets on the center of the Weyl algebra). Bavula [10] has obtained a shorter proof, and also obtained a positive solution of an analog of the Dixmier Conjecture for integro differential operators, cf. [8]. He also proved that every monomorphism of the Lie algebra of triangular polynomial derivations is an automorphism [9] (an analog of Dixmier's conjecture).

The Jacobian Conjecture is closely related to many questions of affine algebraic geometry concerning affine space, such as the Cancellation Conjecture (see Sect. 3.4). If we replace the variety of commutative associative algebras (and the accompanying affine spaces) by an arbitrary algebraic variety,[1] one easily gets a counterexample to the JC. So, strategically these questions deal with some specific

[1]Algebraic geometers use the word *variety*, roughly speaking, for objects whose local structure is obtained from the solution of system of algebraic equations. In the framework of universal algebra, this notion is used for subcategories of algebras defined by a given set of identities. A deep analog of these notions is given in [12].

properties of affine space which we do not yet understand, and for which we do not have the appropriate formulation apart from these very difficult questions.

It seems that these properties do indicate some sort of quantization. From that perspective, noncommutative analogs of these problems (in particular, the Jacobian Conjecture and the analog of the Cancellation Conjecture) become interesting for free associative algebras, and more generally, for arbitrary varieties of algebras.

We work in the language of universal algebra, in which an algebra is defined in terms of a set of operators, called its *signature*. This approach enhances the investigation of the Yagzhev correspondence between endomorphisms and algebras. We work with deformations and the so-called *packing properties* to be introduced in Sects. 3 and 3.2.1, which denote specific noncommutative phenomena which enable one to solve the JC for the free associative algebra.

From the viewpoint of universal algebra, the Jacobian conjecture becomes a problem of "Burnside type," by which we mean the question of whether a given finitely generated algebraic structure satisfying given periodicity conditions is necessarily finite, cf. Zelmanov [107]. Burnside originally posed the question of the finiteness of a finitely generated group satisfying the identity $x^n = 1$. (For odd $n \geq 661$, counterexamples were found by Novikov and Adian, and quite recently Adian reduced the estimate from 661 to 101.) Another class of counterexamples was discovered by Ol'shanskij [60]. Kurosh posed the question of local finiteness of algebras whose elements are algebraic over the base field. For algebraicity of bounded degree, the question has a positive solution, but otherwise there are the Golod–Shafarevich counterexamples.

Burnside-type problems play an important role in algebra. Their solution in the associative case is closely tied to Specht's problem of whether any set of polynomial identities can be deduced from a finite subset. The JC can be formulated in the context of whether one system of identities implies another, which also relates to Specht's problem.

In the Lie algebra case there is a similar notion. An element $x \in L$ is called *Engel of degree n* if $[\dots[[y,x],x]\dots,x] = 0$ for any y in the Lie algebra L. Zelmanov's result that any finitely generated Lie algebra of bounded Engel degree is nilpotent yielded his solution of the Restricted Burnside Problem for groups. Yagzhev introduced the notion of *Engelian* and *weakly nilpotent* algebras of arbitrary signature (see Definitions 5 and 4), and proved that the JC is equivalent to the question of weak nilpotence of algebras of Engel type satisfying a system of Capelli identities, thereby showing the relation of the JC with problems of Burnside type.

A Negative Approach. Let us mention a way of constructing counterexamples. This approach, developed by Gizatullin, Kulikov, Shafarevich, Vitushkin, and others, is related to decomposing polynomial mappings into the composition of σ-processes [41, 48, 70, 89–91]. It allows one to solve some polynomial automorphism problems, including tameness problems, the most famous of which is *Nagata's Problem* concerning the wildness of Nagata's automorphism

$$(x,y,z) \mapsto (x - 2(xz + y^2)y - (xz + y^2)^2 z,\ y + (xz + y^2)z,\ z),$$

cf. [57]. Its solution by Shestakov and Umirbaev [73] is the major advance in this area in the last decade. The Nagata automorphism can be constructed as a product of automorphisms of $K(z)[x, y]$, some of them having non-polynomial coefficients (in $K(z)$). The following theorem of Abhyankar–Moh–Suzuki [2, 53, 78] can be viewed in this context:

AMS Theorem. If f and g are polynomials in $K[z]$ of degrees n and m for which $K[f, g] = K[z]$, then n divides m or m divides n.

Degree estimate theorems are polynomial analogs to Liouville's approximation theorem in algebraic number theory [23,50,51,54]. T. Kishimoto has proposed using a program of Sarkisov, in particular for Nagata's Problem. Although difficulties remain in applying "σ-processes" (decomposition of birational mappings into standard blow-up operations) to the affine case, these may provide new insight. If we consider affine transformations of the plane, we have relatively simple singularities at infinity, although for bigger dimensions they can be more complicated. Blow-ups provide some understanding of birational mappings with singularities. Relevant information may be provided in the affine case. The paper [21] contains some deep considerations about singularities.

2 The Jacobian Conjecture and Burnside-Type Problems, via Algebras

In this section we translate the Jacobian Conjecture to the language of algebras and their identities. This can be done at two levels: at the level of the algebra obtained from a polynomial mapping, leading to the notion of *weak nilpotence* and *Yagzhev algebras* and at the level of the differential and the algebra arising from the Jacobian, leading to the notion of *Engel type*. The Jacobian Conjecture is the link between these two notions.

2.1 *The Yagzhev Correspondence*

2.1.1 Polynomial Mappings in Universal Algebra

Yagzhev's approach is to pass from algebraic geometry to universal algebra. Accordingly, we work in the framework of a universal algebra A having signature Ω. $A^{(m)}$ denotes $A \times \cdots \times A$, taken m times.

We fix a commutative, associative base ring C, and consider C-modules equipped with extra *operators* $A^{(m)} \to A$, which we call *m-ary*. Often one of these operators will be (binary) multiplication. These operators will be multilinear, i.e., linear with respect to each argument. Thus, we can define the *degree* of an operator to be its number of arguments. We say an operator $\Psi(x_1, \ldots, x_m)$ is *symmetric* if $\Psi(x_1, \ldots, x_m) = \Psi(x_{\pi(1)}, \ldots, x_{\pi(m)})$ for all permutations π.

Definition 1. A *string* of operators is defined inductively. Any operator $\Psi(x_1, \dots, x_m)$ is a string of degree m, and if s_j are strings of degree d_j, then $\Psi(s_1, \dots, s_m)$ is a string of degree $\sum_{j=1}^{m} d_j$. A mapping

$$\alpha : A^{(m)} \to A$$

is called *polynomial* if it can be expressed as a sum of strings of operators of the algebra A. The *degree* of the mapping is the maximal length of these strings.

Example. Suppose an algebra A has two extra operators: a binary operator $\alpha(x, y)$ and a tertiary operator $\beta(x, y, z)$. The mapping $F : A \to A$ given by $x \to x + \alpha(x, x) + \beta(\alpha(x, x), x, x)$ is a polynomial mapping of A, having degree 4. Note that if A is finite dimensional as a vector space, not every polynomial mapping of A as an affine space is a polynomial mapping of A as an algebra.

2.1.2 Yagzhev's Correspondence Between Polynomial Mappings and Algebras

Here we associate an algebraic structure with each polynomial map. Let V be an n-dimensional vector space over the field $\mathbf{k}$, and $F : V \to V$ be a polynomial mapping of degree m. Replacing F by the composite TF, where T is a translation such that $TF(0) = 0$, we may assume that $F(0) = 0$. Given a base $\{\vec{e}_i\}_{i=1}^{n}$ of V, and for an element v of V written uniquely as a sum $\sum x_i \vec{e}_i$, for $x_i \in \mathbf{k}$, the coefficients of $\vec{e}_i$ in $F(v)$ are (commutative) polynomials in the x_i. Then F can be written in the following form:

$$x_i \mapsto F_{0i}(\vec{x}) + F_{1i}(\vec{x}) + \cdots + F_{mi}(\vec{x})$$

where each $F_{\alpha i}(\vec{x})$ is a homogeneous form of degree α, i.e.,

$$F_{\alpha i}(\vec{x}) = \sum_{j_1+\cdots+j_n=\alpha} \kappa_J x_1^{j_1} \cdots x_n^{j_n},$$

with $F_{0i} = 0$ for all i, and $F_{1i}(\vec{x}) = \sum_{k=1}^{n} \mu_{ki} x_k$.

We are interested in invertible mappings that have a nonsingular Jacobian matrix (μ_{ij}). In particular, this matrix is nondegenerate at the origin. In this case $\det(\mu_{ij}) \neq 0$, and by composing F with an affine transformation, we arrive at the situation for which $\mu_{ki} = \delta_{ki}$. Thus, the mapping F may be taken to have the following form:

$$x_i \to x_i - \sum_{k=2}^{m} F_{ki}. \tag{1}$$

Suppose we have a mapping as in (1). Then the Jacobi matrix can be written as $E - G_1 - \cdots - G_{m-1}$ where G_i is an $n \times n$ matrix with entries which are homogeneous polynomials of degree i. If the Jacobian is 1, then it is invertible with inverse a polynomial matrix (of homogeneous degree at most $(n-1)(m-1)$, obtained via the adjoint matrix).

If we write the inverse as a formal power series, we compare the homogeneous components and get:

$$\sum_{j_i m_{j_i} = s} M_J = 0, \tag{2}$$

where M_J is the sum of products $a_{\alpha_1} a_{\alpha_q}$ in which the factor a_j occurs m_j times, and J denotes the multi-index $(j_1, \dots, j_q)$.

Yagzhev considered the cubic homogeneous mapping $\vec{x} \to \vec{x} + (\vec{x}, \vec{x}, \vec{x})$, whereby the Jacobian matrix becomes $E - G_3$. We return to this case in Remark 1. The slightly more general approach given here presents the Yagzhev correspondence more clearly and also provides tools for investigating deformations and packing properties (see Sect. 3.5.1) i.e., when the mapping has the form

$$x_i \to x_i + P_i(x_1, \dots, x_n);\ i = 1, \dots, n,$$

with P_i cubic homogenous polynomials), but the more general situation of arbitrary degree.

For any ℓ, the set of (vector valued) forms $\{F_{\ell,i}\}_{i=1}^{n}$ can be interpreted as a homogeneous mapping $\Phi_\ell : V \to V$ of degree ℓ. When Char(**k**) does not divide ℓ, we take instead the polarization of this mapping, i.e. the multilinear symmetric mapping

$$\Psi_\ell : V^{\otimes \ell} \to V$$

such that

$$(F_{\ell,i}(x_1), \dots, F_{\ell,i}(x_n)) = \Psi_\ell(\vec{x}, \dots, \vec{x}) \cdot \ell!$$

Then (1) can be rewritten as

$$\vec{x} \to \vec{x} - \sum_{\ell=2}^{m} \Psi_\ell(\vec{x}, \dots, \vec{x}). \tag{3}$$

We define the algebra $(A, \{\Psi_\ell\})$, where A is the vector space V and the Ψ_ℓ are viewed as operators $A^\ell \to A$.

Definition 2. The **Yagzhev correspondence** is the correspondence from the polynomial mapping (V, F) to the algebra $(A, \{\Psi_\ell\})$.

2.2 *Translation of the Invertibility Condition to the Language of Identities*

The next step is to bring in algebraic varieties, defined in terms of identities.

Definition 3. A *polynomial identity* (PI) of A is a polynomial mapping of A, all of whose values are identically zero.

The *algebraic variety* generated by an algebra A, denoted as $\mathrm{Var}(A)$, is the class of all algebras satisfying the same PIs as A.

Now we come to a crucial idea of Yagzhev:

The invertibility of F and the invertibility of the Jacobian of F can be expressed via (2) in the language of polynomial identities.

Namely, let $y = F(x) = x - \sum_{\ell=2}^{m} \Psi_\ell(x)$. Then

$$F^{-1}(x) = \sum_t t(x), \tag{4}$$

where each t is a *term*, a formal expression in the mappings $\{\Psi_\ell\}_{\ell=2}^{m}$ and the symbol x. Note that the expressions $\Psi_2(x, \Psi_3(x,x,x))$ and $\Psi_2(\Psi_3(x,x,x),x)$ are different although they represent same element of the algebra. Denote by $|t|$ the number of occurrences of variables, including multiplicity, which are included in t.

The invertibility of F means that, for all $q \geq q_0$,

$$\sum_{|t|=q} t(a) = 0, \quad \forall a \in A. \tag{5}$$

Thus we have translated invertibility of the mapping F to the language of identities. (Yagzhev had an analogous formula, where the terms only involved Ψ_3.)

Definition 4. An element $a \in A$ is called *nilpotent* of index $\leq n$ if

$$M(a, a, \dots, a) = 0$$

for each monomial $M(x_1, x_2, \dots)$ of degree $\geq n$. The algebra A is *weakly nilpotent* if each element of A is nilpotent. A is *weakly nilpotent of class k* if each element of A is nilpotent of index k. (Some authors use the terminology *index* instead of *class*.) Equation (5) means A is weakly nilpotent.

To stress this fundamental notion of Yagzhev, we define a *Yagzhev algebra* of *order* q_0 to be a weakly nilpotent algebra, i.e., satisfying the identities (5), also called the *system of Yagzhev identities* arising from F.

Summarizing, we get the following fundamental translation from conditions on the endomorphism F to identities of algebras.

Theorem 1. *The endomorphism F is invertible if and only if the corresponding algebra is a Yagzhev algebra of high enough order.*

2.2.1 Algebras of Engel Type

The analogous procedure can be carried out for the differential mapping. We recall that Ψ_ℓ is a symmetric multilinear mapping of degree ℓ. We denote the mapping $y \to \Psi_\ell(y, x, \dots, x)$ as $\mathrm{Ad}_{\ell-1}(x)$.

Definition 5. An algebra A is of *Engel type* s if it satisfies a system of identities

$$\sum_{\ell m_\ell = s} \quad \sum_{\alpha_1+\cdots+\alpha_q=m_\ell} \mathrm{Ad}_{\alpha_1}(x)\cdots \mathrm{Ad}_{\alpha_q}(x) = 0. \tag{6}$$

A is of *Engel type* if A has Engel type s for some s.

Theorem 2. *The endomorphism F has Jacobian* 1 *if and only if the corresponding algebra has Engel type s for some s.*

Proof. Let $x' = x + dx$. Then

$$\begin{aligned} \Psi_\ell(x') = {} & \Psi_\ell(x) + \ell\Psi_\ell(dx, x, \dots, x) \\ & + \text{ forms containing more than one occurrence of } dx. \end{aligned} \tag{7}$$

Hence the differential of the mapping

$$F : \vec{x} \mapsto \vec{x} - \sum_{\ell=2}^{m} \Psi_\ell(\vec{x}, \dots, \vec{x})$$

is

$$\left(E - \sum_{\ell=2}^{m} \ell\, \mathrm{Ad}_{\ell-1}(x) \right) \cdot dx$$

The identities (2) are equivalent to the system of identities (6) in the signature $\Omega = (\Psi_2, \dots, \Psi_m)$, taking $a_{\alpha_j} = \mathrm{Ad}_{\alpha_j}$ and $m_j = \deg \Psi_\ell - 1$.

Thus, we have reformulated the condition of invertibility of the Jacobian in the language of identities.

As explained in [35], it is well known from [5, 98] that the Jacobian Conjecture can be reduced to the cubic homogeneous case; i.e., it is enough to consider mappings of type

$$x \to x + \Psi_3(x, x, x).$$

In this case the Jacobian assumption is equivalent to the *Engel condition*—nilpotence of the mapping $\mathrm{Ad}_3(x)[y]$ (i.e., the mapping $y \to (y, x, x)$). Invertibility, considered in [5], is equivalent to weak nilpotence, i.e., to the identity $\sum_{|t|=k} t = 0$ holding for all sufficiently large k.

Remark 1. In the cubic homogeneous case, $j = 1$, $\alpha_j = 2$ and $m_j = s$, and we define the linear map

$$\mathrm{Ad}_{xx} : y \to (x, x, y)$$

and the index set $T_j \subset \{1, \ldots, q\}$ such that $i \in T_j$ if and only if $\alpha_i = j$.

Then the equality (6) has the following form:

$$\mathrm{Ad}_{xx}^{s/2} = 0.$$

Thus, for a ternary symmetric algebra, Engel type means that the operators Ad_{xx} for all x are nilpotent. In other words, the mapping

$$\mathrm{Ad}_3(x) : y \to (x, x, y)$$

is nilpotent. Yagzhev called this the *Engel condition.* (For Lie algebras the nilpotence of the operator $\mathrm{Ad}_x : y \to (x, y)$ is the usual Engel condition. Here we have a generalization for arbitrary signature.)

Here are Yagzhev's original definitions, for edification. A binary algebra A is *Engelian* if for any element $a \in A$ the subalgebra $\langle R_a, L_a \rangle$ of vector space endomorphisms of A generated by the left multiplication operator L_a and the right multiplication operator R_a is nilpotent, and *weakly Engelian* if for any element $a \in A$ the operator $R_a + L_a$ is nilpotent.

This leads us to the *Generalized Jacobian Conjecture*:

Conjecture. Let A be an algebra with symmetric $\mathbf{k}$-linear operators Ψ_ℓ, for $\ell = 1, \ldots, m$. In any variety of Engel type, A is a Yagzhev algebra.

By Theorem 2, this conjecture would yield the Jacobian Conjecture.

2.2.2 The Case of Binary Algebras

When A is a binary algebra, *Engel type* means that the left and right multiplication mappings are both nilpotent.

A well-known result of Wang [5] shows that the Jacobian Conjecture holds for quadratic mappings

$$\vec{x} \to \vec{x} + \Psi_2(\vec{x}, \vec{x}).$$

If two different points $(x_1, \ldots, x_n)$ and $(y_1, \ldots, y_n)$ of an affine space are mapped to the same point by $(f_1, \ldots, f_n)$, then the fact that the vertex of a parabola is in the middle of the interval whose endpoints are at the roots shows that all $f_i(\vec{x})$ have gradients at this midpoint $P = (\vec{x} + \vec{y})/2$ perpendicular to the line segment $[\vec{x}, \vec{y}]$. Hence the Jacobian is zero at the midpoint P. This fact holds in any characteristic $\neq 2$.

In Sect. 2.3 we prove the following theorem of Yagzhev, cf. Definition 6 below.

Theorem 3 (Yagzhev). *Every symmetric binary Engel-type algebra of order k satisfying the system of Capelli identities of order n is weakly nilpotent, of weak nilpotence index bounded by some function $F(k, n)$.*

Remark 2. Yagzhev formulates his theorem in the following way:

Every binary weakly Engel algebra of order k satisfying the system of Capelli identities of order n is weakly nilpotent, of index bounded by some function $F(k, n)$.

We obtain this reformulation, by replacing the algebra A by the algebra A^+ with multiplication given by $(a, b) = ab + ba$.

The following problems may help us understand the situation.

Problem. Obtain a straightforward proof of this theorem and deduce from it the Jacobian Conjecture for quadratic mappings.

Problem (Generalized Jacobian Conjecture for Quadratic Mappings). Is every symmetric binary algebra of Engel type k, a Yagzhev algebra?

2.2.3 The Case of Ternary Algebras

As we have observed, Yagzhev reduced the Jacobian Conjecture over a field of characteristic zero to the question:

Is every finite dimensional ternary Engel algebra a Yagzhev algebra?

Drużkowski [33, 34] reduced this to the case when all cubic forms Ψ_{3i} are cubes of linear forms. Van den Essen and his school reduced the JC to the symmetric case; see [37, 38] for details. Bass et al. [5] use other methods including inversions. Yagzhev's approach matches that of [5], but using identities instead.

2.2.4 An Example in Nonzero Characteristic of an Engel Algebra That Is Not a Yagzhev Algebra

Now we give an example, over an arbitrary field $\mathbf{k}$ of characteristic $p > 3$, of a finite dimensional Engel algebra that is not a Yagzhev algebra, i.e., not weakly nilpotent. This means that the situation for binary algebras differs intrinsically from that for ternary algebras, and it would be worthwhile to understand why.

Theorem 4. *If* $\mathrm{Char}(\mathbf{k}) = p > 3$, *then there exists a finite dimensional* $\mathbf{k}$*-algebra that is Engel but not weakly nilpotent.*

Proof. Consider the noninvertible mapping $F : \mathbf{k}[x] \to \mathbf{k}[x]$ with Jacobian 1:

$$F : x \to x + x^p.$$

We introduce new commuting indeterminates $\{y_i\}_{i=1}^n$ and extend this mapping to $k[x, y_1, \ldots, y_n]$ by sending $y_i \mapsto y_i$. If n is big enough, then it is possible to find tame automorphisms G_1 and G_2 such that $G_1 \circ F \circ G_2$ is a cubic mapping $\vec{x} \to \vec{x} + \Psi_3(\vec{x})$, as follows:

Suppose we have a mapping

$$F : x_i \to P(x) + M$$

where $M = t_1t_2t_3t_4$ is a monomial of degree at least 4. Introduce two new commuting indeterminates z, y and take $F(z) = z$, $F(y) = y$.

Define the mapping G_1 via $G_1(z) = z + t_1t_2$, $G_1(y) = y + t_3t_4$ with G_1 fixing all other indeterminates; define G_2 via $G_2(x) = x - yz$ with G_2 fixing all other indeterminates.

The composite mapping $G_1 \circ F \circ G_2$ sends x to $P(x) - yz - yt_1t_2 - zt_3t_4$, y to $y + t_3t_4$, z to $z + t_1t_2$ and agrees with F on all other indeterminates.

Note that we have removed the monomial $M = t_1t_2t_3t_4$ from the image of F, but instead have obtained various monomials of smaller degree (t_1t_2 , t_3t_4, zy, zt_3t_4, yt_1t_2). It is easy to see that this process terminates.

Our new mapping $H(x) = x + \Psi_2(x) + \Psi_3(x)$ is noninvertible and has Jacobian 1. Consider its blowup

$$R : x \mapsto x + T^2 y + T\Psi_2(x),\ y \mapsto y - \Psi_3(x),\ T \mapsto T.$$

This mapping R is invertible if and only if the initial mapping is invertible, and has Jacobian 1 if and only if the initial mapping has Jacobian 1, by [98, Lemma 2]. This mapping is also cubic homogeneous. The corresponding ternary algebra is Engel, but not weakly nilpotent. □

This example shows that a direct combinatorial approach to the Jacobian Conjecture encounters difficulties, and in working with related Burnside-type problems (in the sense of Zelmanov [107], dealing with nilpotence properties of Engel algebras, as indicated in the introduction), one should take into account specific properties arising in characteristic zero.

Definition 1. An algebra A is *nilpotent* of *class* $\leq n$ if $M(a_1, a_2, \ldots) = 0$ for each monomial $M(x_1, x_2, \ldots)$ of degree $\geq n$. An ideal I of A is *strongly nilpotent* of *class* $\leq n$ if $M(a_1, a_2, \ldots) = 0$ for each monomial $M(x_1, x_2, \ldots)$ in which indeterminates occurring in strings of total degree $\geq n$ have been substituted to elements of I.

Although the notions of nilpotent and strongly nilpotent coincide in the associative case, they differ for ideals of nonassociative algebras. For example, consider the following algebra suggested by Shestakov: A is the algebra generated by a, b, z

satisfying the relations $a^2 = b$, $bz = a$ and all other products 0. Then $I = Fa + Fb$ is nilpotent as a subalgebra, satisfying $I^3 = 0$ but not strongly nilpotent (as an ideal), since

$$b = ((a(bz))z)a \neq 0,$$

and one can continue indefinitely in this vein. Also, [45] contains an example of a finite dimensional non-associative algebra without any ideal which is maximal with respect to being nilpotent as a subalgebra.

In connection with the Generalized Jacobian Conjecture in characteristic 0, it follows from results of Yagzhev [105], also cf. [42], that there exists a 20-dimensional Engel algebra over $\mathbb{Q}$, not weakly nilpotent, satisfying the identities

$$x^2y = -yx^2, \quad (((yx^2)x^2)x^2)x^2 = 0,$$

$$(xy + yx)y = 2y^2x, \quad x^2y^2 = 0.$$

However, this algebra can be seen to be Yagzhev (see Definition 4).

For associative algebras, one uses the term "nil" instead of "weakly nilpotent." Any nil subalgebra of a finite dimensional associative algebra is nilpotent, by Wedderburn's Theorem [92]). Jacobson generalized this result to other settings, cf. [68, Theorem 15.23]; and Shestakov [71] generalized it to a wide class of Jordan algebras (not necessarily commutative).

Yagzhev's investigation of weak nilpotence has applications to the Koethe Conjecture, for algebras over uncountable fields. He reproved:

* *In every associative algebra over an uncountable field, the sum of every two nil right ideals is a nil right ideal* [A.V. Yagzhev, *On the Koethe problem*, unpublished (in Russian)].

(This was proved first by Amitsur [3]. Amitsur's result is for affine algebras, but one can easily reduce to the affine case.)

2.2.5 Algebras Satisfying Systems of Capelli Identities

Definition 6. The *Capelli polynomial* C_k of order k is

$$C_k := \sum_{\sigma \in S_k} (-1)^\sigma x_{\sigma(1)} y_1 \cdots x_{\sigma(k)} y_k.$$

It is obvious that an associative algebra satisfies the Capelli identity c_k iff, for any monomial $M(x_1, \dots, x_k, y_1, \dots, y_r)$ multilinear in the x_i, the following equation holds identically in A:

$$\sum_{\sigma \in S_k} (-1)^\sigma M(v_{\sigma(1)}, \dots, v_{\sigma(k)}, y_1, \dots, y_r) = 0. \tag{8}$$

However, this does not apply to nonassociative algebras, so we need to generalize this condition.

Definition 7. The algebra A satisfies a **system of Capelli identities** of order k, if (8) holds identically in A for any monomial $M(x_1, \dots, x_k, y_1, \dots, y_r)$ multilinear in the x_i.

Any algebra of dimension $< k$ over a field satisfies a system of Capelli identities of order k. Algebras satisfying systems of Capelli identities behave much like finite dimensional algebras. They were introduced and systematically studied by Rasmyslov [64, 65].

Using Rasmyslov's method, Zubrilin [113], also see [66, 111], proved that if A is an arbitrary algebra satisfying the system of Capelli identities of order n, then the chain of ideals defining the *solvable radical* stabilizes at the nth step. More precisely, we utilize a Baer-type radical, along the lines of Amitsur [4].

Given an algebra A, we define $\mathrm{Solv}_1 := \mathrm{Solv}_1(A) = \sum\{$Strongly nilpotent ideals of $A\}$, and inductively, given Solv_k, define Solv_{k+1} by $\mathrm{Solv}_{k+1} / \mathrm{Solv}_k = \mathrm{Solv}_1(A/\mathrm{Solv}_k)$. For a limit ordinal α, define

$$\mathrm{Solv}_\alpha = \cup_{\beta<\alpha} \mathrm{Solv}_\beta .$$

This must stabilize at some ordinal α, for which we define $\mathrm{Solv}(A) = \mathrm{Solv}_\alpha$.

Clearly $\mathrm{Solv}(A/\mathrm{Solv}(A)) = 0$; i.e., $A/\mathrm{Solv}(A)$ has no nonzero strongly nilpotent ideals. Actually, Amitsur [4] defines $\zeta(A)$ as built up from ideals having trivial multiplication, and proves [4, Theorem 1.1] that $\zeta(A)$ is the intersection of the prime ideals of A.

We shall use the notion of *sandwich*, introduced by Kaplansky and Kostrikin, which is a powerful tool for Burnside-type problems [107]. An ideal I is called a *sandwich ideal* if, for any k,

$$M(z_1, z_2, x_1, \dots, x_k) = 0$$

for any $z_1, z_2 \in I$, any set of elements $x_1, \dots, x_k$, and any multilinear monomial M of degree $k + 2$. (Similarly, if the operations of an algebra have degree $\leq \ell$, then it is natural to use ℓ-sandwiches, which by definition satisfy the property that

$$M(z_1, \dots, z_\ell, x_1, \dots, x_k) = 0$$

for any $z_1, \dots, z_\ell \in I$, any set of elements $x_1, \dots, x_k$, and any multilinear monomial M of degree $k + \ell$.)

The next useful lemma follows from a result of [113]:

Lemma 1. *If an ideal I is strongly nilpotent of class ℓ, then there exists a decreasing sequence of ideals $I = I_1 \supseteq \cdots \supseteq I_{l+1} = 0$ such that I_s/I_{s+1} is a sandwich ideal in A/I_{s+1} for all $s \leq l$.*

Definition 8. An algebra A is *representable* if it can be embedded into an algebra finite dimensional over some extension of the ground field.

Remark 3. Zubrilin [113], properly clarified, proved the more precise statement, that if an algebra A of arbitrary signature satisfies a system of Capelli identities C_{n+1}, then there exists a sequence $B_0 \subseteq B_1 \subseteq \cdots \subseteq B_n$ of strongly nilpotent ideals such that:

- The natural projection of B_i in A/B_{i-1} is a strongly nilpotent ideal.
- A/B_n is representable.
- If $I_1 \subseteq I_2 \subseteq \cdots \subseteq I_n$ is any sequence of ideals of A such that I_{j+1}/I_j is a sandwich ideal in A/I_j, then $B_n \supseteq I_n$.

Such a sequence of ideals will be called a *Baer–Amitsur sequence*. In affine space the Zariski closure of the radical is radical, and hence the factor algebra is representable. (Although the radical coincides with the linear closure if the base field is infinite (see [18]), this assertion holds for arbitrary signatures and base fields.) Hence in representable algebras, the Baer–Amitsur sequence stabilizes after finitely many steps. Lemma 1 follows from these considerations.

Our next main goal is to prove Theorem 5 below, but first we need another notion.

2.2.6 The Tree Associated with a Monomial

Effects of nilpotence have been used by different authors in another language. We associate a *rooted labelled tree* with any monomial: Any branching vertex indicates the symbol of an operator, whose outgoing edges are the terms in the corresponding symbol. Here is the precise definition.

Definition 9. Let $M(x_1, \ldots, x_n)$ be a monomial in an algebra A of arbitrary signature. One can associate the tree T_M by an inductive procedure:

1. If M is a single variable, then T_M is just the vertex $\bullet$.
2. Let $M = g(M_1, \ldots, M_k)$, where g is a k-ary operator. We assume inductively that the trees T_i, $i = 1, \ldots, k$, are already defined. Then the tree T_M is the disjoint union of the T_i, together with the root $\bullet$ and arrows starting with $\bullet$ and ending with the roots of the trees T_i.

Remark 4. Sometimes one labels T_M according to the operator g and the positions inside g.

If the outgoing degree of each vertex is 0 or 2, the tree is called *binary*. If the outgoing degree of each vertex is either 0 or 3, the tree is called *ternary*. If each operator is binary, T_M will be binary; if each operator is ternary, T_M will be ternary.

2.3 Lifting Yagzhev Algebras

Recall Definitions 4 and 5.

Theorem 5. *Suppose A is an algebra of Engel type, and let I be a sandwich ideal of A. If A/I is Yagzhev, then A is Yagzhev.*

Proof. The proof follows easily from the following two propositions.

Let k be the class of weak nilpotence of A/I. We call a branch of the tree *fat* if it has more than k entries.

Proposition 1. *(a) The sum of all monomials of any degree $s > k$ belongs to I.*
(b) Let $x_1, \ldots, x_n$ be fixed indeterminates, and M be an arbitrary monomial, with $s_1, \ldots, s_\ell > k$. Then

$$\sum_{|t_1|=s_1,\ldots,|t_\ell|=s_\ell} M(x_1, \ldots, x_n, t_1, \ldots, t_\ell) \equiv 0. \tag{9}$$

(c) The sum of all monomials of degree s, containing at least ℓ non-intersecting fat branches, is zero.

Proof. (a) is just a reformulation of the weak nilpotence of A/I; (b) follows from (a) and the sandwich property of an ideal I. To get (c) from (b), it is enough to consider the highest non-intersecting fat branches.

Proposition 2 (Yagzhev). *The linearization of the sum of all terms with a fixed fat branch of length n is the complete linearization of the function*

$$\sum_{\sigma \in S_n} \prod (\mathrm{Ad}_{k_{\sigma(i)}})(z)(t).$$

Theorem 2, Lemma 1, and Zubrilin's result give us the following major result.

Theorem 6. *In characteristic zero, the Jacobian conjecture is equivalent to the following statement.*

Any algebra of Engel type satisfying some system of Capelli identities is a Yagzhev algebra.

This theorem generalizes the following result of Yagzhev.

Theorem 7. *The Jacobian conjecture is equivalent to the following statement:*

Any ternary Engel algebra in characteristic 0 satisfying a system of Capelli identities is a Yagzhev algebra.

The Yagzhev correspondence and the results of this section (in particular, Theorem 6) yield the proof of Theorem 3.

2.3.1 Sparse Identities

Generalizing Capelli identities, we say that an algebra satisfies a system of *sparse identities* when there exist k and coefficients α_σ such that for any monomial $M(x_1, \ldots, x_k, y_1, \ldots, y_r)$ multilinear in x_i, the following equation holds:

$$\sum_\sigma \alpha_\sigma M(c_1 v_{\sigma(1)} d_1, \ldots, c_k v_{\sigma(k)} d_k, y_1, \ldots, y_r) = 0. \tag{10}$$

Note that one needs only to check (10) for monomials. The system of Capelli identities is a special case of a system of sparse identities (when $\alpha_\sigma = (-1)^\sigma$). This concept ties in with the following "few long branches" lemma [112], concerning the structure of trees of monomials for algebras with sparse identities:

Lemma 2 (Few Long Branches). *Suppose an algebra A satisfies a system of sparse identities of order m. Then any monomial is linearly representable by monomials such that the corresponding tree has not more than $m - 1$ disjoint branches of length $\geq m$.*

Lemma 2 may be useful in studying nilpotence of Engel algebras.

2.4 *Inversion Formulas and Problems of Burnside Type*

We have seen that the JC relates to problems of "Specht type" (concerning whether one set of polynomial identities implies another), as well as problems of Burnside type.

Burnside-type problems become more complicated in nonzero characteristic; cf. Zelmanov's review article [107].

Bass et al. [5] attacked the JC by means of inversion formulas. D. Wright [93] wrote an inversion formula for the symmetric case and related it to a combinatorial structure called the *Grossman–Larson Algebra*. Namely, write $F = X - H$, and define $J(H)$ to be the Jacobian matrix of H. Wright proved the JC for the case where H is homogeneous and $J(H)^3 = 0$, and also for the case where H is cubic and $J(H)^4 = 0$; these correspond in Yagzhev's terminology to the cases of Engel types 3 and 4, respectively. Also, the so-called *chain vanishing theorem* in [93] follows from Engel type. Similar results were obtained earlier by Singer [75] using tree formulas for formal inverses. The inversion formula, introduced in [5], was investigated by D. Wright and his school. Many authors use the language of the so-called *tree expansion* (see [75, 93] for details). In view of Theorem 4, the tree expansion technique should be highly nontrivial.

The Jacobian Conjecture can be formulated as a question of quantum field theory (see [1]), in which tree expansions are seen to correspond to Feynmann diagrams.

In the papers [75,93] (see also [94]), trees with one label correspond to elements of the algebra A built by Yagzhev, and 2-labelled trees correspond to the elements of the operator algebra $D(A)$ (the algebra generated by operators $x \to M(x, \vec{y})$, where M is some monomial). These authors deduce weak nilpotence from the Engel conditions of degree 3 and 4. The inversion formula for automorphisms of tensor product of Weyl algebras and the ring of polynomials was studied intensively in the papers [7, 10]. Using techniques from [14], this yields a slightly different proof of the equivalence between the JC and DC, by an argument similar to one given in Yagzhev [Invertibility criteria of a polynomial mapping, unpublished (in Russian)]. Yagzhev's approach makes the situation much clearer, and the known approaches to the Jacobian Conjecture using inversion formulas can be explained from this viewpoint.

Remark 5. The most recent inversion formula (and probably the most algebraically explicit one) was obtained by Bavula [6]. The coefficient q_0 can be made explicit in (5), by means of the Gabber Inequality, which says that if

$$f : K^n \to K_n; \quad x_i \to f_i(\vec{x})$$

is a polynomial automorphism, with $\deg(f) = \max_i \deg(f_i)$, then $\deg(f^{-1}) \leq \deg(f)^{n-1})$

In fact, we are working with *operads*, cf. the classical book [55]. A review of operad theory and its relation with physics and PI-theory, in particular Burnside-type problems, will appear in Piontkovsky [62]; see also [47, 63]. Operad theory provides a supply of natural identities and varieties, but they also correspond to geometric facts. For example, the Jacobi identity corresponds to the fact that the altitudes of a triangle are concurrent. M. Dehn's observations that the Desargue property of a projective plane corresponds to associativity of its coordinate ring, and Pappus' property to its commutativity, can be considered as a first step in operad theory. Operads are important in mathematical physics, and formulas for the famous Kontsevich quantization theorem resemble formulas for the inverse mapping. The operators considered here are operads.

3 The Jacobian Conjecture for Varieties and Deformations

In this section we consider analogs of the JC for other varieties of algebras, partially with the aim on throwing light on the classical JC (for the commutative associative polynomial algebra). See [110] for background on nonassociative algebras.

3.1 Generalization of the Jacobian Conjecture to Arbitrary Varieties

Birman [22] already proved the JC for free groups in 1973. The JC for free associative algebras (in two generators) was established in 1982 by Dicks and Levin [27, 28], utilizing Fox derivatives, which we describe later on. Their result was reproved by Yagzhev [96], whose ideas are sketched in this section. Schofield [69] proved the full version. Yagzhev then applied these ideas to other varieties of algebras [103, 105] including nonassociative commutative algebras and anti-commutative algebras; Umirbaev [82] generalized these to "Schreier varieties," defined by the property that every subalgebra of a free algebra is free. The JC for free Lie algebras was proved by Reutenauer [67], Shpilrain [74], and Umirbaev [81].

The Jacobian Conjecture for varieties generated by finite dimensional algebras is closely related to the Jacobian Conjecture in the usual commutative associative case, which is the most important.

Let $\mathfrak{M}$ be a variety of algebras of some signature Ω over a given field $\mathbf{k}$ of characteristic zero, and $\mathbf{k}_{\mathfrak{M}}\langle\vec{x}\rangle$ the relatively free algebra in $\mathfrak{M}$ with generators $\vec{x} = \{x_i : i \in I\}$. We assume that $|\Omega|, |I| < \infty$, $I = 1, \ldots, n$.

Take a set $\vec{y} = \{y_i\}_{i=1}^n$ of new indeterminates. For any $f(\vec{x}) \in k_{\mathfrak{M}}\langle\vec{x}\rangle$ one can define an element $\hat{f}(\vec{x}, \vec{y}) \in k_{\mathfrak{M}}\langle\vec{x}, \vec{y}\rangle$ via the equation

$$f(x_1 + y_1, \ldots, x_n + y_n) = f(\vec{y}) + \hat{f}(\vec{x}, \vec{y}) + R(\vec{x}, \vec{y}) \tag{11}$$

where $\hat{f}(\vec{x}, \vec{y})$ has degree 1 with respect to $\vec{x}$, and $R(\vec{x}, \vec{y})$ is the sum of monomials of degree ≥ 2 with respect to $\vec{x}$; $\hat{f}$ is a generalization of the differential.

Let $\alpha \in \mathrm{End}(k_{\mathfrak{M}} < \vec{x} >)$, i.e.,

$$\alpha : x_i \mapsto f_i(\vec{x}); \; i = 1, \ldots, n. \tag{12}$$

Definition 1. Define the Jacobi endomorphism $\hat{\alpha} \in \mathrm{End}(\mathbf{k}_{\mathfrak{M}}\langle\vec{x}, \vec{y}\rangle)$ via the equality

$$\hat{\alpha} : \begin{cases} x_i \to \hat{f}_i(\vec{x}), \\ y_i \to y_i. \end{cases} \tag{13}$$

The Jacobi mapping $f \mapsto \hat{f}$ satisfies the chain rule, in the sense that it preserves composition.

Remark 1. It is not difficult to check (and is well known) that if $\alpha \in \mathrm{Aut}(\mathbf{k}_{\mathfrak{M}} < \vec{x} >)$ then $\hat{\alpha} \in \mathrm{Aut}(\mathbf{k}_{\mathfrak{M}} < \vec{x}, \vec{y} >)$.

The inverse implication is called the *Jacobian Conjecture for the variety* $\mathfrak{M}$. Here is an important special case.

Definition 2. Let $A \in \mathfrak{M}$ be a finite dimensional algebra, with base $\{\vec{e}_i\}_{i=1}^N$. Consider a set of commutative indeterminates $\vec{\nu} = \{\nu_{si} | s = 1, \ldots, n; i = 1, \ldots, N\}$. The elements

$$z_j = \sum_{i=1}^N \nu_{ji}\vec{e}_i; \quad j = 1, \ldots, n$$

are called *generic elements of* A.

Usually in the matrix algebra $\mathbb{M}_m(\mathbf{k})$, the set of matrix units $\{e_{ij}\}_{i,j=1}^m$ is taken as the base. In this case $e_{ij}e_{kl} = \delta_{jk}e_{il}$ and $z_l = \sum_{ij} \lambda_{ij}^l e_{ij}$, $l = 1, \ldots, n$.

Definition 3. A *generic matrix* is a matrix whose entries are distinct commutative indeterminates, and the so-called *algebra of generic matrices of order* m is generated by associative generic $m \times m$ matrices.

The algebra of generic matrices is prime, and every prime, relatively free, finitely generated associative PI-algebra is isomorphic to an algebra of generic matrices. If we include taking traces as an operator in the signature, then we get the *algebra of generic matrices with trace*. That algebra is a Noetherian module over its center.

Define the $\mathbf{k}$-linear mappings

$$\Omega_i : \mathbf{k}_{\mathfrak{M}}\langle\vec{x}\rangle \to \mathbf{k}[\nu]; \quad i = 1, \ldots, n$$

via the relation

$$f\left(\sum_{i=1}^{N} \nu_{1i}e_i, \ldots, \sum_{i=1}^{N} \nu_{ni}e_i\right) = \sum_{i=1}^{N} (f\Omega_i)e_i.$$

It is easy to see that the polynomials $f\Omega_i$ are uniquely determined by f.

One can define the mapping

$$\varphi_A : \mathrm{End}(k_{\mathfrak{M}} < \vec{x} >) \to \mathrm{End}(k[\vec{\nu}])$$

as follows: If

$$\alpha \in \mathrm{End}(k_{\mathfrak{M}} < \vec{x} >) : x_s \to f_s(\vec{x}) \quad s = 1, \ldots, n,$$

then $\varphi_A(\alpha) \in \mathrm{End}(k[\vec{\nu}])$ can be defined via the relation

$$\varphi_A(\alpha) : \nu_{si} \to P_{si}(\vec{\nu}); \quad s = 1, \ldots, n; \quad i = 1, \ldots, n,$$

where $P_{si}(\vec{\nu}) = f_s\Omega_i$.

The following proposition is well known.

Proposition 1 ([105]). *Let* $A \in \mathfrak{M}$ *be a finite dimensional algebra, and* $\vec{x} = \{x_1, \ldots, x_n\}$ *be a finite set of commutative indeterminates. Then the mapping* φ_A *is a semigroup homomorphism, sending 1 to 1, and automorphisms to automorphisms. Also the mapping* φ_A *commutes with the operation* $\hat{}$ *of taking the Jacobi endomorphism, in the sense that* $\widehat{\varphi_A(\alpha)} = \varphi_A(\hat{\alpha})$. *If* φ *is invertible, then* $\hat{\varphi}$ *is also invertible.*

This proposition is important, since as noted after Remark 1, the opposite direction is the JC.

3.2 Deformations and the Jacobian Conjecture for Free Associative Algebras

Definition 4. A *T-ideal* is a completely characteristic ideal, i.e., stable under any endomorphism.

Proposition 2. *Suppose A is a relatively free algebra in the variety $\mathfrak{M}$, I is a T-ideal in A, and $\mathfrak{M}' = \mathrm{Var}(A/I)$. Any polynomial mapping $F : A \to A$ induces a natural mapping $F' : A/I \to A/I$, as well as a mapping $\hat{F}'$ in $\mathfrak{M}'$. If F is invertible, then F' is invertible; if $\hat{F}$ is invertible, then $\hat{F}'$ is also invertible.*

For example, let F be a polynomial endomorphism of the free associative algebra $k < \vec{x} >$, and I_n be the T-ideal of the algebra of generic matrices of order n. Then $F(I_n) \subseteq I_n$ for all n. Hence F induces an endomorphism F_{I_n} of $k < \vec{x} > /I_n$. In particular, this is a semigroup homomorphism. Thus, if F is invertible, then F_{I_n} is invertible, but not vice versa.

The Jacobian mapping $\widehat{F_{I_n}}$ of the reduced endomorphism F_{I_n} is the reduction of the Jacobian mapping of F.

3.2.1 The Jacobian Conjecture and the Packing Property

This subsection is based on the *packing property* and deformations. Let us illustrate the main idea. It is well known that the composite of *all* quadratic extensions of $\mathbb{Q}$ is infinite dimensional over $\mathbb{Q}$. Hence all such extensions cannot be embedded ("packed") into a single commutative finite dimensional $\mathbb{Q}$-algebra. However, all of them can be packed into $\mathbb{M}_2(\mathbb{Q})$. We formalize the notion of packing in Sect. 3.5.1. Moreover, for *any* elements *not* in $\mathbb{Q}$ there is a parametric family of embeddings (because it embeds non-centrally and thus can be deformed via conjugation by a parametric set of matrices). Uniqueness thus means belonging to the center. Similarly, adjoining noncommutative coefficients allows one to decompose polynomials, as will be elaborated below.

This idea allows us to solve equations via a finite dimensional extension, and to find a parametric sets of solutions if some solution does not belong to the original algebra. That situation contradicts local invertibility.

Let F be an endomorphism of the free associative algebra having invertible Jacobian. We suppose that $F(0) = 0$ and

$$F(x_i) = x_i + \sum \quad \text{terms of order} \quad \geq 2.$$

We intend to show how the invertibility of the Jacobian implies invertibility of the mapping F.

Let $Y_1, \dots, Y_k$ be generic $m \times m$ matrices. Consider the system of equations

$$\{F_i(X_1, \dots, X_n) = Y_i; \quad i = 1, \dots, k\}.$$

This system has a solution over some finite extension of order m of the field generated by the center of the algebra of generic matrices *with trace*.

Consider the set of block diagonal $mn \times mn$ matrices:

$$A = \begin{pmatrix} A_1 & 0 & & \dots 0 \\ 0 & A_2 & 0 & \dots 0 \\ \vdots & & \ddots & \vdots \\ 0 & & \dots & A_n \end{pmatrix}, \tag{14}$$

where the A_j are $m \times m$ matrices.

Next, we consider the system of equations

$$\{F_i(X_1, \dots, X_n) = Y_i; \quad i = 1, \dots, k\}, \tag{15}$$

where the $mn \times mn$ matrices Y_i have the form (14) with the A_j generic matrices.

Any m-dimensional extension of the base field $\mathbf{k}$ is embedded into $\mathbb{M}_m(\mathbf{k})$. But $\mathbb{M}_{mn}(\mathbf{k}) \simeq \mathbb{M}_m(\mathbf{k}) \otimes \mathbb{M}_n(\mathbf{k})$. It follows that for appropriate m, the system (15) has a unique solution in the matrix ring with traces. (Each is given by a matrix power series where the summands are matrices whose entries are homogeneous forms, seen by rewriting $Y_i = X_i +$ terms of order 2 as $X_i = Y_i +$ terms of order 2, and iterating.) The solution is unique since their entries are distinct commuting indeterminates.

If F is invertible, then this solution must have block diagonal form. However, if F is not invertible, this solution need not have block diagonal form. Now we translate invertibility of the Jacobian to the language of **parametric families** or **deformations**.

Consider the matrices

$$E_\lambda^\ell = \begin{pmatrix} E & 0 & & \dots & 0 \\ 0 & \ddots & & \dots & 0 \\ 0 & \dots & \lambda \cdot E & & 0 \\ \vdots & & \dots & \ddots & \vdots \\ 0 & & \dots & & E \end{pmatrix}$$

where E denotes the identity matrix. (The index ℓ designates the position of the block $\lambda \cdot E$.) Taking X_j not to be a block diagonal matrix, then for some ℓ we obtain a non-constant parametric family $E_\lambda^\ell X_j (E_\lambda^\ell)^{-1}$ dependent on λ.

On the other hand, if Y_i has form (14), then $E_\lambda^\ell Y_i (E_\lambda^\ell)^{-1} = Y_i$ for all $\lambda \neq 0$; $\ell = 1, \dots, k$.

Hence, if F_{I_n} is not an automorphism, then we have a **continuous parametric set of solutions**. But if the Jacobian mapping is invertible, it is locally 1:1, a contradiction. This argument yields the following result.

Theorem 1. *For $F \in \mathrm{End}(k\langle\vec{x}\rangle)$, if the Jacobian of F is invertible, then the reduction F_{I_n} of F, modulo the T-ideal of the algebra of generic matrices, is invertible.*

See [103] for further details of the proof. Because any relatively free affine algebra of characteristic 0 satisfies the set of identities of some matrix algebra, it is the quotient of the algebra of generic matrices by some T-ideal J. But J maps into itself after any endomorphism of the algebra. We conclude:

Corollary 3.1.1. *If $F \in \mathrm{End}(k < \vec{x} >)$ and the Jacobian of F is invertible, then the reduction F_J of F modulo any proper T-ideal J is invertible.*

In order to get invertibility of $\vec{F}$ itself, Yagzhev used the additional ideas:

- The block diagonal technique works equally well on skew fields.
- The above algebraic constructions can be carried out on Ore extensions, in particular for the *Weyl algebras* $W_n = \mathbf{k}[x_1, \ldots, x_n; \partial_1, \ldots, \partial_n]$.
- By a result of L. Makar-Limanov, the free associative algebra can be embedded into the ring of fractions of the Weyl algebra. This provides a nice presentation for mapping the free algebra.

Definition 5. Let A be an algebra, $B \subset A$ a subalgebra, and $\alpha : A \to A$ a polynomial mapping of A (and hence $\alpha(B) \subset B$, see Definition 1). B is a *test algebra for* α, if $\alpha(A \backslash B) \neq A \backslash B$.

The next theorem shows the universality of the notion of a test algebra. An endomorphism is called *rationally invertible* if it is invertible over Cohn's skew field of fractions [24] of $\mathbf{k}\langle\vec{x}\rangle$.

Theorem 2 (Yagzhev). *For any $\alpha \in \mathrm{End}(k < \vec{x} >)$, one of the two statements holds:*

- *α is rationally invertible, and its reduction to any finite dimensional factor also is rationally invertible.*
- *There exists a test algebra for some finite dimensional reduction of α.*

This theorem implies the Jacobian conjecture for free associative algebras. We do not go into details, referring the reader to the papers [103, 105].

Remark 2. The same idea is used in quantum physics. The polynomial $x^2 + y^2 + z^2$ cannot be decomposed for any commutative ring of coefficients. However, it can be decomposed using noncommutative ring of coefficients (Pauli matrices). The Laplace operator in three-dimensional space can be decomposed in such a manner.

3.2.2 Reduction to Nonzero Characteristic

One can work with deformations equally well in nonzero characteristic. However, the naive Jacobian condition does not give us parametric families, because of consequences of inseparability. Hence it is interesting using deformations to get a reasonable version of the JC for characteristic $p > 0$, especially because of recent progress in the JC related to the reduction of holonomic modules to the case of characteristic p and investigation of the p-curvature or Poisson brackets on the center [14, 15, 79].

In his very last paper [A.V. Yagzhev, Invertibility criteria of a polynomial mapping, unpublished (in Russian)], A.V. Yagzhev approached the JC using positive characteristics. He noticed that the existence of a counterexample is equivalent to the existence of an Engel, but not Yagzhev, finite dimensional ternary algebra in each positive characteristic $p \gg 0$. (This fact is also used in the papers [14, 15, 79].)

If a counterexample to the JC exists, then such an algebra A exists even over a finite field, and hence can be finite. It generates a locally finite variety of algebras that are of Engel type, but not Yagzhev. This situation can be reduced to the case of a locally semiprime variety. Any relatively free algebra of this variety is semiprime, and the centroid of its localization is a finite direct sum of fields. The situation can be reduced to one field, and he tried to construct an embedding which is not an automorphism. This would contradict the finiteness property.

Since a reduction of an endomorphism as a mapping on points of finite height may be an automorphism, the issue of injectivity also arises. However, this approach looks promising, and may involve new ideas, such as in the papers [14, 15, 79]. Perhaps different infinitesimal conditions (like the Jacobian condition in characteristic zero) can be found.

3.3 *The Jacobian Conjecture for Other Classes of Algebras*

Although the Jacobian Conjecture remains open for commutative associative algebras, it has been established for other classes of algebras, including free associative algebras, free Lie algebras, and free metabelian algebras. See Sect. 3.1 for further details.

An algebra is *metabelian* if it satisfies the identity $[x, y][z, t] = 0$.

The case of free metabelian algebras, established by Umirbaev [84], involves some interesting new ideas that we describe now. His method of proof is by means of co-multiplication, taken from the theory of Hopf algebras and quantization. Let A^{op} denote the opposite algebra of the free associative algebra A, with generators t_i. For $f \in A$ we denote the corresponding element of A^{op} as f^*. Put $\lambda : A^{\mathrm{op}} \otimes A \to A$ be the mapping such that $\lambda(\sum f_i^* \otimes g_i) = \sum f_i g_i$. $I_A := \ker(\lambda)$ is a free A bimodule with generators $t_i^* \otimes 1 - 1 \otimes t_i$. The mapping $d_A : A \to I_A$ such that $d_A(a) = a^* \otimes 1 - 1 \otimes a$ is called the *universal derivation* of A. The *Fox derivatives* $\partial a/\partial t_i \in A^{\mathrm{op}} \otimes A$ [40] are defined via $d_A(a) = \sum_i (t_i^* \otimes 1 - 1 \otimes t_i)\partial a/\partial t_i$, cf. [28] and [84].

Let $C = A/\operatorname{Id}([A, A])$, the free commutative associative algebra, and let $B = A/\operatorname{Id}([A, A])^2$, the free metabelian algebra. Let

$$\partial(a) = (\partial a/\partial t_1, \ldots, \partial a/\partial t_n).$$

One can define the natural derivations

$$\bar{\partial} : A \to (A' \otimes A)^n \to (C' \otimes C)^n,$$
$$\tilde{\partial} : A \to (C' \otimes C)^n \to C^n. \qquad (16)$$

where the mapping $(C' \otimes C)^n) \to C^n$ is induced by λ. Then $\ker(\bar{\partial}) = \mathrm{Id}([A, A])^2 + F$ and $\bar{\partial}$ induces a derivation $B \to (C' \otimes C)^n$, whereas $\tilde{\partial}$ induces the usual derivation $C \to C^n$. Let $\Delta : C \to C' \otimes C$ be the mapping induced by d_A, i.e., $\Delta(f) = f^* \otimes 1 - 1 \otimes f$, and let $z_i = \Delta(x_i)$. The *Jacobi matrix* is defined in the natural way and provides the formulation of the JC for free metabelian algebras. One of the crucial steps in proving the JC for free metabelian algebras is the following homological lemma from [84]:

Lemma 1. *Let* $\vec{u} = (u_1, \dots, u_n) \in (C^{op} \otimes C)^n$. *Then* $\vec{u} = \bar{\partial}(\bar{w})$ *for some* $w \in \mathrm{Id}([A, A])$ *iff*

$$\sum z_i u_i = 0.$$

The proof also requires the following theorem:

Theorem 3. *Let* $\varphi \in \mathrm{End}(C)$. *Then* $\varphi \in \mathrm{Aut}(C)$ *iff* $\mathrm{Id}(\Delta(\varphi(x_i)))_{i=1}^n = \mathrm{Id}(z_i)_{i=1}^n$.

The paper [84] also includes the following result:

Theorem 4. *Any automorphism of* C *can be extended to an automorphism of* B, *using the* JC *for the free metabelian algebra* B.

This is a nontrivial result, unlike the extension of an automorphism of B to an automorphism of $A/\,\mathrm{Id}([A, A])^n$ for any $n > 1$.

3.4 *Questions Related to the Jacobian Conjecture*

Let us turn to other interesting questions which can be linked to the Jacobian Conjecture. The quantization procedure is a bridge between the commutative and noncommutative cases and is deeply connected to the JC and related questions. Some of these questions also are discussed in the paper [30].

Relations between the free associative algebra and the classical commutative situation are very deep. In particular, Bergman's theorem that any commutative subalgebra of the free associative algebra is isomorphic to a polynomial ring in one indeterminate is the noncommutative analog of Zak's theorem [106] that any integrally closed subring of a polynomial ring of Krull dimension 1 is isomorphic to a polynomial ring in one indeterminate.

For example, Bergman's theorem is used to describe the automorphism group $\mathrm{Aut}(\mathrm{End}(\mathbf{k}\langle x_1, \ldots, x_n\rangle))$ [13]; Zak's theorem is used in the same way to describe the group $\mathrm{Aut}(\mathrm{End}(\mathbf{k}[x_1, \ldots, x_n]))$ [16]. Also see [77].

Question. *Can one prove Bergman's theorem via quantization?*

Quantization could be a key idea for understanding Jacobian-type problems in other varieties of algebras.

1. Cancellation problems.

We recall three classical problems.

1. *Let K_1 and K_2 be affine domains for which $K_1[t] \simeq K_2[t]$. Is it true that $K_1 \simeq K_2$?*
2. *Let K_1 and K_2 be an affine fields for which $K_1(t) \simeq K_2(t)$. Is it true that $K_1 \simeq K_2$? In particular, if $K(t)$ is a field of rational functions over the field $\mathbf{k}$, is it true that K is also a field of rational functions over $\mathbf{k}$?*
3. *If $K[t] \simeq \mathbf{k}[x_1, \ldots, x_n]$, is it true that $K \simeq \mathbf{k}[x_1, \ldots, x_{n-1}]$?*

The answers to Problems 1 and 2 are "No" in general (even if $\mathbf{k} = \mathbb{C}$); see the fundamental paper [11], as well as [17] and the references therein. However, Problem 2 has a positive solution in low dimensions. Problem 3 is currently called the *Cancellation Conjecture*, although Zariski's original cancellation conjecture was for fields (Problem 2). See [26, 44, 56, 76] for Zariski's conjecture and related problems. For $n \geq 3$, the Cancellation Conjecture (Problem 3) remains open, to the best of our knowledge, and it is reasonable to pose the Cancellation Conjecture for free associative rings and ask the following:

Question. *If $K * \mathbf{k}[t] \simeq \mathbf{k}{<}x_1, \ldots, x_n{>}$, then is $K \simeq \mathbf{k}{<}x_1, \ldots, x_{n-1}{>}$?*

This question was solved for $n = 2$ by Drensky and Yu [32].

2. The Tame Automorphism Problem.

Yagzhev utilized his approach to study the tame automorphism problem. Unfortunately, these papers are not preserved.

It is easy to see that every endomorphism ϕ of a commutative algebra can be lifted to some endomorphism of the free associative algebra, and hence to some endomorphism of the algebra of generic matrices. However, it is not clear that any automorphism ϕ can be lifted to an automorphism.

We recall that an automorphism of $\mathbf{k}[x_1, \ldots, x_n]$ is *elementary* if it has the form

$$x_1 \mapsto x_1 + f(x_2, \ldots, x_n), \quad x_i \mapsto x_i, \quad \forall i \geq 2.$$

A *tame automorphism* is a product of elementary automorphisms, and a non-tame automorphism is called *wild*. The "tame automorphism problem" asks whether any automorphism is tame. Jung [43] and van der Kulk [49] proved this for $n = 2$, (also see [58, 59] for free groups, [24] for free Lie algebras, and [25, 52] for free associative algebras), so one takes $n > 2$.

Elementary automorphisms can be lifted to automorphisms of the free associative algebra; hence, every tame automorphism can be so lifted. If an automorphism φ cannot be lifted to an automorphism of the algebra of generic matrices, it cannot be tame. This gives us an approach to the tame automorphism problem.

We can lift an automorphism of $\mathbf{k}[x_1, \dots, x_n]$ to an endomorphism of $\mathbf{k}\langle x_1, \dots, x_n\rangle$ in many ways. Then replacing $x_1, \dots, x_n$ by $N \times N$ generic matrices induces a polynomial mapping $F_{(N)} : \mathbf{k}^{nN^2} \to \mathbf{k}^{nN^2}$.

For each automorphism φ, the invertibility of this mapping can be transformed into compatibility of some system of equations. For example, Theorem 10.5 of [61] says that the Nagata automorphism is wild, provided that a certain system of five equations in 27 unknowns has no solutions. Whether Peretz' method can effectively attack tameness questions remains to be seen. The wildness of the Nagata automorphism was established by Shestakov and Umirbaev [73]. One important ingredient in the proof is *degree estimates* of an expression $p(f, g)$ of algebraically independent polynomials f and g in terms of the degrees of f and g, provided neither leading term is proportional to a power of the other, initiated by Shestakov and Umirbaev [72]. An exposition based on their method is given in Kuroda [50].

One of the most important tools is the degree estimation technique, which in the multidimensional case becomes the analysis of leading terms, and is more complicated. We refer to the deep papers [23, 46, 50]. Several papers of Kishimoto, although containing gaps, also provide deep insights.

One can also ask the weaker question of "coordinate tameness": Is the image of (x, y, z) under the Nagata automorphism the image under some (other) tame automorphism? This also fails, by [88].

An automorphism φ is called *stably tame* if, when several new indeterminates $\{t_i\}$ are adjoined, the extension of φ given by $\varphi'(t_i) = t_i$ is tame; otherwise, it is called *stably wild*. Stable tameness of automorphisms of $\mathbf{k}[x, y, z]$ fixing z is proved in [21]; similar results for $\mathbf{k}\langle x, y, z\rangle$ are given in [20].

Yagzhev tried to construct wild automorphisms via polynomial automorphisms of the Cayley–Dickson algebra with base $\{\vec{e}_i\}_{i=1}^8$, and the set $\{\nu_i, \xi_i, \varsigma_i\}_{i=1}^8$ of commuting indeterminates. Let

$$x = \sum \nu_i \vec{e}_i,\ y = \sum \xi_i \vec{e}_i,\ z = \sum \varsigma_i \vec{e}_i.$$

Let (x, y, z) denote the associator $(xy)z - x(yz)$ of the elements x, y, z, and write

$$(x, y, z)^2 = \sum f_i(\vec{\nu}, \vec{\xi}, \vec{\varsigma})\vec{e}_i.$$

Then the endomorphism G of the polynomial algebra given by

$$G: \nu_i \to \nu_i + f_i(\vec{\nu}, \vec{\xi}, \vec{\varsigma}), \quad \xi_i \to \xi_i, \ \varsigma_i \to \varsigma_i,$$

is an automorphism, which likely is stably wild.

In the free associative case, perhaps it is possible to construct an example of an automorphism, the wildness of which could be proved by considering its Jacobi endomorphism (Definition 1). Yagzhev tried to construct examples of algebras $R = A \otimes A^{\mathrm{op}}$ over which there are invertible matrices that cannot decompose as products of elementary ones. Yagzhev conjectured that the automorphism

$$x_1 \to x_1 + y_1(x_1y_2 - y_1x_2), \ x_2 \to x_2 + (x_1y_2 - y_1x_2)y_2, \ y_1 \to y_1, \ y_2 \to y_2$$

of the free associative algebra is wild.

Umirbaev [83] proved in characteristic 0 that the *Anick automorphism* $x \to x + y(xy - yz)$, $y \to y$, $z \to z + (zy - yz)y$ is wild, by using metabelian algebras. The proof uses description of the defining relations of 3-variable automorphism groups [85–87]. Drensky and Yu [29, 31] proved in characteristic 0 that the image of x under the Anick Automorphism is not the image of any tame automorphism.

Stable Tameness Conjecture *Every automorphism of the polynomial algebra* $\mathbf{k}[x_1, \dots, x_n]$, *resp. of the free associative algebra* $\mathbf{k}\langle x_1, \dots, x_n \rangle$, *is stably tame.*

Lifting in the free associative case is related to quantization. It provides some light on the similarities and differences between the commutative and noncommutative cases. Every tame automorphism of the polynomial ring can be lifted to an automorphism of the free associative algebra. There was a conjecture that *any wild* z*-automorphism of* $\mathbf{k}[x, y, z]$ *(i.e., fixing* z*) over an arbitrary field* $\mathbf{k}$ *cannot be lifted to a* z*-automorphism of* $\mathbf{k}\langle x, y, z \rangle$. In particular, the Nagata automorphism cannot be so lifted [30]. This conjecture was solved by Belov and Yu [19] over an arbitrary field. However, the general lifting conjecture is still open. In particular, it is not known whether the Nagata automorphism can be lifted to an automorphism of the free algebra. (Such a lifting could not fix z.)

The paper [19] describes all the z-automorphisms of $\mathbf{k}\langle x, y, z \rangle$ over an arbitrary field $\mathbf{k}$. Based on that work, Belov and Yu [20] proved that every z-automorphism of $\mathbf{k}\langle x, y, z \rangle$ is stably tame, for all fields $\mathbf{k}$. A similar result in the commutative case is proved by Berson, van den Essen, and Wright [21]. These are the important first steps towards solving the stable tameness conjecture in the noncommutative and commutative cases.

The free associative situation is much more rigid than the polynomial case. Degree estimates for the free associative case are the same for prime characteristic [51] as in characteristic 0 [54]. The methodology is different from the commutative case, for which degree estimates (as well as examples of wild automorphisms) are not known in prime characteristic.

J.-T. Yu found some evidence of a connection between the lifting conjecture and the Embedding Conjecture of Abhyankar and Sathaye. Lifting seems to be "easier."

3.5 *Reduction to Simple Algebras*

This subsection is devoted to finding test algebras.

Any prime algebra B satisfying a system of Capelli identities of order $n+1$ (n minimal such) is said to have *rank* n. In this case, its operator algebra is PI. The localization of B is a simple algebra of dimension n over its centroid, which is a field. This is the famous *rank theorem* [65].

3.5.1 Packing Properties

Definition 6. Let $\mathcal{M} = \{\mathfrak{M}_i : i \in I\}$ be an arbitrary set of varieties of algebras. We say that $\mathcal{M}$ satisfies the *packing property*, if for any $n \in \mathbb{N}$ there exists a prime algebra A of rank n in some $\mathfrak{M}_j$ such that any prime algebra in any $\mathfrak{M}_i$ of rank n can be embedded into some central extension $K \otimes A$ of A.

$\mathcal{M}$ satisfies the *finite packing property* if, for any finite set of prime algebras $A_j \in \mathfrak{M}_i$, there exists a prime algebra A in some $\mathfrak{M}_k$ such that each A_j can be embedded into A.

The set of proper subvarieties of associative algebras satisfying a system of Capelli identities of some order k satisfies the packing property (because any simple associative algebra is a matrix algebra over field).

However, the varieties of alternative algebras satisfying a system of Capelli identities of order >8, or of Jordan algebras satisfying a system of Capelli identities of order >27, do not even satisfy the finite packing property. Indeed, the matrix algebra of order 2 and the Cayley–Dickson algebra cannot be embedded into a common prime alternative algebra. Similarly, $\mathbb{H}_3$ and the Jordan algebra of symmetric matrices cannot be embedded into a common Jordan prime algebra. (Both of these assertions follow easily by considering their PIs.)

It is not known whether or not the packing property holds for Engel algebras satisfying a system of Capelli identities; knowing the answer would enable us to resolve the JC, as will be seen below.

Theorem 5. *If the set of varieties of Engel algebras (of arbitrary fixed order) satisfying a system of Capelli identities of some order satisfies the packing property, then the Jacobian Conjecture has a positive solution.*

Theorem 6. *The set of varieties from the previous theorem satisfies the finite packing property.*

Most of the remainder of this section is devoted to the proof of these two theorems.

Problem. *Using the packing property and deformations, give a reasonable analog of the* JC *in nonzero characteristic.* (The naive approach using only the determinant of the Jacobian does not work.)

3.5.2 Construction of Simple Yagzhev Algebras

Using the Yagzhev correspondence and composition of elementary automorphisms it is possible to construct a new algebra of Engel type.

Theorem 7. *Let A be an algebra of Engel type. Then A can be embedded into a prime algebra of Engel type.*

Proof. Consider the mapping $F : V \to V$ (cf. (1)) given by

$$F : x_i \mapsto x_i + \sum_j \Psi_{ij}; \quad i = 1, \dots, n$$

(where the Ψ_{ij} are forms of homogenous degree j). Adjoining new indeterminates $\{t_i\}_{i=0}^n$, we put $F(t_i) = t_i$ for $i = 0, \dots, n$.

Now we take the transformation

$$G : t_0 \mapsto t_0, \quad x_i \mapsto x_i, \quad t_i \mapsto t_i + t_0 x_i^2, \quad \text{for} \quad i = 1, \dots, n.$$

The composite $F \circ G$ has invertible Jacobian (and hence the corresponding algebra has Engel type) and can be expressed as follows:

$$F \circ G : x_i \mapsto x_i + \sum_j \Psi_{ij}, \quad t_0 \mapsto t_0, \quad t_i \mapsto t_i + t_0 x_i^2 \quad \text{for} \quad i = 1, \dots, n.$$

It is easy to see that the corresponding algebra $\hat{A}$ also satisfies the following properties:

- $\hat{A}$ contains A as a subalgebra (for $t_0 = 0$).
- If A corresponds to a cubic homogenous mapping (and thus is Engel), then $\hat{A}$ also corresponds to a cubic homogenous mapping (and thus is Engel).
- If some of the forms Ψ_{ij} are not zero, then A does not have nonzero ideals with product 0, and hence is prime (but its localization need not be simple!).

Any algebra A with operators can be embedded, using the previous construction, to a prime algebra with nonzero multiplication. The theorem is proved. □

Embedding via the previous theorem preserves the cubic homogeneous case, but does not yet give us an embedding into a simple algebra of Engel type.

Theorem 8. *Any algebra A of Engel type can be embedded into a simple algebra of Engel type.*

Proof. We start from the following observation:

Lemma 2. *Suppose A is a finite dimensional algebra, equipped with a base $\vec{e}_1, \dots, \vec{e}_n, \vec{e}_{n+1}$. If for any $1 \le i, j \le n+1$ there exist operators ω_{ij} in the signature $\Omega(A)$ such that $\omega_{ij}(\vec{e}_i, \dots, \vec{e}_i, \vec{e}_{n+1}) = \vec{e}_j$, with all other values on the base vectors being zero, then A is simple.*

This lemma implies:

Lemma 3. *Let F be a polynomial endomorphism of $\mathbb{C}[x_1, \ldots, x_n; t_1, t_2]$, where*

$$F(x_i) = \sum_j \Psi_{ij}.$$

For notational convenience we put $x_{n+1} = t_1$ and $x_{n+2} = t_2$. Let $\{k_{ij}\}_{i=1,j}^{s}$ be a set of natural numbers such that

- *For any x_i there exists k_{ij} such that among all Ψ_{ij} there is* exactly one *term of degree k_{ij}, and it has the form $\Psi_{i,k_{ij}} = t_1 x_j^{k_{ij}-1}$.*
- *For t_2 and any x_i there exists k_{iq} such that among all Ψ_{ij} there is* exactly one *term of degree k_{iq}, and it has the form $\Psi_{n+2,k_{iq}} = t_1 x_j^{k_{iq}-1}$.*
- *For t_1 and any x_i there exists k_{iq} such that among all Ψ_{ij} there is* exactly one *term of degree k_{iq}, and it has the form $\Psi_{n+1,k_{iq}} = t_2 x_j^{k_{iq}-1}$.*

Then the corresponding algebra is simple.

Proof. Adjoin the term $t_\ell x_i^{k-1}$ to the x_i, for $\ell = 1, 2$. Let e_i be the base vector corresponding to x_i. Take the corresponding k_{ij}-ary operator

$$\omega : \omega(\vec{e}_i, \ldots, \vec{e}_i, \vec{e}_{n+\ell})) = \vec{e}_j,$$

with all other products zero. Now we apply the previous lemma. □

Remark 3. In order to be flexible with constructions via the Yagzhev correspondence, we are working in the general, not necessary cubic, case.

Now we can conclude the proof of Theorem 8. Let F be the mapping corresponding to the algebra A:

$$F : x_i \mapsto x_i + \sum_j \Psi_{ij}, \quad i = 1, \ldots, n,$$

where Ψ_{ij} are forms of homogeneous degree j. Let us adjoin new indeterminates $\{t_1, t_2\}$ and put $F(t_i) = t_i$, for $i = 1, 2$.

We choose all $k_{\alpha,\beta} > \max(\deg(\Psi_{ij}))$ and assume that these numbers are sufficiently large. Then we consider the mappings

$$G_{k_{ij}} : x_i \mapsto x_i + x_j^{k_{ij}-1} t_1,\ i \leq n; \quad t_1 \mapsto t_1; \quad t_2 \mapsto t_2; \quad x_s \mapsto x_s \text{ for } s \neq i.$$

$$G_{k_{i(n+2)}} : t_2 \mapsto x_i^{k_{ij}-1} t_1; \quad t_1 \mapsto t_1; \quad x_s \mapsto x_s \text{ for } 1 \leq s \leq n.$$

$$G_{k_{i(n+1)}} : t_1 \mapsto x_i^{k_{ij}-1} t_2; \quad t_2 \mapsto t_2; \quad x_s \mapsto x_s \text{ for } 1 \leq s \leq n.$$

These mappings are elementary automorphisms.

Consider the mapping $H = \circ_{k_{ij}} G_{k_{ij}} \circ F$, where the composite is taken in order of ascending $k_{\alpha\beta}$, and then with F. If the $k_{\alpha\beta}$ grow quickly enough, then the terms obtained in the previous step do not affect the lowest term obtained at the next step, and this term will be as described in the lemma. The theorem is proved. □

Proof of Theorem 6. The direct sum of Engel-type algebras is also of Engel type, and by Theorem 8 can be embedded into a simple algebra of Engel type. □

The Yagzhev correspondence and algebraic extensions.

For notational simplicity, we consider a cubic homogeneous mapping

$$F : x_i \mapsto x_i + \Psi_{3i}(\vec{x}).$$

We shall construct the Yagzhev correspondence of an algebraic extension.

Consider the equation

$$t^s = \sum_{p=1}^{s} \lambda_p t^{s-p},$$

where the λ_p are formal parameters. If $m \geq s$, then for some λ_{pm}, which can be expressed as polynomials in $\{\lambda_p\}_{p=1}^{s-1}$, we have

$$t^m = \sum_{p=1}^{s} \lambda_{pm} t^{s-p}.$$

Let A be the algebra corresponding to the mapping F. Consider

$$A \otimes \mathbf{k}[\lambda_1, \dots, \lambda_s]$$

and its finite algebraic extension $\hat{A} = A \otimes \mathbf{k}[\lambda_1, \dots, \lambda_s, t]$. Now we take the mapping corresponding (via the Yagzhev correspondence) to the ground ring $R = \mathbf{k}[\lambda_1, \dots, \lambda_s]$ and algebra $\hat{A}$.

For $m = 1, \dots, s-1$, we define new formal indeterminates, denoted as $T^m x_i$. Namely, we put $T^0 x_i = x_i$ and for $m \geq s$, we identify $T^m x_i$ with $\sum_{p=1}^{s} \lambda_{pm} T^{s-p} x_i$, where $\{\lambda_p\}_{p=1}^{s-1}$ are formal parameters in the centroid of some extension $R \otimes A$. Now we extend the mapping F, by putting

$$F(T^m x_i) = T^m x_i + T^{3m} \Psi_{3i}(\vec{x}), \quad m = 1, \dots, s-1.$$

We get a natural mapping corresponding to the algebraic extension.

Now we can take more symbols T_j, $j = 1, \dots, s$, and equations

$$T_j^s = \sum_{p=1}^{s} \lambda_{pj} T_j^{s-p}$$

and a new set of indeterminates $x_{ijk} = T_j^k x_i$ for $j = 1, \dots, s$ and $i = 1, \dots, n$. Then we put

$$x_{ijm} = T_j^m x_i = \sum_{p=1}^{s} \lambda_{jpm} T_j^{s-p} x_i$$

and

$$F(x_{ijm}) = x_{ijm} + T_j^{3m} \Psi_{3i}(\vec{x}), \quad m = 1, \dots, s-1.$$

This yields an "algebraic extension" of A.

Deformations of algebraic extensions.

Let $m = 2$. Let us introduce new indeterminates y_1, y_2, put $F(y_i) = y_1$, $i = 1, 2$, and compose F with the automorphism

$$G : T_1^1 x_i \mapsto T_1^1 x_i + y_1 x_i, \quad T_1^1 x_i \mapsto T_2^1 x_i + y_1 x_i, \quad x_i \mapsto x_i, \quad i = 1, 2,$$

$$y_1 \mapsto y_1 + y_2^2 y_1, \quad y_2 \mapsto y_2.$$

(Note that the $T_1^1 x_i$ and $T_2^1 x_i$ are *new* indeterminates and not proportional to x_i!) Then compose G with the automorphism $H : y_2 \mapsto y_2 + y_1^2$, where H fixes the other indeterminates. Let us call the corresponding new algebra $\hat{A}$. It is easy to see that $\mathrm{Var}(A) \neq \mathrm{Var}(\hat{A})$.

Define an *identity of the pair* (A, B), for $A \subseteq B$ to be a polynomial in two sets of indeterminates x_i, z_j that vanishes whenever the x_i are evaluated in A and z_j in B.) The *variety of the pair* (A, B) is the class of pairs of algebras satisfying the identities of (A, B).

Recall that by the rank theorem, any prime algebra A of rank n can be embedded into an n-dimensional simple algebra $\hat{A}$. We consider the variety of the pair $(A, \hat{A})$.

Considerations of deformations yield the following:

Proposition 3. *Suppose for all simple n-dimensional pairs there exists a universal pair in which all of them can be embedded. Then the Jacobian Conjecture has a positive solution.*

We see the relation with

The Razmyslov–Kushkulei Theorem [65]. *Over an algebraically closed field, any two finite dimensional simple algebras satisfying the same identities are isomorphic.*

The difficulty in applying this theorem is that the identities may depend on parameters. Also, the natural generalization of the Rasmyslov–Kushkulei theorem for a variety and subvariety does not hold: Even if $\mathrm{Var}(B) \subset \mathrm{Var}(A)$, where B and A are simple finite dimensional algebras over some algebraically closed field, B need not be embeddable to A.

Acknowledgements The first and third authors are supported by the Israel Science Foundation grant No. 1207/12. The research of Jie-Tai Yu was partially supported by an RGC-GRF Grant.

Yagzhev was a doctoral student of the second author, L. Bokut.

We thank I.P. Shestakov for useful comments, and also thank the referees for many helpful suggestions in improving the exposition.

We are grateful to Yagzhev's widow G.I. Yagzheva, and also to Jean-Yves Sharbonel, for providing some unpublished materials.

References

1. A. Abdesselam, The Jacobian Conjecture as a problem of Perturbative Quantum Field theory. Ann. Henri Poincare **4**(2), 199–215 (2003)
2. S. Abhyankar, T. Moh, Embedding of the line in the plane. J. Reine Angew. Math. **276**, 148–166 (1975)
3. S.A. Amitsur, Algebras over infinite fields. Proc. Am. Math. Soc. **7**, 35–48 (1956)
4. S.A. Amitsur, A general theory of radicals: III. Applications. Am. J. Math. **75**, 126–136 (1954)
5. H. Bass, E.H. Connell, D. Wright, The Jacobian Conjecture: reduction of degree and formal expansion of the inverse. Bull. Am. Math. Soc. (N.S.) **7**(2), 287–330 (1982)
6. V.V. Bavula, The inversion formulae for automorphisms of Weil algebras and polynomial algebras. J. Pure Appl. Algebra **210**, 147–159 (2007)
7. V.V. Bavula, The inversion formulae for automorphisms of polynomial algebras and rings of differential operators in prime characteristic. J. Pure Appl. Algebra **212**(10), 2320–2337 (2008)
8. V.V. Bavula, An analogue of the conjecture of Dixmier is true for the algebra of polynomial integro-differential operators. J. Algebra **372**, 237–250 (2012)
9. V.V. Bavula, Every monomorphism of the Lie algebra of unitriangular polynomial derivations is an authomorphism. C. R. Acad. Sci. Paris Ser. 1 **350**(11–12), 553–556 (2012)
10. V.V. Bavula, The JacobianConjecture$_{2n}$ implies the Dixmier Problem$_n$. arXiv:math/0512250
11. A. Beauville, J.-L. Colliot-Thelene, J-J Sansuc, P. Swinnerton-Dyer, Varietes stablement rationnelles non rationnelles. Ann. Math. **121**, 283–318 (1985)
12. A. Belov, Local finite basis property and local representability of varieties of associative rings. Izvestia Russ. Acad. Sci. **74**(1), 3–134 (2010). English transl.: Izvestiya Math. **74**, 1–126 (2010)
13. A. Belov, A. Berzins, R. Lipianskii, Automorphisms of the endomorphism group of the free associative algebra. Int. J. Algebra Comput. **17**(5/6), 923–939 (2007)
14. A. Belov, M.L. Kontsevich, Jacobian and Dixmier Conjectures are stably equivalent. Moscow Math. J. **7**(2), 209–218 (2007) (A special volume dedicated to the 60-th anniversary of A.G.Khovanskii)
15. A. Belov, M.L. Kontsevich, Automorphisms of Weyl algebras. Lett. Math. Phys. **74**(3), 181–199 (2005) (A special volume dedicated to the memory of F.A. Berezin)
16. A. Belov, R. Lipyanskii, Automorphisms of the endomorphism group of the free associative-commutative algebra over an arbitrary field. J. Algebra **333**, 40–54 (2011)
17. A. Belov, L. Makar-Limanov, J.T. Yu, On the generalised cancellation conjecture. J. Algebra **281**, 161–166 (2004)
18. A. Belov, L.H. Rowen, U. Vishne, Structure of Zariski-closed algebras. Trans. Am. Math. Soc. **362**, 4695–4734 (2012)

19. A. Belov, J.-T. Yu, On the lifting of the Nagata automorphism. Selecta Math. (N.S.) **17**, 935–945 (2011)
20. A. Belov-Kanel, J.-T. Yu, Stable tameness of automorphisms of $F\langle x, y, z\rangle$ fixing z. Selecta Math. (N.S.) **18**, 799–802 (2012)
21. J. Berson, A. van den Essen, D. Wright, Stable tameness of two-dimensional polynomial automorphisms over a regular ring, 2007 (rev. 2010). Adv. Math. **230**, 2176–2197 (2012)
22. J. Birman, An inverse function theorem for free groups. Proc. Am. Math. Soc. **41**, 634–638 (1973)
23. P. Bonnet, S. Vénéreau, Relations between the leading terms of a polynomial automorphism. J. Algebra **322**(2), 579–599 (2009)
24. P.M. Cohn, *Free Rings and Their Relations*, 2nd edn. (Academic, London, 1985)
25. A.J. Czerniakiewicz, Automorphisms of a free associative algebra of rank 2, II. Trans. Am. Math. Soc. **171**, 309–315 (1972)
26. W. Danielewski, On the cancellation problem and automorphism groups of affine algebraic varieties, Warsaw. Preprint (1989)
27. W. Dicks, Automorphisms of the free algebra of rank two. Group actions on rings. (Brunswick, Maine, 1984). Contemp. Math. **43**, 63–68 (1985)
28. W. Dicks, J. Levin, Jacobian conjecture for free associative algebras. Commun. Alg. **10**(12), 1285–1306 (1982)
29. V. Drensky, J.-T. Yu, The strong Anick conjecture. Proc. Natl. Acad. Sci. USA **103**, 4836–4840 (2006)
30. V. Drensky, J.-T. Yu, Coordinates and automorphisms of polynomial and free associative algebras of rank three. Front. Math. China **2**(1), 13–46 (2007)
31. V. Drensky, J.-T. Yu, The strong Anick conjecture is true. J. Eur. Math. Soc. **9**, 659–679 (2007)
32. V. Drensky, J.-T. Yu, A cancellation conjecture for free associative algebras. Proc. Am. Math. Soc. **136**, 3391–3394 (2008)
33. L.M. Druźkowski, The Jacobian conjecture: symmetric reduction and solution in the symmetric cubic linear case. Ann. Polon. Math. **87**, 83–92 (2005)
34. L.M. Druźkowski, New reduction in the Jacobian conjecture. Effective methods in algebraic and analytic geometry, 2000 (Krakw). Univ. Iagel. Acta Math. **39**, 203–206 (2001)
35. A. Van den Essen, *Polynomial Automorphisms and the Jacobian Conjecture*. Progress in Mathematics, vol. 190 (Birkhauser, Basel, 2000), pp. xviii, 329
36. A. Van den Essen, The Amazing Image Conjecture. arXiv:1006.5801
37. A. Van den Essen, M. de Bondt, in *Recent Progress on the Jacobian Conjecture*. Proc. of the Int. Conf. Singularity Theory in honour of S. Lojawiewicz, Cracow, 22–26 March 2004. Ann. Polon. Math., vol. 87 (2005), pp. 1–11
38. A. Van den Essen, M. de Bondt, The Jacobian Conjecture for symmetric Druźkowski mappings. Ann. Polon. Math. **86**(1), 43–46 (2005)
39. A. Van den Essen, D. Wright, W. Zhao, *On the Image Conjecture*. J. Algebra **340**, 211—224 (2011)
40. R.H. Fox, Free differential calculus, I. Derivation in the free group ring. Ann. Math. **57**(2), 547–560 (1953)
41. M.H. Gizatullin, Automorphisms of affine surfaces, I, II. Math. USSR-Izvestiya **11**(1), 54–103 (1977)
42. G. Gorni, G. Zampieri, Yagzhev polynomial mappings: on the structure of the Taylor expansion of their local inverse. Polon. Math. **64**, 285–290 (1996)
43. H.W.E. Jung, Uber ganze birationale Transformationen der Ebene. J. Reine Angew. Math. **184**, 161–174 (1942)
44. S. Kaliman, M. Zaidenberg, Families of affine planes: the existence of a cylinder. Michigan Math. J. **49**, 353–367 (2001)
45. E. Kuzmin, I.P. Shestakov, Non-associative structures (English). Algebra VI. Encycl. Math. Sci. **57**, 197–280 (1995). Translation from Itogi Nauki Tekh. Ser. Sovrem. Probl. Mat. Fundam. Napravleniya **57**, 179–266 (1990)
46. M. Karas', Multidegrees of tame automorphisms of C^n. Dissertationes Math. **477**, 55 (2011)

47. A. Khoroshkin, D. Piontkovski, On generating series of finitely presented operads. Preprint. arXiv:1202.5170 (2012)
48. V.S. Kulikov, The Jacobian conjecture and nilpotent mappings, in *Complex Analysis in Modern Mathematics* (FAZIS, Moscow, 2001 in Russian), pp. 167–179. Eng. Trans. J. Math. Sci. **106**, 3312–3319 (2001)
49. W. van der Kulk, On polynomial rings in two variables. Nieuw Arch. Wiskunde **1**, 33–41 (1953)
50. S. Kuroda, Shestakov-Umirbaev reductions and Nagata's conjecture on a polynomial automorphism. Tohoku Math. J. **62**, 75–115 (2010)
51. Y.C. Li, J.T. Yu, Degree estimate for subalgebras. J. Algebra **362**, 92–98 (2012)
52. L. Makar-Limanov, The automorphisms of the free algebra with two generators. Funkcional. Anal. i Priloen. **4**(3), 107–108 (1970 in Russian)
53. L. Makar-Limanov, A new proof of the Abhyankar-Moh-Suzuki Theorem, 18 pp. arXiv:1212.0163
54. L. Makar-Limanov, J.-T. Yu, Degree estimate for subalgebras generated by two elements. J. Eur. Math. Soc. **10**, 533–541 (2008)
55. M. Markl, S. Shnider, J. Stasheff, *Operads in Algebra, Topology and Physics*. Math. Surveys and Monographs, vol. 96 (AMS, Providence, 2002)
56. M. Miyanishi, T. Sugie, Affine surfaces containing cylinderlike open sets. J. Math. Kyoto Univ. **20**, 11–42 (1980)
57. M. Nagata, *On the Automorphism Group of* $k[x, y]$, Department of Mathematics, Kyoto University, Lectures in Mathematics, vol. 5 (Kinokuniya Book-Store Co., Ltd., Tokyo, 1972), pp. v, 53
58. J. Nielsen, Die Isomorphismen der allgemeinen, undendlichen Gruppen mit zwei Eerzeugenden. Math. Ann. **78**, 385–397 (1918)
59. J. Nielsen, Die Isomorphismengruppe der freien Gruppen. Math. Ann. **91**, 169–209 (1924)
60. A.Yu. Ol'shanskij, Groups of bounded period with subgroups of prime order. Algebra i Logika **21**, 553–618 (1982). Translation in Algebra Log. **21**, 369–418 (1983)
61. R. Peretz, Constructing Polynomial Mappings Using Non-commutative Algebras. In Aff. Algebr. Geometry. Contemporary Mathematics, vol. 369 (American Mathematical Society, Providence, 2005), pp. 197–232
62. D. Piontkovski, Operads versus Varieties: a dictionary of universal algebra. Preprint (2011)
63. D. Piontkovski, On Kurosh problem in varieties of algebras. Translated from Proceedings of Kurosh conference (Fund. Prikl. Matematika **14**, 5, 171–184 (2008)). J. Math. Sci. **163**(6), 743–750 (2009)
64. Yu.P. Razmyslov, Algebras satisfying identity relations of Capelli type. Izv. Akad. Nauk SSSR Ser. Mat. **45**, 143–166, 240 (1981 in Russian)
65. Yu.P. Razmyslov, *Identities of Algebras and Their Representations*. Sovremennaya Algebra [Modern Algebra] (Nauka, Moscow 1989), p. 432. Translations of Mathematical Monographs, vol. 138 (American Mathematical Society, Providence, 1994), pp. xiv, 318
66. Yu.P. Razmyslov, K.A. Zubrilin, Nilpotency of obstacles for the representability of algebras that satisfy Capelli identities, and representations of finite type. Uspekhi Mat. Nauk **48**, 171–172 (1993 in Russian); Translation in Russ. Math. Surv. **48**, 183–184 (1993)
67. C. Reutenauer, Applications of a noncommutative Jacobian matrix. J. Pure Appl. Algebra **77**, 634–638 (1992)
68. L.H. Rowen, Graduate algebra: Noncommutative view. Graduate Studies in Mathematics, vol. 91 (AMS, Providence, 2008)
69. A.H. Schofield, *Representations of Rings Over Skew Fields*. London Mathematical Society Lecture Note Series, vol. 92 (Cambridge University Press, Cambridge, 1985)
70. I.R. Shafarevich, On some infinite dimensional groups. Recondidi di Matematica **25**, 208–212 (1966)
71. I.P. Shestakov, Finite-dimensional algebras with a nil basis. Algebra i Logika **10**, 87–99 (1971 in Russian)

72. I.P. Shestakov, U.U. Umirbaev, Poisson brackets and two-generated subalgebras of rings of polynomials. J. Am. Math. Soc. **17**(1), 181–196 (2004)
73. I.P. Shestakov, U.U. Umirbaev, The tame and the wild automorphisms of polynomial rings in three variables. J. Am. Math. Soc. **17**(1), 197–227 (2004)
74. V. Shpilrain, On generators of L/R^2 Lie algebras. Proc. Am. Math. Soc. **119**, 1039–1043 (1993)
75. D. Singer, On Catalan trees and the Jacobian conjecture. Electron. J. Combin. **8**(1), Research Paper 2, 35 (2001) (electronic)
76. V. Shpilrain, J.-T. Yu, Affine varieties with equivalent cylinders. J. Algebra **251**(1), 295–307 (2002)
77. V. Shpilrain, J.-T. Yu, Factor algebras of free algebras: on a problem of G. Bergman. Bull. Lond. Math. Soc. **35**, 706–710 (2003)
78. M. Suzuki, Propiétés topologiques des polynômes de deux variables complexes, et automorphismes algébraique de l'espace C^2. J. Math. Soc. Jpn. **26**, 241–257 (1974)
79. Y. Tsuchimoto, Preliminaries on Dixmier conjecture. Mem. Fac. Sci. Kochi Univ. Ser. A Math. **24**, 43–59 (2003)
80. Y. Tsuchimoto, Endomorphisms of Weyl algebra and p-curvatures. Osaka J. Math. **42**(2), 435–452 (2005)
81. U.U. Umirbaev, On Jacobian matrices of Lie algebras, in *6th All-Union Conference on Varieties of Algebraic Systems*, Magnitogorsk, 1990, pp. 32–33
82. U.U. Umirbaev, Shreer varieties of algebras. Algebra i Logika **33**(3), 317–340, 343 (1994 in Russian). Translation in Algebra Log. **33**, 180–193 (1994)
83. U.U. Umirbaev, *Tame and Wild Automorphisms of Polynomial Algebras and Free Associative Algebras* (Max-Planck-Institute für Mathematics, Bonn). Preprint MPIM 2004-108
84. U.U. Umirbaev, On an extension of automorphisms of polynomial rings. Sibirsk. Mat. Zh. **36**, 911–916 (1995 in Russian). Translation in Siberian Math. J. **36**, 787–791 (1995)
85. U.U. Umirbaev, The Anick automorphism of free associative algebras. J. Reine Angew. Math. **605**, 165–178 (2007)
86. U.U. Umirbaev, Defining relations of the tame automorphism group of polynomial algebras in three variables. J. Reine Angew. Math. **600**, 203–235 (2006)
87. U.U. Umirbaev, Defining relations for automorphism groups of free algebras. J. Algebra **314**, 209–225 (2007)
88. U.U. Umirbaev, J.T. Yu, The strong Nagata conjecture. Proc. Natl. Acad. Sci. USA **101**, 4352–4355 (2004)
89. A.G. Vitushkin, Computation of the Jacobian of a rational transformation of C^2 and some applications. Mat. Zametki **66**(2), 308–312 (1999 in Russian). Transl. in Math. Notes **66**(1–2), 245–249 (1999)
90. A.G. Vitushkin, A criterion for the representability of a chain of σ-processes by a composition of triangular chains. Mat. Zametki **65**(5), 643–653 (1999 in Russian). Transl. in Math. Notes **65**(5–6), 539–547 (1999)
91. A.G. Vitushkin, Description of the homology of a ramified covering of C^2. Mat. Zametki **64**(6), 839–846 (1998 in Russian). Transl. in Math. Notes **64**, 726–731 (1999)
92. J.H.M. Wedderburn, Note on algebras. Ann. Math. **38**, 854–856 (1937)
93. D. Wright, *The Jacobian Conjecture as a Problem in Combinatorics*, in the monograph Affine Algebraic Geometry, in honor of Masayoshi Miyanishi,edited by Takayuki Hibi (Osaka University Press, 2007). ArXiv: math.Co/0511214 v2, 22 Mar 2006
94. D. Wright, The Jacobian Conjecture: ideal membership questions and recent advances. Affine algebraic geometry. Contemp. Math. **369**, 261–276 (2005)
95. A.V. Yagzhev, The generators of the group of tame automorphisms of an algebra of polynomials. Sibirsk. Mat. Ž. **18**(1), 222–225, 240 (1977 in Russian)
96. A.V. Yagzhev, On endomorphisms of free algebras. Sibirsk. Mat. Zh. **21**(1), 181–192, 238 (1980 in Russian)

97. A.V. Yagzhev, On the algorithmic problem of recognition of automorphisms among the endomorphisms of free associate algebras of finite rank. Sibirsk. Mat. Zh. **21**(1), 193–199, 238 (1980 in Russian)
98. A.V. Yagzhev, On a problem of Keller. Sibirsk. Mat. Zh. **21**(5), 141–150, 191 (1980 in Russian)
99. A.V. Yagzhev, Finiteness of the set of conservative polynomials of a given degree. Mat. Zametki **41**(2), 148–151, 285 (1987 in Russian)
100. A.V. Yagzhev, Nilpotency of extensions of an abelian group by an abelian group. Mat. Zametki **43**(3), 424–427, 431 (1988 in Russian); translation in Math. Notes **43**(3–4), 244–245 (1988)
101. A.V. Yagzhev, Locally nilpotent subgroups of the holomorph of an abelian group. Mat. Zametki **46**(6), 118 (1989 in Russian)
102. A.V. Yagzhev, A sufficient condition for the algebraicity of an automorphism of a group. Algebra i Logika **28**(1), 117–119, 124 (1989 in Russian). Translation in Algebra and Logic **28**(1), 83–85 (1989)
103. A.V. Yagzhev, Invertibility of endomorphisms of free associative algebras. Mat. Zametki **49**(4), 142–147, 160 (1991 in Russian). Translation in Math. Notes **49**(3–4), 426–430 (1991)
104. A.V. Yagzhev, Engel algebras satisfying Capelli identities, in *Proceedings of Shafarevich Seminar*, Moscow, 2000, pp. 83–88 (in Russian)
105. A.V. Yagzhev, Endomorphisms of polynomial rings and free algebras of different varieties, in *Proceedings of Shafarevich Seminar*, Moscow, 2000, pp. 15–47 (in Russian)
106. A. Zaks, Dedekind subrings of $K[x_1, \ldots, x_n]$ are rings of polynomials. Israel J. Math. **9**, 285–289 (1971)
107. E. Zelmanov, On the nilpotence of nilalgebras. Lect. Notes Math. **1352**, 227–240 (1988)
108. W. Zhao, New Proofs for the Abhyankar-Gurjar Inversion Formula and the Equivalence of the Jacobian Conjecture and the Vanishing Conjecture. Proc. Am. Math. Soc. **139**, 3141–3154 (2011)
109. W. Zhao, Mathieu subspaces of associative algebras. J. Algebra **350**, 245–272 (2012). arXiv:1005.4260
110. K.A. Zhevlakov, A.M. Slin'ko, I.P. Shestakov, A.I. Shirshov, *Nearly Associative Rings* (Nauka, Moscow, 1978 in Russian)
111. K.A. Zubrilin, Algebras that satisfy the Capelli identities. Mat. Sb. 186(3), 53–64 (1995 in Russian). Translation in Sb. Math. **186**(3), 359–370 (1995)
112. K.A. Zubrilin, On the class of nilpotence of obstruction for the representability of algebras satisfying Capelli identities. Fundam. Prikl. Mat. **1**(2), 409–430 (1995 in Russian)
113. K.A. Zubrilin, On the Baer ideal in algebras that satisfy the Capelli identities. Mat. Sb. **189**, 73–82 (1998 in Russian). Translation in Sb. Math. **189**, 1809–1818 (1998)

Equivariant Triviality of Quasi-Monomial Triangular $\mathbb{G}_a$-Actions on $\mathbb{A}^4$

Adrien Dubouloz, David R. Finston, and Imad Jaradat

Abstract We give a direct and self-contained proof of the fact that additive group actions on affine four-space generated by certain types of triangular derivations are translations whenever they are proper. The argument, which is based on explicit techniques, provides an illustration of the difficulties encountered and an introduction to the more abstract methods which were used recently by the authors to solve the general triangular case.

Subject Classification: 14R20; 14L30

1 Introduction

It was established recently in [6] that a proper $\mathbb{G}_a$-action $\sigma : \mathbb{G}_a \times \mathbb{A}^4 \to \mathbb{A}^4$ on the affine space $\mathbb{A}^4 = \mathrm{Spec}(k[x_1, x_2, x_3, x_4])$ over an algebraically closed field k of characteristic zero generated by a triangular derivation

A. Dubouloz (✉)
Institut de Mathématiques de Bourgogne, UMR 5584 du CNRS, Université de Bourgogne, 9 av. A. Savary, BP 47870, 21078 Dijon, France
e-mail: Adrien.Dubouloz@u-bourgogne.fr

D.R. Finston
Mathematics Department, Brooklyn College CUNY, 2900 Bedford Avenue, Brooklyn, NY 11210, USA
e-mail: dfinston@brooklyn.cuny.edu

I. Jaradat
Department of Mathematical Sciences, New Mexico State University, Las Cruces, NM 88003, USA
e-mail: imad_jar@nmsu.edu

I. Cheltsov et al. (eds.), *Automorphisms in Birational and Affine Geometry*, Springer Proceedings in Mathematics & Statistics 79,
DOI 10.1007/978-3-319-05681-4_16,

$$\partial = r_1(x_1)\partial_{x_2} + q(x_1, x_2)\partial_{x_3} + p(x_1, x_2, x_3)\partial_{x_4}$$

is globally equivariantly trivial with geometric quotient $\mathbb{A}^4/\mathbb{G}_a$ isomorphic to $\mathbb{A}^3$. That is, $\mathbb{A}^4$ is equivariantly isomorphic to the trivial $\mathbb{G}_a$-bundle over $\mathbb{A}^3$, a property which is usually abbreviated by saying that σ is a translation.

In contrast with the situation in lower dimensions, there exist fixed point free yet improper triangular actions on $\mathbb{A}^4$, see e.g. [15]. So the properness assumption on σ, which by definition means the properness of the morphism $\Phi = (\sigma, \mathrm{pr}_2) : \mathbb{G}_a \times \mathbb{A}^4 \to \mathbb{A}^4 \times \mathbb{A}^4$, is crucial. Nevertheless, the proof given in [6] exploits this assumption in a rather abstract fashion, with the consequence that the relation between the properness of a given triangular action and the properties of its corresponding derivation remains quite elusive.

The aim of this note is to provide a more explicit treatment for a particular class of triangular derivations that we call quasi-monomial: these are k-derivations of $k[x_1, x_2, x_3, x_4]$ of the form $\partial = \tilde{\partial} + p(x_1, x_2, x_3)\partial_{x_4}$ where $p(x_1, x_2, x_3) \in k[x_1, x_2, x_3]$ and where $\tilde{\partial}$ denotes the natural extension to $k[x_1, x_2, x_3, x_4]$ of a nonzero monomial $k[x_1]$-derivation $ax_1^{n'}\partial_{x_2} + bx_1^q x_2^m \partial_{x_3}$ of $k[x_1, x_2, x_3]$, where $a, b \in k$ and $n', m, q \in \mathbb{Z}_{\geq 0}$. Showing that a proper action generated by a derivation of this type is a translation is already nontrivial, and the methods we present here were actually a source of inspiration for the general argument developed in [6]. The strategy consists in building on an earlier result [5] which asserts that if $p(x_1, x_2, x_3)$ is a polynomial in x_1 and x_2 only, then the properness of the corresponding $\mathbb{G}_a$-action is indeed equivalent to its equivariant triviality. Derivations $\partial = \tilde{\partial} + p(x_1, x_2)\partial_{x_4}$ were dubbed twin-triangular in *loc. cit.* and [2].

A proper action being in particular fixed point free, we may safely restrict to quasi-monomial triangular derivations ∂ generating fixed point free actions, a condition which is equivalent to $\partial x_2 = ax_1^{n'}$, $\partial x_3 = bx_1^q x_2^m$ and $\partial x_4 = p(x_1, x_2, x_3)$ generating the unit ideal in $k[x_1, x_2, x_3]$. Next we can dispense with all cases in which such derivations are already known to have a slice. Since this holds in particular if ∂ contains at least two variables of $k[x_1, x_2, x_3, x_4]$ in its kernel [1], we may thus assume that $a, b \in k^*$ and then that $n' > 0$ since otherwise $a^{-1}x_2$ is a slice for ∂. Up to a triangular change of the coordinate x_3, we may assume next that $0 \leq q < n'$ and set $n' = n + q$ where $n > 0$. Then we may suppose that $m > 0$ since otherwise $x_2 - ab^{-1}x_1^n x_3$ is a second variable of $k[x_1, x_2, x_3, x_4]$ contained in the kernel of ∂. Summing up, we are reduced after a last linear coordinate change to considering derivations of the form

$$\partial = x_1^{n+q}\partial_{x_2} + (m+1)x_1^q x_2^m \partial_{x_3} + p(x_1, x_2, x_3)\partial_{x_4}$$

where $m, n > 0$, $q \geq 0$ and where x_1^{n+q}, $x_1^n x_2^m$ and $p(x_1, x_2, x_3)$ generate the unit ideal in $k[x_1, x_2, x_3]$.[1]

Now the argument proceeds in three steps: we first exhibit in the case $q = 0$ certain restrictions on the form of the derivation imposed by the condition that the

[1] The factor $(m + 1)$ in $\partial(x_3)$ is chosen to simplify calculations in the next sections.

corresponding $\mathbb{G}_a$-action is proper by means of the study of a suitable invariant fibration on $\mathbb{A}^4$. We use this information to infer the existence of a coordinate system on $\mathbb{A}^4$ in which every derivation ∂ as above takes again the form $\delta_0 = x_1^n\partial_{x_2} + (m+1)x_2^m\partial_{x_3} + p_0(x_1, x_2, x_3)\partial_{x_4}$, but with additional very particular numerical restrictions on the exponents of the monomials occurring in the polynomial p_0. This additional information is then exploited through a variant of the valuative criterion for properness which renders the conclusion that non twin-triangular δ_0 as above generate improper $\mathbb{G}_a$-actions. It also provides a complete characterization of quasi-monomial twin-triangular derivations generating proper actions, which enables a direct proof that such proper actions are indeed translations. The general case is finally obtained in an indirect fashion by comparing the $\mathbb{G}_a$-actions σ_q and σ generated by the derivations

$$\partial_q = x_1^{n+q}\partial_{x_2} + (m+1)x_1^q x_2^m\partial_{x_3} + p(x_1, x_2, x_3)\partial_{x_4}$$

and

$$\partial = x_1^n\partial_{x_2} + (m+1)x_2^m\partial_{x_3} + p(x_1, x_2, x_3)\partial_{x_4}.$$

We show that the properness of σ_q implies that of σ and that in this case the two actions have isomorphic geometric quotients.

Notation 1. Hereafter we write $\underline{r}$ for an element $(r_1, r_2, r_3) \in \mathbb{Z}^3_{\geq 0}$ and we denote a monomial $cx_1^{r_1}x_2^{r_2}x_3^{r_3} \in k[x_1, x_2, x_3]$ by $c\underline{x}^{\underline{r}}$. Given two fixed integers $m, n \in \mathbb{Z}_{\geq 0}$, we set for every $l_1, l_2, l_3 \in \mathbb{Z}$:

$$\nu(l_1, l_2) = l_1 - nl_2,\ \eta(l_1, l_2) = (m+1)l_2 + l_1,\ \mu(l_1, l_2, l_3) = nml_3 + nl_2 + l_1.$$

We also set formally for every triple of elements τ, y_1, y_2 in a ring:

$$H_\tau(y_1, y_2) = y_1 + \tau y_2 \text{ and } G_\tau(y_1, y_2) = H_\tau(y_1, y_2)^{m+1} - y_1^{m+1}.$$

For an $(m+1)$-th root of unity $\xi \in k$ and integers $l_1, l_2 \in \mathbb{Z}_{\geq 0}$, we let

$$\Theta_{l_1,l_2}(\xi) = \int_1^\xi z^{l_1}(z^{m+1} - 1)^{l_2} dz = F(\xi) - F(1),$$

where $F(z)$ is any formal antiderivative of $f(z) = z^{l_1}(z^{m+1} - 1)^{l_2}$. Then a straightforward induction on l_2 and integration by parts yields

$$\Theta_{l_1,l_2}(\xi) = \frac{(-1)^{l_2} l_2!\,(m+1)^{l_2}}{\prod_{j=0}^{l_2} \eta(l_1+1, j)}(\xi^{l_1+1} - 1).$$

Note that $\Theta_{l_1,l_2}(\xi) = 0$ if and only if $\xi = 1$ or $l_1 + 1 \in (m+1)\mathbb{Z}$.

2 Invariant Hypersurfaces

In this section, we consider quasi-monomial triangular derivations

$$\partial = x_1^n \partial_{x_2} + (m+1)x_2^m \partial_{x_3} + p(x_1, x_2, x_3)\partial_{x_4}$$

where $n, m > 0$ and where x_1^n, x_2^m and $p(x_1, x_2, x_3)$ generate the unit ideal in $k[x_1, x_2, x_3]$. Up to a linear change of coordinates preserving the first three variables we may thus assume that $p = 1 + \sum_{\underline{r} \in M} c_{\underline{r}} \underline{x}^{\underline{r}}$ where for every $\underline{r} = (r_1, r_2, r_3) \in M \subset \mathbb{Z}_{\geq 0}^3$ either $r_1 > 0$ or $r_2 > 0$.

Since a derivation as above annihilates the polynomial $f = x_2^{m+1} - x_1^n x_3$, the corresponding $\mathbb{G}_a$-action on $\mathbb{A}^4$ restricts to an action on the level surfaces $V_\lambda = f^{-1}(\lambda)$, $\lambda \in k$, of f. We will exploit the fact that the properness of the action generated by ∂ implies that of its restriction on every V_λ to get a hint at the structure of derivations generating proper actions.

Since the defining equation for V_λ does not involve the last variable x_4, $V_\lambda \simeq S_\lambda \times \mathbb{A}^1$ where $S_\lambda \subset \mathbb{A}^3 = \mathrm{Spec}(k[x_1, x_2, x_3])$ is a surface stable under the $\mathbb{G}_a$-action on $\mathbb{A}^3$ generated by the derivation $x_1^n \partial_{x_2} + (m+1)x_2^m \partial_{x_3}$. We note that S_λ, hence V_λ, is smooth provided that $\lambda \neq 0$ and that in this case, the induced $\mathbb{G}_a$-action on S_λ is fixed point free. The projection $\mathrm{pr}_1 : V_\lambda \to S_\lambda$ is equivariant, leading to a commutative diagram

$$\begin{array}{ccc} V_\lambda & \longrightarrow & V_\lambda/\mathbb{G}_a \\ \downarrow & & \downarrow \\ S_\lambda & \longrightarrow & S_\lambda/\mathbb{G}_a \end{array}$$

which expresses the geometric quotient $V_\lambda/\mathbb{G}_a$ as the total space of a $\mathbb{G}_a$-bundle over the geometric quotient $S_\lambda/\mathbb{G}_a$.

Those quotients exists a priori as smooth algebraic spaces of dimension 2 and 1 respectively [10], but in fact, as shown in the proof of Lemma 1 below, $S_\lambda/\mathbb{G}_a$ is a scheme isomorphic to the affine line $\tilde{\mathbb{A}}^1$ with an $(m+1)$-fold origin, obtained by gluing $m+1$ copies $X_i = \mathrm{Spec}(k[x_1])$, $i = 1, \ldots, m+1$, of $\mathbb{A}^1$ by the identity outside their respective origins. This enables a description of the $\mathbb{G}_a$-bundle $V_\lambda/\mathbb{G}_a \to S_\lambda/\mathbb{G}_a \simeq \tilde{\mathbb{A}}^1$ in terms of Čech 1-cocycles for the covering $\mathcal{U} = \{X_i\}_{i=1,\ldots,m+1}$ of $\tilde{\mathbb{A}}^1$. Noting that $C^1(\mathcal{U}, \mathcal{O}_{\tilde{\mathbb{A}}^1}) \simeq k[x_1^{\pm 1}]^{(m+1)^2}$, we have:

Lemma 1. *Let $\lambda \neq 0$ and let $\lambda_1, \ldots, \lambda_{m+1} \in k$ be the distinct $(m+1)$-th roots of λ. Then $V_\lambda/\mathbb{G}_a \to S_\lambda/\mathbb{G}_a \simeq \tilde{\mathbb{A}}^1$ is a locally trivial $\mathbb{G}_a$-bundle with isomorphy class in $H^1(\tilde{\mathbb{A}}^1, \mathcal{O}_{\tilde{\mathbb{A}}^1})$ represented by the Čech 1-cocycle*

$$\left\{h_{ij}(\lambda;x_1) = x_1^{-n}\int_{\lambda_j}^{\lambda_i} p\left(x_1,\tau,x_1^{-n}(\tau^{m+1}-\lambda)\right)d\tau\right\}_{i,j=1}^{m+1} \in C^1(\mathcal{U},\mathcal{O}_{\tilde{\mathbb{A}}^1}).$$

Proof. Let us first recall the description of the $\mathbb{G}_a$-bundle $S_\lambda \to S_\lambda/\mathbb{G}_a \simeq \tilde{\mathbb{A}}^1$. The surface S_λ is covered by $\mathbb{G}_a$-invariant open subsets $U_i = S\setminus\bigcup_{j\neq i}C_j$, $i = 1,\ldots,m+1$, where the curves $C_j = \{x_1 = x_2-\lambda_j = 0\} \subset S_\lambda$ are $\mathbb{G}_a$-orbits. Noting that the rational function $t_i = x_1^{-n}(x_2-\lambda_i) = \prod_{j\neq i}(x_2-\lambda_j)^{-1}x_3$ on S_λ is a regular slice for the $\mathbb{G}_a$-action induced on U_i, i.e., $t_i \in \Gamma(U_i,\mathcal{O}_{S_\lambda})$ and $\partial\mid_{U_i}(t_i) = 1$, we obtain a collection of $\mathbb{G}_a$-equivariant isomorphisms $U_i \simeq \mathrm{Spec}(k[x_1][t_i]) \simeq X_i\times\mathbb{G}_a$, $i = 1,\ldots,m+1$, where $\mathbb{G}_a$ acts by translations on the second factor. It follows that $S_\lambda/\mathbb{G}_a$ is isomorphic to the scheme $\tilde{\mathbb{A}}^1$ obtained by gluing the $X_i \simeq U_i/\mathbb{G}_a$ by the identity along the open subsets $X_i\setminus\{0\} \simeq (U_i\setminus C_i)/\mathbb{G}_a$, and that $g_{ij} = (t_i-t_j)\mid_{U_i\cap U_j} = x_1^{-n}(\lambda_j-\lambda_i) \in k[x_1^{\pm1}]$, $i,j = 1,\ldots,m+1$, is a Čech 1-cocycle representing the isomorphy class of $S_\lambda \to S_\lambda/\mathbb{G}_a \simeq \tilde{\mathbb{A}}^1$.

To determine the structure of the induced bundle $V_\lambda/\mathbb{G}_a \to S_\lambda/\mathbb{G}_a$, we observe that $V_\lambda \simeq S_\lambda\times\mathbb{A}^1$ is covered by the $\mathbb{G}_a$-invariant open subsets $W_i = U_i\times\mathbb{A}^1 \simeq \mathrm{Spec}(k[x_1,t_i][x_4])$, $i = 1,\ldots,m+1$. Since $x_2\mid_{U_i} = x_1^n t_i+\lambda_i$ and $x_3\mid_{U_i} = x_1^{-n}\left((x_1^n t_i+\lambda_i)^{m+1}-\lambda\right)$, we have,

$$\partial(x_4)\mid_{W_i} = p\left(x_1, x_1^n t_i+\lambda_i, x_1^{-n}((x_1^n t_i+\lambda_i)^{m+1}-\lambda)\right) = \Phi_i(x_1,t_i)$$

and, since t_i is again a slice for the induced $\mathbb{G}_a$-action on W_i, we conclude that $W_i \simeq \mathrm{Spec}(k[x_1,u_i][t_i]) \simeq \mathbb{A}^2\times\mathbb{G}_a$ where $u_i = x_4\mid_{W_i} - \int_0^{t_i}\Phi_i(x_1,\tau)d\tau \in k[x_1,t_i,x_4]^{\mathbb{G}_a}$. By construction, we have

$$\begin{aligned} u_i-u_j\mid_{W_i\cap W_j} &= \left(\int_0^{t_j}\Phi_j(x_1,z_j)dz_j - \int_0^{t_i}\Phi_i(x_1,z_i)dz_i\right)\mid_{W_i\cap W_j} \\ &= x_1^{-n}\int_{\lambda_j}^{\lambda_i} p\left(x_1,\tau,x_1^{-n}(\tau^{m+1}-\lambda)\right)d\tau, \end{aligned}$$

the second equality being obtained by making the respective change of variables $z_\ell = x_1^{-n}(\tau-\lambda_\ell)$, $\ell = i,j$ in the integrals. So $V_\lambda/\mathbb{G}_a \to S_\lambda/\mathbb{G}_a$ is a $\mathbb{G}_a$-bundle represented by the advertised Čech 1-cocycle. □

Knowing the structure of the induced $\mathbb{G}_a$-bundle $V_\lambda/\mathbb{G}_a \to S_\lambda/\mathbb{G}_a$ leads to a first efficient criterion to test the properness of the $\mathbb{G}_a$-action on $\mathbb{A}^4$ generated by a derivation ∂ as above. Indeed, since a fixed point free $\mathbb{G}_a$-action on a variety X is proper if and only if its geometric quotient $X/\mathbb{G}_a$ is a separated algebraic space [11, Lecture 3], the properness of the $\mathbb{G}_a$-action generated by ∂ implies the separatedness of $V_\lambda/\mathbb{G}_a$ for every $\lambda \neq 0$.

By virtue of Fieseler's criterion [7], this holds if and only if each of the rational functions $h_{ij}(\lambda;x_1) \in k[x_1^{\pm1}]$, $i\neq j$, of the corresponding Čech cocycle has a pole at 0. This leads to the following criterion:

Proposition 1. *If* $\partial = x_1^n \partial_{x_2} + (m+1)x_2^m \partial_{x_3} + p(\underline{x})\partial_{x_4}$ *generates a proper* $\mathbb{G}_a$*-action, then for every* $\lambda \neq 0$ *and every pair of distinct indices* $i, j \in \{1, \ldots, m+1\}$*, the rational function*

$$h_{ij}(\lambda; x_1) = x_1^{-n} \int_{\lambda_j}^{\lambda_i} p\big(x_1, \tau, x_1^{-n}(\tau^{m+1} - \lambda)\big) d\tau$$

lies in $k[x_1^{\pm}] \setminus k[x_1]$*.*

For a polynomial $p = 1 + \sum_{\underline{r} \in M} c_{\underline{r}} \underline{x}^{\underline{r}}$, we find more explicitly that

$$h_{ij}(\lambda; x_1) = \frac{(\lambda_i - \lambda_j)}{x_1^n} \left[1 + \sum_{\underline{r} \in M} c_{\underline{r}} \lambda^{r_3} \Sigma_{r_2,r_3} \frac{\lambda_i^{r_2+1} - \lambda_j^{r_2+1}}{\lambda_i - \lambda_j} x_1^{\nu(r_1,r_3)} \right]$$

where $\Sigma_{r_2,r_3} = \sum_{\ell=0}^{r_3} \binom{r_3}{\ell} \frac{(-1)^{r_3-\ell}}{\eta(r_2+1,\ell)} \neq 0$. This reveals that the only monomials $c_{\underline{r}} \underline{x}^{\underline{r}}$ which could lead to the improperness of the induced $\mathbb{G}_a$-action on some V_λ are those with $\nu(r_1, r_3) = r_1 - nr_3 \geq 0$. Note also that monomials for which $r_2 \equiv m \mod (m+1)$ do not contribute to the above formula.

3 Normalized Derivations

The previous observations suggest a consideration of the following natural further normalization of quasi-monomial derivations

$$\partial = x_1^n \partial_{x_2} + (m+1)x_2^m \partial_{x_3} + p(x_1, x_2, x_3)\partial_{x_4},$$

which is a particular case of the $\sharp$-reduction introduced in [6]:

Lemma 2. *Every quasi-monomial derivation* ∂ *as above is conjugate to one*

$$\delta_0 = x_1^{n+q} \partial_{x_2} + (m+1)x_1^q x_2^m \partial_{x_3} + p_0(\underline{x})\partial_{x_4}$$

where $p_0(\underline{x}) = 1 + \sum_{\underline{r} \in M} c_{\underline{r}} \underline{x}^{\underline{r}}$ *has the property that for each* $\underline{r} = (r_1, r_2, r_3) \in M$ *either* $(r_1 < nr_3$ and $0 \leq r_2 < m)$ *or* $(r_3 = 0$ and $r_2 \not\equiv m \mod (m+1))$.

Proof. The image by ∂ of a monomial $bx_1^\alpha x_2^\beta x_3^\gamma$ is equal to $b\beta x_1^{\alpha+n} x_2^{\beta-1} x_3^\gamma + b(m+1)\gamma x_1^\alpha x_2^{\beta+m} x_3^{\gamma-1}$. It follows that triangular coordinate changes of the form $\tilde{x}_4 = x_4 + bx_1^\alpha x_2^\beta x_3^\gamma$ allow to replace a monomial $cx_1^{r_1} x_2^{r_2} x_3^{r_3}$ of p by the monomial

$$c' x_1^{r_1'} x_2^{r_2'} x_3^{r_3'} = \begin{cases} -\frac{cr_3(m+1)}{r_2+1} x_1^{r_1-n} x_2^{r_2+(m+1)} x_3^{r_3-1} & \text{if } r_1 \geq n \\ -\frac{c(r_2-m)}{(m+1)(r_3+1)} x_1^{r_1+n} x_2^{r_2-(m+1)} x_3^{r_3+1} & \text{if } r_2 \geq m. \end{cases}$$

Note that in both cases we have $r_2' \equiv r_2 \bmod (m+1)$ and $r_1' - nr_3' = r_1 - nr_3$. A suitable sequence of triangular changes of coordinates of the second type reduces to the situation where $0 \leq r_2 < m$ for every monomial of p. Then monomials for which $r_1 - nr_3 \geq 0$ can be reduced using the first type of coordinate change to monomials with $r_3 = 0$ and $r_2 \not\equiv m \bmod (m+1)$. □

Remark 1. For a derivation δ_0 normalized as in Lemma 2, the polynomial p_0 has the property that for a given pair of integers $(\eta, \mu) \in \mathbb{Z}^2_{\geq 0}$ there exists at most one monomial $cx_1^{r_1}x_2^{r_2}x_3^{r_3}$ of p_0 such that $\eta(r_2, r_3) = (m+1)r_3 + r_2 = \eta$ and $\mu(r_1, r_2, r_3) = n\eta(r_2, r_3) + r_1 - nr_3 = \mu$. Indeed, for two monomials $cx_1^{r_1}x_2^{r_2}x_3^{r_3}$ and $c'x_1^{r_1'}x_2^{r_2'}x_3^{r_3'}$ satisfying $\eta(r_2, r_3) = \eta(r_2', r_3')$ and $\mu(r_1, r_2, r_3) = \mu(r_1', r_2', r_3')$, if both $r_2, r_2' < m$ or $r_3 = r_3' = 0$ then clearly $r_1 = r_1'$ and $r_2 = r_2'$. Otherwise, if say $r_2 < m$ and $r_3' = 0$, then the definition of $\mu(r_2, r_2, r_3)$ forces $r_3 = 0$ as well.

Example 1. Consider the twin-triangular quasi-monomial derivations

$$d_\ell = x_1\partial_{x_2} + 2x_2\partial_{x_3} + (1 + x_2^\ell)\partial_{x_4}, \quad \ell \geq 1,$$

generating fixed point free $\mathbb{G}_a$-actions on $\mathbb{A}^4$. If ℓ is even, then d_ℓ is normalized and it was established in [3] that the corresponding action is improper. On the other hand, if ℓ is odd, then the normalization of d_ℓ as in Lemma 2 is equal to the derivation $\delta_0 = x_1\partial_{x_2} + 2x_2\partial_{x_3} + \partial_{x_4}$ which has x_4 as an obvious slice. The corresponding $\mathbb{G}_a$-action is thus a translation.

By construction, a derivation δ_0 normalized as in Lemma 2 is either twin-triangular, i.e., $p_0(\underline{x}) \in k[x_1, x_2] \subset k[x_1, x_2, x_3]$, or there exists at least one nonzero monomial $c_{\underline{r}}\underline{x}^{\underline{r}}$ of $p_0(\underline{x})$ for which $\nu(r_1, r_3) = r_1 - nr_3 < 0$. In the first case, the so generated $\mathbb{G}_a$-action is known to be proper if and only if it is a translation [5], and an effective criterion for its properness will be given in the next section. In contrast, the following example illustrates the general fact, which will also be established in the next section, that in the second case the corresponding $\mathbb{G}_a$-action on $\mathbb{A}^4$ is always improper.

Example 2. Let $\sigma : \mathbb{G}_a \times \mathbb{A}^4 \to \mathbb{A}^4$ be the $\mathbb{G}_a$-action generated by a non twin-triangular normalized derivation

$$\delta_0 = x_1^n\partial_{x_2} + (m+1)x_2^m\partial_{x_3} + \left(1 + c\underline{x}^{(r_1,r_2,r_3)}\right)\partial_{x_4}.$$

Note that the cocycle $h_{ij}(\lambda; x_1)$ associated with ∂ has a pole at 0 for every $\lambda \neq 0$ and every pair of distinct indices $i, j \in \{1, \ldots, m+1\}$ so that by virtue of Proposition 1, the induced $\mathbb{G}_a$-action on every hypersurface V_λ, $\lambda \neq 0$, is proper. Nevertheless, assuming further that $k = \mathbb{C}$, we will show that $\Phi = (\sigma, \mathrm{pr}_2) : \mathbb{G}_a \times \mathbb{A}^4 \to \mathbb{A}^4 \times \mathbb{A}^4$ is not proper when considered as a holomorphic map between the corresponding varieties equipped with their respective underlying structures of analytic manifolds, hence not a proper morphism of algebraic varieties [12]. The non properness of Φ will follow from the existence of sequences of points $y_\ell = \left(y_{1,\ell}, y_{2,\ell}, y_{3,\ell}, y_{4,\ell}\right) \in$

$\mathbb{A}^4$ and $t_\ell \in \mathbb{G}_a$, $\ell \in \mathbb{N}$, such that y_ℓ and $\sigma(t_\ell, y_\ell)$ converge when $\ell \to \infty$ while $\lim_{\ell\to\infty} |t_\ell| = \infty$: namely, the closure of the set $\{\Phi(t_\ell, y_\ell)\}_{\ell\in\mathbb{N}} \in \mathbb{A}^4 \times \mathbb{A}^4$ will be a compact subset whose inverse image by Φ is unbounded.

With $r_1 < nr_3$, the desired sequences can be produced as follows: let $y_\ell = (a\ell^{-1}, b\ell^{-\beta}, \ell^{-\gamma}, 0)$ and $t_\ell = \ell^\omega$, where $a \in \mathbb{C}^*$, $b = a^n(\xi - 1)^{-1}$ for a primitive $(m+1)$-th root of unity ξ, and where $\beta, \gamma, \omega \in \mathbb{Q}_{>0}$ satisfy the relations $\beta = n - \omega$ and $\gamma = m\beta$. Note that $\lim_{\ell\to\infty} y_\ell = (0,0,0,0)$.

Using the identity

$$\sigma^*(f)(x_1,x_2,x_3,x_4,t) = f(x_1,x_2,x_3,x_4) + \int_0^t \sigma^*(\delta_0(f))(x_1,x_2,x_3,x_4,\tau)d\tau$$

which holds in $k[x_1,x_2,x_3,x_4][t]$ for every $f \in k[x_1,x_2,x_3,x_4]$ due to the fact that δ_0 is locally nilpotent, we obtain that the morphism $\sigma(t,\cdot)$ maps a point $x = (x_1,x_2,x_3,x_4) \in \mathbb{A}^4$ to the point

$$\sigma(t,x) = \begin{pmatrix} x_1 \\ H_t(x_2, x_1^n) \\ x_3 + x_1^{-n} G_t(x_2, x_1^n) \\ x_4 + t + cx_1^{r_1} \int_0^t H_\tau(x_2,x_1^n)^{r_2}\big(x_3 + x_1^{-n}G_\tau(x_2,x_1^n)\big)^{r_3} d\tau \end{pmatrix}^T$$

where H_τ and G_τ are defined in Notation 1. It follows in particular that $\sigma(t_\ell, y_\ell) = \big(a\ell^{-1}, a^n\xi(\xi-1)^{-1}\ell^{-\beta}, \ell^{-\gamma}, z_{4,\ell}\big)$ where

$$\begin{aligned} z_{4,\ell} &= \ell^\omega + ca^{r_1}\ell^{-r_1} \int_0^{\ell^\omega} H_\tau(b\ell^{-\beta}, a^n\ell^{-n})^{r_2}(\ell^{-\gamma} + a^{-n}\ell^n G_\tau(b\ell^{-\beta}, a^n\ell^{-n}))^{r_3} d\tau \\ &= \ell^\omega + ca^{r_1}b^{\eta(r_2,r_3)}\ell^\kappa \int_0^1 H_\tau(1, a^nb^{-1})^{r_2}(\ell^{-\omega}b^{-m-1} + a^{-n}G_\tau(1,a^nb^{-1}))^{r_3} d\tau \end{aligned}$$

for $\kappa = \omega - \nu(r_1,r_3) + (\omega - n)\eta(r_2,r_3)$. So the first three coordinates of $\sigma(t_\ell, y_\ell)$ converge to 0 when $\ell \to \infty$, and it remains to show that we can choose the parameters $a \in \mathbb{C}^*$ and $\omega \in \mathbb{Q}_+$ in such a way that the sequence $z_{4,\ell}$ converges. To that end, we let

$$\omega = n + \frac{\nu(r_1,r_3)}{\eta(r_2,r_3)} = n + \frac{r_1 - nr_3}{(m+1)r_3 + r_2} \in (0,n)$$

to obtain that $z_{4,\ell} = (1 + R_1)\ell^\omega + O(1)$ where $R_1 \in \mathbb{C}$ is given by the formula

$$\begin{aligned} R_1 &= ca^{\nu(r_1,r_3)}b^{\eta(r_2,r_3)} \int_0^1 H_\tau\big(1, a^nb^{-1}\big)^{r_2} G_\tau\big(1,a^nb^{-1}\big)^{r_3} d\tau \\ &= ca^{\nu(r_1,r_3)-n}b^{\eta(r_2+1,r_3)}\Theta_{r_2,r_3}(\xi). \end{aligned}$$

Since $r_2 \not\equiv m \bmod (m+1)$, we have $\Theta_{r_2,r_3-1}(\xi) \neq 0$ (see Notation 1), and so the equation $R_1 = -1$ can be solved with respect to a to obtain the convergence of the sequence $z_{4,\ell}$.

4 An Effective Criterion for Properness

This section is devoted to the proof of the following result:

Proposition 2. *Let $\delta_0 = x_1^n \partial_{x_2} + (m+1)x_2^m \partial_{x_3} + p_0(\underline{x})\partial_{x_4}$ be a derivation normalized as in Lemma 2. If δ_0 generates a proper $\mathbb{G}_a$-action on $\mathbb{A}^4$, then $(p_0(\underline{x}) - 1) \in x_1 k[x_1, x_2]$, i.e., δ_0 is twin-triangular and every nonconstant monomial of p_0 is divisible by x_1.*

The topological criterion for properness used in the last example was of course limited in that it applies only to varieties defined over $\mathbb{C}$. Moreover, the technicality in the construction of the sequences $\{y_\ell\}_{\ell\in\mathbb{N}}$ and $\{t_\ell\}_{\ell\in\mathbb{N}}$ presents daunting challenges to extending the argument to the more general situation considered in Proposition 2. To prove this proposition, we will use an alternative approach through the valuative criterion for properness.

Note that since $\mathbb{G}_a$ has no nontrivial finite subgroup, the properness of an action $\sigma : \mathbb{G}_a \times \mathbb{A}^4 \to \mathbb{A}^4$ implies that the proper morphism $\Phi = (\sigma, \mathrm{pr}_2) : \mathbb{G}_a \times \mathbb{A}^4 \to \mathbb{A}^4 \times \mathbb{A}^4$ is a monomorphism, hence a closed immersion [8, 8.11.5]. Equivalently, the comorphism

$$\Phi^* = \mathrm{id} \otimes \sigma^* : k[x_1, x_2, x_3, x_4] \otimes_k k[x_1, x_2, x_3, x_4] \to k[x_1, x_2, x_3, x_4][t]$$

is surjective, which, because $k[x_1, x_2, x_3, x_4][t]$ is integrally closed, is in turn equivalent to the fact that every discrete valuation ring of $k(x_1, x_2, x_3, x_4)(t)$ which contains all x_i and $\sigma^*(x_i)$ also contains t. The strategy for the proof of Proposition 2 is to exploit the existence of a nonconstant monomial $c_{\underline{r}}\underline{x}^{\underline{r}}$ having either $r_1 = r_3 = 0$ in case δ_0 is twin-triangular or $r_1 < nr_3$ otherwise to construct a discrete valuation υ of $k(x_1, x_2, x_3, x_4)(t)$ such that $\upsilon(x_i)$ and $\upsilon(\sigma^*(x_i))$ are nonnegative for every $i = 1, \ldots, 4$, while $\upsilon(t) < 0$.

Proof. We write $p_0(\underline{x}) = 1 + \sum_{\underline{r}\in M} c_{\underline{r}}\underline{x}^{\underline{r}} \in k[x_1, x_2, x_3]$. Letting $\xi \neq 1$ be a primitive $(m+1)$-th root of unity, we find similarly as in Example 2 above that $\sigma^*(x_1) = x_1, \sigma^*(x_2) = H_t(x_2, x_1^n), \sigma^*(x_3) = x_3 + x_1^{-n} G_t(x_2, x_1^n)$ and

$$\sigma^*(x_4) = x_4 + t + \sum_{\underline{r}\in M} c_{\underline{r}} x_1^{r_1} \int_0^t H_\tau(x_2, x_1^n)^{r_2} \left(x_3 + x_1^{-n} G_\tau(x_2, x_1^n)\right)^{r_3} d\tau.$$

By definition of $G_t(x_2, x_1^n)$, $\sigma^*(x_3) \equiv x_3$ modulo the ideal $\left(x_2 - tx_1^n(\xi - 1)^{-1}\right)$ and the residue class of $\sigma^*(x_4)$ modulo the ideal $\left(x_2 - tx_1^n(\xi - 1)^{-1}, x_3\right)$ is equal to

$$\begin{aligned}\overline{\sigma^*(x_4)} &= x_4 + t + \sum_{\underline{r}\in M} c_{\underline{r}} x_1^{\mu(r_1,r_2,r_3)} \int_0^t H_\tau(t(\xi-1)^{-1},1)^{r_2}\, G_\tau(t(\xi-1)^{-1},1)^{r_3} d\tau \\ &= x_4 + t + \sum_{\underline{r}\in M} c_{\underline{r}} (\xi-1)^{-\eta(r_2+1,r_3)} \Theta_{r_2,r_3}(\xi)\, x_1^{\mu(r_1,r_2,r_3)} t^{\eta(r_2+1,r_3)} \\ &= x_4 + t(1 + \sum_{\underline{r}\in M} \beta_{\underline{r}}\, x_1^{\mu(r_1,r_2,r_3)} t^{\eta(r_2,r_3)})\end{aligned}$$

where, for every $\underline{r} = (r_1, r_2, r_3) \in M$, $\beta_{\underline{r}} = c_{\underline{r}}(\xi-1)^{-\eta(r_2+1,r_3)}\Theta_{r_2,r_3}(\xi) \neq 0$ as $r_2 \not\equiv m \bmod (m+1)$. We view

$$Y(x_1,t) = \frac{\overline{\sigma^*(x_4)} - x_4}{t} = 1 + \sum_{\underline{r}\in M} \beta_{\underline{r}} x_1^{\mu(r_1,r_2,r_3)} t^{\eta(r_2,r_3)} \in k[x_1,t]$$

as a polynomial in t with coefficients in the algebraically closed field $\mathbb{F}$ of Puiseux series in the variable x_1.

Now suppose that either $p_0 \in k[x_1,x_2,x_3] \setminus k[x_1,x_2]$ or $(p_0 - 1) \in k[x_1,x_2] \setminus x_1k[x_1,x_2]$. In each case, it follows from Remark 1 that $Y(x_1,t)$ is nonconstant and our goal is to check that there exists a root $\Psi(x_1) \in \mathbb{F}$ of $Y(x_1,t)$ for which the factor $\big(t - \Psi(x_1)\big)$ of $Y(x_1,t)$ can be assigned a value large enough to ensure that the value of $tY(x_1,t)$ will be nonnegative.

Write $\epsilon(\underline{r}) = \frac{\mu(r_1,r_2,r_3)}{\eta(r_2,r_3)} = n + \frac{r_1 - nr_3}{\eta(r_2,r_3)}$, let $\epsilon_0 := \min_{\underline{r}\in M}\{\epsilon(\underline{r})\}_{\underline{r}}$ and let M_0 be the subset of M on which the minimum occurred. Note that $\varepsilon_0 = n$ if $p_0 \in k[x_1,x_2]$ and that otherwise $\varepsilon_0 < n$. Furthermore, every $\underline{r} \in M_0$ has $r_2 \neq 0$ in the first case and $r_3 \neq 0$ in the second. Because of Remark 1, there exists a unique $\underline{r}_0 \in M_0$ such that $\eta(r_2,r_3)$ is minimal. Denote this value of $\eta(r_2,r_3)$ by η_0 and by μ_0 the corresponding value of $\mu(r_1,r_2,r_3)$. Since ϵ_0 is minimal, all the points of the Newton polygon for the set $\{(0,0)\} \cup \{(\eta(r_2,r_3), \mu(r_1,r_2,r_3))\} \subset \mathbb{R}^2$ lie above or on the line L that passes through $(0,0)$ and (η_0,μ_0). The segment $[(0,0),(\eta_0,\mu_0)]$ is thus an edge of the polygon and hence, by virtue of the Newton-Puiseux Theorem (see e.g. [14, Theorem 3.1]), there exists a root $\Psi(x_1) \in \mathbb{F}$ of $Y(x_1,t)$ whose Puiseux expansion has the form $\Psi(x_1) = a_1 x_1^{\gamma_1}(1 + O(x_1^{\gamma_2}))$ where $a_1 \in k$, $\gamma_1 = -\mathrm{Slope}\,(L) = -\epsilon_0 < 0$ and $\gamma_2 > 0$. Substituting $\Psi(x_1)$ into the Eq. $Y(x_1,t) = 0$, we get

$$1 + \sum_{\underline{r}\in M} \beta_{\underline{r}} \Big(a_1 x_1^{\gamma_1}(1 + O(x_1^{\gamma_2}))\Big)^{\eta(r_2,r_3)} x_1^{\mu(r_1,r_2,r_3)} = 0.$$

The constant term $1 + \sum_{\underline{r}\in M_0} \beta_{\underline{r}} a_1^{\eta(r_2,r_3)}$ of the last equation must be zero, and since $\eta(r_2,r_3) \neq 0$ for every $\underline{r} \in M_0$, we conclude that $a_1 \neq 0$.

Now using the algebraic independence of $t - \Psi(x_1)$, $x_2 - tx_1^n(\xi-1)^{-1}$, x_3 and x_4 over $\mathbb{F}$, we can define a discrete valuation on $\mathbb{F}(x_2,x_3,x_4)(t)$ as follows: first we let $\upsilon(x_1) = \eta_0 > 0$. By construction, $\upsilon(\Psi(x_1)) = \gamma_1\eta_0 = -\mu_0 < 0$ and so choosing $\upsilon(t - \Psi(x_1))$ to be nonnegative will force $\upsilon(t) = -\mu_0 < 0$. Our choice of the root $\Psi(x_1)$ of $Y(x_1, t)$ forces $\upsilon(t - \tilde{\Psi}(x_1)) = -\mu_0$ for every other root $\tilde{\Psi}$ of $Y(x_1, t)$ in $\mathbb{F}$ and so, it is enough to choose $\upsilon(t - \Psi(x_1))$ large enough to obtain $\upsilon(tY(x_1, t)) \geq 0$. Since for the triple $\underline{r}_0 = (r_1, r_2, r_3)$ we have either $r_1 = r_3 = 0$ or $r_1 - nr_3 < 0$, it follows in turn that $\upsilon\left(tx_1^n\right) = \upsilon(t) + n\upsilon(x_1) = -\mu_0 + n\eta_0 \geq 0$, with strict inequality in the case where $p_0 \in k[x_1, x_2, x_3] \setminus k[x_1, x_2]$. So choosing $\upsilon(x_2 - tx_1^n(\xi - 1)^{-1})$ to be nonnegative forces $\upsilon(x_2) \geq 0$ whence $\upsilon(\sigma^*(x_2)) \geq 0$. Writing

$$\sigma^*(x_3) = x_3 + t(x_2 - tx_1^n(\xi - 1)^{-1})R(x_2, tx_1^n)$$

$$\sigma^*(x_4) = x_4 + tY(x_1, t) + (x_2 - tx_1^n(\xi - 1)^{-1})S(x_1, x_2, t) + \sum T_i(x_1, x_2, t)x_3^i$$

for suitable polynomials R, S, T_i with coefficients in k, we see that choosing $\upsilon(x_3) \geq 0$ and $\upsilon(x_2 - tx_1^n(\xi - 1)^{-1}))$ sufficiently large is enough to obtain the nonnegativity of $\upsilon(\sigma^*(x_3))$ and $\upsilon((x_2 - tx_1^n(\xi - 1)^{-1})S(x_1, x_2, t))$. Finally, choosing $\upsilon(x_4) \geq 0$ and $\upsilon(x_3)$ sufficiently large guarantees that $\upsilon(\sigma^*(x_4)) \geq 0$. The restriction of υ to $k(x_1, x_2, x_3, x_4)(t)$ is the required valuation. □

5 Applications

We conclude with our main result:

Theorem 1. *For each $q \in \mathbb{Z}_{\geq 0}$ the $\mathbb{G}_a$-action on $\mathbb{A}^4$ generated by a quasi-monomial derivation $\partial_q = x_1^{n+q}\partial_{x_2} + (m+1)x_1^q x_2^m \partial_{x_3} + p(\underline{x})\partial_{x_4}$ is either improper or a translation.*

Proof. We first consider the case where $q = 0$. We may assume that ∂_0 is normalized as in Lemma 2. If ∂_0 generates a proper $\mathbb{G}_a$-action, then by virtue of Proposition 2, $p = 1 + x_1 q(x_1, x_2)$ for some polynomial $q(x_1, x_2) \in k[x_1, x_2]$. So ∂_0 is an extension to $k[x_1, x_2, x_3, x_4]$ of the triangular $k[x_1]$-derivation $\overline{\partial}_0 = x_1^n\partial_{x_2} + (1 + x_1 q(x_1, x_2))\partial_{x_4}$ of $k[x_1, x_2, x_4]$. Since the latter generates a fixed point free action on $\mathbb{A}^3$, it is a translation by virtue of [13]. Any slice $s \in k[x_1, x_2, x_4]$ for $\overline{\partial}_0$ is then also a slice for ∂_0, and we conclude that σ_0 is globally equivariantly trivial, with geometric quotient isomorphic to $\mathrm{Spec}(k[x_1, x_2, x_4]/(s)[x_3]) \simeq \mathbb{A}^2 \times \mathbb{A}^1 \simeq \mathbb{A}^3$.

Now assume that $q > 0$ and let $\partial = x_1^n\partial_{x_2} + (m+1)x_2^m\partial_{x_3} + p(\underline{x})\partial_{x_4}$. It is enough to show that ∂_q and ∂ have isomorphic kernels and that the properness of the $\mathbb{G}_a$-action σ_q generated by ∂_q implies the properness of the $\mathbb{G}_a$-action σ generated by ∂. Indeed, if so, the previous case implies that $\mathrm{Ker}\partial_q \simeq \mathrm{Ker}\partial$ is a polynomial ring in three variables over k, in particular a regular ring. Thus σ_q is locally equivariantly trivial by [3], whence a translation by [4].

To check that ∂_q and ∂ have isomorphic kernels, we note that since x_1 belongs to the kernel of ∂, the derivation $x_1^q\partial$ is again locally nilpotent. Letting $\rho : k[x_1, x_2, x_3][x_4] \to k[x_1, x_2, x_3][x_4]$ be the $k[x_1, x_2, x_3]$-algebra endomorphism defined by $\rho(x_4) = x_1^q x_4$, we have $\rho \circ x_1^q \partial = \partial_q \circ \rho$ and by induction $\rho \circ x_1^{\ell q} \partial^{(\ell)} = \partial_q^{(\ell)} \circ \rho$ for every $\ell \in \mathbb{Z}_{>0}$. The kernel algorithm [16] applied to ∂ has as initial data the polynomial invariants $c_j = x_1^{n_j} \tilde{c}_j$ where $\tilde{c}_j = \exp(-\frac{x_2}{x_1^n}\partial)(x_j)$ for $j = 1, \ldots, 4$, and where n_j is the least power for which $x_1^{n_j}\tilde{c}_j \in k[x_1, x_2, x_3, x_4]$. Extending ρ to $k[x_1^{\pm 1}, x_2, x_3][x_4]$ we have

$$\rho_i(\tilde{c}_j) = \begin{cases} \exp(-\frac{x_2}{x_1^{n+q}}\partial_q)(x_j) & 1 \le j \le 3 \\ x_1^q \exp(-\frac{x_2}{x_1^{n+q}}\partial_q)(x_j) & j = 4. \end{cases}$$

It follows that the polynomial invariants c_j, $j = 1, \ldots, 4$, are also the initial data for the kernel algorithm for ∂_q and we conclude that $\ker\partial_q \simeq \ker\partial$.

Now suppose that σ is improper. Letting $f_i(\underline{x}, t) = \sigma^*(x_i) - x_i$ and $g_i(\underline{x}, t) = \sigma_q^*(x_i) - x_i$, $i = 1, \ldots, 4$, considered as elements of $k[x_1, x_2, x_3][t]$, it follows from the definition of ∂ and ∂_q that $f_i(\underline{x}, t) = g_i(\underline{x}, x_1^q t)$ for $i = 1, 2, 3$ while $f_4(\underline{x}, t) = x_1^{-q} g_4(\underline{x}, x_1^q t)$. The construction given in the proof of Proposition 2 implies the existence of valuations $\upsilon : k(x_1, x_2, x_3, x_4, t) \to \mathbb{Z} \cup \{\infty\}$ negative on t, nonnegative on all x_i and $\sigma^*(x_i)$ and with the property that $\upsilon(g_4(\underline{x}, t))$ is arbitrarily bigger than $\upsilon(x_1)$. In particular, we may choose such a valuation υ for which $\upsilon(g_4(\underline{x}, t)) \ge q\upsilon(x_1)$. Then it is straightforward to check using the above relations between the f_i and g_i that the valuation υ_q defined by $\upsilon_q(x_i) = \upsilon(x_i) \ge 0$, $i = 1, \ldots, 4$ and $\upsilon_q(t) = \upsilon(t) - q\upsilon(x_1) < 0$ is nonnegative on all $\sigma_q^*(x_i)$. So σ_q is improper, which completes the proof. □

Remark 2. The conclusion of Theorem 1 holds more generally for derivations $\partial = x_1^{n+q}\partial_{x_2} + x_1^q g(x_2)\partial_{x_3} + p(\underline{x})\partial_{x_4}$. A more elaborate version of Lemma 2 provides a normalized form for such ∂. The valuative criterion for properness applies again here, but the reduction of $\sigma^*(x_3)$ is carried out over $k[[tx_1^n]][x_2]$ rather than over $k[tx_1^n][x_2]$. For the details we refer to the thesis [9]. As a consequence, the conclusion of Theorem 1 also holds for instance for $\mathbb{G}_a$-actions generated by triangular derivations $\partial = x_1\partial_{x_2} + q(x_1, x_2)\partial_{x_3} + p(\underline{x})\partial_{x_4}$. Indeed, writing $q(x_1, x_2) = \sum_{i=1}^{l} a_i x_2^{m_i} + \sum_\ell q_\ell(x_2) x_1^{n_\ell}$ where $n_\ell \ge 1$ for all ℓ, the triangular change of coordinates $\tilde{x}_3 = x_3 - \sum_\ell x_1^{n_\ell - 1} \int_0^{x_2} q_\ell(u)du$ brings the derivation to the form covered by the aforementioned version.

Acknowledgements The authors wish to thank the anonymous referees for their very careful reading and valuable comments which helped to correct some inaccuracies in a previous version and contributed to improve the quality of the presentation.

Research supported in part by National Science Foundation Grant 0936691 and ANR Grant "BirPol" ANR-11-JS01-004-01.

References

1. D. Daigle, G. Freudenburg, Locally nilpotent derivations over a UFD and an application to rank two locally nilpotent derivations of $k[x_1, \ldots, x_n]$. J. Algebra **204**(2), 353–371 (1998)
2. J.K. Deveney, D.R. Finston, Twin triangular derivations. Osaka. J. Math. **39**(2), 275–282 (2002)
3. J. Deveney, D.R. Finston, M. Gehrke, $\mathbb{G}_a$-actions on $\mathbb{C}^n$. Commun. Algebra **22**, 4977–4988 (1994)
4. J.K. Deveney, D.R. Finston, P. van Rossum, Triangular $\mathbb{G}_a$-actions on $\mathbb{C}^4$. Proc. Am. Math. Soc. **132**(10), 2841–2848 (2004)
5. A. Dubouloz, D. Finston, Proper twin triangular $\mathbb{G}_a$-actions on $\mathbb{A}^4$ are translations. Proc. Am. Math. Soc. **142**, 1513–1526. arXiv:1109.6302. (2014)
6. A. Dubouloz, D. Finston, I. Jaradat, Proper triangular $\mathbb{G}_a$-actions on $\mathbb{A}^4$ are translations. arXiv:1303.1032 (submitted)
7. K-H. Fieseler, On complex affine surfaces with $\mathbb{C}^+$-action. Comment. Math. Helv. **69**(1), 5–27 (1994)
8. A. Grothendieck, Éléments de géométrie algébrique : IV. Étude locale des schémas et des morphismes de schémas, Troisième partie. Publ. Math. IHÉS. **28**, 5–255 (1966)
9. I. Jaradat, *Equivariant triviality of proper $\mathbb{G}_a$-actions on $\mathbb{A}^4_{\mathbb{C}}$*. Ph.D. thesis, New Mexico State University (2013)
10. G. Laumon, L. Moret-Bailly, *Champs algébriques*, Ergebnisse der Mathematik und ihrer Grenzgebiete (Springer, New York, 2000)
11. H. Popp, *Moduli Theory and Classification Theory of Algebraic Varieties*. Lecture Notes in Mathematics, vol. 620 (Springer, Berlin, 1977)
12. J-P. Serre, Géométrie algébrique et géométrie analytique. Ann. Inst. Fourier Grenoble **6**, 1–42 (1955–1956)
13. D. Snow, Triangular actions on $\mathbb{C}^3$. Manuscripta Math. **60**, 407–415 (1988)
14. R. Walker, *Algebraic Curves* (Princeton University Press, Princeton, 1950)
15. J. Winkelmann, On free holomorphic $\mathbb{C}$-actions on $\mathbb{C}^n$ and homogeneous Stein manifolds. Math. Ann. **286**(1–3), 593–612 (1990)
16. A. van den Essen, An algorithm to compute the invariant ring of a $\mathbb{G}_a$-action on an affine variety. J. Symbolic Comput. **16**(6), 551–555 (1993)

Automorphism Groups of Certain Rational Hypersurfaces in Complex Four-Space

Adrien Dubouloz, Lucy Moser-Jauslin, and Pierre-Marie Poloni

Abstract The Russell cubic is a smooth contractible affine complex threefold which is not isomorphic to affine three-space. In previous articles, we discussed the structure of the automorphism group of this variety. Here we review some consequences of this structure and generalize some results to other hypersurfaces which arise as deformations of Koras–Russell threefolds.

Subject Classification: 14L30; 13R20

1 Introduction

In order to prove the linearizability of algebraic actions of $\mathbb{C}^*$ on affine three-space, [9, 10], Koras and Russell studied hyperbolic $\mathbb{C}^*$-actions on more general smooth contractible threefolds. This led them to introduce a set of threefolds which are smooth affine and contractible, however not isomorphic to $\mathbf{A}^3$. These varieties are now known as Koras–Russell threefolds. One of the families of these varieties, called Koras–Russell threefolds of the first kind, is given by hypersurfaces $X_{d,k,\ell}$ in the affine space $\mathbb{A}^4 = \mathrm{Spec}(\mathbb{C}[x,y,z,t])$ defined by equations of the form $x^d y + z^k + t^\ell + x = 0$ where $d \geq 2$ and $2 \leq k < \ell$ with k and ℓ relatively

A. Dubouloz (✉) • L. Moser-Jauslin
Institut de Mathématiques de Bourgogne, UMR 5584 du CNRS, Université de Bourgogne, 9 av. A. Savary, BP 47870, 21078 Dijon, France
e-mail: Adrien.Dubouloz@u-bourgogne.fr; moser@u-bourgogne.fr

P.-M. Poloni
Mathematisches Institut, Universität Basel, Rheinsprung 21, 4051 Basel, Switzerland
e-mail: pierre-marie.poloni@unibas.ch

I. Cheltsov et al. (eds.), *Automorphisms in Birational and Affine Geometry*, Springer Proceedings in Mathematics & Statistics 79,
DOI 10.1007/978-3-319-05681-4_17,

prime. All of these threefolds admit algebraic actions of the complex additive group $\mathbb{G}_a$ and they were originally proven to be not isomorphic to affine space by means of invariants associated with those actions. These invariants, known as the Derksen and Makar-Limanov invariants, are defined respectively for an affine variety $X = \mathrm{Spec}(A)$ admitting nontrivial $\mathbb{G}_a$-actions as the sub-algebra $\mathrm{Dk}(X)$ of A consisting of regular functions invariant under *at least* one nontrivial $\mathbb{G}_a$-action on X and its sub-algebra $\mathrm{ML}(X)$ consisting of regular functions invariants under *all* nontrivial such actions.

These tools have since become important and useful to study affine algebraic varieties. In particular, one of the central elements in the proofs of many existing results concerning Koras–Russell threefolds of the first kind, and some of the generalizations we consider in this chapter is the fact that their Makar-Limanov and Derksen invariants are equal to $\mathbb{C}[x]$ and $\mathbb{C}[x,z,t]$, respectively (see, for example, [6], Lemma 8.3 and [7], Example 9.1. for the Koras–Russell threefolds). This property imposes very strong restrictions on the nature of isomorphisms between such varieties which enable sometimes an explicit description of their isomorphism classes and automorphism groups.

Koras–Russell threefolds of the first kind belong to the more general family of hypersurfaces $X = X(d, r_0, g)$ in $\mathbb{A}^4 = \mathrm{Spec}(\mathbb{C}[x, y, z, t])$ defined by equations of the form

$$x^d y + r_0(z,t) + xg(x,z,t) = 0$$

where $d \geq 2$, $r_0 \in \mathbb{C}[z,t]$, and $g \in \mathbb{C}[x,z,t]$. All varieties of this type share the property that they come equipped with a flat $\mathbb{A}^2$-fibration $\pi = \mathrm{pr}_x : X \to \mathbb{A}^1 = \mathrm{Spec}(\mathbb{C}[x])$ restricting to a trivial bundle over the complement of the origin and with degenerate fiber $\pi^{-1}(0)$ isomorphic to the cylinder $C_0 \times \mathbb{A}^1$ over the plane curve $C_0 \subset \mathrm{Spec}(\mathbb{C}[z,t])$ with equation $r_0 = 0$. In particular, noting that $\pi^{-1}(\mathbf{A}^1 \setminus \{0\})$ is factorial, and that $\pi^{-1}(0) = \mathrm{div}(x)$ is a prime principal divisor if and only if C_0 is reduced and irreducible, we see that a threefold X is factorial whenever the corresponding curve C_0 is reduced and irreducible (see also [12]). A combination of [5, 13] implies that X is isomorphic to $\mathbb{A}^3$ if and only if $\pi^{-1}(0)$ is reduced and isomorphic to $\mathbb{A}^2$, whence, by virtue of [1] if and only if C_0 is isomorphic to the affine line. Furthermore, identifying the coordinate ring A of X with the sub-algebra $\mathbb{C}[x,z,t,x^{-d}(r_0 + xg(x,z,t))]$ of $\mathbb{C}[x,z,t]_x$ via the canonical localization homomorphism with respect to x gives rise to a description of X as the affine modification $\sigma = \mathrm{pr}_{x,z,t}\,|_X \colon X \to \mathbb{A}^3$ of $\mathbb{A}^3 = \mathrm{Spec}(\mathbb{C}[x,z,t])$ with center at the closed subscheme Z with defining ideal $J = (x^d, r_0(z,t) + xg(x,z,t))$ and divisor $D = \{x^d = 0\}$ in the sense of [8]. That is, X is isomorphic to the complement of the proper transform of D in the blow-up of $\mathbb{A}^3$ with center at Z. Noting that the closed subscheme Z of $\mathbb{A}^3$ is supported along the curve $C_0 \subset \mathrm{Spec}(\mathbb{C}[z,t])$, this description implies that a smooth X for which C_0 is irreducible, topologically contractible, but not isomorphic to the affine line, is an exotic $\mathbb{A}^3$ [8, Theorem 3.1]. This holds for instance for smooth deformations of Koras–Russell threefolds of the first kind defined by equations of the form $x^d y + z^k + t^\ell + xg(x, y, z) = 0$, with $k, \ell \geq 2$ relatively prime and $g(0, 0, 0) = 1$, corresponding to the irreducible, singular, topologically contractible plane curves $C_0 = \{z^k + t^\ell = 0\}$.

The present chapter reviews three complementary applications of Derksen and Makar-Limanov invariants to the study of threefolds $X(d, r_0, g)$ as above. First we summarize several properties of automorphism groups of Koras–Russell threefolds of the first kind which appeared separately in previous articles by the authors, and we complete the picture with a characterization of certain natural subgroups of these automorphism groups. Then we turn to the study of non-necessarily smooth deformations of Koras–Russell threefolds defined by equations of the form $x^d y + z^k + t^\ell + xg(x, y, z) = 0$. We explain how to obtain a description of isomorphism classes of these threefolds that is reminiscent of the (mini)-versal deformation of the corresponding singular plane curve $C_0 = \{z^k + t^\ell = 0\}$. Finally, we illustrate on an example of a threefold $X = \{x^d y + r_0(z, t) = 0\}$ with non-connected associated plane curve $C_0 = \{r_0 = 0\}$ a general procedure to construct new types of automorphisms of X which do not admit any extension to automorphisms of the ambient space $\mathbb{A}^4$.

2 A Preliminary Observation

Let $d \in \mathbf{N}$ and $r_0 \in \mathbb{C}[z, t]$ be fixed. For any $g \in \mathbb{C}[x, z, t]$, we denote by

$$J_g = (x^d, r_0 + xg)$$

the ideal of $\mathbb{C}[x, z, t]$ generated by x^d and $r_0 + xg$, and by $A(g)$ the coordinate ring of the hypersurface $X(g)$ of $\mathbf{A}^4 = \mathrm{Spec}(\mathbb{C}[x, y, z, t])$ defined by the equation

$$x^d y + r_0(z, t) + xg(x, z, t) = 0.$$

Corresponding to the presentation of $X(g)$ as the affine modification $\sigma : X(g) \to \mathbb{A}^3$ mentioned in the introduction, we have a chain of inclusions

$$\mathbb{C}[x, z, t] \subset A(g) \subset A(g)[x^{-1}] \simeq \mathbb{C}[x, x^{-1}, z, t].$$

The second inclusion is induced by the localization homomorphism with respect to the regular element $x \in A(g)$, identifying $y \in A(g)$ with $-x^{-d}(r_0 + xg) \in \mathbb{C}[x, x^{-1}, z, t]$.

Given a pair of polynomials $f, g \in \mathbb{C}[x, z, t]$, the universal property of affine modifications [8, Proposition 2.1] implies that every automorphism φ of $\mathbb{C}[x, z, t]$ which fixes the ideal (x) and maps J_g into J_f lifts in a unique way to a morphism $\tilde{\varphi} : A(g) \to A(f)$ restricting to φ on the subring $\mathbb{C}[x, z, t]$. Actually, $\tilde{\varphi}$ is even an isomorphism. Indeed, by hypothesis, there exist $\alpha \in \mathbb{C}^*$ and $a, b \in \mathbb{C}[x, z, t]$

such that $\varphi(x) = \alpha x$ and $\varphi(r_0 + xg) = ax^d + b(r_0 + xf)$. The second equation implies that b is congruent modulo x to a nonzero constant whence that its residue class in $\mathbb{C}[x,z,t]/(x^d)$ is a unit for every $d \geq 1$. Choosing $b' \in \mathbb{C}[x,z,t]$ such that $bb' \equiv 1 \mod (x^d)$ and multiplying the previous equation by it, we conclude that $r_0 + xf \in \varphi(J_g)$. Thus φ maps J_g isomorphically onto J_f.

The following lemma will be used several times throughout this chapter.

Lemma 1. *With the notation above assume further that the Derksen and Makar-Limanov invariants of $X(f)$ and $X(g)$ are equal to $\mathbb{C}[x,z,t]$ and $\mathbb{C}[x]$, respectively. Then the previous construction provides a one-to-one correspondence between isomorphisms from $A(g)$ to $A(f)$ and automorphisms of $\mathbb{C}[x,z,t]$ which fix the ideal (x) and map J_g into J_f.*

Proof. Note first that the hypotheses imply that neither $X(f)$ nor $X(g)$ is isomorphic to $\mathbb{A}^3$ and hence, as a consequence of [13], that the surface $C_0 \times \mathbb{A}^1 = \{r_0 = 0\} \subset \mathrm{Spec}(\mathbb{C}[x,z,t])$ is not isomorphic to $\mathbb{A}^2$. Since an isomorphism between $A(g)$ and $A(f)$ preserves the Makar-Limanov and the Derksen invariants, it restricts to an automorphism φ of $\mathbb{C}[x,z,t]$ and an automorphism of $\mathbb{C}[x]$. That is, $\varphi(x)$ is of the form $ax + b$ where $a \in \mathbb{C}^*$ and $b \in \mathbb{C}$. Actually, $b = 0$ since by the previous remark the zero set of $ax+b$ in $X(f)$ and $X(g)$ is non-isomorphic to $\mathbb{A}^2$ if and only if $b = 0$. This shows that φ fixes the ideal (x). Noting that $J_g = x^d A(g) \cap \mathbb{C}[x,z,t]$ and similarly for J_f, we conclude that $\varphi(J_g) = J_f$, which completes the proof. □

3 Automorphisms of Koras–Russell Threefolds of the First Kind

In this section, we consider Koras–Russell threefolds of the first kind $X = X(d,k,\ell)$ corresponding to the cases where $d \geq 2$, $r_0 = z^k + t^\ell$, and $g = 1$. Since $\mathrm{Dk}(X) = \mathbb{C}[x,z,t]$ and $\mathrm{ML}(X) = \mathbb{C}[x]$, we deduce from Lemma 1 that the projection $\sigma = \mathrm{pr}_{x,z,t}\,|_X\colon X \to \mathbb{A}^3$ gives rise to an isomorphism between the automorphism group of X and the subgroup $\mathcal{A}$ of automorphisms of $\mathbb{C}[x,z,t]$ which preserve the ideals (x) and $(x^d, r_0 + x)$. In particular, $\mathbb{C}^*$ acts linearly on X. In fact, letting $\mathcal{A}_n$, $1 \leq n \leq d$ be the normal subgroup of $\mathcal{A}$ consisting of the automorphisms φ which fix x and which are congruent to the identity modulo (x^n), it was shown more precisely in [2, 11] that

$$\mathrm{Aut}(X) \simeq \mathcal{A}_1 \rtimes \mathbb{C}^* \quad \text{and} \quad \mathcal{A}_n/\mathcal{A}_{n+1} \cong (\mathbb{C}[z,t],+) \quad \text{for all } 1 \leq n \leq d-1.$$

The next proposition summarizes several consequences of this description:

Proposition 1. *Let $X = X_{d,k,\ell} \subset \mathbf{A}^4$ be a Koras–Russell threefold of the first kind. Then the following hold:*

1) Every automorphism of X extends to an automorphism of $\mathbf{A}^4$.

2) *The group* Aut(X) *acts on* X *with exactly four orbits:*
   - *an open orbit* $\{x \neq 0\} \simeq \mathbb{C}^* \times \mathbb{C}^2$,
   - *a copy of* $\mathbb{C}^* \times \mathbb{C}$ *given by* $x = 0$ *and* $z \neq 0$,
   - *the line* $\{x = z = t = 0\}$ *minus the point* $(0, 0, 0, 0)$,
   - *a fixed point* $(0, 0, 0, 0)$.
3) *Every finite subgroup of* Aut(X) *is cyclic.*
4) *Every one-parameter unipotent subgroup of* Aut(X) *is contained in* $\mathcal{A}_d$. *In particular, the subgroup generated by all* $\mathbf{G}_a$*-actions on* X *is strictly smaller than* $\mathcal{A}_1$.

Remark 1. In contrast with Property 2) above, the group of holomorphic automorphisms of X acts with at most three orbits. Indeed, one checks for instance that the holomorphic automorphism Ψ of $\mathbf{A}^4$ defined by

$$\Psi(x, y, z, t) = \left(x, e^{x^{d-1}} y - \frac{1 - e^{x^{d-1}}}{x^{d-1}}, e^{\frac{x^{d-1}}{k}} z, e^{\frac{x^{d-1}}{\ell}} t \right)$$

maps X onto itself. Hence Ψ induces a holomorphic automorphism ψ of X for which $\psi(0, 0, 0, 0) = (0, 1, 0, 0)$, i.e., $(0, 0, 0, 0)$ is no longer a fixed point for the action of holomorphic automorphisms of X. The exact number of orbits under the action this group is not known. In particular, it is still an open question whether any threefold $X_{d,k,\ell}$ is biholomorphic to the affine space.

Proof. Properties 1) and 2) were established in [2] for the Russell cubic and in [11] for the general case.

The third property was originally formulated as a question by V. Popov. Since $\mathrm{Aut}(X) \simeq \mathcal{A}_1 \rtimes \mathbb{C}^*$, it is enough to show that $\mathcal{A}_1$ does not contain nontrivial torsion elements. Indeed, if so, the projection to the second factor $\mathbb{C}^*$ will induce an isomorphism between every finite subgroup of Aut(X) and a subgroup of $\mathbb{C}^*$. So suppose that $\varphi \in \mathcal{A}_1$ is a nontrivial torsion element, say of order $m \geq 2$. By possibly switching z and t, we can further assume that $\varphi(z) \neq z$. Choosing $N \in \mathbf{N}$ minimal with the property that $\varphi(z) \equiv z + f(z, t)x^N \mod (x^{N+1})$ for some $f \in \mathbb{C}[z, t] \setminus \{0\}$, we would have $\varphi^m(z) \equiv z + mf(z, t)x^N \not\equiv z \mod (x^{N+1})$, a contradiction.

For the last property, it follows from Lemma 1 that every additive group action on X is induced by a locally nilpotent derivation D of the coordinate ring of X extending a locally nilpotent $\mathbb{C}[x]$-derivation of $\mathbb{C}[x, z, t]$ which maps the ideal $J = (x^d, z^k + t^\ell + x)$ into itself, the second condition being equivalent to the property that the corresponding exponential automorphisms preserve this ideal. We will show that in fact the image of D is contained in the ideal (x^d), which implies that the corresponding one-parameter unipotent subgroup of Aut(X) is contained in $\mathcal{A}_d$. We prove this by induction, assuming that $D \equiv 0$ modulo (x^k) with $0 \leq k < d$. Since $D(z^k + t^\ell + x) = ax^d + b(z^k + t^\ell + x)$ where $a, b \in \mathbb{C}[x, z, t]$, the hypothesis implies that x^k divides b. On the other hand, $D_1 = x^{-k} D$ is again a locally nilpotent derivation such that $D_1(x) = 0$ and for which we have $D_1(z^k +$

$t^\ell + x) = ax^{d-k} + (b/x^k)(z^k + t^\ell + x)$. Thus D_1 induces a locally nilpotent derivation $\overline{D_1}$ of $\mathbb{C}[x,z,t]/(x) \cong \mathbb{C}[z,t]$ which maps the ideal generated by $z^k + t^\ell$ into itself. So $\overline{D_1}$ is necessarily trivial as there is no nontrivial $\mathbf{G}_a$-action preserving a singular plane curve. Thus $D_1 \equiv 0$ modulo (x) and hence $D \equiv 0 \mod (x^{k+1})$. □

Remark 2. Recall that we always have an exact sequence of groups

$$0 \to \mathrm{Aut}_0(\mathbf{A}^4, X) \to \mathrm{Aut}(\mathbf{A}^4, X) \xrightarrow{\rho} \mathrm{Aut}(X)$$

where $\mathrm{Aut}(\mathbf{A}^4, X)$ denotes the subgroup of $\mathrm{Aut}(\mathbf{A}^4)$ consisting of automorphisms which leave X invariant and where $\mathrm{Aut}_0(\mathbf{A}^4, X)$ denotes the kernel of ρ. The surjectivity of ρ was established in Property 1) of the above proposition by constructing explicit lifts of automorphisms of X to automorphisms of $\mathbf{A}^4$. Nevertheless, this construction was only set-theoretic and it is not clear whether the above sequence splits. Note however that since an element $\varphi \in \mathcal{A}_d$ is the identity modulo (x^d) and preserves the subring $\mathbb{C}[x,z,t]$ of the coordinate ring of X, it lifts in a natural way to an automorphism Φ of $\mathbb{C}[x,y,z,t]$ by letting simply $\Phi(y) = y + (\varphi(r_0 + x) - r_0 - x)/x^d$. This gives rise to group homomorphism $\mathcal{A}_d \to \mathrm{Aut}(\mathbf{A}^4, X)$, $\varphi \mapsto \Phi$, which, combined with the fact that the action of $\mathbb{C}^*$ on X comes as the restriction of a linear action on $\mathbf{A}^4$, extends to a group homomorphism $j : \mathcal{A}_d \rtimes \mathbb{C}^* \to \mathrm{Aut}(\mathbf{A}^4, X)$ such that $\rho \circ j = \mathrm{id}$. We do not know whether j can be extended to a splitting of the above exact sequence.

4 Deformations of Koras–Russell Threefolds

In this section, we consider hypersurfaces $X(g)$ of $\mathbf{A}^4 = \mathrm{Spec}(\mathbb{C}[x,y,z,t])$ defined by equations of the form

$$x^d y + r_0(z,t) + xg(x,z,t) = 0$$

where $r_0 = z^k + t^\ell$ and $d \geq 2$ are fixed, and we let the polynomial $g \in \mathbb{C}[x,z,t]$ vary. The case $r_0 = z^2 + t^3$ was treated in [3], leading to the construction of large families of non-isomorphic smooth affine threefolds that are all biholomorphic to each other and diffeomorphic to the affine space. Here, we show that similar techniques can be applied to find isomorphisms between deformations of hypersurfaces in a more general setting.

Theorem 1 below relies again in a crucial way on the fact that the Derksen and Makar-Limanov invariant of threefolds $X(g)$ are equal to $\mathbb{C}[x,z,t]$ and $\mathbb{C}[x]$, respectively. These properties can be checked using the methods developed in [7], namely via a careful study of homogeneous locally nilpotent derivations on a well-chosen quasi-homogeneous deformation of $X(g)$. The complete proof is quite long and technical but is essentially straightforward using the aforementioned methods. We shall omit it here, in particular since it involves no new ideas or arguments.

Notation 1 *We will be considering derivations of* $\mathbb{C}[z,t]$ *defined by the Jacobian of a polynomial. If* $f \in \mathbb{C}[z,t]$*, we denote by* f_z *the partial derivative* $\partial f/\partial z$*, and by* f_t *the partial derivative* $\partial f/\partial t$*. For* $f, g \in \mathbb{C}[z,t]$*, the Jacobian* $\mathrm{Jac}(f,g)$ *denotes* $f_z g_t - f_t g_z$*. Finally,* $\mathrm{Jac}(f,\cdot)$ *denotes the derivation of* $\mathbb{C}[z,t]$ *defined by* $g \mapsto \mathrm{Jac}(f,g)$*.*

Let $B = \mathbb{C}[a_{i,j}]$ be the polynomial ring in the $(k-1)(\ell-1)$ indeterminates $a_{i,j}$, $0 \leq i \leq k-2, 0 \leq j \leq \ell-2$, let $\mathfrak{m}_0 \subset B$ be the maximal ideal generated by the $a_{i,j}$ and let

$$F = r_0 + \sum_{0 \leq i \leq k-2,\, 0 \leq j \leq \ell-2} a_{i,j} z^i t^j \in B[z,t].$$

Given a homomorphism $\alpha : B \to \mathbb{C}[x]$ such that $\alpha(\mathfrak{m}_0) \subset x\mathbb{C}[x]$, the image of F in $\mathbb{C}[x] \otimes_B B[z,t] \simeq \mathbb{C}[x,z,t]$ has the form $r_0 + xg_\alpha$ for some polynomial g_α belonging to the $\mathbb{C}[x]$-submodule of $\mathbb{C}[x,z,t]$ generated by the monomials $z^i t^j$ with $0 \leq i \leq k-2$ and $0 \leq j \leq \ell-2$. Every such homomorphism α thus determines a threefold $\mathfrak{X}_\alpha = X(g_\alpha)$ defined by the equation $x^d y + r_0 + xg_\alpha = 0$.

Theorem 1. *With the notation above, the following statements hold:*

1) *For every* $g \in \mathbb{C}[x,z,t]$*, there exists a homomorphism* $\alpha : B \to \mathbb{C}[x]$ *such that* $X(g)$ *is isomorphic to* $\mathfrak{X}_\alpha$*.*
2) *Two homomorphisms* $\alpha, \beta : B \to \mathbb{C}[x]$ *determine isomorphic threefolds* $\mathfrak{X}_\alpha$ *and* $\mathfrak{X}_\beta$ *if and only if there exist constants* $\lambda, \mu \in \mathbb{C}^*$ *such that the following diagram commutes*

$$\begin{array}{ccc} B & \xrightarrow{\pi_d \circ \alpha} & \mathbb{C}[x]/(x^d) \\ {\scriptstyle \zeta_\lambda}\downarrow & & \downarrow{\scriptstyle x \mapsto \mu x} \\ B & \xrightarrow{\pi_d \circ \beta} & \mathbb{C}[x]/(x^d) \end{array}$$

where $\pi_d : \mathbb{C}[x] \to \mathbb{C}[x]/(x^d)$ *denotes the natural projection and where* $\zeta_\lambda : B \to B$ *is the linear automorphism defined by* $a_{i,j} \mapsto \lambda^{\ell i + kj - k\ell} a_{i,j}$*.*

Remark 3. Noting that the subvariety $\mathcal{V}$ of $T \times \mathbf{A}^2 = \mathrm{Spec}(B[z,t])$ with equation $F = 0$ is the (mini)-versal deformation of the curve $C_0 \subset \mathbf{A}^2$ with equation $r_0 = 0$ (see e.g. §14.1 in [4]), we can reinterpret the above result as the fact that for fixed $d \geq 2$, isomorphism classes of hypersurfaces of the form $X(g)$ are in one-to-one correspondence with one-parameter infinitesimal embedded deformations of order $d-1$ of C_0, up to the equivalence between such deformations defined in the second assertion of the theorem.

Proof. 1) In view of Lemma 1, to prove the first assertion, it is enough to show that, given a polynomial $g \in \mathbb{C}[x,z,t]$, there exist elements b_m, $m = 1,\dots,d-1$ in the sub-vector space of $\mathbb{C}[z,t]$ generated by the monomials $z^i t^j$, $0 \leq i \leq k-2$, $0 \leq j \leq \ell-2$ and a $\mathbb{C}[x]$-automorphism φ of $\mathbb{C}[x][z,t]$ which maps the ideal $(x^d, r_0 + xg)$ into the ideal $(x^d, r_0 + \sum_{m=1}^{d-1} b_m x^m)$. We may write

$$g = \sum_{m\geq 1} u_m x^{m-1} = \sum_{m\geq 1} (s_m + t_m)x^{m-1}$$

where for every m, s_m belongs to the sub-vector space of $\mathbb{C}[z,t]$ generated by the monomials $z^i t^j$, $0 \leq i \leq k-2$, $0 \leq j \leq \ell-2$, while t_m is in the ideal $(z^{k-1}, t^{\ell-1})\mathbb{C}[z,t]$. Let $t_0 = 0$ and denote by $\nu = \nu(g)$ the maximal integer with the property that $t_m = 0$ for every $m \leq \nu - 1$. If $\nu = d$ then we are done. Otherwise, we will proceed by induction.

Note that the image of the $\mathbb{C}$-derivation of $\mathbb{C}[z,t]/(r_0)$ induced by the Jacobian derivation $\mathrm{Jac}(r_0,\cdot)$ of $\mathbb{C}[z,t]$ is equal to the ideal of $\mathbb{C}[z,t]/(r_0)$ generated by the residue classes of z^{k-1} and $t^{\ell-1}$. Indeed, one checks for example that, given $a \geq k-1$ and $b \geq 0$, we have

$$\mathrm{Jac}(r_0, \lambda z^{a-k+1} t^{b+1}) = z^a t^b - \lambda\ell(a-k+1)z^{a-k} t^b r_0 \equiv z^a t^b \mod (r_0),$$

where $\lambda = (k(b+1) + \ell(a-k+1))^{-1}$. This fact guarantees the existence of polynomials $h, f \in \mathbb{C}[z,t]$ such that $\mathrm{Jac}(h, r_0) = t_\nu + r_0 f$.

Let us consider the $\mathbb{C}[x]/(x^{\nu+1})$-automorphism $\overline{\varphi} = \exp(-x^\nu \mathrm{Jac}(h,\cdot))$ of $\mathbb{C}[x]/(x^{\nu+1})[z,t]$. Since its Jacobian is a nonzero constant, it lifts, by the main theorem in [14], to a $\mathbb{C}[x]$-automorphism φ of $\mathbb{C}[x][z,t]$ such that $\varphi(x) = x$ and $\varphi(a) \equiv a - x^\nu \mathrm{Jac}(h,a) \mod (x^{\nu+1})$ for all $a \in \mathbb{C}[x,z,t]$. Then, we have the following congruences modulo $(x^{\nu+1})$.

$$\begin{aligned}
\varphi(r_0 + xg) &\equiv r_0 + \sum_{m=1}^{\nu-1} s_m x^m + x^\nu(s_\nu + t_\nu) - x^\nu \mathrm{Jac}(h, r_0) && \mod (x^{\nu+1})\\
&\equiv r_0 + \sum_{m=1}^{\nu} s_m x^m - x^\nu r_0 f && \mod (x^{\nu+1})\\
&\equiv (1 - x^\nu f)(r_0 + \sum_{m=1}^{\nu} s_m x^m) && \mod (x^{\nu+1}).
\end{aligned}$$

It follows that there exists a polynomial $R \in \mathbb{C}[x,z,t]$ such that $\varphi(r_0 + xg) \equiv (1-x^\nu f)(r_0 + \sum_{m=1}^{\nu} s_m x^m + x^{\nu+1} R) \mod (x^d)$. Letting $\tilde{g} = \sum_{m=1}^{\nu} s_m x^{m-1} + x^\nu R$, we obtain that φ maps the ideal $(x^d, r_0 + xg)$ into the ideal $(x^d, r_0 + x\tilde{g})$, and since $\nu(\tilde{g}) > \nu(g)$ by construction, we are done by induction.

2) Let us first rephrase the second assertion. If we let $\alpha(a_{i,j}) = x\alpha_{i,j}(x)$ and $\beta(a_{i,j}) = x\beta_{i,j}(x)$ with $\alpha_{i,j}, \beta_{i,j} \in \mathbb{C}[x]$ for $0 \leq i \leq k-2, 0 \leq j \leq \ell-2$, the latter states that the threefolds $\mathfrak{X}_\alpha$ and $\mathfrak{X}_\beta$ are isomorphic if and only if there exist two constants $\lambda, \mu \in \mathbb{C}^*$ such that

$$\mu x\alpha_{i,j}(\mu x) \equiv \lambda^{\ell i+kj-k\ell} x\beta_{i,j}(x) \mod (x^d)$$

for all $0 \leq i \leq k-2$ and all $0 \leq j \leq \ell-2$. If such constants $\lambda, \mu \in \mathbb{C}^*$ exist then the automorphism φ of $\mathbb{C}[x,z,t]$ defined by $\varphi(x) = \mu x$, $\varphi(z) = \lambda^{-\ell} z$, $\varphi(t) = \lambda^{-k} t$ satisfies that $\varphi(r_0 + xg_\alpha) \equiv \lambda^{-k\ell}(r_0 + xg_\beta)$ modulo (x^d). Thus, φ maps the ideal $(x^d, r_0 + xg_\alpha)$ into the ideal $(x^d, r_0 + xg_\beta)$, and $\mathfrak{X}_\alpha$ and $\mathfrak{X}_\beta$ are isomorphic by Lemma 1.

Conversely, suppose that $\mathfrak{X}_\alpha$ and $\mathfrak{X}_\beta$ are isomorphic and let φ be an automorphism of $\mathbb{C}[x,z,t]$ which fixes the ideal (x) and maps the ideal (x^d, r_0+xg_α) into the ideal $(x^d, r_0 + xg_\beta)$. Up to changing $\mathfrak{X}_\beta$ by its image under an isomorphism coming from an automorphism of $\mathbb{C}^3$ of the type $(x,z,t) \mapsto (\mu x, \lambda^{-\ell} z, \lambda^{-k} t)$ as above, we may further assume that φ is a $\mathbb{C}[x]$-automorphism of $\mathbb{C}[x][z,t]$ which is the identity modulo (x) and we are thus reduced to the following statement. "Suppose that there exists a $\mathbb{C}[x]$-automorphism φ of $\mathbb{C}[x][z,t]$ which is congruent to the identity modulo (x) and such that $\varphi(r_0 + xg_\alpha) \in (x^d, r_0 + xg_\beta)$. Then g_α and g_β are congruent modulo (x^{d-1})."

Choose ν maximal such that φ is congruent to the identity modulo (x^ν). If $\nu \geq d$, then we are done. So, suppose that $\nu \leq d-1$. Writing down that the Jacobian of φ is constant equal to 1, we remark that there exists a polynomial $h \in \mathbb{C}[z,t]$ such that $\varphi(z)$ and $\varphi(t)$ are congruent modulo $(x^{\nu+1})$ to $z+x^\nu h_t$ and to $t-x^\nu h_z$, respectively. Consequently, we have $\varphi(r_0+xg_\alpha) \equiv r_0+xg_\alpha+x^\nu \mathrm{Jac}(r_0,h)$ modulo $(x^{\nu+1})$. On the other hand, since $\varphi(r_0 + xg_\alpha) \in (x^d, r_0 + xg_\beta)$ and since, by definition, g_α and g_β contain no monomials of the form $cx^{i_1}z^{i_2}t^{i_3}$ with $i_2 \geq k-1$ or $i_3 \geq \ell-1$, there exists a polynomial $a \in \mathbb{C}[z,t]$ such that $\varphi(r_0 + xg_\alpha) \equiv r_0 + xg_\beta + x^\nu a r_0$ modulo $(x^{\nu+1})$.

Writing $xg_\alpha = \sum_{m\geq 1} s_{\alpha,m}x^m$ and $xg_\beta = \sum_{m\geq 1} s_{\beta,m}x^m$ as before, we have thus

$$r_0 + \sum_{m=1}^{\nu-1} s_{\alpha,m}x^m + x^\nu(s_{\alpha,\nu} + \mathrm{Jac}(r_0,h)) \equiv r_0 + \sum_{m=1}^{\nu-1} s_{\beta,m}x^m + x^\nu(s_{\beta,\nu} + ar_0)$$

modulo $(x^{\nu+1})$. Since $\mathrm{Jac}(r_0,h)$ and r_0 both belong to the ideal $(z^{k-1}, t^{\ell-1})$ of $\mathbb{C}[z,t]$, we conclude that $s_{\alpha,m} = s_{\beta,m}$ for every $m = 1,\ldots,\nu$ and that $\mathrm{Jac}(r_0,h) \in r_0\mathbb{C}[z,t]$. This implies in turn that $h = \gamma r_0 + c$ for some $\gamma \in \mathbb{C}[z,t]$ and $c \in \mathbb{C}$. Now consider the $\mathbb{C}[x]/(x^d)$-automorphism $\overline{\theta} = \exp(\delta)$ of $\mathbb{C}[x]/(x^d)[z,t]$ associated with the derivation $\delta = x^\nu \mathrm{Jac}(\cdot, \gamma(r_0 + xg_\alpha))$. Since $\overline{\theta}$ has Jacobian determinant equal to 1 (see [11]), we deduce again from [14] that it lifts to a $\mathbb{C}[x]$-automorphism θ of $\mathbb{C}[x][z,t]$. By construction $\theta \equiv \varphi$ modulo $x^{\nu+1}$ and since $r_0 + xg_\alpha$ divides $\delta(r_0 + xg_\alpha) = x^\nu(r_0 + xg_\alpha)\mathrm{Jac}(r_0 + xg_\alpha, \gamma)$, it maps the ideal $(x^d, r_0 + xg_\alpha)$ into itself. Thus $\varphi_1 = \varphi \circ \theta^{-1}$ is a $\mathbb{C}[x]$-automorphism of $\mathbb{C}[x][z,t]$ congruent to identity modulo $x^{\nu+1}$ which maps $(x^d, r_0 + xg_\alpha)$ into $(x^d, r_0 + xg_\beta)$ and we can conclude the proof by induction. □

5 An Example of a Non-extendible Automorphism

Most of the algebraic results of the previous two sections can be generalized to the situation where $r_0 \in \mathbb{C}[z,t]$ defines a connected and reduced plane curve, provided that the Makar-Limanov and Derksen invariants of the corresponding threefolds are equal to $\mathbb{C}[x]$ and $\mathbb{C}[x,z,t]$, respectively. Here we illustrate a new phenomenon in a particular case where the zero set of r_0 is not connected.

For this section, we fix $d = 2$, $r_0 = z(zt^2 + 1)$ and we let X be the smooth hypersurface in $\mathbf{A}^4 = \mathrm{Spec}(\mathbb{C}[x,y,z,t])$ defined by the equation

$$P = x^2y + z(zt^2 + 1) = 0.$$

Note that X is neither factorial nor topologically contractible. It is straightforward to check using the methods in [7] that $\mathrm{Dk}(X) = \mathbb{C}[x,z,t]$ and $\mathrm{ML}(X) = \mathbb{C}[x]$. We will use the fact that the plane curve C_0 with equation $r_0 = z(zt^2 + 1) = 0$ has two connected components to construct a particular automorphism $\tilde{\varphi}$ of X which cannot extend to the ambient space $\mathbf{A}^4$.

Starting from the polynomial $h = zt^2 \in \mathbb{C}[z,t]$, we first construct in a similar way as in [11] an automorphism φ of $\mathbb{C}[x,z,t]$ as follows: Noting that the $\mathbb{C}[x]/(x^2)$-automorphism $\overline{\varphi} = \exp(x\mathrm{Jac}(h,\cdot))$ of $\mathbb{C}[x]/(x^2)[z,t]$ has Jacobian equal to 1, we deduce again from [14], that it lifts to a $\mathbb{C}[x]$-automorphism φ of $\mathbb{C}[x,z,t]$. In other words, there exists an automorphism φ of $\mathbb{C}[x,z,t]$ with $\varphi(x) = x$ and such that $\varphi \equiv \overline{\varphi} \mod (x^2)$. One checks further using the explicit formulas that $\overline{\varphi}(z) = (1 - 2tx)z$, $\overline{\varphi}(t) = (1 + tx)t$ and that $\overline{\varphi}(r_0) = (1 - 2xt)r_0$. So φ preserves the ideals (x) and $J = (x^2, z(zt^2 + 1))$ and hence lifts to a $\mathbb{C}[x]$-automorphism $\tilde{\varphi}$ of the coordinate ring $\mathbb{C}[X]$ of X.

Proposition 2. *The automorphism of $X \subset \mathbf{A}^4$ determined by $\tilde{\varphi}$ cannot extend to an automorphism of the ambient space.*

Proof. Suppose by contradiction that there exists an automorphism Φ of $\mathbb{C}[x,y,z,t]$ which extends $\tilde{\varphi}$. Then there exists a constant $\lambda \in \mathbb{C}^*$ such that $\Phi(P) = \lambda P$ and so we obtain a commutative diagram

$$\begin{array}{ccc} \mathbb{C}[x,y,z,t] & \xrightarrow{\Phi} & \mathbb{C}[x,y,z,t] \\ \uparrow & & \uparrow \\ \mathbb{C}[u] & \xrightarrow{u \mapsto \lambda u} & \mathbb{C}[u] \end{array}$$

where $\mathbb{C}[u] \to \mathbb{C}[x, y, z, t]$ maps u onto $P = x^2y + z(zt^2 + 1) \in \mathbb{C}[x, y, z, t]$. Passing to the field of fractions of $\mathbb{C}[u]$, we see that Φ induces an isomorphism between the $\mathbb{C}(u)$-algebras

$$A = \mathbb{C}(u)[x, y, z, t]/(P - u) \quad \text{and} \quad A_\lambda = \mathbb{C}(u)[x, y, z, t]/(P - \lambda^{-1}u).$$

Now the key observation is that the Makar-Limanov and Derksen invariants of A and A_λ are equal to $\mathbb{C}(u)[x]$ and $\mathbb{C}(u)[x, z, t]$, respectively (this follows from an application of Theorem 9.1 of [7] which in fact only requires that the base field has characteristic 0, we do not give the details here). Now since Φ induces an isomorphism between the Makar-Limanov invariants of A and A_λ, it follows that $\Phi(x) = \mu x + \nu$ for some $\mu \in \mathbb{C}(u)^*$ and $\nu \in \mathbb{C}(u)$. From the fact that the ideal (x, P) of $\mathbb{C}[x, y, z, t]$ is not prime, whereas the ideals $(x + c, P)$ are prime for all $c \in \mathbb{C}^*$, we can conclude that $\nu = 0$ and since $\tilde{\varphi}(x) = x$, we eventually deduce that $\mu = 1$. Thus $\Phi(x) = x$. Next, noting that Φ induces an automorphism $\overline{\Phi}$ of $\mathbb{C}[x, y, z, t]/(x) \cong \mathbb{C}[y, z, t]$ with $\overline{\Phi}(z(zt^2+1)) = \lambda z(zt^2+1)$, we deduce that there exists $\alpha \in \mathbb{C}^*$ such that $\overline{\Phi}(z) = \alpha z$. By comparing with $\tilde{\varphi}(z)$, we find that $\alpha = 1$. Thus $\overline{\Phi}(zt^2 + 1) = z\overline{\Phi}(t)^2 + 1 = \lambda(zt^2 + 1)$ and by considering the constant terms in the last equality, we conclude that in fact $\lambda = 1$. Also, since $\overline{\varphi}(t) \equiv t \mod (x)$, we have that $\overline{\Phi}(t) \equiv t \mod (x)$. Thus, Φ is congruent to the identity modulo (x).

So Φ actually induces a $\mathbb{C}(u)[x]$-automorphism of A and the observation made on the Makar-Limanov and Derksen invariants of A implies that the induced automorphism preserves the subring $\mathbb{C}(u)[x, z, t]$ of A and fixes the ideal $(x^2, r_0 - u)$. This implies that there exists $H \in \mathbb{C}[u][z, t]$ such that $\Phi(f) \equiv f + x\mathrm{Jac}(H, f) \mod x^2$ for every $f \in \mathbb{C}(u)[z, t]$. Furthermore, since Φ is an extension of $\tilde{\varphi}$, the residue class of H in $\mathbb{C}[u][z, t]/(u) \simeq \mathbb{C}[z, t]$ coincides with $h + c$ for some $c \in \mathbb{C}$. But on the other hand, one has $\mathrm{Jac}(H, r_0 - u) \in (r_0 - u)$ as the restriction of Φ fixes the ideal $(x^2, r_0 - u)$. Since $r_0 - u$ is irreducible, this implies that H is, up to the addition of a polynomial in $\mathbb{C}[u]$, an element of the ideal $(r_0 - u)\mathbb{C}[u][z, t]$ and so its image in $\mathbb{C}[u][z, t]/(u) \simeq \mathbb{C}[z, t]$ is a regular function on $\mathbf{A}^2$ whose restriction to the curve $C_0 = \{r_0 = 0\}$ is constant. This is absurd since by construction $h + c = zt^2 + c$ is locally constant but not constant on C_0. □

Remark 4. Note that there are many examples of non-extendible automorphisms of hypersurfaces. In this setting, for example, we showed in [2] that the hypersurface in $\mathbf{A}^4$ given by the equation

$$x^2y + z^2 + t^3 + x(1 + x + z^2 + t^3) = 0,$$

which is in fact isomorphic to the Russell cubic as a variety, admits automorphisms that do not extend to the ambient space. However, none of these non-extendible automorphisms fixes x. The present example exhibits a new phenomenon coming from the fact that the curve defined by $r_0 = 0$ is not connected. Note also that viewing X as a subvariety of $\mathbf{A}^1 \times \mathbf{A}^3 = \mathrm{Spec}(\mathbb{C}[x][y, z, t])$ the restriction of the automorphism determined by $\tilde{\varphi}$ to every fiber X_s, $s \in \mathbf{A}^1$ of the first projection

$\mathrm{pr}_x : X \to \mathbf{A}^1$ actually extend to an automorphism of $\mathbf{A}^3_s$: indeed, $\tilde{\varphi}$ restricts to the identity modulo x and therefore the restriction of the corresponding automorphism to X_0 extends. On the other hand, the argument used in the proof of Lemma 1 shows immediately that the induced automorphism of $X \mid_{\mathbf{A}^1\setminus\{0\}}$ extends to an automorphism of $\mathbf{A}^1 \setminus \{0\} \times \mathbf{A}^3$.

Acknowledgements This research was supported in part by the ANR Grant BirPol ANR-11-JS01-004-01.

References

1. S. Abhyankar, P. Eakin, W. Heinzer, On the uniqueness of the coefficient ring in a polynomial ring. J. Algebra **23**, 310–342 (1972)
2. A. Dubouloz, L. Moser-Jauslin, P.-M. Poloni, Inequivalent embeddings of the Koras-Russell cubic threefold. Mich. Math. J. **59**(3), 679–694 (2010)
3. A. Dubouloz, L. Moser-Jauslin, P.-M. Poloni, Non cancellation for smooth contractible affine threefolds. Proc. Am. Math. Soc. **139**, 4273–4284 (2011)
4. R. Hartshorne, *Deformation Theory*. Graduate Texts in Mathematics, vol. 257 (Springer, New York, 2010)
5. S. Kaliman, Polynomials with general C^2-fibers are variables. Pac. J. Math. **203**(1), 161–190 (2002)
6. S. Kaliman, L. Makar-Limanov, On the Russell-Koras contractible threefolds. J. Algebraic Geom. **6**(2), 247–268 (1997)
7. S. Kaliman; L. Makar-Limanov, AK-invariant of affine domains, in *Affine Algebraic Geometry* (Osaka University Press, Osaka, 2007), pp. 231–255
8. S. Kaliman; M. Zaidenberg, Affine modifications and affine varieties with a very transitive automorphism group. Transform. Groups **4**, 53–95 (1999)
9. S. Kaliman, M. Koras, L. Makar-Limanov, P. Russell, $\mathbb{C}^*$-actions on $\mathbb{C}^3$ are linearizable. Electron. Res. Announc. Am. Math. Soc. **3**, 63–71 (1997)
10. M. Koras, P. Russell, $\mathbf{C}^*$-actions on $\mathbf{C}^3$: the smooth locus of the quotient is not of hyperbolic type. J. Algebraic Geom. **8**(4), 603–694 (1999)
11. L. Moser-Jauslin, Automorphism groups of Koras-Russell threefolds of the first kind, in *Proceedings of "Affine Algebraic Geometry : A conference in Honour of Peter Russell"*, Montreal 1–5 June 2009. CRM Proceedings Lecture Notes, vol. 54 (2011), pp. 261–270
12. M. Nagata, A remark on the unique factorization theorem. J. Math. Soc. Jpn. **9**(1), 143–145 (1957)
13. A. Sathaye, Polynomial ring in two variables over a D.V.R.: a criterion. Invent. Math. **74**, 159–168 (1983)
14. A. van den Essen, S. Maubach, S.Vénéreau, The special automorphism group of $R[t]/(t^m)[x_1 \ldots, x_n]$ and coordinates of a subring of $R[t][x_1, \ldots, x_n]$. J. Pure Appl. Algebra **210**(1), 141–146 (2007)

Laurent Cancellation for Rings of Transcendence Degree One Over a Field

Gene Freudenburg

Abstract If R is an integral domain and A is an R-algebra, then A has the *Laurent cancellation property over* R if $A^{[\pm n]} \cong_R B^{[\pm n]}$ implies $A \cong_R B$ ($n \geq 0$ and B an R-algebra). Here, $A^{[\pm n]}$ denotes the ring of Laurent polynomials in n variables over A. Our main result (Theorem 4.1) is that, if R is a field and the transcendence degree of A over R is one, then A has the Laurent cancellation property over R. Two additional cases of Laurent cancellation are given in Theorem 5.1

2010 Mathematics Subject Classification: 13B99,16S34 (primary); 13F60 (secondary)

1 Introduction

If R is an integral domain, then an *R-algebra* will mean an integral domain containing R as a subring. If A is an R-algebra and $n \in \mathbb{Z}$, $n \geq 0$, then $A^{[n]}$ is the polynomial ring in n variables over A, and $A^{[\pm n]}$ is the ring of Laurent polynomials in n variables over A.

In this paper, we consider the following question.

> Let R be an integral domain, let A and B be R-algebras, and let n be a non-negative integer. Does $A^{[\pm n]} \cong_R B^{[\pm n]}$ imply $A \cong_R B$?

We say that A has the *Laurent cancellation property over* R if this question has a positive answer for all pairs (B, n).

G. Freudenburg (✉)
Department of Mathematics, Western Michigan University, Kalamazoo, MI 49008, USA
e-mail: gene.freudenburg@wmich.edu

I. Cheltsov et al. (eds.), *Automorphisms in Birational and Affine Geometry*, Springer Proceedings in Mathematics & Statistics 79,
DOI 10.1007/978-3-319-05681-4_18,

Our main result (Theorem 4.1) is that, if **k** is a field, then any **k**-algebra of transcendence degree one over **k** has the Laurent cancellation property over **k**. This result parallels the well-known theorem of Abhyankar, Eakin, and Heinzer which asserts that if A and B are **k**-algebras of transcendence one over **k**, then the condition $A^{[n]} \cong_{\mathbf{k}} B^{[n]}$ for some $n \geq 0$ implies $A \cong_{\mathbf{k}} B$; see [1]. Note that our main result implies that if X and Y are affine algebraic curves over **k**, and if $\mathbb{T}^n$ is the torus of dimension n over **k** ($n \geq 0$), then the condition $X \times \mathbb{T}^n \cong Y \times \mathbb{T}^n$ implies $X \cong Y$.

In [12], Makar-Limanov gives a proof of the Abhyankar–Eakin–Heinzer theorem for the field $\mathbf{k} = \mathbb{C}$ using the theory of locally nilpotent derivations (LNDs); see also [5, Corollary 3.2]. The proof of our main result uses $\mathbb{Z}$-gradings in a similar way. Where Makar-Limanov uses the subring of elements of degree zero for all LNDs, we use the subring of elements of degree zero for all $\mathbb{Z}$-gradings over R. This subring is called the *R-neutral subalgebra* and is denoted $\mathcal{N}_R(A)$. The other key ingredient in the proof is Theorem 3.2, which is the following characterization of Laurent polynomial rings over a field.

Let **k** *be a field, and let* A *be a* **k***-algebra. The following are equivalent.*

1. $A = \mathbf{k}^{[\pm 1]}$.
2. *The following three conditions hold.*
   (a) **k** *is algebraically closed in* A
   (b) $\mathrm{tr.deg}_{\mathbf{k}} A = 1$
   (c) $A^* \not\subset \mathcal{N}_{\mathbf{k}}(A)$

Again using the neutral subalgebra $\mathcal{N}_R(A)$, two additional cases of Laurent cancellation are established in Sect. 5. Suppose R is an integral domain and A is an R-algebra. For each $u \in A^*$, define:

$$R_u(A) = \{\lambda \in R \mid u - \lambda \in A^*\}$$

We say that u is *R-transtable* if $R_u(A) \neq \{0\}$. The set of R-transtable units of A is denoted by A_R^τ. Theorem 5.1 shows that if either $R[A^*] = R[A_R^\tau]$, or A is algebraic over $R[A_R^\tau]$, then A has the Laurent cancellation property over R.

1.1 Background

In [4], Bhatwadekar and Gupta showed that the Laurent polynomial ring $R^{[\pm n]}$ has the Laurent cancellation property over the integral domain R. This is discussed in Sect. 4.2 below. There seem to be no other results in the literature specific to the Laurent (torus) cancellation problem.

Other versions of the cancellation problem have been investigated since Zariski posed the birational version (see [14, Sect. 5]). Beauville et al. [3] solved Zariski's problem by constructing fields $\mathbf{k} \subset K$ with the property that $K^{(3)} \cong_{\mathbf{k}} \mathbf{k}^{(6)}$ but $K \not\cong_{\mathbf{k}} \mathbf{k}^{(3)}$.

The polynomial cancellation problem asks whether, given $\mathbf{k}$-algebras A and B over a field $\mathbf{k}$, the condition $A^{[n]} \cong_{\mathbf{k}} B^{[n]}$ implies $A \cong_{\mathbf{k}} B$. As mentioned, Abhyankar, Eakin, and Heinzer showed that this implication holds if the transcendence degree of A over $\mathbf{k}$ is one. The first examples where this implication does not hold were given by Hochster and Murthy (unpublished); see [10]. In Hochster's example, A is the coordinate ring of the tangent bundle over the real 2-sphere, which is a stably trivial bundle, but not trivial. Danielewski [6] gave a counterexample using the coordinate rings of affine surfaces over $\mathbb{C}$.

The cancellation problem for affine spaces is the special case when A is a polynomial ring. It was shown by Fujita, Miyanishi, and Sugie (characteristic $\mathbf{k} = 0$) and Russell (positive characteristic) that $A^{[n]} \cong_{\mathbf{k}} \mathbf{k}^{[n+2]}$ implies $A \cong_{\mathbf{k}} \mathbf{k}^{[2]}$ [8,13,15]. In [2], Asanuma constructed a $\mathbf{k}$-algebra A over a field $\mathbf{k}$ of positive characteristic and showed that $A^{[1]} \cong_{\mathbf{k}} \mathbf{k}^{[4]}$. His construction is based on the existence of non-standard embeddings of lines in the $\mathbf{k}$-plane. Subsequently, Gupta [9] showed that $A \not\cong_{\mathbf{k}} \mathbf{k}^{[3]}$. For fields of characteristic zero, the cancellation problem for affine spaces remains open. A nice discussion of this problem is found in [11].

1.2 Terminology and Notation

The group of units of the integral domain A is denoted A^*, and the field of fractions of A is $\mathrm{frac}(A)$. Given $f \in A$, A_f denotes the localization of A at f. Given $z \in A^*$, the notation $z^{\pm 1}$ is used for the set $\{z, z^{-1}\}$.

For $n \geq 0$, the polynomial ring in n variables over A is denoted by $A^{[n]}$. If $A[x_1, \ldots, x_n] = A^{[n]}$, the ring of Laurent polynomials over R is the subring of $\mathrm{frac}(A^{[n]})$ defined and denoted by:

$$A^{[\pm n]} = A[x_1, x_1^{-1}, \ldots, x_n, x_n^{-1}]$$

For any subring $S \subset A$, the transcendence degree of A over S is equal to the transcendence degree of $\mathrm{frac}(A)$ over $\mathrm{frac}(S)$, denoted $\mathrm{tr.deg}_S A$. The set of elements in A algebraic over S is denoted by $\mathrm{Alg}_S A$; we also say that $\mathrm{Alg}_S A$ is the *algebraic closure of S in A*. If $S = \mathrm{Alg}_S A$, then S is *algebraically closed in A*. Any $\mathbb{Z}$-grading of A such that $S \subset A_0$ is a *$\mathbb{Z}$-grading over S*, where A_0 denotes the subring of elements of degree zero.

Over a ground field $\mathbf{k}$, $\mathbb{A}^n$ denotes the affine n-space $\mathrm{Spec}(\mathbf{k}^{[n]})$, and $\mathbb{T}^n$ denotes the n-torus $\mathrm{Spec}(\mathbf{k}^{[\pm n]})$.

Throughout this paper, the term *R-algebra* will mean an integral domain containing R as a subring.

2 $\mathbb{Z}$-Gradings and the Neutral Subalgebra

2.1 $\mathbb{Z}$-Gradings

Assume that R is an integral domain, and A is an R-algebra. The set of $\mathbb{Z}$-gradings of A is denoted $\mathbb{Z}(A)$, and the subset of $\mathbb{Z}$-gradings of A over R is denoted $\mathbb{Z}_R(A)$.

Given $\mathfrak{g} \in \mathbb{Z}(A)$, let $\deg_{\mathfrak{g}}$ denote the induced degree function on A, and let $A = \oplus_{i\in\mathbb{Z}} A_i$ be the decomposition of A into $\mathfrak{g}$-homogeneous summands, where A_i consists of $\mathfrak{g}$-homogeneous elements of degree i. Define $A^{\mathfrak{g}} = A_0$, which is a subalgebra of A. The subalgebra $S \subset A$ is $\mathfrak{g}$-*homogeneous* if S is generated by $\mathfrak{g}$-homogeneous elements.

Given $a \in A$, write $a = \sum_{i\in\mathbb{Z}} a_i$, where $a_i \in A_i$ for each i. The $\mathfrak{g}$-*support* of a is defined by:

$$\mathrm{Supp}_{\mathfrak{g}}(a) = \{i \in \mathbb{Z} \mid a_i \neq 0\}$$

Note that (1) $\mathrm{Supp}_{\mathfrak{g}}(a) = \emptyset$ if and only if $a = 0$, and (2) $\#\mathrm{Supp}_{\mathfrak{g}}(a) = 1$ if and only if a is nonzero and homogeneous.

Lemma 2.1. *Let A be an integral domain, and let $\mathfrak{g} \in \mathbb{Z}(A)$ be given.*

(a) *$A^{\mathfrak{g}}$ is algebraically closed in A.*
(b) *If $H \subset A$ is a $\mathfrak{g}$-homogeneous subalgebra, then* $\mathrm{Alg}_H A$ *is also a $\mathfrak{g}$-homogeneous subalgebra.*

Proof. Let $A = \oplus_{i\in\mathbb{Z}} A_i$ be the decomposition of A into $\mathfrak{g}$-homogeneous summands. Given nonzero $a \in A$, write $a = \sum_{i\in\mathbb{Z}} a_i$, where $a_i \in A_i$, and let $\bar{a}$ denote the highest-degree (nonzero) homogeneous summand of a.

In order to prove part (a), let $v \in A$ be algebraic over $A^{\mathfrak{g}}$. If $v \notin A^{\mathfrak{g}}$, then we may assume that $\deg_{\mathfrak{g}} v > 0$. Suppose that $\sum_{0\le i\le n} c_i v^i = 0$ is a nontrivial dependence relation for v over $A^{\mathfrak{g}}$, where $c_i \in A^{\mathfrak{g}}$ for each i, and $n \geq 1$. Since $\deg_{\mathfrak{g}} \bar{v} > 0$ and $\deg_{\mathfrak{g}} c_i = 0$ for each i, we see that $c_n \bar{v}^n = 0$, a contradiction. Therefore, $v \in A^{\mathfrak{g}}$, and $A^{\mathfrak{g}}$ is algebraically closed in A.

For part (b), given an integer $n \geq 0$, let $H(n)$ denote the ring obtained by adjoining to H all elements $a \in \mathrm{Alg}_H A$ such that $\#\mathrm{Supp}_{\mathfrak{g}}(a) \leq n$. In particular, $H(0) = H$. We show by induction on n that, for each $n \geq 1$:

$$H(n) \subset H(1) \tag{1}$$

This property implies $\mathrm{Alg}_H A = H(1)$, which is a $\mathfrak{g}$-homogeneous subring of A.

Assume that, for some $n \geq 2$, $H(n-1) \subset H(1)$. Let $a \in \mathrm{Alg}_H A$ be given such that $\#\mathrm{Supp}_{\mathfrak{g}}(a) = n$, and let $\sum_{i\geq 0} h_i a^i = 0$ be a nontrivial dependence relation for a over H, where $h_i \in H$ for each i. Define:

$$d = \max_{i\geq 0}\{\deg_{\mathfrak{g}} h_i a^i\} \quad \text{and} \quad I = \{i \in \mathbb{Z} \mid i \geq 0\ ,\ \deg_{\mathfrak{g}} h_i a^i = d\}$$

Then I is nonempty, and $\sum_{i \in I} \bar{h}_i \bar{a}^i = 0$. Since H is homogeneous, $\bar{h}_i \in H$ for each i. Therefore, $\bar{a}$ is algebraic over H. Since $a = (a - \bar{a}) + \bar{a}$, it follows that $a \in H(n-1) + H(1)$. Since $H(n-1) \subset H(1)$, we see that $a \in H(1)$, thus proving by induction the equality claimed in (1). □

2.2 *The Neutral Subalgebra*

Assume that R is an integral domain, and A is an R-algebra.

Definition 2.1. The *neutral R-subalgebra* of A is:

$$\mathcal{N}_R(A) = \cap_{\mathfrak{g} \in \mathbb{Z}_R(A)} A^{\mathfrak{g}}$$

A is a *neutral R-algebra* if $\mathcal{N}_R(A) = A$.

Lemma 2.2. *Let R be an integral domain, and let A be an R-algebra.*

(a) *$\mathcal{N}_R(A)$ is algebraically closed in A*
(b) *$\mathcal{N}_R(A^{[n]}) \subset \mathcal{N}_R(A)$ and $\mathcal{N}_R(A^{[\pm n]}) \subset \mathcal{N}_R(A)$ for each $n \geq 0$*
(c) *If A is algebraic over $R[A^*]$, then $\mathcal{N}_R(A^{[n]}) = \mathcal{N}_R(A)$ and $\mathcal{N}_R(A^{[\pm n]}) = \mathcal{N}_R(A)$ for each $n \geq 0$*

Proof. Part (a) is implied by Lemma 2.1(a).

For part (b), let $C = A[y_1^{\pm 1}, \ldots, y_n^{\pm 1}] = A^{[\pm n]}$. For each i, $1 \leq i \leq n$, let C_i be the subring of C generated over A by y_j for all $j \neq i$. Given i, define $\mathfrak{g} \in \mathbb{Z}_{C_i}(C)$ by declaring that y_i is homogeneous of degree one. The subring of elements of degree 0 is C_i. Therefore, $\mathcal{N}_R(C) \subset C_i$ for each i, which implies $\mathcal{N}_R(C) \subset A$. Suppose that there exist $\mathfrak{h} \in \mathbb{Z}_R(A)$ and $f \in \mathcal{N}_R(C)$ such that $\deg_{\mathfrak{h}} f \neq 0$. Then $\mathfrak{h}$ extends to C by setting $\deg_{\mathfrak{h}} y_i = 0$ for each i, meaning $f \notin \mathcal{N}_R(C)$, a contradiction. Therefore, $f \in \mathcal{N}_R(A)$. The argument is the same if $C = A^{[n]}$.

For part (c), let $C = A[y_1^{\pm 1}, \ldots, y_n^{\pm 1}] = A^{[\pm n]}$, and let $f \in A$ be given. Suppose that $\mathfrak{g} \in \mathbb{Z}_R(C)$ has $\deg_{\mathfrak{g}} f \neq 0$. Since every element of A^* is $\mathfrak{g}$-homogeneous, it follows from Lemma 2.1(b) that A is a $\mathfrak{g}$-homogeneous subring of C. Therefore, $\mathfrak{g}$ restricts to an element of $\mathbb{Z}_R(A)$ for which the degree of f is nonzero, meaning that f is not in $\mathcal{N}_R(A)$. The argument is the same if $C = A^{[n]}$. □

Example 2.1. Let R be an integral domain, and define $A = R[x, y]/(x^2 - y^3 - 1)$. Then $\mathcal{N}_R(A) = A$. To see this, let $\mathfrak{g} \in \mathbb{Z}_R(A)$ be given. Set $K = \text{frac}(R)$ and define $A_K = K \otimes_R A$. Then $\mathfrak{g}$ extends to A_K, which is the coordinate ring of the plane curve $C : x^2 - y^3 = 1$ over K. This $\mathbb{Z}$-grading induces an action of the torus $\mathbb{T} = \text{Spec}(K[t, t^{-1}])$ on A_K, namely, if $a \in A_K$ is homogeneous of degree d, then $t \cdot a = t^d a$. If this were a nontrivial action, then C would contain $\mathbb{T}$ as a dense open orbit, implying that C is K-rational, which it is not. Therefore, $\mathfrak{g}$ must be the trivial $\mathbb{Z}$-grading.

3 Laurent Polynomial Rings

3.1 Units and Automorphisms

Lemma 3.1. *Let R be an integral domain, and let $A = R[y_1^{\pm 1}, \dots, y_n^{\pm 1}] = R^{[\pm n]}$. Then:*

$$A^* = R^* \cdot \{y_1^{d_1} \cdots y_n^{d_n} \mid d_i \in \mathbb{Z}, i = 1, \dots, n\} = R^* \cdot \mathbb{Z}^n$$

Proof. By induction, it suffices to prove the case $n = 1$.

Suppose $A = R[y, y^{-1}] = R^{[\pm 1]}$, and let $u \in A^*$ be given. Write $u = p(y)/y^k$ and $u^{-1} = q(y)/y^l$, where $p, q \in R[y] = R^{[1]}$; $p(0) \neq 0$ and $q(0) \neq 0$; and $k, l \geq 0$. We thus have $p(y)q(y) = y^{k+l}$. If $k + l > 0$, then $p(0)q(0) = 0$, contradicting the fact that A is an integral domain. Therefore, $k + l = 0$, meaning that $p(y)q(y) = 1$ in $R[y]$. Since $R[y]^* = R^*$, we see that $p(y) \in R^*$, which proves the lemma. □

We next consider the group of R-automorphisms of $R^{[\pm n]}$, which is a well-understood group. Its main features, which are easy to check, are summarized as follows.

1. Given $E \in GL_n(\mathbb{Z})$, the R-morphism $\phi_E : A \to A$ given by

$$\phi_E(y_i) = \prod_{1 \leq j \leq n} y_j^{e_{ij}}$$

   where $E = (e_{ij})$, defines a faithful action of $GL_n(\mathbb{Z})$ on A by R-automorphisms.
2. Given $a = (a_1, \dots, a_n) \in (R^*)^n$, the R-morphism $\psi_a : A \to A$ given by

$$\psi_a(y_i) = a_i y_i$$

   defines a faithful action of $(R^*)^n$ on A by R-automorphisms.
3. The group of R-automorphisms of $R^{[\pm n]}$ is a semi-direct product of the subgroups $GL_n(\mathbb{Z})$ and $(R^*)^n$.

3.2 A Criterion for Cancellation

Proposition 3.1. *Let R be an integral domain, and let A and B be R-algebras. Assume that, for some $n \geq 0$, there exists an R-isomorphism*

$$F : A^{[\pm n]} \to B^{[\pm n]}$$

such that $F(A^*) \subset B$. Then $A \cong_R B$.

Proof. Let

$$C = A[y_1^{\pm 1}, \dots, y_n^{\pm 1}] = A^{[\pm n]} \quad \text{and} \quad D = B[z_1^{\pm 1}, \dots, z_n^{\pm 1}] = B^{[\pm n]}$$

By Lemma 3.1, we have:

$$C^* = A^* \cdot \{y_1^{d_1} \cdots y_n^{d_n} \mid d_i \in \mathbb{Z}, i = 1, \dots, n\} \quad \text{and}$$
$$D^* = B^* \cdot \{z_1^{e_1} \cdots z_n^{e_n} \mid e_i \in \mathbb{Z}, i = 1, \dots, n\}$$

Thus, given i with $1 \le i \le n$, there exist $b_i \in B^*$ and $e_{ik} \in \mathbb{Z}$ such that:

$$F(y_i) = b_i \prod_{1 \le k \le n} z_k^{e_{ik}}$$

Likewise, there exist $a_i \in A^*$ and $d_{ij} \in \mathbb{Z}$ such that:

$$F^{-1}(z_i) = a_i \prod_{1 \le j \le n} y_j^{d_{ij}}$$

Therefore, given i, we have:

$$\begin{aligned} z_i &= FF^{-1}(z_i) \\ &= F\left(a_i \prod_k y_k^{d_{ik}}\right) \\ &= F(a_i) \prod_{kF} (y_k)^{d_{ik}} \\ &= F(a_i) \prod_k \left(b_k \prod_j z_j^{e_{kj}}\right)^{d_{ik}} \\ &= \left(F(a_i) \prod_k b_k^{d_{ik}}\right) \prod_k \prod_j z_j^{d_{ik} e_{kj}} \\ &= \left(F(a_i) \prod_k b_k^{d_{ik}}\right) \prod_j z_j^{(\sum_k d_{ik} e_{kj})} \end{aligned}$$

Since $F(a_i) \prod_k b_k^{d_{ik}} \in B$ for each i, we conclude that, for each i, j $(1 \le i, j \le n)$:

$$F(a_i) \prod_k b_k^{d_{ik}} = 1 \quad \text{and} \quad \sum_k d_{ik} e_{kj} = \delta_{ij}$$

It follows that, if E is the $n \times n$ matrix $E = (e_{ij})$, then $E \in GL_n(\mathbb{Z})$.

Define $b = (b_1, \ldots, b_n) \in (B^*)^n$. Then for each $i = 1, \ldots, n$ we see that:

$$Z_i := F(y_i) = \phi_E \psi_b(z_i)$$

On one hand:

$$D = \phi_E \psi_b(D) = B[Z_1^{\pm 1}, \ldots, Z_n^{\pm 1}]$$

On the other hand:

$$D = F(C) = F(A)[F(y_1)^{\pm 1}, \ldots, F(y_n)^{\pm 1}] = F(A)[Z_1^{\pm 1}, \ldots, Z_n^{\pm 1}]$$

Therefore, if $I \subset D$ is the ideal $I = (Z_1 - 1, \ldots, Z_n - 1)$, then:

$$A \cong_R F(A) \cong_R D/I \cong_R B$$

□

Note that, if $A^* = R^*$, then Proposition 3.1 implies that A has the Laurent cancellation property over R. In particular, every polynomial ring $A = R^{[n]}$ has the Laurent cancellation property over R.

3.3 A Characterization of Laurent Polynomial Rings over a Field

Theorem 3.1. *Let R be an integral domain, and let A be an R-algebra such that R is algebraically closed in A, $\text{tr.deg}_R A = 1$, and $A^* \not\subset \mathcal{N}_R(A)$.*

(a) *There exists $u \in A^*$ such that $R[A^*] = R[u, u^{-1}] = R^{[\pm 1]}$.*
(b) *There exist $r \in R$ and $w \in A_r^*$ such that $A_r = R_r[w, w^{-1}] = R_r^{[\pm 1]}$.*

Proof. Let $u \in A^*$ be given such that $u \notin \mathcal{N}_R(A)$. By Lemma 2.2(a), we see that $R = \mathcal{N}_R(A)$ and $R[u] = R^{[1]}$.

Let $\mathfrak{g} \in \mathbb{Z}_R(A)$ be such that $u \notin A^{\mathfrak{g}}$, and let $A = \oplus_{i \in \mathbb{Z}} A_i$ be the decomposition of A into $\mathfrak{g}$-homogeneous summands. Since R and $A^{\mathfrak{g}}$ are algebraically closed in A, it must be that either $A^{\mathfrak{g}} = R$ or $A^{\mathfrak{g}} = A$, and therefore $A^{\mathfrak{g}} = R$.

Part (a): If $R[A^*] = R[u, u^{-1}]$, there is nothing further to prove. So assume that $R[A^*]$ is strictly larger than $R[u, u^{-1}]$. Let $v \in A^*$ be such that $v \notin R[u, u^{-1}]$. Then v is $\mathfrak{g}$-homogeneous, and $v \notin A^{\mathfrak{g}} = R$.

Set $d = \gcd(\deg_{\mathfrak{g}} u, \deg_{\mathfrak{g}} v)$, and let $a, b \in \mathbb{Z}$ be such that:

$$\deg_{\mathfrak{g}} u = ad \quad \text{and} \quad \deg_{\mathfrak{g}} v = bd$$

If $ab < 0$, replace u by u^{-1}; in this way we can assume that $ab > 0$. If $a < 0$ and $b < 0$, replace the grading $\mathfrak{g}$ by $(-\mathfrak{g})$; in this way we can assume that $a > 0$ and $b > 0$.

Since v is algebraic over $R[u, u^{-1}]$, there exists a nontrivial dependence relation $P(u, v) = 0$, where $P \in R[x, y] = R^{[2]}$. Define a $\mathbb{Z}$-grading $\mathfrak{h}$ of $R[x, y]$ over R by setting $\deg_{\mathfrak{h}} x = a$ and $\deg_{\mathfrak{h}} y = b$. Then it suffices to assume that $P(x, y)$ is $\mathfrak{h}$-homogeneous.

Let K be the algebraic closure of frac(R). Consider $P(x, y)$ as an element of $K[x, y]$, and view A as a subring of $K \otimes_R A$. By Lemma 4.6 of [7], P has the form:[1]

$$P = x^i y^j \prod_{k=1}^{N} (\alpha_k x^b + \beta_k y^a) \ , \ i, j \geq 0 \ , \ N \geq 1 \ , \ \alpha_k, \beta_k \in K^*$$

Since $P(u, v) = 0$, it follows that $ru^b + v^a = 0$ for some $r \in K^*$. We see that $r = -u^{-b}v^a \in R^*$. Moreover, $a > 1$, since otherwise $v \in R[u, u^{-1}]$.

Let $m, n \in \mathbb{Z}$ be such that $am + bn = 1$. Set $w = u^m v^n$, noting that $w \in A^*$ and $\deg_{\mathfrak{g}} w = d \neq 0$. Then:

$$w^a = (-r)^n u \quad \text{and} \quad w^b = (-r)^{-m} v$$

It follows that:

$$R[u^{\pm 1}, v^{\pm 1}] = R[w^{\pm a}, w^{\pm b}] = R[w, w^{-1}]$$

If $R[A^*] = R[w, w^{-1}]$, the desired result holds. Otherwise, replace u by w and repeat the argument above. Since

$$\deg_{\mathfrak{g}} u = ad > d = \deg_{\mathfrak{g}} w > 0$$

this process must terminate in a finite number of steps. This completes the proof of part (a).

Part (b): The proof of part (b) is a continuation of the algorithm used in the proof of part (a), where units are adjoined where needed.

Suppose that $R[A^*] = R[u, u^{-1}]$. If $R[A^*] = A$, there is nothing further to prove. So assume that $R[A^*] \neq A$, and choose $\mathfrak{g}$-homogeneous $v \in A$ not in $R[u, u^{-1}]$. As before, we obtain an equation $ru^b + v^a = 0$, where $r \in R$, and a, b are relatively prime positive integers with $a > 1$. However, in this case $r \notin R^*$, since otherwise v is a unit.

---

[1]The lemma is stated for fields of characteristic zero, but the proof is valid over any field.

In order to continue the algorithm, we extend $\mathfrak{g}$ to the ring A_r, noting that $v \in A_r^*$. As above, there exists $w \in A_r^*$ such that $0 < \deg_{\mathfrak{g}} w < \deg_{\mathfrak{g}} u$ and $R_r[A_r^*] = R_r[w, w^{-1}]$.

If $A_r = R_r[w, w^{-1}]$, the desired result holds. Otherwise, replace u by w and repeat the argument. As before, since a strict decrease in degrees takes place, the process must terminate in a finite number of steps. This completes the proof of part (b). □

As a consequence of this theorem, we obtain the following characterization of Laurent polynomial rings over a field.

Theorem 3.2. *Let* $\mathbf{k}$ *be a field, and let* A *be a* $\mathbf{k}$*-algebra. The following are equivalent.*

1. $A = \mathbf{k}^{[\pm 1]}$.
2. *The following three conditions hold.*
   (a) $\mathbf{k}$ *is algebraically closed in* A
   (b) $\mathrm{tr.deg}_{\mathbf{k}} A = 1$
   (c) $A^* \not\subset \mathcal{N}_{\mathbf{k}}(A)$

Corollary 3.1. *Let* $\mathbf{k}$ *be a field, and let* A *be a* $\mathbf{k}$*-algebra. Assume that* $\mathbf{k}$ *is algebraically closed in* A*. Given* $u \in A^*$*, if* $u \notin \mathcal{N}_{\mathbf{k}}(A)$*, then there exists* $w \in A^*$ *such that:*

$$\mathrm{Alg}_{\mathbf{k}[u]} A = \mathbf{k}[w, w^{-1}] = \mathbf{k}^{[\pm 1]}$$

Proof. By hypothesis, there exists $\mathfrak{g} \in \mathbb{Z}_{\mathbf{k}}(A)$ such that $\deg_{\mathfrak{g}} u \neq 0$. Set $B = \mathrm{Alg}_{\mathbf{k}[u]} A$. Since u is a unit, u is $\mathfrak{g}$-homogeneous, and $\mathbf{k}[u]$ is a $\mathfrak{g}$-homogeneous subring. By Lemma 2.1(b), it follows that B is $\mathfrak{g}$-homogeneous, meaning that $\mathfrak{g}$ restricts to B. Since $\deg_{\mathfrak{g}} u \neq 0$, we see that $u \notin \mathcal{N}_{\mathbf{k}}(B)$. The result now follows from Theorem 3.2. □

4 Laurent Cancellation for R-Algebras of Transcendence Degree One Over R

4.1 A Reduction

Let R be an integral domain, and let A be an R-algebra. If $n \geq 0$, then since A is algebraically closed in $A^{[\pm n]}$ we have:

$$\mathrm{Alg}_R(A^{[\pm n]}) = \mathrm{Alg}_R(A)$$

Let $\alpha : A^{[\pm n]} \to B^{[\pm n]}$ be an isomorphism of R-algebras. If $S = \mathrm{Alg}_R(A)$, then $\alpha(S) = \mathrm{Alg}_R(B)$, since B is algebraically closed in $B^{[\pm n]}$. Therefore, identifying S and $\alpha(S)$, we can view A and B as S-algebras, and α as an S-isomorphism. In considering the question of Laurent cancellation, it thus suffices to assume R is algebraically closed in A. Note that this condition implies the group A^*/R^* is torsion free.

4.2 Cancellation for Laurent Polynomial Rings

In [4], Lemma 4.5, Bhatwadekar and Gupta showed the following: Let R be an integral domain, and let A be an R-algebra with R algebraically closed in A. Suppose that m, n are nonnegative integers such that:

$$A^{[\pm n]} \cong_R R^{[\pm(m+n)]}$$

Then $A \cong_R R^{[\pm m]}$. This fact will be used in the next section.

4.3 Main Theorem

We show that Laurent cancellation holds for several classes of rings that are of transcendence degree one over a subalgebra. For fields in particular, we obtain the analogue of the Abhyankar–Eakin–Heinzer Theorem for Laurent polynomial rings.

Theorem 4.1. *Let R be an integral domain, and let A be an R-algebra with* $\mathrm{tr.deg}_R A = 1$. *Then A has the Laurent cancellation property over R if any one of the following conditions holds.*

(a) $R[A^*]$ *is algebraic over* R
(b) $A^* \subset \mathcal{N}_R(A)$
(c) R *is a field*

Proof. By Sect. 4.1, it suffices to assume that R is algebraically closed in A. Let $F : A^{[\pm n]} \to B^{[\pm n]}$ be an isomorphism of R-algebras.

Part (a): If $R[A^*]$ is algebraic over R, then $A^* = R^*$, and $F(A^*) = F(R^*) = R^* \subset B$. By Proposition 3.1, it follows that $A \cong_R B$ in this case.

Part (b): Assume that $A^* \subset \mathcal{N}_R(A)$. By part (a), we may also assume that $R[A^*]$ is transcendental over R, meaning that A is algebraic over $R[A^*]$. By Lemma 2.2(c), $\mathcal{N}_R(A^{[\pm n]}) = \mathcal{N}_R(A)$. Therefore:

$$F(A^*) \subset F(\mathcal{N}_R(A)) = F\left(\mathcal{N}_R\left(A^{[\pm n]}\right)\right) = \mathcal{N}_R\left(B^{[\pm n]}\right) \subset \mathcal{N}_R(B) \subset B$$

By Proposition 3.1, it follows that $A \cong_R B$ in this case.

Part (c): Assume that R is a field. By part (b), we may assume that $A^* \not\subset \mathcal{N}_R(A)$. Then, by Corollary 3.2, we have that $A \cong_R R^{[\pm 1]}$. By the theorem of Bhatwadekar and Gupta (see Sect. 4.2), it follows that $A \cong_R B$ in this case. □

5 Two Additional Cases of Laurent Cancellation

Let R be an integral domain, and let A be an R-algebra. We say that A has the *strong Laurent cancellation property over* R if, whenever $\alpha : A^{[\pm n]} \to B^{[\pm n]}$ is an R-isomorphism for some R-algebra B and some $n \geq 0$, we have $\alpha(A) = B$.

Given $u \in A^*$, define the set:

$$R_u(A) = \{\lambda \in R \mid u - \lambda \in A^*\}$$

Then u is said to be an *R-transtable* unit if $R_u(A) \neq \{0\}$. The set of R-transtable units of A is denoted A_R^τ.

Lemma 5.1. *Let R be an integral domain, and let A be an R-algebra.*

(a) *If $\alpha : A \to B$ is an isomorphism of R-algebras, then $\alpha(A_R^\tau) = B_R^\tau$.*
(b) $A_R^\tau \subset \mathcal{N}_R(A)$
(c) $(A^{[\pm n]})_R^\tau = A_R^\tau$ *for all integers $n \geq 0$.*

Proof. Part (a): Given $u \in A_R^\tau$, choose nonzero $\lambda \in R_u(A)$. Then $\alpha(u - \lambda) = \alpha(u) - \lambda \in B^*$. Therefore, $\alpha(u) \in B_R^\tau$.

Part (b): Let $\mathfrak{g} \in \mathbb{Z}_R(A)$ be given, and let $A = \oplus_{i \in \mathbb{Z}} A_i$ be the decomposition of A into $\mathfrak{g}$-homogeneous summands. Note that any element of A^* is $\mathfrak{g}$-homogeneous. In particular, given $u \in A_R^\tau$ and non-zero $\lambda \in R_u(A)$, there exist $i, j \in \mathbb{Z}$ such that $u \in A_i$ and $u - \lambda \in A_j$. We therefore have non-zero elements $a = \lambda \in A_0$, $b = u \in A_i$, and $c = u - \lambda \in A_j$ such that $a + b + c = 0$ in $A_0 + A_i + A_j$, and this can only happen if $i = j = 0$.

Part (c): Let $C = A[y_1^{\pm 1}, \ldots, y_n^{\pm 1}] = A^{[\pm n]}$, and let $f \in C_R^\tau$ be given. By part (b), and by Lemma 2.2(b), we see that:

$$f \in C_R^\tau \subset \mathcal{N}_R(C) \subset \mathcal{N}_R(A) \subset A$$

Therefore, given non-zero $\lambda \in R_f(C)$, it follows that $f - \lambda \in C^* \cap A = A^*$, and $f \in A_R^\tau$.

Conversely, if $g \in A_R^\tau$, then $g - \mu \in A^* \subset C^*$ for some nonzero $\mu \in R$. So $g \in C_R^\tau$. □

Theorem 5.1. *Let R be an integral domain, and let A be an R-algebra.*

(a) *If $R[A^*] = R[A_R^\tau]$, then A has the Laurent cancellation property over R.*
(b) *If A is algebraic over $R[A_R^\tau]$, then A has the strong Laurent cancellation property over R.*

Proof. Let B be an R-algebra such that, for some $n \geq 0$, $A^{[\pm n]} \cong_R B^{[\pm n]}$. Let $C = A^{[\pm n]}$ and $D = B^{[\pm n]}$, and let $\alpha : C \to D$ be an R-isomorphism. Assume that:

$$C = A[y_1^{\pm 1}, \ldots, y_n^{\pm 1}] \quad \text{and} \quad D = B[z_1^{\pm 1}, \ldots, z_n^{\pm 1}]$$

Part (a): Assume that $R[A^*] = R[A_R^\tau]$. Using Lemma 5.1, we see that:

$$\alpha(A^*) \subset R[\alpha(A^*)] = \alpha(R[A^*]) = \alpha(R[A_R^\tau]) = \alpha(R[C_R^\tau]) = R[\alpha(C_R^\tau)]$$
$$= R[D_R^\tau] = R[B_R^\tau] \subset B$$

By Proposition 3.1, it follows that $A \cong_R B$.

Part (b): Assume that A is algebraic over $R[A_R^\tau]$. By Lemma 5.1(a), $\alpha(A_R^\tau) = B_R^\tau$. Therefore, given $a \in A$, $\alpha(a)$ is algebraic over B_R^τ. Since B is algebraically closed in D, it follows that $\alpha(a) \in B$. Therefore, if $B_0 = \alpha(A)$, then $B_0 \subset B$, and B_0 is algebraically closed in B.
If $B_0 \neq B$, then there exists $t \in B$ transcendental over B_0. In this case,

$$B_0^{[n+1]} = B_0[t, z_1, \ldots, z_n] \subset D = B[z_1^{\pm 1}, \ldots, z_n^{\pm 1}] = \alpha(C)$$
$$= \alpha(A)[\alpha(y_1)^{\pm 1}, \ldots, \alpha(y_n)^{\pm 1}] = B_0^{[\pm n]}$$

which is absurd: On one hand, $B_0^{[n+1]} \subset D$ shows $\text{tr.deg}_{B_0} D \geq n+1$. On the other hand, $D = B_0^{[\pm n]}$ shows $\text{tr.deg}_{B_0} D = n$. Therefore, $B_0 = B$. □

Example 5.1. Assume that $\mathbf{k}$ is an algebraically closed field, and let $C_1, \ldots, C_m$ ($m \geq 1$) be factorial affine curves over $\mathbf{k}$ other than $\mathbb{A}^1$ or $\mathbb{T}^1$. Form the affine $\mathbf{k}$-variety $X = C_1 \times \cdots \times C_n$, which is smooth, rational, and factorial. Theorem 5.1(b) implies that, if Y is an algebraic variety such that $X \times \mathbb{T}^n \cong_{\mathbf{k}} Y \times \mathbb{T}^n$ for some $n \geq 0$, then $X \cong_{\mathbf{k}} Y$. To see this, we show that the coordinate ring $A = \mathcal{O}(X)$ has $A = \mathbf{k}[A_{\mathbf{k}}^\tau]$. By hypothesis, each ring $A_i = \mathcal{O}(C_i)$, $i = 1, \ldots, m$, is a one-dimensional affine UFD other than $\mathbf{k}^{[1]}$ or $\mathbf{k}^{[\pm 1]}$. For such rings, it is known that $A_i = \mathbf{k}[(A_i)_{\mathbf{k}}^\tau]$; see the proof of Lemma 2.8 of [7]. It follows that $A = \mathbf{k}[A_{\mathbf{k}}^\tau]$.

6 Remarks

Remark 6.1. For the R-algebra A, the question of Laurent cancellation over R is closely related to the question whether $\mathcal{N}_R(A^{[\pm n]}) = \mathcal{N}_R(A)$. In particular, we ask if this equality holds in the case $\text{tr.deg}_R A = 1$. A related question is the following: While a $\mathbb{Z}$-grading of $A^{[\pm n]}$ may not restrict to A, it does give a $\mathbb{Z}$-filtration of both

$A^{[\pm n]}$ and A. Let $\mathrm{Gr}(A^{[\pm n]})$ and $\mathrm{Gr}(A)$ denote the associated graded rings. Does $\mathrm{Gr}(A^{[\pm n]}) = \mathrm{Gr}(A)^{[\pm n]}$?

Remark 6.2. We are not aware of an example of an integral domain R and an R-algebra A such that A fails to have the Laurent cancellation property over R.

Acknowledgements The author wishes to thank the organizers of the GABAG2012 Conference in Levico, Italy. The stimulating environment of that conference gave impetus to the work represented in this paper.

References

1. S. Abhyankar, P. Eakin, W. Heinzer, On the uniqueness of the coefficient ring in a polynomial ring. J. Algebra **23**, 310–342 (1972)
2. T. Asanuma, Polynomial fibre rings of algebras over noetherian rings. Invent. Math. **87**, 101–127 (1987)
3. A. Beauville, J.-L. Colliot-Thélène, J.-J. Sansuc, P. Swinnerton-Dyer, Variétés stablement rationnelles non rationnelles. Ann. Math. **121**(2), 283–318 (1985)
4. S.M. Bhatwadekar, N. Gupta, The structure of a Laurent polynomial fibration in n variables. J. Algebra **353**, 142–157 (2012)
5. A. Crachiola, L. Makar-Limanov, On the rigidity of small domains. J. Algebra **284**, 1–12 (2005)
6. W. Danielewski, On the cancellation problem and automorphism groups of affine algebraic varieties, Warsaw. Preprint (1989)
7. G. Freudenburg, *Algebraic Theory of Locally Nilpotent Derivations*. Encyclopaedia of Mathematical Sciences, vol. 136 (Springer, Berlin, Heidelberg, New York, 2006)
8. T. Fujita, On Zariski problem. Proc. Jpn. Acad. **55A**, 106–110 (1979)
9. N. Gupta, On the cancellation problem for the affine space $\mathbb{A}^3$ in characteristic p. Invent. Math. **195**, 279–288 (2014)
10. M. Hochster, Non-uniqueness of coefficient rings in a polynomial ring. Proc. Am. Math. Soc. **34**, 81–82 (1972)
11. H. Kraft, Challenging problems on affine n-space. Semináire Bourbaki **802**, 295–317 (1995)
12. L. Makar-Limanov, *Locally Nilpotent Derivations, a New Ring Invariant and Applications*. Lecture Notes (Bar-Ilan University, 1998). http://www.math.wayne.edu/~lml/
13. M. Miyanishi, T. Sugie, Affine surfaces containing cylinderlike open sets. J. Math. Kyoto Univ. **20**, 11–42 (1980)
14. M. Nagata, A theorem on valuation rings and its applications. Nagoya Math. J. **29**, 85–91 (1967)
15. K.P. Russell, On affine-ruled rational surfaces. Math. Ann. **255**, 287–302 (1981)

Deformations of $\mathbb{A}^1$-Fibrations

Rajendra V. Gurjar, Kayo Masuda, and Masayoshi Miyanishi

Abstract Let B be an integral domain which is finitely generated over a subdomain R and let D be an R-derivation on B such that the induced derivation $D_{\mathfrak{m}}$ on $B \otimes_R R/\mathfrak{m}$ is locally nilpotent for every maximal ideal $\mathfrak{m}$. We ask if D is locally nilpotent. Theorem 2.1 asserts that this is the case if B and R are affine domains. We next generalize the case of G_a-action treated in Theorem 2.1 to the case of $\mathbb{A}^1$-fibrations and consider the log deformations of affine surfaces with $\mathbb{A}^1$-fibrations. The case of $\mathbb{A}^1$-fibrations of affine type behaves nicely under log deformations, while the case of $\mathbb{A}^1$-fibrations of complete type is more involved [see Dubouloz–Kishimoto (Log-uniruled affine varieties without cylinder-like open subsets, arXiv: 1212.0521, 2012)]. As a corollary, we prove the generic triviality of $\mathbb{A}^2$-fibration over a curve and generalize this result to the case of affine pseudo-planes of ML_0-type under a suitable monodromy condition.

2000 Mathematics Subject Classification: Primary: 14R20; Secondary: 14R25

R.V. Gurjar
School of Mathematics, Tata Institute of Fundamental Research, Homi Bhabha Road, Mumbai 400001, India
e-mail: gurjar@math.tifr.res.in

K. Masuda
School of Science & Technology, Kwansei Gakuin University, 2-1 Gakuen, Sanda 669-1337, Japan
e-mail: kayo@kwansei.ac.jp

M. Miyanishi (✉)
Research Center for Mathematical Sciences, Kwansei Gakuin University, 2-1 Gakuen, Sanda 669-1337, Japan
e-mail: miyanisi@kwansei.ac.jp

I. Cheltsov et al. (eds.), *Automorphisms in Birational and Affine Geometry*, Springer Proceedings in Mathematics & Statistics 79,
DOI 10.1007/978-3-319-05681-4_19,

1 Introduction

An $\mathbb{A}^1$-fibration $\rho : X \to B$ on a smooth affine surface X to a smooth curve B is given as the quotient morphism of a G_a-action if the parameter curve B is an affine curve (see [8]). Meanwhile, it is not so if B is a complete curve. When we deform the surface X under a suitable setting (log deformation), our question is if the neighboring surfaces still have $\mathbb{A}^1$-fibrations of affine type or of complete type according to the type of the $\mathbb{A}^1$-fibration on X being affine or complete. Assuming that the neighboring surfaces have $\mathbb{A}^1$-fibrations, the propagation of the type of $\mathbb{A}^1$-fibration is proved in Lemma 3.2, whose proof reflects the structure of the boundary divisor at infinity of an affine surface with $\mathbb{A}^1$-fibration. The stability of the boundary divisor under small deformations, e.g., the stability of the weighted dual graphs has been discussed in topological methods (e.g., [26]). Furthermore, if such property is inherited by the neighboring surfaces, we still ask if the ambient threefold has an $\mathbb{A}^1$-fibration or equivalently if the generic fiber has an $\mathbb{A}^1$-fibration.

The answer to this question is subtle. We consider first in Sect. 2 the case where each of the fiber surfaces of the deformation has an $\mathbb{A}^1$-fibration of affine type induced by a global vector field on the ambient threefold. This global vector field is in fact given by a locally nilpotent derivation (Theorem 2.1). If the $\mathbb{A}^1$-fibrations on the fiber surfaces are of affine type, we can show (Theorem 3.8), with the absence of monodromies of boundary components that there exists an $\mathbb{A}^1$-fibration on the ambient threefold such that the $\mathbb{A}^1$-fibration on each general fiber surface is induced by the global one up to an automorphism of the fiber surface. The proof of Theorem 3.8 depends on Lemma 3.2 which we prove by observing the behavior of the boundary rational curves. This is done by the use of Hilbert scheme (see [21]) and by killing monodromies by étale finite changes of the base curve.

As a consequence, we can prove the generic triviality of an $\mathbb{A}^2$-fibration over a curve. Namely, if $f : Y \to T$ is a smooth morphism from a smooth affine threefold to a smooth affine curve such that the fiber over every *closed* point of T is isomorphic to the affine plane $\mathbb{A}^2$, then the generic fiber of f is isomorphic to $\mathbb{A}^2$ over the function field $k(T)$ of T and f is an $\mathbb{A}^2$-bundle over an open set of T (see Theorem 3.10). This fact, together with a theorem of Sathaye [29], shows that f is an $\mathbb{A}^2$-bundle over T in the Zariski topology.

The question on the generic triviality is also related to a question on the triviality of a k-form of a surface with an $\mathbb{A}^1$-fibration (see Problem 3.13). In the case of an $\mathbb{A}^1$-fibration of complete type, the answer is negative by Dubouloz–Kishimoto [3] (see Theorem 6.1).

Theorem 3.10 was proved by our predecessors Kaliman–Zaidenberg [16] in a more comprehensive way and without assuming that the base is a curve. The idea in our first proof of Theorem 3.10 is of more algebraic nature and consists of using the existence of a locally nilpotent derivation on the coordinate ring of Y and the second proof of using the Ramanujam–Morrow graph of the normal minimal completion of $\mathbb{A}^2$ was already used in [16]. The related results are also discussed in the article

[23, 28]. We cannot still avoid the use of a theorem of Kambayashi [13] on the absence of separable forms of the affine plane.

Some of the algebro-geometric arguments using Hilbert scheme in Sect. 3 can be replaced by topological arguments using Ehresmann's theorem which might be more appreciated than the use of the Hilbert scheme. But they are restricted to the case of small deformations. This is done in Sect. 4.

In Sect. 5, we extend the above result on the generic triviality of an $\mathbb{A}^2$-fibration over a curve by replacing $\mathbb{A}^2$ by an affine pseudo-plane of ML_0-type which has properties similar to $\mathbb{A}^2$, e.g., the boundary divisor for a minimal normal completion is a linear chain of rational curves. But we still need a condition on the monodromy. An affine pseudo-plane, not necessarily of ML_0-type, is a $\mathbb{Q}$-homology plane, and we note that Flenner–Zaidenberg [5] made a fairly exhaustive consideration for the log deformations of $\mathbb{Q}$-homology planes.

In the final section six, we observe the case of $\mathbb{A}^1$-fibration of complete type and show by an example of Dubouloz–Kishimoto [3] that the ambient threefold does not have an $\mathbb{A}^1$-fibration. But it is still plausible that the ambient threefold is affine-uniruled in the stronger sense that the fiber product of the ambient deformation space by a suitable lifting of the base curve has a global $\mathbb{A}^1$-fibration. But this still remains open.

We use two notations for the intersection of (not necessarily irreducible) subvarieties A, B of codimension one in an ambient threefold. Namely, $A \cap B$ is the intersection of two subvarieties, and $A \cdot B$ is the intersection of effective divisors. In most cases, both are synonymous.

As a final remark, we note that a preprint of Flenner–Kaliman–Zaidenberg [6] recently uploaded on the web treats also deformations of surfaces with $\mathbb{A}^1$-fibrations.

The referees pointed out several flaws in consideration of monodromies in the preliminary versions of the article. In particular, we are indebted for Example 3.6 to one of the referees. The authors are very grateful to the referees for their comments and advice.

2 Triviality of Deformations of Locally Nilpotent Derivations

Let k be an algebraically closed field of characteristic zero which we fix as the ground field. Let $Y = \operatorname{Spec} B$ be an irreducible affine algebraic variety. We define the tangent sheaf $\mathcal{T}_{Y/k}$ as $\mathcal{H}om_{\mathcal{O}_Y}(\Omega^1_{Y/k}, \mathcal{O}_Y)$. A regular vector field on Y is an element of $\Gamma(Y, \mathcal{T}_{Y/k})$. A regular vector field Θ on Y is identified with a derivation D on B via isomorphisms

$$\Gamma(Y, \mathcal{T}_{Y/k}) \cong \operatorname{Hom}_B(\Omega^1_{B/k}, B) \cong \operatorname{Der}_k(B, B).$$

We say that Θ is *locally nilpotent* if so is D. In the first place, we are interested in finding a necessary and sufficient condition for D to be locally nilpotent. Suppose that Y has a fibration $f : Y \to T$. A natural question is to ask whether D is locally nilpotent if the restriction of D on each closed fiber of f is locally nilpotent. The following result shows that this is the case.[1]

Theorem 2.1. *Let $Y = \text{Spec } B$ and $T = \text{Spec } R$ be irreducible affine varieties defined over k and let $f : Y \to T$ be a dominant morphism such that general fibers are irreducible and reduced. We consider R to be a subalgebra of B. Let D be an R-trivial derivation of B such that, for each closed point $t \in T$, the restriction $D_t = D \otimes_R R/\mathfrak{m}$ is a locally nilpotent derivation of $B \otimes_R R/\mathfrak{m}$, where $\mathfrak{m}$ is the maximal ideal of R corresponding to t. Then D is locally nilpotent.*

We need some preliminary results. We retain the notations and assumptions in the above theorem.

Lemma 2.2. *There exist a finitely generated field extension k_0 of the prime field $\mathbb{Q}$ which is a subfield of the ground field k, geometrically integral affine varieties $Y_0 = \text{Spec } B_0$ and $T_0 = \text{Spec } R_0$, a dominant morphism $f_0 : Y_0 \to T_0$ and an R_0-trivial derivation D_0 of B_0 such that the following conditions are satisfied:*

(1) *Y_0, T_0, f_0 and D_0 are defined over k_0.*
(2) *$Y = Y_0 \otimes_{k_0} k, T = T_0 \otimes_{k_0} k, f = f_0 \otimes_{k_0} k$ and $D = D_0 \otimes_{k_0} k$.*
(3) *D_0 is locally nilpotent if and only if so is D.*

Proof. Since B and R are integral domains finitely generated over k, write B and R as the residue rings of certain polynomial rings over k modulo the finitely generated ideals. Write $B = k[x_1, \ldots, x_r]/I$ and $R = k[t_1, \ldots, t_s]/J$. Furthermore, the morphism f is determined by the images $f^*(\eta_j) = \varphi_j(\xi_1, \ldots, \xi_r)$ in B, where $\xi_i = x_i \pmod I$ and $\eta_j = t_j \pmod J$. Adjoin to $\mathbb{Q}$ all coefficients of the finite generators of I and J as well as the coefficients of the φ_j to obtain a subfield k_0 of k. Let $B_0 = k_0[x_1, \ldots, x_r]/I_0$ and $R_0 = k_0[t_1, \ldots, t_s]/J_0$, where I_0 and J_0 are, respectively, the ideals in $k_0[x_1, \ldots, x_r]$ and $k_0[t_1, \ldots, t_s]$ generated by the same generators of I and J. Furthermore, define the homomorphism f_0^* by the assignment $f_0^*(\eta_j) = \varphi_j(\xi_1, \ldots, \xi_r)$. Let $Y_0 = \text{Spec } B_0$, $T_0 = \text{Spec } R_0$ and let $f_0 : Y_0 \to T_0$ be the morphism defined by f_0^*. The derivation D corresponds to a B-module homomorphism $\delta : \Omega^1_{B/R} \to B$. Since $\Omega^1_{B/R} = \Omega^1_{B_0/R_0} \otimes_{k_0} k$, we can enlarge k_0 so that there exists a B_0-homomorphism $\delta_0 : \Omega^1_{B_0/R_0} \to B_0$ satisfying $\delta = \delta_0 \otimes_{k_0} k$. Let $D_0 = \delta_0 \cdot d_0$, where $d_0 : B_0 \to \Omega^1_{B_0/R_0}$ is the standard differentiation. Then we have $D = D_0 \otimes_{k_0} k$.

Let $\Phi_0 : B_0 \to B_0[[u]]$ be the R_0-homomorphism into the formal power series ring in u over B_0 defined by

$$\Phi_0(b_0) = \sum_{i \geq 0} \frac{1}{i!} D_0^i(b_0) u^i \ .$$

[1]The result is also remarked in [3, Remark 13].

Let $\Phi : B \to B[[u]]$ be the R-homomorphism defined in a similar fashion. Then Φ_0 and Φ are determined by the images of the generators of B_0 and B. Since the generators of B_0 and B are the same, we have $\Phi = \Phi_0 \otimes_{k_0} k$. The derivation D_0 is locally nilpotent if and only if Φ_0 splits via the polynomial subring $B_0[u]$ of $B_0[[u]]$. This is the case for D as well. Since Φ_0 splits via $B_0[u]$ if and only if Φ splits via $B[u]$, D_0 is locally nilpotent if and only if so is D. □

Lemma 2.3. *Let k_1 be the algebraic closure of k_0 in k. Let $Y_1 = \operatorname{Spec} B_1$ with $B_1 = B_0 \otimes_{k_0} k_1$, $T_1 = \operatorname{Spec} R_1$ with $R_1 = R_0 \otimes_{k_0} k_1$ and $f_1 = f_0 \otimes_{k_0} k_1$. Let $D_1 = D_0 \otimes_{k_0} k_1$. Then the following assertions hold.*

(1) *Let t_1 be a closed point of T_1. Then the restriction of D_1 on the fiber $f_1^{-1}(t_1)$ is locally nilpotent.*
(2) *D_1 is locally nilpotent if and only if so is D.*

Proof. (1) Let t be the unique closed point of T lying over t_1 by the projection morphism $T \to T_1$, where $R = R_1 \otimes_{k_1} k$. (If $\mathfrak{m}_1$ is the maximal ideal of R_1 corresponding to t_1, $\mathfrak{m}_1 \otimes_{k_1} k$ is the maximal ideal of R corresponding to t.) Then $F_t = f^{-1}(t) = f_1^{-1}(t_1) \otimes_{k_1} k$ and the restriction D_t of D onto F_t is given as $D_{1,t_1} \otimes_{k_1} k$, where D_{1,t_1} is the restriction of D_1 onto $f_1^{-1}(t_1)$. We consider also the R-homomorphism $\Phi : B \to B[[u]]$ and the R_1-homomorphism $\Phi_1 : B_1 \to B_1[[u]]$. As above, let $\mathfrak{m}$ and $\mathfrak{m}_1$ be the maximal ideals of R and R_1 corresponding to t and t_1. Then D_t gives rise to the $R/\mathfrak{m}$-homomorphism $\Phi \otimes_R R/\mathfrak{m} : B \otimes_R R/\mathfrak{m} \to (B \otimes_R R/\mathfrak{m})[[u]]$. Similarly, D_{1,t_1} gives rise to the $R_1/\mathfrak{m}_1$-homomorphism $\Phi_1 \otimes_{R_1} R_1/\mathfrak{m}_1 : B_1 \otimes_{R_1} R_1/\mathfrak{m}_1 \to (B_1 \otimes_{R_1} R_1/\mathfrak{m}_1)[[u]]$, where $R/\mathfrak{m} = k$ and $R_1/\mathfrak{m}_1 = k_1$. Then $\Phi \otimes_R R/\mathfrak{m} = (\Phi_1 \otimes_{R_1} R_1/\mathfrak{m}_1) \otimes_{k_1} k$. Hence $\Phi \otimes_R R/\mathfrak{m}$ splits via $(B \otimes_R R/\mathfrak{m})[u]$ if and only if $\Phi_1 \otimes_{R_1} R_1/\mathfrak{m}_1$ splits via $(B_1 \otimes_{R_1} R_1/\mathfrak{m}_1)[u]$. Hence $D_{1,t}$ is locally nilpotent as so is D_t.
(2) The same argument as above using the homomorphism Φ can be applied.

□

The field k_0 can be embedded into the complex field $\mathbb{C}$ because it is a finitely generated field extension of $\mathbb{Q}$. Hence we can extend the embedding $k_0 \hookrightarrow \mathbb{C}$ to the algebraic closure k_1. Thus k_1 is viewed as a subfield of $\mathbb{C}$. Then Lemma 2.3 holds if one replaces the extension k/k_1 by the extension $\mathbb{C}/k_1$. Hence it suffices to prove Theorem 2.1 with an additional hypothesis $k = \mathbb{C}$.

Lemma 2.4. *Theorem 2.1 holds if k is the complex field $\mathbb{C}$.*

Proof. Let $Y(\mathbb{C})$ be the set of closed points which we view as a complex analytic space embedded into a complex affine space $\mathbb{C}^N$ as a closed set. Consider the Euclidean metric on $\mathbb{C}^N$ and the induced metric topology on $Y(\mathbb{C})$. Then $Y(\mathbb{C})$ is a complete metric space.

Let b be a nonzero element of B. For a positive integer m, define a Zariski closed subset $Y_m(b)$ of $Y(\mathbb{C})$ by

$$Y_m(b) = \{Q \in Y(\mathbb{C}) \mid D^m(b)(Q) = 0\}\,.$$

Since Q lies over a closed point t of $T(\mathbb{C})$ and D_t is locally nilpotent on $f^{-1}(t)$ by the hypothesis, we have

$$f^{-1}(t) \subset \bigcup_{m>0} Y_m(b)\,.$$

This implies that $Y(\mathbb{C}) = \bigcup_{m>0} Y_m(b)$. We claim that $Y(\mathbb{C}) = Y_m(b)$ for some $m > 0$. In fact, this follows by Baire category theorem, which states that if the $Y_m(b)$ are all proper closed subsets, its countable union cannot cover the uncountable set $Y(\mathbb{C})$. If $Y(\mathbb{C}) = Y_m(b)$ for some $m > 0$, then $D^m(b) = 0$. This implies that D is locally nilpotent on B.

One can avoid the use of Baire category theorem in the following way. Suppose that $Y_m(b)$ is a proper closed subset for every $m > 0$. Let H be a general hyperplane in $\mathbb{C}^N$ such that the section $Y(\mathbb{C}) \cap H$ is irreducible, $\dim Y(\mathbb{C}) \cap H = \dim Y(\mathbb{C}) - 1$, and $Y(\mathbb{C}) \cap H = \bigcup_{m>0} (Y_m(b) \cap H)$ with $Y_m(b) \cap H$ a proper closed subset of $Y(\mathbb{C}) \cap H$ for every $m > 0$. We can further take hyperplane sections and find a general linear subspace L in $\mathbb{C}^N$ such that $Y(\mathbb{C}) \cap L$ is an irreducible curve and $Y(\mathbb{C}) \cap L = \bigcup_{m>0} (Y_m(b) \cap L)$, where $Y_m(b) \cap L$ is a proper Zariski closed subset. Hence $Y_m(b) \cap L$ is a finite set, and $\bigcup_{m>0} (Y_m(b) \cap L)$ is a countable set, while $Y(\mathbb{C}) \cap L$ is not a countable set. This is a contradiction. Thus $Y(\mathbb{C}) = Y_m(b)$ for some $m > 0$. □

Let D be a k-derivation on a k-algebra B. It is called *surjective* if D is so as a k-linear mapping. The following result is a consequence of Theorem 2.1

Corollary 2.5. *Let $Y = \operatorname{Spec} B$, $T = \operatorname{Spec} R$ and $f : Y \to T$ be the same as in Theorem 2.1. Let D be an R-derivation of B such that D_t is a surjective k-derivation for every closed point $t \in T$. Assume further that the relative dimension of f is one. Then D is a locally nilpotent derivation and f is an $\mathbb{A}^1$-fibration.*

Proof. Let t be a closed point of T such that the fiber $f^{-1}(t)$ is irreducible and reduced. By [9, Theorem 1.2 and Proposition 1.7], the coordinate ring $B \otimes R/\mathfrak{m}$ of $f^{-1}(t)$ is a polynomial ring $k[x]$ in one variable and $D_t = \partial/\partial x$, where $\mathfrak{m}$ is the maximal ideal of R corresponding to t. Then D_t is locally nilpotent. Taking the base change of $f : Y \to T$ by $U \hookrightarrow T$ if necessary, where U is a small open set of T, we may assume that D_t is locally nilpotent for every closed point t of T. By Theorem 2.1, the derivation D is locally nilpotent and hence f is an $\mathbb{A}^1$-fibration. □

3 Deformations of $\mathbb{A}^1$-Fibrations of Affine Type

In the present section, we assume that the ground field k is the complex field $\mathbb{C}$. Let X be an affine algebraic surface which is normal. Let $p : X \to C$ be an $\mathbb{A}^1$-fibration, where C is an algebraic curve which is either affine or projective and p is surjective. We say that the $\mathbb{A}^1$-fibration p is of *affine type* (resp. *complete type*) if C is affine (resp. complete). The $\mathbb{A}^1$-fibration on X is the quotient morphism of a G_a-action on X if and only if it is of affine type (see [8]). We consider the following result on deformations. For the complex analytic case, one can refer to [19] and also to [12, p. 269].

Lemma 3.1. *Let $\overline{f} : \overline{Y} \to T$ be a smooth projective morphism from a smooth algebraic threefold $\overline{Y}$ to a smooth algebraic curve T. Let C be a smooth rational complete curve contained in $\overline{Y}_0 = \overline{f}^{-1}(t_0)$ for a closed point t_0 of T.*[2] *Then the following assertions hold.*

(1) *The Hilbert scheme* $\mathrm{Hilb}(\overline{Y})$ *has dimension less than or equal to $h^0(C, \mathrm{N}_{C/\overline{Y}})$ in the point $[C]$. If $h^1(C, \mathrm{N}_{C/\overline{Y}}) = 0$, then the equality holds and* $\mathrm{Hilb}(\overline{Y})$ *is smooth at $[C]$. Here $\mathrm{N}_{C/\overline{Y}}$ denotes the normal bundle of C in $\overline{Y}$.*
(2) *Let $n = (C^2)$ on $\overline{Y}_0$. Then $\mathrm{N}_{C/\overline{Y}} \cong \mathcal{O}_C \oplus \mathcal{O}_C(n)$ provided $n \geq -1$.*
(3) *Suppose $n = 0$. Then there exists an étale finite covering $\sigma_2 : T' \to T$ such that the morphism $\overline{f}_{T'}$ splits as*

$$\overline{f}_{T'} : \overline{Y} \times_T T' \xrightarrow{\varphi} V \xrightarrow{\sigma_1} T' ,$$

where φ is a $\mathbb{P}^1$-fibration with C contained as a fiber and σ_1 makes V a smooth T'-scheme of relative dimension one with irreducible fibers. Assume further that every smooth rational complete curve C' in $\overline{Y}_0$ satisfies $(C' \cdot C) = 0$ provided C' is algebraically equivalent to C in $\overline{Y}$. Then the covering $\sigma_2 : T' \to T$ is trivial, i.e., σ_2 is the identity morphism.
(4) *Suppose $n = -1$. Then C does not deform in the fiber $\overline{Y}_0$ but deforms along the morphism $\overline{f}$ after an étale finite base change. Namely, there are an étale finite morphism $\sigma : T' \to T$ and an irreducible subvariety Z of codimension one in $\overline{Y}' := \overline{Y} \times_T T'$ such that Z can be contracted along the fibers of $\overline{f}' : \overline{Y}' \to T'$, where T' is an irreducible smooth affine curve and $\overline{f}'$ is the second projection of $\overline{Y} \times_T T'$ to T'.*
(5) *Assume that there are no (-1)-curves E and E' in $\overline{Y}_0$ such that $E \cap E' \neq \emptyset$ and E is algebraically equivalent to E' as 1-cycles on $\overline{Y}$. Then, after shrinking T to a smaller open set if necessary, we can take Z in the assertion (4) above*

[2] When we write $t \in T$, we tacitly assume that t is a closed point of T.

as a subvariety of $\overline{Y}$. *The contraction of* Z *gives a factorization* $\overline{f}|_Z : Z \xrightarrow{g} T' \xrightarrow{\sigma} T$, *where* g *is a* $\mathbb{P}^1$*-fibration,* C *is a fiber of* g *and* σ *is as above.*

Proof. (1) The assertion follows from Grothendieck [7, Corollary 5.4].

(2) We have an exact sequence

$$0 \to \mathrm{N}_{C/\overline{Y}_0} \to \mathrm{N}_{C/\overline{Y}} \to \mathrm{N}_{\overline{Y}_0/\overline{Y}}|_C \to 0\,,$$

where $\mathrm{N}_{C/\overline{Y}_0} \cong \mathcal{O}_C(n)$ and $\mathrm{N}_{\overline{Y}_0/\overline{Y}}|_C \cong \mathcal{O}_C$. The obstruction for this exact sequence to split lies in $\mathrm{Ext}^1(\mathcal{O}_C, \mathcal{O}_C(n)) \cong \mathrm{H}^1(C, \mathcal{O}_C(n))$, which is zero if $n \geq -1$.

(3) Suppose $n = 0$. Then $\dim_{[C]} \mathrm{Hilb}(\overline{Y}) = 2$ and $[C]$ is a smooth point of $\mathrm{Hilb}(\overline{Y})$. Let H be a relatively ample divisor on $\overline{Y}/T$ and set $P(n) := P_C(n) = h^0(C, \mathcal{O}_C(nH))$ the Hilbert polynomial in n of C with respect to H. Then $\mathrm{Hilb}^P(\overline{Y})$ is a scheme which is projective over T. Let V be the irreducible component of $\mathrm{Hilb}^P(\overline{Y})$ containing the point $[C]$. Then V is a T-scheme with a morphism $\sigma : V \to T$, $\dim V = 2$ and V has relative dimension one over T. Furthermore, there exists a subvariety Z of $\overline{Y} \times_T V$ such that the fibers of the composite morphism

$$g : Z \hookrightarrow \overline{Y} \times_T V \xrightarrow{p_2} V$$

are curves on $\overline{Y}$ parametrized by V. For a general point $v \in V$, the corresponding curve $C' := C_v$ is a smooth rational complete curve because $P_{C'}(n) = P(n)$ and $(C')^2 = 0$ on $\overline{Y}_t = \overline{f}^{-1}(t)$ with $t = \sigma(v)$ because $\dim_{[C']} \mathrm{Hilb}(\overline{Y}) = 2$. In fact, if $(C')^2 \leq -1$, then the exact sequence of normal bundles in (2) implies $h^0(C', \mathrm{N}_{C'/\overline{Y}}) \leq 1$, which contradicts $\dim_{[C']}$ $\mathrm{Hilb}(\overline{Y}) = 2$. If $(C')^2 > 0$, then $\dim_{[C']} \mathrm{Hilb}(\overline{Y}) > 2$, which is again a contradiction. So, $(C')^2 = 0$. Hence $\overline{Y}_t$ has a $\mathbb{P}^1$-fibration $\varphi_t : \overline{Y}_t \to \overline{B}_t$ such that C' is a fiber and $\overline{B}_t$ is a smooth complete curve. By the universality of the Hilbert scheme, there are an open set U of $\overline{B}_t$ and a morphism $\alpha_t : U \to V_t$ such that $\varphi_t^{-1}(U) = Z \times_V U$. Since V is smooth over T, α_t induces an isomorphism from $\overline{B}_t$ to a connected component of $V_t := \sigma^{-1}(t)$. This is the case if we take $v \in V$ from a different connected component of V_t. Let $\sigma : V \xrightarrow{\sigma_1} T' \xrightarrow{\sigma_2} T$ be the Stein factorization of σ. Then σ_2 is an étale finite morphism and $\sigma_1 : V \to T'$ is a smooth morphism of relative dimension one with irreducible fibers. Furthermore, the morphism g above factors as a composite of T'-morphsims

$$g : Z \hookrightarrow (\overline{Y} \times_T T') \times_{T'} V \xrightarrow{p_2} V\,,$$

where Z is identified with $\overline{Y} \times_T T'$ by the above construction. Hence g induces a T'-morphism $\varphi : \overline{Y} \times_T T' \to V$ such that the composite $\sigma_1 \cdot \varphi : \overline{Y} \times_T T' \to V \to T'$ is the pull-back $\overline{f}_{T'} : \overline{Y} \times_T T' \to T'$ of the morphism $\overline{f}$.

In the above argument, we take two curves C, C' corresponding two points v, v' in V_{t_0}. Then C is algebraically equivalent to C' in $\overline{Y}$, and hence $(C \cdot C') = 0$ by the hypothesis. So, $C = C'$ or $C \cap C' = \emptyset$. This implies that C' and C are the fibers of the same $\mathbb{P}^1$-fibration $\varphi_{t_0} : \overline{Y}_{t_0} \to \overline{B}_{t_0}$ and hence that V_{t_0} is irreducible. Namely, $\sigma_2^{-1}(t_0)$ consists of a single point. Hence $\deg \sigma_2 = 1$, i.e., σ_2 is the identity morphism.

(4) Suppose $n = -1$. Then $h^0(C, \mathrm{N}_{C/\overline{Y}}) = 1$ and $h^1(C, \mathrm{N}_{C/\overline{Y}}) = 0$. Hence $\mathrm{Hilb}^P(\overline{Y})$ has dimension one and is smooth at $[C]$, where $P(n) = P_C(n)$ is the Hilbert polynomial of C with respect to H. Let T' be the irreducible component of $\mathrm{Hilb}^P(\overline{Y})$ containing $[C]$. Note that $\dim T' = 1$. Then we find a subvariety Z in $\overline{Y} \times_T T'$ such that C is a fiber of g and every fiber of the T-morphism $g = p_2|_{T'} : Z \to T'$ is a (-1) curve in the fiber $\overline{Y}_t$. In fact, the nearby fibers of C are (-1) curves as a small deformation of C by [19]. Hence, by covering T' by small disks, we know that every fiber of g is a (-1) curve. Further, the projection $\sigma : T' \to T$ is a finite morphism as it is projective and T' is smooth because each fiber is a (-1) curve in $\overline{Y}$ (see the above argument for $[C]$). Furthermore, σ is étale since $\overline{f}$ is locally a product of the fiber and the base in the Euclidean topology. Hence σ induces a local isomorphism between T' and T. This implies that $\overline{Y} \times_T T'$ is a smooth affine threefold and the second projection $\overline{f}' : \overline{Y} \times_T T' \to T'$ is a smooth projective morphism. Now, after an étale finite base change $\sigma : T' \to T$, we may assume that Z is identified with a subvariety of $\overline{Y}$. Since C is a (-1) curve in $\overline{Y}_0$, it is an extremal ray in the cone $\overline{NE}(\overline{Y}_0)$. Since C is algebraically equivalent to the fibers of $g : Z \to T'$, it follows that C is an extremal ray in the relative cone $\overline{NE}(\overline{Y}/T)$. Then it follows from [22, Theorem 3.25] that Z is contracted along the fibers of g in $\overline{Y}$ and the threefold obtained by the contraction is smooth and projective over T.

(5) Let $\sigma^{-1}(t_0) = \{u_1, \ldots, u_d\}$ and let $Z_{u_i} = Z \cdot (\overline{Y} \times \{u_i\})$ for $1 \le i \le d$. Then the Z_{u_i} are the (-1) curves on $\overline{Y}_0$ which are algebraically equivalent to each other as 1-cycles on $\overline{Y}$. By the assumption, $Z_{u_i} \cap Z_{u_j} = \emptyset$ whenever $i \neq j$. This property holds for all $t \in T$ if one shrinks to a smaller open set of t_0. Then we can identify Z with a closed subvariety of $\overline{Y}$. In fact, the projection $p : Z \hookrightarrow Y \times_Y T' \to Y$ is a T-morphism. For the point $t_0 \in T$, the morphism $p \otimes_{\mathcal{O}_{T,t_0}} \hat{\mathcal{O}}_{T,t_0}$ with the completion $\hat{\mathcal{O}}_{T,t_0}$ of $\mathcal{O}_{T,t_0}$ is a direct sum of the closed immersions from $Z \otimes_{\mathcal{O}_{T,t_0}} \hat{\mathcal{O}}_{T',u_i}$ into $Y \otimes_{\mathcal{O}_{T,t_0}} \hat{\mathcal{O}}_{T',u_i}$ for $1 \le i \le r$. So, $p \otimes_{\mathcal{O}_{T,t_0}} \hat{\mathcal{O}}_{T,t_0}$ is a closed immersion. Hence p is a closed immersion locally over t_0 because $\hat{\mathcal{O}}_{T,t_0}$ is faithfully flat over $\mathcal{O}_{T,t_0}$. The rest is the same as in the proof of the assertion (4).

□

Let Y_0 be a smooth affine surface and let $\overline{Y}_0$ be a smooth projective surface containing Y_0 as an open set in such a way that the complement $\overline{Y}_0 \setminus Y_0$ supports a reduced effective divisor D_0 with simple normal crossings. We call $\overline{Y}_0$ a *normal completion* of Y_0 and D_0 the boundary divisor of Y_0. An irreducible component of D_0 is called a (-1) *component* if it is a smooth rational curve with self-intersection number -1. We say that $\overline{Y}_0$ is a *minimal* normal completion if the contraction of a (-1) component of D_0 (if any) results the image of D_0 losing the condition of simple normal crossings.

Let $\overline{f} : \overline{Y} \to T$ be a smooth projective morphism from a smooth algebraic threefold $\overline{Y}$ to a smooth algebraic curve T and let $S = \sum_{i=1}^{r} S_i$ be a reduced effective divisor on $\overline{Y}$ with simple normal crossings. Let $Y = \overline{Y} \setminus S$ and let $f = \overline{f}|_Y$. We assume that for every point $t \in T$, the intersection cycle $D_t = \overline{f}^{-1}(t) \cdot S$ is a reduced effective divisor of $\overline{Y}_t = \overline{f}^{-1}(t)$ with simple normal crossings[3] and $Y_t = Y \cap \overline{Y}_t$ is an affine open set of $\overline{Y}_t$. For a point $t_0 \in T$, we assume that $\overline{Y}_{t_0} = \overline{Y}_0, D_{t_0} = D_0$ and $Y_{t_0} = Y_0$. A collection $(Y, \overline{Y}, S, \overline{f}, t_0)$ is called a *family of logarithmic deformations* of a triple $(Y_0, \overline{Y}_0, D_0)$. We call it simply a *log deformation* of the triple $(Y_0, \overline{Y}_0, D_0)$. Since f is smooth and S is a divisor with simple normal crossings, $(Y, \overline{Y}, S, \overline{f}, t_0)$ is a family of logarithmic deformations in the sense of Kawamata [17, 18]. (See also [20].)

From time to time, we have to make a base change by an étale finite morphism $\sigma : T' \to T$ with irreducible T'. Let $\overline{Y}' = \overline{Y} \times_T T', \overline{f}' = \overline{f} \times_T T', S' = S \times_T T'$ and $Y' = Y \times_T T'$. Since the field extension $k(\overline{Y})/k(T)$ is a regular extension, $\overline{Y}'$ is an irreducible smooth projective threefold, and S' is a divisor with simple normal crossings. Hence $(Y', \overline{Y}', S', \overline{f}', t_0')$ is a family of logarithmic deformations of the triple $(Y_0', \overline{Y}_0', D_0') \cong (Y_0, \overline{Y}_0, D_0)$, where $t_0' \in T'$ with $\sigma(t_0') = t_0$.

We have the following result on logarithmic deformations of affine surfaces with $\mathbb{A}^1$-fibrations.

Lemma 3.2. *Let $(Y, \overline{Y}, S, \overline{f}, t_0)$ be a log deformation of the triple $(Y_0, \overline{Y}_0, D_0)$. Then the following assertions hold.*

(1) *Assume that Y_0 has an $\mathbb{A}^1$-fibration. Then Y_t has an $\mathbb{A}^1$-fibration for every $t \in T$.*
(2) *If Y_0 has an $\mathbb{A}^1$-fibration of affine type (resp. of complete type), then Y_t has also an $\mathbb{A}^1$-fibration of affine type (resp. of complete type) for every $t \in T$.*

Proof. (1) Note that $K_{\overline{Y}_t} = (K_{\overline{Y}} + \overline{Y}_t) \cdot \overline{Y}_t = K_{\overline{Y}} \cdot \overline{Y}_t$ because $\overline{Y}_t$ is algebraically equivalent to $\overline{Y}_{t'}$ for $t' \neq t$. Then $K_{\overline{Y}_t} + D_t = (K_{\overline{Y}} + S) \cdot \overline{Y}_t$. By the hypothesis, $h^0(\overline{Y}_0, \mathcal{O}(n(K_{\overline{Y}_0} + D_0))) = 0$ for every $n > 0$. Then the semicontinuity theorem [11, Theorem 12.8] implies that $h^0(\overline{Y}_t, \mathcal{O}(n(K_{\overline{Y}_t} + D_t))) = 0$ for

[3]In order to avoid the misreading, it is better to specify our definition of simple normal crossings in the case of dimension three. We assume that every irreducible component S_i of S and every fiber $\overline{Y}_t$ are smooth and that analytic-locally at every intersection point P of $S_i \cap S_j$ (resp. $S_i \cap S_j \cap S_k$ or $S_i \cap \overline{Y}_t$), S_i and S_j (resp. S_i, S_j and S_k, or S_i and $\overline{Y}_t$) behave like coordinate hypersurfaces. Hence $S_i \cap S_j$ or $S_i \cap \overline{Y}_t$ are smooth curves at the point P.

every $n > 0$. Hence $\overline{\kappa}(Y_t) = -\infty$. Since Y_t is affine, this implies that Y_t has an $\mathbb{A}^1$-fibration.

(2) Suppose that Y_0 has an $\mathbb{A}^1$-fibration $\rho_0 : Y_0 \to B_0$ which is of affine type. Then ρ_0 defines a pencil Λ_0 on $\overline{Y}_0$.

Suppose first that Λ_0 has no base points and hence defines a $\mathbb{P}^1$-fibration $\overline{\rho}_0 : \overline{Y}_0 \to \overline{B}_0$ such that $\overline{\rho}_0|_{Y_0} = \rho_0$ and $\overline{B}_0$ is a smooth completion of B_0. If $\overline{\rho}_0$ is not minimal, let E be a (-1) curve contained in a fiber of $\overline{\rho}_0$, which is necessarily not contained in Y_0. By Lemma 3.1, E extends along the morphism $\overline{f}$ if one replaces the base T by a suitable étale finite covering T' and can be contracted simultaneously with other (-1) curves contained in the fibers $\overline{Y}_t$ $(t \in T)$. Note that this étale finite change of the base curve does not affect the properties of the fiber surfaces. Hence we may assume that all simultaneous blowing-ups and contractions as applied below are achieved over the base T.

The contraction is performed either within the boundary divisor S or the *simultaneous half-point detachments* in the respective fibers Y_t for $t \in T$. (For the definition of half-point detachment (resp. attachment), see, for example, [4].) Hence the contraction does not change the hypothesis on the simple normal crossing of S and the intersection divisor $S \cdot \overline{Y}_t$. Thus we may assume that $\overline{\rho}_0$ is minimal. Since $B_0 \subsetneqq \overline{B}_0$, a fiber of $\overline{\rho}_0$ is contained in a boundary component, say S_1. Then the intersection $S_1 \cdot \overline{Y}_0$ as a cycle is a disjoint sum of the fibers of $\overline{\rho}_0$ with multiplicity one. Hence $(S_1^2 \cdot \overline{Y}_0) = ((S_1 \cdot \overline{Y}_0)^2)_{\overline{Y}_0} = 0$. Since $\overline{Y}_t$ and $\overline{Y}_0$ are algebraically equivalent, we have $(S_1^2 \cdot \overline{Y}_t) = 0$ for every $t \in T$. Note that $\overline{Y}_t$ is also a ruled surface by Iitaka [12] and minimal by the same reason as for $\overline{Y}_0$. Considering the deformations of a fiber of $\overline{\rho}_0$ appearing in $S_1 \cdot \overline{Y}_0$, we know by Lemma 3.1 that $S_1 \cdot \overline{Y}_t$ is a disjoint sum of smooth rational curves with self-intersection number zero. Namely, $S_1 \cdot \overline{Y}_t$ is a sum of the fibers of a $\mathbb{P}^1$-fibration. Here we may have to replace the $\mathbb{P}^1$-fibration $\overline{\rho}_t$ by the second one if $\overline{Y}_t \cong \mathbb{P}^1 \times \mathbb{P}^1$. In fact, if a smooth complete surface has two different $\mathbb{P}^1$-fibrations and is minimal with respect to one fibration, then the surface is isomorphic to $\mathbb{P}^1 \times \mathbb{P}^1$ and two $\mathbb{P}^1$-fibrations are the vertical and horizontal fibrations. This implies that Y_t has an $\mathbb{A}^1$-fibration of affine type.

Suppose next that Λ_0 has a base point, say P_0, and that the $\mathbb{A}^1$-fibration ρ_0 is of affine type. Then all irreducible components of $D_0 := S \cdot \overline{Y}_0$ are contained in the members of Λ_0. Since the boundary divisor D_0 of $\overline{Y}_0$ is assumed to be a connected divisor with simple normal crossings, there are at most two components of $S \cdot \overline{Y}_0$ passing through P_0, and if there are two of them, they lie on different components of S and P_0 lies on their intersection curve. In particular, if S_1 is a component of S containing P_0, then $S_1 \cdot \overline{Y}_0$ is a disjoint sum of smooth rational curves. Let C_1 be the component of $S_1 \cdot \overline{Y}_0$ passing through P_0 and let F_0 be the member of Λ_0 which contains C_1. We may assume that F_0 is supported by the boundary divisor D_0. If F_0 contains a (-1) curve E such that $P_0 \notin E$, then E extends along the morphism $\overline{f}$ and can be contracted simultaneously along $\overline{f}$ after the base change by an étale finite covering $T' \to T$. So, we may assume that every irreducible component of F_0 not passing P_0 has self-intersection number ≤ -2 on $\overline{Y}_0$. Then we may assume that $(C_1^2)_{\overline{Y}_0} \geq 0$. In fact, if there are two irreducible components of $S \cdot \overline{Y}_0$ passing through P_0 and belonging to the same member F_0 of Λ_0, one of them must have self-intersection number ≥ 0, for otherwise all the components of the member of Λ_0, after the elimination of base points, would have self-intersection number ≤ -2, which is a contradiction. So, we may assume that the one on S_1, i.e., C_1, has self-intersection number ≥ 0. Then the proper transform of C_1 is the unique (-1) curve with multiplicity > 1 in the fiber corresponding to F_0 after the elimination of base points of Λ_0.

On the other hand, $S_1 \cdot \overline{Y}_0$ (as well as $S_i \cdot \overline{Y}_0$ if it is non-empty) is a disjoint sum of smooth rational curves, one of which is C_1. Let $n := (C_1^2)_{\overline{Y}_0} \geq 0$. Then $\mathrm{Hilb}^P(\overline{Y})$ has dimension $n+2$ and is smooth at the point $[C_1]$. Since $C_1 \cong \mathbb{P}^1$ and $\mathrm{N}_{C_1/\overline{Y}} \cong \mathcal{O}(n) \oplus \mathcal{O}$, C_1 extends along the morphism $\overline{f}$. Namely, $\overline{f}|_{S_1} : S_1 \to T$ is a composite of a $\mathbb{P}^1$-fibration $\sigma_1 : S_1 \to T'$ and an étale finite morphism $\sigma_2 : T' \to T$, where C_1 is a fiber of σ_1. By the base change by σ_2, we may assume that $S_1 \cdot \overline{Y}_0 = C_1$. In particular, $(C_1^2)_{S_1} = 0$.

Suppose that C_2 is a component of F_0 meeting C_1. Then C_2 is contained in a different boundary component, say S_2, which intersects S_1. Since $(H \cdot S_2 \cdot \overline{Y}_0) > 0$, we have $(H \cdot S_2 \cdot \overline{Y}_t) > 0$ for every $t \in T$, where H is a relatively ample divisor on $\overline{Y}$ over T. Furthermore, $S_2 \cdot \overline{Y}_0$ is algebraically equivalent to $S_2 \cdot \overline{Y}_t$. Note that $S_2 \cdot \overline{Y}_0$ is a disjoint sum of smooth rational curves, one of which is the curve C_2 connected to C_1. By considering the factorization of $\overline{f}|_{S_2} : S_2 \to T$ into a product of a $\mathbb{P}^1$-fibration and an étale finite morphism as in the case for $S_1 \cdot \overline{Y}_0$ and taking the base change by an étale finite morphism, we may assume that $S_2 \cdot \overline{Y}_0 = C_2$. Hence we have

$$(S_1 \cdot S_2 \cdot \overline{Y}_t) = (S_1 \cdot S_2 \cdot \overline{Y}_0) = (C_1 \cdot (S_2 \cdot \overline{Y}_0))_{\overline{Y}_0} = (C_1 \cdot C_2)_{\overline{Y}_0} = 1.$$

This implies that $S_2 \cdot \overline{Y}_t$ is irreducible for a general point $t \in T$. For otherwise, by the Stein factorization of the morphism $\overline{f}|_{S_2} : S_2 \to T$, the fiber $S_2 \cdot \overline{Y}_t$ is a disjoint sum $A_1 + \cdots + A_s$ of distinct irreducible curves which are algebraically equivalent to each other on S_2. Since

$$\begin{aligned} 1 &= (S_1 \cdot S_2 \cdot \overline{Y}_t) = ((S_1 \cdot S_2) \cdot (S_2 \cdot \overline{Y}_t))_{S_2} \\ &= ((S_1 \cdot S_2) \cdot (A_1 + \cdots + A_s))_{S_2} = s((S_1 \cdot S_2) \cdot A_1), \end{aligned}$$

we have $s = 1$ and $(S_1 \cdot S_2 \cdot A_1) = 1$. So, $\overline{f}|_{S_2} : S_2 \to T$ is now a $\mathbb{P}^1$-bundle and $(C_2^2)_{S_2} = 0$. This implies that $\mathrm{N}_{C_2/\overline{Y}} \cong \mathcal{O}(m) \oplus \mathcal{O}$ with $m = (C_2^2)_{\overline{Y}_0} \leq -2$ and that C_2 extends along the morphism $\overline{f}$. We can argue in the same way as above with irreducible components of F_0 other than C_1.

Assume that no members of Λ_0 except F_0 have irreducible components outside of Y_0. If C_i is shown to move on the component S_i along the morphism $\overline{f}$, we consider a component C_{i+1} anew which meets C_i. Each of them is contained in a distinct irreducible boundary component of S and extends along the morphism

$\overline{f}$. Let $S_1, S_2, \ldots, S_r$ be all the boundary components which meet $\overline{Y}_0$ along the irreducible components of F_0. Then $\overline{Y}_t$ intersects $S_1 + S_2 + \cdots + S_r$ in an effective divisor which has the same form as F_0. Furthermore, we have

$$((S_i \cdot \overline{Y}_t)^2)_{\overline{Y}_t} = (S_i^2 \cdot \overline{Y}_t) = (S_i^2 \cdot \overline{Y}_0) = ((S_i \cdot \overline{Y}_0)^2)_{\overline{Y}_0}$$

for $1 \leq i \leq r$. Namely, the components $S_i \cdot \overline{Y}_t$ $(1 \leq i \leq r)$ with the same multiplicities as $S_i \cdot \overline{Y}_0$ in F_0 is a member F_t of the pencil Λ_t lying outside of Y_t. This implies that Λ_t has a base point P_t and at least one member of Λ_t lies outside of Y_t. So, the $\mathbb{A}^1$-fibration ρ_t on Y_t is of affine type.

If the pencil Λ_0 contains two members F_0, F_0' such that the components C_1, C_1' of F_0, F_0' lie outside of Y_0 and pass through the point P_0, we may assume that F_0 is supported by the boundary components, while F_0' may not. Then no other members of Λ_0 have irreducible components outside of Y_0 because $\overline{Y}_0 \setminus Y_0$ is connected. We can argue as above to show, after a suitable étale finite base change, that the member F_0 moves along the morphism $\overline{f}$, and further that every boundary component of F_0' moves on a boundary component, say S_j', as a fiber of $f|_{S_j'} : S_j' \to T$. Hence the pencil Λ_t has the member F_t corresponding to F_0 whose all components lie outside of Y_t and the member F_t' corresponding to F_0'. In fact, the part of F_t' lying outside of Y_t is determined as above, but since $Y_t \cap F_t'$ is a disjoint union of the $\mathbb{A}^1$ which correspond to the (-1) components of F_t' (the *half-point attachments*), the member F_t' is determined up to its weighted graph. This proof also implies that if ρ_0 is of complete type then ρ_t is of complete type for every $t \in T$. □

Remark 3.3. (1) In the above proof of Lemma 3.2, the case where the pencil Λ_0 has a base point P_0 on one of the connected components $S_1 \cap \overline{Y}_0$, say C_1, there might exist a monodromy on $\overline{Y}$ which transform Λ_0 to a pencil Λ_0' on $\overline{Y}_0$ having a base point P_0' on a different connected component C_1' of $S_1 \cap \overline{Y}_0$. However, we have $(C_1^2)_{\overline{Y}_0} \geq 0$ as shown in the proof, and $(C_1^2) = (C_1'^2)$. Since C_1' is contained in a member of Λ_0, whence $(C_1'^2) < 0$. This is a contradiction. So, $S_1 \cap \overline{Y}_0$ is irreducible.

(2) In the step of the above proof of Lemma 3.2 where we assume that no members of Λ_0 except F_0 have irreducible components outside of Y_0, let P_t' be a point on $C_{1,t} := S_1 \cdot \overline{Y}_t$ other than P_t which is the base point of the given pencil Λ_t. Then there is a pencil Λ_t' on $\overline{Y}_t$ which is similar to Λ_t. In fact, note first that $\overline{Y}_t$ is a rational surface. Perform the same blowing-ups with centers at P_t' and its infinitely near points as those with centers at P_t and its infinitely near points which eliminate the base points of Λ_t. Then we find an effective divisor $\tilde{F}_t'$ supported by the proper transforms of $S_i \cdot \overline{Y}_t$ $(1 \leq i \leq r)$ and the exceptional curves of the blowing-ups such that $\tilde{F}_t'$ has the same form and multiplicities as the corresponding member $\tilde{F}_t$ in the proper transform $\tilde{\Lambda}_t$ of Λ_t after the elimination of base points. Then $(\tilde{F}_t')^2 = 0$ and hence $\tilde{F}_t'$ is a fiber of an $\mathbb{P}^1$-fibration on the blown-up surface of $\overline{Y}_t$. Then the fibers of the $\mathbb{P}^1$-fibration form the pencil Λ_t' on $\overline{Y}_t$ after the reversed contractions. In fact,

the surface $Y_t = \overline{Y}_t \setminus D_t$ is the affine plane with two systems of coordinate lines given as the fibers of Λ_t and Λ'_t. Hence the $\mathbb{A}^1$-fibrations induced by Λ_t and Λ'_t are transformed by an automorphism of $\overline{Y}_t$.

□

The following is one of the simplest examples of our situation.

Example 3.4. Let C be a smooth conic and let S be the subvariety of codimension one in $\mathbb{P}^2 \times C$ defined by

$$S = \{(P, Q) \mid P \in L_Q,\ Q \in C\},$$

where L_Q is the tangent line of C at Q. Let $Y = (\mathbb{P}^2 \times C) \setminus S$ and let $f : Y \to C$ be the projection onto C. We set $T = C$ to fit the previous notations. Set $\overline{Y} = \mathbb{P}^2 \times C$. Then $\overline{f} : \overline{Y} \to T$ is the second projection and the boundary divisor S is irreducible. For every point $Q \in C$, $Y_Q := \mathbb{P}^2 \setminus L_Q$ has a linear pencil Λ_Q generated by C and $2L_Q$, which induces an $\mathbb{A}^1$-fibration of affine type. The restriction $\overline{f}|_S : S \to T$ is a $\mathbb{P}^1$-bundle. Let C be defined by $X_0X_2 = X_1^2$ with respect to a system of homogeneous coordinates (X_0, X_1, X_2) of $\mathbb{P}^2$ and let $\eta = (1, t, t^2)$ be the generic point of C with t an inhomogeneous coordinate on $C \cong \mathbb{P}^1$. Then L_η is defined by $t^2X_0 - 2tX_1 + X_2 = 0$. The generic fiber Y_η of f has an $\mathbb{A}^1$-fibration induced by the linear pencil Λ_η whose general members are the conics defined by $(X_0X_2 - X_1^2) + u(t^2X_0 - 2tX_1 + X_2)^2 = 0$, where $u \in \mathbb{A}^1$. Indeed, the conics are isomorphic to $\mathbb{P}^1_{k(t)}$ since they have the $k(t)$-rational point $(1, t, t^2)$, and Y_η is isomorphic to $\mathbb{A}^2_{k(t)}$. This implies that the affine threefold Y itself has an $\mathbb{A}^1$-fibration. □

In the course of the proof of Lemma 3.2, we frequently used the base change by a finite étale morphism $\sigma : T' \to T$, where T' is taken in such a way that for every $t \in T$, the points $\sigma^{-1}(t)$ correspond bijectively to the connected components of $S_i \cap \overline{Y}_t$, where S_i is an irreducible component of S. Suppose that $\deg \sigma > 1$. Let $\overline{Y}' = \overline{Y} \times_T T'$ and $\overline{f}' = \overline{f} \times_T T'$. Note that $\overline{Y}'$ is smooth because σ is étale. The morphism $\sigma : T' \to T$ gives the Stein factorization $\overline{f}|_{S_i} : S_i \xrightarrow{\varphi} T' \xrightarrow{\sigma} T$. Then the subvariety S_i is considered to be a subvariety of $\overline{Y}'$ via a closed immersion $(\mathrm{id}_{S_i}, \varphi) : S_i \to S_i \times_T T' \hookrightarrow \overline{Y} \times_T T'$. We denote it by S'_i. Let t_1, t_2 be points of T' such that they correspond to the connected components A, B of $S_i \cap \overline{Y}_t$, whence $\sigma(t_1) = \sigma(t_2) = t$. Then A, B are the fibers of S'_i over the points t_1, t_2 of T'. Hence A and B are algebraically equivalent in $\overline{Y}$. Since T' is étale over T, we say more precisely that they are *étale-algebraically equivalent*. We have $(A^2)_{\overline{Y}_t} = (B^2)_{\overline{Y}_t}$. In fact, noting that $\overline{Y}'_{t_1}$ and $\overline{Y}'_{t_2}$ are algebraically equivalent in $\overline{Y}'$ and that $\overline{Y}'_{t_1}$ and $\overline{Y}'_{t_2}$ are isomorphic to $\overline{Y}_t$, we have

$$\begin{aligned}(A^2)_{\overline{Y}_t} &= (A^2)_{\overline{Y}'_{t_1}} = (S'_i \cdot S'_i \cdot \overline{Y}'_{t_1}) \\ &= (S'_i \cdot S'_i \cdot \overline{Y}'_{t_2}) = (B^2)_{\overline{Y}'_{t_2}} = (B^2)_{\overline{Y}_t}\end{aligned}$$

Let C be an irreducible curve in $\overline{Y}_0 \cap S$, say a connected component of $\overline{Y}_0 \cap S_1$ with an irreducible component S_1 of S. We say that C has *no monodromy* in $\overline{Y}$ if $\overline{f}|_{S_1} : S_1 \to T$ has no splitting $\overline{f}|_{S_1} : S_1 \xrightarrow{\sigma_1} T' \xrightarrow{\sigma_2} T$, where σ_2 is an étale finite morphism with $\deg \sigma_2 > 1$. Note that, after a suitable étale finite base change $\overline{Y} \times_T T'$, this condition is fulfilled. Namely, *the monodromy is killed*. Concerning the extra hypothesis in Lemma 3.1(3) and the possibility of achieving the contractions over the base curve T in Lemma 3.1(5), we have the following result.

Lemma 3.5. *Let $(Y, \overline{Y}, S, \overline{f}, t_0)$ be a family of logarithmic deformation of the triple $(Y_0, \overline{Y}_0, D_0)$. Assume that Y_0 has an $\mathbb{A}^1$-fibration of affine type. Let Λ_0 be the pencil on $\overline{Y}_0$ whose general members are the closures of fibers of the $\mathbb{A}^1$-fibration. Suppose that Λ_0 defines a $\mathbb{P}^1$-fibration $\varphi_0 : \overline{Y}_0 \to \overline{B}_0$. Suppose further that the section of φ_0 in $S \cap \overline{Y}_0$ has no monodromy in $\overline{Y}$. Then the following assertions hold.*

(1) *If C is a fiber of φ_0 with $C \cap Y_0 \neq \emptyset$ and C' is a smooth rational complete curve which is algebraically equivalent to C in $\overline{Y}$, then $(C \cdot C') = 0$.*
(2) *There are no two (-1) curves E_1 and E_2 such that they belong to the same connected component of the Hilbert scheme* $\mathrm{Hilb}(\overline{Y})$*, E_1 is an irreducible component of a fiber of φ_0 and $E_1 \cap E_2 \neq \emptyset$.*

Proof. (1) Let S_0 be an irreducible component of S such that $(S_0 \cdot F) = 1$ for a general fiber F of φ_0. Then $S_0 \cap \overline{Y}_0$ contains a cross-section of φ_0. The assumption on the absence of the monodromy implies that $S_0 \cap \overline{Y}_0$ is irreducible and is the section of φ_0. Note that φ_0 contains a fiber F_∞ at infinity which is supported by the intersection of $\overline{Y}_0$ with the boundary divisor S in $\overline{Y}$. Such a fiber exists by the assumption that the $\mathbb{A}^1$-fibration on Y_0 is of affine type. Since $S_0 \cap \overline{Y}_0$ gives the cross-section, F_∞ is supported by $S \setminus S_0$. Since $C \cap (S \setminus S_0) = \emptyset$ and C' is algebraically equivalent to C in $\overline{Y}$, C' does not meet the components of $S \setminus S_0$. Hence $C' \cap F_\infty = \emptyset$, and C' is a component of a fiber of φ_0. So, $(C \cdot C') = 0$.[4]

(2) Suppose that such E_1 and E_2 exist. Since E_1 and E_2 are algebraically equivalent 1-cycles on $\overline{Y}$, E_1 and E_2 have the same intersections with subvarieties of codimension one in $\overline{Y}$. We consider possible cases separately.
   (i) Suppose that both E_1 and E_2 are contained in the fiber at infinity F_∞. Since $E_1 \cap E_2 \neq \emptyset$, it follows that $F_\infty = E_1 + E_2$ with $(E_1 \cdot E_2) = 1$. If E_1 meets the section $S_0 \cap \overline{Y}_0$, then $(E_1 \cdot S_0) = 1$, whence $(E_2 \cdot S_0) = 1$ because E_1 and E_2 are algebraically equivalent in $\overline{Y}$. This is a contradiction. Hence $E_1 \cap E_2 = \emptyset$.
   (ii) Suppose that only E_1 is contained in the fiber at infinity F_∞. Take a smooth fiber F_0 of φ_0 with $F_0 \cap Y_0 \neq \emptyset$ and consider a deformation of F_0 in $\overline{Y}$. Then there exist an étale finite morphism $\sigma_2 : T' \to T$ and

[4] We note here that without the condition on the absence of the monodromy of the cross-section, the assertion fails to hold. See Example 3.6.

a decomposition of $\overline{f}_{T'} : \overline{Y} \times_T T' \xrightarrow{\varphi} V \xrightarrow{\sigma_1} T'$ such that F_0 is a fiber of φ (see Lemma 3.1(3)). Let B be an irreducible curve on V such that $\varphi(F_\infty) \notin B$ and let $W = \varphi^{-1}(B)$. Note that E_1 and E_2 are also algebraically equivalent in $\overline{Y} \times_T T'$. Since $(E_1 \cdot W) = 0$ by the above construction, it follows that $(E_2 \cdot W) = 0$. This implies that E_2 is contained in a fiber of φ_0. Hence $E_1 \cap E_2 = \emptyset$.

(iii) Suppose that E_1 and E_2 are not contained in the fiber F_∞. Then E_1 and E_2 are the fiber components of φ_0 because $(E_i \cdot F_\infty) = 0$ for $i = 1, 2$. If they belong to the same fiber, we obtain a contradiction by the same argument as in the case (i). If they belong to different fibers, then $E_1 \cap E_2 = \emptyset$.

□

The following example, which is due to one of the referees of this article, shows that Lemma 3.5(1) does not hold without the monodromy condition on the section of φ_0.

Example 3.6. Let $Q = \mathbb{P}^1 \times \mathbb{P}^1$ and $T' = \mathbb{A}^1_*$ which is the affine line minus one point and hence is the underlying scheme of the multiplicative group G_m. We denote by ℓ (resp. M) a general fiber of the first projection $p_1 : Q \to \mathbb{P}^1$ (resp. the second projection $p_2 : Q \to \mathbb{P}^1$). Let x (resp. y) be an inhomogeneous coordinate on the first (resp. the second) factor of Q. Set $\ell_\infty = p_1^{-1}(\infty)$ and $M_\infty = p_2^{-1}(\infty)$. We consider an involution ι on $Q \times T'$ defined by $(x, y, z) \mapsto (y, x, -z)$, where z is a coordinate of $\mathbb{A}^1_*$. Let Q' be the blowing-up of Q with center $P_\infty := \ell_\infty \cap M_\infty$ and let E be the exceptional curve. Then the involution ι extends to the threefold $Q' \times T'$ in such a way that $E \times T'$ is stable under ι. Let $\overline{Y}$ be the quotient threefold of $Q' \times T'$ by this $\mathbb{Z}_2$-action induced by the involution ι. Since the projection $p_2 : Q' \times T' \to T'$ is $\mathbb{Z}_2$-equivariant, it induces a morphism $\overline{f} : \overline{Y} \to T$, where $T = T'//\mathbb{Z}_2 \cong \mathbb{A}^1_*$. Let $S_1 = ((\ell_\infty \cup M_\infty) \times T')//\mathbb{Z}_2$, $S_2 = (E \times T')//\mathbb{Z}_2$ and $S = S_1 + S_2$. Further, we let $Y = \overline{Y} \setminus S$ and $f : Y \to T$ the restriction of $\overline{f}$ onto Y. Then the following assertions hold:

(1) The surfaces S_1 and S_2 are smooth irreducible surfaces intersecting normally.
(2) Fix a point $t_0 \in T$ and denote the fibers over t_0 with the subscript 0. Then the collection $(Y, \overline{Y}, S, \overline{f}, t_0)$ is a family of logarithmic deformations of the triple $(Y_0, \overline{Y}_0, D_0)$, where $D_0 = S \cdot \overline{Y}_0$.
(3) For every $t \in T$, $S_1 \cap \overline{Y}_t$ is a disjoint union of two smooth curves C_{1t}, C'_{1t} and $S_2 \cap \overline{Y}_t$ is a smooth rational curve C_{2t}, where $(C_{1t} \cdot C_{2t}) = (C'_{1t} \cdot C_{2t}) = 1$ and $(C_{1t}^2) = ({C'_{1t}}^2) = (C_{2t}^2) = -1$. In particular, C_{1t} is étale-algebraically equivalent to C'_{1t}, and hence has a nontrivial monodromy.
(4) Each fiber $\overline{Y}_t$ is isomorphic to Q' with C_{1t}, C'_{1t} and C_{2t} identified with the proper transforms of M_∞, ℓ_∞ on Q' and E.
(5) Let $\varphi_t : \overline{Y}_t \to \mathbb{P}_1$ be the $\mathbb{P}^1$-fibration induced by the first projection $p_1 : Q \to \mathbb{P}^1$. Then a general fiber $\ell = p_1^{-1}(x)$ is algebraically equivalent to $M = p_2^{-1}(x)$ for $x \in T$.
(6) For every $t \in T$, the affine surface Y_t has an $\mathbb{A}^1$-fibration of affine type.

Proof. (1) Since $(Q \setminus (\ell_\infty \cup M_\infty) = \text{Spec } k[x, y]$, the quotient threefold $V = (Q \times T')//\mathbb{Z}_2$ contains an open set $(\mathbb{A}^2 \times T')//\mathbb{Z}_2$, which has the coordinate ring over k generated by elements $X = x + y$, $U = xy$, $Z = z^2$ and $W = (x - y)z$. Hence the open set is a hypersurface $W^2 = Z(X^2 - 4U)$. The quotient threefold V has a similar open neighborhood of the image of the curve $\{P_\infty\} \times T'$. This can be observed by taking inhomogeneous coordinates x', y' on Q such that $x' = 1/x$ and $y' = 1/y$, where $\ell_\infty \cup M_\infty$ is given by $x'y' = 0$. If we put $W' = x' + y'$, $U' = x'y'$ and $W' = (x' - y')z$, the open neighborhood is defined by a similar equation $W'^2 = Z(X'^2 - 4U')$. Then the image of $(\ell_\infty \cup M_\infty) \times T'$ is given by $U' = 0$. Hence it has an equation $W'^2 = ZX'^2$. So, this is a smooth irreducible surface. The curve E has inhomogeneous coordinate x'/y' (or y'/x'). Hence E is stable under the involution ι. Note that the involution ι has no fixed point because there are no fixed points on the factor T'. The surface S_1 is simultaneously contracted along T, and by the contraction, $\overline{Y}$ becomes a $\mathbb{P}^2$-bundle and the surface S_2 becomes an immersed $\mathbb{P}^1$-bundle. Then the assertion (1) follows easily.

(2) The threefold $\overline{Y}$ is smooth and $\overline{f}$ is a smooth morphism. In fact, every closed fiber of $f = \overline{f}|_Y : Y \to T$ is isomorphic to the affine plane.

(3) If $t = z^2$, C_{1t} (resp. C'_{1t}) is identified with M'_∞ (resp. ℓ'_∞) in $Q' \times \{z\}$ and ℓ'_∞ (resp. M'_∞) in $Q' \times \{-z\}$ under the identification $\overline{Y}_t \cong Q' \times \{z\} \cong Q' \times \{-z\}$, where ℓ'_∞ and M'_∞ are the proper transforms of ℓ_∞ and M_∞ on Q'. Now the rest of the assertions are easily verified.

□

A sufficient condition on the absence of the monodromy in Lemma 3.5 is given by the following result.

Lemma 3.7. *Let the notations and the assumptions be the same as in Lemma 3.5 and its proof. Let $S_0 \cap \overline{Y}_0 = C_{01} \cup \cdots \cup C_{0m}$. Suppose that C_{01} is a section of the $\mathbb{P}^1$-fibration φ_0. If $(C_{01}^2) \geq 0$, then C_{01} has no monodromy in $\overline{Y}$. Namely, $m = 1$ and $S_0 \cap \overline{Y}_0$ is irreducible.*

Proof. Suppose that $m > 1$. Note that $C_{02}, \ldots, C_{0m}$ are mutually disjoint and do not meet a general fiber of φ_0 because they lie outside Y_0 and a general fiber meets only C_{01} in the boundary at infinity. This implies that $C_{02}, \ldots, C_{0m}$ are rational curves and the fiber components of φ_0. By the remark given before Lemma 3.5, we have $(C_{0i}^2) = (C_{01}^2) \geq 0$ for $2 \leq i \leq m$. Then $C_{02}, \ldots, C_{0m}$ are full fibers of φ_0 and hence they meet the section C_{01}. This is a contradiction. □

We prove one of our main theorems.

Theorem 3.8. *Let $f : Y \to T$ be a morphism from a smooth affine threefold onto a smooth curve T with irreducible general fibers. Assume that general fibers of f have $\mathbb{A}^1$-fibrations of affine type. Then, after shrinking T if necessary and taking an étale finite morphism $T' \to T$, the fiber product $Y' = Y \times_T T'$ has an $\mathbb{A}^1$-fibration which factors the morphism $f' = f \times_T T'$. Furthermore, suppose that there is a* *relative normal completion* $\overline{f} : \overline{Y} \to T$ *of* $f : Y \to T$ *satisfying the following conditions:*

(1) $(Y, \overline{Y}, S, \overline{f}, t_0)$ *with* $t_0 \in T$ *and* $S = \overline{Y} \setminus Y$ *is a family of logarithmic deformations of* $(Y_0, \overline{Y}_0, D_0)$ *as above, where* $Y_0 = f^{-1}(t_0), \overline{Y}_0 = \overline{f}^{-1}(t_0)$ *and* $D_0 = S \cdot \overline{Y}_0$.
(2) *The given* $\mathbb{A}^1$*-fibration of affine type on each fiber* Y_t *extends to a* $\mathbb{P}^1$*-fibration* $\varphi_t : \overline{Y}_t \to \overline{B}_t$.
(3) *A section of* φ_0 *in the fiber* $\overline{Y}_0$ *lying in* D_0 *has no monodromy in* $\overline{Y}$.

Then the given morphism $f : Y \to T$ *is factored by an* $\mathbb{A}^1$*-fibration.*

Proof. Embed Y into a smooth threefold $\overline{Y}$ in such a way that f extends to a projective morphism $\overline{f} : \overline{Y} \to T$. We may assume that the complement $S := \overline{Y} \setminus Y$ is a reduced divisor with simple normal crossings. Let $S = S_0 + S_1 + \cdots + S_r$ be the irreducible decomposition of S. For a general point $t \in T$, let Y_t be the fiber $f^{-1}(t)$ and let $\rho_t : Y_t \to B_t$ be the given $\mathbb{A}^1$-fibration on Y_t. By the assumption, B_t is an affine curve. We may assume that Y_t is smooth and hence B_t is smooth. Let $\overline{Y}_t$ be the closure of Y_t in $\overline{Y}$ which we may assume to be a smooth projective surface with t a general point of T. By replacing T by a smaller Zariski open set, we may assume that $\overline{f}$ is a smooth morphism and that $S \cdot \overline{Y}_t$ is a divisor with simple normal crossings for every $t \in T$. Hence we may assume that the condition (1) above is realized.

For each $t \in T$, let Λ_t be the pencil generated by the closures (in $\overline{Y}_t$) of the fibers of the $\mathbb{A}^1$-fibration ρ_t. If Λ_t has a base point, we can eliminate the base points by simultaneous blowing ups on the boundary at infinity after an étale finite base change of T. In this step, we may have to replace, for some $t \in T$, the pencil Λ_t by another pencil Λ'_t which also induces an $\mathbb{A}^1$-fibration of affine type on Y_t (see the proof of Lemma 3.2). So, we may assume that the condition (2) above is also satisfied.

If $S_0 \cap \overline{Y}_0$ contains a section of φ_0, we may assume by an étale finite base change that $S_0 \cap \overline{Y}_0$ is irreducible (see the remark before Lemma 3.5). So, we may assume that the condition (3) is satisfied as well.

Hence, we may assume from the beginning that three conditions are satisfied. The fibration ρ_t extends to a $\mathbb{P}^1$-fibration $\varphi_t : \overline{Y}_t \to \overline{B}_t$ for every $t \in T$, where $\overline{B}_t$ is a smooth completion of B_t. For $t_0 \in T$, we consider the fibration $\varphi_0 : \overline{Y}_0 \to \overline{B}_0$. A general fiber of φ_0 meets one of the irreducible components, say S_0, of S in one point. Then so does every fiber of φ_0 because $S_0 \cdot \overline{Y}_0$ is an irreducible divisor on $\overline{Y}_0$ and the fibers of φ_0 are algebraically equivalent to each other on $\overline{Y}_0$. Hence $S_0 \cdot \overline{Y}_0$ is a section. We claim that

(1) $\overline{Y}_t$ meets the component S_0 for every $t \in T$.
(2) After possibly switching the $\mathbb{A}^1$-fibrations if some Y_t has two $\mathbb{A}^1$-fibrations, we may assume that for every $t \in T$, the fibers of the $\mathbb{P}^1$-fibration φ_t on $\overline{Y}_t$ meet S_0 along a curve $\overline{A}_t$ such that $\overline{A}_t$ is a cross-section of φ_t and hence φ_t induces an isomorphism between $\overline{A}_t$ and $\overline{B}_t$.

In fact, for a relatively ample divisor H of $\overline{Y}$ over T, we have $(H \cdot S_0 \cdot \overline{Y}_0) > 0$, whence $(H \cdot S_0 \cdot \overline{Y}_t) > 0$ for every $t \in T$ because $\overline{Y}_t$ is algebraically equivalent to $\overline{Y}_0$. This implies the assertion (1). To prove the assertion (2), we consider the deformation of a smooth fiber C of φ_0 in $\overline{Y}_0$. Since general fibers Y_t of f have $\mathbb{A}^1$-fibrations of affine type, by Lemma 3.1(3) and Lemma 3.5(1), there is a $\mathbb{P}^1$-fibration $\varphi : \overline{Y} \to V$ such that C is a fiber of φ. Then the restriction $\varphi|_{\overline{Y}_0}$ is the $\mathbb{P}^1$-fibration φ_0. For every $t \in T$, the restriction $\varphi|_{\overline{Y}_t}$ is a $\mathbb{P}^1$-fibration on $\overline{Y}_t$. If it is different from φ_t, we replace φ_t by $\varphi|_{\overline{Y}_t}$. Then $(S_0 \cdot C') = (S_0 \cdot C) = 1$ for a general fiber C' of φ_t because C' is algebraically equivalent to C. The assertion follows immediately.

With the notations in the proof of Lemma 3.1, the isomorphisms $\overline{A}_t \xrightarrow{\sim} V_t := \sigma^{-1}(t) \cong \overline{B}_t$ show that the morphism

$$S_0 \hookrightarrow \overline{Y} \xrightarrow{\varphi} V \xrightarrow{\sigma} T$$

induces a birational T-morphism $S_0 \to V$ and S_0 is a cross-section of φ. It is clear that the boundary divisor S contains no other components which are horizontal to φ. Hence Y has an $\mathbb{A}^1$-fibration. □

As a consequence of Theorem 3.8, we have the following result.

Corollary 3.9. *Let $f : Y \to T$ be a smooth morphism from a smooth affine threefold Y to a smooth affine curve T. Assume that f has a relative projective completion $\overline{f} : \overline{Y} \to T$ which satisfies the same conditions on the boundary divisor S and the intersection of each fiber $\overline{Y}_t$ with S as set in Lemma 3.2. If a fiber Y_0 has a G_a-action, then there exists an étale finite morphism $T' \to T$ such that the threefold $Y' = Y \times_T T'$ has a G_a-action as a T'-scheme. Furthermore, if the relative completion $\overline{f} : \overline{Y} \to T$ is taken so that the three conditions in Theorem 3.8 are satisfied, the threefold Y itself has a G_a-action as a T-scheme.*

Proof. By Lemma 3.2, every fiber Y_t has an $\mathbb{A}^1$-fibration of affine type $\rho_t : Y_t \to B_t$, where B_t is an affine curve. As in the proof of Theorem 3.8, we may assume that the three conditions therein are satisfied. By the same theorem, Y has an $\mathbb{A}^1$-fibration $\rho : Y \to U$ such that f is factored as

$$f : Y \xrightarrow{\rho} U \xrightarrow{\sigma} T\,,$$

where $U_t := \sigma^{-1}(t) \cong B_t$ for every $t \in T$. Then U is an affine scheme after restricting T to a Zariski open set. Then Y has a G_a-action by [8]. □

Given a smooth affine morphism $f : Y \to T$ from a smooth algebraic variety Y to a smooth curve T such that every *closed* fiber is isomorphic to the affine space $\mathbb{A}^n$ of fixed dimension, one can ask if the generic fiber of f is isomorphic to $\mathbb{A}^n$ over the function field $k(T)$. If this is the case with f, we say that *the generic triviality* holds for f. In the case $n = 2$, this holds by the following theorem. If the generic triviality for $n = 2$ holds for $f : Y \to T$ in the setup of Theorem 3.10, a theorem of Sathaye [29] shows that f is an $\mathbb{A}^2$-bundle in the sense of Zariski topology.

Theorem 3.10. *Let $f : Y \to T$ be a smooth morphism from a smooth affine threefold Y to a smooth affine curve T. Assume that the fiber Y_t is isomorphic to $\mathbb{A}^2$ for every closed point of T. Then the generic fiber Y_η of f is isomorphic to the affine plane over the function field of T. Hence $f : Y \to T$ is an $\mathbb{A}^2$-bundle over T after replacing T by an open set if necessary.*

Before giving a proof, we prepare two lemmas where an integral k-scheme is a reduced and irreducible algebraic k-scheme and where a separable K-form of $\mathbb{A}^2$ over a field K is an algebraic variety X defined over K such that $X \otimes_K K'$ is K'-isomorphic to $\mathbb{A}^2$ for a separable algebraic extension K' of K.

Lemma 3.11. *Let $p : X \to T$ be a dominant morphism from an integral k-scheme X to an integral k-scheme T. Assume that the fiber X_t is an integral k-scheme for every closed point t of T. Then the generic fiber $X_\eta = X \times_T \operatorname{Spec} k(T)$ is geometrically integral $k(T)$-scheme.*

Proof. We have only to show that the extension of the function fields $k(X)/k(T)$ is a regular extension. Namely, $k(X)/k(T)$ is a separable extension, i.e., a separable algebraic extension of a transcendental extension of $k(T)$ and $k(T)$ is algebraically closed in $k(X)$. Since the characteristic of k is zero, it suffices to show that $k(T)$ is algebraically closed in $k(X)$. Suppose the contrary. Let K be the algebraic closure of $k(T)$ in $k(X)$, which is a finite algebraic extension of $k(T)$. Let T' be the normalization of T in K. Let $\nu : T' \to T$ be the normalization morphism which is a finite morphism. Then $p : X \to T$ splits as $p : X \xrightarrow{p'} T' \xrightarrow{\nu} T$, which is the Stein factorization. Then the fiber X_t is not irreducible for a general closed point $t \in T$, which is a contradiction to the hypothesis. □

The following result is due to Kambayashi [13].

Lemma 3.12. *Let X be a separable K-form of $\mathbb{A}^2$ for a field K. Then X is isomorphic to $\mathbb{A}^2$ over K.*

The following proof of Theorem 3.10 uses a locally nilpotent derivation and hence is of purely algebraic nature.

Proof of Theorem 3.10. Every closed fiber Y_t has an $\mathbb{A}^1$-fibration of affine type and hence a G_a-action. By Corollary 3.9, there exists an étale finite morphism $T' \to T$ such that $Y' = Y \times_T T'$ has a G_a-action as a T'-scheme. Suppose that the generic fiber $Y'_{\eta'}$ of $f_{T'} : Y' \to T'$ is isomorphic to $\mathbb{A}^2$ over the function field $k(T')$. Since $Y'_{\eta'} = Y_\eta \otimes_{k(T)} k(T')$, it follows by Lemma 3.12 that Y_η is isomorphic to $\mathbb{A}^2$ over $k(T)$. Hence, we may assume from the beginning that Y has a G_a-action which induces $\mathbb{A}^1$-fibrations on general closed fibers Y_t. The G_a-action on a T-scheme Y is induced by a locally nilpotent derivation δ on the coordinate ring B of Y, i.e., $Y = \operatorname{Spec} B$. Let $T = \operatorname{Spec} R$. Here δ is an R-trivial derivation on B. Let A be the kernel of δ. Since B is a smooth k-algebra of dimension 3, A is a finitely generated, normal

k-algebra of dimension 2. The derivation δ induces a locally nilpotent derivation δ_t on $B_t = B \otimes_R R/\mathfrak{m}_t$, where $\mathfrak{m}_t$ is the maximal ideal of R corresponding to a general point t of T. We assume that $\delta_t \neq 0$. Since B_t is a polynomial k-algebra of dimension 2 by the hypothesis, $A_t := \operatorname{Ker} \delta_t$ is a polynomial ring of dimension 1.

Claim 1. $A_t = A \otimes_R R/\mathfrak{m}_t$ if δ_t is nonzero.

Proof. Let $\varphi : B \to B[u]$ be the k-algebra homomorphism defined by

$$\varphi(b) = \sum_{i \geq 0} \frac{1}{i!} \delta^i(b) u^i .$$

Then $\operatorname{Ker} \delta = \operatorname{Ker}(\varphi - \mathrm{id})$. Hence we have an exact sequence of R-modules

$$0 \to A \to B \xrightarrow{\varphi - \mathrm{id}} B[u] .$$

Let $\mathcal{O}_t$ be the local ring of T at t, i.e., the localization of R with respect to $\mathfrak{m}_t$, and let $\hat{\mathcal{O}}_t$ be the $\mathfrak{m}_t$-adic completion of $\mathcal{O}_t$. Since $\hat{\mathcal{O}}_t$ is a flat R-module, we have an exact sequence

$$0 \to A \otimes_R \hat{\mathcal{O}}_t \to B \otimes_R \hat{\mathcal{O}}_t \to (B \otimes_R \hat{\mathcal{O}}_t)[u] . \qquad (*)$$

The completion $\hat{\mathcal{O}}_t$ as a k-module decomposes as $\hat{\mathcal{O}}_t = k \oplus \hat{\mathfrak{m}}_t$, where $\hat{\mathfrak{m}}_t = \mathfrak{m}_t \hat{\mathcal{O}}_t$, the above exact sequence splits as a direct sum of exact sequences of k-modules

$$0 \to A \otimes_R k \to B \otimes_R k \to (B \otimes_R k)[u]$$
$$0 \to A \otimes_R \hat{\mathfrak{m}}_t \to B \otimes_R \hat{\mathfrak{m}}_t \to (B \otimes_R \hat{\mathfrak{m}}_t)[u] .$$

The first one is, in fact, equal to

$$0 \to A \otimes_R R/\mathfrak{m}_t \to B_t \xrightarrow{\varphi_t - \mathrm{id}} B_t[u] ,$$

where φ_t is defined by δ_t in the same way as φ by δ. Hence $\operatorname{Ker} \delta_t = A \otimes_R R/\mathfrak{m}_t = A_t$. □

Let $X = \operatorname{Spec} A$ and let $p : X \to T$ be the morphism induced by the inclusion $R \hookrightarrow A$. Thus $f : Y \to T$ splits as

$$f : Y \xrightarrow{q} X \xrightarrow{p} T ,$$

where q is the quotient morphism by the induced G_a-action on Y.

Claim 2. Suppose that $\delta_t \neq 0$ for every $t \in T$. Then X is a smooth surface with $\mathbb{A}^1$-bundle structure over T.

Proof. Note that R is a Dedekind domain and A is an integral domain. Hence p is a flat morphism. Since f is surjective, p is also surjective. Hence p is a faithfully flat morphism. Further, by Claim 1, $X_t = \text{Spec}\,(A \otimes_R R/\mathfrak{m}_t)$ is equal to Spec A_t for every t, which is isomorphic to $\mathbb{A}^1$. In fact, the kernel of a nontrivial locally nilpotent derivation on a polynomial ring of dimension 2 is a polynomial ring of dimension 1. The generic fiber of p is geometrically integral by Lemma 3.11. Hence, by [14, Theorem 2], X is an $\mathbb{A}^1$-bundle over T. In particular, X is smooth. □

Let $K = k(T)$ be the function field of T. The generic fiber $X_K = X \times_T \text{Spec}\, K$ is geometrically integral as shown in the above proof of Claim 2.

Claim 3. The generic fiber $Y_K = Y \times_T \text{Spec}\, K$ is isomorphic to $\mathbb{A}^2_K$.

Proof. We consider $q_K : Y_K \to X_K$, where $X_K \cong \mathbb{A}^1_K$. We prove the following two assertions.

(1) For every closed point x of X_K, the fiber $Y_K \times_{X_K} \text{Spec}\, K(x)$ is isomorphic to $\mathbb{A}^1_{K(x)}$.
(2) The generic fiber of q_K is geometrically integral.

Note that $K(x)$ is a finite algebraic extension of K. Let T' be the normalization of T in $K' := K(x)$. We consider $Y' := Y \times_T T'$ instead of Y. Then the G_a-action on Y lifts to Y' and the quotient variety is $X' = X \times_T T'$. Indeed, the normalization R' of R in K' is the coordinate ring of T' and is a flat R-module. Then the sequence of R'-modules

$$0 \to A \otimes_R R' \to B \otimes_R R' \xrightarrow{\varphi' - \text{id}} (B \otimes_R R')[u]$$

is exact, where $\varphi' = \varphi \otimes_R R'$. Hence $q_{K'} : Y'_{K'} \to X'_{K'}$, which is the base change of q_K with respect to the field extension K'/K, is the quotient morphism by the G_a-action on $Y'_{K'}$ induced by δ. Since $X'_{K'} = X \times_T \text{Spec}\, K'$, there exists a K'-rational point x' on $X'_{K'}$ such that x is the image of x' by the projection $X'_{K'} \to X_K$. If the fiber of $q_{K'}$ over x', i.e., $Y'_{K'} \times_{X'_{K'}} (\text{Spec}\, K', x')$, is isomorphic to $\mathbb{A}^1_{K'}$, then $Y_K \times_{X_K} \text{Spec}\, K'$ is isomorphic to $\mathbb{A}^1_{K'}$ because $Y'_{K'} \times_{X'_{K'}} \text{Spec}\, K' = Y_K \times_{X_K} \text{Spec}\, K'$. Thus we may assume that x is a K-rational point. Let C be the closure of x in X. Then C is a cross-section of $p : X \to T$. Let $Z := Y \times_X C$. Then $q_C : Z \to C$ is a faithfully flat morphism such that the fiber $q_C^{-1}(w)$ is isomorphic to $\mathbb{A}^1$ for every closed point $w \in C$. In fact, $q_C^{-1}(w)$ is the fiber of $Y_t \to X_t$ over the point $w \in C$, where $t = p(w)$, $Y_t \cong \mathbb{A}^2$, $X_t \cong \mathbb{A}^1$ and $X_t = Y_t /\!/ G_a$. By Lemma 3.11 (which is extended to a non-closed field K), the generic fiber of q_C is geometrically integral, and the generic fiber of q_C, which is $Y_K \times_{X_K} \text{Spec}\, K(x)$, is isomorphic to $\mathbb{A}^1_{K'}$ by [14, Theorem 2]. This proves the first assertion.

The generic point of X_K corresponds to the quotient field $L := Q(A)$. Then it suffices to show that $B \otimes_A Q(A)$ is geometrically integral over $Q(A)$. Meanwhile, $B \otimes_A Q(A)$ has a locally nilpotent derivation $\delta \otimes_A Q(A)$ such that $\text{Ker}\,(\delta \otimes_A Q(A)) = Q(A)$. Hence $B \otimes_A Q(A)$ is a polynomial ring $Q(A)[u]$ in one variable over $Q(A)$ because $\delta \otimes_A Q(A)$ has a slice. So, $B \otimes_A Q(A)$ is geometrically integral over $Q(A)$. Now, by [15, Theorem], Y_K is an $\mathbb{A}^1$-bundle over $X_K \cong \mathbb{A}^1_K$. Hence Y_K is isomorphic to $\mathbb{A}^2_K$. We have to replace T by an open set $T \setminus F$, where $F = \{t \in T \mid \delta_t = 0\}$. This completes the proof of Theorem 3.10. □

We can prove Theorem 3.10 in a more geometric way by making use of a theorem of Ramanujam–Morrow on the boundary divisor of a minimal normal completion of the affine plane [25, 27]. The proof given below is explained in more precise and explicit terms in [16, Lemma 3.2]. In particular, the step to show that $\overline{Y}_K \cong \mathbb{P}^2_K$ and $Y_K \cong \mathbb{A}^2_K$ is due to [*loc. cit.*].

The second proof of Theorem 3.10. Let $f : Y \to T$ be as in Theorem 3.10. Let $\overline{Y}$ be a relative completion such that $\overline{Y}$ is smooth and f extends to a smooth projective morphism $\overline{f} : \overline{Y} \to T$ with the conditions in Lemma 3.2 being satisfied together with $S := \overline{Y} \setminus Y$. To obtain this setting, we may have to shrink T to a smaller open set of T. As in the first proof and the proof of Lemma 3.2, we can apply an étale finite base change $T' \to T$ by which the intersection $S_i \cap \overline{Y}_t$ is irreducible for every irreducible component S_i of S and every $t \in T$. In particular, we assume that $\overline{Y}_t$ is a smooth normal completion of Y_t for every $t \in T$, where Y_t is isomorphic to $\mathbb{A}^2$. Fix one such completed fiber, say $\overline{Y}_0 = \overline{f}^{-1}(t_0)$, and consider the reduced effective divisor $\overline{Y}_0 \setminus Y_0$ with $Y_0 = f^{-1}(t_0) \cong \mathbb{A}^2$. Namely, $(Y, \overline{Y}, S, \overline{f}, t_0)$ is a log deformation of $(Y_0, \overline{Y}_0, D_0)$. If the dual graph of this divisor is not linear, then it contains a (-1)-curve meeting at most two other components of D_0 by a result of Ramanujam [27]. By (4) of Lemma 3.1, such a (-1)-curve deforms along the fibers of $\overline{f}$ and we get an irreducible component, say S_1, of $S = \sum_{i=0}^r S_i$ which can be contracted. Repeating this argument, we can assume that all the dual graphs for $\overline{Y}_t \setminus Y_t$, as t varies on the set of closed points of T, are linear chains of smooth rational curves. By [25], at least one of these curves is a (0)-curve. Fix such a (0)-curve C_1 in $\overline{Y}_0 \setminus Y_0$. Then C_1 deforms along the fibers of $\overline{f}$ and forms an irreducible component, say S_1, of S by abuse of the notations. By the argument in the proof of Lemma 3.2, if C_2 is a component of $\overline{Y}_0 \setminus Y_0$ meeting C_1, it deforms along the fibers of $\overline{f}$ on an irreducible component, say S_2, of S. Repeating this argument, we know that all irreducible components of $\overline{Y}_0 \setminus Y_0$ extend along the fibers of $\overline{f}$ to form the irreducible components of S and that the dual graphs of $\overline{Y}_t \setminus Y_t$ are the same for every $t \in T$. Now let K be the function field of T over k. We consider the generic fibers $\overline{Y}_K$ and Y_K of $\overline{f}$ and f. Then the dual graph of $\overline{Y}_K \setminus Y_K$ is the same linear chain of smooth rational curves as the closed fibers $\overline{Y}_t \setminus Y_t$. Write $\overline{Y}_0 \setminus Y_0 = \sum_{i=1}^r C_i$. If C_i and C_j meet for $i \neq j$, then the intersection point $C_i \cap C_j$ moves on the intersection curve $S_i \cdot S_j$. Since any minimal normal completion of $\mathbb{A}^2$ can be brought to $\mathbb{P}^2$ by blowing ups and downs with centers on the boundary divisor, we can blow up simultaneously the intersection curves and blow down the proper transforms of the S_i according to the blowing ups and downs on $\overline{Y}_0$. Here we note that the beginning center of blowing up is a point on a (0)-curve C_1. In this case, we choose a suitable cross-section on the irreducible component S_1 which is a $\mathbb{P}^1$-bundle in the Zariski topology because $\dim T = 1$. Note that if T is irrational, then the chosen cross-section may meet the intersection curves on S_1 with other components of S. Then we shrink T so that the cross-section does not meet the intersection curves. If T is rational, S_1 is a trivial $\mathbb{P}^1$-bundle, hence we do not need the procedure of shrinking T. Thus we may assume that, for every $t \in T$, $\overline{Y}_t$ is isomorphic to $\mathbb{P}^2$ and $\overline{Y}_t \setminus Y_t$ is a single curve C_t with $(C_t)^2 = 1$. This implies that $\overline{Y}_K \cong \mathbb{P}^2_K$ and $Y_K \cong \mathbb{A}^2_K$. □

In connection with Theorem 3.10, we can pose the following:

Problem 3.13. Let K be a field of characteristic zero and let X be a smooth affine surface defined over K. Suppose that $X \otimes_K \overline{K}$ has an $\mathbb{A}^1$-fibration of affine type, where $\overline{K}$ is an algebraic closure of K. Does X then have an $\mathbb{A}^1$-fibration of affine type?

If we consider an $\mathbb{A}^1$-fibration of complete type, an example of Dubouloz–Kishimoto gives a counterexample to a similar problem for the complete type (see Theorem 6.1). In view of Example 3.6 and Theorem 3.8, we need perhaps some condition for a positive answer in the case of affine type which guarantees the absence of monodromy of a cross-section of a given $\mathbb{A}^1$-fibration.

4 Topological Arguments Instead of Hilbert Schemes

In this section we will briefly indicate topological proofs of some of the results in Sect. 3. The use of topological arguments would make the cumbersome geometric arguments more transparent for the people who do not appreciate the heavy machinery like Hilbert scheme.

We will use the following basic theorem due to Ehresmann [30, Chapter V, Proposition 6.4].

Theorem 4.1. *Let M be a connected differentiable manifold, S a closed submanifold, $f : M \to N$ a proper differentiable map such that the tangent maps corresponding to f and $f|_S : S \to N$ are surjective at any point in M and S. Then $f|_{M \setminus S} : M \setminus S \to N$ is a locally trivial fiber bundle with respect to the base N.*

Note that the normal bundle of any fiber of f is trivial. We can give a proof of Ehresmann's theorem using this observation, and the well-known result from differential topology that given a compact submanifold S of a C^∞ manifold X there are arbitrarily small tubular neighborhoods of S in X which are diffeomorphic to neighborhoods of S in the total space of normal bundle of S in X [1, Chapter II, Theorem 11.14].

Now let $\overline{f} : \overline{Y} \to T$ be a smooth projective morphism from a smooth algebraic threefold onto a smooth algebraic curve T. Let $\overline{Y}_t = \overline{f}^{-1}(t)$ be the fiber over $t \in T$. Let S be a simple normal crossing divisor on $\overline{Y}$ such that $D_t := S \cap \overline{Y}_t$ is a simple normal crossing divisor for each $t \in T$ and $Y_t := \overline{Y}_t \setminus D_t$ is affine for each t.

We can assume that $\overline{f} : \overline{Y} \to T$ has the property that the tangent map is surjective at each point. It follows from Ehresmann's theorem that all the surfaces $\overline{Y}_t$ are mutually diffeomorphic. In particular, they have the same topological invariants like the fundamental group π_1 and the Betti number b_i. By shrinking T if necessary, we will assume that the restricted map $\overline{f} : S_i \to T$ is smooth for each i. For fixed i and t_0 the intersection $S_i \cap \overline{Y}_{t_0}$ is a disjoint union of smooth, compact, irreducible curves. Let $C_{t_0,i}$ be one of these irreducible curves. Then for each t which is close to t_0, there is an irreducible curve $C_{t,i}$ in $S_i \cap \overline{Y}_t$ and suitable tubular neighborhoods of $C_{t_0,i}$, $C_{t,i}$ in $\overline{Y}_{t_0}$, $\overline{Y}_t$, respectively, are diffeomorphic by Ehresmann's theorem. This implies that $C_{t_0,i}^2$ in $\overline{Y}_{t_0}$ and $C_{t,i}^2$ in $\overline{Y}_t$ are equal. This proves that the weighted dual graphs of the curves D_t in $\overline{Y}_t$ are the same for each t.

Recall that if X is a smooth projective surface with a smooth rational curve $C \subset X$ such that $C^2 = 0$ then C is a fiber of a $\mathbb{P}^1$-fibration on X. If the irregularity $q(X) > 0$, then the Albanese morphism $X \to \mathrm{Alb}(X)$ gives a $\mathbb{P}^1$-fibration on X with C as a fiber. By the above discussion the fiber surfaces $\overline{Y}_t$ have the same irregularity.

Suppose that $\overline{Y}_0$ has an $\mathbb{A}^1$-fibration of affine type $f : Y_0 \to B$. If $\overline{f} : \overline{Y}_0 \to \overline{B}$ is an extension of f to a smooth completion of Y_0 then, after simultaneous blowing ups and downs along the fibers of $\overline{f}$, we may assume that $D_0 := \overline{Y}_0 \setminus Y_0$ contains at least one (0)-curve which is a tip, i.e., the end component of a maximal twig of D_0. Since D_t and D_0 have the same weighted dual graphs D_t also contains a (0)-curve which is a tip of D_t. Hence, Y_t also has an $\mathbb{A}^1$-fibration of affine type. This proves the assertion (2) in Lemma 2.2.

We can also shorten the part of showing the invariance of the boundary weighted graphs in the second proof of Theorem 3.10. Suppose now that $f : Y \to T$ is a fibration on a smooth affine threefold Y onto a smooth curve T such that every scheme-theoretic fiber of f is isomorphic to $\mathbb{A}^2$. We can embed Y in a smooth projective threefold $\overline{Y}$ such that f extends to a morphism $\overline{f} : \overline{Y} \to T$. By shrinking T we can assume that $\overline{f}$ is smooth, each irreducible component S_i of $\overline{Y} \setminus Y$ intersects each $\overline{Y}_t$ transversally, etc. By the above discussions, each $D_t := \overline{Y}_t \setminus Y_t$ has the same weighted dual graph. Since Y_t is isomorphic to $\mathbb{A}^2$, we can argue as in the second proof of Theorem 3.10 using the result of Ramanujam–Morrow to conclude that f is a trivial $\mathbb{A}^2$-bundle on a nonempty Zariski-open subset of T. This observation applies also to the proof of Theorem 5.6.

5 Deformations of ML_0 Surfaces

For $i = 0, 1, 2$, an ML_i surface is by definition a smooth affine surface X such that the Makar-Limanov invariant $\mathrm{ML}(X)$ has transcendence degree i over k [10]. In this section, we assume that the ground field k is the complex field $\mathbb{C}$. Let $\mathcal{F} = (Y, \overline{Y}, S, \overline{f}, t_0)$ be a family satisfying the conditions of Lemma 3.2. Let $D_0 = S \cap \overline{Y}_0$.

Lemma 5.1. *Let* $\mathcal{F} = (Y, \overline{Y}, S, \overline{f}, t_0)$ *be a log deformation of* $(Y_0, \overline{Y}_0, D_0)$. *Assume that* D_0 *is a tree of smooth rational curves satisfying one of the following conditions.*

(i) D_0 *contains an irreducible component* C_1 *such that* $(C_1^2) \geq 0$.
(ii) D_0 *contains a* (-1) *curve which meets more than two other components of* D_0.

Then the following assertions hold after changing T *by an étale finite covering of an open set of* T *if necessary.*

(1) *Every irreducible component of* D_0 *deforms along the fibers of* $\overline{f}$. *Namely, if* $D_0 = \sum_{i=1}^r C_i$ *is the irreducible decomposition, then, for every* $1 \leq i \leq r$, *there exists an irreducible component* S_i *of* S *such that* $\overline{f}|_{S_i} : S_i \to T$ *has the fiber* $(\overline{f}|_{S_i})^{-1}(t_0) = C_i$. *Furthermore,* $S = \sum_{i=1}^r S_i$.
(2) *For* $t \in T$, *let* $C_{i,t} = (\overline{f}|_{S_i})^{-1}(t)$. *Then* $D_t = \sum_{i=1}^r C_{i,t}$ *and* D_t *has the same weighted graph on* $\overline{Y}_t$ *as* D_0 *does on* $\overline{Y}_0$.
(3) *For every* i, $\overline{f}|_{S_i} : S_i \to T$ *is a trivial* $\mathbb{P}^1$*-bundle over* T.

Proof. By a suitable étale finite base change of T, we may assume that $S_i \cap \overline{Y}_0$ is irreducible for every irreducible component S_i of S. Then the argument is analytic locally almost the same as in the proof for the assertion (2) of Lemma 3.2. Consider the deformation of C_1 along the fibers of $\overline{f}$, which moves along the fibers because $(C_1^2) \geq -1$. Then the components of D_0 which are adjacent to C_1 also move along the fibers of $\overline{f}$. Once these components of D_0 move, then the components adjacent to these components move along the fibers of $\overline{f}$. Since D_0 is connected because Y_0 is affine, all the components of D_0 move along the fibers of $\overline{f}$. If S contains an irreducible component which does not intersect $\overline{Y}_0$, it is a fiber component of $\overline{f}$. Then we remove the fiber by shrinking T. This proves the assertion (1).

Let $S = \sum_{i=1}^r S_i$ be the irreducible decomposition of S. As shown in (1), $S_i \cap \overline{Y}_0 \neq \emptyset$ for every i. Then $S_i \cap \overline{Y}_t \neq \emptyset$ as well by the argument in the proof of Lemma 3.2.

Note that $((S_i \cdot \overline{Y}_t)^2)_{\overline{Y}_t} = (S_i^2 \cdot \overline{Y}_t) = (S_i^2 \cdot \overline{Y}_0) = ((S_i \cdot \overline{Y}_0)^2)_{\overline{Y}_0}$ because $\overline{Y}_t$ is algebraically equivalent to $\overline{Y}_0$. Hence D_0 and D_t have the same dual graphs. □

In order to prove the following result, we use Ehresmann's theorem, which is Theorem 4.1.

Lemma 5.2. *Let* $\mathcal{F} = (Y, \overline{Y}, S, \overline{f}, t_0)$ *be a log deformation of* $(Y_0, \overline{Y}_0, D_0)$ *which satisfies the same conditions as in Lemma 5.1. Assume further that* $p_g(\overline{Y}_0) = q(\overline{Y}_0) = 0$. *Then the following assertions hold:*

(1) Pic $(Y_t) \cong$ Pic (Y_0) *for every* $t \in T$.
(2) $\Gamma(Y_t, \mathcal{O}_{Y_t}^*) \cong \Gamma(Y_0, \mathcal{O}_{Y_0}^*)$ *for every* $t \in T$.

Proof. Since p_g and q are deformation invariants, we have $p_g(\overline{Y}_t) = q(\overline{Y}_t) = 0$ for every $t \in T$. The exact sequence

$$0 \longrightarrow \mathbb{Z} \longrightarrow \mathcal{O}_{\overline{Y}_t} \overset{\exp}{\longrightarrow} \mathcal{O}^*_{\overline{Y}_t} \longrightarrow 0$$

induces an exact sequence

$$H^1(\overline{Y}_t, \mathcal{O}_{\overline{Y}_t}) \to H^1(\overline{Y}_t, \mathcal{O}^*_{\overline{Y}_t}) \to H^2(\overline{Y}_t; \mathbb{Z}) \to H^2(\overline{Y}_t, \mathcal{O}_{\overline{Y}_t})$$

Since $p_g(\overline{Y}_t) = q(\overline{Y}_t) = 0$, we have an isomorphism

$$H^1(\overline{Y}_t, \mathcal{O}^*_{\overline{Y}_t}) \cong H^2(\overline{Y}_t; \mathbb{Z}) .$$

Now consider the canonical homomorphism $\theta_t : H_2(D_t; \mathbb{Z}) \to H_2(\overline{Y}_t; \mathbb{Z})$, where $H_2(\overline{Y}_t; \mathbb{Z}) \cong H^2(\overline{Y}_t; \mathbb{Z}) = \mathrm{Pic}\ (\overline{Y}_t)$ by the Poincaré duality. Then Coim $\theta_t = \mathrm{Pic}\ (Y_t)$ and Ker $\theta_t = \Gamma(Y_t, \mathcal{O}^*_{Y_t})/k^*$.

Let N be a nice tubular neighborhood of S with boundary in $\overline{Y}$. The smooth morphism $\overline{f} : \overline{Y} \to T$ together with its restriction on the $(N, \partial N)$ gives a proper differential mapping which is surjective and submersive. By Theorem 4.1, it is differentiably a locally trivial fibration. Namely, there exists a small disc U of t_0 in T and a diffeomorphism $\varphi_0 : \overline{Y}_0 \times U \overset{\approx}{\longrightarrow} (\overline{f})^{-1}(U)$ such that its restriction induces a diffeomorphism

$$\varphi_0 : (N \cap \overline{Y}_0) \times U \overset{\approx}{\longrightarrow} (\overline{f}|_N)^{-1}(U) .$$

For $t \in U$, noting that U is contractible and hence $H_2(\overline{Y}_0 \times U; \mathbb{Z}) = H_2(\overline{Y}_0; \mathbb{Z})$ and $H_2((N \cap \overline{Y}_0) \times U; \mathbb{Z}) = H_2(N \cap \overline{Y}_0; \mathbb{Z})$, the inclusions $\overline{Y}_t \hookrightarrow (\overline{f})^{-1}(U)$ and $N \cap \overline{Y}_0 \hookrightarrow (\overline{f}|_N)^{-1}(U)$ induces compatible isomorphisms

$$p_t : H_2(\overline{Y}_t; \mathbb{Z}) \to H_2((\overline{f})^{-1}(U); \mathbb{Z}) \overset{(\varphi^{-1})_*}{\longrightarrow} H_2(\overline{Y}_0 \times U; \mathbb{Z}) = H_2(\overline{Y}_0; \mathbb{Z})$$

and its restriction $q_t : H_2(N \cap \overline{Y}_t; \mathbb{Z}) \overset{\sim}{\longrightarrow} H_2(N \cap \overline{Y}_0; \mathbb{Z})$. Since S and hence D_t are strong deformation retracts of N and $N \cap \overline{Y}_t$ respectively, the isomorphism q_t induces an isomorphism $r_t : H_2(D_t; \mathbb{Z}) \overset{\sim}{\longrightarrow} H_2(D_0; \mathbb{Z})$ such that the following diagram

$$\begin{array}{ccc} H_2(D_t; \mathbb{Z}) & \overset{\theta_t}{\longrightarrow} & H_2(\overline{Y}_t; \mathbb{Z}) \\ {\scriptstyle r_t}\downarrow & & \downarrow{\scriptstyle p_t} \\ H_2(D_0; \mathbb{Z}) & \underset{\theta_0}{\longrightarrow} & H_2(\overline{Y}_0; \mathbb{Z}) \end{array}$$

This implies that Pic $(Y_t) \cong$ Pic (Y_0) and $\Gamma(Y_t, \mathcal{O}^*_{Y_t}) \cong \Gamma(Y_0, \mathcal{O}^*_{Y_0})$. If t is an arbitrary point of T, we choose a finite sequence of points $\{t_0, t_1, \ldots, t_n = t\}$ such that t_i is in a small disc U_{i-1} around t_{i-1} $(1 \leq i \leq n)$ for which we can apply the above argument. □

Remark 5.3. By a result of Neumann [26, Theorem 5.1], if X is a normal affine surface, D an SNC divisor at infinity of X which does not contain any (-1)-curve meeting at least three other components of D and all whose maximal twigs are smooth rational curves with self-intersections ≤ -2, then the boundary 3-manifold of a nice tubular neighborhood N of D determines the dual graph of D. If we use the local differentiable triviality of a tubular neighborhood N, this result of Neumann shows that the weighted dual graph of D_t is deformation invariant.

According to [10, Lemmas 1.2 and 1.4], we have the following property and characterization of ML_0-surface.

Lemma 5.4. *Let X be a smooth affine surface and let V be a minimal normal completion of X. Then the following assertions hold:*

(1) *X is an ML_0-surface if and only if $\Gamma(X, \mathcal{O}_X^*) = k^*$ and the dual graph of the boundary divisor $D := V - X$ is a linear chain of smooth rational curves.*
(2) *If X is an ML_0-surface, X has an $\mathbb{A}^1$-fibration, and any $\mathbb{A}^1$-fibration $\rho : X \to B$ has base curve either $B \cong \mathbb{P}^1$ or $B \cong \mathbb{A}^1$. If $B \cong \mathbb{P}^1$, ρ has at most two multiple fibers, and if $B \cong \mathbb{A}^1$, it has at most one multiple fiber.*

The following result is a direct consequence of the above lemmas.

Theorem 5.5. *Let $\mathcal{F} = (Y, \overline{Y}, S, \overline{f}, t_0)$ be a log deformation of $(Y_0, \overline{Y}_0, D_0)$, where Y_0 is an ML_0-surface. Then Y_t is an ML_0-surface for every $t \in T$.*

Proof. If $S \cap \overline{Y}_t$ contains a (-1) curve, then it deforms along the fibers of $\overline{f}$ after an étale finite base change of T, and these (-1) curves are contracted simultaneously by Lemma 3.1. Hence we may assume that $\overline{Y}_t$ is a minimal normal completion of Y_t for every $t \in T$. By Lemma 5.4, $D_0 := S \cap \overline{Y}_0$ is a linear chain of smooth rational curves. Hence $D_t := S \cap \overline{Y}_t$ is also a linear chain of smooth rational curves. By Lemma 5.2, $\Gamma(Y_t, \mathcal{O}_{Y_t}^*) = k^*$ for every $t \in T$ because $\Gamma(Y_0, \mathcal{O}_{Y_0}^*) = k^*$. So, Y_t is an ML_0-surface by Lemma 5.4. □

A smooth affine surface X is, by definition, an *affine pseudo-plane* if it has an $\mathbb{A}^1$-fibration of affine type $p : X \to \mathbb{A}^1$ admitting at most one multiple fiber of the form $m\mathbb{A}^1$ as a singular fiber (see [24] for the definition and relevant results). An affine pseudo-plane is a $\mathbb{Q}$-homology plane, its Picard group is a cyclic group $\mathbb{Z}/m\mathbb{Z}$, and there are no non-constant invertible elements. An ML_0-surface is an affine pseudo-plane if the Picard number is zero.

If $\overline{X}$ is a minimal normal completion of an affine pseudo-plane X, the boundary divisor $D = \overline{X} - X$ is a tree of smooth rational curves, which is not necessarily a linear chain. By blowing-ups and blowing-downs with centers on the boundary divisor D, we can make the completion $\overline{X}$ satisfy the following conditions [24, Lemma 1.7].

(i) There is a $\mathbb{P}^1$-fibration $\overline{p} : \overline{X} \to \mathbb{P}^1$ which extends the $\mathbb{A}^1$-fibration $p : X \to \mathbb{A}^1$.
(ii) The weighted dual graph of D is

$$\underset{\ell}{(0)} \;—\; \underset{M}{(0)} \;—\; A$$

(iii) There is a (-1) curve F_0 (called *feather*) such that $F_0 \cap X \cong \mathbb{A}^1$ and the union F_0 — A is contractible to a smooth rational curve meeting the image of the component M.

Note that X is an ML_0-surface if and only if A is a linear chain. We then call X an *affine pseudo-plane of* ML_0*-type*.

If we are given a log deformation $(Y, \overline{Y}, S, \overline{f}, t_0)$ of the triple $(\overline{Y}_0, D_0, Y_0)$, it follows by Ehresmann's fibration theorem that p_g and the irregularity q of the fiber $\overline{Y}_t$ is independent of t. Furthermore, by Lemma 3.2, Y_t has an $\mathbb{A}^1$-fibration if Y_0 has an $\mathbb{A}^1$-fibration. So, we can expect that Y_t is an affine pseudo-plane if so is Y_0. Indeed, we have the following result.

Theorem 5.6. *Let* $\mathcal{F} = (Y, \overline{Y}, S, \overline{f}, t_0)$ *be a log deformation of* $(\overline{Y}_0, D_0, Y_0)$*. Assume that* Y_0 *is an affine pseudo-plane. Then the following assertions hold:*

(1) *Y_t is an affine pseudo-plane for every point $t \in T$.*
(2) *Assume that Y_0 is an affine pseudo-plane of* ML_0*-type. Assume further that the boundary divisor D_0 in $\overline{Y}_0$ has the same weighted dual graph as above. Then $f : Y \to T$ is a trivial bundle with fiber Y_0 after shrinking T if necessary.*

Proof. (1) We have only to show that Y_t is an affine pseudo-plane for a small deformation of Y_0. After replacing T by an étale finite covering, we may assume that $\overline{Y}_t$ is a minimal normal completion of Y_t for every $t \in T$. Then, by Lemma 5.1, the boundary divisor $D_t = S \cap \overline{Y}_t$ has the same weighted dual graph as shown above for D_0. Hence Y_t has an $\mathbb{A}^1$-fibration of affine type. By Lemma 5.2, $\mathrm{Pic}\,(Y_t) \cong \mathrm{Pic}\,(Y_0)$ which is a finite cyclic group. This implies that Y_t is an affine pseudo-plane.

(2) Consider the completion $\overline{Y}_0$ of Y_0. We may assume that $\overline{Y}_0$ is a minimal normal completion of Y_0. In fact, a (-1)-curve contained in the boundary divisor D_0 which meets at most two other components of D_0 deforms to the nearby fibers and contracted simultaneously over the same T by Lemma 3.5(2). Note that every fiber Y_t has an $\mathbb{A}^1$-fibration of affine type by Lemma 3.2. As in the proof of Lemma 3.2(2), by performing simultaneous (i.e., along the fibers of $\overline{f}$) blowing-ups and blowing-downs on the boundary S, we may assume that Y_0 has an $\mathbb{A}^1$-fibration which extends to a $\mathbb{P}^1$-fibration on $\overline{Y}_0$ and that the boundary divisor D_0 has the weighted dual graph ℓ–M–A as specified in the condition (ii) above, where A is a linear chain by the hypothesis. To perform a simultaneous blowing-up, we may have to choose as the center a cross-section

on an irreducible component S_i which is a $\mathbb{P}^1$-bundle over T. If such a cross-section happens to intersect the curve $S_i \cap S_j$ with another component S_j, we shrink T to avoid this intersection (see the remark in the second proof of Theorem 3.10). Note that the interior Y (more precisely, the inverse image of f of the shrunken T) is not affected under these operations. Then the (0) curve ℓ defines a $\mathbb{P}^1$-fibration $\varphi : \overline{Y} \to V$ (see Lemma 3.5(1)). In particular, ℓ moves in an irreducible component, say S_{-1}, of S. The (0) curve M moves along the fibers of $\overline{f}$ in an irreducible component, say S_0, of S. By Lemma 5.1, the curves in A move along the fibers of $\overline{f}$ and fill out the irreducible components $S_1, \ldots, S_r$ of S. Hence $S = S_{-1} \cup S_0 \cup S_1 \cup \cdots \cup S_r$ and $D_t = S \cdot \overline{Y}_t$ has the same weighted dual graph as D_0.

Now consider a (-1) curve F_0 on $\overline{Y}_0$. By Lemma 3.1, F_0 moves along the fibers of $\overline{f}$ and fills out a smooth irreducible divisor F which meets transversally an irreducible component S_i $(1 \le i \le r)$. In fact, the feather F_0 is unique on Y_0 and $(S_i \cdot F \cdot \overline{Y}_t) = (S_i \cdot F \cdot \overline{Y}_0) = 1$. Let S_1 be the component of S meeting S_0. Let $F_t = F \cap \overline{Y}_t$ and $S_{j,t} = S_j \cap \overline{Y}_t$ for every $t \in T$. Then $F_t + \sum_{j=2}^{r} S_{j,t}$ is contractible to a smooth point P_t lying on $S_{1,t}$. After performing simultaneous elementary transformations on the fiber ℓ which is the fiber at infinity of the $\mathbb{A}^1$-fibration of the affine pseudo-plane Y_t, we may assume that P_t is the intersection point $S_{0,t} \cap S_{1,t}$. By applying Lemma 3.5(2) repeatedly, we can contract F and the components $S_2, \ldots, S_r$ simultaneously. Let $\overline{Z}$ be the threefold obtained from $\overline{Y}$ by these contractions. Then $\overline{Z}$ has a $\mathbb{P}^1$-fibration $\psi : \overline{Z} \to V$ and the image of S_0 is a cross-section. Let $g = \sigma \cdot \psi : \overline{Z} \xrightarrow{\psi} V \xrightarrow{\sigma} T$ (see Lemma 3.1(3) for the notations). For every $t \in T$, $\overline{Z}_t := g^{-1}(t)$ is a minimal $\mathbb{P}^1$-bundle with a cross-section $S_{0,t}$. Since $(S_{0,t})^2 = 0$, $\overline{Z}_t$ is isomorphic to $\mathbb{P}^1 \times \mathbb{P}^1$. Then $\overline{Z}$ is a trivial $\mathbb{P}^1 \times \mathbb{P}^1$-bundle over T after shrinking T if necessary. In fact, $\overline{Z}$ with the images of S_0 and S_{-1} removed is a deformation of $\mathbb{A}^2$, which is locally trivial in the Zariski topology by Theorem 3.10. We may assume that $\psi : \overline{Z} \to V$ is the projection of $\mathbb{P}^1 \times \mathbb{P}^1 \times T$ onto the second and the third factors. Choose a section $\overline{S}_0'$ of ψ which is disjoint from the image $\overline{S}_0$ of S_0. Then there is a nontrivial G_m-action on $\overline{Z}$ along the fibers of ψ which has $\overline{S}_0$ and $\overline{S}_0'$ as the fixed point locus.

Now reverse the contractions $\overline{Y} \to \overline{Z}$. The center of the first simultaneous blowing-up with center $S_0 \cap S_1$ and the centers of the consecutive simultaneous blowing-ups except for the blowing-up which produces the component F are G_m-fixed because the blowing-ups are fiberwise sub-divisional. Only the center Q_t of the last blowing-up on $\overline{Y}_t$ is non-subdivisional. Let

$$\varphi : \overline{Y} \xrightarrow{\sigma} \overline{Y}_1 \xrightarrow{\sigma_1} \overline{Z}$$

be the factorization of φ where σ is the last non-subdivisional blowing-up. By the construction, the natural T-morphism $\overline{f}_1 : \overline{Y}_1 \to T$ is a trivial fibration with fiber $(\overline{Y}_1)_0 = \overline{f}_1^{-1}(t_0)$. Then there exists an element $\{\rho_t\}_{t \in T}$ of $G_m(T)$

such that $\rho_t(Q_{t_0}) = Q_t$ for every $t \in T$ after shrinking T if necessary. Here note that the G_m-action is nontrivial on the component with the point Q_t thereon, for otherwise the G_m-action is trivial from the beginning. Then these $\{\rho_t\}_{t \in T}$ extend to a T-isomorphism $\tilde{\rho} : \overline{Y}_0 \times T \to \overline{Y}$, which induces a T-isomorphism $Y_0 \times T \to Y$. Hence Y is trivial.

□

6 Deformations of $\mathbb{A}^1$-Fibrations of Complete Type

In the setting of Theorem 3.8, if the $\mathbb{A}^1$-fibration of a general fiber Y_t is of complete type, we do not have the same conclusion. This case is treated in a recent work of Dubouloz and Kishimoto [3]. We consider this case by taking the same example of cubic surfaces in $\mathbb{P}^3$ and explain how it is affine-uniruled.

Taking a cubic hypersurface as an example, we first observe the behavior of the log Kodaira dimension for a flat family of smooth affine surfaces. Let $\overset{\vee}{\mathbb{P}^3}$ be the dual projective 3-space whose points correspond to the hyperplanes of $\mathbb{P}^3$. We denote it by T. Let S be a smooth cubic hypersurface in $\mathbb{P}^3$ and let $\mathcal{W} = S \times T$ which is a codimension one subvariety of $\mathbb{P}^3 \times T$. Let $\mathcal{H}$ be the universal hyperplane in $\mathbb{P}^3 \times T$, which is defined by $\xi_0 X_0 + \xi_1 X_1 + \xi_2 X_2 + \xi_3 X_3 = 0$, where (X_0, X_1, X_2, X_3) and $(\xi_0, \xi_1, \xi_2, \xi_3)$ are, respectively, the homogeneous coordinates of $\mathbb{P}^3$ and T. Let $\mathcal{D}$ be the intersection of $\mathcal{W}$ and $\mathcal{H}$ in $\mathbb{P}^3 \times T$. Let $\pi : \mathcal{W} \to T$ be the projection and let $\pi_{\mathcal{D}} : \mathcal{D} \to T$ be the restriction of π onto $\mathcal{D}$. Then π and $\pi_{\mathcal{D}}$ are the flat morphism. For a closed point $t \in T$, $\mathcal{W}_t = \pi^{-1}(t)$ is identified with S and $\mathcal{D}_t = \pi_{\mathcal{D}}^{-1}(t)$ is the hyperplane section $S \cap \mathcal{H}_t$ in $\mathbb{P}^3$, where $\mathcal{H}_t$ is the hyperplane $\tau_0 X_0 + \tau_1 X_1 + \tau_2 X_2 + \tau_3 X_3 = 0$ with $t = (\tau_0, \tau_1, \tau_2, \tau_3)$. Let $\mathcal{X} = \mathcal{W} \setminus \mathcal{D}$ and $p : \mathcal{X} \to T$ be the restriction of π onto $\mathcal{X}$. Then $\mathcal{X}_t = p^{-1}(t)$ is an affine surface $S \setminus (S \cap \mathcal{H}_t)$.

Since S is smooth, the following types of $S \cap \mathcal{H}_t$ are possible. In the following, $F = 0$ denotes the defining equation of S and $H = 0$ does the equation for $\mathcal{H}_t$.

(1) A smooth irreducible plane curve of degree 3.
(2) An irreducible nodal curve, e.g., $F = X_0(X_1^2 - X_2^2) - X_2^3 + X_0^2 X_3 + X_3^3$ and $H = X_3$.
(3) An irreducible cuspidal curve, e.g., $F = X_0 X_1^2 - X_2^3 + X_3(X_0^2 + X_1^2 + X_2^2 + X_3^2)$ and $H = X_3$.
(4) An irreducible conic and a line which meets in two points transversally or in one point with multiplicity two. In fact, let ℓ and D be, respectively, a line and an irreducible conic in $\mathbb{P}^2$ meeting in two points Q_1, Q_2, where Q_1 is possibly equal to Q_2. Let C be a smooth cubic meeting ℓ in three points P_i $(1 \le i \le 3)$ and D in six points P_i $(4 \le i \le 9)$, where the points P_i are all distinct and different from Q_1, Q_2. Choose two points P_1, P_2 on ℓ and four points P_i $(4 \le i \le 7)$ on D. Let $\sigma : S \to \mathbb{P}^2$ be the blowing-up of these six points. Let ℓ', D'

and C' be the proper transforms of ℓ, D and C'. Then S is a cubic hypersurface in $\mathbb{P}^3$ and $K_S \sim -C'$. Since $\ell' + D' \sim C'$, it is a hyperplane section of S with respect to the embedding $\Phi_{|C'|} : S \hookrightarrow \mathbb{P}^3$.

(5) Three lines which are either meeting in one point or not. Let ℓ_i $(1 \le i \le 3)$ be the lines. Let $Q_1 = \ell_1 \cap \ell_3$ and $Q_2 = \ell_2 \cap \ell_3$. In the setting of (4) above, we consider $\ell = \ell_3$ and $D = \ell_1 + \ell_2$. So, if $Q_1 = Q_2$, three lines meet in one point. Choose a smooth cubic C meeting three lines in nine distinct points P_i $(1 \le i \le 9)$ other than Q_1, Q_2. Choose six points from the P_i, two points lying on each line. Then consider the blowing-up in these six points. The rest of the construction is the same as above.

Note that if S is smooth $S \cap \mathcal{H}_t$ cannot have a non-reduced component. In fact, the non-reduced component is a line in $\mathcal{H}_t$. Hence we may write the defining equation of S as

$$F = X_0^2(aX_1 + X_0) + X_3 G(X_0, X_1, X_2, X_3) = 0,$$

where $G = G(X_0, X_1, X_2, X_3)$ is a quadratic homogeneous polynomial and $a \in k$. We understand that $a = 0$ if the non-reduced component has multiplicity three. By the Jacobian criterion, it follows that S has singularities at the points $G = X_0 = X_3 = 0$.

The affine surface $\mathcal{X}_t$ has log Kodaira dimension 0 in the cases (1), (2), (4) with the conic and the line meeting in two distinct points and (5) with non-confluent three lines, and $-\infty$ in the rest of the cases. Although $p : \mathcal{H} \to T$ is a flat family of affine surfaces, the log Kodaira dimension drops to $-\infty$ exactly at the points $t \in T$ where the boundary divisor $S \cap \mathcal{H}_t$ is not a divisor with normal crossings. This accords with a result of Kawamata concerning the invariance of log Kodaira dimension under deformations (cf. [18]).

If $\overline{\kappa}(\mathcal{X}_t) = -\infty$, then $\mathcal{X}_t$ has an $\mathbb{A}^1$-fibration. We note that if $\overline{\kappa}(\mathcal{X}_t) = 0$ then $\mathcal{X}_t$ has an $\mathbb{A}^1_*$-fibration. In fact, we consider the case where the boundary divisor $\mathcal{D}_t$ is a smooth cubic curve. Then S is obtained from $\mathbb{P}^2$ by blowing up six points P_i $(1 \le i \le 6)$ on a smooth cubic curve C. Choose four points P_1, P_2, P_3, P_4 and let Λ be a linear pencil of conics passing through these four points. Let $\sigma : S \to \mathbb{P}^2$ be the blowing-up of six points P_i $(1 \le i \le 6)$. The proper transform $\sigma'\Lambda$ defines a $\mathbb{P}^1$-fibration $f : S \to \mathbb{P}^1$ for which the proper transform $C' = \sigma'(C)$ is a 2-section. Since $\mathcal{X}_t$ is isomorphic to $S \setminus C'$, $\mathcal{X}_t$ has an $\mathbb{A}^1_*$-fibration.

Looking for an $\mathbb{A}^1$-fibration in the case $\overline{\kappa}(\mathcal{X}_t) = -\infty$ is not an easy task. Consider, for example, the case where $X = \mathcal{X}_t$ is obtained as $S \setminus (Q \cup \ell)$, where Q is a smooth conic and ℓ is a line in $\mathbb{P}^2$ which meet in one point with multiplicity two. As explained in the above, such an X is obtained from $\mathbb{P}^2$ by blowing up six points $P_1, \ldots, P_6$ such that P_1, P_2 lie on a line $\tilde{\ell}$ and P_3, P_4, P_5, P_6 are points on a conic $\tilde{Q}$. Then the proper transforms on S of $\tilde{\ell}, \tilde{Q}$ are ℓ, Q. Consider the linear pencil $\tilde{\Lambda}$ on $\mathbb{P}^2$ spanned by $2\tilde{\ell}$ and $\tilde{Q}$. Then a general member of Λ is a smooth conic meeting $\tilde{Q}$ in one point $\tilde{Q} \cap \tilde{\ell}$ with multiplicity four. The proper transform Λ of $\tilde{\Lambda}$ on S defines an $\mathbb{A}^1$-fibration on X.

The following result of Dubouloz–Kishimoto except for the assertion (4) was orally communicated to one of the authors (see [3]).

Theorem 6.1. *Let S be a cubic hypersurface in $\mathbb{P}^3$ with a hyperplane section $S \cap H$ which consists of a line and a conic meeting in one point with multiplicity two. Let $Y = \mathbb{P}^3 \setminus S$ which is a smooth affine threefold. Then the following assertions hold:*

(1) $\overline{\kappa}(Y) = -\infty$.
(2) *Let $f : Y \to \mathbb{A}^1$ be a fibration induced by the linear pencil on $\mathbb{P}^3$ spanned by S and $3H$. Then a general fiber Y_t of f is a cubic hypersurface S_t minus $Q \cup \ell$, where Q is a conic and ℓ is a line which meet in one point with multiplicity two. Hence $\overline{\kappa}(Y_t) = -\infty$ and Y_t has an $\mathbb{A}^1$-fibration.*
(3) *Y has no $\mathbb{A}^1$-fibration.*
(4) *There is a finite covering T' of $\mathbb{A}^1$ such that the normalization of $Y \times_{\mathbb{A}^1} T'$ has an $\mathbb{A}^1$-fibration.*

Proof. (1) Since $K_{\mathbb{P}^3} + S \sim -4H + 3H = -H$, it follows that $\overline{\kappa}(Y) = -\infty$.

(2) The pencil spanned by S and $3H$ has base locus $Q \cup \ell$ and its general member, say S_t, is a cubic hypersurface containing $Q \cup \ell$ as a hyperplane section. It is clear that $S_t \setminus (Q \cup \ell) = Y_t$. Hence, as explained above, Y_t has an $\mathbb{A}^1$-fibration.
(3) Let $\tau : \tilde{S} \to \mathbb{P}^3$ be the cyclic triple covering of $\mathbb{P}^3$ ramified totally over the cubic hypersurface S. Then $\tilde{S}$ is a cubic hypersurface in $\mathbb{P}^4$ and $\tau^*(S) = 3\tilde{H}$, where $\tilde{H}$ is a hyperplane in $\mathbb{P}^4$. The restriction of τ onto $Z := \tilde{S} \setminus \tilde{S} \cap \tilde{H}$ induces a finite étale covering $\tau_Z : Z \to Y$. Suppose that Y has an $\mathbb{A}^1$-fibration $\varphi : Y \to T$. Then T is a rational surface. Since τ_Z is finite étale, this $\mathbb{A}^1$-fibration φ lifts up to an $\mathbb{A}^1$-fibration $\tilde{\varphi} : Z \to \tilde{T}$. By [2], $\tilde{S}$ is unirational and irrational. Hence $\tilde{T}$ is a rational surface. This implies that Z is a rational threefold. This is a contradiction because $\tilde{S}$ is irrational.
(4) There is an open set T of $\mathbb{A}^1$ such that the restriction of f onto $f^{-1}(T)$ is a smooth morphism onto T. By abuse of the notations, we denote $f^{-1}(T)$ by Y anew and the restriction of f onto $f^{-1}(T)$ by f. Hence $f : Y \to T$ is a smooth morphism. Let $K = k(t)$ be the function field of T and let Y_K be the generic fiber. Let $\overline{K}$ be an algebraic closure of K. Then $Y_{\overline{K}} := Y_K \otimes_K \overline{K}$ is identified with $S_{\overline{K}} \setminus (Q \cup \ell)$, where $S_{\overline{K}}$ is a cubic hypersurface in $\mathbb{P}^3_{\overline{K}}$ defined by $F_K = F_0 + tX_3^3 = 0$. Here t is a coordinate of $\mathbb{A}^1$ and (X_0, X_1, X_2, X_3) is a system of homogeneous coordinates of $\mathbb{P}^3$ such that $F_0(X_0, X_1, X_2, X_3) = 0$ is the defining equation of the cubic hypersurface S and the hyperplane H is defined by $X_3 = 0$. Then $Y_{\overline{K}}$ is obtained from $\mathbb{P}^2_{\overline{K}}$ by blowing up six $\overline{K}$-rational points in general position (two points on the image of ℓ and four points on the image of Q). As explained earlier, there is an $\mathbb{A}^1$-fibration on $Y_{\overline{K}}$ which is obtained from conics on $\mathbb{P}^2_{\overline{K}}$ belonging to the pencil spanned by Q and 2ℓ. This construction involves six points on $\mathbb{P}^2_{\overline{K}}$ to be blown up to obtain the cubic hypersurface $S_{\overline{K}}$ and four points (the point $Q \cap \ell$ and its three infinitely near points). Hence there exists a finite algebraic extension K'/K such that all these

points are rational over K'. Let T' be the normalization of T in K'. Let $Y' = Y \otimes_K K'$. Then Y' has an $\mathbb{A}^1$-fibration.

□

Based on the assertion (4) above, we propose the following conjecture.

Conjecture 6.2. Let $f : Y \to T$ be a smooth morphism from a smooth affine threefold Y onto a smooth affine curve T such that every closed fiber Y_t has an $\mathbb{A}^1$-fibration of complete type. Then there exists a finite covering T' of T such that the normalization of $Y \times_T T'$ has an $\mathbb{A}^1$-fibration.

Remark 6.3. The Conjecture 6.2 is true if Theorem 3.8 holds after an étale finite base change of T in the case where the general fibers of f have $\mathbb{A}^1$-fibrations of complete type. A main obstacle in trying to extend the proof in the case of $\mathbb{A}^1$-fibrations of affine type is to show that, with the notations in the proof of Theorem 3.8, the locus of base points P_t of the linear pencil Λ_t on $\overline{Y}_t$ with t varying in an open neighborhood of $t_0 \in T$ (resp. the loci of infinitely near base points) is a cross-section of the $\mathbb{P}^1$-bundle $\overline{f}|_{S_1} : S_1 \to T$ (resp. the exceptional $\mathbb{P}^1$-bundle).

Acknowledgements The second and third authors are supported by Grant-in-Aid for Scientific Research (C), No. 22540059 and (B), No. 24340006, JSPS.

References

1. G. Bredon, *Topology and Geometry*. Graduate Texts in Mathematics, vol. 139 (Springer, New York, 1993)
2. C.H. Clemens, P.A. Griffiths, The intermediate Jacobian of the cubic threefold. Ann. Math. **95**(2), 281–356 (1972)
3. A. Dubouloz, T. Kishimoto, Log-uniruled affine varieties without cylinder-like open subsets. arXiv: 1212.0521 (2012)
4. T. Fujita, On the topology of noncomplete algebraic surfaces. J. Fac. Sci. Univ. Tokyo Sect. IA Math. **29**(3), 503–566 (1982)
5. H. Flenner, M. Zaidenberg, *$\mathbb{Q}$-Acyclic Surfaces and Their Deformations, Classification of Algebraic Varieties* (L'Aquila, 1992). Contemporary Mathematics, vol. 162 (American Mathematical Society, Providence, 1994), pp. 143–208
6. H. Flenner, Sh. Kaliman, M. Zaidenberg, Deformation equivalence of affine ruled surfaces. arXiv:1305.5366v1, 23 May 2013
7. A. Grothendieck, Techniques de construction et théorèmes d'existence en géométrie algébrique IV: Les schémas de Hilbert, Séminaire Bourbaki, 13e année, 1960/61, no. 221, 1–28
8. R.V. Gurjar, K. Masuda, M. Miyanishi, $\mathbb{A}^1$-fibrations on affine threefolds. J. Pure Appl. Algebra **216**, 296–313 (2012)
9. R.V. Gurjar, K. Masuda, M. Miyanishi, Surjective derivations in small dimensions. J. Ramanujan Math. Soc. 28A(Special Issue), 221–246 (2013)
10. R.V. Gurjar, K. Masuda, M. Miyanishi, P. Russell, Affine lines on affine surfaces and the Makar-Limanov invariant. Can. J. Math. **60**(1), 109–139 (2008)
11. R. Hartshorne, *Algebraic Geometry* (Springer, New York, 1977)
12. S. Iitaka, *Deformations of Compact Complex Surfaces*. Global Analysis (Papers in Honor of K. Kodaira) (Univ. Tokyo Press, Tokyo, 1969), pp. 267–272

13. T. Kambayashi, On the absence of nontrivial separable forms of the affine plane. J. Algebra **35**, 449–456 (1975)
14. T. Kambayashi, M. Miyanishi, On flat fibrations by the affine line. Illinois J. Math. **22**, 662–671 (1978)
15. T. Kambayashi, D. Wright, Flat families of affine lines are affine line bundles. Illinois J. Math. **29**, 672–681 (1985)
16. Sh. Kaliman, M. Zaidenberg, Families of affine planes: the existence of a cylinder. Michigan Math. J. **49**(2), 353–367 (2001)
17. Y. Kawamata, On deformations of compactifiable complex manifolds. Math. Ann. **235**(3), 247–265 (1978)
18. Y. Kawamata, On the classification of non-complete algebraic surfaces, in *Algebraic Geometry* (Proc. Summer Meeting, Univ. Copenhagen, Copenhagen, 1978). Lecture Notes in Mathematics, vol. 732 (Springer, Berlin, 1979), pp. 215–232
19. K. Kodaira, On the stability of compact submanifolds of complex manifolds. Am. J. Math. **85**, 79–94 (1963)
20. K. Kodaira, *Complex Manifolds and Deformation of Complex Structures*. Classics in Mathematics (Springer, Berlin, 1986)
21. J. Kollar, *Rational Curves on Algebraic Varieties*. Ergebnisse der Mathematik und ihrer Grenzgebiete, vol. 32 (Springer, Berlin, 1996)
22. J. Kollar, S. Mori, *Birational Geometry of Algebraic Varieties*. Cambridge Tracts in Mathematics, vol. 134 (Cambridge University Press, Cambridge, 1998)
23. M. Miyanishi, An algebro-topological characterization of the affine space of dimension three. Am. J. Math. **106**, 1469–1486 (1984)
24. M. Miyanishi, Affine pseudo-coverings of algebraic surfaces. J. Algebra **294**, 156–176 (2005)
25. J.A. Morrow, Minimal normal compactifications of $\mathbb{C}^2$, in *Complex Analysis (Proc. Conf. Rice Univ., Houston, Tex., 1972), Vol. I: Geometry of Singularities*. Rice Univ. Studies, vol. 59, no. 1 (1973), pp. 97–112
26. W.D. Neumann, On the topology of curves in complex surfaces, in *Topological Methods in Algebraic Transformation Groups*. Progress in Mathematics, vol. 80 (Birkhauser, Boston, 1989), pp. 117–133
27. C.P. Ramanujam, A topological characterization of the affine plane as an algebraic variety. Ann. Math. **94**, 69–88 (1971)
28. P. Russell, Embeddings problems in affine algebraic geometry, in *Polynomial Automorphisms and Related Topics* (Publishing House for Science and Technology, Hanoi, 2007), pp. 113–135
29. A. Sathaye, Polynomial ring in two variables over a DVR: a criterion. Invent. Math. **74**(1), 159–168 (1983)
30. R.O. Wells, *Differential Analysis on Complex Manifolds*. Graduate Texts in Mathematics, vol. 65 (Springer, New York, 2008)

Remark on Deformations of Affine Surfaces with $\mathbb{A}^1$-Fibrations

Takashi Kishimoto

Abstract The possible structure of singular fibers of an $\mathbb{A}^1$-fibration on a smooth affine surface is well understood, in particular, any such fiber is a disjoint union of affine lines (possibly with multiplicities). This paper lies in a three-dimensional generalization of this fact, i.e., properties concerning a fiber component of a given fibration $f : X \to B$ from a smooth affine algebraic threefold X onto a smooth algebraic curve B whose general fibers are affine surfaces admitting $\mathbb{A}^1$-fibrations. The phenomena differ according to the type of $\mathbb{A}^1$-fibrations on general fibers of f (namely, of affine type, or of complete type). More precisely, in case of affine type, each irreducible component of every fiber of $f : X \to B$ admits an effective $\mathbb{G}_a$-action provided $\mathrm{Pic}(X) = (0)$ with some additional conditions concerning a compactification, whereas for the complete type, there exists an example in which a special fiber of $f : \mathbb{A}^3 \to \mathbb{A}^1$ possesses no longer an $\mathbb{A}^1$-fibration.

MSC2010: 14R25, 14R20, 14D06, 14J30.

1 Introduction

1.1. All varieties treated in this chapter are defined over the field of complex numbers $\mathbb{C}$ otherwise mentioned. An affine algebraic variety X is said to be *affine ruled* if X possesses an $\mathbb{A}^1$-fibration $f : X \to Y$, where Y is an algebraic variety of $\dim(Y) = \dim(X) - 1$. This terminology is fitting to the geometric fact that an affine ruled variety X contains an $\mathbb{A}^1$-cylinder, i.e., an open affine subset $U \subseteq X$ of

T. Kishimoto (✉)
Faculty of Science, Department of Mathematics, Saitama University, Saitama 338-8570, Japan
e-mail: tkishimo@rimath.saitama-u.ac.jp

I. Cheltsov et al. (eds.), *Automorphisms in Birational and Affine Geometry*, Springer Proceedings in Mathematics & Statistics 79,
DOI 10.1007/978-3-319-05681-4_20,

the form $U \cong Z \times \mathbb{A}^1$ with a suitable affine variety Z (cf. [18]). It is obvious that an affine ruled affine curve is isomorphic to the affine line $\mathbb{A}^1$. Meanwhile, there are infinitely many affine ruled smooth affine surfaces, hence it is not possible to list all of such surfaces concretely. Nevertheless, there is fortunately a very useful invariant, so-called log Kodaira dimension $\overline{\kappa}$, in order to characterize affine ruled smooth affine surfaces (see [12, 13] for the definition of log Kodaira dimension). More precisely, for a smooth affine surface Y, the following three conditions are equivalent to each other (cf. [19, 25]):

1. $\overline{\kappa}(Y) = -\infty$,
2. Y admits an $\mathbb{A}^1$-fibration,
3. for a general point $y \in Y$, there exists an algebraic curve $C_y \subseteq Y$ passing through y such that the normalization of C_y is isomorphic to $\mathbb{A}^1$. In other words, there exists a Zariski dense subset of Y which is covered by images of the affine line $\mathbb{A}^1$.

Suppose that Y is an affine ruled affine surface, and let $f : Y \to B$ be an $\mathbb{A}^1$-fibration, where B is a smooth algebraic curve. Then there exist two possibilities about the type of f according to the base curve B as follows:

(a) B is an affine curve (in this case f is said to be *of affine type*),
(b) B is a projective curve (in this case f is said to be *of complete type*).

Remark 1.1. It is worthwhile to note that the difference of type of an $\mathbb{A}^1$-fibration $f : Y \to B$ on an affine surface Y as above (i.e., of affine type or of complete type) seems to be tiny geometrically, whereas from the viewpoint of the coordinate ring it is crucial. In fact, in case of (a), it follows that f is realized as a quotient map with respect to a suitable effective $\mathbb{G}_a$-action on Y, which is in turn translated in terms of a locally nilpotent derivation on the coordinate ring $\Gamma(\mathcal{O}_Y)$ of Y, which is a purely algebraic object (cf. [10]). Meanwhile, in case of (b), f is never obtained as a quotient map of an effective $\mathbb{G}_a$-action on Y, so that we are obliged to consider more geometrically in this case.

Remark 1.2. As an immediate consequence of a theorem of Abhyankar–Moh–Suzuki (cf. [1, 26], see also [22, Chapter 2, §1]), every $\mathbb{A}^1$-fibration on the affine plane $\mathbb{A}^2$ is of affine type. But, in general, the type of an $\mathbb{A}^1$-fibration (a) or (b) is not intrinsic on a given smooth affine surface Y. For instance, let us consider the complement $Y := (\mathbb{P}^1 \times \mathbb{P}^1)\backslash\Delta$, where Δ is a diagonal. For a given point $Q \in \Delta$, let us denote by l_i the fiber of the ruling $\pi_i : \mathbb{P}^1 \times \mathbb{P}^1 \to \mathbb{P}^1$ passing through Q $(i = 1, 2)$. Then the restriction of the rational map determined by the linear pencil $\mathscr{L}$ spanned by Δ and $l_1 + l_2$ to Y yields an $\mathbb{A}^1$-fibration over the affine line $\mathbb{A}^1$:

$$\Phi_{\mathscr{L}}|_Y : Y \longrightarrow \mathbb{A}^1,$$

hence $\Phi_{\mathscr{L}}|_Y$ is of affine type. On the other hand, the restrictions $\pi_i|_Y$ $(i = 1, 2)$ give rise to those of complete type.

1.2. Assume that $f : Y \to B$ is an $\mathbb{A}^1$-fibration on a smooth affine surface Y. The structure of appearing singular fibers of f is well known (cf. [24]), namely, for both cases of affine type and of complete type, a fiber $f^*(b)$ $(b \in B)$ is described as follows:

$$f^*(b) = \sum_i m_i l_i \qquad (m_i \geqq 1),$$

where $l_i \cong \mathbb{A}^1$ $(\forall\, i)$ and $l_i \cap l_i = \emptyset$ $(i \neq j)$.

1.3. In consideration of the facts mentioned above in 1.1 and 1.2 about $\mathbb{A}^1$-fibrations on smooth affine surfaces, we propose the following question:

Problem 1.3. Let X be a smooth affine algebraic variety of $\dim(X) = n$. Suppose that X admits a morphism $f : X \to B$ onto a smooth algebraic curve B whose general fibers are affine ruled. Then what kinds of properties does each irreducible component of a fiber of f possess? For example, is every component of its fiber also affine ruled? Can f be factored by means of an $\mathbb{A}^1$-fibration, say $f = h \circ g : X \to Y \to B$, where g is an $\mathbb{A}^1$-fibration over a normal variety Y of $\dim(Y) = n - 1$ (if necessary by shrinking the base curve B)?

1.4. As recalled in 1.2, Problem 1.3 holds true in case of $n = 2$. Meanwhile, even in case of $n = 3$, the satisfactory answer to Problem 1.3 is not known so far. In this chapter, we investigate mainly the case of $n = 3$ according to the type of $\mathbb{A}^1$-fibrations of general fibers of f (i.e., being of affine type or of complete type), separately, by paying attention to the result due to Gurjar, Masuda, Miyanishi (cf. Theorem 1.4), and we obtain Theorems 1.6 and 1.8 below.

1.5. At first, we deal with the case in which general fibers possess $\mathbb{A}^1$-fibrations of affine type. Before stating our results, we need to mention the following result due to Gurjar, Masuda, and Miyanishi (cf. [11]), which plays an essential role to prove Theorem 1.6.

Theorem 1.4 (cf. [11, Theorem 2.8]). *Let $f : X \to B$ be a morphism from a smooth affine algebraic threefold X onto a smooth algebraic curve B such that a general fiber of f is a smooth affine surface with an $\mathbb{A}^1$-fibration of affine type. Then we have the following:*

(1) After shrinking the base curve B if necessary, say $B_0 \subseteq B$, and after taking an étale finite morphism $\tilde{B}_0 \to B_0$, the resulting morphism from f on the fiber product $\tilde{X}_0 := f^{-1}(B_0) \times_{B_0} \tilde{B}_0$, say $\tilde{f} : \tilde{X}_0 \to \tilde{B}_0$ is factored in such a way that:

$$\tilde{f} = \tilde{h} \circ \tilde{g} : \tilde{X}_0 \xrightarrow{\tilde{g}} \tilde{Y}_0 \xrightarrow{\tilde{h}} \tilde{B}_0,$$

where $\tilde{g}$ is an $\mathbb{A}^1$-fibration over a surface $\tilde{Y}_0$.

(2) *If we suppose additionally that there exists a relative completion of* $f : X \to B$, *say* $\overline{f} : \overline{X} \to B$, *which satisfies the following conditions: (Notation: Let* $\Delta := \overline{X} \setminus X$, *and for a point* $b \in B$, *let us put* $\overline{X}_b := \overline{f}^*(b)$, $X_b := f^*(b)$ *and* $\Delta_b := \Delta \cdot \overline{X}_b$.*)*

   (i) $(X, \overline{X}, \Delta, \overline{f}, 0)$ *with a fixed point* $0 \in B$ *is a family of logarithmic deformations of the triple* $(X_0, \overline{X}_0, \Delta_0)$,

   (ii) *A given* $\mathbb{A}^1$*-fibration of affine type on a general fiber* X_b *is extended to a* $\mathbb{P}^1$*-fibration on* $\overline{X}_b$, *say* φ_b, *and*

   (iii) *A section of the* $\mathbb{P}^1$*-fibration* φ_0 *found in* Δ_0 *has no monodromy in* $\overline{X}$,

   then after shrinking the base curve B *if necessary, say* $0 \in B_0 \subseteq B$, *the restricted morphism* $f|_{f^{-1}(B_0)}$ *is factored by an* $\mathbb{A}^1$*-fibration (without the necessity to take an étale finite morphism as in the assertion (1)).*

Remark 1.5. As a special case of such a deformation $f : X \to B$, if a general fiber of f is isomorphic to the affine plane $\mathbb{A}^2$, then there exists an open dense subset $B_0 \subseteq B$ such that the inverse image $f^{-1}(B_0)$ is isomorphic to the fiber product $f^{-1}(B_0) \cong B_0 \times \mathbb{A}^2$ (cf. [17]), in particular, for such an f, the factorization property asked in Problem 1.3 holds true. By experience, such a factorization becomes to be more subtle when the Picard group of a general fiber of f is bigger.

1.6. By making use of Theorem 1.4, we can obtain the following result:

Theorem 1.6. *Let* X *be a smooth affine algebraic threefold with* $\mathrm{Pic}(X) = (0)$, *and let* $f : X \to B$ *be a morphism onto a smooth algebraic curve* B *such that general fibers are equipped with* $\mathbb{A}^1$*-fibrations of affine type. Suppose that there exists a relative completion* $\overline{f} : \overline{X} \to B$ *of* f *which satisfies conditions (i)–(iii) in (2) of Theorem 1.4. Then each irreducible component of every fiber of* f *is an affine surface admitting an effective* $\mathbb{G}_a$*-action, in particular, it possesses an* $\mathbb{A}^1$*-fibration of affine type.*

Remark 1.7. As stated in Theorem 1.4, if a morphism $f : X \to B$ in question, whose general fibers possess $\mathbb{A}^1$-fibrations of affine type, admits a relative completion as in Theorem 1.4 (2), then we have only to shrink the base curve B in order to obtain a factorization of f by an $\mathbb{A}^1$-fibration. However, we need in general to take a suitable étale finite covering of the base curve after shrinking without an assumption about an existence of a relative completion. In fact, in [11, Example 2.6], they construct an example in which we are obliged to take an étale finite covering of degree two of the base to reach a decomposition by means of an $\mathbb{A}^1$-fibration.

1.7. In the case where a general fiber of f has an $\mathbb{A}^1$-fibration of complete type only, the special fibers of $f : X \to B$ behave often in a more complicated manner. In fact, the following result (Theorem 1.8) asserts that the answer to Problem 1.3 is negative in general even if X is isomorphic to the affine 3-space $\mathbb{A}^3$.

Theorem 1.8. *Let* $f(x, y, z) = x^3 + y^3 + z(z + 1) \in \mathbb{C}[x, y, z]$ *and let us denote by the same notation* f *the polynomial map:*

$$f : \mathbb{A}^3 = \mathrm{Spec}(\mathbb{C}[x, y, z]) \ni (a, b, c) \mapsto f(a, b, c) \in \mathbb{A}^1 = \mathrm{Spec}(\mathbb{C}[f]),$$

defined by $f(x, y, z)$. Then we have the following:

(1) All fibers $f^*(\alpha)$ of f except for $f^*(-\frac{1}{4})$ are smooth affine surfaces admitting $\mathbb{A}^1$-fibrations of complete type only.
(2) The fiber $f^*(-\frac{1}{4})$ is an irreducible normal affine surface which is not affine ruled.

Remark 1.9. As mentioned in [11] (see also [2]), the behavior of a deformation of surfaces equipped with $\mathbb{A}^1$-fibrations of complete type only is more involved to understand than those having $\mathbb{A}^1$-fibrations of affine type. For instance, let $S \subseteq \mathbb{P}^3$ be a smooth cubic hypersurface and let $H \subseteq \mathbb{P}^3$ be a hyperplane such that the intersection $H|_S$ is either of the form (i) $l + C$ or (ii) $l_1 + l_2 + l_3$, where l and C are a line and a smooth conic meeting at a point tangentially in case of (i), on the other hand, l_1, l_2, and l_3 are lines meeting at a point (an Eckardt point) in case of (ii). Notice that for any smooth cubic surface S, we can find a suitable hyperplane H satisfying (i) or (ii) (cf. [2]). In any case, the morphism from $\mathbb{A}^3 \cong \mathbb{P}^3 \backslash H$ onto $\mathbb{A}^1$ obtained as the restriction of the rational map on $\mathbb{P}^3$ determined by the linear pencil spanned by S and $3H$ yields a family of affine surfaces with $\mathbb{A}^1$-fibrations of complete type only (cf. [2]). Note that the fibration in Theorem 1.8 is regarded as the special one realized as in the above-mentioned fashion in case of (ii). The strange looking behavior of a deformation of surfaces of complete type brings us something interesting, for instance, letting $S \subseteq \mathbb{P}^3$ be the closure of the hypersurface $\mathbb{V}_{\mathbb{A}^3}(f) \subseteq \mathbb{A}^3$ with $f \in \mathbb{C}[x, y, z]$ as in Theorem 1.8, the complement $\mathbb{P}^3 \backslash (S \cup \overline{f^*(-1/4)})$ is covered by mutually disjoint affine lines, notwithstanding, this complement admits neither an $\mathbb{A}^1$-fibration nor an effective $\mathbb{G}_a$-action (cf. Remark 4.1 and Proposition 4.2).

This chapter is organized in the following way. In Sect. 2, we shall yield several examples of affine ruled smooth affine surfaces some of which admit only $\mathbb{A}^1$-fibrations of complete type. Then we say about deformations of affine ruled surfaces in consideration of Theorem 1.4 due to Gurjar, Masuda, and Miyanishi. Roughly speaking, a given morphism $f : X \to B$ from a smooth affine algebraic threefold onto a smooth algebraic curve with general fibers possessing $\mathbb{A}^1$-fibrations of affine type can be factored by means of an $\mathbb{A}^1$-fibration up to shrinking and taking an étale finite morphism of the base curve B. However under the condition about an existence of a relative completion $\overline{f} : \overline{X} \to B$ of f as in Theorem 1.4 (2), we have only to shrink the base B to reach a factorization of f by an $\mathbb{A}^1$-fibration. Then it is not difficult to see the assertion of Theorem 1.6, which is done in Sect. 3. On the other hand, we prove Theorem 1.8 in Sect. 4, in addition, we give the proof for Proposition 4.2. In the final section (Sect. 5), we shall mention some remarks and relevant problems.

2 Preliminaries

2.1. In this section, we shall recall at first some typical examples of affine ruled surfaces, some of which admit only $\mathbb{A}^1$-fibrations of complete type, then we observe a deformation $f : X \to B$ of affine ruled smooth affine surfaces by taking works [2, 11] (cf. Theorem 1.4) into account.

2.2. We begin with several examples of smooth affine surfaces admitting $\mathbb{A}^1$-fibrations of affine type or of complete type, respectively (see Sect. 1 for the definition of being of affine type and of complete type). In fact, affine surfaces with $\mathbb{A}^1$-fibrations of affine type are more handy to treat compared with those having only $\mathbb{A}^1$-fibrations of complete type because such $\mathbb{A}^1$-fibrations are obtained as quotients of effective $\mathbb{G}_a$-actions (cf. [10]), which is in turn translated in terms of locally nilpotent derivations. For instance, the following surfaces are typical examples that admit $\mathbb{A}^1$-fibrations of affine type.

Example 2.1. (1) It is well known that any $\mathbb{A}^1$-fibration on the affine plane $\mathbb{A}^2$ is of affine type with the affine line $\mathbb{A}^1$ as the base curve. Furthermore, all closed fibers are scheme-theoretically isomorphic to $\mathbb{A}^1$ (cf. [1, 26], see also [22, Chapter II]).

(2) Let Y be a $\mathbb{Q}$-homology plane with log Kodaira dimension $\overline{\kappa} = -\infty$, where we recall that Y is said to be a *$\mathbb{Q}$-homology plane*, by definition, if $H_i(Y; \mathbb{Q}) = (0)$ ($\forall\, i > 0$). It is known that an $\mathbb{A}^1$-fibration $g : Y \to B$ on such an Y should be of affine type, more precisely the base curve B is isomorphic to $\mathbb{A}^1$ by noting that every $\mathbb{Q}$-homology plane is rational (cf. [6–8]). Moreover, letting $g^*(b_i)$ ($1 \leqq i \leqq r$) exhaust all of singular fibers of g (if there exist at all), they are of the form $g^*(b_i) = m_i l_i$, where $m_i \geqq 2$ and $l_i \cong \mathbb{A}^1$. In terms of these multiplicities m_i, the first integral homology group and the Picard group are expressed as $H_1(Y; \mathbb{Z}) \cong \bigoplus_{i=1}^r (\mathbb{Z}/m_i\mathbb{Z})$ and $\mathrm{Pic}(Y) \cong \bigoplus_{i=1}^r \mathbb{Z}[l_i] \cong \bigoplus_{i=1}^r (\mathbb{Z}/m_i\mathbb{Z})$ (see [24, Chapter 3] for more informations on $\mathbb{Q}$-homology planes). Furthermore, provided that the Makar-Limanov invariant $\mathrm{ML}(Y)$ of Y, which is defined to be the intersection of kernels of all locally nilpotent derivations on the coordinate ring $\Gamma(\mathcal{O}_Y)$, is trivial, i.e., $\mathrm{ML}(Y) = \mathbb{C}$, Masuda and Miyanishi [21, Theorem 3.1] show that $g : Y \to B$ possesses exactly one multiple fiber $g^*(b_1) = m_1 l_1$ with $m_1 = |H_1(Y; \mathbb{Z})|$, and its universal covering is isomorphic to the Danielewski surface Y_m in the next example (3), further Y is realized as the quotient of Y_m with respect to a suitable $(\mathbb{Z}/m_1\mathbb{Z})$-action on Y_m.

(3) Let us consider a Danielewski surface:

$$Y_m := \big(xy + z^m + 1 = 0\big) \subseteq \mathbb{A}^3 = \mathrm{Spec}\,(\mathbb{C}[x, y, z]).$$

Then the restriction of the projection $\mathrm{pr}_x : \mathbb{A}^3 \to \mathbb{A}^1 = \mathrm{Spec}(\mathbb{C}[x])$ onto Y_m yields an $\mathbb{A}^1$-fibration $g = \mathrm{pr}_x|_{Y_m}$ over the affine line $\mathbb{A}^1$. Moreover, any fiber $g^*(b)$ distinct from the central fiber $g^*(0)$ is scheme-theoretically isomorphic to $\mathbb{A}^1$, whereas the central one $g^*(0)$ is a disjoint union of m affine lines with the respective multiplicities one. By the same fashion, the restriction $\mathrm{pr}_y|_{Y_m}$ is an $\mathbb{A}^1$-fibration over $\mathbb{A}^1$ with the same properties as $\mathrm{pr}_x|_{Y_m}$.

Meanwhile, as examples of affine surfaces that admit only $\mathbb{A}^1$-fibrations of complete type, we have the following:

Example 2.2. Let B be a smooth projective curve of positive genus. Let $\mathscr{E}$ be a vector bundle of $\mathrm{rank}(\mathscr{E}) = 2$ on B, and $\pi : \mathbb{P}_B(\mathscr{E}) \to B$ the associated $\mathbb{P}^1$-bundle. Letting $S \subseteq \mathbb{P}(\mathscr{E})$ be an ample section with respect to π, the affine surface $Y := \mathbb{P}(\mathscr{E})\backslash S$ is equipped with the $\mathbb{A}^1$-bundle $\pi|_Y$ over B. It is easy to see that $\pi|_Y$ is the unique $\mathbb{A}^1$-fibration on Y (up to automorphisms of the base curve B), in particular, Y does not admit an $\mathbb{A}^1$-fibration of affine type.

Example 2.3. Let $S \subseteq \mathbb{P}^3$ be a smooth cubic hypersurface with an Eckardt point, and let $H \subseteq \mathbb{P}^3$ be a hyperplane such that $H|_S$ is composed of three lines meeting at an Eckardt point, say $H|_S = l_1 + l_2 + l_3$. Then we can verify by a straightforward computation that the log Kodaira dimension of the affine surface $S_0 := S\backslash(l_1 \cup l_2 \cup l_3)$ is equal to $\overline{\kappa}(S_0) = -\infty$. Thus S_0 admits an $\mathbb{A}^1$-fibration $g : S_0 \to B$ by [25]. In fact, as investigated explicitly in [2], it follows that the base curve B of g is isomorphic to the projective line $B \cong \mathbb{P}^1$, i.e., S_0 is an affine ruled affine surface having only $\mathbb{A}^1$-fibrations of complete type.

Example 2.4. Even in the case in which a smooth cubic hypersurface $S \subseteq \mathbb{P}^3$ does not possess an Eckardt point, we can find a suitable hyperplane $H \subseteq \mathbb{P}^3$ such that $H|_S$ consists of a line l and a conic C that meet to each other at a single point tangentially (cf. [2]). In this case also, a direct computation says that $\overline{\kappa}(S\backslash(l \cup C)) = -\infty$ to deduce that an affine surface $S\backslash(l \cup C)$ admits an $\mathbb{A}^1$-fibration, which is indeed defined over $\mathbb{P}^1$, in particular, it is of complete type.

2.3. We consider a deformation of affine ruled smooth affine surfaces, say $f : X \to B$, where X is a smooth affine algebraic threefold and B is an algebraic curve by taking Problem 1.3 into account. The properties of X differ according to the type of an $\mathbb{A}^1$-fibration found on a general fiber of f. If a general fiber of $f : X \to B$ is an affine surface possessing an $\mathbb{A}^1$-fibration of complete type only, then we know already several examples in which there does not exist an open subset $U \subseteq X$ such that the restriction $f|_U$ is decomposed by an $\mathbb{A}^1$-fibration over a surface (see e.g. [2] for such examples. Notice that the morphism $f : \mathbb{A}^3 \to \mathbb{A}^1$ found in the assertion of Theorem 1.8 is also one of the examples). On the other hand, if a general fiber of f admits an $\mathbb{A}^1$-fibration of affine type, then Theorem 1.4 gives rise to an information of f and X, which plays an essential role to prove Theorem 1.6. Roughly speaking, such an f can be factorized by means of an $\mathbb{A}^1$-fibration after shrinking and taking an étale finite covering of the base eventually. For an actual application, it is hopeful that we do not have to take an étale finite covering, but we are in general obliged to take an étale covering without a hypothesis about an existence of a relative completion as requested in Theorem 1.4 (2). For instance, [11, Example 2.6] yields such an example.

Remark 2.5. In some case, a deformation $f : X \to B$ of affine ruled surfaces can be decomposed by an $\mathbb{A}_1^*$-fibration over a surface instead of an $\mathbb{A}^1$-fibration itself as in the following example. Let S and H be respectively a smooth cubic surface and a hyperplane in $\mathbb{P}^3$ such that the hyperplane section $H|_S$ is of type (ii) in Remark 1.9, and let $f : \mathbb{A}^3 \to \mathbb{A}^1$ be the morphism obtained as the restriction onto $\mathbb{A}^3 \cong \mathbb{P}^3 \backslash H$ of the rational map defined by the linear pencil spanned by S and $3H$. Then a general fiber of f is a smooth affine surface possessing $\mathbb{A}^1$-fibrations of complete type only, furthermore, f cannot be factored by means of $\mathbb{A}^1$-fibration even if we restrict f onto an open dense subset of X however as remarked in 2.3. Nevertheless, it is instead decomposed in such a way that:

$$f = h \circ g : \mathbb{A}^3 \xrightarrow{g} \mathbb{A}^2 \xrightarrow{h} \mathbb{A}^1,$$

where g is a twisted $\mathbb{A}^1_*$-fibration and h is a trivial $\mathbb{A}^1$-bundle (cf. [3]). It is worthwhile to note that this factorization $f = h \circ g$ is obtained geometrically by use of minimal model program. More precisely, we can embed $\mathbb{A}^3$ into a suitable normal projective threefold V with at worst $\mathbb{Q}$-factorial, terminal singularities, which is equipped with a Mori conic bundle structure $p : V \to W$, i.e., p is obtained as the contraction of an extremal ray in $\overline{\mathrm{NE}}(V)$ with $\dim(W) = 2$ such that the restriction of p onto $\mathbb{A}^3$ gives rise to g. Furthermore, we note that p does not admit any birational section. Indeed, there exists an irreducible component in the boundary $V \backslash \mathbb{A}^3$ which meets general fibers of p twice (cf. [3]). Provided that p admits a birational section, say Θ, the restriction $\Theta|_{\mathbb{A}^3}$ gives rise to a 2-torsion element of $\mathrm{Pic}(\mathbb{A}^3)$, which is absurd. This particular phenomenon in higher dimension is interesting since, for any smooth affine algebraic surface with the factorial coordinate ring, an $\mathbb{A}^1_*$-fibration on it should be untwisted (cf. [24]).

3 Proof of Theorem 1.6

3.1. In this section, we shall prove Theorem 1.6. Let X be a smooth affine algebraic threefold with $\mathrm{Pic}(X) = (0)$, which possesses a morphism $f : X \to B$ onto a smooth algebraic curve B such that general fibers of f are affine surfaces equipped with $\mathbb{A}^1$-fibrations of affine type. Suppose that there exists a relative completion $\overline{f} : \overline{X} \to B$ of f that satisfies conditions (i)–(iii) in (2) of Theorem 1.4. Then, by virtue of Theorem 1.4, we can find an open dense subset $B_0 \subseteq B$ such that the restriction of f onto $X_0 := f^{-1}(B_0) \subseteq X$ can be factored as in the following fashion:

$$(*) \quad f_0 := f|_{X_0} : X_0 \xrightarrow{g_0} Y_0 \xrightarrow{h_0} B_0,$$

where g_0 is an $\mathbb{A}^1$-fibration over a surface. By [18], there exists an open affine subset V_0 of Y_0 such that the inverse image $U_0 := {g_0}^{-1}(V_0) \subseteq X_0$ is isomorphic to the fiber product with $\mathbb{A}^1$, i.e., $U_0 \cong V_0 \times \mathbb{A}^1$. Let δ_0 be a locally nilpotent derivation on $\Gamma(\mathcal{O}_{U_0})$ corresponding to translations along the second factor on $U_0 \cong V_0 \times \mathbb{A}^1$. Notice that the complement $X \backslash U_0$ is purely of codimension one, hence triviality of $\mathrm{Pic}(X)$ implies that there exists a regular function a in $A := \Gamma(\mathcal{O}_X)$ such that the principal divisor $\mathrm{div}_X(a)$ coincides with $X \backslash U_0$, i.e., $A[a^{-1}] = \Gamma(\mathcal{O}_{U_0})$. Hence $\delta := a^N \delta_0$ becomes a locally nilpotent derivation on A with $N \geqq 0$ suitably chosen because A is a finitely generated $\mathbb{C}$-algebra. Letting $R := \mathrm{Ker}(\delta)$ be the kernel of δ, which is known to be a finitely generated $\mathbb{C}$-algebra of dimension two, let us denote by $\pi : X \to Y = \mathrm{Spec}(R)$ the morphism associated with the inclusion $R \subseteq A$. Note that π is nothing but the quotient map with respect to an effective $\mathbb{G}_a$-action on X arising from δ, and $\pi : X \to Y$ contains $g_0|_{U_0} : U_0 \to V_0$ by construction.

3.2. Let $F = \sum_{j=1}^{s} m_j F_j$ be a fiber of f, where F_j is an irreducible component of F and $m_j \geqq 1$. There exists $a_j \in A$ such that the principal divisor $\mathrm{div}_X(a_j)$ coincides with F_j for $1 \leqq j \leqq s$ because of the hypothesis $\mathrm{Pic}(X) = (0)$. Then we have the following:

Claim. $a_j \in R \ \ (1 \leqq j \leqq s)$.

Proof of Claim. Assume to the contrary, for instance, that $a_1 \in A \backslash R$. Then F_1 dominates Y with respect to π. Further, it follows then that a general fiber of f also dominates Y. Indeed, if a general fiber of f does not dominate Y via π, then it does not intersect a general $\mathbb{G}_a$-orbit, say $l := \mathbb{G}_a.[x]$ on X. Let us embed X into a smooth projective threefold Z in such a way that Z possesses a projective fibration $p : Z \to C$, which is an extension of the given morphism $f : X \to B$. Since l meets a fiber component F_1, we have $(p^*(c) \cdot \bar{l}) > 0$ for every point $c \in C$, where $\bar{l}$ is the closure of l in Z. Meanwhile, as l does not meet general fibers of f, the closure $\bar{l}$ should be found in the boundary part $Z \backslash X$, whereas $l = \bar{l} \cap X$ meets F_1, a contradiction. Now let Q be a general point of B_0. Then we have $f^*(Q) = g_0^*(h_0^*(Q))$, which means that none of components of the fiber $f^*(Q)$ does not dominate Y via π, which is a contradiction. □

3.3. Recall that $a_j \in A$ is a prime element vanishing along F_j, that is $\mathrm{div}_X(a_j) = F_j$. Then we are able to ascertain that F_j admits an effective $\mathbb{G}_a$-action by the same argument as in [10, Lemma 3.10] (see also [15, Lemma 2.2]). Indeed, after dividing δ by a_j^n with $n \geqq 0$ appropriately chosen, the resulting $a_j^{-n}\delta$ becomes to be nontrivial along F_j. Note that $a_j^{-n}\delta$ is again locally nilpotent because of $a_j \in R$ (see Claim above). Hence it yields a nontrivial locally nilpotent derivation on the coordinate ring $\Gamma(\mathcal{O}_{F_j})$ of F_j to see that F_j admits an effective $\mathbb{G}_a$-action as desired.

Thus we complete the proof for Theorem 1.6.

Remark 3.1. We notice that the argument in 3.3 cannot be applicable directly if we assume only that $\mathrm{Pic}(X)$ is finite instead of the triviality of $\mathrm{Pic}(X)$. In fact, provided that $\mathrm{Pic}(X) \otimes_{\mathbb{Z}} \mathbb{Q} = (0)$, there exists a regular function $a_j \in A$ such that the support of the principal divisor $\mathrm{div}_X(a_j)$ coincides with F_j, i.e., $\mathrm{div}_X(a_j) = d_j F_j$ for some $d_j \geqq 1$. As seen in 3.3, we can show that $A/a_j A \cong \Gamma(\mathcal{O}_{d_j F_j})$ admits a locally nilpotent derivation $a_j^{-n}\delta$ with $n \geqq 0$ adequately chosen. But, this does not always give rise to a *nontrivial* locally nilpotent derivation on $A/\sqrt{a_j A}$, which is the coordinate ring $\Gamma(\mathcal{O}_{F_j})$ of the reduced one. In fact, such an obstacle occurs already in case of dimension two as seen in the following example:

Example 3.2. Let $X := (xz - y^2 = 0) \subseteq \mathbb{A}^3$, which is an affine cone over the smooth conic $xz - y^2 = 0$, and let us consider a locally nilpotent derivation δ on its coordinate ring $A := \mathbb{C}[x, y, z]/(xz - y^2)$ determined by:

$$\delta = x\frac{\partial}{\partial y} + 2y\frac{\partial}{\partial z}.$$

Notice that δ yields an effective $\mathbb{G}_a$-action on X, which is nothing but the projection onto the x-axis, say $p_x : X \to \mathbb{A}^1 = \mathrm{Spec}(\mathbb{C}[x])$. It is easy to see that p_x has the unique singular fiber $p_x^*(0)$, which is a multiple fiber $p_x^*(0) = 2l$, where l is the line defined by $x = y = 0$. Furthermore, it follows that $\mathrm{Pic}(X) \cong \mathbb{Z}[l] \cong \mathbb{Z}/2\mathbb{Z}$. Since $2l$ is a principal divisor on X, indeed $2l$ is defined by xA, it follows that δ descends to a nontrivial locally nilpotent derivation on the coordinate ring of $2l$, i.e., on $A/xA \cong \mathbb{C}[x, y, z]/(x, y^2)$, however it becomes to be trivial when we restrict to $A/\sqrt{xA} \cong \mathbb{C}[x, y, z]/(x, y)$. This fact can be also verified in terms of geometry by noticing that the singular point $(0, 0, 0) \in X$ belongs to l, hence $\mathbb{G}_a$-action on l should be trivial.

4 Proof of Theorem 1.8

4.1. This section is devoted to the proof of Theorem 1.8. Thus let us set:

$$f(x, y, z) := x^3 + y^3 + z(z + 1) \in \mathbb{C}[x, y, z],$$

and we denote tacitly by the same notation f the polynomial map:

$$f : \mathbb{A}^3 = \mathrm{Spec}(\mathbb{C}[x, y, z]) \ni (a, b, c) \mapsto f(a, b, c) \in \mathbb{A}^1 = \mathrm{Spec}(\mathbb{C}[f]),$$

associated with the inclusion $\mathbb{C}[f] \subseteq \mathbb{C}[x, y, z]$. Further, let us denote by S_α° the fiber $f^*(\alpha)$, i.e., $S_\alpha^\circ := \big(f(x, y, z) - \alpha = 0\big) \subseteq \mathbb{A}^3$ for $\alpha \in \mathbb{C}$. First of all, we shall observe singularities on the fiber S_α°. In fact, we see the following:

Claim 1. S_α° is a smooth affine surface for any $\alpha \in \mathbb{C}\backslash\{-\frac{1}{4}\}$.

Proof of Claim 1. The proof of the assertion depends on a straightforward calculation. Letting S_α be the closure of $S_\alpha^\circ \subseteq \mathbb{A}^3$ in $\mathbb{P}^3$, the surface S_α is defined by the following cubic homogeneous polynomial:

$$F_\alpha(x, y, z, u) := x^3 + y^3 + z(z + u)u - \alpha u^3 \in \mathbb{C}[x, y, z, u],$$

where $u = 0$ corresponds to the hyperplane at infinity H_∞ with respect to the canonical embedding $\mathbb{A}^3{}_{(x,y,z)} \hookrightarrow \mathbb{P}^3{}_{[x:y:z:u]}$. By looking at the Jacobian of F_α:

$$J(F_\alpha) = \left(\frac{\partial F_\alpha}{\partial x}, \frac{\partial F_\alpha}{\partial y}, \frac{\partial F_\alpha}{\partial z}, \frac{\partial F_\alpha}{\partial u}\right) = \left(3x^2, 3y^2, 2zu + u^2, z^2 + 2zu - 3\alpha u^2\right),$$

it follows that S_α is smooth for $\alpha \in \mathbb{C}\backslash\{-\frac{1}{4}\}$, in particular, S°_α is smooth. □

4.2. For a while, we shall observe an affine surface S°_α for $\alpha \in \mathbb{C}\backslash\{-\frac{1}{4}\}$. By the proof of Claim 1 above, the surface S_α, which is the closure of S_α in $\mathbb{P}^3$, is smooth. Meanwhile, we have $-K_{S_\alpha} = H_\infty|_{S_\alpha} = l_1 + l_2 + l_3$ by adjunction, where l_1, l_2 and l_3 are three lines intersecting only at $[0 : 0 : 1 : 0]$, which is an Eckardt point of S_α. Then a straightforward computation shows that $\overline{\kappa}(S^\circ_\alpha) = -\infty$, which implies that S_α possesses an $\mathbb{A}^1$-fibration (cf. [25]). Moreover, it follows that an $\mathbb{A}^1$-fibration on S°_α is defined over $\mathbb{P}^1$, i.e., it is an $\mathbb{A}^1$-fibration of complete type (cf. [2]).

4.3. In the subsequent argument, we investigate the remaining fiber $S^\circ_{(-1/4)}$, which is singular. In fact, we can readily confirm that $\mathrm{Sing}(S_{(-1/4)}) = \{P := [0 : 0 : 1 : -2]\}$. More precisely, we see:

$$f(x, y, z) + \frac{1}{4} = x^3 + y^3 + z(z+1) + \frac{1}{4} = x^3 + y^3 + \left(z + \frac{1}{2}\right)^2,$$

thence the point $P = (0, 0, -1/2) \in S^\circ_{(-1/4)}$ is a Du Val singularity of D_4-type. In particular, P is not a cyclic quotient singularity.

Claim 2. $S^\circ_{(-1/4)}$ does not admit any $\mathbb{A}^1$-fibration.

Proof of Claim 2. Assume to the contrary that $S^\circ_{(-1/4)}$ is equipped with an $\mathbb{A}^1$-fibration. Then by virtue of [23] (see also [5]), the surface $S^\circ_{(-1/4)}$ has at most cyclic quotient singularities. This is absurd as $S^\circ_{(-1/4)}$ has a singularity of D_4-type. □

Thus we complete the proof of Theorem 1.8.

Remark 4.1. Certainly, the fiber $S^\circ_{(-1/4)}$ does not admit any $\mathbb{A}^1$-fibration, whereas S°_α does admit for all $\alpha \in \mathbb{C}\backslash\{-\frac{1}{4}\}$ as seen above. Let

$$g_\alpha : S^\circ_\alpha \longrightarrow \mathbb{P}^1$$

be one of such $\mathbb{A}^1$-fibrations on S°_α. Notice that g_α is defined over $\mathbb{P}^1$, namely an $\mathbb{A}^1$-fibration on S°_α is of complete type (cf. [2]). It follows that any fiber of g_α is a disjoint union of affine lines (cf. [24, Chapter 3, 1.4.2. Lemma]). Put $X := \mathbb{P}^3\backslash(S_{(-1/4)} \cup S_0)$. Then by the argument just mentioned, there exists a two-dimensional family $\mathscr{F} = \{C_\gamma\}$ of affine lines $C_\gamma \cong \mathbb{A}^1$ such that for any point $x \in X$ we can find the unique member $C_\gamma \in \mathscr{F}$ passing through x. In other words, X is covered by mutually disjoint affine lines belonging to $\mathscr{F}$. Nevertheless, we claim that:

Proposition 4.2. *With the above notation, X is not affine ruled.*

Proof. Suppose on the contrary that X is affine ruled. Hence, by definition, we can find an open affine subset $U \subseteq X$, which is a fiber product $U \cong V \times \mathbb{A}^1$ with V a suitable affine algebraic variety. Recall that X is an open subset of $\mathbb{P}^3 \backslash S_0$, thence $\mathbb{P}^3 \backslash S_0$ is also affine ruled. Then $\mathbb{P}^3 \backslash S_0$ admits an effective $\mathbb{G}_a$-action (cf. Remark 4.4), which is a contradiction to [2]. □

Remark 4.3. We do not know so far whether or not the complement $\mathbb{P}^3 \backslash S_{(-1/4)}$ itself is affine ruled.

Remark 4.4. Let Z be a normal affine algebraic threefold with $\mathrm{Pic}(Z) \otimes_{\mathbb{Z}} \mathbb{Q} = (0)$. Then the following three conditions on Z are equivalent to each other:

(1) Z is affine ruled in the sense of 1.1,
(2) Z admits an effective $\mathbb{G}_a$-action,
(3) Z admits an $\mathbb{A}^1$-fibration.

Indeed, the implication (1) $\Rightarrow$ (2) can proceed by the similar argument as in 3.1. Suppose that Z satisfies (1), so that Z contains an open affine subset U of the form $U \cong V \times \mathbb{A}^1$ by definition. Then a locally nilpotent derivation on $\Gamma(\mathcal{O}_U)$, say δ, corresponding to translations along the second factor of $U \cong V \times \mathbb{A}^1$ defines a locally nilpotent derivation on $\Gamma(\mathcal{O}_Z)$ after multiplying a suitable regular function on Z vanishing along the complement $Z \backslash U$. In fact, the existence of a regular function $a \in \Gamma(\mathcal{O}_Z)$ such that $\mathrm{Supp}\ (\mathrm{div}_Z(a)) = Z \backslash U$ is guaranteed by virtue of the finiteness of $\mathrm{Pic}(Z)$. Then $a^N \delta$ becomes a locally nilpotent derivation on $\Gamma(\mathcal{O}_Z)$ with $N \geqq 0$ adequately chosen as $\Gamma(\mathcal{O}_Z)$ is a finitely generated $\mathbb{C}$-algebra, which gives rise to an effective $\mathbb{G}_a$-action on Z as desired. As for (2) $\Rightarrow$ (3), if Z possesses an effective $\mathbb{G}_a$-action, then its quotient map is an $\mathbb{A}^1$-fibration. Notice that the kernel of a corresponding locally nilpotent derivation on $\Gamma(\mathcal{O}_Z)$ is finitely generated over $\mathbb{C}$ (cf. [22, Chapter 1]). Finally as for (3) $\Rightarrow$ (1), assuming that Z has an $\mathbb{A}^1$-fibration $p : Z \to W$ over a surface, there exists an open affine subset $V \subseteq W$ with the property that $p^{-1}(V) \cong V \times \mathbb{A}^1$ by [18], hence Z is affine ruled.

5 Relevant Remarks and Problems

5.1. In this section, we shall mention relevant remarks and questions. It is well known that the defining polynomial $g(x, y) \in \mathbb{C}[x, y]$ of an irreducible curve C in the affine plane $\mathbb{A}^2$, which is isomorphic to the affine line $C \cong \mathbb{A}^1$, gives rise to a trivial $\mathbb{A}^1$-bundle structure:

$$g : \mathbb{A}^2 = \mathrm{Spec}(\mathbb{C}[x, y]) \ni (a, b) \mapsto g(a, b) \in \mathbb{A}^1 = \mathrm{Spec}(\mathbb{C}[g]),$$

by virtue of Abhyankar–Moh–Suzuki's theorem (cf. [1, 26], see also [22, Chapter 2] for an alternative geometric proof). Moreover, as a generalization of Abhyankar–Moh–Suzuki's theorem, Gurjar, Masuda, Miyanishi, and Russell investigate affine lines on a smooth affine surface S with the trivial Makar-Limanov invariant $\mathrm{ML}(S) = \mathbb{C}$ and they prove in [9] that any affine line $C \cong \mathbb{A}^1$ on such an S becomes a fiber component of a suitable $\mathbb{A}^1$-fibration on S.[1] Whereas, if we work with the other smooth affine surfaces, the existence of an affine line $C \cong \mathbb{A}^1$ is not enough to guarantee that there exists an $\mathbb{A}^1$-fibration there which contains C as a fiber component, in general.

Example 5.1. Let S be a $\mathbb{Q}$-homology plane with $\overline{\kappa}(S) = -\infty$ (see e.g. [24, Chapter 3] for relevant results on $\mathbb{Q}$-homology planes). In [20, Theorem 1.1], they have classified affine lines $C \cong \mathbb{A}^1$ on S with the property $\overline{\kappa}(S \backslash C) \geqq 0$. In particular, the existence of such an affine line implies that the Makar-Limanov invariant $\mathrm{ML}(S)$ of S is not trivial, i.e., $\mathrm{ML}(S) \neq \mathbb{C}$ (see [20, Corollary 1.2]), and any fibration on S with C as a fiber component is not an $\mathbb{A}^1$-fibration.

Example 5.2. Let S be a homology plane with $\overline{\kappa}(S) = 1$. Notice that there exist such smooth surfaces (cf. [24, 4.8.3. Theorem]), for instance,

$$\begin{aligned} S_{a,b} : &= \left\{ \frac{(xz+1)^a - (yz+1)^b}{z} = 0 \right\} \subseteq \mathbb{A}^3 \\ &= \mathrm{Spec}(\mathbb{C}[x,y,z]),\ a > b \geqq 1,\ \gcd(a,b) = 1 \end{aligned}$$

yields an example of such surfaces. It is known that there is exactly one affine line on S, say $C \cong \mathbb{A}^1$ (cf. [24, 4.10.1. Theorem]), indeed, in case of $S = S_{a,b}$ above, $\{x = y = 0\}$ is the unique affine line contained in $S_{a,b}$. As the coordinate ring $\Gamma(\mathcal{O}_S)$ of S is UFD (cf. [24, 4.2.1. Lemma]), we can find a regular function $g \in \Gamma(\mathcal{O}_S)$ such that the principal divisor $\mathrm{div}_S(g)$ defined by g coincides with C scheme-theoretically. Then the map defined by g:

$$g : S \ni s \mapsto g(s) \in \mathbb{A}^1 = \mathrm{Spec}(\mathbb{C}[g])$$

is not an $\mathbb{A}^1$-fibration, meanwhile the central fiber $g^*(0) = C$ is isomorphic to $\mathbb{A}^1$.

Example 5.3. Let $g(x,y) := x^a - y^b \in \mathbb{C}[x,y]$, where $a, b \geqq 2$ such that $\gcd(a,b) = 1$. We take a fiber $g^*(\alpha)$ ($\alpha \neq 0$) of the polynomial map $g : \mathbb{A}^2 = \mathrm{Spec}(\mathbb{C}[x,y]) \to \mathbb{A}^1 = \mathrm{Spec}(\mathbb{C}[g])$ associated with the inclusion $\mathbb{C}[g] \subseteq \mathbb{C}[x,y]$, and let us take furthermore a point $Q \in g^*(\alpha)$ on it. We apply then an affine modification of $\mathbb{A}^2$ with the center ($Q \in g^*(\alpha)$), say $\sigma : S \to \mathbb{A}^2$ (see [16] for the definition of *affine modifications*). Then the resulting affine surface S is equipped with a fibration $h := g \circ \sigma$ induced by g. By construction, the fiber $h^*(\alpha)$ is isomorphic to $\mathbb{A}^1$, whereas the other fibers are not isomorphic to $\mathbb{A}^1$.

---

[1]Recall that $\mathrm{ML}(S)$ is defined to be an intersections in $\Gamma(\mathcal{O}_S)$ of $\mathrm{Ker}(\partial)$'s when ∂ ranges over all of locally nilpotent derivations of $\Gamma(\mathcal{O}_S)$.

5.2. As seen in Examples 5.1–5.3 above, the existence of the affine line on a smooth affine surface S is not sufficient in general to guarantee the existence of an $\mathbb{A}^1$-fibration there, even if the Picard group of S is finite, i.e., $\mathrm{Pic}(S) \otimes_{\mathbb{Z}} \mathbb{Q} = (0)$. It is reasonable to think that the similar phenomena would occur in case of higher dimension also, e.g., of dimension three. For example, Abhyankar-Sathaye embedding problem in dimension three asks whether or not an irreducible hypersurface S in the affine 3-space $\mathbb{A}^3$, which is isomorphic to $S \cong \mathbb{A}^2$, is a coordinate plane, namely, all of the fibers of the polynomial map:

$$g : \mathbb{A}^3 = \mathrm{Spec}(\mathbb{C}[x, y, z]) \ni (a, b, c) \mapsto g(a, b, c) \in \mathbb{A}^1 = \mathrm{Spec}(\mathbb{C}[g]),$$

determined by the polynomial $g \in \mathbb{C}[x, y, z]$ defining $S = \mathbb{V}(g) \subseteq \mathbb{A}^3$ are isomorphic to $\mathbb{A}^2$, and the generic one of g is isomorphic to the affine plane $\mathbb{A}^2_{\mathbb{C}(g)}$ over the function field $\mathbb{C}(g)$ of the base curve. Notice that it suffices actually to confirm that *general* fibers of g are isomorphic to $\mathbb{A}^2$ in order to settle out this problem by virtue of [14]. Therefore, the essential of this embedding problem in dimension three consists in how to observe whether or not one hypersurface $S = \mathbb{V}_{\mathbb{A}^3}(g) \subseteq \mathbb{A}^3$, which is isomorphic to the affine plane $\mathbb{A}^2$, is enough to see that fibers in a neighborhood of the origin of the base curve $0 \in \mathbb{A}^1 = \mathrm{Spec}(\mathbb{C}[g])$ are still isomorphic to $\mathbb{A}^2$.

5.3. Instead of the affine plane $\mathbb{A}^2$, more generally, instead of affine ruled rational surfaces in the affine 3-space $\mathbb{A}^3$, we consider affine ruled irrational surfaces in $\mathbb{A}^3$, namely, we propose the following problem:

Problem 5.4. Let $S \subseteq \mathbb{A}^3$ be an irreducible smooth affine hypersurface, which is irrational and affine ruled. Letting $g(x, y, z) \in \mathbb{C}[x, y, z]$ be the polynomial defining S in the affine 3-space $\mathbb{A}^3$, is it true that general fibers of the polynomial map:

$$g : \mathbb{A}^3 = \mathrm{Spec}(\mathbb{C}[x, y, z]) \ni (a, b, c) \mapsto g(a, b, c) \in \mathbb{A}^1 = \mathrm{Spec}(\mathbb{C}[g]),$$

are still (irrational) affine ruled?

Remark 5.5. In [4], we show that if *general* fibers $g^*(\alpha)$ of the map $g : \mathbb{A}^3 \to \mathbb{A}^1$ are irrational and affine ruled, then it follows that the type of an $\mathbb{A}^1$-fibration on $g^*(\alpha)$ is *automatically* of affine type. This is remarkable because, as in Theorem 1.8 or the results in [2], there are examples of polynomials $f(x, y, z) \in \mathbb{C}[x, y, z]$ in three variables such that the polynomial maps determined by f yield fibrations whose general fibers are *rational* affine surfaces having only $\mathbb{A}^1$-fibrations of complete type. More precisely, in the work [4], we show that if general fibers of a given fibration:

$$f : X \longrightarrow B,$$

from a normal affine algebraic threefold X with at most $\mathbb{Q}$-factorial, terminal singularities (but without any condition about $\mathrm{Pic}(X)$) onto a smooth algebraic curve are irrational and affine ruled, then there exists an open affine subset $U \subseteq X$ such that the restriction $f|_U$ can be factored as follows:

$$f|_U = h_0 \circ g_0 : U \xrightarrow{g_0} V \xrightarrow{h_0} W,$$

where $g_0 : U \to V$ is a trivial $\mathbb{A}^1$-bundle, i.e., $U \cong V \times \mathbb{A}^1$ and $W \subseteq B$ is an open subset of B (compare with Theorem 1.4, a result due to Gurjar, Masuda, and Miyanishi). Further, under the additional condition that $\mathrm{Pic}(X) \otimes_{\mathbb{Z}} \mathbb{Q} = (0)$, the above-mentioned factorization can be extended to that on whole X:

$$f = h \circ g : X \xrightarrow{g} Y \xrightarrow{h} B,$$

where g is a quotient map with respect to an effective $\mathbb{G}_a$-action on X. Anyway, it is worthwhile to recognize that the argument in [4] to obtain these results consists in an application of minimal model program in a relative setting.

5.4. Let X be a normal affine algebraic threefold with $\mathrm{Pic}(X) \otimes_{\mathbb{Z}} \mathbb{Q} = (0)$ possessing an effective $\mathbb{G}_a$-action, and let us denote by $\pi : X \to Y$ the corresponding quotient map. In an algebraic viewpoint, letting δ be a locally nilpotent derivation on $A = \Gamma(\mathscr{O}_X)$, which corresponds to the given $\mathbb{G}_a$-action on X, the inclusion $R := \mathrm{Ker}(\delta) \hookrightarrow A$ gives rise to π. It is well known that R is normal, finitely generated over $\mathbb{C}$ and an inert sub-algebra of A (cf. [22, Chapter 1]), furthermore it follows that $\mathrm{Pic}(Y) \otimes_{\mathbb{Z}} \mathbb{Q} = (0)$ (cf. [10, Lemma 1.14]). In particular, for any irreducible curve $C \subseteq Y$, there exists an element $a \in R$ vanishing along C, i.e., $\mathbb{V}_Y(\sqrt{aR}) = C$, and the ideal $\sqrt{aA}$ of A defines a prime divisor $S := \mathbb{V}_X(\sqrt{aA}) \subseteq X$, which coincides with the set-theoretic inverse image $\pi^{-1}(C)$, i.e., $\mathrm{div}_X(a) = dS$ for some $d \geqq 1$. Then as seen in 3.3, we can confirm that δ descends to a nontrivial locally nilpotent derivation on $\Gamma(\mathscr{O}_S)$ after dividing by a^{-n} with $n \geqq 0$ suitably chosen to see that S also admits an effective $\mathbb{G}_a$-action arising from that on X provided $d = 1$. Whereas in case of $d \geqq 2$, δ does not necessarily descend to a nontrivial one on $\Gamma(\mathscr{O}_S)$ (see Example 3.2, which is in fact an example of dimension two, but if we consider X in the affine 4-space $\mathrm{Spec}(\mathbb{C}[x, y, z, u])$ with the same defining equation $xz - y^2 = 0$ instead of the affine 3-space $\mathrm{Spec}(\mathbb{C}[x, y, z])$, then δ, that is the same as in Example 3.2, gives a nontrivial locally nilpotent derivation on the principal divisor $2L$ on X, where L is the (z, u)-plane. However, δ yields in turn a trivial one on L). We note however that in such an example X is not smooth. Instead if we work with *smooth* affine algebraic threefolds with finite Picard groups, we do not know so far the answer to the following problem:

Problem 5.6. Let X be a *smooth* affine algebraic threefold with $\mathrm{Pic}(X) \otimes_{\mathbb{Z}} \mathbb{Q} = (0)$, and let $\pi : X \to Y$ be the quotient morphism with respect to a given effective $\mathbb{G}_a$-action on X. Letting δ be a locally nilpotent derivation on $\Gamma(\mathscr{O}_X)$, which corresponds to π, is it true that for any irreducible curve $C \subseteq Y$, the restriction of δ onto the set-theoretic inverse image $\pi^{-1}(C) \subseteq X$ becomes a nontrivial locally nilpotent derivation on $\Gamma(\mathscr{O}_{\pi^{-1}(C)})$ (after dividing by a suitable regular function on X which vanishes along $\pi^{-1}(C)$ if necessary)? In particular, is $\pi^{-1}(C)$ affine ruled?

Acknowledgements The author would like to thank the referees, Professors Mikhail Zaidenberg and Adrien Dubouloz, for giving him useful and constructive comments consistently. Especially, in the previous version, the author has got wrong with the usage of the crucial result [11, Theorem 2.8], and the referees advised to apply the result appropriately and further persuaded him not to go in a wrong direction with an example. Without their advices and patience, this chapter can never be completed. The part of this work was done during the author's stay at l'Institut de Mathématiques de Bourgogne (Dijon). The author would like to express his hearty thanks to the institute for giving him a splendid research environment. The author was supported by a Grant-in-Aid for Scientific Research of JSPS No. 24740003.

References

1. S. Abhyankar, T.T. Moh, Embeddings of the line in the plane. J. Reine Angew. Math. **276**, 148–166 (1975)
2. A. Dubouloz, T. Kishimoto, Log-uniruled affine varieties without cylinder-like open subsets. Preprint (2012). arXiv:1212.0521v1 [math AG]
3. A. Dubouloz, T. Kishimoto, Minimal model program after resolution of pencils of cubic surfaces and compactifications of $\mathbb{A}^3$. Preprint (2014)
4. A. Dubouloz, T. Kishimoto, Deformations of irrational affine ruled surfaces. Preprint (2014)
5. H. Flenner, M. Zaidenberg, Rational curves and rational singularities. Math. Z. **244**, 549–575 (2003)
6. R.V. Gurjar, A.R. Shastri, On the rationality of complex 2-cells. I. J. Math. Soc. Jpn. **41**, 37–56 (1989)
7. R.V. Gurjar, A.R. Shastri, On the rationality of complex 2-cells. II. J. Math. Soc. Jpn. **41**, 175–212 (1989)
8. R.V. Gurjar, C.R. Pradeep, A.R. Shastri, $\mathbb{Q}$-homology planes are rational. III. Osaka J. Math. **36**, 259–335 (1999)
9. R.V. Gurjar, K. Masuda, M. Miyanishi, P. Russell, Affine lines on affine surfaces and the Makar-Limanov invariant. Can. J. Math. **60**, 109–239 (2008)
10. R.V. Gurjar, K. Masuda, M. Miyanishi, $\mathbb{A}^1$-fibrations on affine threefolds. J. Pure Appl. Algebra **216**, 296–313 (2012)
11. R.V. Gurjar, K. Masuda, M. Miyanishi, Deformations of $\mathbb{A}^1$-fibrations. In this volume.
12. S. Iitaka, On logarithmic Kodaira dimension of algebraic varieties, in *Complex Analysis and Algebraic Geometry*, ed. by W.L. Baily, Jr., T. Shioda (Iwanami Shoten, Tokyo, 1977), pp. 175–189
13. S. Iitaka, *Algebraic Geometry, An introduction to Birational Geometry of Algebraic Varieties*. Graduate Texts in Mathematics, vol. 76 (Springer, New York/Berlin, 1982)
14. S. Kaliman, Polynomials with general $\mathbb{C}^2$-fibers are variables. Pac. J. Math. **203**, 161–189 (2002)
15. S. Kaliman, Free $\mathbb{C}^+$-actions on $\mathbb{C}^3$ are translations. Invent. Math. **156**, 163–173 (2004)
16. S. Kaliman, M. Zaidenberg, Affine modifications and affine hyper surfaces with a very transitive automorphism group. Transform. Groups **4**, 53–95 (1999)
17. S. Kaliman, M. Zaidenberg, Families of affine planes: the existence of a cylinder. Mich. Math. J. **49**, 353–367 (2001)
18. T. Kambayashi, M. Miyanishi, On flat fibrations by the affine line. Illinois J. Math. **22**, 662–671 (1978)

19. S. Keel, J. M[c]Kernan, *Rational Curves on Quasi-Projective Surfaces*. Memoirs of AMS, No. 669 (American Mathematical Society, Providence, 1999)
20. T. Kishimoto, H. Kojima, Affine lines on $\mathbb{Q}$-homology planes with logarithmic Kodaira dimension $-\infty$. Transform. Groups **11**, 659–672 (2006)
21. K. Masuda, M. Miyanishi, The additive group actions on $\mathbb{Q}$-homology planes. Ann. Inst. Fourier Grenoble **53**, 429–464 (2003)
22. M. Miyanishi, *Lectures on Curves on Rational and Unirational Surfaces* (Springer, Berlin, 1978) [Published for Tata Inst. Fund. Res., Bombay]
23. M. Miyanishi, Singularities on normal affine surfaces containing cylinderlike open sets. J. Algebra **68**, 268–275 (1981)
24. M. Miyanishi, *Open Algebraic Surfaces*. CRM Monograph Series, vol. 12 (American Mathematical Society, Providence, 2001)
25. M. Miyanishi, T. Sugie, Affine surfaces containing cylinderlike open sets. J. Math. Kyoto Univ. **20**, 11–42 (1980)
26. M. Suzuki, Propriétés topologiques des polynômes de deux variables complexes et automorphismes algébriques de l'espace $\mathbb{C}^2$. J. Math. Soc. Jpn. **26**, 241–257 (1974)

How to Prove the Wildness of Polynomial Automorphisms: An Example

Shigeru Kuroda

Abstract Nagata (On Automorphism Group of $k[x, y]$, Lectures in Mathematics, Department of Mathematics, Kyoto University, vol. 5. Kinokuniya Book-Store Co. Ltd., Tokyo, 1972) conjectured the wildness of a certain automorphism of the polynomial ring in three variables. This famous conjecture was solved by Shestakov–Umirbaev (J. Am. Math. Soc., 17, 181–196, 197–227, 2004) in the affirmative. Although the Shestakov–Umirbaev theory is powerful and applicable to various situations, not so many researchers seem familiar with this theory due to its technical difficulty.

In this paper, we explain how to prove the wildness of polynomial automorphisms using this theory practically. First, we recall a useful criterion for wildness which is derived from the generalized Shestakov–Umirbaev theory. Then, we demonstrate how to use this criterion effectively by showing the wildness of the exponential automorphisms for some well-known locally nilpotent derivations of rank three.

2010 Mathematics Subject Classification: Primary 14R10; Secondary 13F20,

1 Introduction

Let k be a field, $k[\boldsymbol{x}] = k[x_1, \ldots, x_n]$ the polynomial ring in n variables over k, and $\mathrm{Aut}_k k[\boldsymbol{x}]$ the automorphism group of the k-algebra $k[\boldsymbol{x}]$. We say that $\phi \in \mathrm{Aut}_k k[\boldsymbol{x}]$ is *elementary* if there exists $1 \leq l \leq n$ such that $\phi(x_i) = x_i$ for all $i \neq l$.

S. Kuroda (✉)
Department of Mathematics and Information Sciences, Tokyo Metropolitan University, 1-1 Minami-Osawa, Hachioji, Tokyo 192-0397, Japan
e-mail: kuroda@tmu.ac.jp

I. Cheltsov et al. (eds.), *Automorphisms in Birational and Affine Geometry*, Springer Proceedings in Mathematics & Statistics 79,
DOI 10.1007/978-3-319-05681-4_21,

Note that, if this is the case, we have $\phi(x_l) = \alpha x_l + f$ for some $\alpha \in k^\times$ and $f \in k[\{x_i \mid i \neq l\}]$. The subgroup $\mathrm{T}(n,k)$ of $\mathrm{Aut}_k k[\boldsymbol{x}]$ generated by all the elementary automorphisms of $k[\boldsymbol{x}]$ is called the *tame subgroup* of $\mathrm{Aut}_k k[\boldsymbol{x}]$. We call elements of $\mathrm{Aut}_k k[\boldsymbol{x}] \setminus \mathrm{T}(n,k)$ *wild automorphisms*. When $n \geq 2$, a question arises whether $\mathrm{Aut}_k k[\boldsymbol{x}] = \mathrm{T}(n,k)$. Due to Jung [2] and van der Kulk [12], the answer is yes if $n = 2$. Nagata [9] conjectured that the answer is no when $n = 3$, and gave $\psi \in \mathrm{Aut}_k k[\boldsymbol{x}]$ defined by

$$\psi(x_1) = x_1 - 2(x_1x_3 + x_2^2)x_2 - (x_1x_3 + x_2^2)^2 x_3, \quad \psi(x_2) = x_2 + (x_1x_3 + x_2^2)x_3$$

and $\psi(x_3) = x_3$ as a candidate of wild automorphism. This famous conjecture was solved in the affirmative by Shestakov–Umirbaev [10,11] in the case of $\mathrm{char}(k) = 0$. The question remains open when $n = 3$ and $\mathrm{char}(k) > 0$, and when $n \geq 4$.

It is 10 years since Nagata's conjecture was solved, but not so many researchers seem familiar with the Shestakov–Umirbaev theory because of the technical difficulty. In fact, to decide the wildness of $\phi \in \mathrm{Aut}_k k[\boldsymbol{x}]$ by means of this theory, one needs precise information on the polynomials $\phi(x_1)$, $\phi(x_2)$ and $\phi(x_3)$, which is sometimes quite difficult. For example, let D be a *locally nilpotent derivation* of $k[\boldsymbol{x}]$, i.e., a derivation of $k[\boldsymbol{x}]$ such that, for each $f \in k[\boldsymbol{x}]$, we have $D^l(f) = 0$ for some $l \geq 1$. If $\mathrm{char}(k) = 0$, then we can define the *exponential automorphism* $\exp D \in \mathrm{Aut}_k k[\boldsymbol{x}]$ by

$$(\exp D)(f) = \sum_{l=0}^{\infty} \frac{D^l(f)}{l!} \quad \text{for each } f \in k[\boldsymbol{x}].$$

Various interesting automorphisms are obtained as $\exp D$ for some D. In general, however, it is not easy to describe $(\exp D)(x_i)$'s.

The purpose of this paper is to explain how to prove the wildness of polynomial automorphisms in three variables practically. In Sect. 2, we recall a useful criterion for wildness which is derived from the generalized Shestakov–Umirbaev theory [3,4]. Then, in Sect. 3, we demonstrate how to use the criterion effectively by showing the wildness of the exponential automorphisms for some well-known locally nilpotent derivations of rank three. Here, the *rank* of a derivation D of $k[\boldsymbol{x}]$ is defined to be the minimal number r for which there exists $\tau \in \mathrm{Aut}_k k[\boldsymbol{x}]$ such that $D(\tau(x_i)) = 0$ for $i = 1, \ldots, n - r$. Rank three locally nilpotent derivations of $k[x_1, x_2, x_3]$ are rather complicated and difficult to handle, so the exponential automorphisms for such locally nilpotent derivations are interesting test case to apply the criterion.

2 Monomial Orders

Let $\preceq$ be a *monomial order* on $k[\boldsymbol{x}]$, i.e., a total order on $\mathbf{Z}^n$ such that $\boldsymbol{a} \preceq \boldsymbol{b}$ implies $\boldsymbol{a} + \boldsymbol{c} \preceq \boldsymbol{b} + \boldsymbol{c}$ for each $\boldsymbol{a}, \boldsymbol{b}, \boldsymbol{c} \in \mathbf{Z}^n$, and that the coordinate unit vectors $\boldsymbol{e}_1, \ldots, \boldsymbol{e}_n$ of $\mathbf{R}^n$ are greater than the zero vector. Take any $f \in k[\boldsymbol{x}] \setminus \{0\}$ and write

$f = \sum_{a} \lambda_{a} x^{a}$, where $\lambda_{a} \in k$ and $x^{a} := x_1^{a_1} \cdots x_n^{a_n}$ for each $a = (a_1, \ldots, a_n)$. Then, we define $\deg_{\preceq} f := \max_{\preceq}\{a \mid \lambda_{a} \neq 0\}$ and $\mathrm{in}_{\preceq} f := \lambda_{b} x^{b}$, where $b := \deg_{\preceq} f$.

In what follows, we assume that $\mathrm{char}(k) = 0$ and $n = 3$. The following useful criterion is implicit in the generalized Shestakov–Umirbaev theory, where $\mathbf{Z}_{\geq 0}$ denotes the set of nonnegative integers.

Theorem 1. *$\phi \in \mathrm{Aut}_k k[x]$ is wild if there exists a monomial order $\preceq$ on $k[x]$ for which the following conditions hold:*

(M1) $\deg_{\preceq} \phi(x_1)$, $\deg_{\preceq} \phi(x_2)$ and $\deg_{\preceq} \phi(x_3)$ are linearly dependent over $\mathbf{Z}$, and are pairwise linearly independent over $\mathbf{Z}$.
(M2) $\deg_{\preceq} \phi(x_{i_1})$ is not equal to $p \deg_{\preceq} \phi(x_{i_2}) + q \deg_{\preceq} \phi(x_{i_3})$ for any $p, q \in \mathbf{Z}_{\geq 0}$ and $(i_1, i_2, i_3) = (1,2,3), (2,3,1), (3,1,2)$.

For example, let ψ be Nagata's automorphism mentioned above, and $\preceq_{\mathrm{lex}}$ the *lexicographic order* on $k[x]$ with $e_1 \preceq_{\mathrm{lex}} e_2 \preceq_{\mathrm{lex}} e_3$, i.e., the total order on $\mathbf{Z}^3$ defined by $(a_1, a_2, a_3) \preceq_{\mathrm{lex}} (b_1, b_2, b_3)$ if $a_i = b_i$ for all i, or $a_l < b_l$ for $l := \max\{i \mid a_i \neq b_i\}$. Then, we have

$$\deg_{\preceq_{\mathrm{lex}}} \psi(x_1) = (2,0,3), \quad \deg_{\preceq_{\mathrm{lex}}} \psi(x_2) = (1,0,2), \quad \deg_{\preceq_{\mathrm{lex}}} \psi(x_3) = (0,0,1).$$

Since these three vectors satisfy (M1) and (M2), we can conclude that ψ is wild.

Let us briefly explain how Theorem 1 follows from the generalized Shestakov–Umirbaev theory. Recall that a monomial order $\preceq$ on $k[x]$ induces a structure of totally ordered additive group on $\mathbf{Z}^n$. Set $\mathbf{w} = (e_1, \ldots, e_n)$. Then, in the notation of [5], we have $\deg_{\mathbf{w}} f = \deg_{\preceq} f$ and $f^{\mathbf{w}} = \mathbf{in}_{\preceq} f$ for each $f \in k[x]$. In [5] (after Theorem 5.1.3), we derived a wildness criterion from the generalized Shestakov–Umirbaev theory (see also [6, Theorem 2.4]). According to this criterion, $\phi \in \mathrm{Aut}_k k[x]$ is wild if the following conditions hold for some monomial order $\preceq$:

(1) $\phi(x_1)^{\mathbf{w}}$, $\phi(x_2)^{\mathbf{w}}$ and $\phi(x_3)^{\mathbf{w}}$ are algebraically dependent over k, and are pairwise algebraically independent over k.
(2) $\phi(x_i)^{\mathbf{w}}$ does not belong to $k[\{\phi(x_l)^{\mathbf{w}} \mid l \neq i\}]$ for $i = 1, 2, 3$.

Since $\phi(x_i)^{\mathbf{w}} = \mathbf{in}_{\preceq} \phi(x_i)$ is a monomial for each i, we see that (1) and (2) are equivalent to (M1) and (M2), respectively. Therefore, ϕ is wild if (M1) and (M2) hold for some monomial order $\preceq$ on $k[x]$.

We note that the result in [5] used to derive Theorem 1 is a consequence of [4, Theorem 2.1] which is a generalization of Shestakov–Umirbaev [11, Theorem 1]. As in the case of [11, Theorem 1], the proof of [4, Theorem 2.1] is long and technical (see [7] for the main idea behind the proofs of these theorems; see also [13] for a survey of the theory of Shestakov–Umirbaev).

3 Example

Let l and m be positive integers. We set $f = x_1x_3 - x_2^2$, $r = f^l x_2 + x_1^m$ and

$$g = \frac{f^{2l+1} + r^2}{x_1} = \frac{f^{2l}(x_1x_3 - x_2^2) + (f^l x_2 + x_1^m)^2}{x_1} = f^{2l}x_3 + 2f^l x_1^{m-1}x_2 + x_1^{2m-1}.$$

Using f and g, we define a derivation Δ of $k[\boldsymbol{x}]$ by

$$\Delta(p) = \left|\frac{\partial(f, g, p)}{\partial(x_1, x_2, x_3)}\right| \qquad \text{for each } p \in k[\boldsymbol{x}].$$

In this notation, the following proposition holds.

Proposition 1 ([1,5]). *Δ is an irreducible locally nilpotent derivation of $k[\boldsymbol{x}]$ such that* $\ker \Delta = k[f, g]$ *and $\Delta(r) = -f^l g$. If $m \geq 2$, then Δ is of rank three.*

Here, a derivation D of $k[\boldsymbol{x}]$ is said to be *irreducible* if $D(k[\boldsymbol{x}])$ is contained in no proper principal ideal of $k[\boldsymbol{x}]$.

This result is due to Freudenburg [1] when $m = 2l + 1$. The general case follows from a more general result of the author [5, Theorems 7.1.5(i) and 7.1.6(iii)] (see the discussion at the end of Sect. 7.1 of [5]).

In this section, we prove the following theorem using Theorem 1.

Theorem 2. *$\phi := \exp h\Delta$ is wild for any integers $l, m \geq 1$ and $h \in \ker \Delta \setminus \{0\}$.*

We note that Theorem 2 is a special case of [5, Theorem 7.1.5(ii)]. However, since the proof for the general statement is rather technical and complicated, it is worthwhile to give a direct proof for this simple example.

Let $\preceq_{\text{lex}}$ be the lexicographic order on $k[\boldsymbol{x}]$ with $\boldsymbol{e}_1 \preceq_{\text{lex}} \boldsymbol{e}_2 \preceq_{\text{lex}} \boldsymbol{e}_3$, and put $\deg := \deg_{\preceq_{\text{lex}}}$. Then, we have

$$\deg f = (1, 0, 1) \quad \text{and} \quad \deg g = \deg f^{2l}x_3 = (2l, 0, 2l + 1).$$

Hence, $\deg f$ and $\deg g$ are linearly independent over $\mathbf{Z}$. Thus, $\deg f^i g^j$'s are different for different (i, j)'s. Since h is a nonzero element of $\ker \Delta = k[f, g]$, we may write $\deg h = a \deg f + b \deg g$, where $a, b \in \mathbf{Z}_{\geq 0}$. Since $\Delta(f) = \Delta(g) = 0$, $\Delta(r) = -f^l g$ by Proposition 1, and $x_1 g = f^{2l+1} + r^2$ by the definition of g, we have

$$\Delta(x_1)g = \Delta(x_1 g) = \Delta(f^{2l+1} + r^2) = 2r\Delta(r) = -2rf^l g,$$

and so $\Delta(x_1) = -2rf^l$. Hence, we get

$$\phi(x_1) = x_1 + h\Delta(x_1) + \frac{h^2\Delta^2(x_1)}{2!} + \cdots = x_1 - 2rf^l h + f^{2l}gh^2.$$

Since $\deg r = (l, 1, l)$ is less than $\deg g$, it follows that

$$\deg \phi(x_1) = \deg f^{2l} g h^2 = 2(a+l) \deg f + (2b+1) \deg g. \tag{1}$$

Thus, $\deg \phi(x_1)$ and $\deg f$ are linearly independent over $\mathbf{Z}$. Since

$$f^l \phi(x_2) + \phi(x_1)^m = \phi(f^l x_2 + x_1^m) = \phi(r) = r + h\Delta(r) + \cdots = r - f^l gh$$

and $\deg(r - f^l gh) = \deg f^l gh < \deg \phi(x_1)^m$, we have $\deg f^l \phi(x_2) = \deg \phi(x_1)^m$. Hence, we get

$$\deg \phi(x_2) = m \deg \phi(x_1) - l \deg f. \tag{2}$$

Since $\Delta(f) = 0$, we have

$$\phi(x_1)\phi(x_3) - \phi(x_2)^2 = \phi(x_1 x_3 - x_2^2) = \phi(f) = f,$$

in which $\deg \phi(x_1)\phi(x_3) \geq \deg \phi(x_1) > \deg f$. Thus, $\deg \phi(x_1)\phi(x_3) = \deg \phi(x_2)^2$, and so

$$\begin{aligned}\deg \phi(x_3) &= 2 \deg \phi(x_2) - \deg \phi(x_1) = 2(m \deg \phi(x_1) - l \deg f) - \deg \phi(x_1) \\ &= (2m-1) \deg \phi(x_1) - 2l \deg f.\end{aligned} \tag{3}$$

Now, observe that $(1, 0)$, (m, l) and $(2m-1, 2l)$ are linearly dependent over $\mathbf{Z}$, and are pairwise linearly independent over $\mathbf{Z}$. Moreover, none of these three vectors belongs to the additive subsemigroup of $\mathbf{Z}^2$ generated by the other two vectors. Since $\deg \phi(x_1)$ and $-\deg f$ are linearly independent over $\mathbf{Z}$, we see from (1), (2) and (3) that ϕ satisfies (M1) and (M2). Therefore, ϕ is wild by Theorem 1.

Note: Recently, the author [8] showed that the automorphism group $\mathrm{Aut}_{k[f,g]} k[\boldsymbol{x}]$ of the $k[f, g]$-algebra $k[\boldsymbol{x}]$ is equal to $\{\exp h\Delta \mid h \in k[f, g]\}$. Therefore, we know that $\mathrm{Aut}_{k[f,g]} k[\boldsymbol{x}] \cap \mathrm{T}(3, k) = \{\mathrm{id}_{k[\boldsymbol{x}]}\}$ thanks to Theorem 2.

Acknowledgements The author is partly supported by the Grant-in-Aid for Young Scientists (B) 24740022, Japan Society for the Promotion of Science. He also thanks the referees for the comments.

References

1. G. Freudenburg, Local slice constructions in $k[X, Y, Z]$. Osaka J. Math. **34**, 757–767 (1997)
2. H. Jung, Über ganze birationale Transformationen der Ebene. J. Reine Angew. Math. **184**, 161–174 (1942)

3. S. Kuroda, A generalization of the Shestakov-Umirbaev inequality. J. Math. Soc. Jpn. **60**, 495–510 (2008)
4. S. Kuroda, Shestakov-Umirbaev reductions and Nagata's conjecture on a polynomial automorphism. Tohoku Math. J. **62**, 75–115 (2010)
5. S. Kuroda, Wildness of polynomial automorphisms in three variables. arXiv:math.AC/1110.1466v1
6. S. Kuroda, On the Karaś type theorems for the multidegrees of polynomial automorphisms (2011). arXiv:math.AC/1303.3703v1
7. S. Kuroda, Recent developments in polynomial automorphisms: the solution of Nagata's conjecture and afterwards, Sugaku expositions (to appear)
8. S. Kuroda, The automorphism group of an integral domain over the kernel of a locally nilpotent derivation (in preparation)
9. M. Nagata, in *On Automorphism Group of* $k[x, y]$. Lectures in Mathematics, Department of Mathematics, Kyoto University, vol. 5 (Kinokuniya Book-Store Co. Ltd., Tokyo, 1972)
10. I. Shestakov, U. Umirbaev, Poisson brackets and two-generated subalgebras of rings of polynomials. J. Am. Math. Soc. **17**, 181–196 (2004)
11. I. Shestakov, U. Umirbaev, The tame and the wild automorphisms of polynomial rings in three variables. J. Am. Math. Soc. **17**, 197–227 (2004)
12. W. van der Kulk, On polynomial rings in two variables. Nieuw Arch. Wisk. **1**(3), 33–41 (1953)
13. A. van den Essen, The solution of the tame generators conjecture according to Shestakov and Umirbaev. Colloq. Math. **100**, 181–194 (2004)

Flexibility Properties in Complex Analysis and Affine Algebraic Geometry

Frank Kutzschebauch

Abstract In the last decades affine algebraic varieties and Stein manifolds with big (infinite-dimensional) automorphism groups have been intensively studied. Several notions expressing that the automorphisms group is big have been proposed. All of them imply that the manifold in question is an Oka–Forstnerič manifold. This important notion has also recently merged from the intensive studies around the homotopy principle in Complex Analysis. This homotopy principle, which goes back to the 1930s, has had an enormous impact on the development of the area of Several Complex Variables and the number of its applications is constantly growing. In this overview chapter we present three classes of properties: (1) density property, (2) flexibility, and (3) Oka–Forstnerič. For each class we give the relevant definitions, its most significant features and explain the known implications between all these properties. Many difficult mathematical problems could be solved by applying the developed theory, we indicate some of the most spectacular ones.

2010 *Mathematics Subject Classification.* Primary 32M05. Secondary 14L24, 14L30, 32E10, 32M17, 32Q28.

F. Kutzschebauch was partially supported by Schweizerischer Nationalfond grant 200021-140235/1.

F. Kutzschebauch (✉)
Institute of Mathematics, University of Bern, Sidlerstrasse 5, CH-3012 Bern, Switzerland
e-mail: frank.kutzschebauch@math.unibe.ch

I. Cheltsov et al. (eds.), *Automorphisms in Birational and Affine Geometry*, Springer Proceedings in Mathematics & Statistics 79,
DOI 10.1007/978-3-319-05681-4_22,

1 Introduction

This is a survey of recent developments in Complex Analysis and Affine Algebraic Geometry which emphasize on objects described as elliptic, in the opposite to hyperbolic in the sense of Kobayashi or more general in the sense of Eisenman (all Eisenman measures on these objects vanish identically.)

Here is the scheme of properties we are going to discuss, together with the known implications between them. Although the properties do not require the manifolds to be Stein or affine algebraic, some of the implications do. We therefore assume the manifold to be Stein in the upper row and to be a smooth affine algebraic variety in the lower row.

$$\begin{array}{ccccc} \text{density property (DP)} & \Longrightarrow & \text{holomorphic flexible} & \Longrightarrow & \text{Oka–Forstneric} \\ \Uparrow & & \Uparrow & & \\ \text{algebraic density property (ADP)} & & \text{algebraic flexible} & & \end{array} \tag{1}$$

In each of the following three sections we present one class of properties together with main features. We hope that researchers from the algebraic and from the holomorphic side can join their efforts to enlarge the class of examples and to find out which of the reverse implications between these properties hold.

In the last section we briefly recall that in the presence of a volume form there is a similar property to (algebraic) density property, called (algebraic) volume density property, which if replacing DP and ADP in Scheme (1) by these properties (AVDP, VDP) gives another scheme with the same implications true. Also we elaborate on the reverse implications in our Scheme (1).

We sincerely thank the referee for very carefully reading the chapter and making many valuable comments. He helped a lot to improve the presentation. Many thanks to Finnur Lárusson for careful reading and catching an inaccuracy in a previous version of the text.

2 Density Property

2.1 Definition and Main Features

Considering a question promoted by Walter Rudin, Andersén and Lempert in 1989 [1, 3] proved a remarkable fact about the affine n-space $n \geq 2$, namely that the group generated by shears (maps of the form $(z_1, \ldots, z_n) \mapsto (z_1, \ldots, z_{n-1}, z_n + f(z_1, \ldots, z_{n-1}))$ where $f \in \mathcal{O}(\mathbb{C}^{n-1})$ is a holomorphic function and any linear conjugate of such a map) and overshears (maps of the form $(z_1, \ldots, z_n) \mapsto (z_1, \ldots, z_{n-1}, z_n g(z_1, \ldots, z_{n-1}))$ where $g \in \mathcal{O}^*(\mathbb{C}^{n-1})$ is a nowhere vanishing holomorphic function and any linear conjugate of such a map) are dense in holomorphic automorphism group of $\mathbb{C}^n$, endowed with compact-open topology. The main importance of their work was not the mentioned result but the proof itself which implies, as observed by Forstnerič and Rosay in [17] for $X = \mathbb{C}^n$, the remarkable Andersén–Lempert theorem, see below. The natural generalization from $\mathbb{C}^n$ to arbitrary manifolds X was made by Varolin [40] who introduced the following important property of a complex manifold:

Definition 1. A complex manifold X has the density property if in the compact-open topology the Lie algebra generated by completely integrable holomorphic vector fields on X is dense in the Lie algebra of all holomorphic vector fields on X.

Here a holomorphic vector field Θ on a complex manifold X is called completely integrable if the ODE

$$\frac{d}{dt}\varphi(x,t) = \Theta(\varphi(x,t))$$

$$\varphi(x,0) = x$$

has a solution $\varphi(x,t)$ defined for all complex times $t \in \mathbb{C}$ and all starting points $x \in X$. It gives a complex one-parameter subgroup in the holomorphic automorphism group $\mathrm{Aut_{hol}}(X)$.

The density property is a precise way of saying that the automorphism group of a manifold is big, in particular for a Stein manifold this is underlined by the main result of the theory (see [17] for $\mathbb{C}^n$, [40], a detailed proof can be found in the Appendix of [36] or in [14]).

Theorem 2 (Andersén–Lempert Theorem). *Let X be a Stein manifold with the density property and let Ω be an open subset of X. Suppose that $\Phi : [0,1]\times\Omega \to X$ is a C^1-smooth map such that*

(1) $\Phi_t : \Omega \to X$ *is holomorphic and injective for every* $t \in [0,1]$,
(2) $\Phi_0 : \Omega \to X$ *is the natural embedding of* Ω *into* X, *and*
(3) $\Phi_t(\Omega)$ *is a Runge subset*[1] *of* X *for every* $t \in [0,1]$.

Then for each $\epsilon > 0$ and every compact subset $K \subset \Omega$ there is a continuous family, $\alpha : [0,1] \to \mathrm{Aut_{hol}}(X)$ of holomorphic automorphisms of X such that

$$\alpha_0 = id \text{ and } |\alpha_t - \Phi_t|_K < \epsilon \text{ for every } t \in [0,1]$$

Philosophically one can think of the density property as a tool for realizing local movements by global maps (automorphisms). In some sense it is a substitute

[1]Recall that an open subset U of X is Runge if any holomorphic function on U can be approximated by global holomorphic functions on X in the compact-open topology. Actually, for X Stein by Cartan's Theorem A this definition implies more: for any coherent sheaf on X its section over U can be approximated by global sections.

for cutoff functions which in the differentiable category are used for globalizing local movements. In the holomorphic category we of course lose control on automorphism outside the compact set K. This makes constructions more complicate but still constructing sequences of automorphisms by iterated use of the Andersén–Lempert theorem has led to remarkable constructions.

Let us further remark that the implications of the density property for manifolds which are not Stein have not been explored very much yet. If the manifold is compact all (holomorphic) vector fields are completely integrable, the density property trivially hold and thus cannot give any further information on the manifold.

Remark 3. Andersén and Lempert proved that every algebraic vector field on $\mathbb{C}^n$ is a finite sum of algebraic shear fields (fields of form $p(z_1, \dots z_{n-1}) \frac{\partial}{\partial z_n}$ for a polynomial $p \in \mathbb{C}[\mathbb{C}^{n-1}]$ and their linear conjugates, i.e., fields who's one-parameter subgroups consist of shears) and overshear fields (fields of form $p(z_1, \dots z_{n-1}) z_n \frac{\partial}{\partial z_n}$ for a polynomial $p \in \mathbb{C}[\mathbb{C}^{n-1}]$ and their linear conjugates, i.e., fields whose one-parameter subgroups consist of overshears). Together with the fact that any holomorphic automorphism of $\mathbb{C}^n$ can be joined to the identity by a smooth pat, this shows how the Andersén–Lempert theorem implies that the group generated by shears and overshears is dense in the holomorphic automorphism group of $\mathbb{C}^n$

The algebraic density property can be viewed as a tool to prove the density property, whereas the ways of proving it are purely algebraic work.

Definition 4. An affine algebraic manifold X has the algebraic density property if the Lie algebra $\text{Lie}_{\text{alg}}(X)$ generated by completely integrable algebraic vector fields on it coincides with the Lie algebra $\text{VF}_{\text{alg}}(X)$ of all algebraic vector fields on it.

An algebraic vector field is an algebraic section of the tangent bundle, for example on $\mathbb{C}^n$ it can be written as $\sum_{i=1}^n p_i(z_1, \dots, z_n) \frac{\partial}{\partial z_i}$ with polynomials $p_i \in \mathbb{C}[\mathbb{C}^n]$. If it is completely integrable, its flow gives a one-parameter subgroup in the holomorphic automorphism group not necessarily in the algebraic automorphism group. For example, a polynomial shear field of the form $p(z_1, \dots, z_{n-1}) z_n \frac{\partial}{\partial z_n}$ has the flow map $\gamma(t, z) = (z_1, \dots, z_{n-1}, \exp(t p(z_1, \dots, z_{n-1})) z_n)$. This is the reason that algebraic density property is in the intersection of affine algebraic geometry and complex analysis. It is an algebraic notion, proven using algebraic methods but has implications for the holomorphic automorphism group.

2.2 Applications and Examples

A first application we like to mention is to the notoriously difficult question whether every open Riemann surface can be properly holomorphically embedded into $\mathbb{C}^2$. This is the only dimension for which the conjecture of Forster [12], saying that every Stein manifold of dimension n can be properly holomorphically embedded into $\mathbb{C}^N$ for $N = [\frac{n}{2}] + 1$, is still unsolved. The conjectured dimension is sharp by examples of Forster [12] and has been proven by Eliashberg, Gromov [10] and Schürmann [37] for all dimensions $n \geq 2$. Their methods of proof fail in dimension $n = 1$. But Fornaess Wold invented a clever combination of a use of shears (nice projection property) and Theorem 2 which led to many new embedding theorems for open Riemann surfaces. As an example we like to mention the following two recent results of Forstnerič and Fornaess Wold [18, 19] the first of them being the most general one for open subsets of the complex line:

Theorem 5. *Every domain in the Riemann sphere with at least one and at most countably many boundary components, none of which are points, admits a proper holomorphic embedding into* $\mathbb{C}^2$.

Theorem 6. *If* $\bar{\Sigma}$ *is a (possibly reducible) compact complex curve in* $\mathbb{C}^2$ *with boundary* $\partial\Sigma$ *of class* C^r *for some* $r > 1$*, then the inclusion map* $i : \Sigma = \bar{\Sigma} \setminus \Sigma \to \mathbb{C}^2$ *can be approximated, uniformly on compacts in* Σ*, by proper holomorphic embeddings* $\Sigma \to \mathbb{C}^2$.

Many versions of embeddings with interpolation are also known and proven using the same methods invented by Fornaess Wold in [42].

Another application is to construct non-straightenable holomorphic embeddings of $\mathbb{C}^k$ into $\mathbb{C}^n$ for all pairs of dimensions $0 < k < n$, a fact which is contrary to the situation in affine algebraic geometry, namely contrary to the famous Abhyankar-Moh-Suzuki theorem for $k = 1, n = 2$ and also to work of Kaliman [26] or $2k + 1 < n$, whereas straightenability for the other dimension pairs is still unknown in algebraic geometry. The most recent and quite striking result in this direction says that there are even holomorphic families of pairwise non-equivalent holomorphic embeddings (referring to holomorphic automorphisms of the source and target in the definition below). Here non-straightenable for an embedding $\mathbb{C}^k$ into $\mathbb{C}^n$ means to be not equivalent to the standard embedding.

Definition 7. Two embeddings $\Phi, \Psi \colon X \hookrightarrow \mathbb{C}^n$ are *equivalent* if there exist automorphisms $\varphi \in \mathrm{Aut}(\mathbb{C}^n)$ and $\psi \in \mathrm{Aut}(X)$ such that $\varphi \circ \Phi = \Psi \circ \psi$.

Theorem 8. *see [33]. Let* n, l *be natural numbers with* $n \geq l + 2$*. There exist, for* $k = n - l - 1$*, a family of holomorphic embeddings of* $\mathbb{C}^l$ *into* $\mathbb{C}^n$ *parametrized by* $\mathbb{C}^k$*, such that for different parameters* $w_1 \neq w_2 \in \mathbb{C}^k$ *the embeddings* $\psi_{w_1}, \psi_{w_2} : \mathbb{C}^l \hookrightarrow \mathbb{C}^n$ *are non-equivalent.*

We would like to mention a nice application of Theorem 8 to actions of compact (or equivalently complex reductive, see [31]) groups on $\mathbb{C}^n$. It was a long-standing problem, whether all holomorphic actions of such groups on affine space are linear after a change of variables (see for example the overview article [24]). The first counterexamples to that (Holomorphic Linearization) problem were constructed by Derksen and the first author in [9]. The method from [9] is holomorphic in a parameter and therefore applied to our parametrized situation leads to the following result ([33])

Theorem 9. *For any $n \geq 5$ there is a holomorphic family of $\mathbb{C}^*$-actions on $\mathbb{C}^n$ parametrized by $\mathbb{C}^{n-4}$*

$$\mathbb{C}^{n-4} \times \mathbb{C}^* \times \mathbb{C}^n \to \mathbb{C}^n, \quad (w, \theta, z) \mapsto \theta_w(z)$$

so that for different parameters $w_1 \neq w_2 \in \mathbb{C}^{n-4}$ there is no equivariant isomorphism between the actions θ_{w_1} and θ_{w_2}.

The linearization problem for holomorphic $\mathbb{C}^*$-actions on $\mathbb{C}^n$ is thus solved to the positive for $n = 2$ by Suzuki [39] and still open for $n = 3$. For $n = 4$ there are uncountably many actions (non-linearizable ones among them) [8] and for $n \geq 5$ Theorem 9 implies that there are families. Moreover, there are families including a linear action as a single member of the family;

Theorem 10. *For any $n \geq 5$ there is a holomorphic family of $\mathbb{C}^*$-actions on $\mathbb{C}^n$ parametrized by $\mathbb{C}$*

$$\mathbb{C} \times \mathbb{C}^* \times \mathbb{C}^n \to \mathbb{C}^n \quad (w, \theta, z) \mapsto \theta_w(z)$$

so that for different parameters $w_1 \neq w_2 \in \mathbb{C}$ there is no equivariant isomorphism between the actions θ_{w_1} and θ_{w_2}. Moreover, the action θ_0 is linear.

Open Problem: Suppose X is a Stein manifold with density property and $Y \subset X$ is a closed submanifold. Is there always another proper holomorphic embedding $\varphi : Y \hookrightarrow X$ which is not equivalent to the inclusion $i : Y \hookrightarrow X$?

We should remark that an affirmative answer to this question is stated in [41], but the author apparently had another (weaker) notion of equivalence in mind.

Here comes the essentially complete list of examples of Stein manifolds known to have the density property:

List of examples of Stein manifolds known to have the density property:

1. $X = G/R$ where G is linear algebraic and R a reductive subgroup has ADP and thus DP (defined on p. 2), except for $X = \mathbb{C}$ and $X = (\mathbb{C}^*)^n$. (this includes all examples known from the work of Andersén–Lempert and Varolin and Varolin–Toth and Kaliman–Kutzschebauch, the final result is proven by Donzelli–Dvorsky–Kaliman [7]);
2. The manifolds X given as a submanifold in $\mathbb{C}^{n+2}$ with coordinates $u \in \mathbb{C}$, $v \in \mathbb{C}$, $z \in \mathbb{C}^n$ by the equation $uv = p(z)$, where the zero fiber of the polynomial $p \in \mathbb{C}[\mathbb{C}^n]$ is smooth (otherwise X is not smooth), have ADP [28].
3. The only known non-algebraic example with DP are the manifolds X given as a submanifold in $\mathbb{C}^{n+2}$ with coordinates $u \in \mathbb{C}$, $v \in \mathbb{C}$, $z \in \mathbb{C}^n$ by the equation $uv = f(z)$, where the zero fiber of the holomorphic function $f \in \mathcal{O}(\mathbb{C}^n)$ is smooth (otherwise X is not smooth) [28].
4. Danilov–Gizatullin surfaces have ADP [6].

A variant of density property for (normal) singular varieties (considering vector fields vanishing on a subvariety in particular on the singular locus $\mathrm{Sing}(X)$) was introduced in [34]. A version of Andersén–Lempert theorem holds in this situation which allows to approximate local movements taking place in the smooth part of X by automorphisms fixing the singular locus. It is proven in [34] that normal affine toric varieties have this property. Another version of this generalization considering holomorphic automorphisms of $\mathbb{C}^n$ fixing a codimension two subvariety can be found in [27]. For more information on the density property we refer to the overview article [29].

3 Flexibility

3.1 Definition and Main Features

The notion of flexibility is the most recent among the described properties. It was defined in [4]. First the algebraic version:

Definition 11. Let X be a reduced algebraic variety defined over $\mathbb{C}$ (any algebraically closed field would do). We let $\mathrm{SAut}(X)$ denote the subgroup of $\mathrm{Aut}_{\mathrm{alg}}(X)$ generated by all algebraic one-parameter unipotent subgroups of $\mathrm{Aut}_{\mathrm{alg}}(X)$, i.e., algebraic subgroups isomorphic to the additive group $\mathbb{G}_a$ (usually denoted $\mathbb{C}^+$ in complex analysis). The group $\mathrm{SAut}(X)$ is called the *special automorphism group* of X; this is a normal subgroup of $\mathrm{Aut}_{\mathrm{alg}}(X)$.

Definition 12. We say that a point $x \in X_{\mathrm{reg}}$ is *algebraically flexible* if the tangent space $T_x X$ is spanned by the tangent vectors to the orbits $H.x$ of one-parameter unipotent subgroups $H \subseteq \mathrm{Aut}_{\mathrm{alg}}(X)$. A variety X is called *algebraically flexible* if every point $x \in X_{\mathrm{reg}}$ is.

Clearly, X is algebraically flexible if one point of X_{reg} is and the group $\mathrm{Aut}_{\mathrm{alg}}(X)$ acts transitively on X_{reg}.

The main feature of algebraic flexibility is the following result from [4] (whose proof mainly relies on the Rosenlicht theorem);

Theorem 13. *For an irreducible affine variety X of dimension ≥ 2, the following conditions are equivalent.*

(1) *The group* $\mathrm{SAut}(X)$ *acts transitively on* X_{reg}.
(2) *The group* $\mathrm{SAut}(X)$ *acts infinitely transitively on* X_{reg}.
(3) *X is an algebraically flexible variety.*

The paper [4] also contains versions of simultaneous transitivity (where the space X_{reg} is stratified by orbits of $\mathrm{SAut}(X)$) and versions with jet-interpolation. Moreover, it was recently remarked that the theorem holds for quasi-affine varieties, see Theorem 1.11. in [11].

The holomorphic version of this notion is much less explored, it is obviously implied by the algebraic version in case X is an algebraic variety.

Definition 14. We say that a point $x \in X_{\text{reg}}$ is *holomorphically flexible* if the tangent space $T_x X$ is spanned by the tangent vectors of completely integrable holomorphic vector fields, i.e., holomorphic one-parameter subgroups in $\text{Aut}_{\text{hol}}(X)$. A complex manifold X is called *holomorphically flexible* if every point $x \in X_{\text{reg}}$ is.

Clearly, X is holomorphically flexible if one point of X_{reg} is and the group $\text{Aut}_{\text{hol}}(X)$ acts transitively on X_{reg}.

In the holomorphic category it is still open whether an analogue of Theorem 13 holds.

Open Problem: Are the three equivalences from Theorem 13 true for an irreducible Stein space X? More precisely, if an irreducible Stein space X is holomorphically flexible, does the holomorphic automorphism group $\text{Aut}_{\text{hol}}(X)$ act infinitely transitively on X_{reg}?

It is clear that holomorphic flexibility of X implies that $\text{Aut}_{\text{hol}}(X)$ acts transitively on X_{reg}, i.e., the implication (3) $\Rightarrow$ (1) is true. Indeed, let $\theta_i, i = 1, 2, \ldots, n$ be completely integrable holomorphic vector fields which span the tangent space $T_x X$ at some point $x \in X_{\text{reg}}$, where $n = \dim X$. If $\psi^i : \mathbb{C} \times X \to X, \quad (t, x) \mapsto \psi^i_t(x)$ denote the corresponding one-parameter subgroups, then the map $\mathbb{C}^n \to X, \quad (t_1, t_2, \ldots, t_n) \mapsto \psi^n_{t_n} \circ \psi^{n-1}_{t_{n-1}} \circ \cdots \circ \psi^1_{t_1}(x)$ is of full rank at $t = 0$ and thus by the Inverse Function Theorem a local biholomorphisms from a neighborhood of 0 to a neighborhood of x. Thus the $\text{Aut}_{\text{hol}}(X)$-orbit through any point of X_{reg} is open. If all orbits are open, each orbit is also closed, being the complement of all other orbits. Since X_{reg} is connected, this implies that it consists of one orbit.

The inverse implication (1) $\Rightarrow$ (3) is also true. For the proof we appeal to the Hermann–Nagano Theorem which states that if $\mathfrak{g}$ is a Lie algebra of holomorphic vector fields on a manifold X, then the orbit $R_{\mathfrak{g}}(x)$ (which is the union of all points z over any collection of finitely many fields $v_1, \ldots v_N \in \mathfrak{g}$ and over all times $(t_1, \ldots, t_N)$ for which the expression $z = \psi^N_{t_N} \circ \psi^{N-1}_{t_{N-1}} \circ \cdots \circ \psi^1_{t_1}(x)$ is defined) is a locally closed submanifold and its tangent space at any point $y \in R_{\mathfrak{g}}(x)$ is $T_y R_{\mathfrak{g}}(x) = \text{span}_{v \in \mathfrak{g}} v(y)$. We consider the Lie algebra $\mathfrak{g}$ generated by completely integrable holomorphic vector fields. Since by the assumption the orbit is X_{reg} we conclude that Lie combinations of completely integrable holomorphic vector fields span the tangent space at each point in X_{reg}. Now suppose at some point x_0 the completely integrable fields do not generate $T_{x_0} X_{\text{reg}}$, i.e., there is a proper linear subspace W of $T_{x_0} X_{\text{reg}}$, such that $v(x_0) \in W$ for all completely integrable holomorphic fields v. Any Lie combination of completely integrable holomorphic fields is a limit (in the compact open topology) of sums of completely integrable holomorphic fields due to the formula $\{v, w\} = \lim_{t \to 0} \frac{\phi_t^*(w) - w}{t}$ for the Lie bracket ($\phi_t^*(w)$ is a completely integrable field pulled back by an automorphism, thus completely integrable!). Therefore all Lie combinations of completely integrable fields evaluated at x_0 are contained in $W \subset T_{x_0} X_{\text{reg}}$, a contradiction.

In order to prove the remaining implication (3) $\Rightarrow$ (2) one would like to find suitable functions $f \in \mathrm{Ker}\theta$ for a completely integrable holomorphic vector field θ, vanishing at one point and not vanishing at some other point of X. In general these functions may not exist, an orbit of θ can be dense in X.

At this point it is worth mentioning that for a Stein manifold DP implies all three conditions from Theorem 13. For flexibility this is lemma 26 below, infinite transitivity (with jet-interpolation) is proved by Varolin in [41].

Also the generalized form of DP for Stein spaces defined in [34] implies all three conditions from Theorem 13.

3.2 Examples

Examples of algebraically flexible varieties are homogeneous spaces of semisimple Lie groups (or extensions of semisimple Lie groups by unipotent radicals), toric varieties without non-constant invertible regular functions, cones over flag varieties, and cones over Del Pezzo surfaces of degree at least 4, normal hypersurfaces of the form $uv = p(\bar{x})$ in $\mathbb{C}^{n+2}_{u,v,\bar{x}}$. Moreover, algebraic subsets of codimension at least 2 can be removed as recently shown by Flenner, Kaliman, and Zaidenberg in [11]

Theorem 15. *Let X be a smooth quasi-affine variety of dimension ≥ 2 and $Y \subset X$ a closed subscheme of codimension ≥ 2. If X is flexible, then so is $X \setminus Y$.*

4 Oka–Forstnerič Manifolds

4.1 Historical Introduction to Oka Theory and Motivational Examples

The notion of Oka–Forstnerič manifolds is quite new (it was introduced by Forstnerič in [13], who called them Oka manifolds following a suggestion of Lárusson who underlined the importance of such a notion already in [35]) but the development merging into this important notion, called Oka theory, has a long history. It started with Oka's theorem from 1939 that the second (multiplicative) Cousin problem on a domain of holomorphy is solvable with holomorphic functions if it is solvable with continuous functions. This implies that a holomorphic line bundle on such a domain is holomorphically trivial if it is topologically trivial.

Let us recall how the generalizations of the classical one variable results of Mittag–Leffler (construct meromorphic functions with prescribed main parts) and Weierstrass (construct meromorphic functions with prescribed zeros and poles) are generalized to several complex variables.

Let us recall the first (additive) Cousin problem which arises from the Mittag–Leffler problem, generalizing the famous Mittag–Leffler theorem from one variable to several variables: Given data $\{(U_i, m_i)\}$, where U_i is an open cover of a complex space X and $m_i \in \mathcal{M}(U_i)$ is a meromorphic function on U_i such that every difference $f_{ij} = m_i|_{U_{ij}} - m_j|_{U_{ij}}$ is holomorphic on $U_{ij} = U_i \cap U_j$, find a global meromorphic function $m \in \mathcal{M}(X)$ on X such that $m|_{U_i} - m_i$ is holomorphic on U_i for all i.

For solving this Mittag–Leffler problem one first solves the associated additive Cousin problem, defined as follows: The collection $f_{ij} \in \mathcal{O}(U_{ij})$ defines a 1-cocycle on the cover U_i with values in the sheaf $\mathcal{O}$ of holomorphic functions, meaning that for each triple i, j, k of indexes we have

$$f_{ij} + f_{jk} + f_{ki} = 0 \quad \text{on} \quad U_{ijk} = U_i \cap U_j \cap U_k.$$

Given such a 1-cocycle $\{f_{ij}\}$, the Cousin I problem asks for a collection of holomorphic functions $f_j \in \mathcal{O}(U_j)$ (a 0-cochain) such that

$$f_i - f_j = f_{ij} \quad \text{on} \quad U_{ij}.$$

One expresses this by saying the cocycle splits or it is a 1-coboundary. From the solution to the additive Cousin problem one obtains by setting $m|_{U_i} = m_i - f_i$ a well-defined (since $m_i - m_j = f_{ij} = f_i - f_j \quad \text{on} \quad U_{ij}$) global meromorphic function $m \in \mathcal{M}(X)$ solving the Mittag–Leffler problem.

The vanishing of the first Cech cohomology group $H^1(X, \mathcal{O})$ with coefficients in the sheaf $\mathcal{O}$ means that every 1-cocycle splits on a refinement of the covering. In other words $H^1(X, \mathcal{O}) = 0$ implies that every 1-cocycle becomes a 1-coboundary on a refinement, so every Mittag–Leffler problem is solvable, in particular by Cartan's Theorem B this is true for any Stein manifold.

The second (multiplicative) Cousin Problem arises from the problem of finding meromorphic functions with prescribed zeros and poles, solved by Weierstrass in one variable. Given data $\{(U_i, m_i)\}$, where U_i is an open cover of a complex space X and $m_i \in \mathcal{M}^*(U_i)$ is an invertible (i.e., not vanishing identically on any connected component) meromorphic function on U_i such that for any pair of indexes the quotient $f_{ij} := g_i g_j^{-1}$ is a nowhere vanishing holomorphic function $f_{ij} \in \mathcal{O}^*(U_{ij})$. Our data defines a divisor D on X and the problem is to find a global meromorphic function $m \in \mathcal{M}(X)$ defining this divisor, meaning, such a function that $m m_i^{-1}$ is a nowhere vanishing holomorphic function on U_i for every i. A solution is obtained by solving the second Cousin problem: Given a collection f_{ij} of nowhere vanishing holomorphic functions $f_{ij} : U_{ij} \to \mathbb{C}^*$ satisfying the 1-cocycle condition

$$f_{ii} = 1 \quad f_{ij} f_{ji} = 1 \quad f_{ij} f_{jk} f_{ki} = 1$$

on U_i, U_{ij}, U_{ijk} respectively, find nowhere vanishing holomorphic functions $f_j : U_j \to \mathbb{C}^*$ such that

$$f_i = f_{ij} f_j \quad \text{on } U_{ij}.$$

If such f_i exist then $g_i f_i^{-1} = g_j f_j^{-1}$ on U_{ij} which defines a solution, a meromorphic function $m \in \mathcal{M}(X)$ representing our divisor.

The following cohomological formulation and proof of Oka's Theorem are standard, see e.g. [14] Theorem 5.2.2..

Theorem 16. *If X is a complex space satisfying $H^1(X, \mathcal{O}) = 0$, then the homomorphism $H^1(X, \mathcal{O}^*) \to H^1(X, \mathbb{C}^*)$ induced by the sheaf inclusion $\mathcal{O}^* \hookrightarrow \mathbb{C}^*$ is injective. In particular if a multiplicative Cousin problem is solvable by continuous functions, then it is solvable by holomorphic functions. If in addition we have $H^2(X, \mathcal{O}) = 0$, then the above map is an isomorphism.*

Proof. Consider the exponential sheaf sequence (where $\sigma(f) = e^{2\pi i f}$).

$$\begin{array}{ccccccccc} 0 & \longrightarrow & \mathbb{Z} & \longrightarrow & \mathcal{O} & \xrightarrow{\sigma} & \mathcal{O}^* & \longrightarrow & 1 \\ & & \downarrow{\scriptstyle id} & & \downarrow & & \downarrow & & \\ 0 & \longrightarrow & Z & \longrightarrow & \mathbb{C} & \xrightarrow{\sigma} & \mathbb{C}^* & \longrightarrow & 1 \end{array}$$

Since due to partition of unity $H^1(X, \mathbb{C}) = H^2(X, \mathbb{C}) = 0$ the relevant portion of long exact cohomology sequence is:

$$\begin{array}{ccccccccc} H^1(X, \mathbb{Z}) & \longrightarrow & H^1(X\mathcal{O}) & \longrightarrow & H^1(X, \mathcal{O}^*) & \xrightarrow{c_1} & H^2(X, \mathbb{Z}) & \longrightarrow & H^2(X, \mathcal{O}) \\ & & \downarrow & & \downarrow & & \| & & \downarrow \\ & & 0 & \longrightarrow & H^1(X, \mathbb{C}^*) & \xrightarrow{c_1} & H^2(X, \mathbb{Z}) & \longrightarrow & 0 \end{array}$$

The map in the bottom row is an isomorphism $H^1(X, \mathbb{C}^*) \cong H^2(X, \mathbb{Z})$. If $H^1(X, \mathcal{O}) = 0$ the (1-st Chern class) map c_1 in the first row is injective $0 \to H^1(X, \mathcal{O}^*) \xrightarrow{c_1} H^2(X, \mathbb{Z}) \cong H^1(X, \mathbb{C}^*)$. If in addition $H^2(X, \mathcal{O}) = 0$ this map is an isomorphism. □

By Oka's theorem on a complex space with $H^1(X, \mathcal{O}) = H^2(X, \mathcal{O}) = 0$ (by Theorem B this holds in particular on a Stein space) the natural map from equivalence classes of holomorphic line bundles into equivalence classes of continuous (complex) line bundles is an isomorphism. For higher rank vector bundles this cohomological proof fails due to non-commutativity of the relevant cohomology groups. Nevertheless, Grauert was able to prove the corresponding statement in higher dimensions. The following theorem is the holomorphic counterpart of Quillen's and Suslin's result that projective modules over affine space are free.

Theorem 17. *For a Stein space X the natural map $\mathrm{Vect}^r_{\mathrm{hol}}(X) \to \mathrm{Vect}^r_{\mathrm{top}}(X)$ of equivalence classes of rank r complex vector bundles is a bijection for every $r \in \mathbb{N}$.*

This theorem follows from the following result, named Grauert's Oka principle by H. Cartan, obtained by Grauert [20], Grauert and Kerner [21], and Ramspott [36] (see Theorem 5.3.2. in [14]).

Theorem 18. *If X is a Stein space and $\pi : Z \to X$ is a holomorphic fiber bundle with a complex homogeneous fiber whose structure group is a complex Lie group acting transitively on the fiber, then the inclusion $\Gamma_{\text{hol}}(X, Z) \hookrightarrow \Gamma_{\text{cont}}(X, Z)$ of the space of global holomorphic sections into the space of global continuous sections is a weak homotopy equivalence. In particular every continuous section is homotopic to a holomorphic section.*

An equivariant version of Grauerts Oka principle with respect to an action of a reductive complex Lie group has been proven by Heinzner and Kutzschebauch [23]. This principle in particular implies that the method of constructing counterexamples to the linearization problem, found by Schwarz in the algebraic setting [38], does not work in the holomorphic category. Moreover, the above-mentioned Oka principle was recently used by Kutzschebauch, Lárusson, and Schwarz [32] to show among others a strong linearization result: A generic holomorphic action, which is locally over a common categorical quotient isomorphic to a linear action on $\mathbb{C}^n$, is in fact globally isomorphic to that linear action.

The next step in Oka theory was made by Gromov in his seminal paper [22], which marks the beginning of modern Oka theory. He introduced the notion of dominating spray and ellipticity (see the last section). The great improvement compared to Grauert's Oka principle is the fact that not the fiber together with the transition maps of the bundle, but only certain properties of the fiber totally independent of transition maps allow to derive the conclusions. In the above-cited classical works, the structure group was indeed assumed to be a complex Lie group. However, in modern Oka theory the structure group is completely irrelevant. Moreover, modern Oka theory allows to consider sections of stratified elliptic submersions, generalizing the case of locally trivial fiber bundles. The emphasis shifted from the cohomological to the homotopy theoretic aspect, focusing on those analytic properties of a complex manifold Y which ensure that every continuous map from a Stein space X to Y is homotopic to a holomorphic map, with natural additions concerning approximation and interpolation of maps that are motivated by the extension and approximation theorems for holomorphic functions on Stein spaces. The approximation and extension are needed for generalizing from maps $X \to Y$ (which can be considered as sections of the trivial bundle $X \times Y \to X$ with fiber Y) to sections of holomorphic submersions $Z \to X$ with Oka–Forstnerič fibers and moreover to stratified elliptic submersions.

4.2 Definition and Main Features

Definition 19. A complex manifold Y is an Oka–Forstnerič manifold if every holomorphic map $f : K \to Y$ from (a neighborhood of) a compact convex set

$K \subset \mathbb{C}^n$ (any dimension n) can be approximated uniformly on K by entire maps $\mathbb{C}^n \to Y$.

The property in this definition is also called Convex Approximation Property (CAP), if the dimension n is fixed we speak of $(\mathrm{CAP})_n$, thus (CAP) means $(\mathrm{CAP})_n$ for all n. By work of Forstnerič (CAP) is equivalent to any of 13 different Oka properties, one of them is mentioned in the following Theorem which includes all versions of the classical Oka–Grauert principle discussed in the Introduction. This theorem answers Gromov's question whether Runge approximation on a certain class of compact sets in Euclidean spaces suffices to infer the Oka property. Since all these 13 Oka properties are equivalent characterizations of the same class of manifolds Forstnerič called them Oka manifolds. In order to honor his work on the equivalence of all the Oka properties the author finds the notation Oka–Forstnerič manifolds more appropriate.

Theorem 20. *Let $\pi : Z \to X$ be a holomorphic submersion of a complex space Z onto a reduced Stein space X. Assume that X is exhausted by a sequence of open subsets $U_1 \subset U_2 \subset \cdots \cup_j U_j = X$ such that each restriction $Z|_{U_j} \to U_j$ is a stratified holomorphic fiber bundle whose fibers are Oka manifolds. Then sections $X \to Z$ satisfy the following*

Parametric Oka property (POP)*: Given a compact $\mathcal{O}(X)$-convex subset K of X, a closed complex subvariety A of X, compact sets $P_0 \subset P$ in a Euclidean space $\mathbb{R}^m$, and a continuous map $f : P \times X \to Z$ such that*

(a) *for every $p \in P$, $f(p,\cdot) : X \to Z$ is a section of $Z \to X$ that is holomorphic on a neighborhood of K (independent of p) and such that $f(p,\cdot)|_A$ is holomorphic on A, and*
(b) *$f(p,\cdot)$ is holomorphic on X for every $p \in P_0$*

there is a homotopy $f_t : P \times X \to Z \quad (t \in [0,1])$, with $f_0 = f$, such that f_t enjoys properties (a) and (b) for all $t \in [0,1]$, and also the following hold:

(i) *$f_1(p,\cdot)$ is holomorphic on X for all $p \in P$*
(ii) *f_t is uniformly close to f on $P \times K$ for all $t \in [0,1]$*
(iii) *$f_t = f$ on $(P_0 \times X) \cup (P \times A)$ for all $t \in [0,1]$*

As a general reference for Oka theory we refer to the monograph [14] and the overview article [15].

4.3 Applications and Examples

The number of applications of the Oka theory is growing, we already indicated the classical Cousin problems and Grauert's classification of holomorphic vector bundles over Stein spaces in the introduction. The only application we would like to mention is a recent solution to a problem posed by Gromov, called the Vaserstein problem. It is a natural question about the K_1-group of the ring of holomorphic functions or in simple terms it is asking whether (and when) in a matrix whose entries are holomorphic functions (parameters) the Gauss elimination process can be performed in a way holomorphically depending on the parameter. This is to our knowledge the only application where a stratified version of an Oka theorem is needed, i.e., no proof using a non-stratified version is known.

Theorem 21. *(see [25]). Let X be a finite dimensional reduced Stein space and $f: X \to SL_m(\mathbb{C})$ be a holomorphic mapping that is null-homotopic. Then there exist a natural number K and holomorphic mappings $G_1, \ldots, G_K: X \to \mathbb{C}^{m(m-1)/2}$ such that f can be written as a product of upper and lower diagonal unipotent matrices*

$$f(x) = \begin{pmatrix} 1 & 0 \\ G_1(x) & 1 \end{pmatrix} \begin{pmatrix} 1 & G_2(x) \\ 0 & 1 \end{pmatrix} \cdots \begin{pmatrix} 1 & G_K(x) \\ 0 & 1 \end{pmatrix}$$

for every $x \in X$.

Here the assumption null-homotopic means that the map is homotopic through continuous maps to a constant map (matrix), which since Grauert's Oka principle, Theorem 18, is equivalent of being null-homotopic through holomorphic maps. This is an obvious necessary condition since multiplying all lower/upper diagonal matrices in the product by $t \in [0, 1]$ yields a homotopy to the (constant) identity matrix. It is a result of Vaserstein that null-homotopic is also sufficient in order to factorize the map as a product with continuous entries. Thus we have the Oka principle. For the existence of a holomorphic factorization there are only topological obstructions, it exists iff a topological factorization exists.

Now we come to examples of Oka–Forstnerič manifolds:

A Riemann surface is an Oka–Forstnerič manifold iff it is non-hyperbolic, i.e., one of $\mathbb{P}^1$ $\mathbb{C}$, $\mathbb{C}^*$, or a compact torus $\mathbb{C}/\Gamma$.

Oka–Forstnerič manifolds enjoy the following functorial properties, for elliptic manifolds (see Definition below) these properties are unknown.

- If $\pi : E \to B$ is a holomorphic covering map of complex manifolds then B is Oka–Forstnerič iff E is ([14] Proposition 5.5.2).
- If E and X are complex manifolds and $\pi : E \to X$ is a holomorphic fiber bundle whose fiber is an Oka–Forstnerič manifold, the X is an Oka–Forstnerič manifold iff E is ([14] Theorem 5.5.4).
- If a complex manifold Y is exhausted by open domains $D_1 \subset D_2 \subset \cdots \subset \cup_{j=1}^{\infty} = Y$ such that every D_j is an Oka–Forstnerič manifold, then Y is an Oka–Forstnerič manifold. In particular every long $\mathbb{C}^n$ is an Oka–Forstnerič manifold. (A manifold is called a long $\mathbb{C}^n$ if all D_j are biholomorphic to $\mathbb{C}^n$. If the inclusion $D_i \subset D_{i+1}$ is not a Runge pair, on those manifolds the ring of holomorphic functions may consist of constants only!!)

The main source of examples are the elliptic manifolds (see Definition 22 below), a notion invented by Gromov. This includes by our scheme (1) of implications all holomorphic flexible manifolds and all manifolds with the density property, in particular complex Lie groups and homogeneous spaces of Lie groups, the manifolds from the classical theorems of Oka and Grauert. For a Stein manifold ellipticity is equivalent to being an Oka–Forstnerič manifold. For general manifolds this is an open question. One possible counterexample is the complement of the ball in $\mathbb{C}^n$, the set $\{z \in \mathbb{C}^n : |z_1|^2 + |z_2|^2 + \ldots + |z_n|^2 > 1\}$. It was recently shown by Andrist and Wold [5] that it is not elliptic for $n \geq 3$, whereas it has two "nearby" properties implied by being an Oka–Forstnerič manifold, strongly dominable and $\mathrm{CAP_{n-1}}$ ([16]).

5 Proof of the Implications from Scheme (1): Ellipticity in the Sense of Gromov

First remark that the two bottom up arrows in Scheme (1) are obvious from the definitions. In order to prove the left–right arrows let's define the notions introduced by Gromov [22] revolutionizing Oka theory (see also [14, Chap. 5]):

Definition 22. Let Y be a complex manifold.

(1) A holomorphic spray on Y is a triple (E, π, s) consisting of a holomorphic vector bundle $\pi : E \to Y$ (a spray bundle) and a holomorphic map $s : E \to Y$ (a spray map) such that for each $y \in Y$ we have $s(0_y) = y$.
(2) A spray (E, π, s) on Y is dominating on a subset $U \subset Y$ if the differential $d_{0_y} s : T_{0_y} E \to T_y Y$ maps the vertical tangent space E_y of $T_{0_y} E$ surjectively onto $T_y Y$ for every $y \in U$, s is dominating if this holds for all $y \in Y$.
(3) A complex manifold Y is elliptic if it admits a dominating holomorphic spray.

The main result of Gromov can now be formulated in the following way.

Theorem 23. *An elliptic manifold is an Oka–Forstnerič manifold.*

Of course Gromov proved the full Oka principle for elliptic manifolds. This proof can now be decomposed in two stages. The main (and the only) use of ellipticity is to prove a homotopy version of Runge (Oka–Weil) theorem, which in particular gives CAP (= Oka–Forstnerič) and the second stage is CAP implies Oka principle.

Gromov's theorem proves our implication

$$\textbf{holomorphically flexible} \Longrightarrow \textbf{Oka--Forstneric manifold}$$

using the following example of a spray given by Gromov and Lemma 25;

Example 24. Given completely integrable holomorphic vector fields $\theta_1, \theta_2, \ldots, \theta_N$ on a complex manifold X such that at each point $x \in X$ they span the tangent space, $\mathrm{span}(\theta_1(x), \theta_2(x), \ldots, \theta_N(x) = T_x X$. Let $\psi^i : \mathbb{C} \times X \to X, \quad (t, x) \mapsto \psi_t^i(x)$ denote the corresponding one-parameter subgroups; Then the map $s : \mathbb{C}^N \times X \to X$

defined by $((t_1, t_2, \ldots, t_n), x) \mapsto \psi^n_{t_n} \circ \psi^{n-1}_{t_{n-1}} \circ \cdots \circ \psi^1_{t_1}(x)$ is of full rank at $t = 0$ for any x. It is therefore a dominating spray map from the trivial bundle $X \times \mathbb{C}^N \to X$.

Lemma 25. *If a Stein manifold X is holomorphically flexible, then there are finitely many completely integrable holomorphic fields which span the tangent space $T_x X$ at every point $x \in X$*

Proof. To prove that there are finitely many completely integrable holomorphic fields that span each tangent space, let us start with n fields $\theta_1, \ldots, \theta_n$ which span the tangent space at some point x_0 and thus outside a proper analytic subset A. The set A may have countably many irreducible components $A_1, A_2, A_3, \ldots$.

It suffices now to find a holomorphic automorphism $\Phi \in \mathrm{Aut_{hol}}(X)$ such that $\Phi(X \setminus A) \cap A_i \neq \emptyset$ for every $i = 1, 2, 3, \ldots$. Indeed, for such an automorphism Φ the completely integrable holomorphic vector fields $\Phi_*(\theta_1), \ldots, \Phi_*(\theta_n)$ span the tangent space at a general point in each A_i, i.e., together with the fields $\theta_1, \ldots, \theta_n$ they span the tangent space at each point outside an analytic subset B of a smaller dimension than A. Then the induction by dimension implies the desired conclusion.

In order to construct Φ consider a monotonically increasing sequence of compacts $K_1 \subset K_2 \subset \ldots$ in X such that $\bigcup_i K_i = X$ and a closed imbedding $\iota : X \hookrightarrow \mathbb{C}^m$. For every continuous map $\varphi : X \to \mathbb{C}^m$ denote by $||\varphi||_i$ the standard norm of the restriction of φ to K_i. Let d be the metric on the space $\mathrm{Aut_{hol}}(X)$ of holomorphic automorphisms of X given by the formula

$$d(\Phi, \Psi) = \sum_{i=1}^{\infty} 2^{-i}(\min(||\Phi - \Psi||_i, 1) + \min(||\Phi^{-1} - \Psi^{-1}||_i, 1) \qquad (4.1)$$

where automorphisms $\Phi^{\pm 1}, \Psi^{\pm 1} \in \mathrm{Aut_{hol}}(X)$ are viewed as continuous maps from X to $\mathbb{C}^m$. This metric makes $\mathrm{Aut_{hol}}(X)$ a complete metric space.

Set $Z_i = \{\Psi \in \mathrm{Aut_{hol}}(X) : \Psi(A_i) \cap (X \setminus A) \neq \emptyset\}$. Note that Z_i is open in $\mathrm{Aut_{hol}}(X)$ and let us show that it is also everywhere dense.

Since completely integrable holomorphic fields generate the tangent space at each point of X, we can choose such a field θ non-tangent to A_i. Then for every $\Psi \in \mathrm{Aut_{hol}}(X)$ its composition with general elements of the flow induced by θ is in Z_i. That is, a perturbation of Ψ belongs to Z_i which proves that Z_i is everywhere dense in $\mathrm{Aut_{hol}}(X)$. By the Baire category theorem the set $\bigcap_{i=1}^{\infty} Z_i$ is not empty which yields the existence of the desired automorphism. □

Since the question whether holomorphic maps are approximable by morphisms is an important issue in algebraic geometry, we would like to remark at this point that there is an application of Oka-theory to this question. Clearly there is an obvious notion of algebraic spray, thus algebraic ellipticity. Also the proof of the above lemma generalizes showing that an algebraically flexible manifold is algebraically elliptic. These algebraically elliptic manifolds satisfy an algebraic version of CAP. However, in the algebraic category, simple examples show that algebraic CAP does not imply the full algebraic Oka principle, but only a weaker statement that being approximable by algebraic morphisms is a homotopy invariant property (at least for maps from affine algebraic manifolds to algebraically elliptic manifolds). For a precise treatment of this question we refer to [14] chapter 7.10.

The implication DP $\Longrightarrow$ holomorphically flexible is contained in the following Lemma.

Lemma 26. *If a Stein manifold X has the density property, the completely integrable holomorphic vector fields span the tangent space at each point $x \in X$.*

Proof. It follows from the density property that Lie combinations of completely integrable holomorphic vector fields span the tangent space $T_x X$ at any given point $x \in X$. Observe that every Lie bracket $[\nu, \mu]$ of completely integrable holomorphic vector fields can be approximated by a linear combination of such fields which follows immediately from the equality $[\nu, \mu] = \lim_{t \to 0} \frac{\phi_t^*(\nu) - \nu}{t}$ where ϕ_t is the flow generated by μ. Thus the completely integrable holomorphic vector fields span $T_x X$ at any $x \in X$. □

6 Concluding Remarks and Open Problems

There is also another property which has similar consequences as the density property for holomorphic automorphisms preserving a volume form.

Definition 27. Let a complex manifold X be equipped with a holomorphic volume form ω (i.e., ω is nowhere vanishing section of the canonical bundle). We say that X has the volume density property (VDP) with respect to ω if in the compact-open topology the Lie algebra generated by completely integrable holomorphic vector fields ν such that $\nu(\omega) = 0$ is dense in the Lie algebra of all holomorphic vector fields that annihilate ω (note that condition $\nu(\omega) = 0$ is equivalent to the fact that ν is of ω-divergence zero). If X is affine algebraic we say that X has the algebraic volume density property (AVDP) with respect to an algebraic volume form ω if the Lie algebra generated by completely integrable algebraic vector fields ν such that $\nu(\omega) = 0$, coincides with the Lie algebra of all algebraic vector fields that annihilate ω.

For Stein manifolds with the volume density property (VDP) an Andersén–Lempert theorem for volume preserving maps holds. The implication **(AVDP)** $\Rightarrow$ **(VDP)** holds but its proof is not trivial (see [30]). Also **(VDP)** $\Rightarrow$ **holomorphic flexibility** is true (see [29]). Thus we can have a scheme of implications like (1) with (DP) replaced by (VDP) and (ADP) replaced by (AVDP).

Volume density property and density property are not implied by each other, if X has density property it may not even admit a holomorphic volume form, if X has volume density property with respect to one volume form it may not have it with respect to another volume form and there is no reason to expect it has density property. For example, $(\mathbb{C}^*)^n$ for $n > 1$ has (algebraic) volume density property with respect to the Haar form, it does not have algebraic density property [2] and it is expected not to have density property. It is a potential counterexample to the reverse of the left horizontal arrow in scheme (1).

Concerning the reverse implications in scheme (1): The variety $(\mathbb{C}^*)^n, n > 1$ is an obvious counterexample to the reverse of the right vertical arrow, the others are more delicate.

Open Problem: Which of the other three implications in scheme (1) are reversible for a Stein manifold (resp. smooth affine algebraic variety for the vertical arrow)?

The main problem here is that no method is known how to classify (meaning exclude the existence of any other than the obvious) completely integrable holomorphic vector fields on Stein manifolds with any of our flexibility properties. There is not even a classification of completely integrable holomorphic vector fields on $\mathbb{C}^2$ available.

References

1. E. Andersén, Volume-preserving automorphisms of $\mathbb{C}^n$. Complex Variables Theory Appl. **14**(1–4), 223–235 (1990)
2. E. Andersén, Complete vector fields on $(\mathbb{C}^*)^n$. Proc. Am. Math. Soc. **128**(4), 1079–1085 (2000)
3. E. Andersén, L. Lempert, On the group of holomorphic automorphisms of $\mathbb{C}^n$. Invent. Math. **110**(2), 371–388 (1992)
4. I. Arzhantsev, H. Flenner, S. Kaliman, F. Kutzschebauch, M. Zaidenberg, Flexible varieties and automorphism groups. Duke Math. J. **162**(4), 767–823 (2013)
5. R. Andrist, E. Fornaess Wold, The complement of the closed unit ball in $\mathbb{C}^3$ is not subelliptic. arXiv:1303.1804 (2013)
6. F. Donzelli, Algebraic density property of Danilov-Gizatullin surfaces. Math. Z. **272**(3–4), 1187–1194 (2012)
7. F. Donzelli, A. Dvorsky, S. Kaliman, Algebraic density property of homogeneous spaces. Transform. Groups **15**(3), 551–576 (2010)
8. H. Derksen, F. Kutzschebauch, Global holomorphic linearization of actions of compact Lie groups on $\mathbb{C}^n$. Contemp. Math. **222**, 201–210 (1999)
9. H. Derksen, F. Kutzschebauch, Nonlinearizable holomorphic group actions. Math. Ann. **311**(1), 41–53 (1998)
10. Y. Eliashberg, M. Gromov, Embeddings of Stein manifolds of dimension n into the affine space of dimension $3n/2 + 1$. Ann. Math. **136**(2)(1), 123–135 (1992)
11. H. Flenner, S. Kaliman, M. Zaidenberg The Gromov-Winkelmann theorem for flexible varieties. arXiv:1305.6417 (2013)
12. O. Forster, Plongements des variétés de Stein. Comment. Math. Helv. **45**, 170–184 (1970)
13. F. Forstnerič, Oka manifolds. C. R. Acad. Sci. Paris Ser. I **347**, 1017–1020 (2009)
14. F. Forstnerič, *Stein Manifolds and Holomorphic Mappings*. Ergebnisse der Mathematik und ihrer Grenzgebiete. 3. Folge, vol. 56 (Springer, Berlin, 2011)
15. F. Forstnerič, F. Lárusson. Survey of Oka theory. New York J. Math. **17a**, 11–38 (2011)
16. F. Forstneri č, T. Ritter, Oka properties of ball complements. arXiv:1303.2239 (2013)
17. F. Forstnerič, J.-P. Rosay, Approximation of biholomorphic mappings by automorphisms of $\mathbb{C}^n$. Invent. Math. **112**(2), 323–349 (1993)
18. F. Forstneric, E.F. Wold, Embeddings of infinitely connected planar domains into C^2. Anal. PDE **6**(2), 499–514 (2013)

19. F. Forstneric, E.F. Wold Bordered Riemann surfaces in $\mathbb{C}^2$. J. Math. Pures Appl. **91**(1), 100–114 (2009)
20. H. Grauert, Analytische Faserungen über holomorph-vollständigen Räumen. Math. Ann. **135**, 263–273 (1958)
21. H. Grauert, H. Kerner, Approximation von holomorphen Schnittflächen in Faserbündeln mit homogener Faser. Arch. Math. **14**, 328–333 (1963)
22. M. Gromov, Oka's principle for holomorphic sections of elliptic bundles. J. Am. Math. Soc. **2** 851–897 (1989)
23. P. Heinzner, F. Kutzschebauch, An equivariant version of Grauert's Oka principle. Invent. Math. **119**, 317–346 (1995)
24. A.T. Huckleberry, Actions of groups of holomorphic transformations. *Several Complex Variables*, VI. Encyclopaedia of Mathematical Sciences, vol. 69 (Springer, 1990), 143–196
25. B. Ivarsson, Kutzschebauch, F. Holomorphic factorization of mappings into $SL_n(\mathbb{C})$. Ann. Math. **175**(1), 45–69 (2012)
26. S. Kaliman, Extensions of isomorphisms between affine algebraic subvarieties of k^n to automorphisms of k^n. Proc. Am. Math. Soc. **113**(2), 325–333 (1991)
27. S. Kaliman, F. Kutzschebauch Criteria for the density property of complex manifolds. Invent. Math. **172**(1), 71–87 (2008)
28. S. Kaliman, F. Kutzschebauch, Density property for hypersurfaces $uv = p(\bar{x})$. Math. Z. **258**(1), 115–131 (2008)
29. S. Kaliman, F. Kutzschebauch, *On the present state of the Andersen-Lempert theory*, In: Affine Algebraic Geometry: The Russell Festschrift. CRM Proceedings and Lecture Notes, vol. 54 (Centre de Recherches Mathématiques, 2011), pp. 85–122
30. S. Kaliman, F. Kutzschebauch, Algebraic volume density property of affine algebraic manifolds, Invent. Math. **181**(3) (2010) 605–647
31. F. Kutzschebauch, Compact and reductive subgroups of the group of holomorphic automorphisms of $\mathbb{C}^n$. Singularities and complex analytic geometry (Kyoto, 1997). Sūrikaisekikenkyūsho Kōkyūroku **1033**, 81–93 (1998)
32. F. Kutzschebauch, F. Larusson, G.W. Schwarz, An Oka principle for equivariant isomorphisms. J. Reine Angew. Math. p. 22. DOI 10.1515/crelle-2013-0064
33. F. Kutzschebauch, S. Lodin, Holomorphic families of non-equivalent embeddings and of holomorphic group actions on affine space. Duke Math. J. **162**, 49–94 (2013)
34. F. Kutzschebauch, A. Liendo, M. Leuenberger, On the algebraic density property for affine toric varieties, preprint (2013)
35. F. Lárusson, Model structures and the Oka principle. J. Pure Appl. Algebra **192**(1–3), 203–223 (2004)
36. K. J. Ramspott, Stetige und holomorphe Schnitte in Bündeln mit homogener Faser. Math. Z. **89**, 234–246 (1965)
37. J. Schürmann, Embeddings of Stein spaces into affine spaces of minimal dimension. Math. Ann. **307**(3), 381–399 (1997)
38. G.W. Schwarz, Exotic algebraic group actions. C. R. Acad. Sci. Paris Sér. I Math. **309**, 89–94 (1989)
39. M. Suzuki, Sur les opérations holomorphes du groupe additif complexe sur l'espace de deux variables complexes. Ann. Sci. Scuola Norm. Sup. **10**(4), 517–546 (1977)
40. D. Varolin, The density property for complex manifolds and geometric structures. J. Geom. Anal. **11**(1), 135–160 (2001)
41. D. Varolin, The density property for complex manifolds and geometric structures. II. Int. J. Math. **11**(6), 837–847 (2000)
42. E.F. Wold, Embedding Riemann surfaces properly into $\mathbb{C}^2$. Int. J. Math. **17**(8), 963–974 (2006)

Strongly Residual Coordinates over $A[x]$

Drew Lewis

Abstract For a commutative ring A, a polynomial $f \in A[x]^{[n]}$ is called a *strongly residual coordinate* if f becomes a coordinate (over A) upon going modulo x, and f becomes a coordinate (over $A[x, x^{-1}]$) upon inverting x. We study the question of when a strongly residual coordinate in $A[x]^{[n]}$ is a coordinate, a question closely related to the Dolgachev–Weisfeiler conjecture. It is known that all strongly residual coordinates are coordinates for $n = 2$ over an integral domain of characteristic zero. We show that a large class of strongly residual coordinates that are generated by elementaries over $A[x, x^{-1}]$ are in fact coordinates for arbitrary n, with a stronger result in the $n = 3$ case. As an application, we show that all Vénéreau-type polynomials are 1-stable coordinates.

MSC: 14R10, 14R25

1 Introduction

Let A (and all other rings) be a commutative ring with one. An *A-coordinate* (if A is understood, we simply say *coordinate*; some authors prefer the term *variable*) is a polynomial $f \in A^{[n]}$ for which there exist $f_2, \ldots, f_n \in A^{[n]}$ such that $A[f, f_2, \ldots, f_n] = A^{[n]}$. It is natural to ask when a polynomial is a coordinate; this question is extremely deep and has been studied for some time. There are several longstanding conjectures giving a criteria for a polynomial to be a coordinate:

D. Lewis (✉)
University of Alabama, Tuscaloosa, AL, USA
e-mail: amlewis@as.ua.edu

I. Cheltsov et al. (eds.), *Automorphisms in Birational and Affine Geometry*, Springer Proceedings in Mathematics & Statistics 79,
DOI 10.1007/978-3-319-05681-4_23,

Conjecture 1 (Abhyankar–Sathaye). Let A be a ring of characteristic zero, and let $f \in A^{[n]}$. If $A^{[n]}/(f) \cong A^{[n-1]}$, then f is an A-coordinate.

Conjecture 2 (Dolgachev–Weisfeiler). Suppose $A = \mathbb{C}^{[r]}$, and let $f \in A^{[n]}$. If $A[f] \hookrightarrow A^{[n]}$ is an affine fibration, then f is an A-coordinate.

Conjecture 3 (Stable coordinate conjecture). Suppose $A = \mathbb{C}^{[r]}$, and let $f \in A^{[n]}$. If f is a coordinate in $A^{[n+m]}$ for some $m > 0$, then f is a coordinate in $A^{[n]}$.

The Abhyankar–Sathaye conjecture was shown for A a field and $n = 2$ by Abhyankar and Moh [1] and Suzuki [15], independently; this was later generalized to A a $\mathbb{Q}$-algebra (still $n = 2$) by van den Essen and van Rossum [17]. The $n = 2$ case of the Dolgachev–Weisfeiler conjecture follows from the results of Asanuma [2] and Hamann [7]. The case where both $n = 3$ and $A = \mathbb{C}$ follows from a theorem of Sathaye [13]; see [5] for more details on the background of the Dolgachev–Weisfeiler conjecture. The stable coordinates conjecture is known for $n = 2$, due to van den Essen and van Rossum [17]; van den Essen [16] also showed the $r = 0$ part of the $n = 3$ stable coordinates conjecture.

There are several examples of polynomials satisfying the hypotheses of these conjectures whose status as a coordinate is unresolved. Many are constructed via a slight variation of the following classical method for constructing exotic automorphisms of $A^{[n]}$: let $x \in A$ be a nonzero divisor. One may easily construct elementary automorphisms (those that fix $n - 1$ variables) of $A_x^{[n]}$; then, one can carefully compose these automorphisms (over A_x) to produce an endomorphism of $A^{[n]}$. It is a simple application of the formal inverse function theorem to see that such maps must, in fact, be automorphisms of $A^{[n]}$. The well-known Nagata map arises in this manner:

$$\begin{aligned}\sigma &= (y + x(xz - y^2), z + 2y(xz - y^2) + x(xz - y^2)^2) \\ &= (y, z + \frac{y^2}{x}) \circ (y + x^2 z, z) \circ (y, z - \frac{y^2}{x}) \qquad (1)\end{aligned}$$

The Nagata map is somewhat atypical, in that it arises as a conjugation. A more typical example is perhaps

$$\begin{aligned}\sigma' &= (y + x(x^2z - y^2), z + 2xyz + 6y^2(x^2z - y^2) \\ &\quad + (1 + 6xy)(x^2z - y^2)^2 + 2(x^2z - y^2)^3) \\ &= (y, z + \frac{2y^3}{x}) \circ (y, z + \frac{y^2}{x^2}) \circ (y + x^3 z, z) \circ (y, z - \frac{y^2}{x^2}) \qquad (2)\end{aligned}$$

These two examples are generalized in Lemma 8 and Theorem 13, respectively. While σ and σ' are generated over $\mathbb{C}[x, x^{-1}]$ by elementary automorphisms, Shestakov and Umirbaev [14] famously proved that they are wild (i.e., not generated by elementary and linear automorphisms) as automorphisms of $\mathbb{C}[x, y, z]$ over $\mathbb{C}$. They are, however, both stably tame ([3]).

When interested in producing exotic polynomials, we may relax the construction somewhat; let y be a variable of $A^{[n]}$, and compose elementary automorphisms of $A_x^{[n]}$ until the resulting map has its y-component in $A^{[n]}$. For example, the Vénéreau polynomial $f = y + x(xz + y(yu + z^2))$ arises as the y-component of the following automorphism over $\mathbb{C}[x, x^{-1}]$

$$\phi = (y + x^2 z, z, u) \circ \left(y, z + \frac{y(yu + z^2)}{x}, u - \frac{2z(yu + z^2)}{x} - y(yu + z^2)^2 \right) \quad (3)$$

This type of construction motivates the following definition:

Definition 1. A polynomial $f \in A[x]^{[n]}$ is called a *strongly residual coordinate* if f is a coordinate over $A[x, x^{-1}]$ and if $\bar{f}$, the image modulo x, is a coordinate over A.

Remark 1. If A is a field, then a strongly residual coordinate is a residual coordinate (i.e., $A[x, f] \hookrightarrow A[x]^{[n]}$ is an affine fibration).

The Vénéreau polynomial is perhaps the most widely known example of a strongly residual coordinate that satisfies the hypotheses of Conjectures 1, 2, and 3 (with $A = \mathbb{C}[x]$), yet it is an open question whether it is a coordinate (see [6, 8, 18], and [11], among others, for more on that particular question).

One may observe that the second automorphism in the above composition (3) is essentially the Nagata map, and is wild over $\mathbb{C}[x, x^{-1}]$. The wildness of this map is a crucial difficulty in resolving the status of the Vénéreau polynomial. Our present goal is to show that a large class of strongly residual coordinates generated by maps that are elementary over $\mathbb{C}[x, x^{-1}]$ are coordinates. Our methods are quite constructive and algorithmic, although the computations can become unwieldy quite quickly. One application is to show that all Vénéreau-type polynomials, a generalization of the Vénéreau polynomial studied by the author in [11], are one-stable coordinates (coming from the fact that the Nagata map is one-stably tame). Additionally, we also very quickly recover a result of Russell (Corollary 6) on coordinates in 3 variables over a field of characteristic zero.

2 Preliminaries

Throughout, we set $R = A[x]$ and $S = R_x = A[x, x^{-1}]$. We adopt the standard notation for automorphism groups of the polynomial ring $A^{[n]} = A[z_1, \ldots, z_n]$:

1. $\mathrm{GA}_n(A)$ denotes the general automorphism group $\mathrm{Aut}_{\mathrm{Spec}\,A}(\mathrm{Spec}\,A^{[n]})$, which is anti-isomorphic to $\mathrm{Aut}_A\,A^{[n]}$ (some authors choose to define it as the latter).
2. $\mathrm{EA}_n(A)$ denotes the subgroup generated by the elementary automorphisms; that is, those fixing $n-1$ variables.
3. $\mathrm{TA}_n(A) = \langle \mathrm{EA}_n(A), \mathrm{GL}_n(A) \rangle$ is the tame subgroup.

4. $D_n(A) \leq GL_n(A)$ is the subgroup of diagonal matrices.
5. $P_n(A) \leq GL_n(A)$ is the subgroup of permutation matrices.
6. $GP_n(A) = D_n(A)P_n(A) \leq GL_n(A)$ is the subgroup of generalized permutation matrices.

We also make one non-standard definition when working over $R = A[x]$:

7. $IA_n(R) = \{\phi \in GA_n(R) \mid \phi \equiv \text{id} \pmod x\}$ is the subgroup of all automorphisms that are equal to the identity modulo x. It is the kernel of the natural map $GA_n(R) \to GA_n(A)$.

Remark 2. In fact, the surjection $GA_n(R) \to GA_n(A)$ splits (by the natural inclusion), so we have $GA_n(R) \cong IA_n(R) \rtimes GA_n(A)$.

Definition 2. Let $f_1, \ldots f_m \in R^{[n]}$.

1. $(f_1, \ldots, f_m)$ is called a *partial system of coordinates* (over R) if there exists $g_{m+1}, \ldots, g_n \in R^{[n]}$ such that $(f_1, \ldots, f_m, g_{m+1}, \ldots, g_n) \in GA_n(R)$.
2. $(f_1, \ldots, f_m)$ is called a *partial system of residual coordinates*[1] if $R[f_1, \ldots, f_m] \hookrightarrow R^{[n]}$ is an affine fibration; that is, $R^{[n]}$ is flat over $R[f_1, \ldots, f_m]$ and for each prime ideal $\mathfrak{p} \in \operatorname{Spec} R[f_1, \ldots, f_m]$, $R^{[n]} \otimes_{R[f_1, \ldots, f_m]} \kappa(\mathfrak{p}) \cong \kappa(\mathfrak{p})^{[n-m]}$.
3. $(f_1, \ldots, f_m)$ is called a *partial system of strongly x-residual coordinates* if $(f_1, \ldots, f_m)$ is a partial system of coordinates over S and $(\bar{f}_1, \ldots, \bar{f}_m)$, the images modulo x, is a partial system of coordinates over $A = \bar{R} = R/xR$. If x is understood, we may, in a slight abuse, simply say *strongly residual coordinate*.

A single polynomial is called a *coordinate* (respectively *residual coordinate*, *strongly residual coordinate*) when $m = 1$ in the above definitions.

The Dolgachev–Weisfeiler conjecture can be stated in this context as

Conjecture 4. Partial systems of residual coordinates are partial systems of coordinates

Similarly, we have

Conjecture 5. Partial systems of strongly residual coordinates are partial systems of coordinates.

Our main focus will be on constructing and identifying strongly residual coordinates that are coordinates, although in some cases our methods will generalize slightly to partial systems of coordinates. While we lose some generality as compared to considering residual coordinates, we are able to use some very constructive approaches. We first give a short, direct proof of the $n = 2$ case (for coordinates) that shows the flavor of our methods:

[1]Some authors have used the term *x-residual coordinate* instead; however, as the definition does not depend on the choice of the variable x, we will stick with *residual coordinate*.

Theorem 1. *Let A be an integral domain of characteristic zero, and $R = A[x]$. Let $f \in R^{[2]}$ be a strongly residual coordinate. Then f is a coordinate.*

Proof. Since $\bar{f}$ is a coordinate in $\bar{R}^{[2]} = \bar{R}[y,z]$, without loss of generality we may assume $f = y + xQ$ for some $Q \in R[y,z]$. Since f is an S-coordinate, perhaps after composing with an element of $GL_2(S)$, we obtain some $\phi = (y + xQ, z + x^{-t}P) \in GA_2(S)$ with $J\phi = 1$ and $P \in R^{[2]} \setminus xR^{[2]}$. We inductively show that such a map ϕ is elementarily (over S) equivalent to a map with $t \leq 0$, which gives an element of $GA_2(R)$. We compute

$$J\phi = J(y,z) + xJ(Q,z) + x^{1-t}J(Q,P) + x^{-t}J(y,P)$$

Since $J\phi = 1$, we have $xJ(Q,z) + x^{1-t}J(Q,P) + x^{-t}J(y,P) = 0$. Thus, comparing x-degrees, we must have $J(y,P) \in xR^{[2]}$. This means $P = P_0(y) + xP_1$ for some $P_1 \in R^{[2]}$. Then we have $(y, z - x^{-t}P_0(y)) \circ \phi = (y + xQ, z + x^{-t+1}P')$ for some $P' \in R^{[2]}$ by Taylor's formula, allowing us to apply the inductive hypothesis. □

Remark 3. Analogous results for residual coordinates are due to Kambayashi and Miyanishi [9] and Kambayashi and Wright [10].

The $n = 3$ case remains open, with the Vénéreau polynomial providing the most widely known example of a strongly residual coordinate whose status as a coordinate has not been determined.

We next describe some notation necessary to state the most general form of our results.

Definition 3. Given $\tau = (t_1, \ldots, t_n) \in \mathbb{N}^n$, define $A_\tau = R^{[m]}[x^{t_1}z_1, \ldots, x^{t_n}z_n]$. We also set $A_\tau[\hat{z}_k] = A_\tau \cap R^{[m+n]}[\hat{z}_k] = R^{[m]}[x^{t_1}z_1, \ldots, \widehat{x^{t_k}z_k}, \ldots, x^{t_n}z_n]$. We will use $\geq$ to denote the product order on $\mathbb{N}^n$; thus, given $\sigma, \tau \in \mathbb{N}^n$, $\sigma \geq \tau$ if and only if $A_\sigma \subset A_\tau$.

Given $\tau \in \mathbb{N}^n$ and $\phi \in GA_n(R^{[m]})$, we will consider the natural action

$$\phi^\tau := (x^{-t_1}z_1, \ldots, x^{-t_n}z_n) \circ \phi \circ (x^{t_1}z_1, \ldots, x^{t_n}z_n)$$

Note that algebraically, the image of this action this gives us the group $\mathrm{Aut}_{R^{[m]}} A_\tau$; we denote the corresponding automorphism group of $\mathrm{Spec}\, A_\tau$ by $GA_n^\tau(R^{[m]}) \leq GA_n(S^{[m]})$. For any subgroup $H \leq GA_n(R^{[m]})$, we analogously define $H^\tau = \{\phi^\tau \mid \phi \in H\} \leq GA_n^\tau(R^{[m]})$. We will concern ourselves mostly with $EA_n^\tau(R^{[m]})$, $GL_n^\tau(R^{[m]})$, $GP_n^\tau(R^{[m]})$, and $IA_n^\tau(R^{[m]})$.

We also define, choosing variables $R[y_1, \ldots, y_m] = R^{[m]}$,

$$\begin{aligned} IA_{m+n}^\tau(R) := IA_{m+n}^{(\mathbf{0},\tau)}(R) = \{&(y_1, \ldots, y_m, x^{-t_1}z_1, \ldots, x^{-t_n}z_n) \\ &\circ\phi \circ (y_1, \ldots, y_m, x^{t_1}z_1, \ldots, x^{t_n}z_n) \mid \phi \in IA_{m+n}(R)\} \end{aligned}$$

where $(\mathbf{0}, \tau) = (0, \ldots, 0, t_1, \ldots, t_n) \in \mathbb{N}^{m+n}$. Note that $\mathrm{IA}^{\tau}_{m+n}(R) \supset \mathrm{IA}^{\tau}_{n}(R^{[m]})$.

Automorphisms in these subgroups can be characterized by the following lemma.

Lemma 2. *Let* $\tau = (t_1, \ldots, t_n) \in \mathbb{N}^n$.

1. *Let* $\alpha \in \mathrm{IA}^{\tau}_{m+n}(R)$. *Then there exist* $F_1, \ldots, F_m, G_1, \ldots, G_n \in A_\tau$ *such that*

$$\alpha = (y_1 + xF_1, \ldots, y_m + xF_m, z_1 + x^{-t_1+1}G_1, \ldots, z_n + x^{-t_n+1}G_n)$$

2. *Let* $\Phi \in \mathrm{EA}^{\tau}_{n}(R^{[m]})$ *be elementary. Then there exists* $P(\hat{z}_k) \in A_\tau[\hat{z}_k]$ *such that*

$$\Phi = (z_1, \ldots, z_{k-1}, z_k + x^{-t_k} P(\hat{z}_k), z_{k+1}, \ldots, z_n)$$

3. *Let* $\gamma \in \mathrm{GL}^{\tau}_{n}(R^{[m]})$. *Then there exists* $a_{ij} \in R^{[m]} \setminus xR^{[m]}$ *such that*

$$\gamma = (a_{11}z_1 + a_{12}x^{t_2-t_1}z_2 + \cdots + a_{1n}x^{t_n-t_1}z_n, \ldots, a_{1n}x^{t_1-t_n}z_1 \\ + \cdots + a_{n-1,n}x^{t_{n-1}-t_n}z_{n-1} + a_{nn}z_n)$$

The rest of the chapter is organized as follows: the most general form of our results is given in Main Theorems 1 and 2 in the next section. Here, we state a couple of less technical versions that are easier to apply. This section concludes with some more concrete applications of these results. The subsequent section consists of a series of increasingly technical lemmas culminating in the two Main Theorems in Sect. 3.3.

Theorem 3. *Let* $\phi \in \mathrm{EA}_n(S^{[m]})$, *and write* $\phi = \Phi_0 \circ \cdots \circ \Phi_q$ *as a product of elementaries. For* $0 \le i \le q$ *define* $\tau_i \in \mathbb{N}^n$ *to be minimal such that* $(\Phi_i \circ \cdots \circ \Phi_q)(A_{\tau_i}) \subset R^{[m+n]}$. *Let* $\alpha \in \mathrm{IA}^{\tau_0}_{n+m}(R)$, *and set* $\theta = \alpha \circ \phi$. *Suppose also that either*

1. *A is an integral domain and* $n = 2$, *or*
2. $\Phi_i \in \mathrm{EA}^{\tau_i}_{n}(R^{[m]})$ *for* $0 \le i \le q$

Then the partial system of strongly residual coordinates $(\theta(y_1), \ldots, \theta(y_m))$ *is a partial system of coordinates over R. Moreover, if A is a regular domain and* $\alpha \in \mathrm{TA}_{m+n}(S)$, *then* $(\theta(y_1), \ldots, \theta(y_m))$ *can be extended to a stably tame automorphism over R.*

Proof. If we assume hypothesis 1, the theorem follows immediately from Main Theorem 2. If we instead assume the second hypothesis, we need only to show that $\tau_0 \ge \cdots \ge \tau_q$, as then the result follows from Main Theorem 1. Let $i < q$. Since $\Phi_i \in \mathrm{EA}^{\tau_i}_{n}(R^{[m]})$, we have $\Phi_i(A_{\tau_i}) = A_{\tau_i}$. Then $(\Phi_i \circ \cdots \circ \Phi_q)(A_{\tau_i}) = (\Phi_{i+1} \circ \cdots \circ \Phi_q)(A_{\tau_i}) \subset R^{[m+n]}$. Then the minimality assumption on τ_{i+1} immediately implies $\tau_i \ge \tau_{i+1}$ as required. □

It is often more practical to rephrase the general ($n > 2$) case in the following way:

Theorem 4. *Let $\phi \in \mathrm{EA}_n(S^{[m]})$, and write $\phi = \Phi_0 \circ \cdots \circ \Phi_q$ as a product of elementaries. Define inductively the sequence $(\sigma_0, \ldots, \sigma_q)$, which we call the induced σ-sequence, by letting $\sigma_{q+1} = \mathbf{0} \in \mathbb{N}^n$ and setting each $\sigma_i \in \mathbb{N}^n$ to be minimal such that $\Phi_i(A_{\sigma_i}) \subset A_{\sigma_{i+1}}$ (for $0 \le i \le q$). Let $\alpha \in \mathrm{IA}_{n+m}^{\sigma_0}(R)$, and set $\theta = \alpha \circ \phi$. Then the partial system of strongly residual coordinates $(\theta(y_1), \ldots, \theta(y_m))$ is a partial system of coordinates over R. Moreover, if A is a regular domain and $\alpha \in \mathrm{TA}_{m+n}(S)$, then $(\theta(y_1), \ldots, \theta(y_m))$ can be extended to a stably tame automorphism over R.*

Proof. We will use the definition of σ_i to show the following two facts:

1. $\Phi_i \in \mathrm{EA}_n^{\sigma_i}(R^{[m]})$
2. $\sigma_0 \ge \cdots \ge \sigma_q$

Once these are shown, we can apply Main Theorem 1 to achieve the result. To see these two facts, write $\sigma_i = (s_{i,1}, \ldots, s_{i,n})$. Without loss of generality, suppose Φ_i is elementary in z_1, and write

$$\Phi_i = (z_1 + x^{-s} P(x^{s_{i+1,2}} z_2, \ldots, x^{s_{i+1,n}} z_n), z_2, \ldots, z_n)$$

for some $P(\hat{z}_1) \in A_{\sigma_{i+1}}[\hat{z}_1] \setminus x A_{\sigma_{i+1}}[\hat{z}_1]$. Clearly, the minimality condition on σ_i guarantees $s_{i,k} = s_{i+1,k}$ for $k = 2, \ldots, n$. Since $\Phi_i(x^{s_{i,1}} z_1) = x^{s_{i,1}} z_1 + x^{s_{i,1}-s} P(\hat{z}_1) \in A_{\sigma_{i+1}} \setminus x A_{\sigma_{i+1}}$, we see $s_{i,1} \ge s_{i+1,1}$ (giving $\sigma_i \ge \sigma_{i+1}$) and $s \le s_{i,1}$. From the latter, one easily sees that $\Phi_i = (x^{-s_{i,1}} z_1, \ldots, x^{-s_{i,n}} z_n) \circ (z_1 + x^{s_{i,1}-s} P(z_2, \ldots, z_n), z_2, \ldots, z_n) \circ (x^{s_{i,1}} z_1, \ldots, x^{s_{i,n}} z_n) \in \mathrm{EA}_n^{\sigma_i}(R^{[m]})$. □

The remainder of this section is devoted to consequences of these three theorems in more concrete settings.

Example 1. Let $m = 1$ and $n = 1$. Set

$$\alpha = (y + x^2 z, z) \qquad \Phi_0 = \left(y, z - \frac{y^2}{x}\right)$$

Theorem 4 implies $(\alpha \circ \Phi_0)(y) = y + x(xz - y^2)$ is a coordinate. The construction produces the Nagata map

$$\sigma = (y + x(xz - y^2), z + 2y(xz - y^2) + x(xz - y^2)^2)$$

Example 2. Let $m = 1$, $n = 1$, and $R = k[x, t]$. Set

$$\alpha = (y + x^2 z, z) \qquad \Phi_0 = \left(y, z + \frac{yt}{x}\right)$$

Theorem 4 implies $y + x(xz + yt)$ is a coordinate. The construction produces Anick's example

$$\beta = (y + x(xz + yt), z - t(xz + yt))$$

In [11], a generalization of the Vénéreau polynomial called Vénéreau-type polynomials were studied by the author. They are polynomials of the form $y + xQ(xz+y(yu+z^2), x^2u-2xz(yu+z^2)-y(yu+z^2)^2) \in \mathbb{C}[x,y,z,u]$ where $Q \in \mathbb{C}[x]^{[2]}$. Many Vénéreau-type polynomials remain as strongly residual coordinates that have not been resolved as coordinates. However, we are able to show them all to be 1-stable coordinates, generalizing Freudenburg's result [6] that the Vénéreau polynomial is a 1-stable coordinate.[2]

Corollary 5. *Every Vénéreau-type polynomial is a 1-stable coordinate.*

Proof. Let $Q \in \mathbb{C}[x][xz, x^2u]$, and set

$$\alpha = (y + xQ, z, u, t)$$

$$\Phi_0 = (y, z + yt, u, t) \qquad \Phi_3 = (y, z - yt, u, t)$$

$$\Phi_1 = (y, z, u - 2zt - yt^2, t) \qquad \Phi_4 = (y, z, u - 2zt + yt^2, t)$$

$$\Phi_2 = (y, z, u, t + \frac{yu + z^2}{x})$$

A direct computation shows that $(\alpha \circ \Phi_0 \circ \cdots \circ \Phi_4)(y) = y + xQ(xz + y(yu + z^2), x^2u - 2xz(yu + z^2) - y(yu + z^2)^2)$ is an arbitrary Vénéreau-type polynomial. We compute the induced σ-sequence $(1,2,1) \geq (0,2,1) \geq (0,0,1) \geq (0,0,0) \geq (0,0,0)$, and note that since $Q \in A_{\sigma_0}$, then $\alpha \in \mathrm{IA}_4^{\sigma_0}(\mathbb{C}[x])$. It then follows immediately from Theorem 4 (with $m = 1$, $n = 3$, $y_1 = y$, $z_1 = z$, $z_2 = u$, and $z_3 = t$) that any Vénéreau-type polynomial is a $\mathbb{C}[x]$-coordinate in $\mathbb{C}[x][y,z,u,t]$. □

The following result is first due to Russell [12] and later appeared also in [4].

Corollary 6. *Let k be a field, and let $P \in k[x,y,z]$ be of the form $P = y + xf(x,y) + \lambda x^s z$ for some $s \in \mathbb{N}$, $\lambda \in k^*$ and $f \in k[x,y]$. Then P is a $k[x]$-coordinate.*

Proof. Here $R = k[x]$ and $S = k[x,x^{-1}]$. Let $\theta = (y + \lambda x^s z, z) \circ (y, z + \lambda^{-1}x^{1-s}f(x,y)) \in \mathrm{EA}_3(S)$. Then Theorem 3 yields $\theta(y)$ is a $k[x]$-coordinate, and one easily checks that $\theta(y) = P$. □

3 Main Results

The two main theorems can be found in Sect. 3.3. Sections 3.1 and 3.2 contain a series of technical results necessary for the proofs of the main theorems. Section 3.1 contains the tools necessary to prove Theorem 27, a slightly stronger version of

[2]Our construction provides a different coordinate system than Freudenburg's.

Main Theorem 1 which is useful in proving Main Theorem 2. Section 3.2 contains additional calculations that are needed to prove Main Theorem 2.

In Sect. 3.1, the reader will recognize Lemma 8 as generalizing the Nagata construction (1), while Theorem 13 is a generalization of the Nagata-like construction (2). We also prove several lemmas allowing us to transpose a composition of elements of two subgroups by modifying only one of the two elements; see Corollaries 9 and 18. The section concludes with some basic results about $\mathrm{GP}_n(S^{[m]})$ which are needed for Theorem 27.

Section 3.2 may be skipped by the reader interested only in Theorem 4 and Main Theorem 1. In it, we develop the necessary tools to handle the $n = 2$ case when the elementary automorphisms are allowed to be any elements of $\mathrm{EA}_2(S^{[m]})$ (as opposed to the general case of Main Theorem 1, where we require $\Phi_i \in \mathrm{GA}_n^{\tau_i}(S^{[m]})$ with $\tau_i \in \mathbb{N}^n$ and $\tau_0 \geq \cdots \geq \tau_q$). The key induction step in Main Theorem 1 relies on the $\tau_i \geq \tau_{i+1}$ hypothesis. To prove Main Theorem 2, the basic idea is to replace the given composition with a different (possibly shorter) composition that does satisfy a hypothesis of the form $\tau_0 \geq \cdots \geq \tau_q$, and then proceed as in Main Theorem 1.

We use Lemmas 22 and 25 to show that the only obstacles to this replacement process are elements of $\mathrm{EA}_2^{\tau_i}(R^{[m]}) \cap \mathrm{GL}_2^{\tau_i}(R^{[m]})$. In fact, the root of the problem is that we could end up with an element of $\mathrm{GP}_n(S^{[m]})$. We thus use Lemma 23 to distill these out of any elements of $\mathrm{EA}_2^{\tau_i}(R^{[m]}) \cap \mathrm{GL}_2^{\tau_i}(R^{[m]})$. We then use the "bookkeeping" results Lemma 19, Corollary 20, and Lemma 21 to keep track of elements of $\mathrm{GP}_n(S^{[m]})$ as we push them out of the way, allowing us to substitute a suitable composition satisfying the hypothesis of Theorem 27.

3.1 The General Case

We begin with an observation that follows immediately from Taylor's formula.

Lemma 7. *Let $\tau \in \mathbb{N}^n$ and let $P \in A_\tau$.*

1. *If $\phi \in \mathrm{GA}_n^\tau(R^{[m]})$, then $\phi(A_\tau) = A_\tau$.*
2. *If $\alpha \in \mathrm{IA}_{m+n}^\tau(R)$, then $\alpha(P) - P \in xA_\tau$.*

Next, we note that $\mathrm{GA}_n^\tau(R^{[m]})$ is contained in the normalizer of $\mathrm{IA}_{m+n}^\tau(R)$ in $\mathrm{GA}_{m+n}(S)$. This is slightly more general than the fact that $\mathrm{IA}_n^\tau(R^{[m]}) \lhd \mathrm{GA}_n^\tau(R^{[m]})$.

Lemma 8. *Let $\tau \in \mathbb{N}^n$. Then $\mathrm{IA}_{m+n}^{(\mathbf{0},\tau)}(R) \lhd \mathrm{GA}_{m+n}^{(\mathbf{0},\tau)}(R)$. In particular, for any $\alpha \in \mathrm{IA}_{m+n}^\tau(R)$ and $\phi \in \mathrm{GA}_n^\tau(R^{[m]})$, we have $\phi^{-1} \circ \alpha \circ \phi \in \mathrm{IA}_{m+n}^\tau(R)$.*

Proof. Simply note that the surjection $R = A[x] \to A$ induces a short exact sequence

$$0 \to \mathrm{IA}_{m+n}^{(\mathbf{0},\tau)}(R) \to \mathrm{GA}_{m+n}^{(\mathbf{0},\tau)}(R) \to \mathrm{GA}_{m+n}^{(\mathbf{0},\tau)}(A) \to 0 \quad (4)$$

Here, we are viewing $\mathrm{GA}_{m+n}(A) \leq \mathrm{GA}_{m+n}(R)$ by extension of scalars, and thus obtaining $\mathrm{GA}_{m+n}^{(\mathbf{0},\tau)}(A) \leq \mathrm{GA}_{m+n}^{(\mathbf{0},\tau)}(R)$. □

As a result, we have the ability to "push" an element of $\mathrm{GA}_n^{\tau}(R^{[m]})$ past an element of $\mathrm{IA}_{m+n}^{\tau}(R)$, at the price of replacing with a different element of $\mathrm{IA}_{m+n}^{\tau}(R)$.

Corollary 9. *Let $\tau \in \mathbb{N}^n$, $\alpha \in \mathrm{IA}_{m+n}^{\tau}(R)$, and $\phi \in \mathrm{GA}_n^{\tau}(R^{[m]})$. Then there exists $\alpha' \in \mathrm{IA}_{m+n}^{\tau}(R)$ such that $\alpha \circ \phi = \phi \circ \alpha'$.*

Next, we show how we can use elementaries to transform elements of $\mathrm{IA}_{m+n}^{\tau}(R)$ into elements of $\mathrm{IA}_{m+n}^{\sigma}(R)$ for any $\sigma \leq \tau$.

Lemma 10. *Let $\tau \in \mathbb{N}^n$ and $\alpha \in \mathrm{IA}_{m+n}^{\tau}(R)$. Then there exists $\phi \in \mathrm{EA}_n^{\tau}(R^{[m]}) \cap \mathrm{IA}_n^{\tau}(R^{[m]})$ such that*

$$\phi \circ \alpha \in \bigcap_{0 \leq \sigma \leq \tau} \mathrm{IA}_{m+n}^{\sigma}(R)$$

Proof. We begin by writing $\tau = (t_1, \ldots, t_n) \in \mathbb{N}^n$ and, using Lemma 2,

$$\alpha = (y_1 + xF_1, \ldots, y_m + xF_m, z_1 + x^{-t_1+1}Q_1, \ldots, z_n + x^{-t_n+1}Q_n) \tag{5}$$

for some $F_1, \ldots, F_m, Q_1, \ldots, Q_n \in A_\tau$. We prove the following by induction.

Claim 11. For any $\sigma' = (s_1, \ldots, s_n) \in \mathbb{N}^n$, there exists $\phi \in \mathrm{EA}_n^{\tau}(R^{[m]}) \cap \mathrm{IA}_n^{\tau}(R^{[m]})$ such that $\phi \circ \alpha \in \mathrm{IA}_{m+n}^{\tau}(R)$ is of the form

$$\phi\circ\alpha = (y_1+xF_1, \ldots, y_m+xF_m, z_1+xz_1G_1+x^{-t_1+s_1+1}H_1, \ldots, z_n+xz_nG_n+x^{-t_n+s_n+1}H_n)$$

for some $G_1, H_1, \ldots, G_n, H_n \in A_\tau$.

Clearly the case $\sigma' = \tau$ proves the lemma, since $\sigma \leq \tau$ implies $A_\tau \subset A_\sigma$. We induct on σ' in the partial ordering of $\mathbb{N}^n$. Our base case of $\sigma' = (0, \ldots, 0)$ is provided by $\phi = \mathrm{id}$ (from (5)).

Suppose the claim holds for $\sigma' \in \mathbb{N}^n$. We will show that this implies the claim for $\sigma' + e_k$, where e_k is the k-th standard basis vector of $\mathbb{N}^n$. Without loss of generality, we take $k = 1$, so $e_1 = (1, 0, \ldots, 0)$. By the inductive hypothesis, we may write

$$\alpha' := \phi \circ \alpha = (y_1 + xF_1, \ldots, y_m + xF_m, z_1 + xz_1G_1 + x^{-t_1+s_1+1}H_1, \ldots, z_n + xz_nG_n + x^{-t_n+s_n+1}H_n)$$

for some $G_i, H_i \in A_\tau$ and $\phi \in \mathrm{EA}_n^{\tau}(R^{[m]}) \cap \mathrm{IA}_n^{\tau}(R^{[m]})$. Write $H_1 = P(\hat{z}_1) + x^{t_1}z_1Q$ for some $Q \in A_\tau$ and $P(\hat{z}_1) \in A_\tau[\hat{z}_1]$. Then we may set $\phi' = (z_1 - x^{-t_1+s_1+1}P(\hat{z}_1), z_2, \ldots, z_n) \in \mathrm{EA}_n^{\tau}(R^{[m]}) \cap \mathrm{IA}_n^{\tau}(R^{[m]})$ and compute

$$
\begin{aligned}
(\phi' \circ \alpha')(z_1) &= z_1 + xz_1G_1 + x^{-t_1+s_1+1}(H_1 - \alpha'(P(\hat{z}_1))) \\
&= z_1 + xz_1G_1 + x^{-t_1+s_1+1}(P(\hat{z}_1) + x^{t_1}z_1Q - \alpha'(P(\hat{z}_1)) \\
&= z_1 + xz_1(G_1 + x^{s_1}Q) + x^{-t_1+s_1+1}(P(\hat{z}_1) - \alpha'(P(\hat{z}_1))) \quad (6)
\end{aligned}
$$

Since $\alpha' \in \mathrm{IA}^{\tau}_{m+n}(R)$, we can write (by Lemma 7) $\alpha'(P(\hat{z}_1)) = P(\hat{z}_1) - xH_1'$ for some $H_1' \in A_\tau$. We also set $G_1' = G_1 + x^{s_1}Q \in A_\tau$, and thus clearly see from (6) that $\phi' \circ \alpha'$ is of the required form:

$$
\begin{aligned}
\phi' \circ \alpha' = (&y_1 + xF_1, \ldots, y_m + xF_m, z_1 + xz_1G_1' + x^{-t_1+(s_1+1)+1}H_1', \\
&z_2 + xz_2G_2 + x^{-t_2+s_2+1}H_2, \ldots, z_n + xz_nG_n + x^{-t_n+s_n+1}H_n)
\end{aligned}
$$

□

Corollary 12. *Let $\sigma \leq \tau \in \mathbb{N}^n$, and let $\alpha \in \mathrm{IA}^{\tau}_{m+n}(R)$. Then there exists $\beta \in \mathrm{IA}^{\sigma}_{m+n}(R)$ and $\phi \in \mathrm{EA}^{\tau}_n(R^{[m]})$ such that $\alpha = \beta \circ \phi$. Moreover, if $\tau - \sigma = (0, \ldots, 0, \delta, 0, \ldots, 0)$, then ϕ can be taken to be elementary.*

Proof. Applying Lemma 10 to α^{-1}, we obtain $\phi \in \mathrm{EA}^{\tau}_n(R^{[m]})$ such that $\phi \circ \alpha^{-1} = \beta^{-1}$ for some $\beta \in \mathrm{IA}^{\sigma}_{m+n}(R)$. Rearranging yields $\alpha = \beta \circ \phi$ as required. □

The next theorem provides the crucial inductive step necessary to prove Theorem 27, and thus Main Theorem 1.

Theorem 13. *Let $\tau \in \mathbb{N}^n$, $\alpha \in \mathrm{IA}^{\tau}_{m+n}(R)$, and let $\phi \in \mathrm{GA}^{\tau}_n(R^{[m]})$. Then there exists $\tilde{\phi} \in \langle \phi, \mathrm{EA}^{\tau}_n(R^{[m]}) \rangle$ such that*

$$\tilde{\phi} \circ \alpha \circ \phi \in \bigcap_{0 \leq \sigma \leq \tau} \mathrm{IA}^{\sigma}_{m+n}(R)$$

In particular, $\tilde{\phi} \circ \alpha \circ \phi \in \mathrm{IA}_{m+n}(R)$; and if $\phi, \alpha \in \mathrm{TA}_{m+n}(S)$, then $\tilde{\phi} \circ \alpha \circ \phi \in \mathrm{TA}_{m+n}(S)$ as well.

Proof. By Lemma 8, $\phi^{-1} \circ \alpha \circ \phi \in \mathrm{IA}^{\tau}_{m+n}(R)$. But then by Lemma 10, there exists $\psi \in \mathrm{EA}^{\tau}_n(R^{[m]})$ such that

$$\psi \circ (\phi^{-1} \circ \alpha \circ \phi) \in \bigcap_{0 \leq \sigma \leq \tau} \mathrm{IA}^{\sigma}_{m+n}(R)$$

So we simply set $\tilde{\phi} = \psi \circ \phi^{-1} \in \langle \phi, \mathrm{EA}^{\tau}_n(R^{[m]}) \rangle$ to obtain the desired result. □

At this point, one could go ahead and directly prove Main Theorem 1. However, it will be useful in proving Main Theorem 2 to have the stronger result of Theorem 27 (which immediately implies Main Theorem 1). The rest of this section is devoted to studying $\mathrm{GP}_n(S^{[m]})$ and its relation with other subgroups, which we will need to prove Theorem 27.

Definition 4. Let A be a connected, reduced ring. Given $\rho \in \mathrm{GP}_n(S^{[m]})$, we can then write $\rho = (\lambda_{\sigma(1)} x^{r_{\sigma(1)}} z_{\sigma(1)}, \ldots, \lambda_{\sigma(n)} x^{r_{\sigma(n)}} z_{\sigma(n)})$ for some permutation $\sigma \in \mathfrak{S}_n$, $\lambda_i \in A^*$, and $r_i \in \mathbb{Z}$. If we are also given $\tau = (t_1, \ldots, t_n) \in \mathbb{N}^n$, we can define $\rho(\tau) = (t_{\sigma^{-1}(1)} + r_1, \ldots, t_{\sigma^{-1}(n)} + r_n) \in \mathbb{Z}^n$.

Remark 4. The condition that A is connected and reduced is essential to obtain $(A^{[m]}[x, x^{-1}])^* = \{\lambda x^r \mid \lambda \in A^*, r \in \mathbb{Z}\}$, which is what allows us to write ρ in the given form.

The definition of $\rho(\tau)$ is chosen precisely so that the following lemma holds.

Lemma 14. *Let A be a connected, reduced ring, let $\rho \in \mathrm{GP}_n(S^{[m]})$ and let $\tau \in \mathbb{N}^n$. Then $\rho(A_\tau) = A_{\rho(\tau)}$.*

Proof. Write $\rho = (\lambda_{\sigma(1)} x^{r_{\sigma(1)}} z_{\sigma(1)}, \ldots, \lambda_{\sigma(n)} x^{r_{\sigma(n)}} z_{\sigma(n)})$ as in Definition 4, and let $\tau = (t_1, \ldots, t_n)$. Then

$$\begin{aligned}
\rho(A_\tau) &= A\left[x^{t_1}\left(\lambda_{\sigma(1)} x^{r_{\sigma(1)}} z_{\sigma(1)}\right), \ldots, x^{t_n}\left(\lambda_{\sigma(n)} x^{r_{\sigma(n)}} z_{\sigma(n)}\right)\right] \\
&= A\left[\lambda_1 x^{r_1 + t_{\sigma^{-1}(1)}} z_1, \ldots, \lambda_n x^{r_n + t_{\sigma^{-1}(n)}} z_n\right] \\
&= A\left[x^{r_1 + t_{\sigma^{-1}(1)}} z_1, \ldots, x^{r_n + t_{\sigma^{-1}(n)}} z_n\right] \\
&= A_{\rho(\tau)}
\end{aligned}$$

□

Recall that for $\tau = (t_1, \ldots, t_n) \in \mathbb{N}^n$, one obtains ϕ^τ from ϕ via conjugation by $(x^{t_1} z_1, \ldots, x^{t_n} z_n) \in \mathrm{GP}_n(S^{[m]})$. Then, recalling that for any subgroup $H \leq \mathrm{GA}_n(R^{[m]})$, $H^\tau = \{\phi^\tau \mid \phi \in H\}$, we immediately see the following.

Lemma 15. *Let A be a connected, reduced ring. Let $H \leq \mathrm{GA}_n(R^{[m]})$ be any subgroup whose normalizer contains $\mathrm{GP}_n(R^{[m]})$. Let $\tau \in \mathbb{N}^n$, $\rho \in \mathrm{GP}_n(S^{[m]})$, and $\phi \in \mathrm{GA}_n(S^{[m]})$. Then $\phi \in H^\tau$ if and only if $\rho^{-1} \circ \phi \circ \rho \in H^{\rho(\tau)}$.*

Proof. Since H is closed under conjugation by $\mathrm{GP}_n(R^{[m]})$, from Definition 4 it suffices to assume $\rho = (x^{r_1} z_1, \ldots, x^{r_n} z_n)$ for some $r_i \in \mathbb{Z}$; then if $\tau = (t_1, \ldots, t_n)$, $\rho(\tau) = (t_1 + r_1, \ldots, t_n + r_n)$. Now the claim is immediate from the definition of H^τ. □

Remark 5. Observe that $\mathrm{IA}_{m+n}(R)$ and $\mathrm{EA}_n(R^{[m]})$ are closed under conjugation by $\mathrm{GP}_n(R^{[m]})$.

The next three corollaries can be thought of as tools for pushing elements of $\mathrm{GP}_n(R^{[m]})$ out of the way.

Corollary 16. *Let A be a connected, reduced ring, let $\tau \in \mathbb{N}^n$, and let $\alpha \in \mathrm{IA}^\tau_{m+n}(R)$ and $\rho \in \mathrm{GP}_n(S^{[m]})$. Then $\rho^{-1} \circ \alpha \circ \rho \in \mathrm{IA}^{\rho(\tau)}_{m+n}(R)$.*

Proof. One need only consider $(\mathbf{0}, \tau) \in \mathbb{N}^{m+n}$ and view $\rho \in \mathrm{GP}_{n+m}(S)$; it then follows from the previous lemma. □

Corollary 17. *Let A be a connected, reduced ring, let $\tau \in \mathbb{N}^n$, let $\Phi \in \mathrm{EA}_n^{\tau}(R^{[m]})$ be elementary, and let $\rho \in \mathrm{GP}_n(S^{[m]})$. Then $\Phi \in \mathrm{EA}_n^{\rho(\tau)}(R^{[m]})$ if and only if $\rho \circ \Phi \circ \rho^{-1} \in \mathrm{EA}_n^{\tau}(R^{[m]})$.*

Corollary 18. *Let A be a connected, reduced ring, let $\phi \in \mathrm{EA}_n^{\tau}(R^{[m]})$ and $\rho \in \mathrm{GP}_n(S^{[m]})$. Then there exists $\phi' \in \mathrm{EA}_n^{\rho(\tau)}(R^{[m]})$ such that $\phi \circ \rho = \rho \circ \phi'$. Moreover, if ϕ is elementary, then so is ϕ'.*

3.2 The $n = 2$ Case

This section contains the additional tools necessary to prove Main Theorem 2. First, we prove a lemma allowing us to shorten a composition of automorphisms in the more complicated $\rho_i(\tau_i) \leq \tau_{i+1}$ case.

Lemma 19. *Let A be a connected, reduced ring. Let $\tau_0, \ldots, \tau_{q+1} \in \mathbb{N}^n$. Let $\rho_i \in \mathrm{GP}_n(S^{[m]})$, $\alpha_i \in \mathrm{IA}_{m+n}^{\tau_i}(R)$, $\phi_i \in \mathrm{EA}_n^{\tau_{i+1}}(R^{[m]})$ for $0 \leq i < q$, and $\phi_q \in \mathrm{GA}_n(S^{[m]})$. Suppose also that $\rho_i(\tau_i) \leq \tau_{i+1}$ for each $0 \leq i \leq q$. Then there exist $\alpha' \in \mathrm{IA}_{m+n}^{\tau_0}(R)$ and, for each $0 \leq i < q$, $\phi_i' \in \mathrm{EA}_n^{(\rho_{i+1} \circ \cdots \circ \rho_q)(\tau_{i+1})}(R^{[m]})$ such that*

$$\alpha_0 \circ \rho_0 \circ \phi_0 \circ \cdots \circ \alpha_q \circ \rho_q \circ \phi_q = \alpha' \circ (\rho_0 \circ \cdots \circ \rho_q) \circ \phi_0' \circ \phi_1' \circ \cdots \circ \phi_{q-1}' \circ \phi_q$$

Proof. The proof is by induction on q. If $q = 0$, the claim is trivial, so we assume $q > 0$. By the inductive hypothesis, we may assume

$$\alpha_1 \circ \rho_1 \circ \phi_1 \circ \cdots \circ \alpha_q \circ \rho_q \circ \phi_q = \alpha_1' \circ \rho_1' \circ \phi_1' \circ \cdots \circ \phi_{q-1}' \circ \phi_q$$

for some $\alpha_1' \in \mathrm{IA}_{m+n}^{\tau_1}(R)$, $\phi_i' \in \mathrm{EA}_n^{(\rho_{i+1} \circ \cdots \circ \rho_q)(\tau_{i+1})}(R^{[m]})$, and $\rho_1' := \rho_1 \circ \cdots \circ \rho_q$. Note that it now suffices to find $\alpha' \in \mathrm{IA}_n^{\tau_0}(R^{[m]})$ and $\phi_0' \in \mathrm{EA}_2^{\rho_1'(\tau_1)}(R^{[m]})$ such that

$$\alpha_0 \circ \rho_0 \circ \phi_0 \circ \alpha_1' \circ \rho_1' = \alpha' \circ (\rho_0 \circ \rho_1') \circ \phi_0'$$

From Corollary 9, since $\phi_0 \in \mathrm{EA}_n^{\tau_1}(R^{[m]})$ and $\alpha_1' \in \mathrm{IA}_{m+n}^{\tau_1}(R)$, there exists $\tilde{\alpha} \in \mathrm{IA}_{m+n}^{\tau_1}(R)$ such that

$$\phi_0 \circ \alpha_1' = \tilde{\alpha} \circ \phi_0 \tag{7}$$

Then, plugging this $\tilde{\alpha}$ into Corollary 16, we see there exists $\alpha'' \in \mathrm{IA}_{m+n}^{\rho_0^{-1}(\tau_1)}(R)$ such that

$$\rho_0 \circ \tilde{\alpha} = \alpha'' \circ \rho_0 \tag{8}$$

In addition, by Corollary 12, since $\tau_0 \le \rho_0^{-1}(\tau_1)$, there exist $\beta \in \mathrm{IA}_{m+n}^{\tau_0}(R)$ and $\tilde{\phi} \in \mathrm{EA}_n^{\rho_0^{-1}(\tau_1)}(R^{[m]})$ such that

$$\alpha'' = \beta \circ \tilde{\phi} \tag{9}$$

Combining (7), (8), and (9), we have

$$\begin{aligned} \alpha_0 \circ \rho_0 \circ \phi_0 \circ \alpha_1' \circ \rho_1' &= \alpha_0 \circ \rho_0 \circ \tilde{\alpha} \circ \phi_0 \circ \rho_1' \\ &= \alpha_0 \circ \alpha'' \circ \rho_0 \circ \phi_0 \circ \rho_1' \\ &= \alpha_0 \circ \beta \circ \tilde{\phi} \circ \rho_0 \circ \phi_0 \circ \rho_1' \end{aligned} \tag{10}$$

Now by Corollary 18, since $\tilde{\phi} \in \mathrm{EA}_n^{\rho_0^{-1}(\tau_1)}(R^{[m]})$, there exists $\phi' \in \mathrm{EA}_2^{\tau_1}(R^{[m]})$ such that

$$\tilde{\phi} \circ \rho_0 = \rho_0 \circ \phi' \tag{11}$$

Then since $\phi_0, \phi' \in \mathrm{EA}_n^{\tau_1}(R^{[m]})$, we again use Corollary 18 to obtain $\phi_0' \in \mathrm{EA}_2^{\rho_1'(\tau_1)}(R)$ such that

$$(\phi' \circ \phi_0) \circ \rho_1' = \rho_1' \circ \phi_0' \tag{12}$$

Then , combining (11) and (12),

$$\begin{aligned} \alpha_0 \circ \beta \circ \tilde{\phi} \circ \rho_0 \circ \phi_0 \circ \alpha_1' \circ \rho_1' &= \alpha_0 \circ \beta \circ \rho_0 \circ \phi' \circ \phi_0 \circ \rho_1' \\ &= \alpha_0 \circ \beta \circ \rho_0 \circ \rho_1' \circ \phi_0' \end{aligned} \tag{13}$$

From (10) and (13), we obtain

$$\alpha_0 \circ \rho_0 \circ \phi_0 \circ \alpha_1' \circ \rho_1' = \alpha_0 \circ \beta \circ \rho_0 \circ \rho_1' \circ \phi_0'$$

So we simply set $\alpha' = \alpha_0 \circ \beta \in \mathrm{IA}_{m+n}^{\tau_0}(R)$ to achieve the desired result. □

In fact, this same proof gives the following, noting that the hypothesis $\tau_2 - \rho_1(\tau_1) = \delta e_k$ is what implies that the resulting Φ is elementary (recall that e_k is the k-th standard basis vector of $\mathbb{N}^n$):

Corollary 20. *Let A be a connected, reduced ring. Suppose $\tau_1, \tau_2 \in \mathbb{N}^n$, $\alpha_1 \in \mathrm{IA}_{m+n}^{\tau_1}(R)$, $\alpha_2 \in \mathrm{IA}_{m+n}^{\tau_2}(R)$, and $\rho_1, \rho_2 \in \mathrm{GP}_2(S^{[m]})$. If $\tau_2 - \rho_1(\tau_1) = \delta e_k$ for some $1 \le k \le n$ and $\delta \in \mathbb{N}$, then there exist $\alpha' \in \mathrm{IA}_{m+n}^{\tau_1}$, $\rho' \in \mathrm{GP}_2(S^{[m]})$, and elementary $\Phi \in \mathrm{EA}_n^{\rho_2(\tau_2)}(R^{[m]})$ such that*

$$\alpha_1 \circ \rho_1 \circ \alpha_2 \circ \rho_2 = \alpha' \circ \rho' \circ \Phi$$

Next we show a bookkeeping lemma, allowing us to push any generalized permutations ρ_i around as is convenient.

Lemma 21. *Suppose A is a connected, reduced ring. Let $\tau_0, \dots, \tau_q \in \mathbb{N}^n$, $\Phi_0, \dots, \Phi_q \in \mathrm{EA}_n(S^{[m]})$ be elementaries, $\alpha_i \in \mathrm{IA}^{\tau_i}_{m+n}(R)$, and $\rho_i \in \mathrm{GP}_n(S^{[m]})$. Set*

$$\omega_i = \alpha_i \circ \rho_i \circ \Phi_i \circ \cdots \circ \alpha_q \circ \rho_q \circ \Phi_q$$

Also set $\Phi'_i = \rho_i \circ \Phi_i \circ \rho_i^{-1}$. Then the following conditions are equivalent:

1. *Each $\tau_i \in \mathbb{N}^n$ is minimal such that $\omega_i(A_{\tau_i}) \subset R^{[m+n]}$*
2. *Each $\tau_i \in \mathbb{N}^n$ is minimal such that $(\Phi_i \circ \omega_{i+1})(A_{\rho_i(\tau_i)}) \subset R^{[m+n]}$.*
3. *Each $\tau_i \in \mathbb{N}^n$ is minimal such that $(\rho_i \circ \Phi_i \circ \omega_{i+1})(A_{\tau_i}) \subset R^{[m+n]}$.*
4. *Each $\tau_i \in \mathbb{N}^n$ is minimal such that $(\Phi'_i \circ \rho_i \circ \omega_{i+1})(A_{\tau_i}) \subset R^{[m+n]}$.*

Moreover, if the above are satisfied, then, writing $\tau_i = (t_{i,1}, \dots, t_{i,n})$,

1. *If Φ'_i is elementary in z_j, then $(\rho_i \circ \omega_{i+1})(x^{t_{i,k}} z_k) \in R^{[m+n]} \setminus xR^{[m+n]}$ for all $k \neq j$.*
2. *If Φ_i is elementary in z_j, then $\rho_i(\tau_i) - \tau_{i+1} = \delta_i e_j$ for some $\delta_i \in \mathbb{Z}$ (recall $e_j = (0, \dots, 0, 1, 0, \dots, 0)$).*

Proof. The equivalence of (2) and (3) is immediate from the fact that $\rho_i(A_{\tau_i}) = A_{\rho_i(\tau_i)}$. Since $\alpha_i \in \mathrm{IA}^{\tau_i}_{m+n}(R) \subset \mathrm{GA}^{\tau_i}_{m+n}(R)$, we have $\alpha_i(A_{\tau_i}) = A_{\tau_i}$ and thus $\omega_i(A_{\tau_i}) = (\rho_i \circ \Phi_i \circ \omega_{i+1})(A_{\tau_i})$, giving the equivalence of (1) and (3). The equivalence of (3) and (4) follows immediately from the definition of Φ'_i.

Suppose now that the four conditions are satisfied. Suppose also that Φ'_i is elementary in z_j, so that $\Phi'_i(z_k) = z_k$ for $k \neq j$. Then (4) immediately implies $(\rho_i \circ \omega_{i+1})(x^{t_{i,k}} z_k) = (\Phi'_i \circ \rho_i \circ \omega_{i+1})(x^{t_{i,k}} z_k) \in R^{[m+n]} \setminus xR^{[m+n]}$. Now suppose (perhaps instead) that Φ_i is elementary in z_j. Then $(\Phi_i \circ \omega_{i+1})(x^s z_k) = \omega_{i+1}(x^s z_k)$ for $k \neq j$. The minimal s such that this lies in $R^{[m+n]}$ is precisely $t_{i+1,k}$, so we see from (2) that $\rho_i(\tau_i) = \tau_{i+1} + \delta_i e_j$ for some $\delta_i \in \mathbb{Z}$. □

One key difference in the statements of the two main theorems is that in Main Theorem 1, we must assume $\Phi_i \in \mathrm{GA}^{\tau_i}_n(R^{[m]})$, whereas in Main Theorem 2 we simply assume more generally $\Phi_i \in \mathrm{EA}_2(S^{[m]})$. To account for this in the latter case, we use the following criteria to find a τ such that $\Phi_i \in \mathrm{EA}^{\tau}_2(R^{[m]})$.

Lemma 22. *Assume A is an integral domain. Let $\tau = (t_1, t_2) \in \mathbb{N}^2$, and let $\omega \in \mathrm{GA}_{m+n}(S)$ such that $\omega(x^{t_1} z_1), \omega(x^{t_2} z_2) \in R^{[m+2]} \setminus xR^{[m+2]}$. Suppose $\Phi \in \mathrm{EA}_2(S^{[m]})$ is of the form $\Phi = (z_1 + \lambda(x^{t_2} z_2)^d, z_2)$ for some $\lambda \in S^{[m]}$ and $d \in \mathbb{N}$. If $(\Phi \circ \omega)(A_\tau) \subset R^{[m+2]}$, then $\Phi \in \mathrm{EA}^{\tau}_2(R^{[m]})$.*

Proof. Write $\lambda = x^{-r}\mu$ for some $r \in \mathbb{Z}$ and $\mu \in R^{[m]} \setminus xR^{[m]}$. It suffices to show that $r \leq t_1$. We compute

$$(\Phi \circ \omega)(x^{t_1} z_1) = \omega(x^{t_1} z_1) + \mu x^{t_1 - r} \left(\omega(x^{t_2} z_2)\right)^d$$

Since $(\Phi\circ\omega)(A_\tau) \subset R^{[m+2]}$, we must have $(\Phi\circ\omega)(x^{t_1}z_1) \in R^{[m+2]}$. But $\omega(x^{t_1}z_1) \in R^{[m+2]}$ by assumption, so we then have $\mu x^{t_1-r}\left(\omega(x^{t_2}z_2)\right)^d \in R^{[m+2]}$. Since R is a domain and $\mu, \omega(x^{t_2}z_2) \in R^{[m+2]}\backslash xR^{[m+2]}$, we must have $\mu\left(\omega(x^{t_2}z_2)\right)^d \in R^{[m+2]}\backslash xR^{[m+2]}$, and thus $r \le t_1$ as required. □

The next lemma is our key tool for distilling out the troublesome elements of $\mathrm{GP}_2(S^{[m]})$.

Lemma 23. *Let $\tau = (t_1, t_2) \in \mathbb{N}^2$. Let $\Phi \in \mathrm{EA}_2^\tau(R^{[m]})$ be elementary, and let $\beta = (az_1 + bx^{t_2-t_1}z_2, dz_2 + cx^{t_1-t_2}z_1) \in \mathrm{GL}_2^\tau(R^{[m]})$.*

1. *If Φ is elementary in z_1 and either $c = 0$ or $d = 0$, then there exists $\rho \in \mathrm{GP}_2^\tau(R^{[m]})$ and elementary $\Phi' \in \mathrm{EA}_2^\tau(R^{[m]})$ such that $\Phi\circ\beta = \rho\circ\Phi'$.*
2. *If Φ is elementary in z_2 and either $a = 0$ or $b = 0$, then there exists $\rho \in \mathrm{GP}_2^\tau(R^{[m]})$ and elementary $\Phi' \in \mathrm{EA}_2^\tau(R^{[m]})$ such that $\Phi\circ\beta = \rho\circ\Phi'$.*

Proof. Suppose Φ is elementary in z_1 and write $\Phi = (z_1 + x^{-t_1}P(x^{t_2}z_2), z_2)$. First, suppose $c = 0$, so

$$\begin{aligned}\Phi\circ\beta &= (az_1 + bx^{t_2-t_1}z_2 + x^{-t_1}P(dx^{t_2}z_2), dz_2)\\ &= (az_1, dz_2)\circ(z_1 + \frac{1}{a}x^{-t_1}(bx^{t_2}z_2 + P(dx^{t_2}z_2)), z_2)\end{aligned}$$

If instead $d = 0$, then

$$\begin{aligned}\Phi\circ\beta &= (az_1 + bx^{t_2-t_1}z_2 + x^{-t_1}P(cx^{t_1}z_1), cx^{t_1-t_2}z_1)\\ &= (bx^{t_2-t_1}z_2, cx^{t_1-t_2}z_1)\circ(z_1, z_2 + \frac{1}{b}x^{-t_2}(ax^{t_1}z_1 + P(cx^{t_1}z_1)), z_2)\end{aligned}$$

These are both precisely in the desired form. The case where Φ is elementary in z_2 follows similarly. □

Our final preliminary result, Lemma 25, can be thought of as zeroing in on the crucial technical obstacle: namely, elements of $\mathrm{GL}_2(S^{[m]})$ (which can then be further refined by the previous lemma). We make the following definition to aid in its proof.

Definition 5. Let $\tau = (t_1, \ldots, t_n) \in \mathbb{N}$. Note that as in (4), we have $\mathrm{GA}_n(A^{[m]}) \le \mathrm{GA}_n(R^{[m]})$. Then, we can consider $\mathrm{EA}_n(A^{[m]}) \le \mathrm{GA}_n(R^{[m]})$, and define

$$\mathrm{EA}_n^\tau(A^{[m]}) := \{(x^{-t_1}z_1, \ldots, x^{-t_n}z_n)\circ\phi\circ(x^{t_1}z_1, \ldots, x^{t_n}z_n) \mid \phi \in \mathrm{EA}_n(A^{[m]})\} \le \mathrm{GA}_n^\tau(R^{[m]})$$

Given $\Phi \in \mathrm{EA}_n^\tau(R^{[m]})$, we will use $\overline{\Phi}$ to denote its image under the quotient map $\mathrm{EA}_n^\tau(R^{[m]}) \to \mathrm{EA}_n^\tau(A^{[m]})$. That is, if $\Phi = (x^{-t_1}z_1, \ldots, x^{-t_n}z_n)\circ\phi\circ(x^{t_1}z_1, \ldots, x^{t_n}z_n)$ for some $\phi \in \mathrm{EA}_n(R^{[m]})$, we set

$$\overline{\Phi} := (x^{-t_1}z_1, \ldots, x^{-t_n}z_n)\circ\overline{\phi}\circ(x^{t_1}z_1, \ldots, x^{t_n}z_n)$$

where $\bar{\phi} \in \mathrm{EA}_n(A^{[m]})$ is the image of ϕ modulo x. We define other subgroups such as $\mathrm{GL}_2^{\tau}(A^{[m]})$ in a similar way.

We observe in the following lemma that a composition in $\mathrm{EA}_n^{\tau}(R^{[m]})$ is determined by a composition in $\mathrm{EA}_n^{\tau}(A^{[m]})$ and an element of $\mathrm{IA}_n^{\tau}(R^{[m]})$.

Lemma 24. *Let $\tau \in \mathbb{N}$, and let $\Phi_1, \ldots, \Phi_q \in \mathrm{EA}_n^{\tau}(R^{[m]})$. Then there exists $\alpha \in \mathrm{IA}_n^{\tau}(R^{[m]})$ such that $\Phi_1 \circ \cdots \circ \Phi_q = \alpha \circ \overline{\Phi}_1 \circ \cdots \circ \overline{\Phi}_q$.*

Proof. First, note that it suffices to assume each Φ_i is elementary. The key observation is that if $\Phi \in \mathrm{EA}_n^{\tau}(R^{[m]})$ is elementary, then $\Phi \circ \overline{\Phi}^{-1} \in \mathrm{IA}_n^{\tau}(R^{[m]})$. We prove the lemma by induction on q: suppose $\Phi_2 \circ \cdots \circ \Phi_q = \alpha \circ \overline{\Phi}_2 \circ \cdots \circ \overline{\Phi}_q$ for some $\alpha \in \mathrm{IA}_n^{\tau}(R^{[m]})$. Then noting that $\Phi_1 = \beta \circ \overline{\Phi}_1$ for some $\beta \in \mathrm{IA}_n^{\tau}(R^{[m]})$, we can write

$$\begin{aligned}\Phi_1 \circ \cdots \Phi_q &= \beta \circ \overline{\Phi}_1 \circ \alpha \circ \overline{\Phi}_2 \circ \cdots \circ \overline{\Phi}_q \\ &= \beta \circ \alpha' \circ \overline{\Phi}_1 \circ \cdots \circ \overline{\Phi}_q\end{aligned}$$

for some $\alpha' \in \mathrm{IA}_n^{\tau}(R^{[m]})$ by Corollary 9 (applied to $\overline{\Phi}_1^{-1}$). The simple observation that $\beta \circ \alpha' \in \mathrm{IA}_n^{\tau}(R^{[m]})$ completes the proof. □

Now, we are ready to state and prove our final lemma.

Lemma 25. *Suppose A is an integral domain. Let $\tau = (t_1, t_2) \in \mathbb{N}^2$, let $\Phi_1, \ldots, \Phi_q \in \mathrm{EA}_2^{\tau}(R^{[m]})$ be elementaries, and let $\omega \in \mathrm{GA}_2(S^{[m]})$. Assume that the following three conditions hold:*

1. *Either $\omega(x^{t_1} z_1) \in xR^{[m+2]}$ and $\omega(x^{t_2} z_2) \in R^{[m+2]} \setminus xR^{[m+2]}$, or $\omega(x^{t_2} z_2) \in xR^{[m+2]}$ and $\omega(x^{t_1} z_1) \in R^{[m+2]} \setminus xR^{[m+2]}$.*
2. *Setting $\omega_i = \Phi_i \circ \cdots \circ \Phi_q \circ \omega$, we have $\omega_i(x^{t_1} z_1), \omega_i(x^{t_2} z_2) \in R^{[m+2]} \setminus xR^{[m+2]}$ for $1 < i \leq q$*
3. *$\omega_1(x^{t_1} z_1) \in xR^{[m+2]}$*

Then there exists $\alpha \in \mathrm{IA}_2^{\tau}(R^{[m]})$, $\rho \in \mathrm{GP}_2^{\tau}(R^{[m]})$ and elementary $\Phi \in \mathrm{EA}_2^{\tau}(R^{[m]}) \cap \mathrm{GL}_2^{\tau}(R^{[m]})$ such that $\Phi_1 \circ \cdots \circ \Phi_q = \alpha \circ \rho \circ \Phi$.

Proof. We first prove the following claim:

Claim 26. There exist $r \leq q$, nonlinear elementaries $\Phi_1', \ldots, \Phi_r' \in \mathrm{EA}_2^{\tau}(A^{[m]})$, and $\beta_1, \ldots, \beta_r \in \mathrm{GL}_2^{\tau}(A^{[m]})$ such that, setting

$$\omega_k' = \Phi_k' \circ \beta_k \circ \cdots \circ \Phi_r' \circ \beta_r \circ \omega$$

for $1 \leq k \leq r$, and writing $\beta_k = (a_k z_1 + b_k x^{t_1 - t_2} z_2, d_k z_2 + c_k x^{t_1 - t_2} z_1)$ for some $a_k, b_k, c_k, d_k \in A^{[m]}$,

1. There exists $1 = i_1 < i_2 < \cdots < i_{r-1} < i_r = q$ such that, for each $1 \leq k \leq r$,

$$\omega_{i_k} = \alpha_k \circ \rho_k \circ \omega'_k$$

for some $\alpha_k \in \mathrm{IA}_2^\tau(R^{[m]})$ and $\rho_k \in \mathrm{GP}_2^\tau(R^{[m]})$
2. If Φ'_k is elementary in z_1, then either
   a. $\beta_k = \mathrm{id}$ and Φ'_{k+1} is elementary in z_2, or
   b. $c_k \neq 0$ and $d_k \neq 0$.
3. If Φ'_k is elementary in z_2, then either
   a. $\beta_k = \mathrm{id}$ and Φ'_{k+1} is elementary in z_1, or
   b. $a_k \neq 0$ and $b_k \neq 0$.

Proof. By Lemma 24, for each $1 \leq k \leq q$, there exists $\alpha_k \in \mathrm{IA}_2^\tau(R^{[m]})$ such that

$$\omega_k = \alpha_k \circ \overline{\Phi_k} \circ \cdots \circ \overline{\Phi_q} \circ \omega$$

By concatenating adjacent linear elements, we obtain nonlinear elementaries $\Phi'_1, \ldots \Phi'_r \in \mathrm{EA}_2^\tau(A^{[m]})$ and $\beta_1, \ldots, \beta_r \in \mathrm{GL}_2^\tau(A^{[m]}) \cap \mathrm{EA}_2^\tau(A^{[m]})$ such that for each $1 \leq k \leq r$, there exists $1 \leq i_k \leq q$ with

$$\omega_{i_k} = \alpha_k \circ \Phi'_k \circ \beta_k \circ \cdots \circ \Phi'_r \circ \beta_r \circ \omega$$

Now we inductively apply Lemma 23. We assume Φ'_k is elementary in z_1 (the z_2 case follows similarly). If $c_k = 0$ or $d_k = 0$, then by Lemma 23 we can write $\Phi'_k \circ \beta_k = \rho_k \circ \Phi''_k$ for some $\rho_k \in \mathrm{GP}_2^\tau(A^{[m]})$ and nonlinear elementary $\Phi''_k \in \mathrm{EA}_2^\tau(A^{[m]})$. Then, if $k > 1$, we set $\beta'_{k-1} = \beta_{k-1} \circ \rho_k \in \mathrm{GL}_2^\tau(A^{[m]})$. Thus, we have

$$\omega_{i_k} = \alpha_k \circ \rho_k \circ \Phi''_k \circ \beta'_k \circ \cdots \circ \Phi''_r \circ \beta'_r \circ \omega$$

and

$$\omega'_{k-1} = \Phi'_{k-1} \circ \beta'_{k-1} \circ \omega'_k$$

Finally, if $\beta_k = \mathrm{id}$ and Φ'_k, Φ'_{k+1} are elementaries in the same variable, then their composition is also an elementary, and we can thus concatenate the two and shorten our sequence. □

Now, let $\Phi'_k, \beta_k, \alpha_k, \rho_k$ be as in the claim. Since $\alpha_k \in \mathrm{IA}_2^\tau(R^{[m]})$ and $\rho_k \in \mathrm{GP}_2^\tau(A^{[m]})$, then $\omega_{i_k}(z_j)$ and $\omega'_k(z_{\sigma(j)})$ (where $\sigma \in \mathfrak{S}_2$ is the permutation induced by ρ_k) have the same x-degrees; thus , setting $F_k = \omega'_k(x^{t_1}z_1)$ and $G_k = \omega'_k(x^{t_2}z_2)$, we may rewrite our three assumptions as

1. Either $\overline{F_{r+1}} = 0$ and $\overline{G_{r+1}} \neq 0$; or $\overline{G_{r+1}} = 0$ and $\overline{F_{r+1}} \neq 0$.
2. For $1 < k \leq r$, $\overline{F_k} \neq 0$ and $\overline{G_k} \neq 0$
3. Either $\overline{F_1} = 0$ or $\overline{G_1} = 0$

We now use a quick induction to show that 1 and 2 imply for $1 \leq k \leq r+1$, if Φ'_k is elementary in z_1, then $\deg \overline{F_k} > \deg \overline{G_k}$; and if Φ'_k is elementary in z_2, then $\deg \overline{F_k} < \deg \overline{G_k}$. Note that assumption 1 yields the base case of $k = r+1$. Let $k < r$, and without loss of generality, assume Φ'_k is elementary in z_1. We may then write $\Phi'_k = (z_1 + x^{-t_1} P_k(x^{t_2} z_2), z_2)$. Then, since $\omega'_k = \Phi'_k \circ \beta_k \circ \omega'_{k+1}$, we see

$$\overline{F_k} = a_k \overline{F_{k+1}} + b_k \overline{G_{k+1}} + P_k \left(c_k \overline{F_{k+1}} + d_k \overline{G_{k+1}} \right)$$

$$\overline{G_k} = c_k \overline{F_{k+1}} + d_k \overline{G_{k+1}}$$

Note that we may assume (from the claim) that $\deg P_k \geq 2$ for $2 \leq k \leq q$. Recall also that $c_k \neq 0$ and $d_k \neq 0$ from the claim; then by the induction hypothesis, $\deg \overline{F_{k+1}} \neq \deg \overline{G_{k+1}}$, thus

$$\begin{aligned} \deg \overline{G_k} &= \max \left\{ \deg \overline{F_{k+1}}, \deg \overline{G_{k+1}} \right\} \\ \deg \overline{F_k} &= (\deg P_k) \left(\deg \overline{G_k} \right) \\ &> \deg \overline{G_k} \end{aligned}$$

Now we can see that assumption 3 implies that $\Phi'_1 = \mathrm{id}$; for if not, and Φ'_1 is elementary in z_1 (the other case follows similarly), the above argument shows $\deg \overline{F_1} > \deg \overline{G_1}$; but as $\overline{G_1}$ is a nontrivial linear combination of two nonzero polynomials of differing degrees (namely $c_1 \overline{F_2} + d_1 \overline{G_2}$), it is nonzero, contradicting assumption 3.

Thus, we may now write

$$\overline{F_1} = a_1 \overline{F_2} + b_1 \overline{G_2}$$

$$\overline{G_1} = c_1 \overline{F_2} + d_1 \overline{G_2}$$

But since $\deg \overline{F_2} \neq \deg \overline{G_2}$, we can only have $\overline{F_1} = 0$ or $\overline{G_1} = 0$ if $\overline{F_2} = 0$ or $\overline{G_2} = 0$. By assumption 2, we must then have $r = 1$; moreover, β_1 must be triangular, since assumption 3 yields exactly one of $\overline{F_2}, \overline{G_2}$ is nonzero. Then by Lemma 23, we must have $\beta_1 = \rho' \circ \Phi'$ for some $\rho' \in \mathrm{GP}_2^{\tau}(A^{[m]})$ and $\Phi' \in \mathrm{EA}_2^{\tau}(A^{[m]}) \cap \mathrm{GL}_2^{\tau}(A^{[m]})$. Then we have

$$\omega_1 = \alpha_1 \circ \rho_1 \circ \omega'_1 = \alpha_1 \circ \rho_1 \circ \beta_1 \circ \omega = \alpha_1 \circ (\rho_1 \circ \rho') \circ \Phi' \circ \omega$$

which yields the lemma. □

3.3 Main Theorems

We can now state and prove our main theorems.

Main Theorem 1 *Let* $\tau_0 \geq \cdots \geq \tau_q \in \mathbb{N}^n$. *For* $0 \leq i \leq q$, *let* $\Phi_i \in \mathrm{GA}_n^{\tau_i}(R^{[m]})$ *and* $\alpha_i \in \mathrm{IA}_{m+n}^{\tau_i}(R)$. *Set*

$$\psi = \alpha_0 \circ \Phi_0 \circ \cdots \circ \alpha_q \circ \Phi_q$$

Then $(\psi(y_1), \ldots, \psi(y_m))$ *is a partial system of coordinates over* R. *Moreover, if* A *is a regular domain, and* $\alpha_i, \Phi_i \in \mathrm{TA}_{m+n}(S)$ *for* $0 \leq i \leq q$, *then* $(\psi(y_1), \ldots, \psi(y_m))$ *can be extended to a stably tame automorphism of* $R^{[m+n]}$.

Rather than prove this directly, we prove a slightly stronger version below that immediately implies the above statement. The inclusion of the permutation maps ρ_i is not necessary to achieve Main Theorem 1, but will help us in our proof of Main Theorem 2. Note that if we assume each ρ_i is of the form in Definition 4, then we may drop the assumption "A is a connected, reduced ring". In particular, we do not need to assume A is connected and reduced in Main Theorem 1, since we set $\rho_i = \mathrm{id}$ for each i to obtain it from Theorem 27.

Theorem 27. *Let* A *be a connected, reduced ring, and let* $\tau_0, \ldots, \tau_q \in \mathbb{N}^n$. *Let* $\rho_i \in \mathrm{GP}_n(S^{[m]})$, $\gamma_i \in \mathrm{GL}_{m+n}(S)$, $\alpha_i \in \mathrm{IA}_{m+n}^{\tau_i}(R)$, *and* $\Phi_i \in \mathrm{GA}_n^{\rho_i(\tau_i)}(R^{[m]})$ *for each* $0 \leq i \leq q$. *Set*

$$\psi_i = \alpha_0 \circ \rho_0 \circ \Phi_0 \circ \cdots \circ \alpha_i \circ \rho_i \circ \Phi_i$$

Suppose $\rho_i(\tau_i) \geq \tau_{i+1}$ *for each* $0 \leq i \leq q$. *Then for each* $0 \leq i \leq q$, *there exists* $\theta_i \in \mathrm{IA}_{m+n}^{\tau_{i+1}}(R) \cap \mathrm{IA}_{m+n}(R)$ *with* $\theta_i(y_j) = \psi_i(y_j)$ *for each* $1 \leq j \leq m$. *Moreover, if* A *is a regular domain and* $\alpha_k, \Phi_k \in \mathrm{TA}_{m+n}(S)$ *for* $0 \leq k \leq i$, *then* θ_i *is stably tame.*

Proof. The proof is by induction on i. Note that we may use a trivial base case of $i = -1$ and $\theta_{-1} = \mathrm{id}$. So we suppose $i \geq 0$. By the induction hypothesis we have $\theta_{i-1} \in \mathrm{IA}_{m+n}^{\tau_i}(R)$. Thus, $(\theta_{i-1} \circ \alpha_i) \in \mathrm{IA}_{m+n}^{\tau_i}(R)$, and by Corollary 16, $\rho_i^{-1} \circ (\theta_{i-1} \circ \alpha_i) \circ \rho_i \in \mathrm{IA}_{m+n}^{\rho_i(\tau_i)}(R)$. Since $\Phi_i \in \mathrm{GA}_n^{\rho_i(\tau_i)}(R)$ and $\tau_{i+1} \leq \rho(\tau_i)$, we can apply Theorem 13 to obtain $\tilde{\Phi} \in \mathrm{GA}_n^{\rho_i(\tau_i)}(R^{[m]})$ such that

$$\theta_i := \tilde{\Phi} \circ (\rho_i^{-1} \circ \theta_{i-1} \circ \alpha_i \circ \rho_i) \circ \Phi_i \in \mathrm{IA}_{m+n}^{\tau_{i+1}}(R) \cap \mathrm{IA}_{m+n}(R)$$

Noting that $\tilde{\Phi}, \rho_i \in \mathrm{GA}_n(S^{[m]})$ and thus fix each y_j, and by the inductive hypothesis $\theta_{i-1}(y_j) = \psi_{i-1}(y_j)$ we have

$$\begin{aligned}\theta_i(y_j) &= (\tilde{\Phi} \circ \rho_i^{-1} \circ \theta_{i-1} \circ \alpha_i \circ \rho_i \circ \Phi_i)(y_j) \\ &= (\theta_{i-1} \circ \alpha_i \circ \rho_i \circ \Phi_i)(y_j)\end{aligned}$$

$$= (\psi_{i-1} \circ \alpha_i \circ \rho_i \circ \Phi_i)(y_j)$$
$$= \psi_i(y_j)$$

for each $1 \leq j \leq m$. Moreover, if $\alpha_0, \Phi_0, \ldots, \alpha_i, \Phi_i \in \mathrm{TA}_{m+n}(S)$, then the inductive hypothesis along with Theorem 13 guarantee $\theta_i \in \mathrm{TA}_{m+n}(S)$ as well. Noting that since $\theta_i \in \mathrm{IA}_{m+n}(R)$ we have $\theta_i \equiv \mathrm{id} \pmod{x}$, the stable tameness assertion follows immediately from the following result of Berson, van den Essen, and Wright:

Theorem 28 ([3], Theorem 4.5). *Let A be a regular domain, and let $\phi \in \mathrm{GA}_n(R)$ with $J\phi = 1$. If $\phi \in \mathrm{TA}_n(S)$ and $\bar{\phi} \in \mathrm{EA}_n(R/xR)$, then ϕ is stably tame.*

□

Main Theorem 2 *Suppose A is an integral domain. Let $\Phi_0, \ldots, \Phi_q \in \mathrm{EA}_2(S^{[m]})$ be elementaries. Let $\alpha_i \in \mathrm{GA}_{m+2}(S)$ and $\rho_i \in \mathrm{GP}_2(S^{[m]})$ for each $0 \leq i \leq q$. Set*

$$\omega_i = \alpha_i \circ \rho_i \circ \Phi_i \circ \cdots \circ \alpha_q \circ \rho_q \circ \Phi_q$$

and define $\tau_i \in \mathbb{N}^2$ to be minimal such that $\omega_i(A_{\tau_i}) \subset R^{[m+2]}$ for $0 \leq i \leq q$. If $\alpha_i \in \mathrm{IA}^{\tau_i}_{m+2}(R)$ for each $0 \leq i \leq q$, then there exists $\theta \in \mathrm{IA}_{m+2}(R)$ such that $\theta(y_j) = \omega_0(y_j)$. Moreover, if A is a regular domain and $\alpha_0, \ldots, \alpha_q \in \mathrm{TA}_{m+2}(S)$, then θ is stably tame.

The theorem follows from following theorem, which allows us to apply Theorem 27 to $\omega_0 = \tilde{\omega}_0$. By convention, we will let $\tau_{q+1} = 0$.

Theorem 29. *Suppose A is an integral domain. Let $\Phi_0, \ldots, \Phi_q \in \mathrm{EA}_2(S^{[m]})$ be elementaries. Let $\alpha_i \in \mathrm{GA}_{m+2}(S)$ and $\rho_i \in \mathrm{GP}_2(S^{[m]})$ for each $0 \leq i \leq q$. Set*

$$\omega_i = \alpha_i \circ \rho_i \circ \Phi_i \circ \cdots \circ \alpha_q \circ \rho_q \circ \Phi_q$$

and define $\tau_i \in \mathbb{N}^2$ to be minimal such that $\omega_i(A_{\tau_i}) \subset R^{[m+2]}$ for $0 \leq i \leq q$. Assume also that $\alpha_i \in \mathrm{IA}^{\tau_i}_{m+2}(R)$ for each $0 \leq i \leq q$. Then there exists $r \leq q$ and

1. *A sequence $0 = j_0 < j_1 < \cdots < j_{r-1} < j_r = q$*
2. *A sequence $\tilde{\tau}_0, \ldots, \tilde{\tau}_r \in \mathbb{N}^2$*
3. *$\tilde{\rho}_0, \ldots, \tilde{\rho}_r \in \mathrm{GP}_2(S^{[m]})$*
4. *For each $0 \leq i \leq r$, $\tilde{\alpha}_i \in \mathrm{IA}^{\tilde{\tau}_i}_{m+2}(R)$ and $\tilde{\Phi}_i \in \mathrm{EA}_2^{\tilde{\rho}_i(\tilde{\tau}_i)}(R^{[m]})$*

such that

1. *$\tilde{\rho}_i(\tilde{\tau}_i) \geq \tilde{\tau}_{i+1}$ for $0 \leq i \leq r$*
2. *Setting $\tilde{\omega}_i = \tilde{\alpha}_i \circ \tilde{\rho}_i \circ \tilde{\Phi}_i \circ \cdots \circ \tilde{\alpha}_q \circ \tilde{\rho}_q \circ \tilde{\Phi}_q$, each $\tilde{\tau}_i$ is minimal such that $\tilde{\omega}_i(A_{\tilde{\tau}_i}) \subset R^{[m+2]}$, for $a \leq i \leq q$*
3. *$\omega_{j_i} = \tilde{\omega}_i$*

Proof. Note that the original composition satisfies all of the conclusions except for $\Phi_i \in \mathrm{EA}_2^{\rho_i(\tau_i)}(R^{[m]})$ and $\rho_i(\tau_i) \geq \tau_{i+1}$. Let $b < q$ be minimal such that $\rho_b(\tau_b) > \tau_{b+1}$, and $\rho_i(\tau_i) \geq \tau_{i+1}$ for $b \leq i \leq q$ (we will see below that this implies $\Phi_i \in \mathrm{EA}_2^{\rho_i(\tau_i)}(R^{[m]})$ for $b \leq i \leq q$). The proof is by induction downwards on b. Let $a < b$ be maximal such that $\rho_a(\tau_a) < \tau_{a+1}$ (if any such a exist). The case where a does not exist will form the base of our induction. If we do have such an a, we will replace our composition with a shorter composition with a smaller b, allowing us to apply the induction hypothesis.

The subsequent claim (with $c = 0$) yields the base case of our induction:

Claim 30. If $\rho_i(\tau_i) \geq \tau_{i+1}$ for $c \leq i \leq q$, then $\Phi_i \in \mathrm{EA}_2^{\rho_i(\tau_i)}(R^{[m]})$ for $c \leq i \leq q$.

Proof. Note that by Corollary 17 this is equivalent to showing that $\Phi_i' := \rho_i \circ \Phi_i \circ \rho_i^{-1} \in \mathrm{EA}_2^{\tau_i}(R)$ for $c \leq i \leq q$. Fix i, and without loss of generality, write $\Phi_i' = (z_1 + \lambda x^{-s}(x^{t_{i,2}}z_2)^d, z_2)$ for some $\lambda \in R^{[m]} \setminus xR^{[m]}$, $s \in \mathbb{Z}$, and $d \in \mathbb{N}$. Then

$$(\rho_i \circ \Phi_i \circ \omega_{i+1})(x^{t_{i,1}}z_1) = (\Phi_i' \circ \rho_i \circ \omega_{i+1})(x^{t_{i,1}}z_1)$$
$$= (\rho_i \circ \omega_{i+1})(x^{t_{i,1}}z_1) + \lambda x^{t_{i,1}-s}\left((\rho_i \circ \omega_{i+1})(x^{t_{i,2}}z_2)\right)^d \tag{14}$$

Since $\rho_i(\tau_i) \geq \tau_{i+1}$, we have $\rho_i(A_{\tau_i}) = A_{\rho_i(\tau_i)} \subset A_{\tau_{i+1}}$. In particular,

$$(\rho_i \circ \omega_{i+1})(x^{t_{i,1}}z_1) \in (\rho_i \circ \omega_{i+1})(A_{\tau_i}) = \omega_{i+1}(A_{\rho_i(\tau_i)}) \subset \omega_{i+1}(A_{\tau_{i+1}}) \subset R^{[m+2]}$$

with the last containment following from the minimality condition on τ_{i+1}. Observe also that, since $\alpha_i(A_{\tau_i}) = A_{\tau_i}$,

$$(\rho_i \circ \Phi_i \circ \omega_{i+1})(x^{t_{i,1}}z_1) \in (\rho_i \circ \Phi_i \circ \omega_{i+1})(A_{\tau_i}) = (\alpha_i \circ \rho_i \circ \Phi_i \circ \omega_{i+1})(A_{\tau_i})$$
$$= \omega_i(A_{\tau_i}) \subset R^{[m+2]}$$

and therefore, from (14) we see $\lambda x^{t_{i,1}-s}\left((\rho_i \circ \omega_{i+1})(x^{t_{i,2}}z_2)\right)^d \in R^{[m+2]}$ as well. Thus since A is a domain, we have $t_{i,1} \geq s$ since $\lambda \notin (x)$ and $(\rho_i \circ \omega_{i+1})(x^{t_{i,2}}z_2) \in R^{[m+2]} \setminus xR^{[m+2]}$ (by Lemma 21). But $t_{i,1} \geq s$ is precisely the condition that $\Phi_i' \in \mathrm{EA}_2^{\tau_i}(R^{[m]})$ as required. ☐

We thus now assume $b > 0$ and a as defined above exists. Then the maximality of a yields $\rho_i(\tau_i) = \tau_{i+1}$ for $a < i < b$. By the preceding claim, we have $\Phi_i \in \mathrm{EA}_2^{\rho_i(\tau_i)}(R^{[m]})$ for $a + 1 \leq i \leq q$.

Observe that since $\rho_a(\tau_a) < \tau_{a+1}$, then

$$(\Phi_a \circ \omega_{a+1})(A_{\tau_{a+1}}) \subset (\Phi_a \circ \omega_{a+1})(A_{\rho_a(\tau_a)}) = (\alpha_a \circ \rho_a \circ \Phi_a \circ \omega_{a+1})(A_{\tau_a})$$
$$= \omega_a(A_{\tau_a}) \subset R^{[m+2]}$$

Then by Lemma 22 (with $\tau = \tau_{a+1}$, and $\omega = \omega_{a+1}$), we must have $\Phi_a \in \mathrm{EA}_2^{\tau_{a+1}}(R^{[m]})$. Also, since $\rho_i(\tau_i) = \tau_{i+1}$ for $a < i < b$, $\Phi_i \in \mathrm{EA}_2^{\tau_i+1}(R^{[m]})$ for $a \leq i < b$. Then by Lemma 19, it suffices to assume that $\alpha_{a+1} = \cdots = \alpha_b = \mathrm{id}$, $\rho_{a+1} = \cdots = \rho_b = \mathrm{id}$, $\rho_a(\tau_a) < \tau_{a+1} = \cdots = \tau_b > \tau_{b+1}$ and $\Phi_i \in \mathrm{EA}_n^{\tau_b}(R^{[m]})$ for $a \leq i \leq b$.

A priori, it seems we may no longer be able to assume the minimality condition on the τ_i when $a \leq i < b$. However, we may simply replace the τ_i by the minimal τ_i such that $\omega_i(A_{\tau_i}) \subset R^{[m+2]}$ (for $a \leq i < b$). Note that our application of Lemma 19 did not change Φ_b; then, while we may need to increase a (but it will not exceed b), we may still assume $\alpha_{a+1} = \cdots = \alpha_b = \mathrm{id}$, $\rho_{a+1} = \cdots = \rho_b = \mathrm{id}$, $\Phi_a, \ldots, \Phi_b \in \mathrm{EA}_n^{\tau_b}(R^{[m]})$, and

$$\rho_a(\tau_a) < \tau_{a+1} = \cdots = \tau_b > \tau_{b+1}$$

We also now see that

$$\omega_a = \alpha_a \circ \rho_a \circ \Phi_a \circ \Phi_{a+1} \circ \cdots \circ \Phi_b \circ \omega_{b+1} \tag{15}$$

Set

$$\tau_b = (t_1, t_2)$$

for some $t_1, t_2 \in \mathbb{N}$. Without loss of generality, assume Φ_a is elementary in z_1. Then since $\rho_a(\tau_a) < \tau_{a+1} = \tau_b = (t_1, t_2)$, the minimality of τ_a implies $(\Phi_a \circ \cdots \circ \Phi_b \circ \omega_{b+1})(x^{t_1} z_1) \in xR^{[m+2]}$. Then, by Lemma 25, we may assume that $\Phi_a \circ \cdots \circ \Phi_b = \alpha \circ \rho \circ \Phi$ for some $\alpha \in \mathrm{IA}_2^{\tau_{a+1}}(R^{[m]})$, $\rho \in \mathrm{GP}_2^{\tau_{a+1}}(R^{[m]})$ and elementary $\Phi \in \mathrm{EA}_2^{\tau_{a+1}}(R^{[m]}) \cap \mathrm{GL}_2^{\tau_{a+1}}(R^{[m]})$. Then from (15), we see

$$\omega_a = \alpha_a \circ \rho_a \circ \alpha \circ \rho \circ \Phi \circ \omega_{b+1}$$

Noting that $\rho_a(\tau_a) < \tau_{a+1}$, by Corollary 20, we have $\alpha_a \circ \rho_a \circ \alpha \circ \rho = \alpha'_a \circ \rho'_a \circ \Phi'$ for some $\alpha'_a \in \mathrm{IA}_{m+2}^{\tau_a}(R)$, $\rho'_a = \rho_a \circ \rho \in \mathrm{GP}_2(S^{[m]})$, and $\Phi' \in \mathrm{EA}_2^{\tau_{a+1}}(R^{[m]})$ (since $\rho(\tau_{a+1}) = \tau_{a+1}$). Thus we have

$$\omega_a = \alpha'_a \circ \rho'_a \circ \Phi' \circ \Phi \circ \omega_{b+1}$$

First, suppose Φ' and Φ are both elementary in the same variable; then we may set $\tilde{\Phi} = \Phi' \circ \Phi$ and $\tilde{\Phi} \in \mathrm{EA}_2^{\tau_{a+1}}(R^{[m]})$ is elementary, and

$$\omega_a = \alpha'_a \circ \rho'_a \circ \tilde{\Phi} \circ \omega_{b+1}$$

Similarly, if we suppose instead that Φ' and Φ are elementary in different variables, then since $\Phi \in \mathrm{GL}_2^{\tau_{a+1}}(R^{[m]})$, by Lemma 23 there exist $\tilde{\rho} \in \mathrm{GP}_2^{\tau_{a+1}}(R^{[m]})$ and $\tilde{\Phi} \in \mathrm{EA}_2^{\tau_{a+1}}(R^{[m]})$ such that $\Phi' \circ \Phi = \tilde{\rho} \circ \tilde{\Phi}$. Then we have

$$\omega_a = \alpha'_a \circ (\rho'_a \circ \tilde{\rho}) \circ \tilde{\Phi} \circ \omega_{b+1}$$

Now set $r = q-(b-a)$. If $i \leq a$, set $j_i = i$, and if $i > a$, set $j_i = i + (b-a)$. Then, for $i \neq a$, set $\tilde{\Phi}_i = \Phi_{j_i}$, $\tilde{\alpha_i} = \alpha_{j_i}$, and $\tilde{\rho_i} = \rho_{j_i}$. Finally, set $\tilde{\Phi}_a = \tilde{\Phi}$, $\tilde{\alpha}_a = \alpha'_a$, and $\tilde{\rho}_a = \rho'_a \circ \tilde{\rho}$. Note that we now have $\tilde{\omega}_i = \omega_{j_i}$ for $0 \leq i \leq r$. Thus, we have a new composition satisfying all of the conclusions for $i > a$. Since $a < b$, the new minimal b has been reduced, and we may apply the induction hypothesis. □

Acknowledgements The author would like to thank David Wright, Brady Rocks, and Eric Edo for helpful discussions and feedback. The author is also indebted to the referee for several helpful suggestions.

References

1. S.S. Abhyankar, T.T. Moh, Embeddings of the line in the plane. J. Reine Angew. Math. **276**, 148–166 (1975)
2. T. Asanuma, Polynomial fibre rings of algebras over Noetherian rings. Invent. Math. **87**, 101–127 (1987)
3. J. Berson, A. van den Essen, D. Wright, Stable tameness of two-dimensional polynomial automorphisms over a regular ring. Adv. Math. **230**, 2176–2197 (2012)
4. S.M. Bhatwadekar, A. Roy, Some results on embedding of a line in 3-space. J. Algebra **142**, 101–109 (1991)
5. D. Daigle, G. Freudenburg, Families of affine fibrations, in *Symmetry and Spaces*. Progress in Mathematics, vol. 278 (Birkhäuser, Boston, 2010), pp. 35–43
6. G. Freudenburg, The Vénéreau polynomials relative to $\mathbb{C}^*$-fibrations and stable coordinates, in *Affine Algebraic Geometry* (Osaka University Press, Osaka, 2007), pp. 203–215
7. E. Hamann, On the R-invariance of $R[x]$. J. Algebra **35**, 1–16 (1975)
8. S. Kaliman, S. Vénéreau, M. Zaidenberg, Simple birational extensions of the polynomial algebra $\mathbb{C}^{[3]}$. Trans. Am. Math. Soc. **356**, 509–555 (2004) (electronic)
9. T. Kambayashi, M. Miyanishi, On flat fibrations by the affine line. Ill. J. Math. **22**, 662–671 (1978)
10. T. Kambayashi, D. Wright, Flat families of affine lines are affine-line bundles. Ill. J. Math. **29**, 672–681 (1985)
11. D. Lewis, Vénéreau-type polynomials as potential counterexamples. J. Pure Appl. Algebra **217**, 946–957 (2013)
12. P. Russell, Simple birational extensions of two dimensional affine rational domains. Compos. Math. **33**, 197–208 (1976)
13. A. Sathaye, Polynomial ring in two variables over a DVR: A criterion. Invent. Math. **74**, 159–168 (1983)
14. I. P. Shestakov, U.U. Umirbaev, The tame and the wild automorphisms of polynomial rings in three variables. J. Am. Math. Soc. **17**, 197–227 (2004) (electronic)
15. M. Suzuki, Propriétés topologiques des polynômes de deux variables complexes, et automorphismes algébriques de l'espace $\mathbf{C}^2$. J. Math. Soc. Jpn. **26**, 241–257 (1974)
16. A. van den Essen, Three-variate stable coordinates are coordinates. Radboud University Nijmegan Technical Report No. 0503 (2005)
17. A. van den Essen, P. van Rossum, Coordinates in two variables over a $\mathbb{Q}$-algebra. Trans. Am. Math. Soc. **356**, 1691–1703 (2004) (electronic)
18. S. Vénéreau, Automorphismes et variables de l'anneau de polynômes $A[y_1, \ldots, y_n]$. Ph.D. thesis, Université Grenoble I, Institut Fourier, 2001

Configuration Spaces of the Affine Line and their Automorphism Groups

Vladimir Lin and Mikhail Zaidenberg

Abstract The configuration space $\mathcal{C}^n(X)$ of an algebraic curve X is the algebraic variety consisting of all n-point subsets $Q \subset X$. We describe the automorphisms of $\mathcal{C}^n(\mathbb{C})$, deduce that the (infinite dimensional) group Aut $\mathcal{C}^n(\mathbb{C})$ is solvable, and obtain an analog of the Mostow decomposition in this group. The Lie algebra and the Makar-Limanov invariant of $\mathcal{C}^n(\mathbb{C})$ are also computed. We obtain similar results for the level hypersurfaces of the discriminant, including its singular zero level.

2010 Mathematics Subject Classification: 14R20, 32M17.

1 Introduction

Let X be an irreducible smooth algebraic curve over the field $\mathbb{C}$. The nth *configuration space* $\mathcal{C}^n(X)$ of X is a smooth affine variety of dimension n consisting of all n-point subsets $Q = \{q_1, \ldots, q_n\} \subset X$ with distinct $q_1, \ldots, q_n$. We would like to study its biregular automorphisms and the algebraic structure of the group Aut $\mathcal{C}^n(X)$.

For a hyperbolic curve X the group Aut $\mathcal{C}^n(X)$ is finite (possibly, trivial for a generic curve). We are interested in the case where X is non-hyperbolic, i.e., one of the curves $\mathbb{C}$, $\mathbb{P}^1 = \mathbb{P}^1_{\mathbb{C}}$, $\mathbb{C}^* = \mathbb{C} \setminus \{0\}$, or an elliptic curve. In the latter two cases the

V. Lin
Technion-Israel institute of Technology, Haifa 32000, Israel
e-mail: vlin@tx.technion.ac.il

M. Zaidenberg (✉)
Institut Fourier de Mathématiques, Université Grenoble I, Grenoble, France
e-mail: Mikhail.Zaidenberg@ujf-grenoble.fr

I. Cheltsov et al. (eds.), *Automorphisms in Birational and Affine Geometry*, Springer Proceedings in Mathematics & Statistics 79,
DOI 10.1007/978-3-319-05681-4_24,

groups Aut $\mathcal{C}^n(X)$ were described in [35] and [10], respectively. Here we investigate automorphisms of the configuration space $\mathcal{C}^n = \mathcal{C}^n(\mathbb{C})$ and of some related spaces.

All varieties in this chapter are defined over $\mathbb{C}$ and reduced; in general, irreducibility is not required. *Morphism* means a regular morphism of varieties. The same applies to the terms *automorphism* and *endomorphism*. The actions of algebraic groups are assumed to be regular. We use the standard notation $\mathcal{O}(\mathcal{Z})$, $\mathcal{O}_+(\mathcal{Z})$, and $\mathcal{O}^\times(\mathcal{Z})$ for the algebra of all regular functions on a variety $\mathcal{Z}$, the additive group of this algebra, and its group of invertible elements, respectively.

For $z \in \mathbb{C}^n$, let $d_n(z)$ denote the discriminant of the monic polynomial

$$P_n(\lambda, z) = \lambda^n + z_1\lambda^{n-1} + \cdots + z_n\,, \quad z = (z_1, \ldots, z_n) \in \mathbb{C}^n = \mathbb{C}^n_{(z)}\,. \tag{1}$$

If $d_n(z) \neq 0$ and $Q \subset \mathbb{C}$ is the set of all roots of $P_n(\cdot, z)$, then $Q \in \mathcal{C}^n$ and

$$D_n(Q) \stackrel{\text{def}}{=\!=} \prod_{\{q',q''\}\subset Q} (q' - q'')^2 = d_n(z)\,. \tag{2}$$

Denoting by $\mathcal{P}^n$ the space of all polynomials (1) with simple roots, we have the natural identification

$$\mathcal{C}^n = \{Q \subset \mathbb{C} \mid \#Q = n\} \cong \mathcal{P}^n = \mathbb{C}^n_{(z)} \setminus \Sigma^{n-1}\,, \quad Q \leftrightarrow z = (z_1, \ldots, z_n)\,, \tag{3}$$

where [1] the *discriminant variety* Σ^{n-1} is defined by

$$\Sigma^{n-1} \stackrel{\text{def}}{=\!=} \{z \in \mathbb{C}^n \mid d_n(z) = 0\}\,. \tag{4}$$

We describe the automorphisms of the configuration space $\mathcal{C}^n$ for $n > 2$, of the discriminant variety Σ^{n-1} for $n > 6$, and, for $n > 4$, of the *special configuration space*

$$\mathcal{SC}^{n-1} \stackrel{\text{def}}{=\!=} \{Q \in \mathcal{C}^n \mid D_n(Q) = 1\} \cong \{z \in \mathbb{C}^n \mid d_n(z) = 1\}\,. \tag{5}$$

This leads to structure theorems for the automorphism groups Aut $\mathcal{C}^n$, Aut $\mathcal{SC}^{n-1}$, and Aut Σ^{n-1}.

The varieties $\mathcal{C}^n$ and Σ^{n-1} can be viewed as the complementary to each other parts of the symmetric power $\operatorname{Sym}^n \mathbb{C} = \mathbb{C}^n/\mathbf{S}(n)$, where $\mathbf{S}(n)$ is the symmetric group permuting the coordinates $q_1, \ldots, q_n$ in $\mathbb{C}^n = \mathbb{C}^n_{(q)}$. We have the natural projections

$$p\colon \mathbb{C}^n \to \operatorname{Sym}^n \mathbb{C} \cong \mathbb{C}^n_{(z)}\,, \quad \Delta^{n-1} \to \Sigma^{n-1}\,, \quad \text{and } \mathbb{C}^n \setminus \Delta^{n-1} \to \mathcal{C}^n\,, \tag{6}$$

[1] The upper index will usually mean the dimension of the variety.

where $\Delta^{n-1} \stackrel{\text{def}}{=\!=} \bigcup_{i\neq j}\{q = (q_1,\ldots,q_n) \in \mathbb{C}^n \mid q_i = q_j\}$ is the big diagonal. The points $z \in \Sigma^{n-1}$ are in one-to-one correspondence with unordered n-term multisets (or corteges) $Q = \{q_1,\ldots,q_n\}$, $q_i \in \mathbb{C}$, with at least one repetition; the regular part $\operatorname{reg}\Sigma^{n-1}$ of Σ^{n-1} consists of all unordered multisets $Q = \{q_1,\ldots,q_{n-2},u,u\} \subset \mathbb{C}$ with $q_i \neq q_j$ for $i \neq j$ and $q_i \neq u$ for all i.

The *barycenter* $\operatorname{bc}(Q)$ of a point $Q \in \operatorname{Sym}^n \mathbb{C} = \mathcal{C}^n \cup \Sigma^{n-1} \cong \mathbb{C}^n_{(z)}$ is defined as

$$\operatorname{bc}(Q) \stackrel{\text{def}}{=\!=} \frac{1}{n}\sum_{q\in Q} q = -z_1/n \tag{7}$$

(if Q is a multiset, the summation takes into account multiplicities).

Let $\mathcal{Z}$ be one of the varieties $\mathcal{C}^n$, $\mathcal{SC}^{n-1}$, or Σ^{n-1}. The corresponding *balanced* variety $\mathcal{Z}_{\text{blc}} \subset \mathcal{Z}$ is defined by

$$\mathcal{Z}_{\text{blc}} = \{Q \in \mathcal{Z} \mid \operatorname{bc}(Q) = 0\}\,, \quad \dim_{\mathbb{C}} \mathcal{Z}_{\text{blc}} = \dim_{\mathbb{C}} \mathcal{Z} - 1\,. \tag{8}$$

A free regular $\mathbb{C}_+$-action τ on $\mathbb{C}^n$ defined by $\tau_\zeta Q = Q + \zeta = \{q_1+\zeta,\ldots,q_n+\zeta\}$ for $\zeta \in \mathbb{C}$ and $Q \in \mathcal{Z}$ gives rise to the projections

$$\pi\colon \mathcal{Z} \to \mathcal{Z}_{\text{blc}},\quad Q \mapsto Q^\circ \stackrel{\text{def}}{=\!=} Q - \operatorname{bc}(Q)\,, \quad \text{and} \quad \pi'\colon \mathcal{Z} \to \mathbb{C},\quad Q \mapsto \operatorname{bc}(Q)\,, \tag{9}$$

where the retraction π yields the orbit map of τ with all fibers isomorphic to $\mathbb{C}$. The corresponding *cylindrical* direct decompositions of our varieties

$$\mathcal{C}^n = \mathcal{C}^{n-1}_{\text{blc}} \times \mathbb{C}\,, \quad \mathcal{SC}^{n-1} = \mathcal{SC}^{n-2}_{\text{blc}} \times \mathbb{C}\,, \quad \text{and} \quad \Sigma^{n-1} = \Sigma^{n-2}_{\text{blc}} \times \mathbb{C} \tag{10}$$

play an important part in the paper.

Our main results related to automorphisms of $\mathcal{C}^n$ are the following two theorems (for more general results see Theorems 4.10, 5.4, 5.5, and Corollary 5.2).

Theorem 1.1. *Assume that $n > 2$. A map $F\colon \mathcal{C}^n \to \mathcal{C}^n$ is an automorphism if and only if it is of the form*

$$F(Q) = s\cdot\pi(Q) + A(\pi(Q))\operatorname{bc}(Q) \ \ \textit{for any}\ Q \in \mathcal{C}^n\,, \tag{11}$$

where $\pi(Q) = Q - \operatorname{bc}(Q)$, $s \in \mathbb{C}^*$, and $A\colon \mathcal{C}^{n-1}_{\text{blc}} \to \operatorname{Aff}\mathbb{C}$ is a regular map.

Theorem 1.2. *If $n \geq 3$, then the following hold.*

(a) *The group $\operatorname{Aut}\mathcal{C}^n$ is solvable. More precisely, it is a semi-direct product*

$$\operatorname{Aut}\mathcal{C}^n \cong \left(\mathcal{O}_+(\mathcal{C}^{n-1}_{\text{blc}}) \rtimes (\mathbb{C}^*)^2\right) \rtimes \mathbb{Z}\,.$$

(b) *Any finite subgroup $\Gamma \subset \operatorname{Aut}\mathcal{C}^n$ is Abelian.*

(c) *Any connected algebraic subgroup G of* Aut $\mathcal{C}^n$ *is either Abelian or metabelian of rank* ≤ 2.[2]

(d) *Any two maximal tori in* Aut $\mathcal{C}^n$ *are conjugated.*

Similar facts are established for $\mathcal{SC}^{n-1}$ and Σ^{n-1}, see loc. cit.

Let us overview some results of [10, 19, 21, 24, 26, 27, 35] initiated the present paper and used in the proofs. Given a smooth irreducible non-hyperbolic algebraic curve X, consider the diagonal action of the group Aut X on the configuration space $\mathcal{C}^n(X)$,

$$\begin{aligned} &\operatorname{Aut} X \ni A\colon \mathcal{C}^n(X) \to \mathcal{C}^n(X)\,, \\ &Q = \{q_1, \dots, q_n\} \mapsto AQ \overset{\text{def}}{=\!=} \{Aq_1, \dots, Aq_n\}\,. \end{aligned} \tag{12}$$

To any morphism $T\colon \mathcal{C}^n(X) \to \operatorname{Aut} X$ we assign an endomorphism F_T of $\mathcal{C}^n(X)$ defined by

$$F_T(Q) \overset{\text{def}}{=\!=} T(Q)Q \quad \text{for all } Q \in \mathcal{C}^n(X)\,. \tag{13}$$

Such endomorphisms F_T are called *tame*. A tame endomorphism preserves each (Aut X)-orbit in $\mathcal{C}^n(X)$. A theorem below implies the converse in the following stronger form: an endomorphism of $\mathcal{C}^n(X)$ whose image is not contained in a *single* (Aut X)-orbit is tame and hence preserves each (Aut X)-orbit. If the image of F is contained in a single (Aut X)-orbit, F is called *orbit-like*.

The braid group of X, $B_n(X) = \pi_1(\mathcal{C}^n(X))$, is non-Abelian for any $n \ge 3$. If $X = \mathbb{C}$, then $B_n(X) = B_n$ is the Artin braid group on n strands. An endomorphism F of $\mathcal{C}^n(X)$ is called *non-Abelian* if the image of the induced endomorphism $F_*\colon \pi_1(\mathcal{C}^n(X)) \to \pi_1(\mathcal{C}^n(X))$ is a non-Abelian group. Otherwise, F is said to be *Abelian*. Rather unexpectedly, this evident algebraic dichotomy gives rise to the following analytic one.

Tame Map Theorem. *Let X be a smooth irreducible non-hyperbolic algebraic curve. For $n > 4$ any non-Abelian endomorphism of $\mathcal{C}^n(X)$ is tame, whereas any Abelian endomorphism of $\mathcal{C}^n(X)$ is orbit-like.*

Remarks 1.3. (a) A proof of Tame Map Theorem for $X = \mathbb{C}$ is sketched in [21] and [22]; a complete proof for $X = \mathbb{C}$ or $\mathbb{P}^1$ in the analytic category can be found in [24, 26], and [27]. For $X = \mathbb{C}^*$ the theorem is proved in [35],[3] and for elliptic curves in [10]. The proofs mutatis mutandis apply in the algebraic setting. We use this theorem to describe automorphisms of the balanced spaces $\mathcal{C}^{n-1}_{\text{blc}}$ and $\Sigma^{n-2}_{\text{blc}}$ (see Theorems 5.1(a),(c) and 7.1); its analytic counterpart is involved in the proof of Theorem 8.2.

[2]The rank of an affine algebraic group is the dimension of its maximal tori.

[3]The complex Weyl chamber of type B studied in [35] is isomorphic to $\mathcal{C}^n(\mathbb{C}^*)$.

(b) A morphism $T: \mathcal{C}^n(X) \to \operatorname{Aut} X$ in the tame representation $F = F_T$ is uniquely determined by a non-Abelian endomorphism F. Indeed, if $T_1(Q)Q = T_2(Q)Q$ for all $Q \in \mathcal{C}^n(X)$, then the automorphism $[T_1(Q)]^{-1}T_2(Q)$ is contained in the Aut (X)-stabilizer of Q, which is trivial for general configurations Q. Therefore, $T_1 = T_2$.

(c) According to Tame Map Theorem and Theorem 1.1, the map F in (11), being an automorphism, must be tame. It is indeed so with the morphism

$$T\colon \mathcal{C}^n \to \operatorname{Aff} \mathbb{C}, \qquad T(Q)\zeta = s \cdot (\zeta - \operatorname{bc}(Q)) + A(\pi(Q)) \operatorname{bc}(Q) \ , \qquad (14)$$

where $\zeta \in \mathbb{C}$ and $Q \in \mathcal{C}^n$.

(d) Let $X = \mathbb{C}$. Then Tame Map Theorem holds also for $n = 3$, but not for $n = 4$. However, any *automorphism* of $\mathcal{C}^4(\mathbb{C})$ is tame. The automorphism groups of $\mathcal{C}^1(\mathbb{C}) \cong \mathbb{C}$ and $\mathcal{C}^2(\mathbb{C}) \cong \mathbb{C}^* \times \mathbb{C}$ are well known, so we assume in the sequel that $n > 2$.

Using Tame Map Theorem, Zinde and Feler [loc. cit.] described all automorphisms of $\mathcal{C}^n(X)$ when $\dim_{\mathbb{C}} \operatorname{Aut} X = 1$, i.e., when X is $\mathbb{C}^*$ or an elliptic curve. For $\mathbb{C}$ and $\mathbb{P}^1$, where the automorphism groups Aff $\mathbb{C}$ and $\mathbf{PSL}(2, \mathbb{C})$ have dimension 2 and 3, respectively, the problem becomes more difficult. The group $\operatorname{Aut} \mathcal{C}^n = \operatorname{Aut} \mathcal{C}^n(\mathbb{C})$ is the subject of the present paper; the case $X = \mathbb{P}^1$ remains open.

The content of this chapter is as follows. In Sect. 2 we propose an abstract scheme to study the automorphism groups of cylinders over rigid bases. An irreducible affine variety $\mathcal{X}$ will be called *rigid* if the images of non-constant morphisms $\mathbb{C} \to \operatorname{reg} \mathcal{X}$ do not cover any Zariski open dense subset in the smooth locus $\operatorname{reg} \mathcal{X}$, i.e., if $\operatorname{reg} \mathcal{X}$ is *non-$\mathbb{C}$-uniruled*, see Definition 2.1. Any cylinder $\mathcal{X} \times \mathbb{C}$ over a rigid base $\mathcal{X}$ is *tight*, meaning that its cylinder structure is unique, see Definition 2.2 and Corollary 2.4.

We show in Sect. 2.3 that the bases $\mathcal{C}^{n-1}_{\text{blc}}$, $\mathcal{SC}^{n-2}_{\text{blc}}$, and $\Sigma^{n-2}_{\text{blc}}$ of cylinders (10) are rigid. So, the scheme of Sect. 2.2 applied to the latter cylinders yields that their automorphisms have a triangular form, see (19)–(20).

For any cylinder $\mathcal{X} \times \mathbb{C}$ over a rigid $\mathcal{X}$ we describe in Sect. 3 the special automorphism group $\operatorname{SAut}(\mathcal{X} \times \mathbb{C})$, and in Sect. 4 the neutral component $\operatorname{Aut}_0 (\mathcal{X} \times \mathbb{C})$ of the group $\operatorname{Aut}(\mathcal{X} \times \mathbb{C})$ and its algebraic subgroups. In Theorem 4.10 we establish an analog of Theorem 1.2 for cylinders over rigid bases. Besides, in Sect. 3 we find the locally nilpotent derivations of the structure ring $\mathcal{O}(\mathcal{X} \times \mathbb{C})$ and its Makar-Limanov invariant subring. In Sect. 4.5 we study the Lie algebra $\operatorname{Lie}(\operatorname{Aut}_0 (\mathcal{X} \times \mathbb{C}))$. These results are used in the subsequent sections in the concrete setting of the varieties $\mathcal{C}^n$, $\mathcal{SC}^{n-1}$, Σ^{n-1}, and their automorphism groups.

Theorems 1.1 and 1.2 are proven in Sect. 5, see Theorems 5.5 and 5.4, respectively. We provide analogs of our main results for the automorphism groups of the special configuration space $\mathcal{SC}^{n-1}$, the discriminant variety Σ^{n-1}, and the pair $(\mathbb{C}^n, \Sigma^{n-1})$. All these groups are solvable; we also find presentations of their Lie algebras. In Sect. 6 we show that all these groups are centerless and describe their commutator series, semisimple and torsion elements. In Sect. 7 we complete the description of the automorphism group of Σ^{n-1}. Finally, in Sect. 8, using the analytic counterpart of Tame Map Theorem, we obtain its analog for the space

$\mathcal{C}^{n-1}_{\text{blc}}$, describe the proper holomorphic self-maps of this space and the group of its biholomorphic automorphisms $\text{Aut}_{\text{hol}}\mathcal{C}^{n-1}_{\text{blc}}$.

2 Cylinders over Rigid Bases

2.1 *Triangular Automorphisms*

(a) Let $\mathfrak{C}$ be a category of sets admitting direct products $\mathcal{X} \times \mathcal{Y}$ of its objects with the standard projections to $\mathcal{X}$ and $\mathcal{Y}$ being morphisms. Every endomorphism $F \in \text{End}(\mathcal{X} \times \mathcal{Y})$ of the form

$$F(x, y) = (Sx, A(x)y) \text{ for any } (x, y) \in \mathcal{X} \times \mathcal{Y} \tag{15}$$

with some $S \in \text{Aut}\,\mathcal{X}$ and a map $A\colon \mathcal{X} \to \text{Aut}\,\mathcal{Y}$ is, in fact, an automorphism of $\mathcal{X} \times \mathcal{Y}$, and all such automorphisms form a subgroup $\text{Aut}_\Delta(\mathcal{X} \times \mathcal{Y}) \subset \text{Aut}\,(\mathcal{X} \times \mathcal{Y})$. Indeed, for F, F' we have $F'F(x, y) = (S'Sx, A'(Sx)A(x)y)$, and the inverse F^{-1} of F corresponds to $S' = S^{-1}$ and $A'(x) = (A(S^{-1}x))^{-1}$. We call such automorphisms F of $\mathcal{X} \times \mathcal{Y}$ *triangular* (with respect to the given product structure).

(b) Suppose, in addition, that $\text{Aut}\,\mathcal{Y}$ is an object in $\mathfrak{C}$. Then $\text{Mor}(\mathcal{X}, \text{Aut}\,\mathcal{Y})$ with the pointwise multiplication of morphisms can be embedded in $\text{Aut}_\Delta(\mathcal{X} \times \mathcal{Y})$ as a normal subgroup consisting of all F of the form $F(x, y) = (x, A(x)y)$, $A \in \text{Mor}(\mathcal{X}, \text{Aut}\,\mathcal{Y})$. The corresponding quotient group is isomorphic to $\text{Aut}\,\mathcal{X}$. So we have the exact sequence

$$1 \to \text{Mor}(\mathcal{X}, \text{Aut}\,\mathcal{Y}) \xrightarrow{i} \text{Aut}_\Delta(\mathcal{X} \times \mathcal{Y}) \to \text{Aut}\,\mathcal{X} \to 1$$

with the splitting monomorphism $\text{Aut}\,\mathcal{X} \xrightarrow{j} \text{Aut}_\Delta(\mathcal{X} \times \mathcal{Y})$, $S \mapsto F$, where $F(x, y) \stackrel{\text{def}}{=\!=} (Sx, y)$, and the semi-direct product decomposition

$$\text{Aut}_\Delta(\mathcal{X} \times \mathcal{Y}) \cong \text{Mor}(\mathcal{X}, \text{Aut}\,\mathcal{Y}) \rtimes \text{Aut}\,\mathcal{X}\,. \tag{16}$$

The second factor acts on the first one via $S.A = A \circ S^{-1}$, where $S \in \text{Aut}\,\mathcal{X}$ and $A \in \text{Mor}(\mathcal{X}, \text{Aut}\,\mathcal{Y})$; this is the action by conjugation in $\text{Aut}_\Delta(\mathcal{X} \times \mathcal{Y})$.

2.2 *Automorphisms of Cylinders over Rigid Bases*

We are interested in the case where $\mathfrak{C}$ is the category of complex algebraic varieties and their morphisms, and $\mathcal{Y} = \mathbb{C}$. Thus, in the sequel we deal with *cylinders* $\mathcal{X} \times \mathbb{C}$. Since $\text{Aut}\,\mathcal{Y} = \text{Aff}\,\mathbb{C} \in \mathfrak{C}$, the assumption in (b) above is fulfilled. Let us introduce the following notions.

Definition 2.1. Recall that an irreducible variety $\mathcal{X}$ is called $\mathbb{C}$-*uniruled* if for some variety $\mathcal{V}$ there is a dominant morphism $\mathcal{V}\times\mathbb{C} \to \mathcal{X}$ non-constant on a general ruling $\{v\} \times \mathbb{C}$, $v \in \mathcal{V}$ (see [17, Definition 5.2 and Proposition 5.1]). We say that $\mathcal{X}$ is *rigid* if its smooth locus $\operatorname{reg}\mathcal{X}$ is non-$\mathbb{C}$-uniruled. For such $\mathcal{X}$, the variety $\mathcal{X} \times \mathbb{C}$ is said to be a *cylinder over a rigid base*.

Definition 2.2. For an irreducible $\mathcal{X}$, we call the cylinder $\mathcal{X}\times\mathbb{C}$ *tight* if its cylinder structure over $\mathcal{X}$ is unique, that is, if for any automorphism $F \in \operatorname{Aut}(\mathcal{X} \times \mathbb{C})$ there is a (unique) automorphism $S \in \operatorname{Aut}\mathcal{X}$ that fits in the commutative diagram

$$\begin{array}{ccc} \mathcal{X}\times\mathbb{C} & \xrightarrow{F} & \mathcal{X}\times\mathbb{C} \\ {\scriptstyle \mathrm{pr}_1}\downarrow & & \downarrow{\scriptstyle \mathrm{pr}_1} \\ \mathcal{X} & \xrightarrow[S]{} & \mathcal{X} \end{array} \tag{17}$$

Thus, $\mathcal{X} \times \mathbb{C}$ is tight if and only if every $F \in \operatorname{Aut}(\mathcal{X} \times \mathbb{C})$ is triangular, so that

$$\operatorname{Aut}(\mathcal{X} \times \mathbb{C}) = \operatorname{Aut}_\Delta(\mathcal{X} \times \mathbb{C}). \tag{18}$$

For a cylinder $\mathcal{X} \times \mathbb{C}$ formula (15) takes the form

$$F(x, y) = (Sx, A(x)y) = (Sx, ay + b) \text{ for any } (x, y) \in \mathcal{X} \times \mathbb{C} \tag{19}$$

with $a \in \mathcal{O}^\times(\mathcal{X})$ and $b \in \mathcal{O}_+(\mathcal{X})$. If $\mathcal{X}\times\mathbb{C}$ is tight, then, by (18) and (19), we have

$$\begin{aligned} &\operatorname{Aut}(\mathcal{X} \times \mathbb{C}) \cong \operatorname{Mor}(\mathcal{X}, \operatorname{Aff}\mathbb{C}) \rtimes \operatorname{Aut}\mathcal{X} \text{ and} \\ &\operatorname{Mor}(\mathcal{X}, \operatorname{Aff}\mathbb{C}) \cong \mathcal{O}_+(\mathcal{X}) \rtimes \mathcal{O}^\times(\mathcal{X}), \end{aligned} \tag{20}$$

where $\mathcal{O}^\times(\mathcal{X})$ acts on $\mathcal{O}_+(\mathcal{X})$ by multiplication $b \mapsto ab$ for $a \in \mathcal{O}^\times(\mathcal{X})$ and $b \in \mathcal{O}_+(\mathcal{X})$. The group $\operatorname{Aut}(\mathcal{X} \times \mathbb{C})$ of a tight cylinder is solvable as soon as $\operatorname{Aut}\mathcal{X}$ is.

The following *strong cancellation property* of an irreducible $\mathcal{X}$ implies the tightness of $\mathcal{X} \times \mathbb{C}$: *for any* $m > 0$, *any variety* $\mathcal{Y}$, *and any isomorphism* $F: \mathcal{X} \times \mathbb{C}^m \xrightarrow{\cong} \mathcal{Y} \times \mathbb{C}^m$ *there is an isomorphism* $S: \mathcal{X} \xrightarrow{\cong} \mathcal{Y}$ *that fits in the commutative diagram*

$$\begin{array}{ccc} \mathcal{X}\times\mathbb{C}^m & \xrightarrow{F} & \mathcal{Y}\times\mathbb{C}^m \\ {\scriptstyle \mathrm{pr}_1}\downarrow & & \downarrow{\scriptstyle \mathrm{pr}_1} \\ \mathcal{X} & \xrightarrow[S]{} & \mathcal{Y} \end{array} \tag{21}$$

A priori, the tightness is weaker than the strong cancellation property. In the following classical example both these properties fail.

Example 2.3. Consider the Danielewski surfaces $S_n = \{x^n y - z^2 + 1 = 0\}$ in $\mathbb{C}^3$, $n \in \mathbb{N}$. By Danielewski's Theorem, the cylinders $S_n \times \mathbb{C}$ are all isomorphic, whereas S_n is not isomorphic to S_m for $n \neq m$. Thus these surfaces provide counterexamples to cancellation[4]. We show that the cylinders over the Danielewski surfaces are not tight.[5]

By Theorem 3.1 in [2], the surface S_1 is flexible, i.e., the tangent vectors to the orbits of the $\mathbb{C}_+$-actions on S_1 generate the tangent space at any point of S_1. It follows easily that the cylinder $S_1 \times \mathbb{C}$ is also flexible. By Theorem 0.1 in [3], the flexibility implies the k-transitivity of the automorphism group Aut $(S_1 \times \mathbb{C})$ for any $k \geq 1$. In particular, for any $n \geq 1$, there are automorphisms of the cylinder $S_n \times \mathbb{C} \simeq S_1 \times \mathbb{C}$ that do not preserve the cylinder structure, and so send it to another such structure over the same base S_n. Thus, none of the cylinders $S_n \times \mathbb{C}$ is tight.

The following theorem is known (see [6, (I)]).

Dryło's Theorem I. *The strong cancellation holds for any rigid affine variety.*

For the reader's convenience, we provide a short argument for the following corollary.

Corollary 2.4. *If $\mathcal{X}$ is rigid, then $\mathcal{X} \times \mathbb{C}$ is tight, i.e.,* Aut $(\mathcal{X} \times \mathbb{C}) = \mathrm{Aut}_\Delta(\mathcal{X} \times \mathbb{C})$.

Proof. Let us show that any $F \in$ Aut $(\mathcal{X} \times \mathbb{C})$ sends the rulings $\{x\} \times \mathbb{C}$ into rulings. Then the same holds for F^{-1}, and so $S \stackrel{\text{def}}{=\!=} \mathrm{pr}_1 \circ F|_{\mathcal{X} \times \{0\}} \in \mathrm{Aut}\, \mathcal{X}$ fits in diagram (17).

Assuming the contrary, we consider the family $\{F(\{x\} \times \mathbb{C})\}_{x \in \mathrm{reg}\, \mathcal{X}}$. Projecting it to reg $\mathcal{X}$ we get a contradiction with the rigidity assumption. □

Corollary 2.4 and Proposition 2.7 below show that the cylinders (10) are tight.

2.3 *Configuration Spaces and Discriminant Levels as Cylinders over Rigid Bases*

In this section we show (see Proposition 2.7) that the bases $\mathcal{C}^{n-1}_{\mathrm{blc}}$, $\mathcal{SC}^{n-2}_{\mathrm{blc}}$, and $\Sigma^{n-2}_{\mathrm{blc}}$ of the cylinders (10) possess a property, which is stronger than the rigidity and, consequently, all automorphisms of these cylinders are triangular[6] (Corollary 2.8).

---

[4]See [18] for further examples of non-cancellation.

[5]The authors thank S. Kaliman for a useful discussion, where the latter observation appeared.

[6]Since D_n is homogeneous, any hypersurface $D_n(Q) = c \neq 0$ is isomorphic to $\mathcal{SC}^{n-1}$.

Notation 2.5. For any X and any $n \in \mathbb{N}$, let $\mathcal{C}^n_{\mathrm{ord}}(X)$ denote the *ordered configuration space* of X, i.e., $\mathcal{C}^n_{\mathrm{ord}}(X) = \{(q_1, \ldots, q_n) \in X^n \mid q_i \neq q_j \text{ for all } i \neq j\}$. The group $\mathbf{S}(n)$ acts freely on $\mathcal{C}^n_{\mathrm{ord}}(X)$ and, by definition, $\mathcal{C}^n_{\mathrm{ord}}(X)/\mathbf{S}(n) = \mathcal{C}^n(X)$. We let

$$\mathcal{C}^n_{\mathrm{ord}} \stackrel{\mathrm{def}}{=\!=} \mathcal{C}^n_{\mathrm{ord}}(\mathbb{C}) \quad \text{and} \quad \mathcal{C}^{n-1}_{\mathrm{ord,blc}} \stackrel{\mathrm{def}}{=\!=} \{q = \{q_1, \cdots, q_n\} \in \mathcal{C}^n_{\mathrm{ord}} \mid q_1 + \ldots + q_n = 0\}\,.$$

Clearly $\mathcal{C}^{n-1}_{\mathrm{ord,blc}}/\mathbf{S}(n) = \mathcal{C}^{n-1}_{\mathrm{blc}}$.

Recall (see Sect. 1) that the regular part $\operatorname{reg}\Sigma^{n-1}$ of the discriminant variety Σ^{n-1} consists of all unordered n-multisets $Q = \{q_1, \ldots, q_{n-2}, u, u\} \subset \mathbb{C}$ with $q_i \neq q_j$ for $i \neq j$ and $q_i \neq u$ for all i. Since the hyperplane $q_1 + \cdots + q_n = 0$ is transversal to each of the hyperplanes $q_i = q_j$, the regular part $\operatorname{reg}\Sigma^{n-2}_{\mathrm{blc}}$ of $\Sigma^{n-2}_{\mathrm{blc}}$ consists of all multisets $Q = \{q_1, \ldots, q_{n-2}, u, u\}$ as above that satisfy the additional condition $\sum_{i=1}^{n-2} q_i + 2u = 0$. In the proofs of Proposition 2.7 and Theorem 7.1 below we need the following lemma.

Lemma 2.6. *For $n > 2$ the regular part $\operatorname{reg}\Sigma^{n-2}_{\mathrm{blc}}$ of $\Sigma^{n-2}_{\mathrm{blc}}$ is isomorphic to the configuration space $\mathcal{C}^{n-2}(\mathbb{C}^*)$. Consequently, $\operatorname{Aut}(\operatorname{reg}\Sigma^{n-2}_{\mathrm{blc}}) \cong \operatorname{Aut}\mathcal{C}^{n-2}(\mathbb{C}^*)$.*

Proof. An isomorphism $\operatorname{reg}\Sigma^{n-2}_{\mathrm{blc}} \cong \mathcal{C}^{n-2}(\mathbb{C}^*)$ does exist since both these varieties are smooth cross-sections of the $\mathbb{C}_+$-action τ on the cylinder $\operatorname{reg}\Sigma^{n-2} = (\operatorname{reg}\Sigma^{n-2}_{\mathrm{blc}}) \times \mathbb{C}$ (see (10)). To construct such an isomorphism explicitly, for every

$$Q = \{q_1, \ldots, q_{n-2}, u, u\} \in \operatorname{reg}\Sigma^{n-2}_{\mathrm{blc}}\,, \quad \text{where, of course, } u = -\frac{1}{2}\sum_{i=1}^{n-2} q_i\,,$$

we let $\tilde{Q} = \{q_1 - u, \ldots, q_{n-2} - u\}$. Then $\tilde{Q} \in \mathcal{C}^{n-2}(\mathbb{C}^*)$ and we have an epimorphism

$$\varphi\colon \operatorname{reg}\Sigma^{n-2}_{\mathrm{blc}} \to \mathcal{C}^{n-2}(\mathbb{C}^*)\,, \quad \varphi(Q) = \tilde{Q}\,, \tag{22}$$

To show that φ is an isomorphism, for any $Q' = \{q'_1, \ldots, q'_{n-2}\} \in \mathcal{C}^{n-2}(\mathbb{C}^*)$ take

$$v = -\frac{1}{n}\sum_{i=1}^{n-2} q'_i \quad \text{and} \quad Q'' = \{q'_1 + v, \ldots, q'_{n-2} + v, v, v\}\,;$$

notice that $v = u$ for $Q' = \tilde{Q}$ as above. Then $Q'' \in \operatorname{reg}\Sigma^{n-2}_{\mathrm{blc}}$ and the morphism

$$\psi\colon \mathcal{C}^{n-2}(\mathbb{C}^*) \to \operatorname{reg}\Sigma^{n-2}_{\mathrm{blc}}\,, \quad \psi(Q') = Q''\,,$$

is inverse to φ. □

Proposition 2.7. *For $n > 2$, let $\mathcal{X}$ be one of the varieties $\mathcal{C}^{n-1}_{\mathrm{blc}}$, $\mathcal{SC}^{n-2}_{\mathrm{blc}}$, or $\Sigma^{n-2}_{\mathrm{blc}}$. Then any morphism $\mathbb{C} \to \operatorname{reg} \mathcal{X}$ is constant. Consequently, $\mathcal{C}^{n-1}_{\mathrm{blc}} \times \mathbb{C}$, $\mathcal{SC}^{n-2}_{\mathrm{blc}} \times \mathbb{C}$, and $\Sigma^{n-2}_{\mathrm{blc}} \times \mathbb{C}$ in* (10) *are cylinders over rigid bases.*

Proof. Let us show first that any morphism $f: \mathbb{C} \to \mathcal{C}^{n-1}_{\mathrm{blc}}$ is constant. Consider the unramified $\mathbf{S}(n)$-covering $p: \mathcal{C}^{n-1}_{\mathrm{blc,ord}} \to \mathcal{C}^{n-1}_{\mathrm{blc}}$. By the monodromy theorem f can be lifted to a morphism $g = (g_1, \ldots, g_n): \mathbb{C} \to \mathcal{C}^{n-1}_{\mathrm{blc,ord}}$. For any $i \neq j$ the regular function $g_i - g_j$ on $\mathbb{C}$ does not vanish, hence it is constant. In particular, $g_i = g_1 + c_i$, where $c_i \in \mathbb{C}$, $i = 1, \ldots, n$, and so $0 = \sum_{i=1}^{n} g_i = ng_1 + c$, where $c = \sum_{i=1}^{n} c_i$. Thus, $g_1 = \text{const}$, and so $g_i = \text{const}$ for all $i = 1, \ldots, n$. Hence $f = \text{const}$ and the variety $\mathcal{C}^{n-1}_{\mathrm{blc}}$ is rigid.

Since $\mathcal{SC}^{n-2}_{\mathrm{blc}} \subset \mathcal{C}^{n-1}_{\mathrm{blc}}$, any morphism $\mathbb{C} \to \mathcal{SC}^{n-2}_{\mathrm{blc}}$ is constant and $\mathcal{SC}^{n-2}_{\mathrm{blc}}$ is rigid.

It remains to show that any morphism $\mathbb{C} \to \operatorname{reg} \Sigma^{n-2}_{\mathrm{blc}}$ is constant. For $n = 3$ we have $\operatorname{reg} \Sigma^{n-2}_{\mathrm{blc}} \cong \mathbb{C}^*$, hence the claim follows.

For $n \geq 4$, by Lemma 2.6, it suffices to show that any morphism $f: \mathbb{C} \to \mathcal{C}^{n-2}(\mathbb{C}^*)$ is constant. By monodromy theorem f admits a lift $g: \mathbb{C} \to \mathcal{C}^{n-2}_{\mathrm{ord}}(\mathbb{C}^*) \subset (\mathbb{C}^*)^{n-2}$ to the unramified $\mathbf{S}(n-2)$-covering $\mathcal{C}^{n-2}_{\mathrm{ord}}(\mathbb{C}^*) \to \mathcal{C}^{n-2}(\mathbb{C}^*)$. This implies that both g and f are constant, since any morphism $\mathbb{C} \to \mathbb{C}^*$ is. □

The following important result follows from Corollary 2.4 and Proposition 2.7.

Corollary 2.8. *For $n > 2$, all automorphisms of the cylinders*

$$\mathcal{C}^n \cong \mathcal{C}^{n-1}_{\mathrm{blc}} \times \mathbb{C}, \quad \mathcal{SC}^{n-1} \cong \mathcal{SC}^{n-2}_{\mathrm{blc}} \times \mathbb{C}, \quad \textit{and} \quad \Sigma^{n-1} \cong \Sigma^{n-2}_{\mathrm{blc}} \times \mathbb{C}$$

are triangular, and (18)–(20) *hold for the corresponding automorphism groups.*

3 The Special Automorphism Groups

3.1 $\mathbb{C}_+$-Actions and LNDs on Cylinders over Rigid Bases

We start with the following simple lemma.

Lemma 3.1. *If $\mathcal{X}$ is rigid, then any $\mathbb{C}_+$-action on the cylinder $\mathcal{X} \times \mathbb{C}$ preserves each fiber of the first projection $\mathrm{pr}_1: \mathcal{X} \times \mathbb{C} \to \mathcal{X}$.*

Proof. Indeed, since $\mathcal{X}$ is rigid, the induced $\mathbb{C}_+$-action on it is trivial. □

We suppose that this lemma is not true any longer for tight cylinders.

Definition 3.2. Recall that a derivation ∂ of a ring A is *locally nilpotent* if $\partial^n a = 0$ for any $a \in A$ and for some $n \in \mathbb{N}$ depending on a. For any $f \in \ker \partial$ the locally nilpotent derivation $f\partial \in \operatorname{Der} A$ is called a *replica* of ∂ (see [3]).

Consider the locally nilpotent derivation $\partial/\partial y$ on the structure ring $\mathcal{O}(\mathcal{X} \times \mathbb{C}) \cong \mathcal{O}(\mathcal{X})[y]$. The corresponding $\mathbb{C}_+$-action, say τ, on the cylinder $\mathcal{X} \times \mathbb{C}$ acts via translations along the second factor. For arbitrary $\mathbb{C}_+$-actions on the cylinder $\mathcal{X} \times \mathbb{C}$ we have the following description.

Proposition 3.3. *If $\mathcal{X}$ is rigid, then any locally nilpotent derivation ∂ on the coordinate ring $\mathcal{O}(\mathcal{X} \times \mathbb{C})$ is a replica of the derivation $\partial/\partial y$, i.e.,*

$$\partial = f\partial/\partial y, \quad \textit{where} \quad f \in \mathcal{O}^{\tau}(\mathcal{X} \times \mathbb{C}) = \operatorname{pr}_1^*(\mathcal{O}(\mathcal{X})) = \ker \partial/\partial y\,.$$

Consequently, any $\mathbb{C}_+$-action on $\mathcal{X} \times \mathbb{C}$ is of the form

$$(x, y) \mapsto (x, y + \lambda b(x)), \quad \textit{where} \quad \lambda \in \mathbb{C} \quad \textit{and} \quad b \in \mathcal{O}(\mathcal{X})\,.$$

Proof. Indeed, both ∂ and $\partial/\partial y$ can be viewed as regular vector fields on $\mathcal{X} \times \mathbb{C}$, where the latter field is non-vanishing. By Lemma 3.1, these vector fields are proportional. That is, there exists a function $f \in \mathcal{O}^{\tau}(\mathcal{X} \times \mathbb{C})$ such that $\partial = f\partial/\partial y$, which proves the first assertion. Now the second follows. □

3.2 *The Group* SAut$(\mathcal{X} \times \mathbb{C})$

Definition 3.4. Let $\mathcal{Z}$ be an irreducible algebraic variety. A subgroup $G \subset \operatorname{Aut} \mathcal{Z}$ is called *algebraic* if it admits a structure of an algebraic group such that the natural map $G \times \mathcal{Z} \to \mathcal{Z}$ is a morphism. The *special automorphism group* SAut$\mathcal{Z}$ is the subgroup of Aut $\mathcal{Z}$ generated by all algebraic subgroups of Aut $\mathcal{Z}$ isomorphic to $\mathbb{C}_+$ (see e.g. [3]). Clearly, SAut $\mathcal{Z}$ is a normal subgroup of Aut $\mathcal{Z}$.

Assume that $\mathcal{X}$ is rigid. Due to (20) we have the decomposition

$$\operatorname{Aut}(\mathcal{X} \times \mathbb{C}) \cong \left(\mathcal{O}_+(\mathcal{X}) \rtimes \mathcal{O}^{\times}(\mathcal{X})\right) \rtimes \operatorname{Aut} \mathcal{X}\,. \tag{23}$$

In general, the group SAut of an affine variety is not necessarily Abelian. Thus the following corollary emphasizes a special character of the varieties that we are dealing with.

Corollary 3.5. *If $\mathcal{X}$ is rigid, then $G =$ SAut$(\mathcal{X} \times \mathbb{C})$ is an Abelian group with Lie algebra*[7]

[7] See Sect. 4.5 below.

$$L = \mathrm{Lie}\, G = \mathcal{O}_+^{\tau}(\mathcal{X}\times\mathbb{C})\partial/\partial y = \mathcal{O}_+(\mathcal{X})\partial/\partial y\,.$$

Furthermore, the exponential map exp: $L \to G$ *yields an isomorphism of groups*

$$\mathcal{O}_+(\mathcal{X}) \xrightarrow{\cong} \mathrm{SAut}(\mathcal{X}\times\mathbb{C})\,.$$

Proof. This follows easily from Proposition 3.3. □

Consider the standard $\mathbb{C}_+$-action τ on $\mathcal{X}\times\mathbb{C}$ by shifts along the second factor, and let $U = \exp(\mathbb{C}\partial/\partial y)$ be the corresponding one-parameter unipotent subgroup of SAut $(\mathcal{X}\times\mathbb{C})$. Consider also the subgroup $B \overset{\mathrm{def}}{=\!=} U\cdot \mathrm{Aut}\,\mathcal{X} \cong U \rtimes \mathrm{Aut}\,\mathcal{X}$ of Aut $(\mathcal{X}\times\mathbb{C})$, and let $B_0 \cong U \rtimes \mathrm{Aut}_0\,\mathcal{X}$ be its neutral component. More generally, given a character χ of Aut $\mathcal{X}$ (of $\mathrm{Aut}_0\,\mathcal{X}$, respectively) we let

$$B(\chi) = \big\{F \in \mathrm{Aut}\,(\mathcal{X}\times\mathbb{C}) \mid F : (x,y) \mapsto (Sx, \chi(S)y + b),\ S \in \mathrm{Aut}\,\mathcal{X},\ b \in \mathbb{C}\big\}\,.$$

The group $B_0(\chi)$ is defined in a similar way. Thus, $B = B(1)$ and $B_0 = B_0(1)$. Clearly, $B(\chi)$ ($B_0(\chi)$, respectively) is algebraic as soon as Aut $\mathcal{X}$ ($\mathrm{Aut}_0\,\mathcal{X}$, respectively) is.

From Proposition 3.3 we deduce the following result.

Corollary 3.6. *If $\mathcal{X}$ is a rigid affine variety, then the orbits of the automorphism group* Aut $(\mathcal{X}\times\mathbb{C})$ *(of* $\mathrm{Aut}_0\,(\mathcal{X}\times\mathbb{C})$*, respectively) coincide with the orbits of the group $B(\chi)$ ($B_0(\chi)$, respectively), whatever is the character χ of* Aut $\mathcal{X}$ *(of* $\mathrm{Aut}_0\mathcal{X}$*, respectively).*

Proof. We give a proof for the group Aut $(\mathcal{X}\times\mathbb{C})$; that for $\mathrm{Aut}_0\,(\mathcal{X}\times\mathbb{C})$ is similar. Recall that any automorphism F of the cylinder $\mathcal{X}\times\mathbb{C}$ over a rigid base $\mathcal{X}$ can be written as

$$F(x,y) = (Sx, A(x)y) \ \text{ for any } (x,y)\in\mathcal{X}\times\mathbb{C}\,, \tag{24}$$

where $S \in \mathrm{Aut}\,\mathcal{X}$ and $A \in \mathrm{Mor}(\mathcal{X}, \mathrm{Aff}\,\mathbb{C})$. It follows that the $B(\chi)$-orbit $B(\chi)Q$ of a point $Q = (x,y)$ in $\mathcal{X}\times\mathbb{C}$ is $B(\chi)Q = [(\mathrm{Aut}\,\mathcal{X})x]\times\mathbb{C}$. By virtue of Proposition 3.3 the SAut $(\mathcal{X}\times\mathbb{C})$-orbits in $\mathcal{X}\times\mathbb{C}$ coincide with the τ-orbits, that is, with the rulings of the cylinder $\mathcal{X}\times\mathbb{C}$. Now the assertion follows from decomposition (23) and the isomorphism $\mathcal{O}_+(\mathcal{X}) \cong \mathrm{SAut}(\mathcal{X}\times\mathbb{C})$ of Corollary 3.5.
□

Remarks 3.7. (a) It is known that the group $\mathrm{Aut}_0\,\mathcal{X}$ of a rigid affine variety $\mathcal{X}$ is an algebraic torus, see Theorem 4.10(a) below. Hence $B_0(\chi)$ is a metabelian linear algebraic group isomorphic to a semi-direct product $\mathbb{C}_+ \rtimes (\mathbb{C}^*)^r$, where $r \ge 0$ and $(\mathbb{C}^*)^r$ acts on $\mathbb{C}_+$ via multiplication by the character χ of the torus $(\mathbb{C}^*)^r$.

(b) We have the following assertion in the spirit of Tame Map Theorem: *Given a rigid affine variety $\mathcal{X}$ and a character χ of* Aut $\mathcal{X}$*, any automorphism F of the cylinder $\mathcal{X}\times\mathbb{C}$ admits a unique factorization*

$$F:\mathcal{X}\times\mathbb{C}\stackrel{T\times\{\mathrm{id}\}}{\longrightarrow}B(\chi)\times(\mathcal{X}\times\mathbb{C})\stackrel{\alpha}{\longrightarrow}\mathcal{X}\times\mathbb{C}\,,\qquad(25)$$

where α stands for the $B(\chi)$-action on $\mathcal{X}\times\mathbb{C}$, and $T:\mathcal{X}\times\mathbb{C}\to B(\chi)$ is a morphism with a constant Aut $\mathcal{X}$*-component.*

Indeed, by Corollary 3.6 for any point $Q=(x,y)\in\mathcal{X}\times\mathbb{C}$ there exists an element $T(Q)\in B(\chi)$, $T(Q):(x',y')\mapsto(S(Q)x',\chi(S(Q))y'+f(Q))$ for some $f(Q)\in\mathbb{C}$, such that

$$F(Q)=T(Q)Q=(S(Q)x,\chi(S(Q))y+f(Q))\in\mathcal{X}\times\mathbb{C}\,.\qquad(26)$$

On the other hand, according to (24),

$$F(Q)=(Sx,a(x)y+b(x))\in\mathcal{X}\times\mathbb{C}\,,\qquad(27)$$

where $S\in\operatorname{Aut}\mathcal{X}$, $a\in\mathcal{O}^{\times}(\mathcal{X})$, and $b\in\mathcal{O}_{+}(\mathcal{X})$ are uniquely determined by F. Comparing (26) and (27) yields $S(Q)x=Sx$ and $f(Q)=(a(x)-\chi(S))y+b(x)$ for any $Q\in\mathcal{X}\times\mathbb{C}$. Vice versa, the latter equalities define unique $f\in\mathcal{O}(\mathcal{X}\times\mathbb{C})$ and $S\in\operatorname{Aut}\mathcal{X}$ such that $T:\mathcal{X}\times\mathbb{C}\to B(\chi)$, $Q\mapsto(S,z\mapsto\chi(S)z+f(Q))$, fits in (25) i.e., $F=\alpha\circ(T\times\mathrm{id})$, as required. □

Formula (14) corresponds to the particular case $\mathcal{X}\times\mathbb{C}=\mathcal{C}^{n-1}_{\mathrm{blc}}\times\mathbb{C}\cong\mathcal{C}^{n}$. In this case $\operatorname{Aut}\mathcal{X}=\operatorname{Aut}\mathcal{C}^{n-1}_{\mathrm{blc}}=\mathbb{C}^{*}$ (see Theorem 5.1(a) below), and the character $\chi:\mathbb{C}^{*}\to\mathbb{C}^{*}$ is the identity.

3.3 The Makar-Limanov Invariant of a Cylinder

The subalgebra of τ-invariants $\mathcal{O}^{\tau}(\mathcal{X}\times\mathbb{C})\subset\mathcal{O}(\mathcal{X}\times\mathbb{C})$ admits yet another interpretation.

Definition 3.8. Let Y be an affine algebraic variety over $\mathbb{C}$. The ring of invariants $\mathcal{O}(Y)^{\mathrm{SAut}Y}$ is called the *Makar-Limanov invariant* of Y and is denoted by $\mathrm{ML}(Y)$. This ring is invariant under the induced action of the group Aut Y on $\mathcal{O}(Y)$.

Any algebraic subgroup $H\subset\operatorname{Aut}Y$ isomorphic to $\mathbb{C}_{+}$ can be written as $H=\exp(\mathbb{C}\partial)$, where the infinitesimal generator $\partial\in\operatorname{Der}\mathcal{O}(Y)$ is a locally nilpotent derivation on the coordinate ring $\mathcal{O}(Y)$ ([12]). We have $\mathcal{O}(Y)^{H}=\ker\partial$ and so

$$\mathrm{ML}(Y)=\bigcap\ker\partial\,,$$

where ∂ runs over the set $\mathrm{LND}(\mathcal{O}(Y))$ of all locally nilpotent derivations of the algebra $\mathcal{O}(Y)$. The next corollary is immediate from Proposition 3.3.

Corollary 3.9. *If an irreducible affine variety $\mathcal{X}$ is rigid, then*

$$\mathrm{ML}(\mathcal{X}\times\mathbb{C}) = \mathcal{O}^{\tau}(\mathcal{X}\times\mathbb{C}) = \mathcal{O}(\mathcal{X}).$$

Alternatively, this corollary follows from the next more general result (cf. also [4]).

Dryło's Theorem II ([7]). *Let $\mathcal{X}$ and $\mathcal{Y}$ be irreducible affine varieties over an algebraically closed field k. If $\mathcal{X}$ is rigid, then* $\mathrm{ML}(\mathcal{X}\times\mathcal{Y}) = \mathcal{O}(\mathcal{X})\otimes_k \mathrm{ML}(\mathcal{Y})$.

4 Ind-Group Structure and Algebraic Subgroups

4.1 Aut $(\mathcal{X}\times\mathbb{C})$ *as Ind-Group*

Recall the following notions (see [20, 34]).

Definition 4.1. An *ind-group* is a group G equipped with an increasing filtration $G = \bigcup_{i\in\mathbb{N}} G_i$, where the components G_i are algebraic varieties (and not necessarily algebraic groups) such that the natural inclusion $G_i \hookrightarrow G_{i+1}$, the multiplication map $G_i\times G_j \to G_{m(i,j)}$, $(g_i, g_j)\mapsto g_i g_j$, and the inversion $G_i \to G_{k(i)}$, $g_i \mapsto g_i^{-1}$, are morphisms for any $i, j \in \mathbb{N}$ with a suitable choice of $m(i, j)$, $k(i) \in \mathbb{N}$.

Examples 4.2 (a) *Ind-structure on* $\mathcal{O}_+(\mathcal{X})$. Given an affine variety $\mathcal{X}$ we fix a closed embedding $\mathcal{X}\hookrightarrow\mathbb{C}^N$. For $f\in\mathcal{O}_+(\mathcal{X})$ we define its degree $\deg f$ as the minimal degree of a polynomial extension of f to $\mathbb{C}^N$. Letting

$$G_i = \{f\in\mathcal{O}_+(\mathcal{X}) \mid \deg f \le i\} \tag{28}$$

we obtain a filtration of the group $\mathcal{O}_+(\mathcal{X})$ by an increasing sequence of connected Abelian algebraic subgroups G_i ($i\in\mathbb{N}$), hence an ind-structure on $\mathcal{O}_+(\mathcal{X})$.

(b) *Ind-structure on* Aut $\mathcal{X}$. Given again a closed embedding $\mathcal{X}\hookrightarrow\mathbb{C}^N$, any automorphism $F\in$ Aut $\mathcal{X}$ can be written as $F = (f_1,\dots,f_N)$, where $f_j\in\mathcal{O}_+(\mathcal{X})$. Letting

$$\deg F = \max_{j=1,\dots,N}\{\deg f_j\} \quad \text{and} \quad G_i = \{F\in \mathrm{Aut}\,\mathcal{X} \mid \deg F \le i\}$$

we obtain an ind-group structure Aut $\mathcal{X} = \bigcup_{i\in\mathbb{N}} G_i$ compatible with the action of Aut $\mathcal{X}$ on $\mathcal{X}$. The latter means that the maps $G_i\times\mathcal{X}\to\mathcal{X}$, $(F,x)\mapsto F(x)$, are morphisms of algebraic varieties. It is well known that any two such ind-structures on Aut $\mathcal{X}$ are equivalent.

(c) *Ind-structure on* Aut $(\mathcal{X}\times\mathbb{C})$. For a cylinder $\mathcal{X}\times\mathbb{C}$ over a rigid base an ind-structure on the group Aut $(\mathcal{X}\times\mathbb{C})$ can be defined via the ind-structures on the factors $\mathcal{O}_+(\mathcal{X})$, $\mathcal{O}^\times(\mathcal{X})$, and Aut $\mathcal{X}$ in decomposition (23).

4.2 $\mathcal{O}^\times(\mathcal{X})$ as Ind-Group

Any extension of an algebraic group by a countable group is an ind-group. In particular, the group $\mathcal{O}^\times(\mathcal{X})$ is an ind-group due to the following well-known fact (see [33, Lemme 1]).

Lemma 4.3 (Samuel's Lemma). *For any irreducible algebraic variety $\mathcal{X}$ defined over an algebraically closed field k we have*

$$\mathcal{O}^\times(\mathcal{X}) \cong k^* \times \mathbb{Z}^m \text{ for some } m \geq 0\,.$$

If $k = \mathbb{C}$, then $m \leq \operatorname{rank} H^1(\mathcal{X}, \mathbb{Z})$.

We provide an argument for $k = \mathbb{C}$, which follows the sheaf-theoretic proofs of the topological Bruschlinsky [5] and Eilenberg [8, 9] theorems; see also [11, Lemma 1.1] for a proof in the general case and further references.

Proof. The sheaves $\mathcal{Z}_\mathcal{X}$, $\mathcal{C}_\mathcal{X}$, and $\mathcal{C}^*_\mathcal{X}$ of germs of continuous functions with values in $\mathbb{Z}$, $\mathbb{C}$, and $\mathbb{C}^*$, respectively, form the exact sequence $0 \to \mathcal{Z}_\mathcal{X} \stackrel{\cdot 2\pi i}{\longrightarrow} \mathcal{C}_\mathcal{X} \stackrel{\exp}{\longrightarrow} \mathcal{C}^*_\mathcal{X} \to 1$. As $\mathcal{X}$ is connected and paracompact, and the sheaf $\mathcal{C}_\mathcal{X}$ is fine, $H^1(\mathcal{X}, \mathcal{C}_\mathcal{X}) = 0$ and the corresponding exact cohomology sequence takes the form

$$0 \to \mathbb{Z} \stackrel{\cdot 2\pi i}{\longrightarrow} C(\mathcal{X}) \stackrel{\exp}{\longrightarrow} C^*(\mathcal{X}) \stackrel{\rho}{\longrightarrow} H^1(\mathcal{X}, \mathbb{Z}) \to 0\,.$$

Restricting the homomorphism ρ to $\mathcal{O}^\times(\mathcal{X}) \subset C^*(\mathcal{X})$ and taking into account that the conditions $\varphi \in \mathcal{O}(\mathcal{X})$ and $e^\varphi \in \mathcal{O}^\times(\mathcal{X})$ imply $\varphi = \text{const}$, we obtain the exact sequence

$$\mathbb{C} \stackrel{\exp}{\longrightarrow} \mathcal{O}^\times(\mathcal{X}) \stackrel{\rho}{\longrightarrow} H^1(\mathcal{X}, \mathbb{Z})\,.$$

Since $H^1(\mathcal{X}, \mathbb{Z})$ is a free Abelian group of finite rank, the image of ρ is isomorphic to $\mathbb{Z}^m$ with some $m \leq \operatorname{rank} H^1(\mathcal{X}, \mathbb{Z})$. This implies that the Abelian group $\mathcal{O}^\times(\mathcal{X})$ admits the desired direct decomposition. □

Examples 4.4 (The Groups of Units on the Balanced Spaces) (a) The discriminant D_n is the restriction to $\mathcal{C}^n = \mathbb{C}^n_{(z)} \setminus \{z \mid d_n(z) = 0\}$ of the irreducible discriminant polynomial d_n. Since $\mathcal{C}^n = \mathcal{C}^{n-1}_{\text{blc}} \times \mathbb{C}$, the group $H^1(\mathcal{C}^{n-1}_{\text{blc}}, \mathbb{Z}) = H^1(\mathcal{C}^n, \mathbb{Z}) \cong \mathbb{Z}$ is generated by the cohomology class of D_n, all elements of $\mathcal{O}^\times(\mathcal{C}^{n-1}_{\text{blc}})$ are of the form sD^k_n with $s \in \mathbb{C}^*$ and $k \in \mathbb{Z}$, and $\mathcal{O}^\times(\mathcal{C}^{n-1}_{\text{blc}}) \cong \mathbb{C}^* \times \mathbb{Z}$.

(b) The projection $D_n\colon \mathcal{C}^{n-1}_{\text{blc}} \to \mathbb{C}^*$, $Q \mapsto D_n(Q)$, is a locally trivial fiber bundle with fibers isomorphic to $\mathcal{SC}^{n-2}_{\text{blc}}$. Since $\pi_2(\mathbb{C}^*) = 0$, the final segment of the corresponding homotopical exact sequence looks as follows:

$$1 \to \pi_1(\mathcal{SC}^{n-2}_{\text{blc}}) \to \pi_1(\mathcal{C}^{n-1}_{\text{blc}}) \to \pi_1(\mathbb{C}^*) \to 1\,.$$

Now, $\pi_1(\mathcal{C}^{n-1}_{\text{blc}})$ is the Artin braid group B_n and we can rewrite this sequence as

$$1 \to \pi_1(\mathcal{S}\mathcal{C}^{n-2}_{\text{blc}}) \to B_n \to \mathbb{Z} \to 1\,,$$

so that the commutator subgroup B'_n is contained in $\pi_1(\mathcal{S}\mathcal{C}^{n-2}_{\text{blc}})$. Since $B_n/B'_n \cong \mathbb{Z}$ and the torsion of any nontrivial quotient group of $\mathbb{Z}$ is nontrivial, it follows that $\pi_1(\mathcal{S}\mathcal{C}^{n-2}_{\text{blc}}) \cong B'_n$. By [13, Lemma 2.2], $B''_n = B'_n$ for any $n > 4$, and so $\text{Hom}(B'_n, \mathbb{Z}) = 0$. Finally, $H^1(\mathcal{S}\mathcal{C}^{n-2}_{\text{blc}}, \mathbb{Z}) \cong \text{Hom}(\pi_1(\mathcal{S}\mathcal{C}^{n-2}_{\text{blc}}), \mathbb{Z}) = \text{Hom}(B'_n, \mathbb{Z}) = 0$ and $\mathcal{O}^\times(\mathcal{S}\mathcal{C}^{n-2}_{\text{blc}}) \cong \mathbb{C}^*$.

(c) The discriminant d_n and its restriction $d_n|_{z_1=0}$ to the hyperplane $z_1 = 0$ are quasi-homogeneous. So, the zero level sets $\Sigma^{n-1} = \{d_n = 0\}$ and $\Sigma^{n-2}_{\text{blc}} = \Sigma^{n-1} \cap \{z_1 = 0\}$ are contractible, and hence $H^1(\Sigma^{n-2}_{\text{blc}}, \mathbb{Z}) = 0$. By Lemma 4.3, $\mathcal{O}^\times(\Sigma^{n-2}_{\text{blc}}) \cong \mathbb{C}^*$.

4.3 *The Neutral Component* $\text{Aut}_0\,(\mathcal{X} \times \mathbb{C})$

Definition 4.5. The neutral component G_0 of an ind-group $G = \bigcup_{i\in\mathbb{N}} G_i$ is defined as union of those connected components of the G_i that contain the unity e_G of G. In other words, G_0 is the union of all connected algebraic subvarieties of G passing through e_G. Recall that a subset $V \subset G$ is an algebraic subvariety if it is a subvariety of a piece G_i. Clearly, G_0 is a normal ind-subgroup of G.

For an irreducible affine variety $\mathcal{X}$, the neutral component $\text{Aut}_0\,\mathcal{X}$ is the union of all connected algebraic subvarieties of $\text{Aut}\,\mathcal{X}$ which contain the identity. Thus $\text{Aut}_0\,\mathcal{X}$ is as well the neutral component of $\text{Aut}\,\mathcal{X}$ in the sense of [32].

From Corollary 2.4, Lemma 4.3, and decomposition (23) we derive the following.

Theorem 4.6. *For a cylinder $\mathcal{X} \times \mathbb{C}$ over a rigid base we have*

$$\text{Aut}_0\,(\mathcal{X} \times \mathbb{C}) \cong \mathcal{O}_+(\mathcal{X}) \rtimes (\mathbb{C}^* \times \text{Aut}_0\,\mathcal{X})\,. \tag{29}$$

Proof. For a semi-direct product of two ind-groups H and H' we have $(H \rtimes H')_0 = H_0 \rtimes H'_0$. Thus, from (23) we get a decomposition

$$\text{Aut}_0\,(\mathcal{X} \times \mathbb{C}) \cong (\mathcal{O}_+(\mathcal{X}) \rtimes \mathbb{C}^*) \rtimes \text{Aut}_0\,\mathcal{X}\,.$$

It suffices to show that the factors $\mathbb{C}^*$ and $\text{Aut}_0\,(\mathcal{X})$ in this decomposition commute. That is, that $FF' = F'F$ for any two automorphisms $F, F' \in \text{Aut}_0\,(\mathcal{X} \times \mathbb{C})$ of the form $F : (x, y) \mapsto (x, ty)$ and $F' : (x, y) \mapsto (Sx, y)$, where $S \in \text{Aut}\,\mathcal{X}$ and $t \in \mathbb{C}^*$, see (19). However, the latter equality is evident. □

Remark 4.7 (The unipotent radical of $\text{Aut}_0\,(\mathcal{X} \times \mathbb{C})$*).* A rigid variety $\mathcal{X}$ does not admit any nontrivial action of a unipotent linear algebraic group. Thus, any such subgroup of $\mathrm{Aut}_0(\mathcal{X} \times \mathbb{C})$ is contained in the subgroup $\mathrm{SAut}(\mathcal{X} \times \mathbb{C}) \cong \mathcal{O}_+(\mathcal{X})$, see (29), and so it is Abelian. Due to Corollary 3.5, the normal Abelian subgroup $\mathrm{SAut}(\mathcal{X} \times \mathbb{C})$ can be regarded as the unipotent radical of $\mathrm{Aut}_0(\mathcal{X} \times \mathbb{C})$. Notice that $\mathrm{SAut}(\mathcal{X} \times \mathbb{C})$ is a union of an increasing sequence of connected algebraic subgroups, see Example 4.2 (a). We need the following slightly stronger result.

Lemma 4.8. *Let $\mathcal{X} \times \mathbb{C}$ be a cylinder over a rigid base. Then the special automorphism group $\mathrm{SAut}(\mathcal{X} \times \mathbb{C}) \subset \mathrm{Aut}_0(\mathcal{X} \times \mathbb{C})$ is a countable increasing union of connected unipotent algebraic subgroups $U_i \subset \mathrm{SAut}(\mathcal{X} \times \mathbb{C})$, which are normal in $\mathrm{Aut}_0(\mathcal{X} \times \mathbb{C})$.*

Proof. The action of $\mathrm{Aut}_0\,\mathcal{X}$ on the normal subgroup $\mathcal{O}_+(\mathcal{X}) \lhd \mathrm{Aut}_0(\mathcal{X} \times \mathbb{C})$ in (29) is given by $b \mapsto b \circ S$ for $b \in \mathcal{O}_+(\mathcal{X})$ and $S \in \mathrm{Aut}_0\,\mathcal{X}$, cf. the proof of Theorem 4.6. The $\mathbb{C}^*$-subgroup in (29) acts on $\mathcal{O}_+(\mathcal{X})$ via homotheties $b \mapsto t^{-1}b$, where $b \in \mathcal{O}_+(\mathcal{X})$ and $t \in \mathbb{C}^*$. Therefore, the linear representation of the product $\mathbb{C}^* \times \mathrm{Aut}_0\,\mathcal{X}$ on $\mathcal{O}_+(\mathcal{X})$ is locally finite. In particular, the subspace $G_i = \{f \in \mathcal{O}_+(\mathcal{X}) \mid \deg f \le i\}$ as in (28) is of finite dimension and is contained in another finite dimensional subspace, say U_i, which is stable under the action of $\mathbb{C}^* \times \mathrm{Aut}_0\,\mathcal{X}$ and hence normal when regarded as a subgroup of $\mathrm{Aut}_0(\mathcal{X} \times \mathbb{C})$. Since the sequence $(G_i)_{i \in \mathbb{N}}$ is increasing, we can choose the sequence $(U_i)_{i \in \mathbb{N}}$ being also increasing. □

Corollary 4.9. *Let $\mathcal{X}$ be a rigid affine variety such that the group $\mathrm{Aut}_0\,\mathcal{X}$ is algebraic.*[8] *Then $\mathrm{Aut}_0(\mathcal{X} \times \mathbb{C}) = \bigcup_{i \in \mathbb{N}} B_i$, where $(B_i)_{i \in \mathbb{N}}$ is an increasing sequence of connected algebraic subgroups.*

Proof. It is enough to let $B_i = U_i \rtimes (\mathbb{C}^* \times \mathrm{Aut}_0\,\mathcal{X})$. □

4.4 *Algebraic Subgroups of* $\mathrm{Aut}_0(\mathcal{X} \times \mathbb{C})$

In this subsection we keep the assumptions of Corollary 4.9. By this corollary the group $\mathrm{Aut}_0(\mathcal{X} \times \mathbb{C})$ is a union of connected affine algebraic subgroups. Hence the notions of semisimple and unipotent elements, and as well of Jordan decomposition are well defined in $\mathrm{Aut}_0(\mathcal{X} \times \mathbb{C})$ due to their invariance. Moreover, by virtue of Remark 4.7 for any connected affine algebraic subgroup G of $\mathrm{Aut}_0(\mathcal{X} \times \mathbb{C})$ the unipotent radical of G equals $G \cap \mathrm{SAut}(\mathcal{X} \times \mathbb{C})$. So $\mathrm{SAut}(\mathcal{X} \times \mathbb{C})$ is the set of all unipotent elements of $\mathrm{Aut}_0(\mathcal{X} \times \mathbb{C})$. The next result shows that, under the assumptions of Corollary 4.9, decomposition (29) can be viewed as an analog of the Mostow decomposition for algebraic groups. Recall that Mostow's version of the Levi–Malcev Theorem [29] (see also [14] or [31, Chap. 2, Sect. 1, Theorem 3]) states that any connected algebraic group over a field of characteristic zero admits

[8]Due to a result of Iitaka [16, Proposition 5] the latter assumption holds if the regular locus $\mathrm{reg}\,\mathcal{X}$ is of non-negative logarithmic Kodaira dimension. Moreover, in this case $\mathrm{Aut}_0\,\mathcal{X}$ is an algebraic torus.

a decomposition into a semidirect product of its unipotent radical and a maximal reductive subgroup. Any two such maximal reductive subgroups are conjugated via an element of the unipotent radical.

Theorem 4.10. *Let $\mathcal{X}$ be a rigid affine variety such that* $\mathrm{Aut}_0\,\mathcal{X}$ *is an algebraic group. Then the following hold.*

(a) *The group* $\mathrm{Aut}_0\,\mathcal{X}$ *is isomorphic to the algebraic torus* $(\mathbb{C}^*)^r$, $r = \dim \mathrm{Aut}_0\,\mathcal{X}$.
(b) *The group* $\mathrm{Aut}_0\,(\mathcal{X}\times\mathbb{C}) \cong \mathcal{O}_+(\mathcal{X}) \rtimes (\mathbb{C}^*)^{r+1}$ *is metabelian.*
(c) *Any connected algebraic subgroup G of* $\mathrm{Aut}\,(\mathcal{X}\times\mathbb{C})$ *is either Abelian or metabelian of rank* $\le r+1$.
(d) *Any algebraic torus in* $\mathrm{Aut}_0\,(\mathcal{X}\times\mathbb{C})$ *is contained in a maximal torus of rank* $r+1$. *Any two maximal tori are conjugated via an element of* $\mathrm{SAut}(\mathcal{X}\times\mathbb{C})$.
(e) *Any semisimple element of* $\mathrm{Aut}_0\,(\mathcal{X}\times\mathbb{C})$ *is contained in a maximal torus. Any finite subgroup of* $\mathrm{Aut}_0\,(\mathcal{X}\times\mathbb{C})$ *is Abelian and contained in a maximal torus.*

Proof. By our assumptions $\mathrm{Aut}_0\,\mathcal{X}$ is a connected linear algebraic group without any unipotent subgroup. Hence by Lemma 3 in [16] this group is an algebraic torus. This proves (a).

By virtue of (29) and (a) we have

$$\mathrm{Aut}_0\,(\mathcal{X}\times\mathbb{C}) \cong \mathcal{O}_+(\mathcal{X}) \rtimes (\mathbb{C}^*)^{r+1}\,. \tag{30}$$

This proves (b).

By Corollary 4.9 the group $\mathrm{Aut}_0\,(\mathcal{X}\times\mathbb{C})$ is covered by an increasing sequence of connected algebraic subgroups $(B_i)_{i\in\mathbb{N}}$. Any algebraic subgroup $G \subset \mathrm{Aut}_0\,(\mathcal{X}\times\mathbb{C})$ is contained in one of them, say $G \subset B_i$, where $B_i = U_i \rtimes (\mathbb{C}^*)^{r+1}$ is metabelian. This proves (c).

Now (d) follows by the classical Mostow Theorem applied to an appropriate subgroup B_i, which contains the tori under consideration.

The same argument proves (e). Indeed, both assertions of (e) hold for connected solvable affine algebraic groups due to Proposition 19.4(a) in [15, Chap. 7]. □

Remark 4.11. The assertion of (c) holds with $r = \dim\mathcal{X}$ even without the assumption that $\mathrm{Aut}_0\,\mathcal{X}$ is an algebraic group. We wonder whether (a) also holds in this generality.

4.5 *The Lie Algebra of* $\mathrm{Aut}_0\,(\mathcal{X}\times\mathbb{C})$

The Lie algebra of an ind-group is defined in [34], see also [20]. For an ind-group G of type $G = \varinjlim_i G_i$, where $(G_i)_{i\in\mathbb{N}}$ is an increasing sequence of connected algebraic subgroups of G, the Lie algebra $\mathrm{Lie}\,(G)$ coincides with the inductive limit $\varinjlim_i \mathrm{Lie}\,(G_i)$. From Corollary 4.9 and decomposition (29) we deduce the following presentation.

Theorem 4.12. *Under the assumptions of Theorem 4.10 we have*

$$\mathrm{Lie}\,(\mathrm{Aut}_0\,(\mathcal{X}\times\mathbb{C})) = I \rtimes L\,, \tag{31}$$

where

$$I \overset{\text{def}}{=\!=} \{b(x)\partial/\partial y \mid b \in \mathcal{O}_+(\mathcal{X})\} = \mathcal{O}_+(\mathcal{X})\partial/\partial y \tag{32}$$

is the Abelian ideal consisting of all locally nilpotent derivations of the algebra $\mathcal{O}(\mathcal{X}\times\mathbb{C})$, *and the Lie subalgebra*

$$L \cong \mathrm{Lie}\left(\mathbb{C}^* \times \mathrm{Aut}_0\,\mathcal{X}\right) \tag{33}$$

corresponding to the second factor in (29) *is a Cartan subalgebra, i.e., a maximal Abelian subalgebra of* Lie (Aut$_0$ ($\mathcal{X}\times\mathbb{C}$)) *consisting of semisimple elements. Furthermore, we have the presentation*

$$\mathrm{Lie}\,(\mathrm{Aut}_0\,(\mathcal{X}\times\mathbb{C})) = \langle b(x)\partial/\partial y,\ y\partial/\partial y,\ \partial \mid b \in \mathcal{O}_+(\mathcal{X}),\ \partial \in \mathrm{Lie}\,(\mathrm{Aut}_0\,\mathcal{X})\rangle \tag{34}$$

and relations

$$[\partial, y\partial/\partial y] = 0,\ [\partial, b\partial/\partial y] = (\partial b)\partial/\partial y,\ \textit{and}\ [b\partial/\partial y, y\partial/\partial y] = b\partial/\partial y \tag{35}$$

for any $b \in \mathcal{O}_+(\mathcal{X})$ *and any* $\partial \in \mathrm{Lie}\,(\mathrm{Aut}_0\,\mathcal{X})$.

Proof. Decomposition (31) is a direct consequence of (29), and (34) follows from (29) and (31). The first relation in (35) follows from the fact that the factors $\mathbb{C}^*$ and $\mathrm{Aut}_0\,\mathcal{X}$ in (29) commute. To show the other two relations it suffices to verify these on the functions of the form $f(x)y^k \in \mathcal{O}(\mathcal{X}\times\mathbb{C}) = \mathcal{O}(\mathcal{X})[y]$, where $k \geq 0$. The latter computation is easy, and so we omit it. □

5 Automorphisms of Configuration Spaces and Discriminant Levels

5.1 Automorphisms of Balanced Spaces

In view of Corollary 2.8, to compute the automorphism groups of the varieties $\mathcal{C}^n$, $\mathcal{SC}^{n-1}$, and Σ^{n-1} we need to know the automorphism groups of the corresponding balanced spaces $\mathcal{C}^{n-1}_{\mathrm{blc}}$, $\mathcal{SC}^{n-2}_{\mathrm{blc}}$, and $\Sigma^{n-2}_{\mathrm{blc}}$. The latter groups have been already described in the literature. We formulate the corresponding results and provide necessary references. Then we give a short argument for (a) based on Tame Map Theorem. The proof of (c) will be done in Sect. 7.

Theorem 5.1. (a) $\mathrm{Aut}\, \mathcal{C}^{n-1}_{\mathrm{blc}} \cong \mathbb{C}^*$ *for* $n > 2$. *Any automorphism* $S \in \mathrm{Aut}\, \mathcal{C}^{n-1}_{\mathrm{blc}}$ *is of the form* $Q^\circ \mapsto sQ^\circ$, *where* $Q^\circ \in \mathcal{C}^{n-1}_{\mathrm{blc}}$ *and* $s \in \mathbb{C}^*$.

(b) $\mathrm{Aut}\, \mathcal{SC}^{n-2}_{\mathrm{blc}} \cong \mathbb{Z}/n(n-1)\mathbb{Z}$ *for* $n > 4$. *Any automorphism* $S \in \mathrm{Aut}\, \mathcal{SC}^{n-2}_{\mathrm{blc}}$ *is of the form* $Q^\circ \mapsto sQ^\circ$, *where* $Q^\circ \in \mathcal{SC}^{n-2}_{\mathrm{blc}}$, $s \in \mathbb{C}^*$, *and* $s^{n(n-1)} = 1$.

(c) $\mathrm{Aut}\Sigma^{n-2}_{\mathrm{blc}} \cong \mathbb{C}^*$ *for* $n > 6$. *Any automorphism* $S \in \mathrm{Aut}\Sigma^{n-2}_{\mathrm{blc}}$ *is of the form* $Q^\circ \mapsto sQ^\circ$, *where* $s \in \mathbb{C}^*$ *and every point* $Q^\circ \in \Sigma^{n-2}_{\mathrm{blc}}$ *is considered as an unordered multiset* $Q^\circ = \{q_1, \ldots, q_n\} \subset \mathbb{C}$ *with at least one repetition.*

Statement (a) is a simple consequence of Tame Map Theorem. It was stated in [21] and [22, Sect. 8.2.1]; we reproduce a short argument below. In Theorem 8.3(c) we give a more general result in the analytic setting.

A proof of (b) is sketched in [19]. Actually, a theorem of Kaliman ([19]) says that *every non-constant holomorphic self-map of* $\mathcal{SC}^{n-2}_{\mathrm{blc}}$ *is a biregular automorphism of the above form.* A complete proof can be found in [26, Theorem 12.13]. This proof exploits the following property of the Artin braid group B_n (see [23, Theorem 7.7] or [25, Theorem 8.9]): *For* $n > 4$, *the intersection* $J_n = B'_n \cap PB_n$ *of the commutator subgroup* B'_n *of* B_n *with the pure braid group* PB_n *is a completely characteristic subgroup of* B'_n.

Our proof of (c) is based on a part of Tame Map Theorem due to Zinde ([35, Theorems 7 and 8]), which describes the automorphisms of the configuration space $\mathcal{C}^n(\mathbb{C}^*)$. Since by Lemma 2.6 $\mathrm{reg}\, \Sigma^{n-2}_{\mathrm{blc}} \cong \mathcal{C}^{n-2}(\mathbb{C}^*)$, from results in [loc. cit.] it follows that for $n > 6$

$$\mathrm{Aut}(\mathrm{reg}\, \Sigma^{n-2}_{\mathrm{blc}}) \cong \mathrm{Aut}\, \mathcal{C}^{n-2}(\mathbb{C}^*) \cong (\mathrm{Aut}\, \mathbb{C}^*) \times \mathbb{Z} \cong (\mathbb{C}^* \times \mathbb{Z}) \rtimes (\mathbb{Z}/2\mathbb{Z})\,.$$

In Theorem 7.1 below we show that only the elements of the connected component $\mathbb{C}^*$ of $\mathrm{Aut}(\mathrm{reg}\, \Sigma^{n-2}_{\mathrm{blc}})$ can be extended to automorphisms of the whole variety $\Sigma^{n-2}_{\mathrm{blc}}$. This implies both assertions in (c).

Proof of (a). The extension F of S to $\mathcal{C}^n$ defined by $F(Q) = S(Q - \mathrm{bc}(Q))$ for all $Q \in \mathcal{C}^n$ is a non-Abelian endomorphism of $\mathcal{C}^n$ such that $F(\mathcal{C}^n) \subset \mathcal{C}^{n-1}_{\mathrm{blc}}$. By Tame Map Theorem and Remark 1.3 (b), there is a unique morphism $T\colon \mathcal{C}^n \to \mathrm{Aff}\, \mathbb{C}$ such that $F = F_T$. Since F preserves $\mathcal{C}^{n-1}_{\mathrm{blc}}$, we have $T(Q)(0) = 0$ for any $Q \in \mathcal{C}^{n-1}_{\mathrm{blc}}$ and hence also for any $Q \in \mathcal{C}^n$. So, $T(Q)\zeta = a(Q)\zeta$ for all $\zeta \in \mathbb{C}$ and $Q \in \mathcal{C}^n$, where $a \in \mathcal{O}^\times(\mathcal{C}^n)$. According to Example 4.4 (a), $a = sD^k_n$ for some $s \in \mathbb{C}^*$ and $k \in \mathbb{Z}$, so that $S(Q) = sD^k_n(Q) \cdot Q$ on $\mathcal{C}^{n-1}_{\mathrm{blc}}$. Similarly, for the inverse automorphism S^{-1} we obtain that $S^{-1}(Q) = tD^l_n \cdot Q$ on $\mathcal{C}^{n-1}_{\mathrm{blc}}$ with some $t \in \mathbb{C}^*$ and $l \in \mathbb{Z}$. Since D_n is a homogeneous function on $\mathcal{C}^n$ (namely, $D_n(sQ) = s^{n(n-1)}Q$ for all $Q \in \mathcal{C}^n$ and $s \in \mathbb{C}^*$), from the identity $S \circ S^{-1} = \mathrm{id}$ we deduce that $k = l = 0$ and $t = s^{-1}$, as required. □

By Theorem 5.1 in all three cases the automorphism groups of the corresponding balanced spaces are algebraic groups. Hence Theorem 4.10 applies and leads to the following corollary.

Corollary 5.2. *The conclusions (b)–(e) of Theorem 4.10 hold for the groups* $\mathrm{Aut}_0\, \mathcal{C}^n$ $(n > 2)$ *and* $\mathrm{Aut}_0\, \Sigma^{n-1}$ $(n > 6)$ *with* $r = 1$ *and for* $\mathrm{Aut}_0\, \mathcal{SC}^{n-1}$ $(n > 4)$ *with* $r = 0$*, where these varieties are viewed as cylinders over the corresponding balanced spaces.*

Remark 5.3. Recall (see Sect. 1) that $\mathrm{Sym}^n\, \mathbb{C}$ viewed as the space of all unordered multisets $Q = \{q_1, \ldots, q_n\} \subset \mathbb{C}$ is a disjoint union of $\mathcal{C}^n$ and Σ^{n-1}. The tautological (Aff $\mathbb{C}$)-action on $\mathbb{C}$ induces the diagonal (Aff $\mathbb{C}$)-action on $\mathrm{Sym}^n\, \mathbb{C}$; both the spaces $\mathcal{C}^n$ and Σ^{n-1} are invariant under the latter action. It follows from Tame Map Theorem and Remark 1.3 (d) that for $n > 2$ the (Aut $\mathcal{C}^n$)-orbits coincide with the orbits of the diagonal (Aff $\mathbb{C}$)-action on $\mathcal{C}^n$ (see [27, Sect. 2.2]). Corollaries 3.6, 5.2 and Theorems 4.10, 5.1 imply that for $n > 6$ the $(\mathrm{Aut}\Sigma^{n-1})$-orbits coincide with the orbits of the above diagonal (Aff $\mathbb{C}$)-action on Σ^{n-1}. For $n > 4$ the (Aut $\mathcal{SC}^{n-1}$)-orbits coincide with the orbits of the subgroup $\mathbb{C} \rtimes (\mathbb{Z}/n(n-1)\mathbb{Z}) \subset \mathrm{Aff}\, \mathbb{C}$ acting on $\mathcal{SC}^{n-1}$.

5.2 *The Groups* Aut $\mathcal{C}^n$, Aut $\mathcal{SC}^{n-1}$, *and* AutΣ^{n-1}

For our favorite varieties $\mathcal{C}^n$, $\mathcal{SC}^{n-1}$, and Σ^{n-1} we dispose at present all necessary ingredients in decomposition (23). Gathering this information we obtain the following description.

Theorem 5.4. *For* $n > 2$,

$$\mathrm{Aut}\, \mathcal{C}^n \cong \left(\mathcal{O}_+(\mathcal{C}^{n-1}_{\mathrm{blc}}) \rtimes (\mathbb{C}^*)^2\right) \rtimes \mathbb{Z}\,. \tag{36}$$

For $n > 4$,

$$\mathrm{Aut}\, \mathcal{SC}^{n-1} \cong \mathcal{O}_+(\mathcal{SC}^{n-2}_{\mathrm{blc}}) \rtimes \left(\mathbb{C}^* \times (\mathbb{Z}/n(n-1)\mathbb{Z})\right)\,. \tag{37}$$

For $n > 6$,

$$\mathrm{Aut}\Sigma^{n-1} \cong \mathcal{O}_+(\Sigma^{n-1}_{\mathrm{blc}}) \rtimes (\mathbb{C}^*)^2\,. \tag{38}$$

So, the group Aut $\mathcal{C}^n$ *is solvable, whereas* Aut $\mathcal{SC}^{n-1}$ *and* AutΣ^{n-1} *are metabelian. Furthermore, any finite subgroup of each of these automorphism groups is Abelian.*

Proof. As in the proof of Theorem 4.6, one can show that the factor $\mathbb{C}^*$ of the group of units on the corresponding balanced space commutes with the last factor in (23). Taking this into account, the isomorphisms in (36)–(38) are obtained after substitution of the factors in (23) using Examples 4.4 and Theorem 5.1.

For the connected group AutΣ^{n-1} in (38) the last assertion holds due to Theorem 4.10. The same argument applies in the case of Aut $\mathcal{C}^n$. Indeed, the decomposition in (36) provides a surjection $\eta : \operatorname{Aut} \mathcal{C}^n \to \mathbb{Z}$, and any finite subgroup of $\operatorname{Aut} \mathcal{C}^n$ is contained in the kernel $\ker \eta = \operatorname{Aut}_0 \mathcal{C}^n$.

The isomorphism in (37) yields a surjection $\operatorname{Aut} \mathcal{SC}^{n-1} \to \mathbb{C}^* \times (\mathbb{Z}/n(n-1)\mathbb{Z})$ with a torsion free kernel $\operatorname{SAut}\mathcal{SC}^{n-1} \cong \mathcal{O}_+(\mathcal{SC}^{n-2}_{\mathrm{blc}})$. Since any finite subgroup of $\operatorname{Aut} \mathcal{SC}^{n-1}$ meets this kernel just in the neutral element, it injects into the Abelian group $\mathbb{C}^* \times (\mathbb{Z}/n(n-1)\mathbb{Z})$ and so is Abelian. □

5.3 Automorphisms of $\mathcal{C}^n$, $\mathcal{SC}^{n-1}$, and Σ^{n-1}

All three varieties can be viewed as subspaces of the nth symmetric power $\operatorname{Sym}^n \mathbb{C} = \mathbb{C}^n_{(q)}/\mathbf{S}(n) \cong \mathbb{C}^n_{(z)}$. Elements Q of one of the spaces $\mathcal{C}^n$ or $\mathcal{SC}^{n-1} \subset \mathcal{C}^n$ are, as before, n-point configurations in $\mathbb{C}$, whereas the discriminant variety $\Sigma^{n-1} = (\operatorname{Sym}^n \mathbb{C}) \setminus \mathcal{C}^n$ consists of all unordered multisets $Q = \{q_1, \ldots, q_n\} \subset \mathbb{C}$ with at least one repetition (see Sect. 1).

Theorem 5.5. *Let $\mathcal{Z}$ be one of the varieties $\mathcal{C}^n$ $(n > 2)$, $\mathcal{SC}^{n-1}$ $(n > 4)$, or Σ^{n-1} $(n > 6)$, and let $\mathcal{Z}_{\mathrm{blc}}$ be the corresponding balanced space* (see *(8)). A map $F : \mathcal{Z} \to \mathcal{Z}$ is an automorphism if and only if*

$$F(Q) = sQ^\circ + a(Q^\circ)\,\mathrm{bc}(Q) + b(Q^\circ) \ \text{ for all } \ Q \in \mathcal{Z}\,, \qquad (39)$$

where $Q^\circ = Q - \mathrm{bc}(Q) \in \mathcal{Z}_{\mathrm{blc}}$, $s \in \mathbb{C}^*$, $b \in \mathcal{O}_+(\mathcal{Z}_{\mathrm{blc}})$, and

(a) *$a = tD_n^k$ with $t \in \mathbb{C}^*$ and $k \in \mathbb{Z}$, if $\mathcal{Z} = \mathcal{C}^n$;*
(b) *$a \in \mathbb{C}^*$ and $s^{n(n-1)} = 1$, if $\mathcal{Z} = \mathcal{SC}^{n-1}$;*
(c) *$a \in \mathbb{C}^*$, if $\mathcal{Z} = \Sigma^{n-1}$.*

Proof. Let F be an automorphism of the cylinder $\mathcal{Z} = \mathcal{Z}_{\mathrm{blc}} \times \mathbb{C}$ (cf. (10)). According to Corollary 2.8, Theorem 5.1, and Example 4.4, F is triangular of the form

$$F(Q) = (sQ^\circ, a(Q^\circ)\,\mathrm{bc}(Q) + b(Q^\circ)) = sQ^\circ + a(Q^\circ)\,\mathrm{bc}(Q) + b(Q^\circ)\,,$$

where in each of the cases (a), (b), (c) data s, a, and b are as stated above. Conversely, such F with s, a, and b as in one of the cases (a), (b), (c) is a (triangular) automorphism of $\mathcal{Z}_{\mathrm{blc}} \times \mathbb{C} = \mathcal{Z}$ corresponding to the automorphism $S\colon Q^\circ \mapsto sQ^\circ$ of $\mathcal{Z}_{\mathrm{blc}}$ and the morphism $A\colon \mathcal{Z}_{\mathrm{blc}} \to \operatorname{Aff} \mathbb{C}$, $A(Q^\circ)\colon \zeta \mapsto a\zeta + b$ for all $Q^\circ \in \mathcal{Z}_{\mathrm{blc}}$ and $\zeta \in \mathbb{C}$. □

Remarks 5.6. (a) Consider the algebraic torus $\mathcal{T}$ of rank 2 consisting of all transformations

$$\nu(s,t)\colon Q \mapsto s \cdot (Q - \mathrm{bc}(Q)) + t\,\mathrm{bc}(Q)\,, \qquad (40)$$

where $(s,t) \in (\mathbb{C}^*)^2$ and $Q \in \operatorname{Sym}^n \mathbb{C}$.

Both subspaces $\mathcal{C}^n, \Sigma^{n-1} \subset \mathrm{Sym}^n \mathbb{C}$ are invariant under this action. In fact, $\mathcal{T}$ is a maximal torus in each of the corresponding automorphism groups. The subgroup of $\mathcal{T}$ given by $s^{n(n-1)} = 1$ and isomorphic to $(\mathbb{Z}/n(n-1)\mathbb{Z}) \times \mathbb{C}^*$ acts on $\mathcal{SC}^{n-1}$.

(b) As in Theorem 5.5, let $\mathcal{Z}$ be one of the varieties $\mathcal{C}^n$ $(n > 2)$, $\mathcal{SC}^{n-1}$ $(n > 4)$, or Σ^{n-1} $(n > 6)$, and $\mathcal{Z}_{\mathrm{blc}}$ be the corresponding balanced space. Using Proposition 3.3 one can deduce the following: *Any $\mathbb{C}_+$-action on $\mathcal{Z}$ is of the form*

$$Q \mapsto Q + \lambda b(Q - \mathrm{bc}(Q)), \quad \textit{where } Q \in \mathcal{Z}, \ \lambda \in \mathbb{C}_+, \ b \in \mathcal{O}(\mathcal{Z}_{\mathrm{blc}}). \quad (41)$$

The case $b = 1$ corresponds to the τ-action.

(c) It follows from (39) that $F = F_T$ with T as in (14) for any F in one of the above groups.

5.4 *The Group* Aut $(\mathbb{C}^n, \Sigma^{n-1})$

The space $\mathrm{Sym}^n \mathbb{C} \cong \mathcal{C}^n \cup \Sigma^{n-1}$ of all unordered n-multisets $Q = \{q_1, \ldots, q_n\} \subset \mathbb{C}$ can be identified with the space $\mathbb{C}^n_{(z)} \cong \mathbb{C}^n$ of all polynomials (1). The corresponding balanced space $\mathcal{C}^{n-1}_{\mathrm{blc}} \cup \Sigma^{n-2}_{\mathrm{blc}} \cong \mathbb{C}^{n-1}$ consists of all polynomials $\lambda^n + z_2\lambda^{n-2} + \cdots + z_n$. An automorphism F of $\mathcal{C}^n$ as in (39) extends to an endomorphism of the ambient affine space $\mathbb{C}^n$ if and only if the rational functions $a(Q - \mathrm{bc}(Q))$ and $b(Q - \mathrm{bc}(Q))$ on $\mathbb{C}^n$ in (39) are regular, i.e., $a, b \in \mathcal{O}(\mathbb{C}^n_{\mathrm{blc}}) \cong \mathbb{C}[z_2, \ldots, z_n]$. Such an endomorphism F admits an inverse, say F', on $\mathbb{C}^n$ if and only if the corresponding functions a' and b' are also regular i.e. $a', b' \in \mathcal{O}(\mathbb{C}^n_{\mathrm{blc}})$. In particular $a = \mathrm{const} \in \mathbb{C}^*$. This leads to the following description.

Theorem 5.7. *For any $n > 2$ we have*

$$\mathrm{Aut}\,(\mathbb{C}^n, \Sigma^{n-1}) \cong \mathbb{C}[z_2, \ldots, z_n] \rtimes (\mathbb{C}^*)^2,$$

where the 2-torus $(\mathbb{C}^*)^2$ and the group $\mathbb{C}[z_2, \ldots, z_n] \cong \mathcal{O}_+(\mathbb{C}^{n-1})$ act on $\mathbb{C}^n \cong \mathrm{Sym}^n \mathbb{C}$ via (40) *and* (41) *with $\lambda = 1$, respectively.* □

5.5 *The Lie Algebras* Lie $(\mathrm{Aut}_0\, \mathcal{C}^n)$, Lie $(\mathrm{Aut}_0\, \mathcal{SC}^{n-1})$, *and* Lie $(\mathrm{Aut}_0\, \Sigma^{n-1})$

5.5.1 The Lie Algebra Lie $(\mathrm{Aut}_0\, \mathcal{C}^n)$

Let $\partial_\tau \in \mathrm{LND}(\mathcal{O}(\mathcal{C}^n))$ be the infinitesimal generator of the $\mathbb{C}_+$-action τ on $\mathcal{C}^n \subset \mathbb{C}^n_{(z)}$. By (31) and Remark 5.6 (b) for $n > 2$ there is the Levi–Malcev–Mostow decomposition

$$\text{Lie}\,(\text{Aut}_0\,\mathcal{C}^n) = I \oplus \text{Lie}\mathcal{T}$$

with Abelian summands, where $I = \mathcal{O}(\mathcal{C}^{n-1}_{\text{blc}})\partial_\tau$ is as in (32) and the 2-torus $\mathcal{T} \subset \text{Aut}_0\,\mathcal{C}^n$ as in Remark 5.6 (b) consists of all transformations $\nu(s,t)$ as in (40) with $(s,t) \in (\mathbb{C}^*)^2$ and $Q \in \mathcal{C}^n$. Thus $\mathcal{T}$ is the direct product of two its 1-subtori with infinitesimal generators, say ∂_s and ∂_t, respectively. These derivations are locally finite and locally bounded on $\mathcal{O}(\mathcal{C}^n)$ and their sum $\partial_s + \partial_t$ is the infinitesimal generator of the $\mathbb{C}^*$-action $Q \mapsto \lambda Q$ ($\lambda \in \mathbb{C}^*$) on $\mathcal{C}^n$. With this notation we have the following description.

Proposition 5.8. *For $n > 2$ the Lie algebra*

$$\text{Lie}\,(\text{Aut}_0\,\mathcal{C}^n) = \langle I, \partial_s, \partial_t \rangle\,, \quad \textit{where} \quad I = \mathcal{O}(\mathcal{C}^{n-1}_{\text{blc}})\partial_\tau\,, \tag{42}$$

is uniquely determined by the commutator relations

$$[\partial_s, \partial_t] = 0, \quad [\partial_s, b\partial_\tau] = (\partial_s b)\partial_\tau, \quad \textit{and} \quad [b\partial_\tau, \partial_t] = b\partial_\tau\,, \tag{43}$$

where b runs over $\mathcal{O}(\mathcal{C}^{n-1}_{\text{blc}})$. Furthermore, in the coordinates $z_1, \ldots, z_n$ in $\mathbb{C}^n = \mathbb{C}^n_{(z)}$ the derivations ∂_τ, ∂_t, and ∂_s are given by

$$\partial_\tau = \sum_{i=1}^{n}(n-i+1)z_{i-1}\frac{\partial}{\partial z_i}, \quad \partial_t = (-z_1/n)\partial_\tau\,, \quad \textit{and} \quad \partial_s = \sum_{k=1}^{n} kz_k\frac{\partial}{\partial z_k} - \partial_t\,, \tag{44}$$

where $z_0 \stackrel{\text{def}}{=\!=} 1$.

Proof. From (34) and (35) in Theorem 4.12 we obtain (42) and (43), respectively. The diagonal $\mathbb{C}_+$-action $(q_1, \ldots, q_n) \mapsto (q_1 + \lambda, \ldots, q_n + \lambda)$, $\lambda \in \mathbb{C}_+$, on the affine space $\mathbb{C}^n_{(q)}$ has for infinitesimal generator the derivation

$$\partial^{(n)} = \sum_{i=1}^{n}\frac{\partial}{\partial q_i} \in \text{LND}(\mathbb{C}[q_1, \ldots, q_n])\,.$$

This $\mathbb{C}_+$-action on $\mathbb{C}^n_{(q)}$ descends to the $\mathbb{C}_+$-action τ on the base of the Vieta covering

$$p\colon \mathbb{C}^n_{(q)} \to \mathbb{C}^n_{(q)}/\mathbf{S}(n) = \mathbb{C}^n_{(z)}, \quad (q_1, \ldots, q_n) \mapsto (z_1, \ldots, z_n)\,,$$

where $z_i = (-1)^i\sigma_i(q_1, \ldots, q_n)$ with σ_i being the elementary symmetric polynomial of degree i. We have $\partial^{(n)}(\sigma_i) = (n-i+1)\sigma_{i-1}$. Hence in the coordinates $z_1, \ldots, z_n$ on $\mathbb{C}^n_{(z)}$ the infinitesimal generator ∂_τ of τ is given by the first equality in (44).

The derivations ∂_s, ∂_t, and ∂_τ preserve the subring $\mathbb{C}[z_1, \ldots, z_n] \subset \mathcal{O}(\mathcal{C}^n)$ and admit natural extensions from $\mathbb{C}[z_1, \ldots, z_n]$ to $\mathbb{C}[q_1, \ldots, q_n]$ denoted by the same symbols, where

$$\partial_\tau : q_i \mapsto 1, \quad \partial_t : q_i \mapsto \frac{1}{n}\sum_{k=1}^n q_k, \quad \text{and} \quad \partial_s : q_i \mapsto q_i - \frac{1}{n}\sum_{k=1}^n q_k, \quad i = 1, \ldots, n.$$

It follows that $\partial_t = (-z_1/n)\partial_\tau$, which yields the second equality in (44). Applying these derivations to the coordinate functions $z_i = (-1)^i \sigma_i(q_1, \ldots, q_n)$ the last equality in (44) follows as well. □

5.5.2 The Lie Algebra Lie $(\mathrm{Aut}_0\, \Sigma^{n-1})$

We have seen in the proof of Proposition 5.8 that the $\mathbb{C}_+$-action τ and the action of the 2-torus $\mathcal{T}$ on $\mathcal{C}^n$ extend regularly to the ambient affine space $\mathbb{C}^n_{(z)}$, along with the derivations ∂_τ, ∂_t, and ∂_s given by (44). The discriminant d_n on $\mathbb{C}^n_{(z)}$ is invariant under τ. Hence $\partial_\tau d_n = 0$ and so the complete vector field ∂_τ is tangent along the level hypersurfaces of d_n, in particular, along $\mathcal{SC}^{n-1} = \{d_n = 1\}$ and $\Sigma^{n-1} = \{d_n = 0\}$. The induced locally nilpotent derivations of the structure rings $\mathcal{O}(\mathcal{SC}^{n-1})$ and $\mathcal{O}(\Sigma^{n-1})$ will be still denoted by ∂_τ.

The action of the 2-torus $\mathcal{T}$ on $\mathbb{C}^n_{(z)}$ stabilizes Σ^{n-1}. Hence ∂_t and ∂_s generate commuting semisimple derivations of $\mathcal{O}(\Sigma^{n-1})$ denoted by the same symbols. Using these observations and notation we can deduce from Theorem 4.12 and Corollary 5.2 the following description (cf. [28]).

Proposition 5.9. *For $n > 6$ the Lie algebra*

$$\mathrm{Lie}\,(\mathrm{Aut}_0\, \Sigma^{n-1}) = \langle I, \partial_s, \partial_t \rangle, \quad \textit{where} \quad I = \mathcal{O}_+(\Sigma^{n-2}_{\mathrm{blc}})\partial_\tau,$$

is uniquely determined by relations (43), *where b runs over $\mathcal{O}_+(\Sigma^{n-2}_{\mathrm{blc}})$.*

Proof. The proof goes along the same lines as that of Proposition 5.8, and so we leave it to the reader. □

5.5.3 The Lie Algebra Lie $(\mathrm{Aut}_0\, \mathcal{SC}^{n-1})$

Since $\partial_\tau d_n = 0$, for the derivation $\partial_t = (-z_1/n)\partial_\tau$ (see (44)) we have $\partial_t d_n = 0$. Hence the vector field ∂_t is tangent as well to each of the level hypersurfaces of d_n. In particular, ∂_t induces a semisimple derivation of $\mathcal{O}(\mathcal{SC}^{n-1})$ (denoted again by ∂_t) and generates a $\mathbb{C}^*$-action T on $\mathcal{SC}^{n-1}$. So we arrive at the following description.

Proposition 5.10. *For $n > 4$ the Lie algebra*

$$\mathrm{Lie}\,(\mathrm{Aut}_0\, \mathcal{SC}^{n-1}) = \langle I, \partial_t \rangle, \quad \textit{where} \quad I = \mathcal{O}_+(\mathcal{SC}^{n-2}_{\mathrm{blc}})\partial_\tau,$$

is uniquely determined by the relations $[b\partial_\tau, \partial_t] = b\partial_\tau$, *where* b *runs over* $\mathcal{O}_+(\mathcal{SC}^{n-2}_{\text{blc}})$.

Proof. This follows from Theorem 4.12 and Corollary 5.2 in the same way as before. We leave the details to the reader. □

6 More on the Group Aut $(\mathcal{X} \times \mathbb{C})$

6.1 The Center of Aut $(\mathcal{X} \times \mathbb{C})$

The following lemma provides a formula for the commutator of two triangular automorphisms of a product $\mathcal{X} \times \mathbb{C}$. We let

$$F = F(S, A) \colon (x, y) \mapsto (Sx, A(x)y)\,,$$

where $(x, y) \in \mathcal{X} \times \mathbb{C}$, $S \in \operatorname{Aut} \mathcal{X}$, and $A : \mathcal{X} \to \operatorname{Aff} \mathbb{C}$ (cf. (24)).

Lemma 6.1. *If the group* Aut $\mathcal{X}$ *is Abelian, then for any* $F = F(S, A)$ *and* $F' = F(S', A')$ *in* $\operatorname{Aut}_\Delta (\mathcal{X} \times \mathbb{C})$ *and any* $(x, y) \in \mathcal{X} \times \mathbb{C}$ *we have*

$$[F', F](x, y) = (x, (A'(x))^{-1}(A(S'x))^{-1}A'(Sx)A(x))y)\,. \tag{45}$$

Consequently, F *and* F' *commute if and only if*

$$A(S'x)A'(x) = A'(Sx)A(x) \quad \textit{for any} \quad x \in \mathcal{X}\,. \tag{46}$$

Proof. The proof is straightforward. □

Applying this lemma to general cylinders we deduce the following facts.

Proposition 6.2. *Let* $\mathcal{X}$ *be an affine variety. If the group* Aut $\mathcal{X}$ *is Abelian, then the center of the group* $\operatorname{Aut}_\Delta (\mathcal{X} \times \mathbb{C})$ *is trivial. The same conclusion holds for the groups* Aut $\mathcal{C}^n$ $(n > 2)$, Aut $\mathcal{SC}^{n-1}$ $(n > 4)$, *and* AutΣ^{n-1} $(n > 6)$.

Proof. Consider two elements $F = F(S, A)$ and $F' = F(S', A')$ in $\operatorname{Aut}_\Delta (\mathcal{X} \times \mathbb{C})$, where $S, S' \in \operatorname{Aut} \mathcal{X}$ and $A \colon y \mapsto ay + b$, $A' \colon y \mapsto a'y + b'$ with $a, a' \in \mathcal{O}^\times(\mathcal{X})$ and $b, b' \in \mathcal{O}_+(\mathcal{X})$. If F and F' commute then (46) is equivalent to the system

$$(a \circ S') \cdot a' = a \cdot (a' \circ S) \quad \text{and} \quad (a' \circ S) \cdot b + b' \circ S = (a \circ S') \cdot b' + b \circ S'\,. \tag{47}$$

Assume that F is a central element, i.e. (47) holds for any F'. Letting in the second relation $S' = \text{id}$, $a' = 2$, and $b' = 0$ yields $b = 0$. Now this relation reduces to

$$b' \circ S = (a \circ S') \cdot b'\,.$$

Letting $b' = 1$ yields $a = 1$ and so $A = \mathrm{id}$ and $b' \circ S = b'$ for any $b' \in \mathcal{O}_+(\mathcal{X})$. If $S \neq \mathrm{id}$ this leads to a contradiction, provided that b' is non-constant on an S-orbit in $\mathcal{X}$. Hence $S = \mathrm{id}$ and so $F = \mathrm{id}$, as claimed.

The last assertion follows now from Corollary 2.4, since the bases of the cylinders $\mathcal{C}^n$, $\mathcal{SC}^{n-1}$, and Σ^{n-1} are rigid varieties with Abelian automorphism groups, see (10), Proposition 2.7, and Theorem 5.1. □

6.2 Commutator Series

Let us introduce the following notation.

Notation. Let $\mathcal{X}$ be a rigid variety, and let $\mathcal{D} \subseteq \mathrm{Aut}\,(\mathcal{X} \times \mathbb{C})$ be the subgroup consisting of all automorphisms of the form $F = F(\mathrm{id}, A)$, where $A\colon y \mapsto ty + b$ with $t \in \mathbb{C}^*$ and $b \in \mathcal{O}_+(\mathcal{X})$. It is easily seen that

$$\mathrm{SAut}\,(\mathcal{X} \times \mathbb{C}) \lhd \mathcal{D} \lhd \mathrm{Aut}_0\,(\mathcal{X} \times \mathbb{C})\,.$$

Furthermore, $\mathcal{D} \cong \mathcal{O}_+(\mathcal{X}) \rtimes \mathbb{C}^*$ under the isomorphism in (29), with quotient group $\mathrm{Aut}_0\,(\mathcal{X} \times \mathbb{C})/\mathcal{D} \cong \mathrm{Aut}_0\,\mathcal{X}$. In particular, for $\mathcal{X} \times \mathbb{C} \cong \mathcal{SC}^{n-1}$, $n > 4$, we have $\mathcal{D} = \mathrm{Aut}_0\,(\mathcal{X} \times \mathbb{C})$, see Corollary 2.4 and Theorem 5.4.

It is known ([15, Chap. 7, Theorem 19.3(a)]) that for any connected solvable affine algebraic group G the commutator subgroup $[G, G]$ is contained in the unipotent radical G_{u} of G. In our setting a similar result holds.

Theorem 6.3. *Let $\mathcal{X}$ be a rigid affine variety. If the group $\mathrm{Aut}_0\,\mathcal{X}$ is Abelian, then*

$$[\mathrm{Aut}_0\,(\mathcal{X} \times \mathbb{C}), \mathrm{Aut}_0\,(\mathcal{X} \times \mathbb{C})] = [\mathcal{D}, \mathcal{D}] = \mathrm{SAut}\,(\mathcal{X} \times \mathbb{C})\,. \tag{48}$$

Consequently, the commutator series of the group $\mathrm{Aut}_0\,(\mathcal{X} \times \mathbb{C})$ is

$$1 \lhd \mathrm{SAut}(\mathcal{X} \times \mathbb{C}) \lhd \mathrm{Aut}_0\,(\mathcal{X} \times \mathbb{C})\,. \tag{49}$$

The same conclusions hold for the groups $\mathrm{Aut}_0\,\mathcal{C}^n$ $(n > 2)$, $\mathrm{Aut}_0\,\mathcal{SC}^{n-1}$ $(n > 4)$, and $\mathrm{Aut}\,\Sigma^{n-1} = \mathrm{Aut}_0\,\Sigma^{n-1}$ $(n > 6)$.

Proof. By (29), $\mathrm{SAut}\,(\mathcal{X} \times \mathbb{C}) \lhd \mathrm{Aut}_0\,(\mathcal{X} \times \mathbb{C})$ is a normal subgroup with the Abelian quotient

$$\mathrm{Aut}_0\,(\mathcal{X} \times \mathbb{C})/\mathrm{SAut}\,(\mathcal{X} \times \mathbb{C}) \cong \mathbb{C}^* \times \mathrm{Aut}\,\mathcal{X}\,.$$

Hence $[\mathrm{Aut}_0\,(\mathcal{X} \times \mathbb{C}), \mathrm{Aut}_0\,(\mathcal{X} \times \mathbb{C})] \subset \mathrm{SAut}\,(\mathcal{X} \times \mathbb{C})$. To show (48) it suffices to establish the inclusion $\mathrm{SAut}\,(\mathcal{X} \times \mathbb{C}) \subset [\mathcal{D}, \mathcal{D}]$. However, by virtue of (45) any $F = F(\mathrm{id}, A) \in \mathrm{SAut}\,(\mathcal{X} \times \mathbb{C})$, where $A : y \mapsto y + b$ with $b \in \mathcal{O}_+(\mathcal{X})$, can be written as commutator $F = [F', F'']$, where $F' = F(\mathrm{id}, A')$ and $F'' = F(\mathrm{id}, A'')$ with $A' : y \mapsto -y - b/2$ and $A'' : y \mapsto y + b/2$ (in fact, $A = [A', A'']$).

Now (49) follows from (48), since the group SAut $(\mathcal{X}\times\mathbb{C}) \cong \mathcal{O}_+(\mathcal{X})$ is Abelian.

By Theorem 5.1, the groups $\mathrm{Aut}_0\,\mathcal{C}^n$ $(n > 2)$, $\mathrm{Aut}_0\,\mathcal{SC}^{n-1}$ $(n > 4)$, and $\mathrm{Aut}\;\Sigma^{n-1} = \mathrm{Aut}_0\Sigma^{n-1}$ $(n > 6)$ satisfy our assumptions. So, the conclusions hold also for these groups. □

For the group Aut $\mathcal{SC}^{n-1}$ the following hold.

Theorem 6.4. *For* $n > 4$ *we have* $[\mathrm{Aut}\;\mathcal{SC}^{n-1}, \mathrm{Aut}\;\mathcal{SC}^{n-1}] = \mathrm{SAut}\mathcal{SC}^{n-1}$. *Hence the commutator series of* Aut $\mathcal{SC}^{n-1}$ *is*

$$1 \lhd \mathrm{SAut}\mathcal{SC}^{n-1} \lhd \mathrm{Aut}\;\mathcal{SC}^{n-1}$$

with the Abelian normal subgroup $\mathrm{SAut}\mathcal{SC}^{n-1} \cong \mathcal{O}_+(\mathcal{SC}^{n-2}_{\mathrm{blc}})$ *and the Abelian quotient group*

$$\mathrm{Aut}\;\mathcal{SC}^{n-1}/\mathrm{SAut}\mathcal{SC}^{n-1} \cong \mathbb{C}^* \times (\mathbb{Z}/n(n-1)\mathbb{Z})\,. \tag{50}$$

Proof. Theorem 6.3 yields the inclusion $\mathrm{SAut}\mathcal{SC}^{n-1} \subset [\mathrm{Aut}\;\mathcal{SC}^{n-1}, \mathrm{Aut}\;\mathcal{SC}^{n-1}]$. The opposite inclusion follows from (50), which is in turn a consequence of Theorem 5.4. Hence the assertions follow. □

Consider further the group Aut $\mathcal{C}^n$. Notice that the quotient groups

$$(\mathrm{Aut}\;\mathcal{C}^n)/\mathcal{D} \cong \mathbb{C}^* \times \mathbb{Z} \quad \text{and} \quad \mathcal{D}/\mathrm{SAut}\mathcal{C}^n \cong \mathbb{C}^* \tag{51}$$

are Abelian, see Theorem 5.4. Hence

$$[\mathrm{Aut}\;\mathcal{C}^n, \mathrm{Aut}\;\mathcal{C}^n] \subseteq \mathcal{D}, \quad \text{where} \quad [\mathcal{D}, \mathcal{D}] = \mathrm{SAut}\mathcal{C}^n\,, \tag{52}$$

see Theorem 6.3. More precisely, the following holds.

Theorem 6.5. *For* $n > 2$ *we have* $[\mathrm{Aut}\;\mathcal{C}^n, \mathrm{Aut}\;\mathcal{C}^n] = \mathcal{D}$. *Hence the commutator series of the group* Aut $\mathcal{C}^n$ *is*

$$1 \lhd \mathrm{SAut}\mathcal{C}^n \lhd \mathcal{D} \lhd \mathrm{Aut}\;\mathcal{C}^n$$

with Abelian quotient groups, see (51).

Proof. By virtue of (52) to establish the first equality it suffices to prove the inclusion $\mathcal{D} \subseteq [\mathrm{Aut}\;\mathcal{C}^n, \mathrm{Aut}\;\mathcal{C}^n]$. We show below that, moreover, any element $F_0 \in \mathcal{D}$ is a product of two commutators in Aut $\mathcal{C}^n$.

Indeed, choosing as before F and F' in $\mathcal{D}$ such that $[F', F]\colon Q \to Q + b(Q)$ and replacing F_0 by $[F', F]^{-1}F_0$ we may suppose that $F_0 = F(\mathrm{id}, A_0)$, where $A_0\colon y \mapsto ty$ with $t \in \mathbb{C}^*$.

Let $\tilde{F} = F(S, A)$ and $\tilde{F}' = F(S', A')$, where

$$S: Q^\circ \mapsto sQ^\circ, \quad S': Q^\circ \mapsto s'Q^\circ, \quad \text{and}$$
$$A(Q^\circ): y \mapsto sD_n^k(Q^\circ)y, \quad A'(Q^\circ): y \mapsto s'D_n^{k'}(Q^\circ)y$$

for $Q^\circ \in \mathcal{C}_{\text{blc}}^{n-1}$ and $y \in \mathbb{C}$. By (45) we obtain

$$[\tilde{F}', \tilde{F}](Q) = F(\text{id}, A''), \quad \text{where} \quad A'' : y \mapsto (s^{k'} s'^{-k})^{n(n-1)} y$$

does not depend on $Q^\circ \in \mathcal{C}_{\text{blc}}^{n-1}$. Given $t \in \mathbb{C}^*$ we can find $s, s' \in \mathbb{C}^*$ and $k, k' \in \mathbb{Z}$ such that $(s^{k'} s'^{-k})^{n(n-1)} = t$. With this choice, $F_0 = [\tilde{F}', \tilde{F}]$ and we are done. □

6.3 *Torsion in* $\mathrm{Aut}_0\,(\mathcal{X} \times \mathbb{C})$

We let $\mathcal{T}$ denote the maximal torus in $\mathrm{Aut}_0\,(\mathcal{X} \times \mathbb{C})$ which corresponds to the factor $\mathbb{C}^* \times \mathrm{Aut}_0\,\mathcal{X} \cong (\mathbb{C}^*)^{r+1}$ under the isomorphisms as in (29) and (30). From Theorem 4.10 ((d) and (e)) we deduce the following proposition.

Proposition 6.6. *Under the assumptions of Theorem 4.10 any semisimple (in particular, any torsion) element of the group* $\mathrm{Aut}_0\,(\mathcal{X} \times \mathbb{C})$ *is conjugate to an element of the maximal torus* $\mathcal{T}$ *via an element of the unipotent radical* $\mathrm{SAut}(\mathcal{X} \times \mathbb{C})$. *The same conclusion holds for any finite subgroup of* $\mathrm{Aut}_0\,(\mathcal{X} \times \mathbb{C})$.

Using this proposition we arrive at the following description of the semisimple and torsion elements in the automorphism group of a cylinder over a rigid base.

Corollary 6.7. *Under the assumptions of Theorem 4.10 an element* $F \in \mathrm{Aut}_0\,(\mathcal{X} \times \mathbb{C})$ *is semisimple if and only if it can be written as*

$$F: (x, y) \mapsto (Sx, ty + tb(x) - b(Sx)), \quad \textit{where} \quad (x, y) \in \mathcal{X} \times \mathbb{C}, \tag{53}$$

for some triplet (S, t, b) *with* $S \in \mathrm{Aut}_0\,\mathcal{X}$, $t \in \mathbb{C}^*$, *and* $b \in \mathcal{O}(\mathcal{X})$. *Such an element* F *is torsion with* $F^m = \text{id}$ *if and only if* $S^m = \text{id}$ *and* $t^m = 1$.

Let $\mathcal{Z}$ be one of the varieties $\mathcal{C}^n$ $(n > 2)$, $\mathcal{SC}^{n-1}$ $(n > 4)$, or Σ^{n-1} $(n > 6)$, and let $\mathcal{Z}_{\text{blc}}$ be the corresponding balanced variety. Denote by $G_{\mathcal{Z}}$ one of the groups $\mathrm{Aut}\,\mathcal{C}^n$, $\mathrm{Aut}_0\,\mathcal{SC}^{n-1}$, and $\mathrm{Aut}\,\Sigma^{n-1}$. With this notation we have the following results.

Theorem 6.8. *The semisimple elements of the group* $G_{\mathcal{Z}}$ *are precisely the automorphisms of the form*

$$F: Q \mapsto sQ^\circ + t \cdot \text{bc}(Q) + t \cdot b(Q^\circ) - b(sQ^\circ) \quad \textit{for all} \ Q \in \mathcal{Z}, \tag{54}$$

where $Q^\circ = Q - \text{bc}(Q) \in \mathcal{Z}_{\text{blc}}$, $b \in \mathcal{O}_+(\mathcal{Z}_{\text{blc}})$, *and* $s, t \in \mathbb{C}^*$, *with* $s = 1$ *when* $\mathcal{Z} = \mathcal{SC}^{n-1}$. *Such an element* F *is torsion with* $F^m = \text{id}$ *if and only if, in addition,* $s^m = t^m = 1$.

Proof. Since $\operatorname{Aut} \mathcal{C}^n / \operatorname{Aut}_0 \mathcal{C}^n \cong \mathbb{Z}$, the torsion elements of $\operatorname{Aut} \mathcal{C}^n$ are that of the neutral component $\operatorname{Aut}_0 \mathcal{C}^n$. Taking into account that the group $\operatorname{Aut} \Sigma^{n-1}$ ($n > 6$) is connected, in all three cases Corollary 6.7 applies and yields the result after a simple calculation using the description in (40) and (41). □

In Example 6.11 below we construct some particular torsion elements of the group $\operatorname{Aut} \mathcal{C}^n$. We show that for any $b \in \mathcal{O}_+(\mathcal{C}^{n-1}_{\text{blc}})$ there is an element $F \in \operatorname{Tors}(\operatorname{Aut} \mathcal{C}^n)$ of the form

$$F\colon Q \mapsto sQ^\circ + t\,\text{bc}(Q) + b(Q^\circ), \quad \text{where} \quad Q^\circ = Q - \text{bc}(Q). \tag{55}$$

We use the following lemma. Its proof proceeds by induction on m; we leave the details to the reader.

Lemma 6.9. *Let $n > 2$, and let $F \in \operatorname{Aut}_0 \mathcal{C}^n$ be given by* (55)*. Letting $Q^\circ = Q - \text{bc}(Q)$ for any $m \in \mathbb{N}$ we have*

$$F^m(Q) = s^m Q^\circ + t^m \,\text{bc}(Q) + \sum_{j=0}^{m-1} t^{m-j-1} b(s^j Q^\circ).$$

Consequently, $F^m = \text{id}$ if and only if $s^m = t^m = 1$ and the function b satisfies the equation

$$\sum_{j=0}^{m-1} t^{m-j-1} b(s^j Q^\circ) = 0 \ \textit{for any} \ Q^\circ \in \mathcal{C}^{n-1}_{\text{blc}}. \tag{56}$$

Remark 6.10. It follows from Lemma 6.9 and Theorem 6.8 that for $m \geq 2$ a function $b \in \mathcal{O}(\mathcal{C}^{n-1}_{\text{blc}})$ satisfies (56) for a given pair (s, t) of mth roots of unity if and only if it can be written as $b(Q^\circ) = t\tilde{b}(Q^\circ) - \tilde{b}(sQ^\circ)$ for some $\tilde{b} \in \mathcal{O}(\mathcal{C}^{n-1}_{\text{blc}})$. The inversion formula

$$\tilde{b}(Q^\circ) = \sum_{j=0}^{m-1} \frac{m-j}{m} t^{m-j} b(s^{j-1} Q^\circ) \ \text{for any} \ Q^\circ \in \mathcal{C}^{n-1}_{\text{blc}} \tag{57}$$

allows to find such a function $\tilde{b} \in \mathcal{O}(\mathcal{C}^{n-1}_{\text{blc}})$ for a given solution $b \in \mathcal{O}(\mathcal{C}^{n-1}_{\text{blc}})$ of (56).

Examples 6.11 (Automorphisms of Finite Order) (a) For $m > 1$, pick any $b \in \mathcal{O}(\mathcal{C}^{n-1}_{\text{blc}})$ and any mth root of unity $t \neq 1$. Then the automorphism $F\colon Q \mapsto (Q - \text{bc}(Q)) + t\,\text{bc}(Q) + b(Q - bc(Q))$ satisfies $F \neq \text{id}$ and $F^m = \text{id}$.

(b) Let $b \in \mathcal{O}(\mathcal{C}^n)$ be invariant under the diagonal $(\operatorname{Aff} \mathbb{C})$-action on $\mathcal{C}^n$. For instance, $b(Q) = cD_n^{-k}(Q) \sum\limits_{q', q'' \in Q} (q' - q'')^{kn(n-1)}$ is such a function for any $k \in \mathbb{N}$ and $c \in \mathbb{C}$. Take any $m > 2$, and let s and t be two distinct mth roots of unity, where $t \neq 1$. Then the automorphism F as in (55) satisfies $F \neq \mathrm{id}$ and $F^m = \mathrm{id}$.

(c) Any automorphism F of $\mathcal{C}^n$ of the form $F\colon Q \mapsto -Q + b(Q - \mathrm{bc}(Q))$ with $b \in \mathcal{O}(\mathcal{C}^{n-1}_{\mathrm{blc}})$ is an involution. For instance, one can take

$$b(Q) = cD_n^r(Q) \sum_{\{q',q''\}\subset Q} (q' - q'')^{2m},$$

where $r, m \in \mathbb{Z}$, $|r| + |m| > 0$, and $c \in \mathbb{C}$.

7 The Group Aut $\Sigma^{n-2}_{\mathrm{blc}}$

In this section we prove part (c) of Theorem 5.1. Let us recall this assertion.

Theorem 7.1. *For $n > 6$ we have*

$$\mathrm{Aut}\, \Sigma^{n-2}_{\mathrm{blc}} \cong \mathbb{C}^*,$$

where $s \in \mathbb{C}^*$ acts on $Q \in \Sigma^{n-2}_{\mathrm{blc}}$ via $Q \mapsto sQ$.

For the proof we need some preparation.

The function $h_n \in \mathcal{O}^{\times}(\mathcal{C}^n(\mathbb{C}^*))$, $h_n(Q) \stackrel{\mathrm{def}}{=\!=} D_n(Q)/(q_1 \cdot \ldots \cdot q_n)^{n-1}$, is invariant under the diagonal action of Aut $\mathbb{C}^*$ on $\mathcal{C}^n(\mathbb{C}^*)$. For $\varepsilon = \pm 1$ and $Q = \{q_1, \ldots, q_n\} \in \mathcal{C}^n(\mathbb{C}^*)$, we set $Q^{\varepsilon} \stackrel{\mathrm{def}}{=\!=} \{q_1^{\varepsilon}, \ldots, q_n^{\varepsilon}\}$. With this notation, we have the following description of the group Aut $\mathcal{C}^n(\mathbb{C}^*)$.

Zinde's Theorem ([35, Theorem 8][9]). *Let $n > 4$. A map $F\colon \mathcal{C}^n(\mathbb{C}^*) \to \mathcal{C}^n(\mathbb{C}^*)$ is an automorphism if and only if there exist $\varepsilon \in \{1, -1\}$, $s \in \mathbb{C}^*$, and $k \in \mathbb{Z}$ such that*

$$F(Q) = sh_n^k(Q)Q^{\varepsilon} \ \text{ for all } \ Q \in \mathcal{C}^n(\mathbb{C}^*). \tag{58}$$

Hence, Aut $\mathcal{C}^n(\mathbb{C}^*) \cong (\mathbb{C}^* \times \mathbb{Z}) \rtimes (\mathbb{Z}/2\mathbb{Z})$.

Recall (see e.g. [1, 30]) that for $n \geq 4$ the singular locus $\mathrm{sing}\, \Sigma^{n-1} = \Sigma^{n-1} \setminus \mathrm{reg}\, \Sigma^{n-1}$ of Σ^{n-1} is the union[10] of the *Maxwell stratum* $\Sigma^{n-2}_{\mathrm{Maxw}}$ and the *Arnold caustic* $\Sigma^{n-2}_{\mathrm{cau}}$ defined by

$$\begin{aligned} \Sigma^{n-2}_{\mathrm{Maxw}} &= p(\{q_{n-2} = q_{n-1} = q_n\}) \quad \text{and} \\ \Sigma^{n-2}_{\mathrm{cau}} &= p(\{q_{n-3} = q_{n-2},\, q_{n-1} = q_n\}), \end{aligned} \tag{59}$$

[9] In the Arxive version of this paper we provide a proof of this theorem conformal to our notation.

[10] This is not a stratification of $\mathrm{sing}\, \Sigma^{n-1}$ since $\Sigma^{n-2}_{\mathrm{Maxw}} \cap \Sigma^{n-2}_{\mathrm{cau}} \neq \varnothing$.

where p is the projection (6). So, $\Sigma^{n-2}_{\text{Maxw}}$ and Σ_{cau} consist, respectively, of all unordered n-multisets $Q \subset \mathbb{C}$ that can be written as $Q = \{q_1, \ldots, q_{n-3}, u, u, u\}$ and $Q = \{q_1, \ldots, q_{n-4}, u, u, v, v\}$.

Proof of Theorem 7.1. Let an isomorphism $\varphi\colon \text{reg}\, \Sigma^{n-2}_{\text{blc}} \xrightarrow{\simeq} \mathcal{C}^{n-2}(\mathbb{C}^*)$ be defined as in (22). Any automorphism $F \in \text{Aut}\, \mathcal{C}^{n-2}(\mathbb{C}^*)$ as in (58) yields an automorphism $\tilde{F} = \varphi^{-1} \circ F \circ \varphi \in \text{Aut}\,(\text{reg}\, \Sigma^{n-2}_{\text{blc}})$. In particular, the $\mathbb{C}^*$-action $Q \mapsto sQ$ $(s \in \mathbb{C}^*,$ $Q \in \mathcal{C}^{n-2}(\mathbb{C}^*))$ on $\mathcal{C}^{n-2}(\mathbb{C}^*)$ induces a $\mathbb{C}^*$-action on $\text{reg}\, \Sigma^{n-2}_{\text{blc}}$ given again by $Q \mapsto sQ$ $(s \in \mathbb{C}^*)$. The latter $\mathbb{C}^*$-action extends to $\Sigma^{n-2}_{\text{blc}}$ so that the origin $\bar{0} \in \Sigma^{n-2}_{\text{blc}}$ is a unique fixed point. This fixed point lies in the closure of any one-dimensional $\mathbb{C}^*$-orbit.

We have to show that, for F as in (58), the automorphism $\tilde{F}$ extends to an automorphism of $\Sigma^{n-2}_{\text{blc}}$ if and only if $k = 0$ and $\varepsilon = 1$, that is, iff $F \in \text{Aut}\,(\mathcal{C}^{n-2}(\mathbb{C}^*))$ belongs to the identity component $\text{Aut}_0(\mathcal{C}^{n-2}(\mathbb{C}^*)) \cong \mathbb{C}^*$. Since in the latter case $\tilde{F}$ does admit an extension, we may restrict to the case, where $s = 1$ in (58), and so, $F\colon Q \mapsto h^k_{n-2} Q^{\varepsilon}$.

The (Aut $\mathbb{C}^*$)-invariant function $h_{n-2} \in \mathcal{O}^{\times}(\mathcal{C}^{n-2}(\mathbb{C}^*))$ lifts to an invertible regular function $g \overset{\text{def}}{=\!=} h_{n-2} \circ \varphi$ on $\text{reg}\, \Sigma^{n-2}_{\text{blc}}$. An automorphism $F_k\colon Q \mapsto h^k_{n-2} Q$ of $\mathcal{C}^{n-2}(\mathbb{C}^*)$ $(k \in \mathbb{Z})$ induces the automorphism $\tilde{F}_k\colon Q \mapsto g^k Q$ of $\text{reg}\, \Sigma^{n-2}_{\text{blc}}$.

The subgroup $\text{Aut}_0(\mathcal{C}^{n-2}(\mathbb{C}^*)) \cong \mathbb{C}^*$ of $\text{Aut}\,(\mathcal{C}^{n-2}(\mathbb{C}^*))$ being normal, any automorphism $F \in \text{Aut}\,(\mathcal{C}^{n-2}(\mathbb{C}^*))$ sends the $\mathbb{C}^*$-orbits in $\mathcal{C}^{n-2}(\mathbb{C}^*)$ into $\mathbb{C}^*$-orbits of the same dimension. Since the function h_{n-2} is constant along the $\mathbb{C}^*$-orbits, the multiplication $Q \mapsto h^k_{n-2} Q$ preserves each $\mathbb{C}^*$-orbit. Hence the automorphism $F\colon Q \mapsto h^k_{n-2} Q^{\varepsilon}$ sends the $\mathbb{C}^*$-orbits in $\mathcal{C}^{n-2}(\mathbb{C}^*)$ into $\mathbb{C}^*$-orbits. It follows that $\tilde{F}$ also sends the $\mathbb{C}^*$-orbits in $\text{reg}\, \Sigma^{n-2}_{\text{blc}}$ into $\mathbb{C}^*$-orbits.

The involution $Q \mapsto Q^{-1}$ on $\mathcal{C}^{n-2}(\mathbb{C}^*)$ sends any $\mathbb{C}^*$-orbit into another such orbit interchanging the punctures, while the multiplication $Q \mapsto h^k_{n-2} Q$ preserves the punctures. Hence $\tilde{F}$ interchanges as well the punctures of the $\mathbb{C}^*$-orbits in $\text{reg}\, \Sigma^{n-2}_{\text{blc}}$ as soon as $\varepsilon = -1$.

On the other hand, if $\tilde{F} \in \text{Aut}\,(\text{reg}\, \Sigma^{n-2}_{\text{blc}})$ admits an extension, say $\bar{F}$, to an automorphism of $\Sigma^{n-2}_{\text{blc}}$, then $\bar{F}$ should fix the origin. Indeed, $\bar{F}$ normalizes the $\mathbb{C}^*$-action on $\Sigma^{n-2}_{\text{blc}}$, hence it preserves the only $\mathbb{C}^*$-fixed point $0 \in \Sigma^{n-2}_{\text{blc}}$. The origin is a unique common point of the $\mathbb{C}^*$-orbit closures. Hence $\tilde{F}$ cannot interchange the punctures of the $\mathbb{C}^*$-orbits in $\text{reg}\, \Sigma^{n-2}_{\text{blc}}$. This proves that $\varepsilon = 1$ for such an extendable $\tilde{F}$.

The function $h_{n-2} \in \mathcal{O}^{\times}(\mathcal{C}^{n-2}(\mathbb{C}^*))$ can be regarded as the rational function $d_{n-2}(z)/z^{n-3}_{n-2}$ on $\mathbb{C}^{n-2}_{(z)}$, where $z_{n-2} = (-1)^{n-2} \prod_{i=1}^{n-2} q_i$. It has pole along the coordinate hyperplane $z_{n-2} = 0$, and h^{-1}_{n-2} has pole along the discriminant hypersurface $\Sigma^{n-3} = \{d_{n-2} = 0\}$. It follows by (22) that g regarded as a rational function on $\Sigma^{n-2}_{\text{blc}}$ has pole along the caustic $\Sigma^{n-2}_{\text{cau}} \cap \Sigma^{n-2}_{\text{blc}}$, and g^{-1} has pole along the Maxwell stratum $\Sigma^{n-2}_{\text{Maxw}} \cap \Sigma^{n-2}_{\text{blc}}$, see (59). Anyhow, the automorphism $\tilde{F}_k\colon Q \mapsto g^k_n Q$ of $\text{reg}\, \Sigma^{n-2}_{\text{blc}}$ does not admit an extension to an automorphism of $\Sigma^{n-2}_{\text{blc}}$ unless $k = 0$ in (58). Thus, $k = 0$ in the case that $\varepsilon = 1$ and $\tilde{F} \in \text{Aut}\,(\text{reg}\, \Sigma^{n-2}_{\text{blc}})$ admits an extension to an automorphism of $\Sigma^{n-2}_{\text{blc}}$, as stated. □

8 Holomorphic Endomorphisms of the Balanced Configuration Space

In this section, $\mathcal{O}^{\times}_{\mathrm{hol}}(\mathcal{Z})$ stands for the multiplicative group of the algebra $\mathcal{O}_{\mathrm{hol}}(\mathcal{Z})$ of all holomorphic functions on a complex space $\mathcal{Z}$, and $\mathcal{O}_{\mathrm{hol},+}(\mathcal{Z})$ for its additive group.

Any holomorphic endomorphism f of $\mathcal{C}^{n-1}_{\mathrm{blc}}$ extends to a holomorphic endomorphism of $\mathcal{C}^n$. Such an extension is non-Abelian whenever f is non-Abelian (see the definition in the Introduction). The *minimal* extension F given by $F(Q) = f(Q - \mathrm{bc}(Q))$ for all $Q \in \mathcal{C}^n$ maps $\mathcal{C}^n$ to $\mathcal{C}^{n-1}_{\mathrm{blc}} \subset \mathcal{C}^n$, see (9).

Among affine transformations of $\mathbb{C}$ acting diagonally on $\mathcal{C}^n$, only the elements of the multiplicative subgroup $\mathbb{C}^* \subset \mathrm{Aff}\,\mathbb{C}$ fixing the origin $0 \in \mathbb{C}$ preserve the balanced configuration space $\mathcal{C}^{n-1}_{\mathrm{blc}} \subset \mathcal{C}^n$. Let $\mathcal{S}$ denote this $\mathbb{C}^*$-action on each of the spaces $\mathcal{C}^n$ and $\mathcal{C}^{n-1}_{\mathrm{blc}}$, and let $\mathcal{O}^{\mathcal{S}}_{\mathrm{hol}}(\mathcal{C}^{n-1}_{\mathrm{blc}})$ be the subalgebra of $\mathcal{O}_{\mathrm{hol}}(\mathcal{C}^{n-1}_{\mathrm{blc}})$ consisting of all $\mathcal{S}$-invariant functions.

For any configuration $Q^\circ \in \mathcal{C}^{n-1}_{\mathrm{blc}}$ its $\mathbb{C}^*$-stabilizer $\mathrm{St}_{\mathbb{C}^*}(Q^\circ) = \{\zeta \in \mathbb{C}^* \mid \zeta \cdot Q^\circ = Q^\circ\}$ is a cyclic rotation subgroup in $\mathbb{C}^*$ of order $\le n$ permuting elements of Q°. If $n \ge 3$, it follows that the set $\{Q^\circ \in \mathcal{C}^{n-1}_{\mathrm{blc}} \mid \mathrm{St}_{\mathbb{C}^*}(Q^\circ) \ne \{1\}\}$ is a Zariski closed subset in $\mathcal{C}^{n-1}_{\mathrm{blc}}$ of dimension 1 and $\{Q^\circ \in \mathcal{C}^{n-1}_{\mathrm{blc}} \mid \mathrm{St}_{\mathbb{C}^*}(Q^\circ) = \{1\}\}$ is a Zariski open dense subset of $\mathcal{C}^{n-1}_{\mathrm{blc}}$.

Definition 8.1. We say that a holomorphic self-map f of $\mathcal{C}^{n-1}_{\mathrm{blc}}$ is $\mathbb{C}^*$-*tame*, if there is a holomorphic function $h\colon \mathcal{C}^{n-1}_{\mathrm{blc}} \to \mathbb{C}^*$ such that $f(Q^\circ) = h(Q^\circ) \cdot Q^\circ$ for all $Q^\circ \in \mathcal{C}^{n-1}_{\mathrm{blc}}$.

Notice that the cohomology group $H^1(\mathcal{C}^{n-1}_{\mathrm{blc}}, \mathbb{Z}) \cong \mathbb{Z}$ of the Stein manifold $\mathcal{C}^{n-1}_{\mathrm{blc}}$ is generated by the cohomology class of the discriminant $D_n|_{\mathcal{C}^{n-1}_{\mathrm{blc}}}$ (see (2)) restricted to $\mathcal{C}^{n-1}_{\mathrm{blc}}$. Hence any function $h \in \mathcal{O}^{\times}(\mathcal{C}^{n-1}_{\mathrm{blc}})$ can be written as $h = e^{\chi} D_n^m$ with some $\chi \in \mathcal{O}_{\mathrm{hol}}(\mathcal{C}^{n-1}_{\mathrm{blc}})$ and $m \in \mathbb{Z}$.

The results below, stated in [21] and [22, Sect. 8.2.1], are simple consequences of the analytic counterpart of Tame Map Theorem (see [26] or [27] for the proof) and the facts mentioned above.

Theorem 8.2. *For $n > 4$ every non-Abelian holomorphic self-map f of $\mathcal{C}^{n-1}_{\mathrm{blc}}$ is $\mathbb{C}^*$-tame, i.e., it can be given by*

$$f(Q^\circ) = \mathcal{S}_{e^{\chi(Q^\circ)} D_n^m(Q^\circ)} Q^\circ = e^{\chi(Q^\circ)} D_n^m(Q^\circ) \cdot Q^\circ \quad \text{for all} \quad Q^\circ \in \mathcal{C}^{n-1}_{\mathrm{blc}}, \tag{60}$$

where $\chi \in \mathcal{O}_{\mathrm{hol}}(\mathcal{C}^{n-1}_{\mathrm{blc}})$ and $m \in \mathbb{Z}$.

Proof. The map f admits a holomorphic non-Abelian extension $F\colon \mathcal{C}^n \to \mathcal{C}^{n-1}_{\mathrm{blc}} \subset \mathcal{C}^n$ defined by $F(Q) = f(Q - \mathrm{bc}(Q))$ for all $Q \in \mathcal{C}^n$. By the analytic version of Tame Map Theorem, $F(Q) = A(Q)Q + B(Q)$ for all $Q \in \mathcal{C}^n$ with $A \in \mathcal{O}^{\times}_{\mathrm{hol}}(\mathcal{C}^n)$ and $B \in \mathcal{O}_{\mathrm{hol}}(\mathcal{C}^n)$. Since $\mathcal{C}^{n-1}_{\mathrm{blc}} \subset \mathcal{C}^n$ and $\mathrm{bc}(Q^\circ) = 0$ for any $Q^\circ \in \mathcal{C}^{n-1}_{\mathrm{blc}}$, we see that

$f(Q^\circ) = a(Q^\circ)Q^\circ + b(Q^\circ)$ for all $Q^\circ \in \mathcal{C}^{n-1}_{\text{blc}}$, where $a = A|_{\mathcal{C}^{n-1}_{\text{blc}}}$ and $b = B|_{\mathcal{C}^{n-1}_{\text{blc}}}$.

Moreover, $b = 0$, since the condition $\text{bc}(f(Q^\circ)) = \text{bc}(Q^\circ) = 0$ implies that

$$b(Q^\circ) = a(Q^\circ)\,\text{bc}(Q^\circ) + b(Q^\circ) = \text{bc}(a(Q^\circ)Q^\circ + b(Q^\circ)) = \text{bc}(f(Q^\circ)) = 0$$

for all $Q^\circ \in \mathcal{C}^{n-1}_{\text{blc}}$ and

$$a = e^\chi D_n^m \text{ for some } m \in \mathbb{Z} \text{ and } \chi \in \mathcal{O}_{\text{hol}}(\mathcal{C}^{n-1}_{\text{blc}}).$$

This proves (60). □

Theorem 8.3. *Let $n \geq 3$ and let $f = f_{\chi,m}\colon \mathcal{C}^{n-1}_{\text{blc}} \to \mathcal{C}^{n-1}_{\text{blc}}$ be a holomorphic map of the form* (60)*. Then the following hold.*

(a) *The map f is surjective,*[11] *and the set $f^{-1}(Q^\circ)$ is discrete for any $Q^\circ \in \mathcal{C}^{n-1}_{\text{blc}}$. This set consists of all points $\omega \cdot Q^\circ$, where $\omega \in \mathbb{C}^*$ is any root of the system of equations*

$$\omega^{mn(n-1)+1} e^{\chi(\omega \cdot Q^\circ)} D_n^m(Q^\circ) \cdot Q^\circ = Q^\circ\,, \tag{61}$$

which always has solutions.

(b) *The map f is proper (in the complex topology) if and only if $\chi \in \mathcal{O}^S_{\text{hol}}(\mathcal{C}^{n-1}_{\text{blc}})$. In this case $f\colon \mathcal{C}^{n-1}_{\text{blc}} \to \mathcal{C}^{n-1}_{\text{blc}}$ is a finite unramified cyclic holomorphic covering of degree $N = mn(n-1)+1$. The corresponding normal subgroup $f_*(\pi_1(\mathcal{C}^{n-1}_{\text{blc}}))$ of index N in the Artin braid group $B_n = \pi_1(\mathcal{C}^{n-1}_{\text{blc}})$ consists of all $g = \sigma_{i_1}^{m_1} \cdot \ldots \cdot \sigma_{i_q}^{m_q} \in B_n$ such that N divides $m_1 + \ldots + m_q$, where $\{\sigma_1, \ldots, \sigma_{n-1}\}$ is the standard system of generators in B_n. Every two such coverings of the same degree are equivalent.*

(c) *The map f is a biholomorphic automorphism of $\mathcal{C}^{n-1}_{\text{blc}}$ if and only if it is of the form $f(Q^\circ) = e^{\chi(Q^\circ)} \cdot Q^\circ$ for any $Q^\circ \in \mathcal{C}^{n-1}_{\text{blc}}$ and some $\chi \in \mathcal{O}^S_{\text{hol}}(\mathcal{C}^{n-1}_{\text{blc}})$. Every automorphism is isotopic to the identity and $\text{Aut}_{\text{hol}}\mathcal{C}^{n-1}_{\text{blc}} \cong \mathcal{O}^S_{\text{hol},+}(\mathcal{C}^{n-1}_{\text{blc}})/2\pi i\mathbb{Z}$.*

(d) *If f is regular, then $\chi = \text{const}$ and so $f(Q^\circ) = cD_n^m(Q^\circ) \cdot Q^\circ$ for all $Q^\circ \in \mathcal{C}^{n-1}_{\text{blc}}$, where $c \in \mathbb{C}^*$ and $m \in \mathbb{Z}$. Every biregular automorphism f of $\mathcal{C}^{n-1}_{\text{blc}}$ is of the form $f(Q^\circ) = s \cdot Q^\circ$, $Q^\circ \in \mathcal{C}^{n-1}_{\text{blc}}$, where $s \in \mathbb{C}^*$. In particular, the group of all biregular automorphisms $\text{Aut}\,\mathcal{C}^{n-1}_{\text{blc}}$ is isomorphic to $\mathbb{C}^*$.*

Proof. (a) Given a configuration $Q^\circ \in \mathcal{C}^{n-1}_{\text{blc}}$, we set

$$\psi_{Q^\circ}(\omega) \overset{\text{def}}{=\!=} \omega^{mn(n-1)+1} e^{\chi(\omega \cdot Q^\circ)} D_n^m(Q^\circ) \text{ for any } \omega \in \mathbb{C}^*. \tag{62}$$

[11] In view of Theorem 8.2, for $n > 4$ *any* non-Abelian holomorphic endomorphism of $\mathcal{C}^{n-1}_{\text{blc}}$ is surjective.

Clearly $\psi_{Q^\circ} \in \mathcal{O}^\times_{\mathrm{hol}}(\mathcal{C}^{n-1}_{\mathrm{blc}})$ and $\psi_{Q^\circ} \neq \mathrm{const}$, since $mn(n-1)+1 \neq 0$ and $e^{\chi(\omega\cdot Q^\circ)}$ cannot be a non-constant rational function of $\omega \in \mathbb{C}^*$. Hence, by the Picard theorem, $\psi_{Q^\circ}(\mathbb{C}^*) = \mathbb{C}^*$. According to (60), we have

$$f(\omega\cdot Q^\circ) = \omega^{mn(n-1)+1} e^{\chi(\omega\cdot Q^\circ)} D_n^m(Q^\circ)\cdot Q^\circ = \psi_{Q^\circ}(\omega)\cdot Q^\circ\,.$$

Thus, taking $\omega \in \mathbb{C}^*$ such that $\psi_{Q^\circ}(\omega) = 1$, we see that $Q^\circ \in f(\mathcal{C}^{n-1}_{\mathrm{blc}})$. Hence f is surjective. Furthermore, all such ω satisfy the system of equations (61). Since the stabilizer $\mathrm{St}_{\mathbb{C}^*}(Q^\circ)$ is finite, all solutions ω of (61) form a finite union of countable discrete subsets of $\mathbb{C}^*$. Thus the set $f^{-1}(Q^\circ)$ is countable and discrete.

(b) If f as in (60) is proper, then $f^{-1}(Q^\circ)$ is finite for any $Q^\circ \in \mathcal{C}^{n-1}_{\mathrm{blc}}$. This is possible only when the exponent $\chi(\omega\cdot Q^\circ)$ in (61) and (62) does not depend on $\omega \in \mathbb{C}^*$, i.e., the function χ is $\mathcal{S}$-invariant. Then, for any fixed Q°, the function (62) takes the form

$$\psi_{Q^\circ}(\omega) = \tilde{\psi}_{Q^\circ}(\omega) \stackrel{\mathrm{def}}{=} \omega^{mn(n-1)+1} e^{\chi(Q^\circ)} D_n^m(Q^\circ)\,.$$

The latter function is homogeneous of degree $N = mn(n-1)+1$ and the equation $\tilde{\psi}_{Q^\circ}(\omega) = 1$ has precisely N distinct solutions $\omega_1,\dots,\omega_N$. If the stabilizer $\mathrm{St}_{\mathbb{C}^*}(Q^\circ)$ is trivial, then $f^{-1}(Q^\circ)$ consists on N distinct points $\omega_1 Q^\circ,\dots,\omega_N Q^\circ$. If $\mathrm{St}_{\mathbb{C}^*}(Q^\circ) \neq \{1\}$, then, to find the preimage $f^{-1}(Q^\circ)$, we have to solve the inclusion $\tilde{\psi}_{Q^\circ}(\omega) \in \mathrm{St}_{\mathbb{C}^*}(Q^\circ)$. Fix some ω_0 such that $\tilde{\psi}_{Q^\circ}(\omega_0) = 1$, take any solution $\omega \in \mathbb{C}^*$ of the above inclusion, and set $\lambda = \omega/\omega_0$. Then

$$\lambda^N = \left(\frac{\omega}{\omega_0}\right)^N = \frac{\tilde{\psi}_{Q^\circ}(\omega)}{\tilde{\psi}_{Q^\circ}(\omega_0)} = \tilde{\psi}_{Q^\circ}(\omega) \in \mathrm{St}_{\mathbb{C}^*}(Q^\circ)\,. \tag{63}$$

The preimage $f^{-1}(Q^\circ)$ of Q° consists of all configurations $\omega Q^\circ = \omega_0\lambda Q^\circ$, where λ runs over all solutions of the inclusion (63). All such configurations $\omega_0\lambda Q^\circ$ form a periodic sequence $\omega_0\lambda^k Q^\circ$, $k \in \mathbb{Z}_{\geq 0}$, with period N; therefore, this sequence contains precisely N distinct elements. It follows easily from these facts that $f\colon \mathcal{C}^{n-1}_{\mathrm{blc}} \to \mathcal{C}^{n-1}_{\mathrm{blc}}$ is an unramified cyclic covering of degree N.

The proof of the other assertions in (b) is easy and we leave it to the reader.

(c) First, let $\chi \in \mathcal{O}^{\mathcal{S}}_{\mathrm{hol}}(\mathcal{C}^{n-1}_{\mathrm{blc}})$, $f_1(Q^\circ) = e^{\chi(Q^\circ)}Q^\circ$, and $f_2(Q^\circ) = e^{-\chi(Q^\circ)}Q^\circ$. It follows from the $\mathcal{S}$-invariance of χ that $f_1(f_2(Q^\circ)) = f_2(f_1(Q^\circ)) = Q^\circ$ for every $Q^\circ \in \mathcal{C}^{n-1}_{\mathrm{blc}}$. Thus f_1 and f_2 are mutually inverse biholomorphic automorphisms of $\mathcal{C}^{n-1}_{\mathrm{blc}}$. To prove the converse notice that any automorphism is a proper map. According to Theorem 8.2 (formula (60)) and part (b), such a map is of the form $Q^\circ \mapsto e^{\chi(Q^\circ)}Q^\circ$ with $\chi \in \mathcal{O}^{\mathcal{S}}_{\mathrm{hol}}(\mathcal{C}^{n-1}_{\mathrm{blc}})$. The other two assertions of part (c) are clear.

(d) A map as in (60) is regular if and only if $\chi = \mathrm{const}$, i.e.,

$$f(Q^\circ) = sD_n^m(Q^\circ) \cdot Q^\circ \text{ for all } Q^\circ \in \mathcal{C}_{\text{blc}}^{n-1}, \text{ where } s \in \mathbb{C}^* \text{ and } m \in \mathbb{Z}.$$

It is a biregular automorphism of $\mathcal{C}_{\text{blc}}^{n-1}$ if and only if $m = 0$. Hence Aut $\mathcal{C}_{\text{blc}}^{n-1} \cong \mathbb{C}^*$.

□

Remark 8.4 (Dimension of the Image). In what follows we assume that $n > 4$. According to [27, Theorem 14], for $X = \mathbb{C}$ or $\mathbb{P}^1$ and any non-Abelian holomorphic endomorphism F of $\mathcal{C}^n(X)$ we have $\dim_{\mathbb{C}} F(\mathcal{C}^n(X)) \geq n - \dim_{\mathbb{C}}(\text{Aut}X) + 1$. Moreover, by [27, Remark 7] or Theorems 8.2 and 8.3 (a) above, the composition $\pi \circ F$ of any non-Abelian holomorphic endomorphism F of $\mathcal{C}^n$ with the projection $\pi: \mathcal{C}^n \to \mathcal{C}_{\text{blc}}^{n-1}$ is surjective, so that $\dim_{\mathbb{C}} F(\mathcal{C}^n) \geq n - 1$. Clearly, this bound cannot be improved. Seemingly, no examples of F with $\dim_{\mathbb{C}} F(\mathcal{C}^n(\mathbb{P}^1)) < n$ are known. Zinde [35] proved that any non-Abelian holomorphic endomorphism of $\mathcal{C}^n(\mathbb{C}^*)$ is surjective.

References

1. V.I. Arnold, On some topological invariants of algebraic functions, I. Trudy Moskov. Mat. Obshch. **21**, 27–46 (1970) (Russian) [English translation: Trans. Moscow Math. Soc. **21**, 30–52 (1970)]
2. I.V. Arzhantsev, K. Kuyumzhiyan, M. Zaidenberg, Flag varieties, toric varieties, and suspensions: Three instances of infinite transitivity. Sb. Math. **203**, 3–30 (2012)
3. I.V. Arzhantsev, H. Flenner, S. Kaliman, F. Kutzschebauch, M. Zaidenberg, Flexible varieties and automorphism groups. Duke Math. J. **162**(4), 767–823 (2013)
4. T. Bandman, L. Makar-Limanov, *Cylinders over Affine Surfaces*, 14 pp. (1998) [arXiv:math/9807146]
5. N. Bruschlinsky, Stetige Abbildungen und Bettische Gruppen der Dimensionzahl 1 und 3. Math. Ann. **109**, 525–537 (1934)
6. R. Dryło, Non-uniruledness and the cancellation problem, I. Ann. Polon. Math. **87**, 93–98 (2005); II. Ann. Polon. Math. **92**, 41–48 (2007)
7. R. Dryło, A remark on a paper of Crachiola and Makar-Limanov. Bull. Pol. Acad. Sci. Math. **59**(3), 203–206 (2011)
8. S. Eilenberg, Transformations continues en circonférence et la topologie du plan. Fund. Math. **26**, 61–112 (1936)
9. S. Eilenberg, Généralisation du théorème de M. H. Hopf sur les classes des transformations en surfaces sphériques. Compos. Math. **6**, 428–433 (1939)
10. Y. Feler, Configuration spaces of tori. Atti Accad. Naz. Lincei Cl. Sci. Fis. Mat. Natur. Rend. Lincei (9) Mat. Appl. **18**(2), 139–151 (2007)
11. T.J. Ford, *The Group of Units on an Affine Variety*, 22 pp. (2013) [arXiv:1303.5687]
12. G. Freudenburg, *Algebraic Theory of Locally Nilpotent Derivations*. Encyclopaedia of Mathematical Sciences, vol. 136, Invariant Theory and Algebraic Transformation Groups, vol. VII (Springer, Berlin, 2006)
13. E.A. Gorin, V.Ja. Lin, Algebraic equations with continuous coefficients, and certain questions of the algebraic theory of braids. Mat. Sb. (N.S.) **78(120)**(4), 579–610 (1969) (Russian) [English translation: Math. USSR-Sb. **7**(4), 569–596 (1969)]
14. G. Hochschild, Cohomology of algebraic linear groups. Ill. J. Math. **5**(3), 492–519 (1961)

15. J.E. Humphreys, *Linear Algebraic Groups*. Graduate Texts in Mathematics, vol. 21 (Springer, New York, 1975), 247 pp.
16. S. Iitaka, On logarithmic Kodaira dimension of algebraic varieties, in *Complex Analysis and Algebraic Geometry* (Iwanami Shoten, Tokyo, 1977), pp. 175–189
17. Z. Jelonek, Testing sets for properness of polynomial mappings. Math. Ann. **315**, 1–35 (1999)
18. Z. Jelonek, On the cancellation problem. Math. Ann. **344**, 769–778 (2009)
19. S.I. Kaliman, Holomorphic endomorphisms of the manifold of complex polynomials with discriminant 1. Uspehi Mat. Nauk **31**(1), 251–252 (1976) (Russian)
20. S. Kumar, *Kac-Moody Groups, Their Flag Varieties and Representation Theory*. Progress in Mathematics, vol. 204 (Birkhuser, Boston, 2002)
21. V.Ya. Lin, Algebraic functions with universal discriminant manifolds, Funkt. Analiz i Priloz. **6**(1), 81–82 (1972) (Russian) [English translation: Funct. Anal. Appl. **6**(1), 73–74 (1972)]
22. V.Ya. Lin, Artin braids and the groups and spaces connected with them, in *Itogi Nauki i Tekhniki*. Algebra, Topologiya, Geometriya, vol. 17 (VINITI, Moscow, 1979), pp. 159–227 (Russian). [English translation: J. Soviet Math. **18**, 736–788 (1982)]
23. V.Ya. Lin, *Braids, permutations, polynomials-I*, Max-Planck-Institute für Mathematik, Bonn, MPI 96–118, 112 pp. (1996), http://www.mpim-bonn.mpg.de/preblob/3958
24. V.Ya. Lin, *Configuration spaces of* $\mathbb{C}$ *and* $\mathbb{CP}^1$*: Some analytic properties*, Max-Planck-Institute für Mathematik, Bonn, MPI 2003-98, 80 pp. (2003), http://www.mpim-bonn.mpg.de/preblob/2266
25. V.Ya. Lin, *Braids and Permutations*, 112 pp. (2004), http://arxiv.org/abs/math/0404528v1
26. V.Ya. Lin, *Configuration Spaces of* $\mathbb{C}$ *and* $\mathbb{CP}^1$*: Some Analytic Properties*, 89 pp. (2004) [arxiv.org/abs/math/0403120v3]
27. V.Ya. Lin, Algebraic functions, configuration spaces, Teichmüller spaces, and new holomorphically combinatorial invariants. Funkt. Analiz i Priloz. **45**(3), 55–78 (2011) (Russian) [English translation: Funct. Anal. Appl. **45**(3), 204–224 (2011)]
28. O.V. Ljashko, Geometry of bifurcation diagrams. Uspehi Mat. Nauk **34**(3), 205–206 (1979) (Russian)
29. G.D. Mostow, Fully reducible subgroups of algebraic groups. Am. J. Math. **78**, 200–221 (1956)
30. F.Napolitano, Topology of complements of strata of the discriminant of polynomials. C. R. Acad. Sci. Paris Sér. I Math. **327**, 665–670 (1998)
31. V.P. Platonov, A.S. Rapinchuk, *Algebraic Groups and Number Theory* (Nauka, Moscow, 1991) (Russian) [English translation: V. Platonov, A. Rapinchuk, *Algebraic Groups and Number Theory*. Pure and Applied Mathematics, vol. 139 (Academic, Boston, 1994)]
32. C.P. Ramanujam, A note on automorphism groups of algebraic varieties. Math. Ann. **156**, 25–33 (1964)
33. P. Samuel, À propos du théorème des unités. Bull. Sci. Math. **90**(2), 89–96 (1966)
34. I.R. Shafarevich, On some infinite-dimensional groups, II. Math. USSR Izv. **18**, 185–194 (1982)
35. V.M. Zinde, Holomorphic mappings of the spaces of regular orbits of Coxeter groups of series B and D. Sibirsk. Mat. Zh. **18**(5), 1015–1026 (1978) (Russian) [English translation: Siberian Math. J. **18**(5), 716–724 (1978)]. http://link.springer.com/content/pdf/10.1007%2FBF00967010.pdf

On the Newton Polygon of a Jacobian Mate

Leonid Makar-Limanov

To the memory of Shreeram Abhyankar one of the champions of the Jacobian Conjecture

Abstract This note contains an up-to-date description of the "minimal" Newton polygons of the polynomials satisfying the Jacobian condition.

Mathematics Subject Classification (2000): Primary 14R15, 12E05; Secondary 12E12.

1 Introduction

Consider two polynomials $f,\ g \in \mathbb{C}[x, y]$ where $\mathbb{C}$ is the field of complex numbers with the Jacobian $\mathrm{J}(f, g) = 1$ and $\mathbb{C}[f, g] \neq \mathbb{C}[x, y]$, i.e., a counterexample to the JC (Jacobian conjecture) which states that $\mathrm{J}(f, g) = 1$ implies $\mathbb{C}[f, g] = \mathbb{C}[x, y]$ (see [9]). This conjecture occasionally becomes a theorem even for many years but today it is a problem.

One of the approaches to this problem which is still popular is through obtaining information about the Newton polygons of polynomials f and g. It is known for many years that there exists an automorphism ξ of $\mathbb{C}[x, y]$ such that the Newton polygon $\mathcal{N}(\xi(f))$ of $\xi(f)$ contains a vertex $v = (m, n)$ where $n > m > 0$ and is included in a trapezoid with the vertex v, edges parallel to the y axes and to the bisectrix of the first quadrant adjacent to v, and two edges belonging to the coordinate axes (see [1,2,7,10,12–15,17–19,21]). This was improved quite recently

L. Makar-Limanov (✉)
Department of Mathematics, Wayne State University, Detroit, MI 48202, USA

Max-Planck-Institut für Mathematik, 53111 Bonn, Germany

Department of Mathematics & Computer Science, the Weizmann Institute of Science, Rehovot 76100, Israel

Department of Mathematics, University of Michigan, Ann Arbor, MI 48109, USA
e-mail: lml@math.wayne.edu

I. Cheltsov et al. (eds.), *Automorphisms in Birational and Affine Geometry*, Springer Proceedings in Mathematics & Statistics 79,
DOI 10.1007/978-3-319-05681-4_25,

by Pierrette Cassou-Noguès who showed that $\mathcal{N}(f)$ does not have an edge parallel to the bisectrix (see [3]). Here a shorter (and more elementary) version of the proof of this fact is suggested. A proof of the "trapezoid" part based on the work [5] of Dixmier published in 1968 is also included to have all the information on $\mathcal{N}(f)$ in one place with streamlined proofs.

As a byproduct we'll get a proof of the Jung theorem that any automorphism of $\mathbb{C}[x, y]$ is a composition of linear and "triangular" automorphisms.

2 Trapezoidal Shape

In this section, using technique developed by Dixmier in [5], we will check the claim that if $f \in \mathbb{C}[x, y]$ is a Jacobian mate, i.e., when $\mathrm{J}(f, g) = \frac{\partial f}{\partial x}\frac{\partial g}{\partial y} - \frac{\partial f}{\partial y}\frac{\partial g}{\partial x} = 1$ for some $g \in \mathbb{C}[x, y]$, then there exists an automorphism ξ of $\mathbb{C}[x, y]$ such that the Newton polygon $\mathcal{N}(\xi(f))$ of $\xi(f)$ is contained in a trapezoid described in the introduction.

Recall that if $p \in \mathbb{C}[x, y]$ is a polynomial in two variables and each monomial of p is represented by a lattice point on the plane with the coordinate vector equal to the degree vector of this monomial, then the convex hull $\mathcal{N}(p)$ of the points so obtained is called the Newton polygon. For reasons which are not clear to me Newton included the origin (a nonzero constant term) in his definition.

Define a weight degree function on $\mathbb{C}[x, y]$ as follows. First, take weights $w(x) = \alpha$, $w(y) = \beta$ where $\alpha, \beta \in \mathbb{Z}$ and put $w(x^i y^j) = i\alpha + j\beta$. For a $p \in \mathbb{C}[x, y]$ denote the support of p, i.e., the collection of all monomials appearing in p with nonzero coefficients by $\mathrm{supp}(p)$ and define $w(p) = \max(w(x^i y^j) | x^i y^j \in \mathrm{supp}(p))$. Polynomial p can be written as $p = \sum p_i$ where p_i are forms homogeneous relative to w. The leading form p_w of p according to w is the form of the maximal weight in this presentation.

2.1 *Lemma on Independence*

Take any two algebraically independent polynomials $a, b \in \mathbb{C}[x, y]$ and a nonzero weight degree function w on $\mathbb{C}[x, y]$. Then there exists an $h \in \mathbb{C}[a, b]$ for which $\mathrm{J}(a_w, h_w) \neq 0$, i.e., h_w and a_w are algebraically independent.

Proof. A standard proof of this fact would be based on the notion of Gelfand-Kirillov dimension (see [6]) and is rather well known. The proof below uses a deficiency function

$$\mathrm{def}_w(a, h) = w(\mathrm{J}(a, h)) - w(h)$$

(somewhat similar to the one introduced in [11]) and is more question specific. This function is defined and has values in $\mathbb{Z}$ when $\mathrm{J}(a, h) \neq 0$, i.e., def_w is defined for any $h \in \mathbb{C}[a,b]$ which is algebraically independent with a. Observe that $\operatorname{def}_w(a, hr(a)) = \operatorname{def}_w(a,h)$, $r(a) \in \mathbb{C}[a] \setminus 0$; $\operatorname{def}_w(a,h) \leq w(a) - w(xy)$; and that $\operatorname{def}_w(a,h^k) = \operatorname{def}_w(a,h)$ since $\mathrm{J}(a,h^k) = kh^{k-1}\mathrm{J}(a,h)$.

If a_w and b_w are algebraically dependent, then there exists an irreducible nonzero polynomial $q = \sum_{i=0}^{k} q_i(x)y^i \in F[x,y]$ for which $q(a_w, b_w) = 0$ and all monomials with nonzero coefficients have the same degree relative to the weight $W(x) = w(a)$, $W(y) = w(b)$. Elements a, $b' = q(a,b)$ are algebraically independent since a and b are algebraically independent but there is a drop in weight, i.e., $w(b') < w(q_k(a)b^k)$.

We have $\operatorname{def}_w(a,b') = w(J(a,b')) - w(b') = w(\sum_i J(a, q_i(a)b^i)) - w(b') > w(J(a, q_k(a)b^k)) - w(q_k(a)b^k) = \operatorname{def}_w(a,b^k) = \operatorname{def}_w(a,b)$ since $w(b') < w(q_k(a)b^k)$ while $w(J(a,q_k(a)b^k)) = w(kq_k(a)b^{k-1}) + w(J(a,b)) = w(\sum_i iq_i(a)\, b^{i-1}) + w(J(a,b)) = w(\sum_i J(a,q_i(a)b^i))$ because $\sum_i iq_i(a_w)\, b_w^{i-1} \neq 0$ since q is irreducible. If a_w, b'_w are algebraically dependent, we repeat the procedure and obtain a pair a, b'' with $\operatorname{def}_w(a,b'') > \operatorname{def}_w(a,b')$, etc.. Since $\operatorname{def}_w(a,h) \leq w(a) - w(xy)$ for any h and $\operatorname{def}_w(a,h) \in \mathbb{Z}$, the process will stop after a finite number of steps and we will get an element $h \in \mathbb{C}[a,b]$ for which h_w is algebraically independent with a_w. □

Now back to our polynomials f, g with $\mathrm{J}(f,g) = 1$. These two polynomials are algebraically independent. To prove it consider a derivation ∂ given on $\mathbb{C}[x,y]$ by $\partial(h) = \mathrm{J}(f,h)$. When ∂ is restricted to $\mathbb{C}[f,g]$ this is the ordinary partial derivative relative to g. Hence if $p(f,g) = 0$, then $p_g(f,g) = 0$ and a contradiction is reached if we assume that p is an irreducible dependence.

This derivation is locally nilpotent on $\mathbb{C}[f,g]$, i.e., $\partial^d(h) = 0$ for $h \in \mathbb{C}[f,g]$ and $d = \deg_g(h) + 1$. Therefore ∂_w which is given by $\partial_w(h) = \mathrm{J}(f_w, h)$ on the ring $\mathbb{C}[f,g]_w$ generated by the leading w forms of elements in $\mathbb{C}[f,g]$ is also a locally nilpotent derivation. Indeed a straightforward computation shows that $\mathrm{J}(a,b)_w = \mathrm{J}(a_w,b_w)$ if $\mathrm{J}(a_w,b_w) \neq 0$.

Take a weight degree function for which $w(f) \neq 0$ and a w-homogenous form $\chi \in \mathbb{C}[x,y]$ for which $f_w = \chi^d$ where d is maximal possible. Then by Lemma on independence there exists a $\psi \in \mathbb{C}[f,g]_w$ which is algebraically independent with χ, i.e., $\partial_w(\psi) \neq 0$. Take k for which $\partial_w^k(\psi) \neq 0$ and $\partial_w^{k+1}(\psi) = 0$ and denote $\partial_w^{k-1}(\psi)$ by ω. Then $\partial_w^2(\omega) = 0$, $\partial_w(\omega) \neq 0$ and $\partial_w(\omega) = c_1\chi^{d_1}$ since χ and $\partial_w(\omega)$ are homogeneous. Therefore $\mathrm{J}(\chi^d, \omega) = c_1\chi^{d_1}$ and $\mathrm{J}(\chi,\omega) = c_2\chi^{d_1-d+1}$. For computational purposes it is convenient to introduce $\varsigma = \frac{\omega}{c_2\chi^{d_1-d}} \in \mathbb{C}(x,y)$; then $\mathrm{J}(\chi,\varsigma) = \chi$ and $w(\varsigma) = w(xy)$.

If $w(x) = 0$, then $\chi = y^j p(z)$, $\varsigma = yq(z)$ where $z = x$; if $w(x) \neq 0$ we can write $\chi = x^r p(z)$, $\varsigma = x^s q(z)$ where $z = x^{\frac{\beta}{-\alpha}} y$. In both cases $p(z) \in \mathbb{C}[z]$, $q(z) \in \mathbb{C}(z)$. In the second case r, $s \in \mathbb{Q}$ and $w(\chi) = r\alpha$, $w(\varsigma) = s\alpha$. (Recall that $w(x) = \alpha$, $w(y) = \beta$.) In any case the relation $\mathrm{J}(\chi,\varsigma) = \chi$ is equivalent to

$$\tau p' q - \rho p q' = cp \tag{1}$$

where $\rho = w(\chi)$, $\tau = w(\varsigma) = w(xy)$, and $c \in \mathbb{C}^*$.

(1) can be rewritten as $\ln(p^\tau q^{-\rho})' = \frac{c}{q}$ or

$$p^\tau = q^\rho \exp(c \int \frac{dz}{q}). \quad (2)$$

If $\rho\tau > 0$, then $q(z)$ must be a polynomial since a pole of $q(z)$ would induce a pole of $p(z)$ in the same point.

Now we are ready to discuss the shape of $\mathcal{N}(f)$. Let $m = \deg_x(f)$, $n = \deg_y(f)$. Assume that f does not contain a monomial $cx^m y^n$. Then $\mathcal{N}(f)$ has a vertex (m, k) where $k < n$ (and maximal possible) and an edge e with the vertex (m, k) and a negative slope. We can find a weight degree function w so that the Newton polygon of the leading form f_w of f relative to w is e. Since the slope of e is negative $\rho\tau$ is positive and $\varsigma = x^s q(z)$ is a homogeneous polynomial. Indeed, $w(x) \neq 0$ and we checked above that ς is a polynomial in z and therefore a polynomial in y. Since $w(y) \neq 0$ similar considerations show that ς is a polynomial in x.

There are just four options for $\mathcal{N}(\varsigma)$ because $w(\varsigma) = w(xy)$. Here is the list of all possibilities: (1) $\varsigma = cxy$; (2) $\varsigma = cx(y + c_1 x^k)$, $k > 0$; (3) $\varsigma = c(x + c_1 y^k)y$, $k > 0$; (4) $\varsigma = c(x + c_1 y)(y + c_2 x)$, $c_1 c_2 \neq 0$. In each case there is an automorphism of $\mathbb{C}[x, y]$ which transforms ς into cxy and then the image of $\chi = f_e$ under this automorphism is also a monomial ($J(\chi, cxy) = \chi$ is satisfied only by monomials $x^i y^j$ where $c(i - j) = 1$ and these monomials have different weights). Hence in the first case χ is a monomial, in the second case $\chi = c_3 x^a (y + c_1 x^k)^b$, in the third case $\chi = c_3 (x + c_1 y^k)^a y^b$, and in the fourth case $\chi = c_3 (x + c_1 y)^a (y + c_2 x)^b$.

Define $A(f) = \deg_x(f) \deg_y(f)$. In each case there is an automorphism ζ such that $A(\zeta(f)) < A(f)$: in the second and the forth cases we can take $\zeta(x) = x$, $\zeta(y) = y - c_1 x^k$ (indeed, $\zeta(x^a (y + c_1 x^k)^b) = x^a (y - c_1 x^k + c_1 x^k)^b = x^a y^b$ and $\deg_x(\zeta(f)) < \deg_x(f)$, $\deg_y(\zeta(f)) = \deg_y(f)$) and in the third and the forth cases we can take $\zeta(x) = x - c_1 y^k$, $\zeta(y) = y$ (then $\deg_x(\zeta(f)) = \deg_x(f)$, $\deg_y(\zeta(f)) < \deg_y(f)$).

Hence if $x^m y^n \notin \mathrm{supp}(f)$ one of the automorphisms $\zeta(x) = x$, $\zeta(y) = y - c_1 x^k$; $\zeta(x) = x - c_1 y^k$, $\zeta(y) = y$ (usually automorphisms $\zeta(x) = x$, $\zeta(y) = y + \phi(x)$ and $\zeta(x) = x + \phi(y)$, $\zeta(y) = y$ are called *triangular*) decreases $A(f)$. Since A is a nonnegative integer there is an automorphism ξ which is a composition of triangular automorphisms for which $A(\xi(f))$ is minimal possible and $\mathcal{N}(\xi(f))$ contains a vertex $(\deg_x(\xi(f)), \deg_y(\xi(f)))$.

Replace f by $\xi(f)$ for which $A(\xi(f))$ is minimal. The leading form of f, say for a weight $w(x) = 1$, $w(y) = 1$ is $x^m y^n$. The corresponding $\varsigma = cxy$. Since $J(x^m y^n, cxy) = c_1 x^m y^n$ where $c_1 \neq 0$ we cannot have $m = n$ and an assumption that $n > m$ is not restrictive (if $m > n$ apply an automorphism $\alpha(x) = y$, $\alpha(y) = x$).

If $m = 0$, then $f = f(y)$. Since then $J(f, g) = -f_y g_x$ this implies that $\deg_y(f) = 1$, $g = g_0(y) + cx$ where $c \in \mathbb{C}^*$ and $\mathbb{C}[f, g] = \mathbb{C}[x, y]$.

Consider again a weight given by $w(x) = 1$, $w(y) = 1$. Then $f_w = x^m y^n$. As we observed above ∂_w defined by $\partial_w(h) = \mathrm{J}(f_w, h)$ is locally nilpotent on $\mathbb{C}[f, g]_w$. If $\mathbb{C}[f, g] = \mathbb{C}[x, y]$, then $\mathbb{C}[f, g]_w = \mathbb{C}[x, y]_w = \mathbb{C}[x, y]$. Hence if $\mathbb{C}[f, g] = \mathbb{C}[x, y]$, then $\partial(h) = \mathrm{J}(x^m y^n, h)$ is a locally nilpotent derivation on $\mathbb{C}[x, y]$. If $m > 0$, then $\partial^j(y) = \frac{m(m+d)\ldots(m+(j-1)d)}{j!} x^{j(m-1)} y^{j(n-1)+1}$ where $d = n - m > 0$ is never zero and $\mathbb{C}[f, g] \neq \mathbb{C}[x, y]$.

These observations prove a theorem of Jung (see [8]) that any automorphism is a composition of triangular and linear automorphisms. If α is an automorphism of $\mathbb{C}[x, y]$, then $f = \alpha(x)$ is a Jacobian mate since by the chain rule $\mathrm{J}(\alpha(x), \alpha(y)) = c \in \mathbb{C}^*$. As we saw we can apply several triangular automorphisms after which the image of f is a polynomial which is linear in either x or y (since both cases $n > m$ and $m > n$ are possible). After that an additional triangular automorphism reduce (f, g) to either $(c_1 x, c_2 y + g_1(x))$ or $(c_1 y, c_2 x + g_1(y))$ and another triangular automorphism to $(c_1 x, c_2 y)$ or $(c_1 y, c_2 x)$. Finally a linear automorphism reduces the images to (x, y).

From now on assume that $m > 0$. Then there are two edges containing $v = (m, n)$ as a vertex, the edge e which is either horizontal or below the horizontal line and the edge e' which is either vertical or to the left of the vertical line.

Consider the edge e and the weight w for which $\mathcal{N}(f_w) = e$. If the slope of e is less than 1, then $\rho\tau > 0$, ς is a polynomial and $w(\varsigma) = w(xy)$. In the case e is horizontal $\varsigma = yq(x)$ where $q(x)$ is a polynomial and after an appropriate automorphism $x \to x - c$, $y \to y$ we may assume that $q(0) = 0$. If $w(x) \neq 0$ and $w(y) \neq 0$, then $\varsigma(0, 0) = 0$ because of the shape of $\mathcal{N}(\varsigma)$. If $\varsigma = cxy$, then e is a vertex contrary to our assumption. If $\varsigma = c_1 xy + \cdots + c_2 x^i y^j$ where $c_2 \neq 0$ and $i > 1$, then $j = \mu(i - 1) + 1$ where μ is the slope and $\mathrm{J}(x^m y^n, x^i y^j) = (mj - ni)x^{m+i-1} y^{n+j-1} \neq 0$ since $mj - ni = (m\mu - n)(i - 1) + m - n < 0$ (recall that $n > m$ and $0 \leq \mu < 1$). But then $\deg_x(\mathrm{J}(f_w, \varsigma)) > \deg_x(f_w)$ and $\mathrm{J}(f_w, \varsigma) \neq c f_w$, a contradiction.

Therefore the slope of e is at least 1. If slope is 1 we cannot get a contradiction using only $\mathrm{J}(f_w, \varsigma) = f_w$ since $\mathrm{J}(y^k h(xy), xy) = -k y^k h(xy)$.

2.2 Edge with Slope One

Newton introduced the polygon which we call the Newton polygon in order to find a solution y of $f(x, y) = 0$ in terms of x (see [16]). Here is the process of obtaining such a solution. Consider an edge e of $\mathcal{N}(f)$ which is not parallel to the x axes and take a weight $w(x) = \alpha$, $w(y) = \beta$ which corresponds to e (the choice of weight is unique if we assume that $\alpha, \beta \in \mathbb{Z}$, $\alpha > 0$ and $(\alpha, \beta) = 1$). Then the leading form f_w allows to determine the first summand of the solution as follows. Consider an equation $f_w = 0$. Since f_w is a homogeneous form and $\alpha \neq 0$ solutions of this equation are $y = c_i x^{\frac{\beta}{\alpha}}$ where $c_i \in \mathbb{C}$. Choose any c_i and replace $f(x, y)$ by $f_1(x, y) = f(x, c_i x^{\frac{\beta}{\alpha}} + y)$ which is not necessarily a polynomial in x but is a polynomial in y, and consider the Newton polygon of f_1. This polygon contains the *degree* vertex v of e, i.e., the vertex with y coordinate equal to $\deg_y(f_w)$ and an edge e' which is a modification of e (e' may collapse to v). Take the other vertex v_1 of e' (if $e' = v$ take $v_1 = v$). Use the edge e_1 for which v_1 is the degree vertex to determine the next summand and so on. After possibly a countable number of steps we obtain a vertex v_μ and the edge e_μ for which v_μ is not the degree vertex, i.e., either e_μ is horizontal or the degree vertex of e_μ has a larger y coordinate than the y coordinate of v_μ. It is possible only if $\mathcal{N}(f_\mu)$ does not have any vertices on the x axis. Therefore $f_\mu(x, 0) = 0$ and a solution is obtained.

The process of obtaining a solution is more straightforward then it may seem from this description. The denominators of fractional powers of x (if denominators and numerators of these rational numbers are assumed to be relatively prime) do not exceed $\deg_y(f)$. Indeed, for any initial weight there are at most $\deg_y(f)$ solutions while a summand $cx^{\frac{M}{N}}$ can be replaced by $c\varepsilon^M x^{\frac{M}{N}}$ where $\varepsilon^N = 1$ and hence at least N solutions can be obtained (also see [20] for a more elaborate explanation).

If $\mathcal{N}(f)$ has an edge which is parallel to the bisectrix of the first quadrant, i.e., the edge with the slope 1 we can start the resolution process with the weight $w(x) = 1$, $w(y) = -1$. If we choose a nonzero root of the equation $f_w = 0$, then a solution $y = cx^{-1} + \sum_{i=1}^{\infty} c_i x^{\frac{r_i}{N}}$ where $c \in \mathbb{C}^*$ and $-1 < \frac{r_1}{N} < \frac{r_2}{N} < \dots$ will be obtained.

It is time to recall our particular situation. We have two polynomials $f, g \in \mathbb{C}[x, y]$ with $\mathrm{J}(f, g) = 1$ and the Newton polygon of f supposedly contains an edge with slope 1. David Wright observed in [24] that the differential form $ydx - g(x, y)df(x, y)$ is exact if and only if $\mathrm{J}(f, g) = 1$ (a calculus exercise) and therefore

$$ydx - g(x, y)df(x, y) = dH(x, y) \tag{3}$$

where $H \in \mathbb{C}[x, y]$ (see the proof of Theorem 3.3 in [24]). By the chain rule $dH(x, \phi(x)) = \phi(x)dx - g(x, \phi(x))df(x, \phi(x))$ for any expression $\phi(x)$ for which the derivative $\frac{d}{dx}$ is defined.

Take for $\phi(x)$ a solution $y = cx^{-1} + \sum_{i=1}^{\infty} c_i x^{\frac{r_i}{N}}$ for $f(x, y) = 0$.

Then $f(x, \phi(x)) = 0$ and $dH(x, \phi(x)) = \phi(x)dx$ or

$$\frac{dH(x, \phi(x))}{dx} = \phi(x). \tag{4}$$

Since ϕ contains x^{-1} with a nonzero coefficient $H(x, \phi(x))$ should contain $\ln x$ with a nonzero coefficient which is clearly not possible. □

We see that on a smooth curve γ given by $f(x, y) = 0$ the differential form ydx is exact. This is a very strong restriction on γ. If γ is a rational curve and we do not mind logarithms ydx on γ is exact but the exactness of the restriction of ydx on γ does not imply that the genus of γ is zero (even if logarithms are forbidden). E.g. for $\varphi = x^k y^{2k}(y^k - 1)^{k-1}$, $\psi = xy(y^k - 1)$ we have $\mathrm{J}(\varphi, \psi) = k\varphi$ and $ydx - \frac{\psi}{k\varphi}d\varphi = d[xy(2 - y^k)]$. Hence $ydx = d[xy(2 - y^k)]$ on $\varphi = 1$. This curve is birationally equivalent to the kth Fermat curve: $x^k y^{2k}(y^k - 1)^{k-1} = 1$, hence $x^k y^{2k}(y^k - 1)^k = y^k - 1$ and $[xy^2(y^k - 1)]^k = y^k - 1$.

Apparently a description of curves on which the form ydx is exact is not known and possibly is rather complicated. I do not have a conjectural description of these curves but to find one seems to be very interesting.

3 Conclusion

A reader may ask if it is possible to extract more information from (1) and (2). For example when $\rho\tau > 0$ it is easy to observe that all roots of q must be of multiplicity 1; that all roots of p are also roots of q; that $\varsigma = xyh(x^a y^b)$ where a, b are relatively prime integers and h is a polynomial and hence $m = l(1 + ka)$, $n = l(1 + kb)$ (e.g., when the right leading edge is vertical, then $a = 0$ and m divides n); that there is a root of p with multiplicity larger than $\frac{\rho}{\tau}$, this observation was made by Nagata in [14] and Vinberg (private communication); and possibly something else which eludes me. The problem is that there are plenty of polynomial solutions even for a more restrictive Davenport equation $ap'r - bpr' = 1$ where a, b are positive relatively prime integers both larger than 1 (see [4, 22, 23, 25]). Similarly there are plenty of forms which satisfy the Dixmier equation (2) when ρ and τ have different signs. So we cannot eliminate additional edges of $\mathcal{N}(f)$ using only this approach. It is not very surprising, everybody who thought about JC knows of its slippery nature! Clearly a description of curves on which ydx is exact will help, but this question is possibly harder than JC.

Acknowledgements The author is grateful to the Max-Planck-Institut für Mathematik in Bonn, Germany, where he is presently a visitor. While working on this project the author was supported by an NSA grant H98230-09-1-0008, by an NSF grant DMS-0904713, a Fulbright fellowship awarded by the United States-Israel Educational Foundation, and a FAPESP grant 2011/52030-5 awarded by the State of São Paulo, Brazil.

Finally, a suggestion by a friendly referee to include a remark on the Jung theorem, which is accepted with gratitude helped to improve the paper.

References

1. S.S. Abhyankar, *Lectures on Expansion Techniques in Algebraic Geometry* (Tata Institute of Fundamental Research, Bombay, 1977)
2. H. Appelgate, H. Onishi, The Jacobian conjecture in two variables. J. Pure Appl. Algebra **37**(3), 215–227 (1985)
3. P. Cassou-Noguès, Newton trees at infinity of algebraic curves, in *Affine Algebraic Geometry*. CRM Proceedings of Lecture Notes, vol. 54 (American Mathematical Society, Providence, 2011), pp. 1–19 [The Russell Festschrift]
4. H. Davenport, On $f^3(t) - g^2(t)$. Norske Vid. Selsk. Forh. (Trondheim) **38**, 86–87 (1965)
5. J. Dixmier, Sur les algèbres de Weyl. Bull. Soc. Math. Fr. **96**, 209–242 (1968)

6. I. Gelfand, A. Kirillov, Sur les corps liés aux algèbres enveloppantes des algèbres de Lie. (French) Inst. Hautes Études Sci. Publ. Math. **31**, 5–19 (1966)
7. R. Heitmann, On the Jacobian conjecture. J. Pure Appl. Algebra **64**, 35–72 (1990)
8. H.W.E. Jung, Über ganze birationale Transformationen der Ebene. J. Reine Angew. Math. **184**, 161–174 (1942)
9. O.H. Keller, Ganze cremona-transformationen. Monatsh. Math. Physik **47**, 299–306 (1939)
10. J. Lang, Jacobian pairs II. J. Pure Appl. Algebra **74**, 61–71 (1991)
11. L. Makar-Limanov, Commutativity of certain centralizers in the rings $R_{n,k}$. Funk. An. i ego Pril. **4**, 332–333 (1970)
12. J. McKay, S. Wang, A note on the Jacobian condition and two points at infinity. Proc. Am. Math. Soc. **111**(1), 35–43 (1991)
13. T. Moh, On the Jacobian conjecture and the configuration of roots. J. Reine Angew. Math. **340**, 140–212 (1983)
14. M. Nagata, Two-dimensional Jacobian conjecture, in *Algebra and Topology* (Korea Inst. Tech., Taejon, 1988), pp. 77–98
15. M. Nagata, Some remarks on the two-dimensional Jacobian conjecture. Chin. J. Math. **17**(1), 1–7 (1989)
16. I. Newton, De methodis serierum et fluxionum, in *The Mathematical Papers of Isaac Newton*, vol. 3, ed. by D.T. Whiteside (Cambridge University Press, Cambridge, 1967–1981), pp. 32–353, 43–71
17. A. Nowicki, Y. Nakai, On Appelgate-Onishi's lemmas. J. Pure Appl. Algebra **51**(3), 305–310 (1988)
18. A. Nowicki, Y. Nakai, Correction to: "On Appelgate-Onishi's lemmas" [J. Pure Appl. Algebra **51**(3), 305–310 (1988)]; J. Pure Appl. Algebra **58**(1), 101 (1989)
19. M. Oka, On the boundary obstructions to the Jacobian problem. Kodai Math. J. **6**(3), 419–433 (1983)
20. V.A. Puiseux, Recherches sur les fonctions algebriques. J. Math. Pure Appl. **15** 365–480 (1850)
21. R. Rentschler, Opérations du groupe additif sur le plan affine (French). C. R. Acad. Sci. Paris Sér. A-B **267**, A384–A387 (1968)
22. T. Shioda, Elliptic surfaces and Davenport-Stothers triples. Comment. Math. Univ. St. Pauli **54**(1), 49–68 (2005)
23. W. Stothers, Polynomial identities and Hauptmoduln. Quart. J. Math. Oxford Ser. (2) 32(127), 349–370 (1981)
24. D. Wright, Affine surfaces fibered by affine lines over the projective line. Ill. J. Math. **41**(4), 589–605 (1997)
25. U. Zannier, On Davenport's bound for the degree of $f^3 - g^2$ and Riemann's existence theorem. Acta Arith. **71**(2), 107–137 (1995)

Keller Maps of Low Degree over Finite Fields

Stefan Maubach and Roel Willems

Abstract For a finite field $\mathbb{F}_q$, the set of polynomial endomorphisms of $\mathbb{F}_q^n$ of degree d is bounded when n and d are fixed. This makes it possible to compute the set of *all* polynomial automorphisms of degree d or less (while it is still an open problem to determine generators of the group of polynomial automorphisms). In this chapter, we do exactly that: we compute the set of all automorphisms for the dimensions and degrees for which it is computationally feasible. In addition, we study a slightly larger class of endomorphisms, the "mock automorphisms," which are Keller maps inducing bijections of the space $\mathbb{F}_q^n$ (essentially characteristic p counterexamples to the Jacobian Conjecture which are injective) and determine some of their equivalence classes. We also determine equivalence classes of locally finite polynomial endomorphisms of low degree. The results of this chapter are mainly of a computational nature, and the conjectures we can make due to these computations, but we also prove a few theoretical results related to mock automorphisms.

AMS classification: 14R10, 14Q15, 37P05

Research partially done while funded by Veni-grant of council for the physical sciences, Netherlands Organisation for scientific research (NWO).

Funded by Phd-grant of council for the physical sciences, Netherlands Organisation for scientific research (NWO).

S. Maubach (✉)
Jacobs University, 28759 Bremen, Germany
e-mail: s.maubach@jacobs-university.de

R. Willems
Sopra Banking Software, Avenue de Tervuren 226, 1150 Bruxelles, Belgium
e-mail: roelwill@gmail.com

I. Cheltsov et al. (eds.), *Automorphisms in Birational and Affine Geometry*, Springer Proceedings in Mathematics & Statistics 79,
DOI 10.1007/978-3-319-05681-4_26,

1 Introduction

1.1 *Notations and Definitions*

The following notations are mostly standard: throughout this paper, $\mathbb{F}_q$ will be a finite field of characteristic p where $q = p^r$ for some $r \in \mathbb{N}^*$. When $F_1, \ldots, F_n \in k[x_1, \ldots, x_n]$ (k a field), then $F := (F_1, \ldots, F_n)$ is a polynomial endomorphism over k. If there exists a polynomial endomorphism G such that $F(G) = G(F) = (x_1, \ldots, x_n)$, then F is a polynomial automorphism (which is stronger than stating that F induces a bijection on k^n). The polynomial automorphisms in n variables over k form a group, denoted $\mathrm{GA}_n(k)$ (compare the notation $\mathrm{GL}_n(k)$), while the set of polynomial endomorphisms is denoted by $\mathrm{ME}_n(k)$ (the monoid of endomorphisms). If $F \in \mathrm{GA}_n(k)$ such that $\deg(F_i) = 1$ for all $1 \leq i \leq n$, then F is called **affine**. The affine automorphisms form a subgroup of $\mathrm{GA}_n(k)$ denoted by $\mathrm{Aff}_n(k)$. In case $F \in \mathrm{GA}_n(k)$ such that $F_i \in k[x_i, \ldots, x_n]$ for each $1 \leq i \leq n$, then F is called **triangular**, or **Jonquière**. The triangular automorphisms form a subgroup of $\mathrm{GA}_n(k)$, denoted by $\mathrm{J}_n(k)$. The subgroup of $\mathrm{GA}_n(k)$ generated by $\mathrm{Aff}_n(k)$ and $\mathrm{J}_n(k)$ is called the **tame automorphism group**, denoted by $\mathrm{TA}_n(k)$. By $\deg(F)$ we will denote the maximum of $\deg(F_i)$.

If $F, G \in \mathrm{ME}_n(k)$, then F and G are called **equivalent (tamely equivalent)** if there exist $N, M \in \mathrm{GA}_n(k)$ ($N, M \in \mathrm{TA}_n(k)$) such that $NFM = G$. If $F \in \mathrm{ME}_n(k)$ then we say that $(F, x_{n+1}, \ldots, x_{n+m}) \in \mathrm{ME}_{n+m}(k)$ is a **stabilization** of F. We hence introduce the terms **stably equivalent** and **stably tamely equivalent** meaning that a stabilization of F and G are equivalent or tamely equivalent.

We now introduce some notations which are more specific to this chapter.

$$\begin{aligned}
\mathrm{ME}_n^d(k) &= \{F \in \mathrm{ME}_n(k) \mid \deg(F) \leq d\},\\
\overline{\mathrm{ME}}_n(k) &= \{F \in \mathrm{ME}_n(k) \mid \text{affine part of } F \text{ is identity.}\},\\
\overline{\mathrm{ME}}_n^d(k) &= \mathrm{ME}_n^d(k) \cap \overline{\mathrm{ME}}_n(k),
\end{aligned}$$

and of course the corresponding intersections with $\mathrm{GA}_n(k)$:

$$\begin{aligned}
\mathrm{GA}_n^d(k) &:= \mathrm{ME}_n^d(k) \cap \mathrm{GA}_n(k),\\
\overline{\mathrm{GA}}_n(k) &:= \overline{\mathrm{ME}}_n(k) \cap \mathrm{GA}_n(k),\\
\overline{\mathrm{GA}}_n^d(k) &:= \overline{\mathrm{ME}}_n^d(k) \cap \mathrm{GA}_n(k) = \overline{\mathrm{GA}}_n(k) \cap \mathrm{GA}_n^d(k).
\end{aligned}$$

1.2 *Background*

The automorphism group $\mathrm{GA}_n(k)$ is one of the basic objects in (affine) algebraic geometry, and the understanding of its structure a much studied question. If $n = 1$ then $\mathrm{GA}_1(k) = \mathrm{Aff}_1(k)$, and if $n = 2$ then one has the Jung–van der Kulk theorem

[7, 16], stating among others that $GA_2(k) = TA_2(k)$. However, in dimension 3 the structure of $GA_3(k)$ is almost completely unknown. The only strong result is in fact a *negative* result by Umirbaev and Shestakov [13, 14], stating that if $\operatorname{char} k = 0$, then $TA_3(k) \neq GA_3(k)$.

It might be that all types of automorphisms known in $GA_3(k)$ have already surfaced, but the possibility exists that there are some strange automorphisms that have eluded common knowledge so far. (See [9–11].) But, in the case $k = \mathbb{F}_q$, we have an opportunity: one could simply check the finite set of endomorphisms up to a certain degree d, i.e., $ME_n^d(\mathbb{F}_q)$, and determine which ones are automorphisms. Any "new" type of automorphisms *have* to surface in this way. (In this respect, also note the paper [5], which classifies the polynomial automorphisms in $GA_3(\mathbb{C})$ of degree 2.)

Unfortunately, the computations rapidly become unfeasible if the degree d, the number of variables n, or the size of the finite field $\mathbb{F}_q$, are too large. We didn't find any significant shortcuts except the ones mentioned in Sect. 3. In the end, for us scanning through lists of $2^{30} = 8^{10}$ endomorphisms was feasible, but $3^{20} = 9^{10} > 2^{31}$ was barely out of reach.

Nevertheless—these are computations that *have* to be done at some point, and we did it. In fact, with a negligible additional effort we could study special sets of interesting Keller maps:

2 Mock Automorphisms: Keller Maps in Characteristic p

We remind the reader that a **Keller map** is a polynomial map $F \in ME_n(R)$ (R a commutative ring) such that $\det(\operatorname{Jac}(F)) \in R^*$. A useful criterion is that if F is invertible, then $\det(\operatorname{Jac}(F)) \in R^*$. The converse is a notorious problem in characteristic zero:

Jacobian Conjecture. (Short JC) Let k be a field of characteristic zero, $F \in ME_n(k)$, and $\det(\operatorname{Jac}(F)) \in k^*$, then F is a polynomial automorphism.

The JC in $\operatorname{char}(k) = p$ is not true in general, as already in one variable, $F(x_1) := x_1 - x_1^p$ has Jacobian 1, but $F(0) = F(1)$ and so F is not a bijection.

Definition 2.1. Let $F \in ME_n(\mathbb{F}_q)$. We say that F is a **mock polynomial automorphism** if F is a Keller map and $F : \mathbb{F}_q^n \longrightarrow \mathbb{F}_q^n$ is a bijection. (Polynomial automorphisms are also mock polynomial automorphisms.)

There are two reasons to study mock polynomial automorphisms:

One reason is that such maps are interesting for cryptography: they are "multivariate permutation polynomials," but in a completely nontrivial way. (For example, in dimension 1 permutation polynomials are important tools—the multivariate versions show equal promise.)

The second reason is that they *are* counterexamples to the Jacobian Conjecture in characteristic p and can help shed insight in this problem in characteristic zero.

Related to this, we want to discuss the connection between algebraic independence of $F_1, \ldots, F_n$, the determinant of the Jacobian nonzero or even constant, and F being a bijection. These discussions point out why mock automorphisms are the "best" objects to consider over $\mathbb{F}_q$ when searching for Keller maps in characteristic zero (in particular over $\mathbb{Q}$ or $\mathbb{Z}$).

Theorem 2.2. *Suppose $F \in \mathrm{ME}_n(\mathbb{Z})$ is a Keller map (i.e., $\det(\mathrm{Jac}(F)) = \pm 1$), and that $F \mod p$ is a mock automorphism for infinitely many p. Then $F \in \mathrm{GA}_n(\mathbb{Z})$.*

Proof. This follows directly from Theorem 10.3.8 of [15], which we recall in our notation: "Let $F \in \mathrm{ME}_n(\mathbb{Z})$. If $F \mod p$ is injective for all but finitely many primes p, then $d := \det(\mathrm{Jac}(F)) \in \mathbb{Z}\backslash\{0\}$ and $F \in \mathrm{GA}_n(\mathbb{Z}[d^{-1}])$." Since here we assume that $d = \det(\mathrm{Jac}(F)) = \pm 1$, $F \in \mathrm{GA}_n(\mathbb{Z})$. □

We also want to point out here that injectivity over other fields has far stretching consequences:

1. If the field K is algebraically closed, then any injective polynomial map $F : K^n \longrightarrow K^n$ is surjective.
2. Any injective polynomial map $F : \mathbb{R}^n \longrightarrow \mathbb{R}^n$ is surjective.
3. Any injective polynomial map $F : \mathbb{C}^n \longrightarrow \mathbb{C}^n$ is an automorphism.

We refer to [2, 3, 8, 12] for these results.

Conjecture 2.3. Suppose $F = (F_1, \ldots, F_n) \in \mathrm{ME}_n(\mathbb{F}_q)$ and F induces a bijection $\mathbb{F}_q^n \longrightarrow \mathbb{F}_q^n$. Then $F_1, \ldots, F_n$ are algebraically independent.

The point of Conjecture 2.3 is to emphasize that it is reasonable to assume algebraic independence of the elements.

Lemma 2.4. *Let $F_1, \ldots, F_n \in k^{[n]}$ where $\mathrm{char}(k) = p$. Assume that $F_1, \ldots, F_n$ are algebraically dependent. If $Q \neq 0$ is of lowest possible degree such that $Q(F_1, \ldots, F_n) = 0$, then $Q \in k[x_1, \ldots, x_n]\backslash k[x_1^p, \ldots, x_n^p]$.*

Proof. Let Q be a lowest degree nonzero polynomial such that $Q(F_1, \ldots, F_n) = 0$. Assume that $Q \in k[x_1^p, \ldots, x_n^p]$, i.e., $Q = \tilde{Q}(x_1^p, \ldots, x_n^p)$ for some $\tilde{Q} \in k^{[n]}$. Denote $x^\alpha = x_1^{\alpha_1} \cdots x_n^{\alpha_n}$ if $\alpha \in \mathbb{N}^n$ and write $\tilde{Q}(x_1, \ldots, x_n) = \sum \lambda_\alpha x^\alpha$, i.e., the λ_α are the coefficients of $\tilde{Q}$. Then

$$\tilde{Q}(x_1^p, \ldots, x_n^p) = \sum \lambda_\alpha (x^\alpha)^p = \left(\sum \mu_\alpha x^\alpha\right)^p$$

where $\mu_\alpha^p = \lambda_\alpha$ (and thus μ_α is in a finite extension k' of k). Define $G = \sum \mu_\alpha x^\alpha$. Note that $\deg(G) = \frac{1}{p} \deg Q < \deg(Q)$. We can define L to be a finite extension of k such that $\mu_\alpha \in L$ for all $\alpha \in \mathbb{N}^n$. Let $l_1, \ldots, l_m$ be a k-basis of L. Then there exist $\mu_{\alpha,i} \in k$ such that $\sum_{i=1}^m \mu_{\alpha,i} l_i = \mu_\alpha$. Thus, there exist $G_i \in k[x_1, \ldots, x_n]$ satisfying $\deg(G_i) \leq \deg(G)$ such that $G = \sum_{i=1}^m G_i l_i$ and

$$\left(\sum_{i=1}^{m} G_i l_i\right)^p = \left(\sum \mu_\alpha x^\alpha\right)^p = \tilde{Q}(x_1^p, \ldots, x_n^p) = Q.$$

Now $0 = Q(F_1, \ldots, F_n) = \left(\sum_{i=1}^m G_i(F_1, \ldots, F_n) l_i\right)^p$ and thus $\sum_{i=1}^m G_i(F_1, \ldots, F_n) l_i = 0$. Since $l_1, \ldots, l_m$ form a k-basis of L, they also form a $k(x_1, \ldots, x_n)$-basis of $L(x_1, \ldots, x_n)$. Hence, $\sum_{i=1}^m G_i(F_1, \ldots, F_n) l_i = 0$ implies $G_i(F_1, \ldots, F_n) = 0$ for all $1 \le i \le m$. Not all G_i are zero, and since all G_i are of lower degree than Q, we get a contradiction. Thus $Q \notin k[x_1^p, \ldots, x_n^p]$. □

Lemma 2.5. *Let* $F = (F_1, \ldots, F_n) \in \mathrm{ME}_n(k)$ *where* $\mathrm{char}(k) = p$. *Assume* $F_1, \ldots, F_n$ *algebraically dependent. Then* $\det(\mathrm{Jac}(F)) = 0$.

Proof. Let $Q \neq 0$ be of lowest degree such that $Q(F_1, \ldots, F_n) = 0$. By lemma 2.4 we know that $Q \notin k[x_1^p, \ldots, x_n^p]$, which implies that at least one of the derivatives $\partial_i Q \neq 0$, or in other words $\nabla Q \neq 0$. Now $0 = Q(F)$, hence $0 = \mathrm{Jac}(F) \cdot (\nabla Q)(F) = 0$. Note that if $\partial_i Q \neq 0$ then $(\partial_i Q)(F) \neq 0$ since Q is assumed to be of lowest degree such that $Q(F) = 0$. Thus, the vector $(\nabla Q)(F) \neq 0$, and hence $\mathrm{Jac}(F)$ is not invertible over the field $k(x_1, \ldots, x_n)$, meaning its determinant is zero. □

Of course, the converse of the above lemma is not true—but conjecture 2.3 would be a sort-of converse.

We now want to point out that having $\det(\mathrm{Jac}(F))$ constant is even better:

Example 2.6. Let $F = (x + y(x^p - x), y + y(x^p - x))$. Then F induces a bijection of $\mathbb{F}_p^2$ but does not induce a bijection of K^2 for any field extension $[K : \mathbb{F}_p] < \infty$.

Proof. Note that F induces the identity map $\mathbb{F}_p^2 \longrightarrow \mathbb{F}_p^2$ and thus indeed is bijective. Let K be any extension of $\mathbb{F}_p$ and pick $s \in K \backslash \mathbb{F}_p$. Then $s^p - s \neq 0$ and we can define $t = \frac{-s}{s^p - s}$. Then $F(s,t) = (s + t(s^p - s), t + t(s^p - s)) = (0, t - s) = F(0, t - s)$ and thus the map $K^2 \longrightarrow K^2$ induced by F is not injective. □

The above example is not a mock automorphism, and in fact, we conjecture:

Conjecture 2.7. Let $F \in \mathrm{ME}_n(\mathbb{F}_q)$ be a mock automorphism. Then there exist infinitely many finite extensions $[K : \mathbb{F}_q] < \infty$ such that F is a mock automorphism over K, i.e., F induces a bijection $K^n \longrightarrow K^n$.

The above conjecture is already challenging in dimension 1, we will only give a proof for a special case (as we need it later in the chapter):

Lemma 2.8. *Let* $f(x) \in \mathbb{F}_q x + \mathbb{F}_q x^p + \mathbb{F}_q x^{p^2} + \mathbb{F}_q x^{p^3} + \cdots$. *If* $f : \mathbb{F}_q \longrightarrow \mathbb{F}_q$ *is a bijection, then there exist infinitely many extensions* K *of* $\mathbb{F}_q$ *such that* $f : K \longrightarrow K$ *is a bijection.*

More precisely, if $f(x) = x f_1(x) \cdots f_k(x)$ *where* $f_i \in \mathbb{F}_q[x]$ *irreducible, and* $d_i = \deg(f_i)$, *and* $[K : \mathbb{F}_q] = m < \infty$, *then* $f : K \longrightarrow K$ *is a bijection if and only if* $d_1, \ldots, d_k$ *all do not divide* m.

Proof. Assume that K is some extension of $\mathbb{F}_q$ and $a, b \in K$ such that $f(a) = f(b)$. Since $a^{p^k} - b^{p^k} = (a-b)^{p^k}$ for any $k \in \mathbb{N}$, we get $0 = f(a) - f(b) = f(a-b)$. Hence, $a - b$ is a root of $f(x)$, and $a \neq b$ is equivalent to having a nonzero root of $f(x)$. Thus: K is an extension of $\mathbb{F}_q$ for which $f(x)$ only has $x = 0$ as root, if and only if f is injective on $\mathbb{F}_q$. Factoring $f(x) = x f_1(x) \cdots f_d(x)$, we see that we find a nonzero root if and only if $f_i(x)$ has a root in K for some i, which is equivalent to K containing $\mathbb{F}_q[x]/(f_i(x)) \cong \mathbb{F}_{q^{d_i}}$. If $[K : \mathbb{F}_q] = m < \infty$, this is equivalent to d_i dividing m. □

See [1] for other interesting results related to polynomials as in the above Lemma 2.8.

In this chapter, we compute (some of) the (mock) automorphisms for $n = 3$, $d \leq 3$, and $q \leq 5$. In particular, we compute classes up to (tame) equivalence. Note that the set of mock automorphisms over a field $\mathbb{F}_q$ form a monoid (being closed under composition), and that a mock automorphism over $\mathbb{F}_{q^r}$ is obviously a mock automorphism over $\mathbb{F}_q$.

3 Generalities on Polynomial Automorphisms

The following trivial lemma explains why we only study polynomial maps having affine part identity:

Lemma 3.1. *Let* $F \in \mathrm{GA}_n^d(k)$. *Then there exists a unique* $\alpha, \beta \in \mathrm{Aff}_n(k)$ *and* $F', F'' \in \overline{\mathrm{GA}}_n^d(k)$ *such that*

$$F = \alpha F' = F'' \beta.$$

Proof. The first equality is trivial (take α to be the inverse of the affine part of F). The second equation follows by considering $F^{-1} = \alpha G$, and then $F = F''\beta$ where $F'' = G^{-1}, \beta = \alpha^{-1}$. The fact that $F'' \in \mathrm{GA}_n^d(k)$ is easy to check by comparing the highest degrees of F'' and F. □

However, the following two (well-known) lemmas show that the Jacobian conjecture is true for the special case where $\deg(F) = 2$ and $\mathrm{char}(k) \geq 3$.

Lemma 3.2. *Let* $F : k^n \to k^n$ *be a polynomial endomorphism of degree* 2 *with* $\det(\mathrm{Jac}(F))$ *nowhere zero. If* $\mathrm{char}(k) = p \neq 2$, *then* F *is injective. In particular, if* k *is a finite field, then* F *is bijective.*

Proof. The Proof of Proposition 4.3.1 of [15] works as long as the characteristic of k is not 2, even though it is only written for the case that k is algebraically closed and of characteristic zero. □

Corollary 3.3. *Let* $F : k^n \to k^n$ *be a polynomial endomorphism of degree* 2 *with* $\det(\mathrm{Jac}(F)) = 1$. *If* $char(k) \neq 2$, *then* F *is an automorphism.*

Proof. Let K be the algebraic closure of k, and consider F as a polynomial endomorphism of K^n. For every finite extension L of k, we have $F : L^n \to L^n$ is a bijection, by the above lemma. Hence, $F : K^n \to K^n$ is a bijection (as K is the infinite union of all finite extensions of k). But K is algebraically closed so a bijection of K^n is a polynomial automorphism, so it has an inverse F^{-1}. Now Lemma 1.1.8 in [15] states that F^{-1} has coefficients in k, which means that F^{-1} is defined over k, which means that F is a polynomial automorphism over k. □

One remark on the previous result about the difference between $\det(\mathrm{Jac}(F)) = 1$ and $\det(\mathrm{Jac}(F))$ is nowhere zero. If $\det(\mathrm{Jac}(F))$ is nowhere zero over k this does not imply that $\det(\mathrm{Jac}(F))$ over K is nowhere zero, consider the following example (see the warning after Corollary 1.1.35 in [15]):

Example 3.4. Let $F = (x, y + axz, z + bxy) \in k[x, y, z]^3$ with k a finite field of characteristic p and $a, b \neq 0 \in k$, such that ab is not a square. Then $\det(\mathrm{Jac}(F)) = 1 - abx^2$ is nowhere zero, but obviously F not invertible.

In this chapter we also consider so-called locally finite polynomial automorphisms. A motivation for studying these automorphisms is that they might generate the automorphism group in a natural way (see [6] for a more elaborate motivation of studying these maps). The reason that we make computations and classifications on them in this chapter is to have some examples on hand to work with in the future: they can be examples of complicated polynomial automorphisms having nice properties.

Definition 3.5. Let $F \in \mathrm{ME}_n(k)$. Then F is called **locally finite** (short LFPE) if $\{\deg(F^n) \mid n \in \mathbb{N}\}$ is bounded, or equivalently, there exists $n \in \mathbb{N}$ and $a_i \in k$ such that $F^n + a_{n-1}F^{n-1} + \cdots + a_1 F + a_0 I = 0$. We say that $T^n + a_{n-1}T^{n-1} + \cdots + a_1 T + a_0$ is a vanishing polynomial for F. In [6], Theorem 1.1, it is shown that these vanishing polynomials form an ideal of $k[T]$, and that there exists a minimal polynomial, denoted $\mathrm{m}_F(T)$.

When trying to classify LFPEs and their minimal polynomials (i.e., using computer calculations) one can use the following lemmas to reduce computations:

Lemma 3.6. *Let $F \in \mathrm{GA}_n(k)$ and $L \in GL_n(k)$ then F is locally finite iff $L^{-1}FL$ is locally finite. In this case, $\mathrm{m}_F(T) = \mathrm{m}_{L^{-1}FL}(T)$.*

Proof. It is not that hard to prove that $P(T)$ is a vanishing polynomial for F if and only if it is a vanishing polynomial of $L^{-1}FL$. (Note that the linearity of L is essential!) □

When classifying LFPEs, one cannot simply restrict to $\overline{\mathrm{GA}}_n(k)$, as it is very well possible that $F \in \overline{\mathrm{GA}}_n(k)$ is not an LFPE, but αF is where $\alpha \in \mathrm{Aff}_n(k)$. However, we can of course restrict to the conjugacy classes of affine parts under linear maps. So in order to classify the locally finite automorphisms (up to some degree d), it suffices to compute the conjugacy classes of $\mathrm{Aff}_n(k)$ under conjugacy by $\mathrm{GL}_n(k)$, and compose a representative of each class with all the elements of $\overline{\mathrm{GA}}_n^d(k)$ and check if the result is locally finite.

When considering LFPEs over finite fields, we have the additional following lemma:

Lemma 3.7. *Let $F \in \mathrm{GA}_n(\mathbb{F}_q)$ be an LFPE. Then F has finite order (as element of* $\mathrm{GA}_n(\mathbb{F}_q)$*).*

Proof. If F is an LFPE, then there exists a minimum polynomial $\mathrm{m}_F(T)$ generating the ideal of vanishing polynomials for F. There exists some $r \in \mathbb{N}$ such that $\mathrm{m}_F(T) \mid T^{q^r} - T$, yielding the result. □

Another concept that surfaces is the following:

Definition 3.8. Let $F = I + H \in \overline{\mathrm{ME}}_n(k)$ where $H = (H_1, \ldots, H_n)$ is the nonlinear part. Then F is said to satisfy the **dependence criterion** if $(H_1, \ldots, H_n)$ are linearly dependent.

Notice that $F \in \overline{\mathrm{ME}}_n(k)$ satisfying the dependence criterion is equivalent to being able to apply a linear conjugation to isolate one variable, i.e., $L^{-1}FL = (x_1, x_2 + H_2, \ldots, x_n + H_n)$ for some linear map L.

4 Computations on Endomorphisms of Low Degree

Why dimension 3 and not dimension 2? Our original motivation to study solely dimension 3 is that there are no non-tame automorphisms in dimension 2. Also, we note also the result of Drensky–Yu in [4], which counts the number of automorphisms in dimension 2 up to a certain degree. However, we realize that since our interest shifted from automorphisms to mock automorphisms, that the dimension 2 case should be researched more in detail in future research.

Note that $\mathrm{GA}_n^d(k) = \mathrm{Aff}_n(k)\overline{\mathrm{GA}}_n^d(k)$, which allows us to restrict to finding all elements in $\overline{\mathrm{GA}}_n^d(k)$. We will consider mock automorphisms of the following form:

- $F \in \overline{\mathrm{ME}}_3^2(\mathbb{F}_p)$ where $p = 2, 3$,
- $I + H \in \overline{\mathrm{ME}}_3^3(\mathbb{F}_2)$ where H is homogeneous of degree 3,
- $I + H \in \overline{\mathrm{ME}}_3^2(\mathbb{F}_q)$ where $q = 4, 5$ but where $H = (H_1, H_2, H_3)$ satisfying the dependence criterion.

We explicitly mention that the following cases were computationally out of reach: Determining $\overline{\mathrm{GA}}_3^3(\mathbb{F}_2)$ in full, determining $\overline{\mathrm{GA}}_4^2(\mathbb{F}_q)$ for any q, determining $\overline{\mathrm{GA}}_3^2(\mathbb{F}_q)$ in full if $q > 2$. For example, it is unknown if there exist any $I + H \in \overline{\mathrm{GA}}_3^2(\mathbb{F}_3)$ *not* satisfying the dependence criterion. In essence, this points out the limits of current computing power—or perhaps better, shows the sheer impressive *size* of these sets even for low n, q.[1]

For the rest of the chapter, as we stay in 3 dimensions, we will rename our variables x, y, z.

4.1 The Finite Field of Two Elements: $\mathbb{F}_2$

Note that there are $(2^3 - 1)(2^3 - 2)(2^3 - 2^2) = 168$ elements of $\mathrm{GL}_3(\mathbb{F}_2)$, and $2^3 \cdot 168 = 1344$ elements in $\mathrm{Aff}_3(\mathbb{F}_2)$.

4.1.1 Degree 2 over $\mathbb{F}_2$

Over $\mathbb{F}_2$, Corollary 3.3 does not hold so there do exist mock automorphisms which are not automorphisms in $\mathrm{ME}_3^2(\mathbb{F}_2)$.

Theorem 4.1. *If $F \in \mathrm{ME}_3^2(\mathbb{F}_2)$ is a mock automorphism, then F is in one of the following four classes:*

(1) The 176 tame automorphisms, equivalent to (x, y, z).
(2) 48 endomorphisms tamely equivalent to $(x^4 + x^2 + x, y, z)$.
(3) 56 endomorphisms tamely equivalent to $(x^8 + x^2 + x, y, z)$.
(4) 56 endomorphisms tamely equivalent to $(x^8 + x^4 + x, y, z)$.

In particular, all automorphisms of this type are tame, i.e., $\mathrm{GA}_3^2(\mathbb{F}_2) = \mathrm{TA}_3^2(\mathbb{F}_2)$ Furthermore, the equivalence classes are all distinct, except possibly class (3) and (4) (see Conjecture 4.3).

There are in total $1344 \cdot 176 = 236{,}544$ *automorphisms of $\mathbb{F}_2^3$ of degree less or equal to* 2.

Proof. The classification is done by computer, see [17] Chap. 5. We can show how for example $(x^8 + x^4 + x, y, z)$ is tamely equivalent to a polynomial endomorphism of degree 2:

$$(x + y^2, y + z^2, z)(x^8 + x^4 + x, y, z)(x, y + x^4 + x^2, z + x^2) =$$
$$(x + y^2, y + x^2 + z^2, z + x^2)$$

What is left is to show that the classes (1), (2), and (3)+(4) are different. Class (1) consists of automorphisms while (2), (3), (4) are not. Using the below Lemma 4.2, the endomorphisms of type (2) are all bijections of $\mathbb{F}_{2^m}^3$ if $3 \nmid m$, and

[1]Even with many extra factors of computing power, the sets grow in size so fast, that the gain will be minimal.

the endomorphisms of type (3) and (4) are all bijections of $\mathbb{F}_{2^m}^3$ if $7 \not| m$. The last sentence follows since $\# \mathrm{Aff}_3(\mathbb{F}_2) = 1344$. □

Corollary 4.2 (of Lemma 2.8). *$x^4 + x^2 + x$ is a bijection of $\mathbb{F}_{2^r}$ if $3 \not| r$, and $x^8 + x^4 + x$ and $x^8 + x^2 + x$ are bijections of $\mathbb{F}_{2^r}$ if $7 \not| r$.*

Question 4.3. 1. Are $x^8 + x^2 + x$ and $x^8 + x^4 + x$ stably (tamely) equivalent?
2. Are $F = (x^8 + x^2 + x, y)$ and $G = (x^8 + x^4 + x, y)$ tamely equivalent?
3. Let $P, Q \in k[x]$ where k is a field of characteristic $p \neq 0$. Does P, Q stably (tamely) equivalent imply that P, Q are equivalent?

The above question is particular to characteristic p, to consider the following:

Lemma 4.4. *Let $P, Q \in k[x]$. Assume that $F := (P(x), y, z)$ is equivalent to $G := (Q(x), y, z)$. Then P' and Q' are equivalent, in particular $Q'(ax+b) = cP'$ for some $a, b, c \in k$, $ac \neq 0$.*

Proof. Equivalent means there exist $S, T \in \mathrm{GA}_3(k)$ such that $SF = GT$. Write J for det(Jac). Now $\mathrm{J}(S) = \lambda, \mathrm{J}(T) = \mu$ for some $\lambda, \mu \in k^*$. Using the chain rule we have

$$\begin{aligned} \mathrm{J}(SF) &= \mathrm{J}(F) \cdot (\mathrm{J}(S) \circ (F)) = \tfrac{\partial P}{\partial x} \cdot (\lambda \circ (F)) = \lambda \tfrac{\partial P}{\partial x} \\ &= \mathrm{J}(GT) = \mathrm{J}(T) \cdot (\mathrm{J}(G) \circ (T)) = \mu \cdot (\tfrac{\partial Q}{\partial x} \circ T) \end{aligned}$$

so

$$Q'(T) = \frac{\lambda}{\mu} P'(x)$$

which means that $T = (T_1, T_2, T_3)$, T_1 must be a polynomial in x, and thus $T_1 = ax + b$ where $a \in k^*, b \in k$. □

Corollary 4.5. *Assume* $\mathrm{char}(k) = 0$. *Let $P, Q \in k[x]$. Assume that $F := (P(x), y, z)$ is equivalent to $G := (Q(x), y, z)$. Then P and Q are equivalent.*

Proof. Lemma 4.4 shows that $P'(ax + b) = cQ'$ for some $a, b, c \in k$, $ac \neq 0$. In characteristic zero we can now integrate both sides and get $a^{-1} P(ax + b) = cQ$ proving the corollary. □

Note that in the "integrate both sides" part the characteristic zero is used, as $(x + x^2 + x^8)' = (x + x^4 + x^8)'$ in characteristic 2.

Note that all the above one-variable polynomials $x^8 + x^2 + x, x^4 + x^2 + x$ have a stabilization which is tamely equivalent to a polynomial endomorphism of degree 2. (In this respect, note that any polynomial endomorphism is stably equivalent to a polynomial endomorphism of degree 3 or less, see Lemma 6.2.5 from [15].)

4.1.2 Locally Finite in Degree 2 over $\mathbb{F}_2$

We now want to classify the locally finite automorphisms among the 236,544 automorphisms over $\mathbb{F}_2$ of degree 2 (or less), and we want to determine the minimum polynomial of each. Using Lemma 3.6, we may classify up to conjugation by a linear map. We found 262 locally finite classes under linear conjugation, with the following minimum polynomials: In the above table, # denotes the number of

Minimumpolynomial	#	t
$F^5 + F^4 + F + I$	16	8
$F^4 + F^3 + F^2 + I$	8	7
$F^4 + F^3 + F + I$	26	6
$F^4 + I$	12	4
$F^4 + F^2 + F + I$	8	7
$F^3 + F^2 + F + I$	139	4
$F^3 + F^2 + I$	2	7
$F^3 + F + I$	2	7
$F^3 + I$	14	3
$F^2 + I$	34	2
$F + I$	1	1

conjugacy classes (i.e., not elements) having this minimum polynomial, while t denotes the order of the automorphism (see Lemma 3.7). Furthermore, observe that # displayed is the number of conjugacy classes that satisfy this relation, not the total number of automorphisms.

4.1.3 Degree 3 over $\mathbb{F}_2$

We only considered the endomorphisms of the form $F = I + H$, where H is homogeneous of degree 3. The below table describes the set of $F \in \mathrm{ME}_3^3(\mathbb{F}_2)$ having the following criteria:

- F is a mock automorphism,
- $F = I + H$, H homogeneous of degree 3.

We found $1,520$ endomorphisms satisfying the above requirements. The table lists them in 20 classes up to conjugation by linear maps:
The first column gives a representative up to linear conjugation, and the one with bold font gives a representative under tamely equivalence for the classes listed beneath it. The second column lists for which field extensions (from $\mathbb{F}_{2^r}$ where $1 \le r \le 5$) the map is also a bijection of $\mathbb{F}_{2^r}^3$ Class 1 are the 400 automorphisms, all of them are tame and satisfy the dependence conjecture. All classes are tamely

	Representative	Bijection over	#
1.	$\mathbf{(x, y, z)}$		
1a.	(x, y, z)	all	1
1b.	$(x, y, z + x^2y + xy^2)$	all	7
1c.	$(x, y, z + x^3 + x^2y + y^3)$	all	14
1d.	$(x, y + x^3, z + x^3)$	all	21
1e.	$(x, y, z + x^3 + x^2y + xy^2)$	all	21
1f.	$(x, y, z + x^2y)$	all	42
1g.	$(x, y + x^3, z + xy^2)$	all	42
1h.	$(x, y + x^3, z + x^2y + xy^2)$	all	42
1i.	$(x, y + z^3, z + x^2y)$	all	42
1j.	$(x, y + x^3, z + x^2y + y^3)$	all	84
1k.	$(x, y + x^3, z + y^3)$	all	84
2.	$\mathbf{(x, y, z + x^3z^4 + xz^2)}$		
2	$(x, y + x^3 + xz^2, z + xy^2 + xz^2)$	$\mathbb{F}_2, \mathbb{F}_4, \mathbb{F}_{16}, \mathbb{F}_{32}$	56
3.	$\mathbf{(x, y, z + x^3z^2 + x^3z^4)}$		
3a.	$(x, y + xz^2, z + x^2y + xy^2)$	$\mathbb{F}_2, \mathbb{F}_4$	84
3b.	$(x, y + xz^2, z + x^3 + x^2y + xy^2)$	$\mathbb{F}_2, \mathbb{F}_4$	84
4.	$\mathbf{(x, y, z + xz^2 + xz^6)}$		
4a.	$(x, y + x^3 + z^3, z + x^3 + xy^2 + xz^2)$	$\mathbb{F}_2$	168
4b.	$(x, y + z^3, z + xy^2 + xz^2)$	$\mathbb{F}_2$	168
5.	$\mathbf{(x, y, z + x^3z^2 + xy^2z^4 + x^2yz^4 + x^3z^6)}$		
5a.	$(x, y + xz^2, z + xy^2 + y^3)$	$\mathbb{F}_2$	168
5b.	$(x, y + xz^2, z + x^3 + x^2y + y^3)$	$\mathbb{F}_2$	168
6.	$\mathbf{(x, y, z + x^3z^2 + xy^2z^2 + x^2yz^4 + x^3z^6)}$		
6.	$(x, y + xy^2 + xz^2, z + x^3 + x^2y)$	$\mathbb{F}_2$	168
7.	$\mathbf{(x + y^2z, y + x^2z + y^2z, z + x^3 + xy^2 + y^3)}$		
7.	$(x + y^2z, y + x^2z + y^2z, z + x^3 + xy^2 + y^3)$	$\mathbb{F}_2$	56

equivalent to a map of the form $(x, y, P(x, y, z))$, except the last class 7—these maps do not satisfy the Dependence Criterion, which makes them very interesting!

The above table might make one think that any mock automorphism in $\mathrm{ME}_3(\mathbb{F}_2)$ of the form $F = (x, y + H_2, z + H_3)$ where H_2, H_3 are homogeneous of the same degree, then one can tamely change the map into one of the form $(x, y, z + K)$, but the below conjecture might give a counterexample:

Conjecture 4.6. Let $F = (x, y + y^8z^2 + y^2z^8, z + y^6z^4 + y^4z^6) \in \mathrm{ME}_3(\mathbb{F}_2)$, which is a mock polynomial automorphism. Then F is not tamely equivalent to a map of the form $(x, y, z + K)$.

Due to our lack of knowledge of the automorphism group $\mathrm{TA}_3(\mathbb{F}_2)$, this conjecture is a hard one unless one finds a good invariant of maps of the form $(x, y, z + K)$.

4.2 *The Finite Field of Three Elements:* $\mathbb{F}_3$

4.2.1 Degree 2 over $\mathbb{F}_3$

Over $\mathbb{F}_3$, there are $(27-1)(27-3)(27-9) = 11{,}232$ elements of $GL_3(\mathbb{F}_3)$ and $27 \cdot 11{,}232 = 303{,}264$ elements of $\mathrm{Aff}_3(\mathbb{F}_3)$.

From corollary 3.3 it follows that if $\det(\mathrm{Jac}(F)) = 1$ and $\deg(F) \leq 2$, then F is an automorphism—so we will not encounter any mock automorphisms which aren't an automorphism in this class. There are 2,835 automorphisms of degree less or equal to 2 having affine part identity, so there are $2{,}835 \cdot 303{,}264 =$ automorphisms of degree 2 or less. *They all turned out to be tame.*

4.2.2 Locally Finite

We computed all conjugacy classes under linear maps of locally finite automorphisms of $\mathbb{F}_3^3$. There are 80 orbits of affine automorphisms, composing a representative of each class with all of the 2,835 tame automorphisms, gives us 226,800 representatives of "conjugacy classes." We checked for each of them whether it was locally finite or not. It turns out that 25,872 of these conjugacy classes are locally finite. And there are exactly a 100 different minimum polynomials that can appear. On the next page, we have put a selected list of ten minimum polynomials (of a total of 100 different ones).

Of the appearing minimal polynomials in this list, all polynomials of degree 3 appear in this list. The highest minimum polynomials are of degree 10. In the table on the previous page, we have put their order (which is determined by the minimum polynomial), number of *conjugacy classes* with this minimum polynomial, and one example. The reader interested in the complete list we refer to Chap. 6 of the Ph.D. thesis of the second author [17].

4.2.3 Degree 3 over $\mathbb{F}_3$

The number of elements in $\overline{\mathrm{ME}}_3(\mathbb{F}_3)$ of the form $(x, y, z) + (0, H_2, H_3)$ (i.e., satisfying the dependency criterion) where H_2, H_3 are homogeneous of degree 3 is too large: this set has 3^{20} elements which was too large for our system to scan through; however, we think that this case is feasible for someone having a stronger, dedicated system and a little more time.

Note that in the next section we *do* attack this case for the larger fields $\mathbb{F}_4, \mathbb{F}_5$, but here the $\det(\mathrm{Jac}(F)) = 1$ criterion is much stronger—the degree 3 part in characteristic 3 plays its part here.

Minimum polynomial	Order	♯	Example
F^2+2I	2	509	$\begin{pmatrix} 2x^2+xy+xz+2x+y^2+z^2 \\ 2x^2+xy+xz+y^2+2y+z^2 \\ 2x^2+xy+xz+y^2+z^2+2z \end{pmatrix}$
$F^3+F^2+2F+2I$	6	5084	$\begin{pmatrix} x^2+xy+2x+y^2 \\ x^2+xy+y^2+2y \\ 2x^2+2y^2+2z \end{pmatrix}$
$F^4+2F^2+2F+2I$	24	2	$\begin{pmatrix} x^2+xz+2x+y^2+2y+z^2 \\ 2x+y+z \\ x^2+xz+x+y^2+y+z^2+z \end{pmatrix}$
F^4+2F^3+2F+I	9	3804	$\begin{pmatrix} 2x^2+2xy+x+2y^2+1 \\ 2x^2+2xy+2y^2+y+1 \\ 2x^2+xy+2x+2y^2+z+1 \end{pmatrix}$
$F^4+F^3+F^2+2F+I$	8	38	$\begin{pmatrix} x^2+2xy+xz+x+y^2+yz+z^2+2z+2 \\ x^2+2xy+xz+y^2+yz+z^2+2z \\ 2x^2+xy+2xz+2x+2y^2+2yz+2y+2z^2+2 \end{pmatrix}$
$F^5+2F^3+2F^2+F+2I$	8	8	$\begin{pmatrix} 2x^2+xy+xz+y^2+2y+z^2 \\ 2x^2+xy+xz+2x+y^2+2y+z^2+z \\ 2x^2+xy+xz+x+y^2+z^2+z \end{pmatrix}$
$F^6+F^5+2F^4+F^3+2I$	24	16	$\begin{pmatrix} y^2+yz+2y+z^2 \\ 2x+y^2+yz+2y+z^2+z \\ x+y^2+yz+z^2+z \end{pmatrix}$
$F^7+F^6+2F+2I$	18	396	$\begin{pmatrix} 2x^2+2xz+2y^2+2y+2z^2+2z+1 \\ x^2+xz+2y+z^2 \\ 2x^2+2xz+2x+2y^2+2y+2z^2+2 \end{pmatrix}$
$F^{10}+F^8+2F^5+F^2+2F+2I$	26	40	$\begin{pmatrix} y+2z^2+z+1 \\ x^2+2xz+x+z^2+1 \\ x+z+1 \end{pmatrix}$
$F^{10}+F^9+2F^8+F^7+F^6+F^5+2F^3+2F+I$	13	48	$\begin{pmatrix} 2x^2+2xy+2xz+y^2+yz+y+2z^2+2z+1 \\ 2x+y+z+2 \\ 2x^2+2xy+2xz+2x+yz+y+2z^2+2z+1 \end{pmatrix}$

4.3 *The Finite Fields* $\mathbb{F}_4$ *and* $\mathbb{F}_5$

In this section we will only restrict to degree 2, and to the maps which satisfy the dependency conjecture. Thus, in this section we restrict to maps F of the form $(x + H_1, y + H_2, z)$ where H_1, H_2 are of degree 2.

4.3.1 The Finite Field $\mathbb{F}_4$

There are $(64 - 1)(64 - 4)(64 - 16) = 181{,}440$ elements in $GL_3(\mathbb{F}_4)$ and $64 \cdot 181{,}440 = 11{,}612{,}160$ elements in $Aff_3(\mathbb{F}_4)$. We considered the following maps:

- $F \in \overline{\mathrm{ME}}_3^2(\mathbb{F}_4)$,
- F is a mock automorphism,
- F is of the form $(x + H_1(x, y, z), y + H_2(x, y, z), z)$ (i.e. F satisfies the dependency criterion).

and we counted 40, 384 such maps. Under tame equivalence, we have the following classes:

1 (x, y, z) (tame automorphisms)
2 $(x + x^2 + x^4, y, z)$

So, surprisingly, we only find a subset of the classes we found over $\mathbb{F}_2$. Well, not really surprising—the dependency criterion removes the classes 3 and 4 of theorem 4.1 from the list. We conjecture that the four classes of theorem 4.1 are the same for $\mathbb{F}_4$:

Conjecture 4.7. (i) Suppose $F \in \mathrm{ME}_3^2(\mathbb{F}_4)$ is a mock automorphism of $\mathbb{F}_4$. Then F is tamely equivalent to $(P(x), y, z)$ where

$$P = x, P = x^4 + x^2 + x, P = x^8 + x^4 + x, \text{ or } P = x^8 + x^2 + x.$$

(ii) Suppose $F \in \mathrm{ME}_3^2(L)$ is a mock automorphism of L, where $[L : \mathbb{F}_2] < \infty$. Then F is tamely equivalent to $(P(x), y, z)$ where

$$P = x, P = x^4 + x^2 + x, P = x^8 + x^4 + x, \text{ or } P = x^8 + x^2 + x.$$

If $3|[L : \mathbb{F}_2]$ then one should remove the class of $P = x^4 + x^2 + x$, and if $7|[L : \mathbb{F}_2]$ then one should remove the classes of $P = x^8 + x^4 + x$ and $P = x^8 + x^2 + x$.

It would be interesting to see a proof of this conjecture by theoretical means—or a counterexample of course.

4.3.2 The Finite Field $\mathbb{F}_5$

There are $(125 - 1)(125 - 5)(125 - 25) = 1{,}488{,}000$ elements in $\mathrm{GL}_3(\mathbb{F}_5)$ and $125 \cdot 1{,}488{,}000 = 1{,}186{,}000{,}000$ elements in $\mathrm{Aff}_3(\mathbb{F}_5)$. We consider maps of the following form: There are 3,625 endomorphisms satisfying the following:

- $F \in \overline{\mathrm{ME}}_3^2(\mathbb{F}_5)$,
- F is a mock automorphism of $\mathbb{F}_5$,
- F satisfies the dependency criterion (i.e. $F = (x + H_1(x, y, z), y + H_2(x, y, z), z)$).

We counted 3,625 such maps—and because of Corollary 3.3, they are all automorphisms. They all turned out to be tame maps.

5 Conclusions

We can gather some of the results in the below theorem:

Theorem 5.1. *Let* $F \in \mathrm{GA}_3^d(\mathbb{F}_q)$. *If one of the below conditions is met, then* F *is tame:*

- $d = 3, q = 2,$
- $d = 2, q = 3,$
- $d = 2, q = 4$ *or* 5, *and* F *satisfies the dependency criterion.*

This gives rise to the following conjecture:

Conjecture 5.2. If $F = I + H \in \mathrm{GA}_n(k)$ where H is homogeneous of degree 2, then F is tame.

This natural conjecture might have been posed before, but we are unaware. (This chapter proves this conjecture for $n = 3$ and $k = \mathbb{F}_2, \mathbb{F}_3$.) We expect that for $n = 3$ and a generic field a solution is within reach.

Unfortunately, the computations did not allow us to go as far as finding some candidate non-tame automorphisms (though the Nagata automorphism is one, however it is of too high degree). However, one of the interesting conclusions is that the set of *classes* (under tame automorphisms) of mock automorphisms seems to be much smaller than we originally expected. In particular, we are puzzled by the interesting two-dimensional question whether the two endomorphisms in $\mathrm{ME}_2(\mathbb{F}_2)$ described by $(x^8 + x^4 + x, y)$ and $(x^8 + x^2 + x, y)$ are not equivalent, as stated in question 4.3.

Computations. For computations we used the MAGMA computer algebra program. The reader interested in the routines we refer to Chap. 6 of the thesis of the second author, [17]. Also, we possess databases usable in MAGMA, which we freely share upon request.

Acknowledgements The authors would like to thank Joost Berson for some useful discussions. Furthermore, we would like to express our gratitude to the two referees, who have significantly improved the readability of the paper on many fronts. Thank you very much!

References

1. J. Berson, Linearized polynomial maps over finite fields. preprint. (2012) [arXiv:1201.1137]
2. A. Bialynicki-Birula, M. Rosenlicht, Injective morphisms of real algebraic varieties. Proc. Am. Math. Soc. **13**, 200–203 (1962)
3. S. Cynk, K. Rusek, Injective endomorphisms of algebraic and analytic sets. Annales Polonici Mathematici **56**(1), 29–35 (1991)
4. V. Drensky, J. Yu, Automorphisms of polynomial algebras and Dirichlet series. J. Algebra **321**(1), 292–302 (2009)

5. J.E. Fornaess, H. Wu, Classifcation of degree 2 polynomial automorphisms of $\mathbb{C}^3$. Publ. Mat. **42**, 195–210 (1998)
6. J-Ph. Furter, S. Maubach, Locally finite polynomial endomorphisms. J. Pure Appl. Algebra **211**(2), 445–458 (2007)
7. H.W.E. Jung, Über ganze birationale Transformationen der Ebene. (German) J. Reine Angew. Math. **184**, 161–174 (1942)
8. M.C. Kang, Injective morphisms of affine varieties. Proc. Am. Math. Soc. **119**, 1–4 (1993)
9. S. Maubach, The automorphism group over finite fields. Serdica Math. J. **27**(4), 343–350 (2001)
10. S. Maubach, A problem on polynomial maps over finite fields, unpublished preprint (2008) [arXiv:0802.0630]
11. S. Maubach, R. Willems, Polynomial automorphisms over finite fields: Mimicking non-tame and tame maps by the Derksen group. Serdica Math. J. **37**, (305–322) 2011
12. W. Rudin, Injective polynomial maps are automorphisms. Am. Math. Mon. **102**,540–543 (1995)
13. I. Shestakov, U.Umirbaev, The tame and the wild automorphisms of polynomial rings in three variables. J. Am. Math. Soc. **17**(1), 197–227 (2004)
14. I. Shestakov, U. Umirbaev, Poisson brackets and two-generated subalgebras of rings of polynomials. J. Am. Math. Soc. **17**(1), 181–196 (2004)
15. A. van den Essen, in *Polynomial Automorphisms and the Jacobian Conjecture*. Progress in Mathematics, vol. 190 (Birkhäuser, Basel, 2000)
16. W. van der Kulk, On polynomial rings in two variables. Nieuw Arch. Wiskunde **1**(3), 33–41 (1953)
17. R. Willems, Polynomial automorphisms and Mathieu subspaces. Ph.-D. thesis, Radboud University, 2011

Cancellation

Peter Russell

Abstract What follows is a slightly expanded and updated version of lectures I gave in May 2011 during a workshop on "Group actions, generalized cohomology theories and affine algebraic geometry" at the University of Ottawa. Among the participants were young beginning mathematicians as well as seasoned experts in diverse aspects of algebra and geometry. The aim (by order of the organizers) was to give all of them a taste of "cancellation for affine algebraic varieties."

As much as possible I have tried in these notes to maintain the informal style of the lectures. They are a very selective and far from an exhaustive treatment of the subject. Should the reader note a tendency to frequently switch between the algebraic and geometric point of view: this is by design.

2000 Mathematics Subject Classification: Primary: 14-02; Secondary:14R10, 14R25, 14H50, 13B25

1 General Cancellation

In a category,

$$\text{if } A \times C \simeq B \times C, \text{ is } A \simeq B \ ?$$

If yes, we say

$$A \text{ has the } \textbf{cancellation property for } \mathrm{C}.$$

P. Russell (✉)
Department of Mathematics, McGill University, Montreal, QC, Canada
e-mail: russell@math.mcgill.ca

I. Cheltsov et al. (eds.), *Automorphisms in Birational and Affine Geometry*, Springer Proceedings in Mathematics & Statistics 79,
DOI 10.1007/978-3-319-05681-4_27,

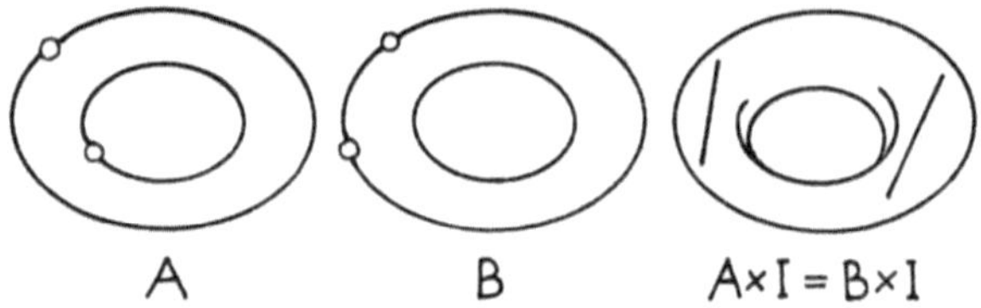

Fig. 1 An example of non-cancellation in topology

Counterexamples exist in the category of **topological spaces**. The nicest, and simplest, example I know (as far as I can remember I learned it from Pavaman Murthy) has $C = I = [0, 1]$, the closed unit interval, A is an annulus in the plane with one puncture on each of the two boundary circles and B an annulus with two punctures on just one of them. That $A \times C \simeq B \times C$ is obvious by play-dough topology (both are a solid torus with two disjoint scratches). That $A \not\simeq B$ is not entirely trivial, but follows from the well-known theorem on invariance of domain. See Fig. 1.

A bit harder to come by are counterexamples with C an open interval, or $C = \mathbb{R}$, but that they exist will be clear from what we do later. It is known to topologists that they exist even with A a contractible open manifold (in dimension ≥ 3).

2 Examples of Cancellation in Algebra

2.1 Uniqueness of the Coefficient Ring

As a first cancellation type question in **algebra**, we consider the problem of the *uniqueness of the coefficient ring in a polynomial ring*.

Let$A^{[n]}$ denote the polynomial ring in n variables over the ring A. The following is a basic

Question 2.1.1. If A and B are rings and $A^{[n]} \simeq B^{[n]}$, is $A \simeq B$? This could depend on n.

If the answer is **yes** with $n = 1$, we say that A is **invariant**.

To fix the ideas let us agree that if nothing is said to the contrary we will be talking about finitely generated commutative algebras over a field k. We can reformulate the question as:

$$\text{does } A \otimes k^{[n]} \simeq B \otimes k^{[n]} \text{ imply } A \simeq B?$$

Using "Spec" we can turn this into a geometric statement: Let $\mathbb{A}^n_k$, or just $\mathbb{A}^n$ if k is understood, denote affine n-space Spec$(k^{[n]})$. Then the question is:

$$\text{if } X = \text{Spec}(A), Y = \text{Spec}(B) \text{ are affine varieties and } X \times \mathbb{A}^n \simeq Y \times \mathbb{A}^n, \text{ is } X \simeq Y?$$

The most intriguing case here is that of $B = k^{[m]}$, i.e., $Y = \mathbb{A}^m$, and $n = 1$:

$$\text{if } X \times \mathbb{A}^1 \simeq \mathbb{A}^{m+1}, \text{ is } X \simeq \mathbb{A}^m?$$

This is **the** cancellation problem. It is open for $m \geq 3$ if $\text{char}(k) = 0$. We will discuss the **yes**-answer for $m = 1, 2$ and a very recently discovered counterexample [19] for $m = 3$ in positive characteristic further below.

More generally we can ask: Does $A \otimes C \simeq B \otimes C$ imply $A \simeq B$, or, for general algebraic varieties X, Y, Z, does $X \times Z \simeq Y \times Z$ imply $X \simeq Y$?

A very interesting variant of the cancellation problem for affine spaces is the original **Zariski cancellation problem**: if $K \supset k$ is a field (say k algebraically closed of characteristic 0), and

$$\text{if } K^{(m)} \simeq k^{(n)}, \text{ is } K \simeq k^{(n-m)}?$$

Here, for a field L, $L^{(n)}$ dentes the purely transcendental extension of degree n. The answer is **no**, a counterexample with $m = 3, n = 6$ is given in [5].

Let me make some comments here on **isomorphic** versus **equal**. Consider

$$D = k[x, y] = k^{[2]} = A^{[1]} = B^{[1]}, A = k[x], B = k[y].$$

We have $A \simeq B$, but $A \neq B$ as subsets of D. We will have to distinguish these two concepts.

Definition 2.1.2. We say that the ring A is **strongly invariant** if

$$\text{whenever } A^{[1]} = D = B^{[1]}, \text{ then } A = B \text{ (as subsets of } D).$$

An equivalent formulation is:

Any isomorphism $\Phi : A^{[1]} \to B^{[1]}$ is induced by an isomorphism $\phi : A \to B$.

Note that it would be sufficient to require $\Phi(A) \subset B$.

Example. Let A be a domain, $D = A^{[n]}$. Then the units of D are in A, so if A is generated by its units, for instance if $A = k[t, t^{-1}]$, it is uniquely determined as coefficient ring and A is strongly invariant.

The geometric version of this is

Definition 2.1.3. X has **strong cancellation** for Z if

any isomorphism $X \times Z \to Y \times Z$ is induced by an isomorphism $X \to Y$.

A note of caution: When trying to prove cancellation, one should avoid any argument that in disguise implies strong cancellation. Such a mistake is usually made by about half the students when I give the problem as an assignment in a topology class.

Here is one of the main theorems of the subject, see [3]. I will assume tacitly in the arguments that k is algebraically closed (it is not always necessary), and that the rings are finitely generated k-algebras.

Theorem 2.1.4. *A domain A of dimension 1 is strongly invariant unless it is $k^{[1]}$, or, an affine curve X has strong cancellation for $\mathbb{A}^1$ unless $X = \mathbb{A}^1$. Moreover, $A = k^{[1]}$ is invariant, or, $X = \mathbb{A}^1$ has cancellation for $\mathbb{A}^1$.*

We first make some general **observations**. Let $D = A^{[n]}$, $X = \mathrm{Spec}(A), Z = \mathrm{Spec}(D)$.

(1) D is a domain $\Longleftrightarrow$ A is a domain.
(2) D is factorial $\Longleftrightarrow$ A is a factorial.
(3) D is regular, i.e., Z is non-singular, $\Longleftrightarrow$ A is regular, i.e., X is non-singular.
(4) The same for normal.
(5) $A^* = D^*$.

Proof of theorem 2.1.4. I will quite intentionally mix geometric and algebraic language in the arguments.

We have the fibrations

$$\pi_X : Z = \mathrm{Spec}(A^{[1]}) \to X = \mathrm{Spec}(A) \textit{ and } \pi_Y : Z = \mathrm{Spec}(B^{[1]}) \to Y = \mathrm{Spec}(B)$$

with all fibers isomorphic to $\mathbb{A}^1 = \mathrm{Spec}(k^{[1]})$. The strategy now is to show that if $A \neq B$, then $X \simeq \mathbb{A}^1$ and $Y \simeq \mathbb{A}^1$.

So suppose $A \neq B$. Then there is a fiber $x_0 \times \mathbb{A}^1$ of π_X that is not a fiber of π_Y and the morphism

$$\eta : \mathbb{A}^1 \to X \times \mathbb{A}^1 = Y \times \mathbb{A}^1 \to Y,$$
$$t \mapsto (x_0, t) = (y(t), s(t)) \mapsto y(t)$$

is not constant and hence dominant, i.e., given by an injective homomorphism $\tilde{\eta} : B \to k^{[1]}$. It follows that the invertible functions on Y are constant.

Now the normalization $\tilde{B}$ of B is isomorphic to a normal subalgebra $R \subset k^{[1]}$, $R \supsetneq k$. One can make a quite elementary algebraic argument for an "affine Lüroth theorem" (exercise, or see [3]), namely that $R \simeq k^{[1]}$. One can also appeal to the usual Lüroth theorem to conclude that $\mathrm{Spec}(R)$ is a rational curve, i.e., the field of quotients of R is $k^{(1)}$. Also, $\mathrm{Spec}(R)$ is non-singular and all invertible functions are constant. Hence $\mathrm{Spec}(R) \simeq \mathbb{A}^1$. It follows that Y_{ns}, the non-singular locus of Y, is isomorphic to an open subset of $\mathbb{A}^1$. If Y is normal, we find $Y \simeq \mathbb{A}^1$. Moreover, X is normal by observation (4) above, and by symmetry $X \simeq \mathbb{A}^1$.

If Y is not normal, then the non-normal locus of Z consists of the fibers $y_1 \times \mathbb{A}^1$ of π_Y with $y_1 \in Y$ a non-normal point. This must also be the set of the fibers $x_1 \times \mathbb{A}^1$ of π_X, where $x_1 \in X$ is a non-normal point. Since distinct fibers of π_X are disjoint,

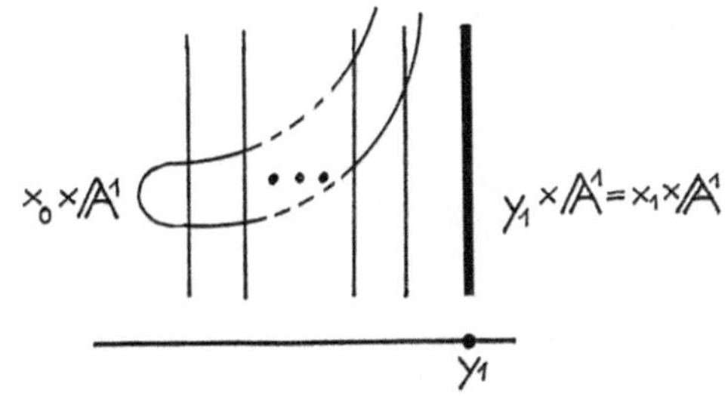

Fig. 2 Non-unique coefficients in a one variable polynomial ring

there is an x_0 as above such that $\mathbb{A}^1 \simeq \eta(\mathbb{A}^1) \subset Y_{ns}$. Since $Y_{ns} \subsetneq \mathbb{A}^1$, $k^{[1]}$ would have nontrivial units and we get a contradiction. See Fig. 2.

There is a considerable literature on the invariance problem for one-dimensional rings, and more generally on cancellation with curves as base. These could be **complete** curves. Fujita and Iitaka [18] give a very general result applicable here that we discuss below. See also [7, 16]. There are examples of elliptic curves E, E_1, E_2 so that $E_1 \times E \simeq E_2 \times E$ and $E_1 \not\simeq E_2$. It turns out that in the category of **abelian varieties** the good notion, giving cancellation, is *isogenous* instead of *isomorphic*. See [8, 35].

2.2 Projective Modules

As a second **algebraic** cancellation situation we consider finitely generated projective modules over an algebra R.

Recall that an R-module P is *projective* if it is a direct summand of a free module, i.e., if there is a module Q so that $P \oplus Q$ is a free module,

$$P \oplus Q = R^n.$$

If we can choose Q to be free, we say P is *stably free*. Then if P is not free, of rank m, say, we have an example of non-cancellation (with "product" the direct sum) in the category of R-modules:

$$P \oplus R^{n-m} \simeq R^m \oplus R^{n-m}.$$

We will see shortly that by taking symmetric algebras this can sometimes be transformed into an example of non-cancellation for rings.

There is a vast literature on projective modules. Good examples can be constructed from **unimodular rows**:

Let R be a domain. $(x_1, \ldots, x_m) \in R^m$ is a *unimodular row* if $(x_1, \ldots, x_m)R$ is the unit ideal, i.e., if

$$\phi : R^m \to R, (y_1, \ldots, y_m) \mapsto y_1x_1 + \ldots y_mx_m$$

is surjective. Write

$$1 = x'_1 x_1 + \cdots x'_m x_m.$$

Then ϕ has a section defined by

$$1 \mapsto (x'_1, \ldots, x'_m),$$

and $R^m \simeq P \oplus R(x'_1, \ldots, x'_m)$, $P = \mathrm{Ker}(\phi)$, $R(x'_1, \ldots, x'_m) \simeq R$.

A favorite **special case** is

$$R = k[x_0, \ldots, x_n] = k[X_0, \ldots, X_n]/(X_0^2 + \cdots + X_n^2 - 1), \; x_i = \overline{X}_i.$$

We obtain the unimodular row $(x_0, \ldots, x_n) \in R^{n+1}$ and $1 \mapsto (x_0, \ldots, x_n)$ defines a section as above (with $x'_i = x_i$). Let P be the corresponding projective module. If $k = \mathbb{R}$, we obtain the real n-sphere $\mathbb{S}^n$ embedded in $\mathbb{R}^{n+1}$ and P is the sheaf of sections of the tangent bundle of $\mathbb{S}^n$: at each point of the sphere the ambient $\mathbb{R}^{n+1}$ splits into **vectors along the position vector** and **those perpendicular to it, so lying in the hyperplane tangent to the sphere**. It is a famous theorem of topology about the parallelizability of spheres (see [6] for instance) that says P is free precisely for $n = 1, 3, 7$. For $n = 2$, I learned this in my undergraduate curves and surfaces course under the heading:"One can't comb the hair on a billiard ball." Anyway, in all other cases, for instance for the two-sphere, you have an example of a stably free, non-free projective module. Let me remark that such examples exist also over algebraically closed fields.

Let R be a ring (commutative) and M an R-module (finitely generated). Let me recall that the **symmetric algebra** $S_R(M)$ is a commutative R-algebra together with a R-module homomorphism

$$i : M \to S_R(M)$$

that is universal for module homomorphisms $M \to A$, where A is a commutative R-algebra. We write $S(M)$ for $S_R(M)$ if R is understood.

Here are some **easy facts**.

1. $S(M)$ is a graded algebra.
2. i is injective and identifies M with the elements of degree 1 in $S(M)$.
3. $S(M)$ is generated by $i(M) = M$.
4. $S(R) = R^{[1]}$.
5. $S(M \oplus N) = S(M) \otimes_R S(N)$, in particular $S(R^n) = R^{[n]}$.

   An **only slightly harder fact**, coming from the graded structure and the universal property, is
6. $S(M)$ is isomorphic to $S(N)$ as R-algebra $\iff$ M, N are isomorphic R-modules.

Now let R be the coordinate ring of the real n-sphere (so we work over $k = \mathbb{R}$) and P the "tangent bundle module" defined above.

Theorem 2.2.1 ([21], see also [13]). *For $n \neq 1, 3, 7$ the rings $S(P)$ and $R^{[n]}$ are non-isomorphic and stably equivalent.*

Proof. We have already established that $S(P)$ and $R^{[n]}$ are stably equivalent and not isomorphic as R-algebras. To finish the proof, i.e., to see that they can not be isomorphic in some other way, we establish

Lemma 2.2.2. *If D is a subalgebra of $R^{[m]}$ such that $R^{[m]} = D[y] \simeq D^{[1]}$, then $R \subset D$.*

An immediate consequence, which proves the theorem, is

Corollary 2.2.3. *R is strongly invariant.*

The key to the *proof of the lemma* is that $R^{[m]}$ is *formally real*, i.e., if a sum of squares vanishes, then each summand vanishes. Let $e = \max(deg_y(x_j))$. Write each x_i as a polynomial in y formally of degree e with highest coefficient a_i. If $e > 0$, $\Sigma x_i^2 = 1$ gives $\Sigma a_i^2 = 0$, so all $a_i = 0$, and we have a contradiction.

Remark 2.2.4. In case $n = 2$ we have

(i) R is strongly invariant.
(ii) $R^{[1]}$ is invariant.
(iii) $R^{[2]}$ is not invariant.

For the *proof of (ii)* use the lemma and the following two facts:

(a) R is factorial.
(b) ([3]) Let $B \subset D \subset B^{[m]}$ be domains and D factorial of transcendence degree 1 over B. Then $D \simeq B^{[1]}$.

Remark. The situation changes considerably if we pass to $k = \mathbb{C}$ from $k = \mathbb{R}$ in 2.2.4. The module $P_{\mathbb{C}}$, or its dual, the module of differentials of $R_{\mathbb{C}}$, is free. We leave this as a not entirely trivial **exercise**. It is best done by writing $R_{\mathbb{C}} = \mathbb{C}[x, y, z]$ with $xy = z^2 - 1$. Note that, in the notation of the next section, $\mathbb{S}^2_{\mathbb{C}}$ is the Danielewski surface Z_1. Hence, as shown there, $R_{\mathbb{C}}$ is not invariant.

2.3 *The Danielewski Examples*

Let $Z_n = \mathrm{Spec}(k[x, y, z]), x^n y = z^2 - 1, n \geq 1$. These are surfaces of a type known as *Danielewski surfaces*.

Theorem 2.3.1 (Danielewski). *For $n \neq m$, $Z_n \not\simeq Z_m$. However $Z_n \times \mathbb{A}^1 \simeq Z_m \times \mathbb{A}^1$.*

Fig. 3 The line with double origin

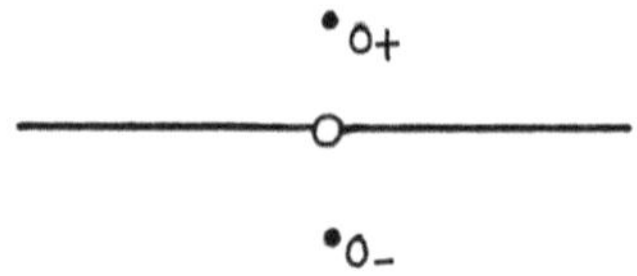

Proof. Consider the projection

$$\pi_n : Z_n \to X = \mathbb{A}^1 = \mathrm{Spec}(k[x]),$$
$$(x, y, z) \mapsto x.$$

We have $\pi_n^{-1}(\{x \neq 0\}) = \mathrm{Spec}(k[x, x^{-1}, z])$ (since $y = \frac{z^2-1}{x^n}$) and $\pi_n^{-1}(\{x = 0\})$ is the disjoint union of two affine lines $\mathbb{A}^1 = \mathrm{Spec}(k[y])$ distinguished by $z = 1$ and $z = -1$. They have multiplicity 1 in the fiber. We can factor π_n as $\delta \circ \pi'_n$,

$$\pi'_n : Z_n \to X', \;\; \delta : X' \to X,$$

where X' is a non-separated scheme, the

affine line with origin 0 doubled into 0_+ and 0_-.

We can think of X' as the union of open sets $U_+ = \mathrm{Spec}(k[x_+]), U_- = \mathrm{Spec}(k[x_-])$ with intersection $U = \mathrm{Spec}(k[x, x^{-1}])$. Here both x_+ and x_- are identified with x on U, not one with x and the other with x^{-1} as in the construction of $\mathbb{P}^1$. See Fig. 3.

Every fiber of $\pi'_n : Z_n \to X'$ and $\pi'_m : Z_m \to X'$ is a line. It will now be clear to the experts that we can make Z_n, Z_m into a *principal homogeneous spaces* for the additive group $\mathbb{G}_a$ over X'. They are then defined by elements $\upsilon_n, \upsilon_m \in H^1(X', \mathcal{O}_{X'})$.

We form the fiber product over X'

$$Z = Z_n \times_{X'} Z_m.$$

The pullback

$$\Pi_m : Z \to Z_n$$

of π'_m by π'_n makes Z into a principal homogeneous space for $\mathbb{G}_a$ over Z_n defined by the pullback of υ_m. Since Z_n is affine, $H^1(Z_n, \mathcal{O}_{Z_n}) = 0$ and

$$Z \simeq Z_n \times \mathbb{A}^1.$$

Similarly

$$Z \simeq Z_m \times \mathbb{A}^1.$$

Concretely, Z_m (resp. Z_n) is defined by a co-cycle Υ_m (resp. Υ_n) on X' that becomes a co-boundary after base extension to Z_n (resp.Z_m).

Consider Z_n. Let $v_+ = \frac{y}{z-1} = \frac{z+1}{x^n}$, $v_- = \frac{y}{z+1} = \frac{z-1}{x^n}$. Note

$$k[x,y,z] \subset k[x,v_+] \subset k[x,x^{-1},z],$$
$$k[x,y,z] \subset k[x,v_-] \subset k[x,x^{-1},z].$$

The sets $\pi_n'^{-1}(U_+) = \mathrm{Spec}(k[x,v_+])$, $\pi_n'^{-1}(U_-) = \mathrm{Spec}(k[x,v_-])$ give an open cover of Z_n, their intersection is $\pi_n'^{-1}(U)$. We have

$$v_+ - v_- = \frac{2}{x^n} \in H^0(U, \mathcal{O}_{X'}) \subset H^0(\pi_n^{-1}(U), \mathcal{O}_{Z_n}),$$

so assigning $\frac{2}{x^n}$ to U gives a Čech co-cycle Υ_n defining υ_n on X'.

It is a nice **exercise** to determine explicitly how Υ_m becomes a co-boundary on Z_n and thereby to explicit construct an isomorphism between $Z_n \times \mathbb{A}^1$ and $Z_m \times \mathbb{A}^1$. See [36] if you don't succeed on your own.

It is important to be aware that the fiber products $Z_n \times_{X'} Z_m$ and $Z_n \times_X Z_m$ are quite different.

We discuss two approaches to **non-isomorphism of Z_n and Z_m**.

1. Foreshadowing his later introduction of the **Makar-Limanov invariant** (see Sect. 6 below), Makar-Limanov shows in [28] that Z_1 and $Z_m, m > 1$, have different automorphism groups. Specifically, Z_m essentially admits a unique $\mathbb{G}_a$-action (its general orbits are fibers of the x-fibration). More precisely,

   x is fixed by any action of $\mathbb{G}_a$ on Z_m.

   This is clearly not the case for Z_1 because of the automorphism interchanging x and y.
2. Let us work over $k = \mathbb{C}$. Then

   the **topology at infinity**, that is the topology of the complements of large compact subsets, is different for Z_n and Z_m. To be precise (see [14,38]), the first homology group at infinity of Z_n has order $2n$. See Sect. 5.3 below.

Hence Z_m, Z_n are not homeomorphic if $m \neq n$. We also obtain an example of topological non-cancellation with $\mathbb{R}$ as factor: Either $Z_m \times \mathbb{R}, Z_n \times \mathbb{R}$ are homeomorphic, or, if not, then $(Z_m \times \mathbb{R}) \times \mathbb{R}, (Z_n \times \mathbb{R}) \times \mathbb{R}$ are homeomorphic.

3 Some (Very Intuitive) Algebraic Geometry

This section is a lightning sprint through about a semester's worth of classical algebraic geometry. At the end we offer some exciting (I think) applications of algebraic geometry to algebra that were developed over the last 40 years or so, but that as far as the geometry is concerned fit very well into the classical framework. I hope that the rewards offered will inspire the non-expert to make a run for the nearest algebraic geometry textbook.

3.1 Divisors and Differentials

Let X be a *complete* (e.g., projective, compact if over $\mathbb{C}$) and *non-singular* algebraic variety of dimension n.

1. A *divisor* $D = \Sigma n_i D_i$ on X is a formal linear combination of closed irreducible co-dimension 1 subvarieties D_i. It is *effective* if all $n_i \geq 0$.
2. If D is effective and $q \in D$, then there is a rational function f on X (element of the function field $k(X)$) that is a *local equation* for D at q,

   i.e., f is defined at q, i.e., is in the local ring of q, and in a neighborhood of q, D is defined (including multiplicities) by the vanishing of f.

   This comes from the fact that the *local ring of* X *at* q is factorial. Splitting a divisor into its positive and negative part we can extend this concept to general divisors.
3. Given $0 \neq f \in k(X)$, there is a unique divisor (f), called the divisor of f, such that, at each $q \in X$, f is a local equation for (f). Divisors D_1, D_2 are *linearly equivalent* if $D_1 - D_2$ is the divisor of a rational function. The **group of divisors modulo linear equivalence**

   $$\mathrm{Pic}(X) = \textit{Divisors/Divisors of rational functions}$$

   is central in the study of X.
4. For a divisor D we define the vector space (it is finite dimensional over k)

   $$\mathcal{L}(D) = \{f \in k(X) | (f) + D \geq 0\} \cup (0)$$

   and put

   $$l(D) = \dim(\mathcal{L}(D)).$$

   If $l(D) > 0$ we obtain a rational map (with a basis of $\mathcal{L}(D)$ as components)

   $$\Phi_D : X \to \mathbb{P} = \mathbb{P}^{l(D)-1}.$$

   We define

   $$|D| = \textit{linear system defined by} D = \{\textit{effective divisors linearly equivalent to } D\}.$$

If $l(D) > 0$, then $|D|$ is the projective space associated to $\mathcal{L}(D)$ and its elements are the pullbacks by Φ_D of the hypersurfaces in $\mathbb{P}$. If $l(D) = 0$, then $|D| = \emptyset$.

6. If f is a rational function we take the engineering point of view that we know what its *differential* df is. (There is, of course, a well-developed algebraic way to introduce differentials into algebraic geometry. Consult a textbook!) Let D be an effective divisor, $q \in D$ and f a local equation at q. Then

$$D \text{ is reduced and non-singular at } q \iff df \neq 0 \text{ at } q.$$

7. Rational functions $x_1, \dots, x_n$ form a *system of parameters at* q provided they are *defined at* q *and the* dx_i *are linearly independent at* q. They then are a separating transcendence base for $k(X)/k$. Then k-derivations of $k(x_1, \cdots, x_n)$, in particular the partial derivatives $\partial/\partial x_i$, extend uniquely to $k(X)$. So it makes sense to take partial derivatives w.r.t. the x_i on all of $k(X)$. Recall that $x_1, \dots, x_n$ is a separating transcendence base if and only if $\partial/\partial x_1, \dots, \partial/\partial x_n$ is a basis of the $k(X)$-vectorspace of k-derivations of $k(X)$, if and only if $dx_1, \dots, dx_n$ is a basis of the $k(X)$-vectorspace V of *rational differential* 1*-forms*

$$\Sigma g_i df_i, \quad f_i, g_i \in k(X).$$

Transition from one transcendence basis to another is, at the level of differential 1-forms, given by the usual *Jacobian matrix.*

8. A rational differential 1-form ω is *regular at* $q \in X$ if it can be expressed as $\omega = \Sigma f_i dx_i$,, where $x_1, \dots, x_n$ is a system of parameters at q and the f_i are regular at q.

3.2 Plurigenera and Kodaira Dimension

1. We form the $\mathcal{O}_X$ module Ω by assigning to each open set U the $\mathcal{O}_X(U)$-module of one-differentials regular at each point of U. It is locally free of rank n. With Ω we can perform various standard linear algebra constructions (open set by open set), in particular take exterior and symmetric powers in various combinations. This leads to a host of new $\mathcal{O}_X$-modules $\Omega^{\clubsuit}$. The numbers $\dim(\Omega^{\clubsuit}(X))$, they are all finite, are collectively known as **plurigenera**.
2. If we apply the highest exterior power operator Λ^n, we obtain the rank one-module Ω^n of *differential* n*-forms*. For these the ideas of 1) can be translated into the language of divisors. If ω is a nonzero n-differential, we define its **divisor** (ω) by: at $q \in X$ write

$$\omega = f dx_1 \wedge \cdots \wedge dx_n,$$

where $x_1, \dots, x_n$ is a system of parameters at q, and decree f to be a local equation of (ω) at q. The usual rules of calculus apply: Transition from one system of parameters to another changes the local equation by a Jacobian determinant which is a nonzero constant at q. All divisors of the form (ω) are linearly equivalent. Any one of them is called a

canonical divisor, usually denoted K_X, or just K.

We observe that if we follow Λ^n by the symmetric power operation S^m we obtain the divisors mK.

3. The number $l(D)$, e.g., $l(K)$, is usually hard to compute. We define a more robust number by observing:

   for large m the numbers $l(mD)$ are the values of a polynomial of degree $\kappa \leq n = \dim(X)$.

   We call $\kappa = \kappa(D)$ the Kodaira dimension of D.
   We write $\kappa(K_X) = \kappa(X)$ and call it the **Kodaira dimension of** X**.**

   Note: $\kappa(D) \geq 0$ if and only if some multiple mD, $m \geq 1$, is linearly equivalent to an effective divisor. In that case $\kappa(D) = \dim(\Phi_{mD}(X))$ for m large. If no such multiple is effective, then the above polynomial is 0 and one puts $\kappa(D) = -\infty$.
4. So far all this is very classical algebraic geometry. Here come the crucial **new definitions** ([22]).

   (a) A divisor Δ is a *divisor with strong normal crossing*, or in brief a **SNC-divisor**, if at each $q \in X$ we can find a system of parameters $x_1, \ldots, x_n$ such that for some $s \leq n$, $f = x_1 \cdots x_s$ is a local equation for Δ.
   (b) We modify the ideas of (1) by allowing *logarithmic singularities* along Δ. To be precise:

   We define an $\mathcal{O}_X$-module $\Omega(log\Delta)$ by decreeing that a one-differential ω is in $\Omega(log\Delta)(U)$ if at each $q \in U$ it can, with notation as in (a), be written as

$$\omega = f_1 \frac{dx_1}{x_1} + \cdots + f_n \frac{dx_s}{x_s} + f_{s+1} dx_{s+1} + \cdots + f_n dx_n,$$

   with $f_1, \ldots, f_n$ regular at q.
   (c) As above, but working with $\Omega^{\clubsuit}(log\Delta)$, we define now **logarithmic pluri-genera** of the pair (X, Δ). The wonderful fact is:

   they are invariants of $U = X \setminus \Delta$.

   This takes some work with resolution of singularities, so we assume $\text{char}(k) = 0$ or restrict the dimension in positive characteristic.
   (d) The role of the canonical divisor is taken on by $K_X + \Delta$ and we can define:

   the logarithmic Kodaira dimension of U **is** $\overline{\kappa}(U) = \kappa(K_X + \Delta)$.

(e) Let $U = \mathrm{Spec}(A)$ be a non-singular affine variety. We can embed U as an open subset in a complete non-singular variety X such that $\Delta = X \setminus U$ is a SNC-divisor. (We again need resolution of singularities.) We will call Δ a divisor at infinity for U. I am repeating myself, but would like to emphasize: If we make sure that Δ is a SNC-divisor, then

the logarithmic plurigenera, so in particular the logarithmic Kodaira dimension, of the pair (X, Δ) are invariants of A.

4 Some Cancellation Rewards

It is known (see [10] for an argument and references) that

a smooth contractible affine variety V over $\mathbb{C}$ of dimension $n \geq 3$ is homeomorphic to $\mathbb{R}^{2n}$.

If such a variety is not isomorphic to $\mathbb{C}^n$ we call it an **exotic affine space**.

The above statement is no longer true in dimension 2. C. P. Ramanujam in a famous paper [32] constructed the first example in 1971. Let us call it $\mathfrak{R}$. This is a smooth, contractible affine surface that is *not simply connected at infinity*, so it is not isomorphic to $\mathbb{C}^2$. See Sect. 5.4 below. Now "not simply connected at infinity" is destroyed by passing from V to $V \times \mathbb{C}$, and it was not known in 1971 whether $\mathfrak{R} \times \mathbb{C}$ is exotic, in fact whether exotic affine spaces exist at all. Here Kodaira dimension came to the rescue.

We need **some easy facts**. Let V be non-singular affine of dimension n.

1. Suppose V is **affine ruled**, i.e., contains an open subvariety U of the form $U' \times \mathbb{A}^1$ (a *cylinderlike open set*). Then $\overline{\kappa}(V) = -\infty$. See Fig. 4.
2. We always have $\overline{\kappa}(V \times \mathbb{A}^1) = -\infty$, but a nonzero logarithmic plurigenus of V survives on $V \times \mathbb{A}^1$. In particular, if $\overline{\kappa}(V) \geq 0$, then some plurigenus coming from n-differentials is nonzero on the $(n+1)$-dimensional variety $V \times \mathbb{A}^1$.
3. All logarithmic plurigenera of $\mathbb{A}^n$ are 0.
4. If $\dim(V) = \dim(V')$ and $V \to V'$ is a dominant separable morphism, then $\overline{\kappa}(V') \leq \overline{\kappa}(V)$.
5. Suppose $\dim(V) = 2$ and $\overline{\kappa}(V) = -\infty$. Then any SNC-divisor at infinity for V is a tree of non-singular rational curves. See [33].

   Let again $\dim(V) = 2$. A hard fact is the following powerful result of Fujita [15]. It is known as **adjunction terminates** in the classical ($\Delta = \emptyset$) situation.
6. Suppose $\kappa(K + \Delta) = -\infty$ and let B be an effective divisor. Then there exists $m \geq 0$ such that $|B + m(K + \Delta)| \neq \emptyset$ and $|B + (m+1)(K + \Delta)| = \emptyset$.

It turns out that effective divisors D so that $D + (K + \Delta)$ (adding $K + \Delta$ is adjunction in the logarithmic case) is not linearly equivalent to an effective divisor have very special properties, see [33]. So (6) is a very strong existence theorem. It is used in many arguments, the proof of Theorem 4.3 below, for instance. See also [27].

Fig. 4 An affine ruling

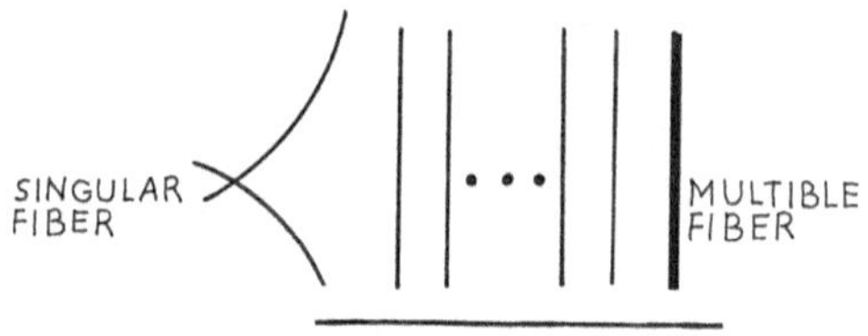

Theorem 4.1. *We have* $\overline{\kappa}(\mathfrak{R}) = 2$ *and hence* $\mathfrak{R} \times \mathbb{C}$ *is an exotic affine three-space.*

We will give more details in Sect. 5.4 below.

Ramanujam also gave this topological characterization of $\mathbb{C}^2 = \mathbb{A}^2_{\mathbb{C}}$.

Theorem 4.2. *A smooth contractible surface over* $\mathbb{C}$ *that is simply connected at infinity is isomorphic to* $\mathbb{C}^2$.

This, however, did not solve **the** cancellation question in dimension 2. For this we need a converse to (1) above.

Theorem 4.3 ([15, 30, 33]). *Let* V *be a smooth affine surface with* $\kappa(V) = -\infty$. *Then* V *is affine ruled.*

It is fairly easy to show

Proposition 4.4. *If* $V = \text{Spec}(A)$ *is a smooth affine surface and*

(i) V *is affine-ruled and*
(ii) $A^* = k^*$ *and* A *is factorial,*
then $V \simeq \mathbb{A}^2$.
Over $\mathbb{C}$ *we can replace (ii) by*
(ii') V *is contractible. (See [17].)*

From this we finally get the cancellation theorem in dimension 2.

Theorem 4.5. $V \times \mathbb{A}^1 \simeq \mathbb{A}^3 \Longrightarrow V \simeq \mathbb{A}^2$, *or,* $k^{[2]}$ *is invariant.*

In fact, for a sufficiently general $\mathbb{A}^2 \subset \mathbb{A}^3$ the projection $\mathbb{A}^3 \to V$ induces a separable dominant morphism $\mathbb{A}^2 \to V$ and we have $\kappa(V) = -\infty$ by (4) above.

Not so long ago a completely algebraic proof of the theorem based on the Makar-Limanov invariant was found by Crachiola and Makar-Limanov [9]. See Sect. 6 below.

Let me mention one more far reaching result. (I am not stating the most general version.)

Theorem 4.6 ([18]). *Let* X *be an algebraic variety and suppose* $\overline{\kappa}(X_{ns}) \geq 0$. *Let* Z *be a non-singular algebraic variety with vanishing logarithmic plurigenera, e.g.,* $Z = \mathbb{A}^n$. *Then* X *has strong cancellation for* Z.

By now the zoo of known exotic affine spaces is quite extensive, and a very active area of research, see the survey paper [39] in particular. They are, of course, prime candidates for testing cancellation questions. For instance, with methods inspired by those of the Danielwski example and using fibrations over *algebraic spaces*,

Dubouloz, Moser-Jauslin, and Poloni [12] recently established non-cancellation (for $\mathbb{A}^1$) for certain exotic threefolds that turned up in the quest of linearizing $\mathbb{C}^*$-actions on $\mathbb{C}^3$. (See [24].)

However, **the** cancellation problem in dimension 3 is open in characteristic 0. This is a main roadblock at present in settling the question whether $(\mathbb{C}^*)^2$-actions on $\mathbb{C}^4$ are linearizable. We will discuss more details and positive characteristic in Sect. 6 below.

5 Some Kodaira Dimension Calculations

5.1 *Kodaira Dimension of Non-singular Curves*

- $\overline{\kappa} = -\infty : \mathbb{P}^1$ $(K_{\mathbb{P}^1} = -2p)$, $\mathbb{A}^1 = \mathbb{P}^1 \setminus \{p\}$ (p one point, $K_{\mathbb{P}^1} + p = -p$).
- $\overline{\kappa} = 0$: Complete non-singular curves C of genus 1 $(K_C = 0)$, $(\mathbb{A}^1)^* = \mathbb{P}^1 \setminus \{q_1, q_2\}$ (q_1, q_2 distinct points, $K_{\mathbb{P}^1} + q_1 + q_2 = 0$).
- $\overline{\kappa} = 1$: the rest, e.g. curves C of genus 1 minus a point, $\mathbb{P}^1 \setminus \{q_1, q_2, q_3\}$ (three distinct points).

5.2 *Some Very Rudimentary Kodaira Dimension on Affine Surfaces*

5.2.1 The Blow-Up

We need preparatory information on producing SNC-divisors at infinity.

(a) The **blow-up**

$$X' \to X$$

of a non-singular surface in a point q is a process that replaces q by a curve

$$E \simeq \mathbb{P}^1$$

(to be thought of as parametrizing the set of directions at q), and that induces an isomorphism

$$X' \setminus E \to X \setminus \{q\}.$$

You get a very good picture of this if you study the blow-up of $\mathbb{A}^2 = \operatorname{Spec}(k[x, y])$ in the origin, which is

$$\operatorname{Spec}(k[x, \frac{y}{x}]) \cup \operatorname{Spec}(k[\frac{x}{y}, y]).$$

For points $(a, b) \neq (0, 0)$ we can use a and the slope b/a as coordinates if $a \neq 0$, and of course b and the other slope a/b if $b \neq 0$. We can use either if $a \neq 0$ and $b \neq 0$. This gives the gluing information.

(b) If D is a reduced divisor (a sum of irreducible curves taken with multiplicity 1) and $q \in D$, let

D' = reduced inverse image of D in X'.

It has E as an irreducible component. We call $D' - E$ the **strict transform** of D. Let

D^* = total inverse image of D in X'.

D^* is the same as D' except that it contains E with multiplicity μ = multiplicity of q on D. The definition of D^* extends in an obvious way to general divisors.

(c) A basic fact is

$$K_{X'} = K_X^* + E.$$

It follows that if D is a reduced divisor and q a point of multiplicity μ on D, then

$$K_{X'} + D' = (K_X + D)^* - (\mu - 2)E. \qquad (*)$$

We can repeat this process with a point on $D' \subset X'$, and so on. This is called "blowing up infinitely near points".

The fundamental theorem about **embedded resolution of curves** says that by blowing up repeatedly over X (ordinary and infinitely near points) we can find a tower

$$X^{(m)} \to \cdots \to X^{(1)} \to X^{(0)} = X. \qquad (**)$$

so that

$D^{(m)}$, the reduced inverse image of D on $X^{(m)}$, is a SNC-divisor.

5.2.2 The Complement of a Curve $D \subset \mathbb{P}^2$

Let $X = \mathbb{P}^2$. Note that $\mathrm{Pic}(X)$ is generated by (the class of) a line L and that $K_X = -3L$.

(a) Consider $D = Q$ = non-singular conic. Then $K + Q = -L$ and $\overline{\kappa}(X \setminus Q) = -\infty$. As we said above, this implies that $U = X \setminus Q$ is affine-ruled. This is easily seen directly here. Given $q \in Q$, there is a pencil of conics four times tangent to Q at q, so meeting Q in q only. This gives a pencil of parallel affine lines $\mathbb{A}^1$ in $U = \mathbb{P}^2 \setminus Q$ which gives an **affine ruling** of U. It has the tangent to Q at q as a fiber of multiplicity 2. Different q give different affine rulings, members of two different rulings meet four times.

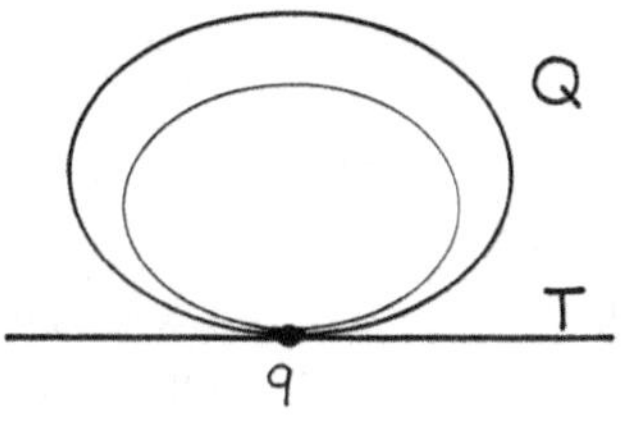

Fig. 5 The pencil of conics tangent in a point

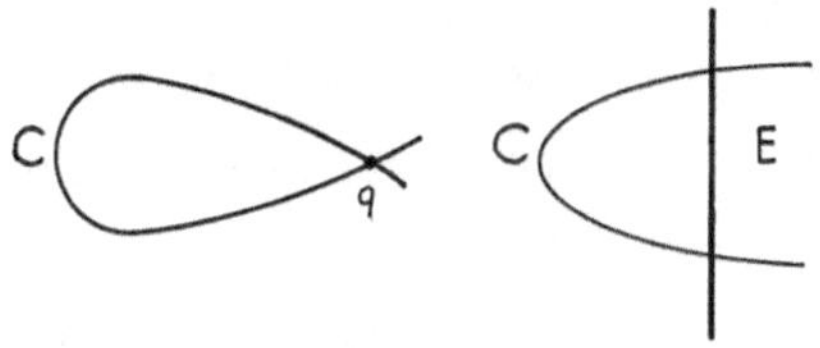

Fig. 6 Normal crossing resolution of a cubic with a node

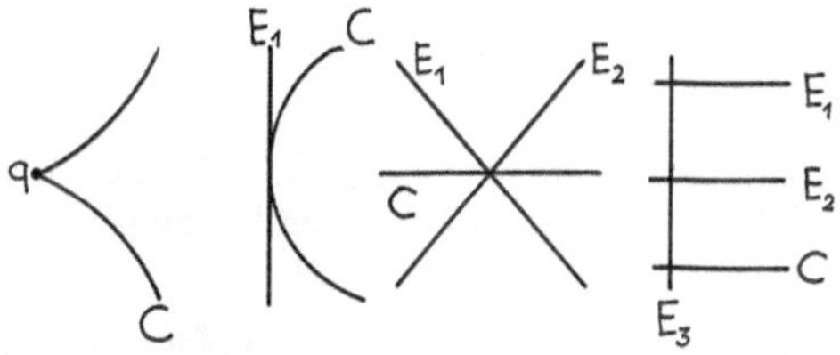

Fig. 7 Normal crossing resolution of a cubic with a cusp

Exercise. Show that $X \setminus Q \not\simeq \mathbb{A}^2$. You may also want to show that $X \setminus Q$ is not isomorphic to a Danielewski surface. See Fig. 5.

(b.1) Consider $D = C$ = non-singular cubic. Then $K + D = 0$ and $\overline{\kappa}(X \setminus C) = 0$.

(b.2) Consider $D = C$ = cubic with a node q. We blow up at q to achieve normal crossing. We have $K_0 + D^{(0)} = 0$ to begin with, and after the blow up $K_1 + D^{(1)} = (K_0 + D^{(0)})^* = 0$ by (*) above. So again $\overline{\kappa}(X \setminus C) = 0$. See Fig. 6.

(b.3) Consider $D = C$= cubic with a cusp q. We now have to blow up three times to achieve normal crossing, twice with multiplicity 2 on $D^{(i)}$, then with multiplicity 3.

We find $K_{i+1} + D^{(i+1)} = (K_i + D^{(i)})^* = 0, i = 0, 1$ and $K_3 + D^{(3)} = -E_3$. Hence $\overline{\kappa}(\mathbb{P}^2 \setminus C) = -\infty$. So the Kodaira dimension is very sensitive to the nature of the singularities we have to resolve. See Fig. 7.

5.2.3 Interlude on the Intersection Pairing

On a complete non-singular surface X we have the **intersection pairing**

$$\mathrm{Pic}(X) \times \mathrm{Pic}(X) \to \mathbb{Z},$$
$$(D, E) \mapsto D \cdot E.$$

For two distinct irreducible curves, $D \cdot E$ is the number of intersection points, counted with multiplicity. We extend the definition linearly to divisors without common component. One finds that

1. $D \cdot E = 0$ if E is linearly equivalent to 0,
2. given a curve C it is possible to find a divisor C' linearly equivalent to C and not having C as a component.

So we can define the pairing unambiguously on $\mathrm{Pic}(X)$. In particular, **self-intersection numbers** are defined for irreducible curves in X.

Exercise. Find $E \cdot E$ for the exceptional locus in the blow-up of a point on a non-singular surface.

The intersection pairing has a central role in the study of surfaces. In our context, if

$$\Delta = \Sigma \Delta_i$$

is a SNC-divisor with irreducible components Δ_i, then the intersection matrix

$$I(\Delta) = (-\Delta_i \cdot \Delta_j)$$

gives a lot of information on the topology at infinity of $U = X \setminus \Delta$. In particular, if Δ is a tree of $\mathbb{P}^1$'s (so simply connected), then

$$|\det(I(\Delta)|$$

is the order of the homology at infinity. There also is a way to get a presentation of the fundamental group at infinity [31, 32]. (The topology at infinity is that of the boundary of a "small" tubular neighborhood of Δ.)

5.3 *More on the Danielewski Surfaces*

It is clear from what we said earlier that the Danielewski surface Z_n is affine ruled. So $\overline{\kappa}(Z_n) = -\infty$. The graph below represents a SNC-divisor Δ completing Z_n. The components are $\mathbb{P}^1$'s meeting normally (if at all), and the labels on the components are the self-intersections. It is not difficult to compute

$$\det(I(\Delta)) = 2n.$$

(See [17], for instance, for relevant formulas and tricks.) See Fig. 8.

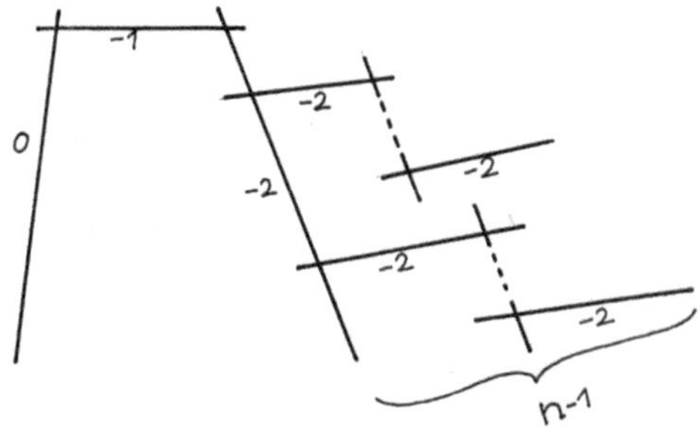

Fig. 8 Divisor at infinity of a Danielewski surface

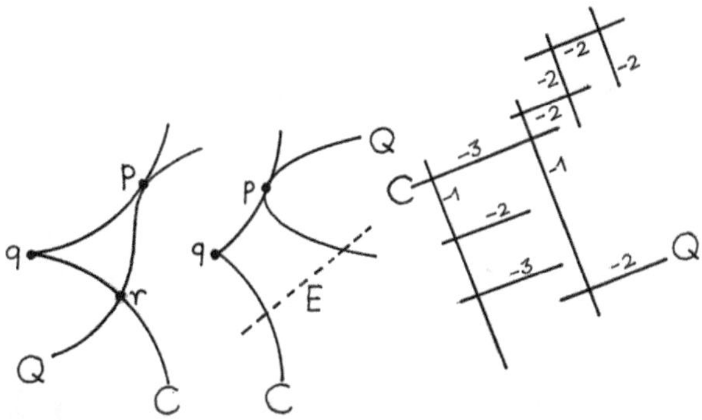

Fig. 9 Divisor at infinity of the Ramanujam surface

5.4 *The Ramanujam Surface* $\mathfrak{R}$

In $\mathbb{P}^2$ take a cubic C with a cusp at q, say. Let Q be an irreducible conic meeting C with multiplicity 5 at a point p and transversally at a point r. (That such a conic exists can be deduced from the group law on the non-singular part of C. It can also be seen directly. One just has to avoid choosing for p a flex-point of C.) Let X be the blow-up of $\mathbb{P}^2$ at r and C_1, Q_1 the strict transforms of C, Q. Then

$$\mathfrak{R} = X \setminus C_1 \cup Q_1.$$

Note that (an open subset of) the exceptional curve E above r becomes part of $\mathfrak{R}$.

To obtain a SNC-completion (X', Δ) of $\mathfrak{R}$ we have to resolve the singularities of $\Delta_1 = C_1 \cup Q_1$ at p and q. We have to blow up over q as in 5.2.2(b.3) and five times over p to separate C and Q. See Fig. 9.

One finds that the intersection matrix of Δ is unimodular and, using [31], that the fundamental group at infinity is an infinite group with trivial abelianization. To calculate the Kodaira dimension we write (on $\mathbb{P}^2$) $2(K + C + Q) = L + L_1 + 2L_2$, where L_1 is the line tangent to C at q, L_2 is the line joining p and r, and L is some line (think of it as not passing through any interesting point). We leave it as an **exercise** to show that $2(K_{X'} + \Delta) = L^* + F$, where F is an effective divisor. (Use (*) above.) Hence $\overline{\kappa}(\mathfrak{R}) \geq \kappa(L^*) = \kappa(L) = 2$.

6 Automorphisms

It will have become apparent by now that it can be a delicate task to decide on isomorphism or non-isomorphism of two algebraic varieties, or algebras, and that one always is on the lookout for new invariants that might help. It is clear that the automorphism group can be such an invariant, but it most of the time is not easy to compute. A more subtle, and it turns out very successful, strategy is to study the way the automorphism group G of a variety X, or certain subgroups of G, act on X. One of the most fertile ideas in this direction, initiated by L. Makar-Limanov, was to measure the abundance, or non-abundance, of actions of the additive group $\mathbb{G}_a$. He made the following

Definition 6.1. Let $X = \mathrm{Spec}(A)$ be an affine variety. Then the **Makar-Limanov Invariant** $\mathrm{ML}(X)$**, or** $\mathrm{ML}(A)$**, of** X**, or** A**,** is the subalgebra of A consisting of the regular functions f on X invariant under **all** actions of the additive group $\mathbb{G}_a$ on X.

He also developed techniques to actually compute this invariant, see [9, 23, 29] for details.

Algebraically, an action $\mathbb{G}_a \times X \to X$ is given by a k-algebra homomorphism $\delta : A \to A[t] \simeq A^{[1]}$. Write

$$\delta(a) = \sum_{i \geq 0} \delta^{(i)}(a) t^i$$

for $a \in A$. Since δ is a k-homomorphism,

(i) each $\delta^{(i)}$ is k-linear,
(ii) for given a we have $\delta(a) \neq 0$ for only finitely many i, and
(iii) we have the Leibniz rule $\delta^{(n)}(ab) = \sum_{i+j=n} \delta^{(i)}(a)\delta^{(j)}(b)$.
The properties of a group action give in addition
(iv) $\delta^{(0)}$ is the identity map, and
(v) the "iteration rule" $\delta^{(i)}\delta^{(j)} = \binom{i+j}{i}\delta^{(i+j)}$.

The algebra of functions on X invariant under the action corresponding to δ is

$$A^{\delta} = \{a \in A | \delta(a) = a\} = \{a \in A | \delta^{(i)}(a) = 0, i = 1, 2, \ldots\}.$$

If $\mathrm{char}(k) = 0$, then $\delta^{(i)} = \frac{1}{i!}(\delta^{(1)})^i$. Hence δ is completely determined by $\delta^{(1)}$, in fact $\delta = \exp(t\delta^{(1)})$. It has therefore become customary to call a δ satisfying (i) through (v) above an

exponential map on A

even if $\mathrm{char}(k) > 0$. It is called **nontrivial** if $A^{\delta} \subsetneq A$, i. e., if $\delta^{(i)} \neq 0$ for some $i > 0$.

Note that $\delta^{(1)}$ is a k-derivation of A by (iii). If $\text{char}(k) = 0$, then given $a \in A$ there exists $n \geq 0$ such that $\delta^{(n)}(a) = 0$. We say that

$\delta^{(1)}$ is a **locally nilpotent derivation (LND) on A**. Note that $A^\delta = \text{Ker}(\delta^{(1)})$.

We denote by $\text{EXP}(A)$ the set of exponential maps of A. Then

$$\text{ML}(A) = \bigcap_{\delta \in \text{EXP}(A)} A^{(\delta)}.$$

If $\text{char}(k) = 0$, we denote by $\text{LND}(A)$ the set of locally nilpotent derivations of A. Then also

$$\text{ML}(A) = \bigcap_{\delta^{(1)} \in \text{LND}(A)} \text{Ker}(\delta^{(1)}).$$

It is clear that

"$\text{ML}(X) = A$" if and only if "there is no nontrivial action of $\mathbb{G}_a$ on X".

In contrast, $\text{ML}(X) = k$ indicates that there is a rich abundance of $\mathbb{G}_a$-actions. It is easily seen that $\text{ML}(\mathbb{A}^n) = k$. One can verify that for the Danielewski surfaces discussed above one has $\text{ML}(Z_1) = k$ and $\text{ML}(Z_n) = k[x]$ if $n > 1$. As an **exercise** show that $\text{ML}(X \setminus Q) = k$ for the example in 5.2.2(a).

Here is a remarkable result proved recently by Neena Gupta [20] using the ML-invariant and generalizing techniques developed in [29]. A somewhat more restricted result was proved earlier in [19]. A key point is that there is no restriction on the characteristic of k. With the complex numbers $\mathbb{C}$ as base field, similar and in some instances more far reaching results can be found in [25].

Theorem 6.2. *Let k be a field and A an integral domain defined by*

$A = k[U, Y, Z, T]/(U^m Y - F(U, Z, T))$ with $F(U, Z, T) \in k[U, Z, T]$ and $m > 1$.

Then the following conditions are equivalent:

(i) $f(Z, T) := F(0, Z, T)$ *is a variable in* $k[Z, T]$.
(ii) $A \simeq k^{[3]}$.

As an application of this result we obtain a counter example to **the** cancellation problem in dimension 3. We have to backtrack a bit. We call $f \in k[Z, T]$ a **line** if $k[Z, T]/(f) \simeq k^{[1]}$ and a **variable** if $k[Z, T] = k[f]^{[1]}$, i.e., if $k[Z, T] = k[f, g]$ for some $g \in k[Z, T]$. Clearly a variable is a line, and by the famous AMS-theorem [4, 37], the converse is true if $\text{char}(k) = 0$. On the other hand, if $\text{char}(k) = p > 0$, then there exist **exotic lines**, i.e., lines that are not variables. An example, due to B. Segre [34], is

$$s(Z,T) = Z^{p^2} - T - T^{p(p+1)}.$$

In fact, a part of the AMS-theorem asserts that if s is a variable, then one of $deg_Z(s), deg_T(s)$ divides the other. So the s above is not a variable. We leave it as an **exercise** to find a polynomial parametrization that exhibits it as a line.

Let u, y, z, t be the images of U, Y, Z, T in A. Gupta shows that $\mathrm{ML}(A) \subset k[u]$, even if $m = 1$ (this is seen easily by exhibiting suitable exponential maps). She also shows that $\mathrm{ML}(A) = k[u]$ if $m > 1$ and $f(Z,T)$ is an exotic line. This is not an easy argument, beginning with the results of [28]. On the other hand, if $m = 1$, it follows that $\mathrm{ML}(A) \subset k[u] \cap k[y] = k$ since we can interchange the role of u and y. So in that case the Makar-Limanov invariant and related techniques do not seem to shed any light on the nature of $X = \mathrm{Spec}(A)$ and cancellation. This is an open problem.

Assume for simplicity that k is algebraically closed and consider what we will call an **Asanuma threefold**, that is, an affine threefold $X = \mathrm{Spec}(A)$ with

a morphism $X \to \mathbb{A}^1$ such that every fiber, including the generic fiber, is an affine plane.

Algebraically this means that we are given $k[u] \subset A$ such that

1. for each $\lambda \in k$ we have $A/(u-\lambda)A \simeq k^{[2]}$, and
2. $A \otimes_{k[u]} k(u) \simeq k(u)^{[2]}$.

Interest in these algebras actually dates back quite some time, see [11].

The following is a very special case of a quite sweeping theorem of Asanuma [1] on the stability of **quasi polynomial algebras**, roughly algebras that are fiber-wise polynomial algebras over a subalgebra.

Theorem 6.3. *If* $\mathrm{Spec}(A)$ *is an Asanuma threefold, then* $A^{[1]} \simeq k^{[4]}$.

It is clear that $k[u] \subset A$ in Theorem 6.2 gives an Asanuma threefold if and only if $f(Z,T)$ is a line. Let us take $F(U,Y,Z,T) = s(Z,T)$, where $s(Z,T)$ is an exotic line. We obtain the promised counter example to cancellation for affine three-space. To be specific:

Theorem 6.4 ([19]). *Let* $\mathrm{char}(k) = p > 0$ *and* $m > 1$. *Let* $A = k[U,Y,Z,T]/(U^mY - (Z^{p^2} - T - T^{p(p-1)}))$. *Then* $A \not\simeq k^{[3]}$ *and* $A^{[1]} \simeq k^{[4]}$.

The program to study the linearizability of $G = \mathbb{C}^*$-actions on $X = \mathbb{C}^3$ outlined in [24] has two quite distinct parts:

1. Determine the quotient $X//G$. It parametrizes the closed orbits. See [27].
2. Determine how the space X is put together from the quotient and the orbits.

The second part, see [26], produced a large family $\mathfrak{F}$ of topologically contractible smooth affine threefolds X with a $\mathbb{C}^*$-action. The X in this family are potentially exotic affine spaces. It was a difficult problem, crucial for the linearizability question, to determine whether those **not equivariantly isomorphic to** $\mathbb{C}^3$ **with**

a linear action are actually exotic, i.e., not isomorphic to $\mathbb{C}^3$. It is fair to say that the Makar-Limanov invariant was invented to provide the solution [23]. The best studied example is the threefold

$$u^2y = u + z^2 + t^3.$$

It was first proved in [29] that it is not isomorphic to $\mathbb{C}^3$. Note that it is exotic also by Gupta's result. The $\mathbb{C}^*$-action is defined by giving u, y, z, t the weights $6, -6, 3, 2$. Note that X here is an Asanuma threefold topologically, every fiber of the u-fibration is homeomorphic to an affine plane.

The X in the family $\mathfrak{F}$ are by definition the contractible smooth affine $\mathbb{C}^*$-threefolds with unique fixpoint q and two-dimensional quotient isomorphic to that for the induced action on the tangent space at q. Can such an $X \not\simeq \mathbb{A}^3$ appear as the quotient of $\mathbb{C}^4$ by a $\mathbb{C}^*$-action? More specifically, can we have $X \times \mathbb{A}^1$ isomorphic to $\mathbb{A}^4$? These are central sub-questions in the investigation of $\mathbb{C}^*$- and $(\mathbb{C}^*)^2$-actions on $\mathbb{C}^4$.

The alert reader will have noted that in positive characteristic the example in Theorem 6.4 provides us with a nonlinearizable $\mathbb{G}_m$-action on $\mathbb{A}^4$. (The action is along the added variable). See [2]. Also, should it turn out that some $\mathrm{Spec}(k[U, Y, T, Z]/(UY - s(Z, T)))$, s an exotic line, is isomorphic to $\mathbb{A}^3$, then we obtain a non-linearizable $\mathbb{G}_m$-action on $\mathbb{A}^3$. The action is by $\tau \cdot (u, y, z, t) = (\tau u, \tau^{-1} y, z, t)$. Then the quotient is $\mathrm{Spec}(k[z, t]) \simeq \mathbb{A}^2$ and the part of the quotient corresponding to fixpoints is the exotic line $s(z, t) = 0$. If the action were linearizable, it would be a coordinate line.

Acknowledgements This research was supported by a grant from NSERC, Canada.

References

1. T. Asanuma, Polynomial fibre rings of algebras over noetherian rings. Invent. Math. **87**, 101–127 (1987)
2. T. Asanuma, Non-linearizable algebraic group actions on $\mathbb{A}^n$. J. Algebra **166**(1), 72–79 (1994)
3. S. Abhyankar, P. Eakin, W. Heinzer, On the uniqueness of the ring of coefficients in a polynomial ring. J. Algebra **23**, 310–342 (1972)
4. S.S. Abhyankar, T.-T. Moh, Embeddings of the line in the plane. J. Reine Angew. Math. **276**, 148–166 (1975)
5. A. Beauville, J.-L. Colliot-Thelene, J.-J. Sansuc, P. Swinnerton-Dyer, Variétés stablement rationnelles non rationnelles. Ann. Math. (2) **121**(2), 283–318 (1985)
6. R. Bott, J. Milnor, On the parallelizability of the spheres. Bull. Am. Math. Soc. **64**, 87–89 (1958)
7. J. Brun, On the cancellation problem for compact analytic manifolds. Proc. Symp. Pure Math. **30**, 245–247 (1976)
8. J. Brun, Sur la simplification dans les ismorphismes analytiques. Ann. Ec. Norm. Sup. **4**(serie 9), 533–538 (1976)
9. A. Crachiola, L. Makar-Limanov, An algebraic proof of a cancellation theorem for surfaces. J. Algebra **320**, 3113–3119 (2008)

10. A. Dimca, in *Topology and Singularities of Hypersurfaces*. Universitext (Springer, New York, 1992)
11. I. Dolgachev, B. Weisfeiler, Unipotent group schemes over integral rings (English translation). Izv. Akad. Nauk USSR Ser. Mat. **38**, 757–799 (1974)
12. A. Dubouloz, L. Moser-Jauslin, P.-M. Poloni, Noncancellation for contractible affine threefolds. Proc. Am. Math. Soc. **139**(12), 4273–4284 (2011)
13. P. Eakin, W. Heinzer, A cancellation problem for rings. in *Conference on Commutative Algebra*. Lecture Notes in Mathematics, vol. 311 (Springer, Berlin, 1973), pp. 61–77
14. K.H. Fieseler, On complex affine surfaces with $\mathbb{C}_+$-action. Math. Helvetici **69**, 5–27 (1994)
15. T. Fujita, On Zariski problem. Proc. Jpn. Acad. Ser. A **55**(3), 106–110 (1979)
16. T. Fujita, Cancellation problem of complete varieties. Invent. Math. **64**, 119–121 (1981)
17. T. Fujita, On the topology of non-complete surfaces. J. Fac. Sci. U. Tokyo **29**, 503–566 (1982)
18. T. Fujita, S. Iitaka, Cancellation theorem for algebraic varieties. J. Fac. Sci. U. Tokyo **24**, 123–127 (1977)
19. N. Gupta, On the cancellation problem for the affine space $\mathbb{A}^3$ in characteristc p. Inventiones Mat. **195**, 279–288 (2014)
20. N. Gupta, On the family of affine theefolds $x^m y = F(x, z, t)$, to appear in Compositio Mat.
21. M. Hochster, Non-uniqueness of the ring of coefficients in a polynomial ring. Proc. Am. Math. Soc. **34**, 81–82 (1972)
22. S. Iitaka, On logarithmic Kodaira dimension of algebraic varieties, in *Complex Analysis and Algebraic Geometry* (Iwanami Shoten, Tokyo, 1977), pp. 175–189
23. S. Kaliman, L. Makar Limanov, On the Russell-Koras contractible threefolds. J. Algebr. Geom. **6**, 247–248 (1997)
24. S. Kaliman, M. Koras, L. Makar-Limanov, P. Russell, $\mathbb{C}^*$-actions on $\mathbb{C}^3$ are linear. Electron. Res. Announc. Am. Math. Soc. **3**, 63–71 (1997)
25. S. Kaliman, S. Vénéreau, M. Zaidenberg, Simple birational extensions of the polynomial algebra $\mathbb{C}^{[3]}$. Trans. Am. Math. Soc. **356**(2), 509–555 (2003)
26. M. Koras, P. Russell, Contractible threefolds and $\mathbb{C}^*$-actions on $\mathbb{C}^3$. J. Algebr. Geom. **6**, 671–695 (1997)
27. M. Koras, P. Russell, $\mathbb{C}^*$-actions on $\mathbb{C}^3$: the smooth locus of the quotient is not of hyperbolic type. J. Algebr. Geom. **8**(4), 603–694 (1999)
28. L. Makar-Limanov, On groups of automorphisms of a class of surfaces. Isr. J. Math. **69**(2), 250–256 (1990)
29. L. Makar-Limanov, On the hypersurface $x + x^2y + z^2 + t^3 = 0$ in $\mathbb{C}^4$ or a $\mathbb{C}^3$-like threefold which is not $\mathbb{C}^3$. Isr. J. Math. **96**(Part B), 419–429 (1996)
30. M. Miyanishi, T. Sugie, Affine surfaces containing cylinderlike open sets. J. Math. Kyoto U. **20**, 11–42 (1980)
31. D. Mumford, The topology of normal singularities of an algebraic surface and a criterion for simplicity. Inst. Hautes tudes Sci. Publ. Math. **9**, 5–22 (1961)
32. C.P. Ramanujam, A topological characterization of the affine plane as an algebraic variety. Ann. Math. **94**, 68–88 (1971)
33. P. Russell, On affine-ruled surfaces. Math. Ann. **255**, 287–302 (1981)
34. B. Segre, Corrispondenze di Möbius e transformazioni cremoniane intere. Atti Acad. Sci. Torino Cl. Fis. Mat. Nat. **91**, 319 (1956/1957)
35. T. Shioda, Some remarks on abelian varieties. J. Fac. Sci. U. Tokyo **24**, 11–21 (1977)
36. V. Shpilrain, J.-T. Yu, Affine varieties with equivalent cylinders. J. Algebra **251**, 295–307 (2002)
37. M. Suzuki, Propriétés topologiques des polynômes de deux variables complexes, et automorphismes algébriques de l'espace $\mathbb{C}^2$. J. Math. Soc. Jpn. **26**, 241–257 (1974)
38. J. Wilkens, On the cancellation problem for surfaces. C. R. Acad. Sci. Paris **326**(Serie I), 1111–1116 (1998)
39. M. Zaidenberg, On exotic algebraic structures on affine spaces (Russian). Algebra i Analiz **19**(5), 3–73 (1999) [English translation in St.Petersburg Math. J. **11**(5), 703–760 (2000)]

FSC
www.fsc.org
MIX
Papier | Fördert
gute Waldnutzung
FSC® C083411

Kolibri 360 GmbH
Ferdinand-Jühlke-Straße 7
99095 Erfurt, Deutschland
produktsicherheit@kolibri360.de